COLLINS
BIRD GUIDE
SECOND EDITION

TEXT & MAPS BY LARS SVENSSON

ILLUSTRATIONS & CAPTIONS BY
KILLIAN MULLARNEY & DAN ZETTERSTRÖM

WITH A SIGNIFICANT CONTRIBUTION BY
PETER J.GRANT

TRANSLATED BY DAVID CHRISTIE & LARS SVENSSON

HarperCollins *Publishers* Ltd
77–85 Fulham Palace Road
London W6 8JB

www.harpercollins.co.uk

Collins is a registered trademark of HarperCollins *Publishers* Ltd.

Originally published in 1999 in Swedish, *Fågelguiden. Europas
och Medelhavsområdets fåglar i fält*, by Bonniers, Stockholm
2nd revised and enlarged edition published in 2009 by Bonnier Fakta, Stockholm

First published in the UK as a hardback edition in 1999
First published in the UK as a paperback edition in 2001

2nd revised and enlarged edition © 2009 HarperCollins *Publishers*, London

© 1999, 2009 Lars Svensson (text and maps),
Killian Mullarney and Dan Zetterström (illustrations and captions)

Translation and English adaptation: David A. Christie & Lars Svensson
Typography: Lars Svensson
Cover: Arctic Tern by Dan Zetterström

09 11 13 15 14 12 10

1 3 5 7 9 10 8 6 4 2

ISBN 978 0 00 726726 2 (Hardback)
ISBN 978 0 00 726814 6 (Paperback)

Reproduction: Fälth & Hässler, Värnamo, Sweden 2009
Printing: Printer, Trento, Italy 2009
Printed in Italy

Contents

Acknowledgements

As with the first edition, a book like this cannot be produced without the help of many, both by direct input and indirectly through new and ground-breaking work made available to us in the ornithological literature and on the net. Our gratitude to a circle of close friends as detailed in the first edition remains, and readers are referred to this for a full list. Here we would like to mention particularly those who have helped us generously while producing the revised edition.

L. S. is indebted to José Luis Copete, Andrew Lassey and Hadoram Shirihai, who all three were not only good company on various trips but also gave freely of their imposing knowledge and experience. Many thanks also to Per Alström, Vladimir Arkhipov, Oleg Belyalov, Martin Collinson, Pierre-André Crochet, Alan Dean, Pete Dunn, David Erterius, Andrew Grieve, Marcel Haas, Magnus Hellström, Guy Kirwan, Hans Larsson, Norbert Lefranc, Klaus Malling Olsen, Urban Olsson, Eugeny Panov, David Parkin, Mike Pearson, George Sangster, Jevgeni Shergalin and Mike Wilson for help, advice and support in various ways. Again, Richard Ranft and the British Library Sound Archive generously provided access to recordings of some of the rarer species. The staff at visited museums have always been welcoming and helpful, for which I am grateful.

K. M. is especially grateful to Mark Constantine for his solid support, invaluable advice and great friendship over so many years. Similarly, friends Paul Archer, Arnoud van den Berg, Richard Crossley, Dick Forsman, Hannu Jännes, Lars Jonsson, Ian Lewington, Pat Lonergan, Aidan Kelly, Dave McAdams, Richard Millington, Colm Moore, René Pop, Magnus Robb, Cornelia Sakali, Alyn Walsh and Pim Wolf have helped greatly in a variety of ways, as well as provide the best of company in the field. Many thanks also to Per Alström, Derek Charles, José Luis Copete, Andrea Corso, Michael Davis, Paul Doherty, Annika Forsten, Magnus Hellström, Paul Holt, Hans Larsson, Antero Lindholm, Bruce Mactavish and Frank Zino for their thoughtful suggestions and/or provision of very useful photographic material. The extraordinary dedication of a number of keen 'larophiles' to elucidating the complexities of large gull ageing and identification for the benefit of all has been of enormous help in the preparation of the new treatments presented here. It is impossible, in the space available, to do any real justice to their work but the assistance of Ruud Altenburg, Chris Gibbins, Hannu Koskinen, Bert-Jan Luijendijk, Ies Meulmeester, Mars Muusse, Theodoor Muusse, Rudy Offereins and especially Visa Rauste, is greatly appreciated.

D. Z. wants to express his gratitude to Ian Andrews, Stefan Asker, Arnoud van den Berg, Christer Brostam, José Luis Copete, Andrea Corso, Göran Ekström, David Fisher, Annika Forsten, Fares Khoury, Markus Lagerqvist, Lars Larsson, Dan Mangsbo, Bill Zetterström and Frank Zino for their kind support.

Last but not least, we thank our severely tried relatives and friends, as well as our publishers, for patience and support.

L. S., K. M., D. Z.

Preface

The *Collins Bird Guide* has, to our joy and satisfaction, had tremendous success since it came out in 1999. It has been published in no fewer than 13 languages and sold a staggering combined 700,000 copies, with nearly a third of these sold in the UK alone. It has been gratifying for us to see it being used so much in the field wherever we travel in Europe, North Africa and the Middle East. The full background and leading principles for the layout of the book can be found in the first edition and will not be repeated here, but the positive reviews and personal feedback indicate that the principles we chose were the right ones, corresponding well to what birders want from a field guide in the 21st century.

This said, there are no grounds for complacency. Shortly after publication of the first edition it was evident that it was already in need of a revision. The reasons for this were twofold. First, the gestation of the first edition took a long time, and with the plan for the book having been laid out in the early and mid 1980s, many things were bound to change or develop in the 15 or so years it took to reach publication, not all of which could be compensated for along the way. New methods to separate similar-looking birds are being developed all the time, and a good field guide needs to be as up to date as possible. Some of this required more space than allocated and had to be left for later inclusion.

The second reason, which was difficult to foresee when we set out to create the book, was that the development of avian taxonomy would take a big leap—after a long period of relative stability—just prior to 1999, and especially after. New research and a partly new approach to taxonomic issues involving genetic methods led to a re-evaluation of the taxonomic status of many taxa formerly regarded as subspecies. These advances in knowledge have had the effect that a number of species formerly regarded as polytypic, comprising several rather distinct subspecies, have now been split into two or more species. These 'new' species are, for very natural reasons, often quite similar to their closest relatives. But they constitute interesting populations with their own life histories, and they deserve their own species accounts in the book, with advice on identification, in both words and pictures.

To accommodate all new species and new information we have had to increase the number of pages by some 10%, but we feel that the book is still a lightweight one so that no-one should be tempted to leave it at home when travelling.

One change will strike the reader familiar with the first edition immediately: the new order of families in the beginning. Newly published genetic research has shown that the two oldest groups of birds are the wildfowl *Anseriformes* and the grouse and their relatives *Galliformes*, these two collectively called *Galloanserae*. Since the arrangement of families in this book is basically the traditional systematic one, with the oldest groups placed first, the book now starts with swans, geese and ducks followed by grouse, pheasants, etc. Only then come the loons (divers), grebes, seabirds, etc., formerly placed first.

Areas where much taxonomic change has taken place lately, and which have affected this revised edition, are the wildfowl, shearwaters, large gulls, thrushes, warblers, flycatchers, shrikes and finches. The revised edition treats no less than 41 new species, 33 of which are the result of the taxonomic changes, and several more subspecies have received better treatment. This has been achieved by the addition of 24 new spreads, and by a redesign of numerous plates and the incorporation of new illustrations.

Obviously, taxonomy is not a static science. New research constantly reveals new relationships and better arrangements. Some proposed changes which have been discussed in recent years have not been adopted in this edition because we have decided to await further research and more universal acceptance. Still, they might well be adopted in future editions. This is particularly true for the passerines, where clearly in a possible future edition readers will have to get used to a profoundly changed sequence. By the look of it now, the natural sequence starting with the oldest families would begin with shrikes and orioles, then group together tits, warblers, bulbuls, larks, reedlings and swallows; thrushes and flycatchers would come close together, while pipits & wagtails would be fitted in between sparrows and finches. Well, let us not cross that bridge until we come to it!

Changes which *have* been incorporated are aimed to improve both text and plates to facilitate identification, even when there is no underlying taxonomic change. With continued travel and through the advances achieved both by others and ourselves, we are keen to let our book mirror the most recent developments in identification. For some groups which were a bit crammed in the previous edition (pigeons & doves, thrushes, warblers, shrikes) we are pleased to have been given the opportunity to afford them more adequate treatment.

All the maps have been revised. Although small, our ambition is that they should be accurate and up to date. New atlases and checklists have been published in recent years, and new references made available, for important countries such as Algeria, the Czech Republic, France, Greece, Morocco, Poland, Spain, Sweden, Turkey, and for the Ural region, and these have all been used to full advantage. It should be noted that very local, rare or recent breeding records are deliberately not always shown on the maps, which aim to show more common and established patterns rather than every recent new breeding site. For several species restricted to a smaller part of the covered range we have introduced maps at a larger scale.

Peter J. Grant was deeply involved in the planning of and preparations for the first edition. His untimely death prevented him from participating fully in the creation of the finished book. We dedicate the second edition to him as a tribute to his many achievements in the field of bird identification, and to his memory.

Lars Svensson, Killian Mullarney, Dan Zetterström
August 2009

Introduction

This book treats all bird species which breed or regularly occur in Europe, North Africa north of 30°N, and Israel, Palestine, Jordan, Syria, Turkey, Armenia, Georgia and Azerbaijan, that is in a large part of the Middle East. Also included are the Canary Islands, Madeira and the Sinai peninsula. (Excluded are the Azores, the Cape Verde Islands, Iraq and Iran.) Europe is bordered in the east by the ridge of the Ural mountains, the Ural river and the Caspian Sea.

713 species are described in the main section. There is also brief mention of 59 occasional visitors. Another 32 species are treated which are either probable escapes from captivity or originally introductions to the area and which now breed in a feral state without human support. An additional 118 very rare stragglers from other continents are merely listed.

Taxonomy and names
With prevailing disagreement both on best taxonomy and on English bird names, resulting in more than one standard available, the solution has been to follow 'author's preference'. Compared to the first edition a few changes have been made to conform better with the list of recommended English names initiated by IOC (Gill & Wright 2006). Standardized vernacular names are of course practical and help communication. But just as taxonomy is continuously developing, so is nomenclature, and asking for total conformity and discipline is unrealistic.

The scientific name of a species is written in Latin (or in latinized form) and consists of two words, the generic name (written with an initial capital letter) and the species epithet (all lowercase letters); these two words together constitute the species name. The scientific name of e.g. the White Wagtail is thus *Motacilla alba*.

To indicate geographical variation within a species, distinct populations—so-called subspecies or races—are designated with a third word, a subspecies epithet (e.g. Pied Wagtail *Motacilla alba yarrellii* of NW Europe, as opposed to *Motacilla alba alba* of the rest of Europe). In this book space permits only the most distinct subspecies to be named and treated. The reader interested in more detail should consult any of the more comprehensive handbooks, such as *BWP* (Cramp *et al*. 1977–94) or Vaurie (1959, 1965).

English names nowadays exist both in short form for everyday use at home, and in a longer form for international use (with added modifiers). A way of conveniently showing both these English names for the same species is practised: at the head of each species entry, bold face is used for the ordinary names and normal type within brackets for the extra modifiers; '(Western) **Jackdaw**' serves as an example.

Although the sequence is mainly based on the so-called natural one, placing the most primitive (old) groups first, we have made a few minor adjustments within genera and, at times, families. The sole purpose of these deviations is to bring together those species which show the greatest resemblance and thereby to facilitate comparisons. We are also aware that recent molecular research has cast new light on relationships and best arrangement, meaning that in possible future editions of this book a rather radical change of sequence will in parts be required.

Abundance symbols
For easy assessment of whether a certain species occurs commonly in Great Britain and Ireland, or whether it is merely a rare vagrant to the isles, abundancy symbols are given to the right of the species name. If you have seen a flock of unknown birds in your garden, it might be a waste of time to read in depth about species which do not even occur in the region. The symbols are explained below.

Status in Great Britain and Ireland

r**B** Resident breeding species.
m**B** Migratory breeding species.
r+m**B** Breeder; some are residents, others migratory.
r(m)**B** Breeding species; mainly resident but a minority are migrants.

To these a qualifier is attached:

1	Very abundant	(estimated > 1 million pairs)
2	Abundant	(estimated > 100,000 pairs)
3	Fairly abundant	(estimated > 10,000 pairs)
4	Scarce or local	(estimated > 100 pairs)
5	Rare	(estimated ≤ 100 pairs)

W Winter visitor (common; many immigrants).
P Passage visitor (common).

Here, too, a qualifier may be added (as under breeding, above); if so, estimates refer to number of birds, not pairs.

V Vagrant.

To this, again, a qualifier is attached:

*	Annual vagrant in some numbers.
**	Only one or a few records a year, or in most.
***	Only one or a very few records per decade. ('Three-star rarity.')
[**V**]	No record considered a result of genuine and unassisted vagrancy, or all records thought to involve escapes from captivity; records which are open to question for other reasons.
—	No record in Great Britain or Ireland.

Distribution maps
Besides the abundancy symbols, distribution maps are included for most species and should help establish quickly whether it is reasonable or not to expect a certain species at a

Breeding range, abandoned in winter.

Present all year around, including when breeding.

Range where the species can be seen on migration.

Winter range.

Main migration direction (not necessarily exact route).

certain site and a given season. Although it was necessary to make the maps small, they should still give a useful summary of the normal occurrence. Unlike in many other books, the maps also show where the birds usually occur during autumn and spring migrations, so far as this is known.

As before, an effort has been made to present as up-to-date and clear maps as possible. All national checklists and atlas surveys published in the last decade covering the treated area have been consulted. Still, modern mapping of the bird fauna is still missing for such important areas as Britain, Finland, Germany, Iceland, Ireland, Libya, Norway and for most of the states which were formerly a part of the Soviet Union.

Size of birds

Each species account opens with the size of the bird, expressed as the length in centimetres from tip of bill to tip of tail measured on the stretched bird (L). For most species which are often seen in flight the wingspan is added (WS).

To indicate the normal size variation within any one species, a size range is always given instead of a single average figure. Many birdwatchers are unaware of the magnitude of this normal size variation, and it is all too easy then to arrive at the wrong conclusion if an identification is based too heavily on a size evaluation.

The length measurements have been taken largely from series of well-prepared skins and in some cases from freshly killed or live birds. (Hardly any have been obtained from the current larger handbooks.)

The wingspans are meant to indicate the largest possible extents which the birds themselves can achieve in flight. (Slightly larger values can be reached if, on a live bird, the wings are stretched tightly by pulling the primaries, but such artificial measurements have been avoided.) Quite a few measurements in the current literature are misleading. Those presented in this book are based to a large extent on original measurements taken on live birds. They have been supplemented by measurements on skins and photographs.

Terminology and symbols

In order to make the book easily accessible to a large public interested in birds and natural history, the specialist jargon has been kept to a minimum. A few technical terms, however, are very useful to know and use, such as the precise terms for different plumages and ages, and these are explained below. It should be noted that most of the terms relating to feather tracts and body parts, the topography of the bird, are also explained separately on the inside of the covers.

PLUMAGES AND AGES

juvenile (juv.) – young, fledged bird wearing its first set of true feathers (the juvenile plumage), but which has not yet moulted any of these feathers.

post-juvenile – all plumages or ages following the juvenile.

young – imprecise term usually referring to juvenile and/or 1st-winter without distinguishing between these two.

immature (imm.) – a bird wearing any plumage other than adult, generally corresponding to the word 'young'.

1st-autumn – bird in its first autumn, 2–5 months old. The term refers to the age, not a particular plumage.

1st-winter (1st-w.) – age category usually referring to the plumage following that of the juvenile, usually gained through a partial (in some species complete) moult in late summer/autumn of 1st calendar-year, and worn until next moult in the spring of 2nd calendar-year.

1st-summer (1st-s.) – age category usually referring to the plumage worn at the approximate age of one year and attained in late winter/spring through partial (in some species complete) moult from 1st-winter plumage, or through abrasion of this plumage; worn until next moult, usually in late summer/autumn in the same year, when replaced by 2nd-winter plumage.

2nd-winter (2nd-w.) – age category usually referring to the plumage following 1st-summer, usually gained through a complete (in some species partial) moult in late summer/autumn of 2nd calendar-year, and worn until next moult in the spring of 3rd calendar-year.

2nd-summer (2nd-s.) – age category usually referring to the plumage worn at the approximate age of two years. It follows the 2nd-winter plumage and is worn until next moult; see also under 1st-summer.

subadult (subad.) – nearly adult, not quite in definitive plumage; imprecise term, often used when exact age is difficult to establish, e.g. among larger gulls or raptors.

adult (ad.) – old, mature bird in definitive plumage. Some species have the same appearance in both summer and winter; others have separate plumages, *adult summer* (*ad. sum.* or *ad. s.*) *and adult winter* (*ad. wint.* or *ad. w.*).

breeding plumage – usually more colourful set of feathers gained by many birds, primarily males, through abrasion or in a spring moult (autumn moult for ducks).

non-breeding plumage – usually equivalent to adult winter plumage, a more cryptic plumage appearing among species with seasonal plumage changes; the term is often used when this plumage is acquired early, even in summer.

eclipse plumage – cryptic, female-like plumage attained by male ducks in summer while moulting the flight-feathers and becoming flightless, and thus needing camouflage. (Cf. *breeding plumage*.)

1st-year, 2nd-year, etc. – age category referring to the entire first, second, etc. year of life, from summer to summer.

calendar-year (cal.-yr) – a bird is in its 1st calendar-year from hatching until 31 Dec of the same year, in its 2nd calendar-year from 1 Jan until next 31 Dec, etc.

GENERAL TERMINOLOGY

albinism – innate lack of pigment, usually in the feathers, leading to partly or completely white plumage.

alula – a group of feathers attached at the digital bone, near the wing-bend on upper forewing. (Syn.: *bastard wing*.)

'arm' – the same as *inner wing*, the part of the wing inside the wing-bend (incl. secondaries and accordant coverts).

axillaries – the feathers covering the 'armpit'.

carpal – the 'wrist', often referred to as the *wing-bend*.

carpal bar – usually dark bar along the leading edge of the inner wing, from the carpal inwards.

cere – bare skin on the base of upper mandible and around nostrils in some groups, e.g. raptors.

eye-stripe – usually contrasting *dark* stripe *through* the eye from near bill-base and backwards to the ear tract.

Fenno-Scandia – Norway, Sweden, Finland and the Kola Peninsula.

'fingers' – the spread tips of the outermost long primaries on broad-winged, soaring birds, e.g. eagles, storks, cranes.

flight-feathers – the long quills on the wing (primaries, secondaries and tertials).

foot projection – the part of the feet (incl. any part of the tarsus) which extends beyond the tail-tip in flight.

forewing – usually the upperwing-coverts (or at least the foremost of these) on the 'arm' of a spread wing.

gape – naked skin at the corner of the mouth.

gonys angle – slight hook on lower mandible of e.g. gulls.

'hand' – the same as *outer wing*, the part of the wing outside the wing-bend (incl. primaries, alula, primary-coverts).

hover – hold position in the air by quick wing movements.

hybrid – a cross between two different species.

inner wing – see '*arm*'.

'jizz' – the shape and movements of a bird, its 'personality'. (Jargon of unclear derivation.)

lateral throat stripe – narrow stripe, often dark, along the side of the throat, below the moustachial stripe.

leucism – innate pigment deficiency leading to pale, washed-out colours of the plumage.

lore – area between eye and base of upper mandible.

mantle – tract of feathers covering the fore-part of the back.

malar – see *lateral throat stripe*.

melanism – innate surplus of dark pigment (notably melanin) causing the plumage to become darker than normal, at times even black.

midwing-panel – area on inner upperwing, often on median and longest lesser coverts, of contrasting colour (lighter on e.g. harriers and petrels, darker on young small gulls).

'mirror' – term used to describe a small white (or light) patch just inside the tip of longest primaries (on gulls).

morph – variant of certain appearance within a species and which is not geographically defined (cf. *subspecies*). The same as formerly often used *phase*.

moult – a natural process of renewal of the plumage, whereby the old feathers are shed and new ones grown.

moustachial stripe – usually narrow stripe, often dark, running from near the base of the lower mandible and along the lower edge of the cheek. (Cf. *lateral throat stripe*.)

nominate subspecies – within a species, the subspecies first to have been named and which therefore, according to the rules of Zoological Nomenclature, is given the same sub-specific epithet as the specific, e.g. *Motacilla alba alba*.

orbital ring – eye-ring of bare, unfeathered skin.

outer wing – see '*hand*'.

Palearctic – ('Old Arctic') the zoogeographical or 'natural' region comprising Europe, North Africa south to the Sahara, often the entire Arabian Peninsula, and Asia south to Pakistan, the Himalayas and central China.

panel – term often used in ornithology to describe an elongated patch of contrasting colour (cf. *wing-panel*).

pelagic – referring to the open sea.

primaries – the quills growing on the 'hand'.

primary patch – pale patch at the base of the primaries, on upperwing or underwing, e.g. on raptors and skuas.

primary projection – the part of the primaries which on the folded wing extends beyond the tip of the tertials.

secondaries – the quills growing on the 'arm', or ulna.

species (*sp.*) – important entity in taxonomy which in brief is a group of natural populations whose members can interbreed freely but which are reproductively isolated from members of other such groups (i.e. other species).

speculum – distinctive, glossy patch on upper secondaries, often on ducks.

submoustachial stripe – a usually light stripe between dark moustachial stripe and dark lateral throat-stripe.

subspecies (*ssp.*) – morphologically (colour, size) on average discernibly different, geographically defined population within a species. The use of quotation marks around a subspecies name indicates that the form is either subtle, or questionable or invalid (e.g. '*omissus*'). (Syn.: race.)

supercilium – usually contrasting light stripe above the eye from near forehead and backwards along side of crown.

taiga – predominantly coniferous, northerly (boreal), wide and largely continuous forest zone in northern Europe.

tail-streamers – elongated outer or central tail-feathers.

tarsus – short for tarsometatarsus, the fusion of the metatarsal bones, in everyday speech the 'leg' of the bird.

tertials – the innermost wing-feathers, usually with somewhat different shape and pattern, serving as cover (protection) for the folded wing. On larger birds tertials are attached to humerus (inner arm), on smaller birds to ulna (and thus are not 'true tertials' but inner secondaries).

tibia – the lower leg, on birds colloquially named 'thigh'.

variety (*var.*) – a variant form within a population (cf. *morph*); also, a domestic breed of constant appearance.

'window' – referring to paler and more translucent inner primaries, e.g. on some raptors and young gulls.

wing-bar – contrastingly coloured, often lighter bars on the wing, often formed by white tips of wing-coverts.

wing-bend – the same as the carpal, the joint between 'hand' and 'arm' of the wing.

wing-panel – imprecise term referring to contrastingly coloured rectangular area on the wing.

wing projection – the part of the wings which extends beyond the tail-tip when the bird is perched.

wingspan (*WS*) – distance from wing-tip to wing-tip.

SYMBOLS AND SIGNS

♂, ♂♂	male, males
♀, ♀♀	female, females
>	more than, larger than
<	less than, smaller than
±	more or less, to a varying degree

Voice and transcriptions

Owing to lack of space, and in order to achieve simplicity and clarity, especially with beginners in mind, only the common and distinct calls are described. Many subtle variations of the normal contact- and alarm-calls are excluded.

Although rendering bird voices in writing inevitably is inexact and personal, a serious effort has been made to convey what is typical for each call by trying to select the letters and style of writing which are most apt. We know by experience that it can be very helpful to read a fitting rendition, particularly after just having heard a certain call. We do not share the opinion that written voice transcriptions are so subjective that they have little value at all.

Whenever you hear an unfamiliar bird voice, do not be afraid to note down what it first brings to mind, be it another bird or something quite unrelated; more than likely this first impression is exactly what you will be reminded of next time you hear it. Your own impressions will thus be important additions to the voice descriptions offered in the book.

Transcriptions of calls are indicated by quotation marks. Syllables in **bold** face are more stressed than the others. Very explosive or fierce calls end with an exclamation mark.

The choice of consonants is meant to show whether the sounds are hard or soft: 'tic' is sharper overall than 'gip', 'kick' has a harder opening than 'bick'. 'z' indicates a sharper and more 'electric', fizzing tone than 'ts'.

The vowels are selected in an attempt to hint somewhat at the relative pitch, although this is not so easy in English. To the common English vowels is added German 'ü' (as in Lübeck or the French word 'mur', pronounced between 'sue' and 'sync') since there is often a need for it. The call 'too-ü' has a rising pitch whereas 'dee-u' is downslurred.

Double vowels denote a longer sound than single: 'viit' is more drawn out than 'vit'. When a vowel at the end of a call is followed by an 'h', as in 'tüh', it is drawn out but somewhat fading at the end, or sounds 'breathed out'.

The way of writing suggests how rapidly the syllables are delivered. 'ki... ki... ki...' is very slow, 'ki, ki, ki,...' denotes a composed rhythm, 'ki ki ki...' is a little quicker, whereas 'ki-ki-ki-...' is rapid like a shuttle, and 'kikikiki...' is very fast (shivering). Even faster is 'kr'r'r'r'r...', indicating a vibrant, rolling sound. Now and then a strict application of these guidelines has had to give way to legibility.

The plates

To help beginners, in particular, to find their way when looking at the plates with their wealth of small images, we have aimed at: (1) similar scale for all main images which are not obvious 'distant views'; (2) constant posture and location on the plate of adult males, adult females, juveniles, etc. (summer males always at lower far right, juveniles at far left, etc.); (3) addition of vignettes and tiny 'distant views' showing characteristic habitat, posture and behaviour, and how each species will appear at long range and in poor light (i.e. as they are so often seen!); and (4) addition of pointers and brief captions highlighting particularly useful clues, which will make the plates easier to 'read'. Fine rules have been inserted between the species in order further to facilitate a quick survey among the many images on the plates.

It should be pointed out that, although great care has been

A Lapwing in bright sunshine (left)—rather as it is shown on the plate on p. 147—and huddled up in dull weather (right)—a totally different creature.

taken to choose typical postures and plumages for the birds portrayed, odd birds which you see live in nature will nevertheless look different from the images on the plates. Some have an abraded or stained plumage, others are in poor condition and adopt uncharacteristic postures, while others again are in good shape but have a posture which is rather odd and which could not be fitted in on the plate. The light, too, can cause all colours to be subdued or the pattern to be less contrasting. It is best therefore to view the illustrations as a guide but not to expect every single detail of posture and plumage to match the bird you have in front of you.

Moult and abrasion of plumage

As soon as a feather is fully grown it becomes a dead part. Sunshine and abrasion bleach and wear the feathers gradually, and in the end their function as a means to achieve flight and to serve as waterproofing or heat regulation is impaired. Unlike fur and nails, feathers are not continuously renewed. Instead, they are replaced in a usually annual process which is called moult.

Another reason for moult, especially of feathers on head and body, is the signalling function which colours and patterns in the plumage of many species have, important in the social life of the bird; different signals may be required at different seasons.

It is disadvantageous to moult heavily during migration (the flight capacity is then often impaired) and, for many species, also while breeding (both breeding and feather growth require much energy). Therefore, most species moult in concentrated periods between these two stages.

The time after the breeding season but before the autumn migration is utilized by many species for a complete moult. Others, especially those wintering in the tropics, undergo their annual complete moult in the winter quarters.

A Herring Gull in autumn approaching the end of the annual complete moult: only the outermost primaries are still growing (making the bird appear confusingly pale-winged, with much of the black on wing-tips still concealed!).

Many of those moulting completely in late summer also have a partial moult in early spring in which some body-feathers and wing-coverts are replaced (but no flight-feathers). And the long-distance migrants with the complete moult in winter often renew some body-feathers either before the autumn migration or prior to the spring migration.

Very large birds which are dependent on undiminished powers of flight and which often soar, e.g. large raptors, storks and pelicans, replace their flight-feathers slowly during a large part of the year. Since a new quill grows only by 5–10 mm a day, each quill requires up to two months for renewal (the longest primaries of large eagles and pelicans even 2½ months). With 10–11 primaries and 15–22 secondaries in each wing of these large species, a complete replacement of the flight-feathers requires 3–4 years in order not to hamper the ability to fly.

Wildfowl shed all their flight-feathers simultaneously. This is done in summer, usually after breeding or near its end. They are flightless for 3–7 weeks (the larger the species is, the longer the time) but survive by keeping in cover in marshes or out of the way on the open sea.

Many ducks have rather peculiar and complex moult strategies. The breeding plumage of the males is often acquired in late autumn, since ducks usually mate in winter. In the summer, when the males of other birds are at their most handsome, male ducks instead change to a female-like so-called eclipse plumage so as to be better camouflaged when the moult renders them temporarily flightless.

The Crane is unique among the landbirds in shedding all flight-feathers simultaneously in summer, like the ducks. The moult takes place apparently only every second year, or at even longer intervals. The moulting bird keeps well hidden on wide, sparsely forested bogs or in reedbeds until capable of flying again, in time for the autumn migration.

As is evident from the examples, the variations in moult are considerable, and almost every species has its own strategy for renewal of the feathers. To master all these variations requires specialist knowledge, but the fundamentals are easy to learn. And they are useful to know since they enable you to age the birds you see closely and sufficiently well, even

The difference in wear and shape of the tail-feathers between winter adult and young Common Redpolls. Note rounder tips and more fresh edges on the adult.

Adult winter   **1st-winter**

with species in which the different age-groups have similar plumages. Correct ageing is often supporting evidence—at times even a prerequisite—for reliable identification.

It is even more common that a means of reliably ageing a bird allows you to sex it as well. In many species the plumages of adult females and young males are very similar (e.g. Horned Lark, some members of the thrush family, starlings, numerous finches and buntings), but if the age is established these two categories can usually be separated.

When young birds become fledged they have fresh wing- and tail-feathers, whereas adults have abraded feathers, these having been grown six months to almost a full year earlier. Soon thereafter, however, the adults of many species start to moult their wing- and tail-feathers in a complete moult, and after a short time the scene is reversed: from about late summer *the wings and tail of adults are fresher than those of young birds*. Since the quality of the juvenile plumage is somewhat inferior and hence is more susceptible to bleaching and abrasion than later plumages, the difference becomes more obvious through autumn and winter.

Juveniles of many species also have body-feathers a little looser in structure and duller, and t*he wing- and tail-feathers and some coverts are often slightly paler and less glossy and are narrower and more pointed*, whereas adults have *slightly darker, glossier and broader feathers with more rounded tips*.

In species with a slower renewal of flight-feathers, e.g. large raptors, juveniles can often be told by having *all wing- and tail-feathers uniformly fresh* (or worn), whereas older birds may be recognized by *slight unevenness in the length, darkness and degree of abrasion among the wing-feathers*.

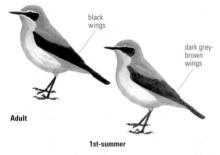

Two male Wheatears in spring. The left-hand one has black wings without light edges and is at least two years old. The one on the right has dark grey-brown wings (the juvenile feathers) with traces of light edges, and is thus one year old.

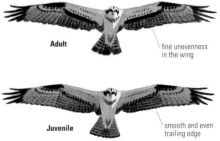

Two Ospreys in autumn. The upper bird has slightly uneven trailing edge of wing, and the flight-feathers are not quite homogeneous; it is adult. The lower bird has uniformly fresh and evenly long flight-feathers finely tipped white; it is juvenile.

In a few species the juveniles, too, have a complete moult in late summer and early autumn, renewing wings and tail at the age of only a few months, e.g. larks, Moustached Warbler, Bearded Reedling, Long-tailed Tit, starlings, sparrows, and Corn Bunting. After this post-juvenile moult there is *no way of separating adults and young birds* of these species.

How birds are identified

One of the thrills of becoming a birdwatcher is the joy of discovery. Therefore you do not want everything on a plate. The perplexities and mistakes experienced as a beginner can be frustrating, but generally only spur you to new efforts. Still, a few general pieces of advice may be appropriate. Bird identification in the field is difficult enough as it is.

The first question to be asked when suspecting that a Black-winged Kite just flew over the local heath or a Citril Finch alighted in the garden: *Is it reasonable?* Is this species supposed to occur in Britain or Ireland, and at this season? Look for a commoner species (e.g. male Hen Harrier and Greenfinch in these examples) if both abundance symbol and range map argue against the suspected rarity!

Is the habitat the right one? There is at least something in the species accounts hinting at the typical environment in which each species thrives. If it does not fit, try other alternatives.

Does the observed behaviour match the description? In many accounts (but not all) there is an indication of characteristic flight and movements, degree of shyness, mating or feeding behaviour, etc. It is of course reassuring if the observed habits match those described under the suspected species.

If everything you have observed matches the selected description so far, now is the time to examine the appearance of the bird more critically. First establish: *What size is it?* If the size is correctly assessed now, much trouble is saved later. Therefore a lot of work has been put in to ensure that accurate measurements of size are given in the book.

It may seem a trivial and easy matter to some to assess size correctly when attempting to identify an unknown bird. But experience shows that both beginners and well-practised birders often make mistakes. And once you have misjudged the size of a bird it is all too easy stubbornly to follow up the wrong clue. 'Oh, yes, it looked like a Red-backed Shrike all right, only this bird was considerably bigger!'

Prejudices about the size of birds can lead you astray. Frequent practice in the field helps, especially if combined with careful study of the sizes given in the book.

against the light   with the light

Whether the light is against or with the observer will affect the impression of the shape of a bird. Seen against the light, this Greenshank appears slimmer ('the light eats the edges') than it does in flat light from behind the observer.

The contrast between bird and background affects the size impression. The dark Arctic Skua appears larger against a light, calm sea than against a dark and stormy one, while the opposite is true for the pale Kittiwake.

A source of error which birdwatchers are likely to encounter is the optical size-illusion occurring when a flock of birds is studied through a telescope with powerful magnification. Since the telescope contracts the field of vision so much, the birds further away *appear* larger than the conspecific birds in the foreground, being in fact of equal size.

The contrast between bird and background also affects the size impression. As a rule, *strong contrast makes a bird look bigger whereas slight contrast makes it look smaller*. A dark bird against a dark sea appears smaller than a white one even if they are equal in size. And a light bird does not look so impressive against a light background as a black one would.

After having tried to assess the size of the bird, a number of important questions follow: *Is the bill small or big, wedge-shaped or thin? If long and slender, is it longer than the length of the head? What is the shape of the body? Has it got long or short neck and tail? Are the legs strikingly long and thin or just 'average'?* Consider such questions on location while observing the bird, not back at home when memory starts to fade and you are unable to check uncertain details further. Ideally, you should take some notes and make a sketch while still with the bird. Even a very rough sketch is usually very helpful later when studying the book or consulting friends.

Now look at colours and patterns and ask yourself similar questions to those regarding the shape: *What are the main colours above and below? Are there any bright and colourful patches? What colours are the legs, eye and bill? Are there any light wing-bars or a supercilium above the eye? Is there any noteworthy pattern on throat, breast, back or crown? Is the rump concolorous with the back or perhaps contrastingly white? Does the tail have white sides or other pattern?*

When examining the plumage of a bird, remember that the *weather and type of light affect the appearance of the colours*. In sunshine at noon the colours of the upperparts certainly look bright, but finer nuances might be subdued or lost and the underparts are in darkest shadow. In overcast but bright weather the contrasts decrease, but on the other hand finer shades are more apparent. On dark and gloomy days, both the true colours and the patterns are difficult to see.

Finally, keep in mind that the voice is an important clue. To distinguish between the many similar species in such families as waders, larks, pipits and warblers, it is very helpful to compare what you hear with the voice descriptions in the book while still in the field. You can also try to memorize or note down your own version of the calls you have heard and compare these notes with recordings at home.

SWANS *Cygnus*

Very large, white waterfowl with long neck used for feeding in shallow waters. Heavily built, and laborious take-off requires running and strong flapping with wings. Gait clumsy, waddling. Food plants, mainly submerged. Nest on ground or on mound of vegetation in marsh or lake or on shore. Monogamous, pair for life. Gregarious in winter.

Mute Swan *Cygnus olor*   r**B**4

L 140–160 cm (body *c.* 80), WS 200–240 cm. Breeds on freshwater lakes, generally with reedbeds, and along coasts. Hardy, requires only open water in winter. Nest a large mound of reed stems etc., or, on coasts, a heap of seaweed. Not shy, and can act aggressively. ♂♂ have territorial fights with wing-splashing rushes and long 'slides' on the water.

IDENTIFICATION *Huge*. Plumage *white*. Neck very long, head small. Tail comparatively long and pointed (useful to know when a swan is upending). *Bill orange-red with prominent black knob* on forehead, black nostrils, cutting edges and nail. *Neck held* either fairly straight (much like Whooper) or, more typically, *smoothly bent in S-shape* when swimming. Threat posture with *wings raised like sails and head lowered* over back diagnostic. Sexes similar (♂ larger; bill-knob of ♂ larger, especially when breeding; colour of bill deeper orange on ♂). Difficult to separate from Whooper in flight at distance; at times, however, head looks more abruptly 'cut off' in front. – Juvenile: Grey-brown with pink tinge; bill first dark grey, then pinkish-grey, without knob. Separated from juvenile Whooper by *darker and browner plumage*. White plumage attained in 1st summer or 2nd autumn. A rare variety is born pure white ('Polish Swan').

VOICE Despite name, has some distinctive calls. Most commonly heard is an explosive, snorting or rumbling 'heeorr!'. Lone swimming birds seeking contact sometimes give rather loud, somewhat gull-like 'ga-oh'. Young and immatures have rather weak, piping 'büi-büi-büi-...'. Aggressive call a 'mean', snake-like hissing. *A loud singing or throbbing sound with each wingbeat in flight* is diagnostic.

Whooper Swan *Cygnus cygnus*   W3 / (m**B**5)

L 140–160 cm (body *c.* 75), WS 205–235 cm. Breeds on tundra pools and small lakes with sheltering vegetation, on damp bogs and in marshes, typically in remote areas in far north, but has spread south recently (linked with less shy, more confiding habits). Migrates to open waters in NW Europe. Hardy, closely following retreat of ice in spring.

IDENTIFICATION *Huge*. Plumage *white*, but many in spring and summer have *head and neck stained brown* (rarely seen in Mute). Neck long, upper part generally held straight when swimming. Best told from Mute and Tundra by *long, wedge-shaped bill, largely yellow* with black tip, yellow forming pointed wedge, reaching in front of nostril. – Juvenile: *Grey* (tinged brownish), clearly *less brown and slightly paler than juvenile Mute*; very similar to juvenile Tundra but (in British Isles) often slightly paler. *Bill pattern of adult discernible* (black of adult is pink, yellow is off-white).

VOICE Highly vocal. *Calls loud and bugling in quality*, similar to those of Tundra Swan but usually louder, slightly lower-pitched on average, *notes more straight*, not with such a marked diphthong, and notes often *given in groups of three or four*, 'kloo-kloo-kloo', instead of one or two as most often in Tundra. *Insignificant, slight hissing sound from wing-beats*, unlike musical throb of Mute.

Tundra Swan (Bewick's Swan)
Cygnus columbianus   W4

L 115–127 cm (body *c.* 60), WS 170–195 cm. Breeds on far NE tundras by the Arctic Ocean, winters in NW Europe.

IDENTIFICATION Very large. A smaller version of Whooper Swan, with *proportionately slightly shorter neck, bigger head and more compact body and bill*; these differences often difficult to appreciate when observing single birds. Safest separation by pattern of bill, Tundra having *slightly less yellow than black* (ssp. *bewickii*, Europe), the *yellow generally a square or rounded patch* (diagnostic), rarely more wedge-shaped (recalling Whooper, but yellow not reaching in front of nostril, and black around gape more solid). Also, with experience, voice is good clue (see below). – Variation: Very rarely N American vagrants (ssp. *columbianus*, 'WHISTLING SWAN') occur, distinguished by practically all-black bill (only a tiny yellow patch in front of eye). – Juvenile: Greyish (often a shade darker than Whooper). *Bill pattern of adult discernible* (black of adult is pink, yellow is off-white).

VOICE Vocal. Calls similar to those of Whooper Swan but *less bugling, more yapping or honking*, and on average *higher-pitched* (large overlap; flight-call rather deep) and softly 'bent' (diphthong impression). Also, notes not repeated in threes or fours, usually in twos, or given singly.

Snow Goose *Anser caerulescens*   V**

L 65–75 cm, WS 133–156 cm. Breeds in extreme NE Siberia (Wrangel Is) and N America. Genuine vagrant to Europe or, more often, escape from captivity.

IDENTIFICATION Of Pink-footed Goose size. Two morphs. – Adult: *Bill and feet reddish*. White morph *white with black primaries*; dark morph ('Blue Goose') has only *head, upper neck and tip of tail white, rest is* various shades of *grey*, palest on upper forewing and on tail-coverts. – Juvenile: *Bare parts dark grey*. Plumage pale greyish with whitish uppertail and darker flight-feathers (white morph) or rather uniform dark grey (dark morph). – Can be confused with smaller Ross's Goose *A. rossii* (p. 422), but apart from smaller size that has much smaller bill.

VOICE Has rather peculiar, soft, cackling notes with an upward-inflected diphthong, 'koeek'. Alarm a deep cackling 'angk-ak-ak-ak'.

Mute Swan

Whooper Swan

Tundra Swan

MUTE SWAN

1st-w.

1st-w.

strong
contrasts
MUTE

subdued pattern
WHOOPER/
TUNDRA

juv.

1st-winter (Feb)

long tail

orange-
red

ad.

WHOOPER SWAN

pair on breeding pool

juv.

1st-winter
(Feb)

Icelandic 1st-
winter birds ac-
quire a whiter
look more rapidly
than birds from
Fenno-Scandia
or Russia

juv.

characteristic
profile, longer
and straighter
than Tundra's

yellow
projects
to below
nostrils

ad.

TUNDRA SWAN

resting on field

juv. can look
surprisingly
long-billed

reduced pale
yellow

juv.

1st-winter (Feb)

shorter neck
than on
Whooper
often
apparent

juv.

amount of
yellow varies
somewhat

***columb-
ianus***
(N America)

less yel-
low on bill
than on
Whooper

ad.

bewickii
(Europe,
Siberia)

SNOW GOOSE

ad. Snow Goose among
Greenland White-
fronts

black primaries help
distinguish true
Snow Geese from
domestic lookalikes

ad.

ad.
('Blue Goose')

ad.

1st-winter

KM

GEESE *Anser, Branta*

Large, heavily built wildfowl, largely specialized on grazing, therefore have strong and fairly long feet centred under body to facilitate walking. Medium-long neck and strong, conical bill with saw-toothed edges of upper mandible. Gregarious, especially during migration and on winter grounds. Shy and wary, but can become remarkably confiding in areas where there is no shooting; in particular, juveniles from arctic breeding sites can be fearless in autumn. Migratory, following traditional pathways, travelling in family parties or large flocks, in flight forming V or undulating, bent line. – Two genera of geese occur, 'grey' (*Anser*) and 'black' (*Branta*). Food vegetable matter. Nest on ground, lined with down. Monogamous, pair for life. Both parents tend the young. Adults moult all flight-feathers simultaneously and become flightless for *c*. 3–4 weeks in summer, usually coinciding with young being unfledged.

IDENTIFYING 'GREY' GEESE

Although, in theory, separating the five species of 'grey' *Anser* geese in the field should present little problem, practice is different. Geese are normally shy birds which do not tolerate close approach, and all too often the interesting individuals are partly hidden by other geese in dense flocks, or by vegetation, or will have their legs stained by mud, or they simply lie down to rest before bill and leg colours are established. Even at moderate range, and against the light, bill and leg colours become surprisingly difficult to assess properly. The orange-red of the Bean Goose, e.g., often appears dull tomato-red at a distance or in overcast weather, and can sometimes even give a pinkish-red impression. Light conditions can be critical when it comes to goose identification.

Size variation is a potential pitfall. ♂♂ are larger than ♀♀, adults larger than juveniles. The extremes may confuse the unwary if they occur in mixed flocks, their size alone implying that they should belong to different species.

Most problems are caused by the trio *Bean Goose, Pinkfooted Goose* and *young White-fronted Goose*. With Bean, it is essential to realize amount of variation in bill (shape, pattern), size and proportions due to geographical variation within wide breeding range; at least two rather distinct races of Bean Goose, and possibly some rare varieties, involved. Young—and even some adult—Pink-footed Geese do not always 'look typical', and angle of light and previous experience are both important for correct identification. Some young White-fronted Geese have not only dark nail but also some dark smudges at the base of the bill and on culmen, and thus resemble Bean Goose at first glance.

Finally, it is important to age geese correctly, the basis for reliable identification. The ageing of 'grey' geese is similar for all species, and involves assessment of degree of neat grooves on the neck ('water-combing' effect), the width and shape at the tips of certain feathers, and the amount of distinct white tips to some of these; the colour of the nail on the bill is a further supportive character in most species. (See figures at lower left.)

(Greater) **White-fronted Goose** *Anser albifrons* **W**3

L 64–78 cm, WS 130–160 cm. Breeds on tundra in far northeast (ssp. *albifrons*), migrating to winter in W, C and SE Europe; birds breeding in Greenland (ssp. *flavirostris*; a separate species according to some) winter largely in Ireland and SW Scotland.

IDENTIFICATION Medium-sized, rather short-necked and compact. Legs orange-red. Adult has *prominent white blaze surrounding base of bill* and *black transverse markings on belly*, confusable only with Lesser White-fronted Goose, but is *larger* (marginal overlap!); *lacks prominent yellow orbital ring* (indistinct, narrow ring on some only); has *heavier bill*; *white blaze does not normally extend to forecrown*, *is fairly straight in side view* (blaze reaches forecrown on Lesser White-fronted, and outline is angled in side view). Ssp. *flavirostris* (see above) is slightly larger, longer-necked and heavier-billed, darker-plumaged (notably breast), with bill orange-yellow (pink restricted to tip), not reddish-pink with just a little yellow at base. – Juvenile: Lacks white blaze and dark belly markings, bill dull pinkish with *dark nail*. Confusable at distance with Bean Goose, but told by *darker grey feathering around base of bill and on forehead*, giving contrast to paler cheek; also, *pale feather edges on upperparts less prominent*, and *base of lower mandible not solidly black* as on Bean Goose.

VOICE Compared with Bean and Greylag Geese, calls are more high-pitched and musical or laughing in quality, not so nasal and raw. Usual call is disyllabic (sometimes tri-), 'kyü-yü' (or 'kyu-yu-yu'). Some calls are deeper, more like those of the larger species.

Lesser White-fronted Goose *Anser erythropus* **V****

L 56–66 cm, WS 115–135 cm. Breeds on marshes and bogs in boreal mountains, generally in willow and birch zone. Has declined, now very rare in Europe. Reintroduction in Fenno-Scandia, (colour-ringed) young programmed to migrate towards southwest in winter (where biocide levels and shooting pressure estimated to be more favourable), having Barnacle Geese as foster parents.

IDENTIFICATION Slightly smaller than similar White-fronted (which see), but marginal overlap. Best told by *promi-*

AGEING OF 'GREY' GEESE

White-fronted Goose   Lesser White-fronted Goose

WHITE-FRONTED GOOSE

dark

BEAN, ad.

note that some
Bean Geese
have white rim
along bill-base

albifrons

juv.

white blaze develops
from end of 1st winter

1st-w.

pinkish

wing-bars not
so marked as
on Bean Goose

albifrons

albifrons

unmarked

ad.

dark
light

albifrons

albifrons

juv.

broad
white

GREYLAG
(to compare)

albifrons
(NE Europe,
N Siberia)

black
patches

long, predominantly orange
(but can look pinkish!)

dark

rather plain

juv.

dark

ad.

light edges

flavirostris

narrow
white

flavirostris
(Greenland)

bold black
patches

resting White-fronted (upper)
and Bean Goose in winter

LESSER WHITE-FRONTED GOOSE

juv. is darker and more uniform
than White-fronted; yellow orbital ring visible at close range

WHITE-
FRONTED
albifrons

note that some White-fronted
Geese have white reaching
forecrown, and may even have
a *narrow* yellow orbital ring

white blaze
reaches crown

juv.

prominent
yellow
orbital
ring in all
plumages

small,
pink

very similar to
White-fronted
Goose in flight; has
slightly shorter
neck and smaller
head and bill

ad.

long-winged,
wing-tips project
beyond tail

usually rather
few belly
markings

three Lesser White-fronted (foreground) and two
White-fronted Geese

K M

nent yellow orbital ring (also on juv.); *white blaze reaching forecrown, white area angled in side view; bill small* ('cute-looking'), always pink; plumage rather *dark*, but *black belly markings fewer*. Wings narrow, flight agile, wingbeats fast. Feet orange-red. – Juvenile: Resembles juvenile White-fronted, but is on average smaller and darker over-all, and nail is usually pale. Still, often inseparable but for presence of *yellow orbital ring*.

VOICE Similar to calls of White-fronted Goose, but even more high-pitched and yelping in character.

Bean Goose *Anser fabalis* **W**4

L 69–88 cm, WS 140–174 cm. Breeds on bogs, marshes and pools in remote taiga (TAIGA BEAN GOOSE, ssp. *fabalis*) or on wet tundra (TUNDRA BEAN GOOSE, *rossicus*), both wintering in W and C Europe. Migrants passing S Fenno-Scandia 2nd half Apr and Sep/Oct. Huge roosting and wintering flocks at favoured places in S Sweden and N Continental Europe; scarce in Britain. Shy and wary.

IDENTIFICATION *Medium-large to large*, rather *dark*, long-winged. Distinct white edges to dark tertials and upper-wing-coverts. Head and neck rather dark. Differs from Grey-lag and Pink-footed in *orange* (or dull red) legs, not pink; also, *in flight upperwing rather dark*, not strikingly pale, and *underwing all dark* (cf. Greylag). Sometimes difficult to separate from Pink-footed, especially at a distance, when orange of bill and legs often looks indeterminably dull red; *back just as dark as flank area* (paler on Pink-footed) helpful, as is narrower white terminal tail-band of Bean. Many have narrow white rim at base of bill; those with most can resemble White-fronted Goose (which see). – Ecologically differentiated subspecies *fabalis* and *rossicus* differ slightly but distinctly. Ssp. *fabalis* about as large as Greylag but has *narrower neck, smaller head, slimmer bill and body*; bill is usually *long*, and base not too deep; amount of orange (or dull red) variable, some with dark restricted to base of lower mandible, tip and a little on culmen, others with a lot of dark at base of bill resembling *rossicus* and Pink-footed. Ssp. *rossicus* is slightly *smaller*, has *shorter neck, darker head and neck* which contrast more markedly with paler body than in *fabalis*, all characters which make it more similar to Pink-footed; bill is short, deep-based and triangular, and *pale* (reddish) *patch almost invariably small*, and base of lower mandible is deeper and more curved. In large flocks of N Europe some birds appear to be intermediates.

VOICE Commonest call is a deep, nasal, trumpeting, disyllabic or trisyllabic, jolting 'ung-unk' or 'yak-ak-ak', at slightly varying pitch when heard from flocks. With practice discernibly deeper in pitch than Pink-footed (otherwise similar).

Pink-footed Goose *Anser brachyrhynchus* **W**2

L 64–76 cm, WS 137–161 cm. Closely related to Bean Goose. Breeds on arctic tundra and mountainsides in Greenland, Iceland and Svalbard, winters NW Europe. Nests on the ground, at times on rocks.

IDENTIFICATION Slightly smaller and more compact than Taiga Bean Goose (*fabalis*), about equal in size and proportions to Tundra Bean (*rossicus*). *Legs pink* (diagnostic), but surprisingly difficult to determine at some distance or in poor light. *Bill* rather *short and triangular*, typically mostly

black with only a pink band across outer part (some have a little more pink). Head rounded, *neck rather short. Head* (and upper neck) *dark* brown-grey, *contrasting with pale* (lower) *neck and body*. A very few have a thin white rim at base of bill (just like many Bean and some Greylag). Lower neck and body pale brown-grey, upperparts often with a 'frosty' blue-grey cast, breast with pinkish-buff suffusion. *Flank area* darkest part of body, *darker than back. In flight, upperwing and back very pale*, distinctly lighter than on Bean and White-fronted. (Greylag has even lighter forewing, but on the other hand darker back.) Also useful to know that Pink-footed has a *wider white trailing edge to tail* than Bean. – Juvenile: For ageing, see p. 16. Resembles adult, but at least some birds are less distinctive, being a little darker and browner, with less contrast between head and body, and lacking blue-grey cast above; leg colour duller and less clean pink than on adult.

VOICE Common calls like Bean Goose but slightly higher-pitched on average, often discern-ibly so with experience. Differ from calls of White-fronted Goose in lower pitch and lack of laughing quality. Softer 'wink-wink' also heard.

Greylag Goose *Anser anser* r**B**3 / **W**3

L 74–84 cm, WS 149–168 cm. Breeds in variety of habitats, mainly wetlands, from shallow lakes with reedbeds and freshwater marshes to islets in larger lakes, coasts (even brackish water), heather, rocky slopes, etc. British breeders mainly resident, most others migratory. Increasing in numbers. Only grey goose to be seen in large numbers in summer in Europe.

IDENTIFICATION Large, equalled in size only by largest Taiga Bean Goose (which see), but *bulkier*, with *thicker neck, larger head and heavier bill*, latter being *all pinkish-orange or pink*. Legs dull pinkish. Wings broad, flight heavy. Plumage rather *plain brown-grey* without strong contrasts, *head and neck* typically *rather pale*. A few have insignificant thin white rim at base of bill, and many have some dark marks on belly. In flight, *upper forewing strikingly pale ash-grey*, contrasting sharply with darker rear parts of wing and with darker back. *Underwing* characteristic, too, otherwise dark wing *having pale grey leading edge* distinctly set off, thus two-coloured, unique among European geese. Greylags flying away from observer can be recognized by size and *pale grey rump*, contrasting with dark brown tertials and back. – Variation: Western birds (ssp. *anser*, most of Europe) comparatively darker, smaller and with more orange tinge to bill; eastern (*rubrirostris*, Russia, Asia) paler, larger and have pinkish bill; E European breeders appear to be intermediates.

VOICE Loud, raw, nasal cackling calls, most characteristic trisyllabic with first higher-pitched (falsetto-like), drawn out (almost disyllabic) and more stressed than following two, 'ki**yaaa**-ga-ga'. Commonly also a similar deep, raw 'ahnk-ang-ang' (thus somewhat more like Bean Goose). Repertoire varied: some calls are deep, others shrill, all being similar to those of domestic goose.

Bean Goose   Pink-footed Goose   Greylag Goose

BEAN GOOSE

note that ssp. *rossicus* resembles Pink-footed Goose, but head and bill distinctly heavier, and feet orange

small orange patch

dark

short

rather short, deep-based

ad.

some *fabalis* have *rossicus*-like bill pattern

white rim on some

long

long

rather dark

distinct white edges

dark

narrow white tip

uniformly dark

long

'TUNDRA BEAN GOOSE' *rossicus* (NW Siberia)

fabalis

dull orange

juv.

ad.

'TAIGA BEAN GOOSE' *fabalis* (N Europe)

grazing family party

PINK-FOOTED GOOSE

some ads. are browner, not so 'frosty' blue-grey above

pinkish-cinnamon tinge

ad.

white rim on some

dark

pale, 'frosty' blue-grey

ad.

darker than back

pink

juv.

pink band

short, stubby

broad white tip

pale blue-grey

pale

uniformly dark

short

GREYLAG GOOSE

paler grey

pinkish

broad pale fringes

rubrirostris (Siberia)

ad.

large pale head

thick

heavy, all pale

very pale

dark

dark

pale

broad white tip

ad.

anser

juv.

dull pinkish

contrastingly pale coverts

anser (Europe)

KM

Canada Goose *Branta canadensis* ⌐**B**3

L 80–105 cm, WS 155–180 cm (referring to feral European birds, mainly ssp. *canadensis*). Breeds on lakes and marshes, along rivers and seashores. Introduced American species; in many areas half-tame or at least not shy. Northern birds migratory. Other races than nominate, some of which are smaller (e.g. *parvipes*), occur rarely as vagrants.

IDENTIFICATION *Large, with very long neck. Black neck and head* with *white patch on head* separates from all except Cackling and Barnacle Geese, but is much larger, longer-necked, and *breast is pale*, not black as in Barnacle; *body brownish*, paler below (Barnacle black, dark grey and white); *the white on head forms 'throat-strap'* (more white on 'face' on Barnacle). For differences from Cackling, see below. Wings long, wingbeats rather slow. – Adult: Buff-white tips on upperparts and flanks form neat bars; head/neck glossy black; 'throat-strap' pure white. – Juvenile: Pale bars on upperparts and flanks less even and well marked; head/neck duller brown-black; white 'throat-strap' tinged pale brown.

VOICE Vocal. Most characteristic call loud, disyllabic, nasal 'awr-**lüt**', second syllable higher-pitched. Variety of other deep, nasal, honking calls, often repeated.

Cackling Goose *Branta hutchinsii* **V***∗∗

L 60–70 cm, WS 125–140 cm. Recently split from Canada Goose. Breeds in Arctic Canada and Alaska on tundra. Rare vagrant to W Europe, apparently only involving ssp. *hutchinsii* from C Canada, in Europe often occurring with Barnacle and Pink-footed Geese.

IDENTIFICATION About *as large as Barnacle Goose.* Compared to Canada Goose a much smaller and *more compact* bird with *much shorter neck* and *shorter legs*, a rather *square head* with *steep forehead* and *small bill* giving similar 'cute' impression as Lesser White-fronted Goose. Often a hint of a whitish neck-collar between black neck and pale brown chest, but this may be missing. *Rather pale brownish-grey above*, paler than average Canada Goose. Ages differ as in Canada Goose.

VOICE Similar to Canada Goose, only more high-pitched.

Barnacle Goose *Branta leucopsis* **W**3

L 58–70 cm, WS 120–142 cm. Breeds colonially, mainly on arctic islands and coasts, preferring rocky coasts and steep slopes to flat tundra. Also, since mid 1970s, on low grassy islands and shores in the Baltic. Migratory, main movements in Oct–early Nov and Apr–May.

IDENTIFICATION Medium-sized, fairly compact, with *thick-set, short neck, rounded head* and *small black bill*. Neck and breast black, *head mostly white*, underparts silvery-white, upperparts grey and barred black and white. In flight, white head surprisingly difficult to see at a distance; *strong contrast between black breast and whitish belly* better character to separate from Brent Goose ssp. *bernicla*; slightly *paler upperwing* sometimes useful clue, too. Wings slightly longer than on Brent, and wingbeat rate roughly as White-fronted, slightly slower than Brent. Flock formation often irregular U-shaped line (like Brent). – Juvenile: Very similar to adult, differing mainly in *less well-marked dark barring on flanks*, and not such solidly black neck and breast; pale tips on upperparts less clean white than on adult, and black and grey

steep fore-head

small bill

almost all 'legitimate' records are of ones and twos that arrive in western extremities of Europe with Barnacles from Greenland

hutchinsii
(also known as 'Richardson's Goose')

ad.

more diffuse. Whitish 'face' on average less tinged yellow than on adult; amount of dark marks around eye varies, however, and is unreliable age character.

VOICE Vocal. Basically only one call, a shrill, monosyllabic bark, 'ka' or 'kaw', at slightly varying pitch when heard in chorus.

Brent Goose *Branta bernicla* **W**2

L 55–62 cm, WS 105–117 cm. Breeds on arctic islands and coasts, preferring low tundra near sea coasts. Migratory, main movements late Sep–Oct and late May, largely keeping to traditional routes; often concentrated passage in spring. Habits duck-like, often resting on open sea (dense flocks) and feeding in shallow waters, upending to reach main food of eel-grass; also grazes on mudflats and grassy fields.

IDENTIFICATION Somewhat smaller than Barnacle Goose, *slimmer-bodied* with slightly longer neck, *smaller head* and *narrower wings*. *Darkish* with *bright white stern. Small white crescent on side of upper neck* visible at close range. In flight appears slim and elegant, with pointed wings swept back a little; wingbeats slower than Eider but somewhat quicker than Barnacle Goose. Migrating flocks often in lines (esp. in spring), flock formation disorderly U-shaped lines or muddle of lines. – Variation: Ssp. *bernicla* (Russia, W Siberia, migrates through Baltic to NW Europe) has *dusky dark grey belly* and slightly paler flanks and dark grey upperparts; ssp. *hrota* (Svalbard, Greenland, passing Iceland and Norway to winter Denmark, England and Ireland) has much *paler underparts*, pale grey-white with rather *clear contrast against black breast*, and buff-tinged upperparts; ssp. *nigricans*, 'BLACK BRANT' (E Siberia, Alaska, NW Canada, rare vagrant; sometimes claimed to be a separate species), has strong contrast between whitish flanks and very dark belly, and the white crescents on neck are large and widest on foreneck (often merging at front). –Adult: Upperparts *plain* dark brown-grey. – Juvenile: *Upperwing-coverts tipped off-white*, giving barred appearance; white neck-crescents lacking at first, developed (late Sep) Oct–Dec. In all three races, juveniles have darker flanks (uniform with belly) than adults, making subspecific identification difficult or even impossible. – 1st-winter: Retains white bars on upperparts. From Oct, paler flanks of ssp. *hrota* develop, facilitating separation from *bernicla*.

VOICE Call is a gargling, guttural 'r'rot' or 'rhut', at slightly varying pitch from flocks.

CANADA GOOSE

ssp. *parvipes* can be very similar to *hutchinsii* race of Cackling Goose, but note shape of head and bill

small, rounded feathers in juv.

juv.

ad.

parvipes – smallest race of Canada Goose, known also as 'Lesser Canada Goose'

light breast

long neck

ad.

BARNACLE GOOSE

light grey

some are tinged yellow on head

diffuse pattern

juv.

neatly barred

ad.

strong contrast

black breast

ad.

BRENT GOOSE

hrota 1st-winter

bernicla ad.

white bars

white

hrota (Svalbard, Greenland) **ad.**

blackish

broad neck patch

dark

hrota **juv.**

hrota 1st-winter

plain grey

bernicla

hrota

bernicla 1st-winter

dark

bernicla (Russia) **ad.**

'BLACK BRANT' *nigricans* (N America) **ad.**

bernicla ad.

KM

Canada Goose

Barnacle Goose

Brent Goose

Red-breasted Goose *Branta ruficollis* V**

L 54–60 cm, WS 110–125 cm. Breeds in small colonies on arctic tundra near coasts or along river mouths, often together with birds of prey as protection against Arctic Fox and other mammal predators. Migratory, spending winter on plains N and W of Black Sea, with decent numbers still occurring in SE Romania and NE Bulgaria Nov–Mar. A very few accompany congeners to winter in W Europe.

IDENTIFICATION A trifle smaller than Brent Goose. *Neck short and thick*, head rounded and *bill very small*. Plumage striking, *deep chestnut-red, black and white*, unmistakable at close range, but at a distance surprisingly dull-looking; then appears *dark* with *broad white flank-stripe*. In flight, note small size, shortish neck, *all-black underwing* and very dark upperwing, and black belly with broad white flank stripe through 'armpit'. – Adult: Only *two* (distinct) *white bars* (tips to wing-coverts) *on closed wing*. Red *cheek patch large, leaving narrow white border*. – Juvenile: *4–5 narrow white bars on closed wing*. (Dull) red *cheek patch small* (virtually absent on some), *leaving broad white border*.

VOICE A shrill, high-pitched 'ki-kwi' or 'kik-yik'.

(Common) Shelduck *Tadorna tadorna* rB3

L 55–65 cm, WS 100–120 cm. Breeds along seashores, at larger lakes and rivers, preferring open, unvegetated areas. Feeds on shallow water, grassy shores and arable fields, latter sometimes far from water. Migratory in N and E. Birds gather in summer from a large part of Europe in huge flocks on German North Sea coast (Waddensee), with smaller parties elsewhere, to moult flight-feathers collectively. Nests in burrow or under dense bush or building.

IDENTIFICATION Medium-sized, goose-like, *long-necked* duck with boldly patterned plumage. Body plump, bill strong with markedly concave culmen, and legs rather long. Wings long and rather pointed, arched in normal flight, wingbeats deep when accelerating. When migrating, usually forms rather small flocks, flying in line low over surface. Plumage largely *white with dark green head, red bill*, a *broad rust-brown belt across breast*, and black scapulars, flight-feathers, tip of tail and stripe on centre of belly. Legs of adults are dull pink, of juvenile greyish. – Adult ♂: *Bright red bill* with *prominent knob*, especially when breeding. Larger general size of ♂ apparent when pair seen together. – Adult ♀: *Bill* often *duller red* with *small knob*. Rust-brown and black marks on chest/belly narrower and less neat. – Juvenile: Forehead, 'face', chin, throat and foreneck white, crown and hindneck brown-grey. No rust-brown chest-band or black stripe on belly. – 1st-summer: Like adult, but distinguished in flight by *white trailing edge of wing*.

VOICE Vocal. In spring courtship, ♂ utters a series of high, whizzing whistles, 'sliss-sliss-sliss-...'. The whizzing sound, often uttered in flight when pursuing ♀, is frequently accompanied by short soft whistles, 'pyu'; also a repeated disyllabic 'pyu-pu'. ♀ has a strong, straight, somewhat nasal, whinnying 'gagagagagaga...', often in flight. Also a disyllabic, growling 'ah-ank', used as alarm, and raucous rolling 'grrah grrah grrah...'.

Ruddy Shelduck *Tadorna ferruginea* V***

L 58–70 cm, WS 110–135 cm. Breeds in variety of inland habitats: on vast steppes at shores of lakes or saltmarshes, along rivers, in hills and even on barren, rocky mountainsides, at times far from nearest water. Nests in hole in cliff, bank, tree or ruin. Largely migratory. Often seen flying at considerable height. Records in W and N Europe mainly involve escapes from captivity, since it is a commonly kept bird in wildfowl collections and parks, but influxes of wild birds sometimes occur (e.g. in 1994).

IDENTIFICATION Roughly the same size as Shelduck (marginally larger), and has similar proportions, with fairly long neck and legs, long and rather narrow wings. *Body bright orange-brown*, head paler cinnamon-buff or creamy-white, especially pale on forehead and 'face'. Rump, tail and flight-feathers black, partly with greenish gloss. *Forewing* (upper and under) *pure white*. Bill black. – Adult ♂: Narrow *black neck-collar*. Slightly darker orange-brown on mantle and breast. – Adult ♀: No neck-collar. Tendency to have more clearly set-off white 'face-mask'. – Juvenile: Resembles ♀, but has grey wash on white of forewing.

VOICE Typical is strongly nasal, 'honking' call. In flight, loud trumpeting 'ang' and disyllabic 'ah-üng' (latter recalling Canada Goose, and—more valid considering range—distant donkey). Hollow, rolling 'ahrrrr' also heard.

Egyptian Goose *Alopochen aegyptiaca* rB4

L 63–73 cm. African species. Feral resident population in S England (introduced in 18th century), Netherlands, France, etc. Found in parkland with rivers, lakes or marshes.

IDENTIFICATION Somewhat larger than Ruddy Shelduck, being similarly *pale brown with large white wing-panels*; told by stocky build, less orange tinge on body, *dark eye-surround*, and, at closer range, *dull pink-red bill* with dark outline. *Long-legged*, legs being dull pinkish-red. Dark brown breast patch. Sexes similar. Rather extensive individual variation. – Juvenile: Duller, lacking dark brown patch on breast and around eye, crown being dull brown, not whitish.

Bar-headed Goose *Anser indicus* [V**]

L 68–78 cm. Central Asian high-altitude species, but escaped feral birds may appear in Europe.

IDENTIFICATION The size of Bean Goose and *very pale* grey (in flight largely whitish-looking). At close range shows distinctive *white head with two black cross-bars*. Hind neck dark grey. Outer parts of flight-feathers dark. Bill and legs yellow, bill comparatively small.

Red-breasted Goose

Shelduck

Ruddy Shelduck

RED-BREASTED GOOSE

surprisingly inconspicuous in flock of Barnacles

narrow, indistinct bars

juv.

two prominent white bars

ad.

wide white flank-stripe

migrating with Brent Geese

fat-necked

ad.

SHELDUCK

bill-knob

ad. ♂

white trailing edge on young

juv.

juv.

ad. ♀

ad. ♂

RUDDY SHELDUCK

ad. summer ♂

♂ loses black neck-ring in winter

juv.

juv.

ad. ♀

ad.

EGYPTIAN GOOSE

dark eye-surround

sexes similar

long, dull red

ad.

BAR-HEADED GOOSE

lacks cross-bars

juv.

ad.

very pale plumage

ad.

KM

DABBLING DUCKS *Anatini*

Surface-feeding small to large-sized ducks of lakes, rivers and shallow water, often upending. Take off from water readily and without prior running. Sexes markedly dissimilar in most. ♂♂ adopt ♀-like, cryptic so-called eclipse plumage in summer during simultaneous moult of flight-feathers, making adults flightless for 3–4 weeks.

Mallard *Anas platyrhynchos* r**B**2 / **P**+**W**2
L 50–60 cm, WS 81–95 cm. Most familiar duck, and ancestor of domestic duck. Breeds in parks, by canals in towns, on eutrophic lakes, woodland marshes, seashores; accepts very small waters (tiny pools, ditches, etc.). Resident in much of Europe except in N and E, where it retreats from the ice in winter. Wide range of nest sites: under bush, tree-hole, artificial nest-basket, near or on buildings.
IDENTIFICATION Large, with stocky build. Head and bill large, tail short. In flight, looks *heavy, wings rather blunt-tipped and broad*, especially at base, wingbeats moderately quick. Legs orange. Often told by size, shape, and *dark blue speculum very prominently bordered white*. – Adult ♂ breeding: *Head metallic green, narrow white neck-collar*, breast purplish-brown, rest of body largely pale grey, stern black, central tail-feathers upcurled; the *bill is uniform dull yellow*. – Adult ♀: Streaked brown, crown and eye-stripe darker, leaving pale supercilium; *bill orange, with blackish culmen* irregularly invading sides of bill at centre (cf. Gadwall); outer tail-feathers dusky buff-white; in flight, note *medium dark belly* and whitish underwing-coverts. – Adult ♂ eclipse: Like ♀, but *bill uniform yellow*, breast tinged rufous and less well marked; head slightly more contrasting. – Juvenile: Very similar to adult ♀.
VOICE Vocal. ♂ has soft, nasal, low 'rhaeb', often repeated when alert on water. Gives a short whistle, 'piu', when courting ♀. Quacking of ♀ loud, a series of hoarse notes, the first one or two being stressed, the following dying off, '**quaek-quaek**-quak-quak-quah-qua-...'; in anxiety, can give prolonged slower series of hard quacks at even pitch.

Gadwall *Anas strepera* r**B**4 / **W**3
L 46–56 cm, WS 78–90 cm. Breeds on variety of fresh (and rarely brackish) waters, mostly on eutrophic lakes or bays with reedbeds and wooded islets. Local breeder in Britain, partly introduced, increasing.
IDENTIFICATION Slightly smaller than Mallard, and of *slimmer build* with *narrower wings* and body. In flight, note *whitish belly* (Mallard pale brownish). Main feature is *small white speculum*, but note that this is prominent only on adult ♂ ('lump of sugar'), and practically nonexistent on juvenile ♀. – Adult ♂ breeding: Head medium brown; *stern*

black (but tail pale brown), rest of *body grey, finely vermiculated*, boldest on breast; *long scapulars pale grey-buff; bill grey-black*. In flight, *prominent white speculum, and black and chestnut on upperwing*. – Adult ♀: Resembles Mallard, but has *whitish belly, small white speculum*, different bill pattern with *orange even stripe along cutting edges*, and *darker brown-grey tail-feathers*. – Adult ♂ eclipse: As ♂ but retains adult ♂ wing; bill with some orange, recalling ♀. – Juvenile: As adult ♀ but body colour brighter buff, contrasting with greyer head.
VOICE ♂ has a low, short, croaking 'ahrk', and high-pitched whistled 'pee' in courtship. ♀ has quacking call like Mallard, but often somewhat harder and more mechanical.

(Northern) Pintail *Anas acuta* r+m**B**5 / **W**3
L 51–62 cm (excl. elongated tail-feathers of ♂ *c.* 10), WS 79–87 cm. Breeds on lakes in lowland and steppe, also mountain and tundra pools. Local breeder in Britain. Partly migratory.
IDENTIFICATION Nearly Mallard-sized but *much more slim and elegant*. Wings long, *narrow and pointed*; tail long and *pointed*, on ♂ markedly so; neck long and narrow; head small; a greyhound among ducks! Flight fast, outer wings swept back a little, long neck and tail obvious. – Adult ♂ breeding: Unmistakable, with long central tail-feathers (cf. Long-tailed Duck); head and upper neck brown, *breast and lower neck white, extending as narrow stripe into brown of head* on each side. Speculum blackish-green, bordered white at rear and pale rufous at front. – Adult ♀: Like a slim ♀ Mallard, scalloped and mottled brown, but: *slim bill dark grey; head almost uniformly brownish,contrasting with rather grey body with more coarse, patchy marks*; and speculum dark brown, at front thinly (hardly visible) but *at rear prominently bordered white* (visible at a mile!). Belly pale brownish-white. – Adult ♂ eclipse: As adult ♀ but retains upperwing of adult ♂, and scapulars are longer and greyer. – Juvenile: As adult ♀ but less neatly scalloped on body, more diffusely spotted and barred. White trailing edge of secondaries less broad, speculum duller brown, tinged greenish on ♂♂.
VOICE ♂ has a short, clear whistle, 'krrü', recalling Teal but lower-pitched; at close range a simultaneous whizzing 'wee-weey' can be heard. During courtship, ♀ has a deep crowing 'cr-r-r-rah', almost like a quarrelling Carrion Crow. Quacking of ♀ intermediate between Mallard and Teal.

(American) Black Duck *Anas rubripes* **V****
L 53–61 cm. American species, proved to be a genuine vagrant to Europe; still, some records could be escapes.
IDENTIFICATION Mallard-sized, and plumage much as very dark ♀ Mallard (beware of melanistic such!). Sexes and ages similar. Head pale brown-grey, contrasting with *very dark* brown-grey and almost unmarked body. *Tail dark; wing speculum bluish-purple* (or turquoise), *bordered narrowly and indistinctly off-white* at rear. *Bill yellowish* on both sexes, ♀ with only a hint of darker culmen.

Mallard

Gadwall

Pintail

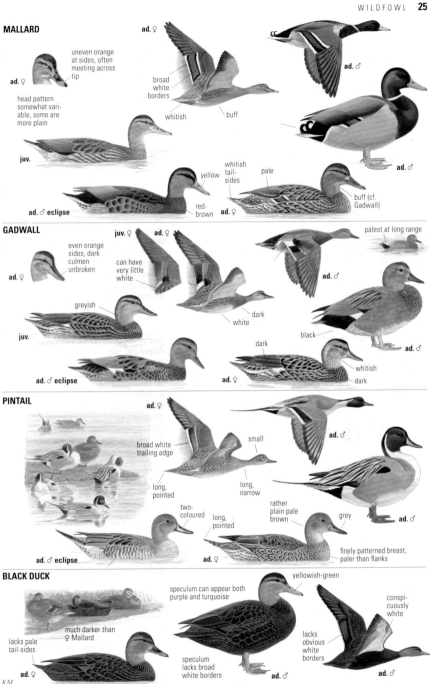

MALLARD

ad. ♀

ad. ♀

uneven orange at sides, often meeting across tip

broad white borders

head pattern somewhat variable, some are more plain

whitish

buff

ad. ♂

juv.

ad. ♂

whitish tail-sides

pale

yellow

ad. ♂ eclipse

red-brown

ad. ♀

buff (cf. Gadwall)

ad. ♀

GADWALL

ad. ♀

even orange sides, dark culmen unbroken

juv. ♀   ad. ♀

palest at long range

can have very little white

greyish

dark

white

ad. ♂

juv.

black

dark

ad. ♂

ad. ♂ eclipse

ad. ♀

whitish

dark

PINTAIL

ad. ♀

broad white trailing edge

small

ad. ♂

long, pointed

long, narrow

rather plain pale brown

grey

two-coloured

long, pointed

ad. ♂

ad. ♂ eclipse

ad. ♀

finely patterned breast, paler than flanks

BLACK DUCK

yellowish-green

speculum can appear both purple and turquoise

conspicuously white

much darker than ♀ Mallard

lacks obvious white borders

lacks pale tail-sides

speculum lacks broad white borders

ad. ♀

ad. ♂

ad. ♂

KM

(Northern) **Shoveler** *Anas clypeata*   m(r)**B**4 / **W**3

L 44–52 cm, WS 73–82 cm. Breeds on shallow eutrophic lakes with rich vegetation, or marshes with sufficient open water. Largely migratory. Although fairly common and widespread, usually seen in pairs or smaller parties only.

IDENTIFICATION Medium-large duck with short neck but *strikingly long and broad bill*, giving *front-heavy* look both when swimming and in flight. Wings rather evenly broad, but appearing slightly narrower than on Mallard. – Adult ♂ breeding: Unmistakable; *head green* (appearing black at distance), *breast white, belly and flanks vividly chestnut*. Upper forewing pale blue, speculum green, bordered white at front. – Adult ♀: Resembles Mallard except for *huge bill*, and *belly dark brown, contrasting with white underwings*; speculum dull green-grey *without white trailing edge* (white bar at front only, being narrower towards body), and upper forewing dull pale grey. – Adult ♂ eclipse: As adult ♀, but especially flanks and belly more rufous-tinged, head darker, and pale blue upper forewing retained. – Juvenile: As adult ♀, but crown and hindneck darker, and belly slightly paler.

VOICE Most often heard is disyllabic, nasal knocking call of ♂ when flushed, 'took-took...took-took...' (sometimes monosyllabic). ♀ has a similar call but wheezy or muffled, 'kerr-aesh'; also a short, hoarse quacking, falling in strength and pitch, like a coarse-voiced ♀ Teal.

(Eurasian) **Wigeon** *Anas penelope*   r(m)**B**4 / **W**2

L 42–50 cm, WS 71–85 cm. Breeds on boreal forest lakes and marshes, tundra pools, also on temperate shallow fresh waters. Migratory in N, wintering largely in marine habitats, also lakes and reservoirs. Gregarious except when nesting. Vegetarian; often grazes on arable fields in winter.

IDENTIFICATION Medium-sized, *short-necked*, with *rounded, comparatively large head, small bill* and *pointed tail*. In flight, these characters apparent as well as *narrow neck, pointed wings*, and outer wing generally swept back. Flight fast. In all plumages, note distinct white belly patch. Axillaries and underwing-coverts dusky pale grey (cf. American Wigeon). – Adult ♂ breeding: Unmistakable; *head and neck chestnut, forehead and crown creamy-yellow*, breast greyish-pink, rest of body grey with white and black stern. In flight, *large white patch on upper forewing* striking, speculum green. – Adult ♀: Rather dull rufous-brown or greyish (some variation, with inclination towards two morphs) with variable diffuse mottling and blotching (some are rather plain), best recognized by: *rather dark plumage; shape; small, pale blue-grey bill with black tip*; and *contrasting white belly*. Speculum dull, dark; innermost secondaries usually edged pale, forming narrow whitish patch (recalling 1st-year ♀ Gadwall). – Adult ♂ eclipse: Resembles adult ♀ but *retains white upper*

forewing, and plumage more *rufous-tinged*. – Juvenile: As adult ♀, but white underparts often faintly mottled, and speculum still duller. – 1st-year ♂: As adult ♂ breeding, but *upper forewing is chiefly grey-brown*, not largely pure white.

VOICE ♂ has characteristic *loud whistling glissando note*, often preceded by low, brief note (and ending with subdued dry trill), 'wu, **wee**-oo(rr)'; at a distance only '**wee**-oo' audible. Also more subdued, conversational 'wip... wee... wip-wü', etc., somewhat recalling ♂ Tufted Duck. In autumn, and at times at other seasons as well, coarse, snorting 'rrah', 'ra-**kaah**' and variants are heard. Flight-call of ♀ is a repeated growling, 'karr karr karr...', recalling Great Crested Grebe in tone.

American Wigeon *Anas americana*   **V**∗

L 48–56 cm. American species; rare straggler to Europe. About 10–20 records annually in Britain.

IDENTIFICATION Roughly of same size as Wigeon; shape much the same, with rounded head and pointed tail, but on average head appears a fraction larger with steeper forehead and fuller nape. – Adult ♂ breeding: *Forehead and centre of crown white, broad stripe over eye and to nape dark green* (looking blackish at distance); rest of head freckled grey and white; both breast and flanks pinkish-brown. – All other plumages: Very similar to Wigeon and at times difficult to distinguish. Note: On average *greyer head and neck* with much more *marked contrast with rufous-brown breast. Axillaries and median underwing-coverts white* (axillaries at the most with faint dusky shafts and indistinct subterminal bars on some), whereas these are pale greyish with brownish-grey shafts and patterning on most Wigeons (may *appear* whitish on Wigeon in strong light, though, and a few *do have* paler underwing). Paler forehead and crown and darker mottling around eye give impression of *dark eye patch* at distance (only a hint of this on some Wigeons). White bases to greater upperwing-coverts on adult ♀ give *pale bar along midwing*.

Marbled Duck *Marmaronetta angustirostris*   —

L 39–42 cm, WS 63–70 cm. Breeds rarely and locally (declining; e.g. now probably <100 pairs left in Spain) in lowland, shallow, well-vegetated fresh waters. Lives much as a dabbling duck, but also shows some affinities to the 'pochards' (smaller diving ducks).

IDENTIFICATION Rather small and slim with long wings and tail and *slender dark bill* (recalling ♀ Pintail), head appearing fairly large and oblong owing to crest-like feathers at nape. *Plumage pale sandy-brown*, diffusely blotched off-white; *eye-surround dusky*. In flight, *wings look pale without marked pattern* (does not have a speculum), but *secondaries palest*, and leading edge shows two pale spots in carpal area when seen head-on, much as on Grey Heron. Sexes and ages similar, ♂♂ tending to have more crest and blacker bill, juvs. having more diffuse and buff-white blotches.

VOICE Usually silent. During courtship, a high-pitched and squeaky whistle, 'veeveeh', is given by ♂.

Shoveler   Wigeon   Marbled Duck

SHOVELER

MALLARD ♀

SHOVE-
LER ♀

can look
black

no white
edge

grey

paler

dark

resting pair

light eye

ad. ♀

ad. ♀

ad. ♂

juv.

ad. ♂ eclipse
(late summer)

ad. ♂

WIGEON

imm. ♂

rounded
head,
short bill

large
white
patch

♀ Wigeon (centre) with ♀ Pintail
(left) and ♀ Gadwall (right)

imm. ♀

1st-years have
subdued whit-
ish tips to
wing-coverts

pointed

white

ad. ♀

ad. ♂

white

buff

ad. ♀

ad. ♂ eclipse

ad. ♂

AMERICAN WIGEON

beware:
'armpits' can
look whitish
on Wigeon!

broad
whitish bar

dark eye patch
often stands out

WIGEON ♀

♂ American Wigeon
among Wigeons

AMERICAN ♀

white axillaries
of American often
visible during
wing-flap

ad. ♀

ad. ♂

white

more grey
than rufous

green

cream-
white

ad. ♀

ad. ♂ eclipse

ad. ♂

MARBLED DUCK

rather pale
in flight

dark tips

pale spot

dark eye-
surround

palest part
of wing

white

three
birds
with a
♀ Pintail

ad.

K M

(Eurasian) Teal *Anas crecca* r(m)**B**4 / **W**2

L 34–38 cm, WS 53–59 cm. Breeds on variety of fresh and brackish waters, preferring lakes and ponds (even quite small ones) in forests, pools in taiga bogs or mountain willows, also along rivers and shallow, well-vegetated seashores, and on eutrophic lakes if near forests, where nest is placed. A common bird, forming large flocks on coastal bays or shallow lakes outside breeding season. Birds from N Europe winter in Britain, but also in Holland, France, etc.

IDENTIFICATION *Smallest* duck, with *narrow, pointed wings, short neck, fast and agile flight* and *dense flock formation*; takes off from water easily and rises steeply, will twist and turn readily in flight. At distance in flight, appears *dark with short, broad white bar along centre of upperwing* (both sexes). Speculum glossy green. – Adult ♂ breeding: Head chestnut with green sides, green colour thinly bordered yellow. *Sides of undertail pale yellow, bordered black*, visible even at distance. *Horizontal white line along grey body*. Told from N American Green-winged Teal (p. 46) by this horizontal white line (Green-winged has vertical band across breast-side), and by complete thin yellow lines on sides of head. – Adult ♀: Brown, streaked and mottled dark. Swimming bird resembles Garganey, but note: *small size*; *small bill*, often a little *orange at base*; moderately dark eye-stripe and *lack of dark stripe over cheeks* make head look fairly plain; *a pale patch or streak along base of tail-side*; chin/throat not so clean and pale buff-white. In flight told by wing pattern; belly diffusely paler at centre; underwing has whitish centre and dark leading and trailing edges. – Eclipse ♂ and juvenile very similar to adult ♀, juveniles generally slightly darker (also on belly), and flanks more 'striped' dark (adult ♀ has more delicate scaly pattern).

VOICE Vocal. ♂ has a clear, ringing whistle, 'treel', confusable only with Pintail's, but higher-pitched. ♀ quacks feeble, high-pitched with a nasal voice, first note higher, following descending, '**peeht** pat pat'; also a hard, fast 'krek-ekekek' when pursuit by ♂ is causing stress.

Garganey *Anas querquedula* m**B**5

L 37–41 cm, WS 59–67 cm. Breeds on shallow, eutrophic fresh waters in lowland and steppe habitats. Nests in tussock on lakeside meadow. Summer visitor (mainly Apr–Sep); strictly migratory, wintering in Africa. Never seen in large flocks in northern part of range (though often in southern).

IDENTIFICATION Slightly larger than Teal, with longer and straighter bill, and on average slightly longer tail. In flight, *wings are slightly paler* (especially outer wing). *Only rarely upends*, prefers to skim surface ('dabble') or dip head (Teal frequently upends). – Adult ♂ breeding: Head purple-brown ('dark' at distance) with *white crescent over eye to side*

of nape; breast dark brown, flanks pale grey; scapulars black and white, pointed. In flight, *pale, dull blue-grey upper forewing* and *sharp contrast between white belly and dark breast*. – Adult ♀: Resembles Teal, but told by slightly *larger size*; *longer, straighter all-grey bill* (no orange at base); hint of *darker stripe across cheeks* and of *pale loral patch* at base of bill, making *head look more striped*; lack of pale patch at base of tail-side. Further, ♀ Garganeys (and imm./eclipse ♂♂) often have quite *clean, pale chin/throat* compared with ♀-type Teals. In flight, *grey-brown forewing* much as on Teal, but slightly paler outer wing, and lacks broad white midwing-bar (has narrow only), and has *broad white trailing edge to secondaries* in ♀ Pintail fashion. – Adult ♂ eclipse: Like adult ♀, but retains adult ♂ wing with *pale blue-grey forewing*. – Juvenile: Like adult ♀, but belly is less pale, and white trailing edge to secondaries is narrower.

VOICE ♂ has characteristic display-call, a dry, wooden rattle, swaying a little in pitch, 'prrerrorrer', like running a fingernail across a comb. ♀ rather silent; has nasal 'ga, ga-ga, ga...', and a feeble, short, high-pitched quacking.

Blue-winged Teal *Anas discors* **V**∗

L 37–41 cm. American species. 1–13 records annually in Britain (may involve escapes among genuine vagrants).

IDENTIFICATION Breeding ♂ distinctive (but beware that eclipse ♂ Shoveler often has white half-moon bordering base of bill). ♀-type plumages resemble Garganey, Teal and, especially, Cinnamon Teal (American species, sometimes kept in European wildfowl collections and may escape from these; see p. 423). Note *prominent pale loral spot* but *lack of dark horizontal, diffuse streak from bill across cheek*; chin and belly pale, finely mottled brown (but not white); greenish or dull yellow feet; *bright pale blue upper forewing*, almost as blue as on adult ♂; *lack of white trailing edge to secondaries*; no pale patch at base of tail-side.

Baikal Teal *Anas formosa* [**V**∗∗∗]

L 39–43 cm. N and E Siberian species. Declining. Very rarely seen in Europe, and arguably doubtful whether any record involves a genuine straggler.

IDENTIFICATION Breeding ♂ unmistakable, with *striking head pattern*; *white vertical narrow line at side of breast*. Adult ♀ and juveniles brownish and streaked like other teals, and wing and head patterns recall both ♀ Teal (often has a pale patch at base of tail-side) and ♀ Garganey (well-patterned head; rather similar wing pattern). Slightly larger and longer-tailed than Teal; bill thin and all grey; *speculum is bordered pale rufous in front*, not whitish, and *white rear border always broader*. – Adult ♀: Head pattern distinctive, with small but *very prominent and circular white loral spot* (enhanced by dark surround); *dark eye-stripe only behind eye*; on many, a *whitish bar* or wedge runs from the white throat *vertically up across the cheek* towards eye; and sometimes there is a *dusky kidney-shaped mark on the upper cheek*. – Adult ♂ eclipse: Like adult ♀, but plumage much richer rufous. – Juvenile: Slightly duller brown-grey than adult ♀. Most similar to Teal, but usually told by more prominent pale loral spot with darker surround (not so conspicuous as on adult ♀, still usually more prominent than on any Teal), all-grey bill, and lack of white midwing-bar.

Teal

Garganey

TEAL

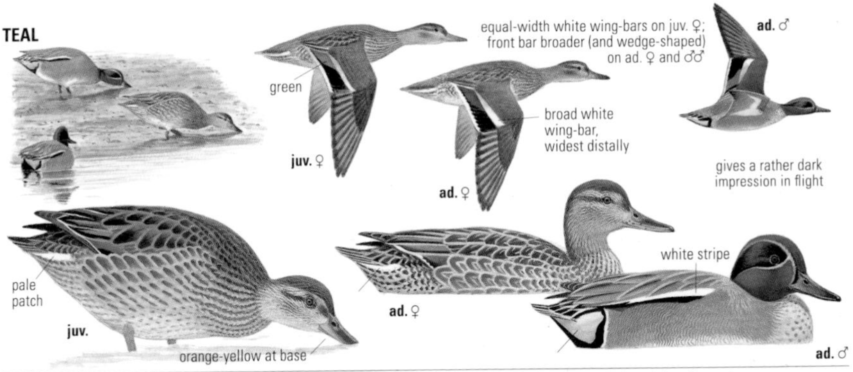

green

juv. ♀

equal-width white wing-bars on juv. ♀;
front bar broader (and wedge-shaped)
on ad. ♀ and ♂♂

broad white
wing-bar,
widest distally

ad. ♂

ad. ♀

gives a rather dark
impression in flight

white stripe

pale
patch

juv.

orange-yellow at base

ad. ♀

ad. ♂

GARGANEY

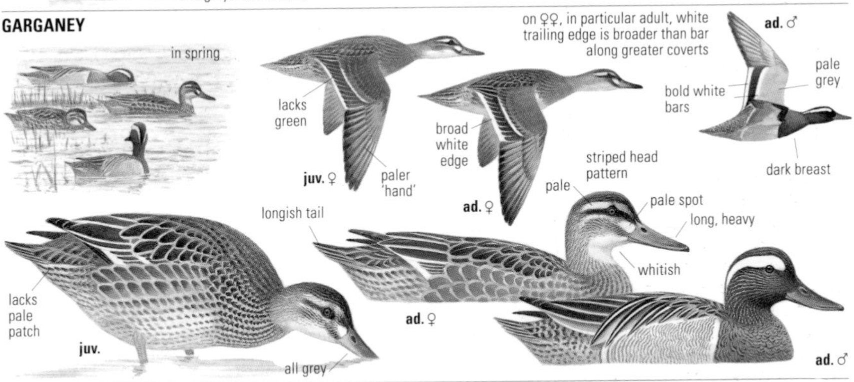

in spring

lacks
green

broad
white
edge

juv. ♀

paler
'hand'

ad. ♀

on ♀♀, in particular adult, white
trailing edge is broader than bar
along greater coverts

ad. ♂

pale
grey

bold white
bars

dark breast

striped head
pattern

pale

pale spot

long, heavy

whitish

longish tail

lacks
pale
patch

juv.

all grey

ad. ♀

ad. ♂

BLUE-WINGED TEAL

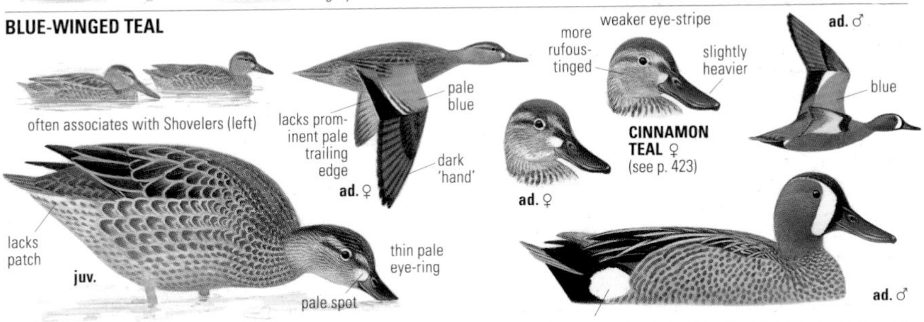

often associates with Shovelers (left)

pale
blue

lacks prom-
inent pale
trailing
edge

dark
'hand'

ad. ♀

weaker eye-stripe

more
rufous-
tinged

slightly
heavier

ad. ♂

blue

CINNAMON
TEAL ♀
(see p. 423)

ad. ♀

lacks
patch

juv.

thin pale
eye-ring

pale spot

ad. ♂

BAIKAL TEAL

compare with ♀ Teal (right)

broad
white
edge

dark
'hand'

ad. ♀

TEAL,
moulting
♂(Oct)

ad. ♂

dark-bordered
light spot

ad. ♀

BAIKAL TEAL

narrow
pale
patch

juv.

red-brown

ad. ♂

KM

DIVING DUCKS *Aythya, Somateria* et al.

A rather loosely applied group name which designates all species, fairly small to large, which feed mainly by diving, only rarely practising upending. Vegetarian or omnivorous. Body rather heavy, wings somewhat shorter (and in some cases blunter) than on dabbling ducks, hence take off from water generally only after prior running on surface. One group ('pochards') found mainly on shallow, eutrophic lakes, another (eiders, scoters, sawbills) in primarily marine habitats. Nest on ground, near water, a scrape lined with down. ♀ alone tends young. Moult as in dabbling ducks (p. 24). Beware of confusing hybrids, though rare, occurring particularly among 'pochards'(p. 34).

(Common) **Pochard** *Aythya ferina*   r(m)**B**4 / **W**3

L 42–49 cm, WS 67–75 cm. Breeds on eutrophic lakes and marshes with sufficient open water (depth > 1 m). Birds in N and E Europe winter in W and S, ♂♂ often already moving in summer. Gregarious. ♂♂ appear more numerous than ♀♀. Often dives for food, but also practises upending and dabbling for food from surface. Nests in vegetation near water.

IDENTIFICATION Medium-sized, short-tailed (rear end sloping down on swimming bird), with long neck and *long bill with concave culmen* running without step into *sloping forehead*, giving fairly distinctive profile with *smoothly peaked crown*. In all plumages, wings have *indistinct greyish wing-bars*. – Adult ♂ breeding: *Head bright chestnut; bill blackish with pale grey band across outer part*; eye reddish; *breast black* with gloss; *flanks and back pale ash-grey*, appearing whitish in strong light; *stern black*. In flight, medium grey upperwing-coverts and pale grey wing-bar give the bird a rather washed-out, pale appearance. – Adult ♀: Along with ♀ Wigeon, most nondescript duck: grey-brown, flanks and back tinged greyish, breast, crown and neck darker and tinged brownish. Diffuse pale and dark head marks, along with head/bill profile, often best clue: diffuse pale loral patch, eye-ring and line behind eye, and diffuse dark patch below eye which reaches lower base of bill; bill with a narrow, dull pale band across outer part (winter), or appearing all dark (summer). Eye rufous-brown. – Adult ♂ eclipse: Differs from breeding in having breast and stern dark brown-grey, and head duller rufous. Eye remains reddish. – Juvenile: Similar to adult ♀; typically more uniform above, and lacks pale line behind eye. Eye yellowish-olive.

VOICE Rather silent. Display-call of ♂ a characteristic wheezing abruptly cut off by a nasal, short note, 'aaa**ooo**-**chaa**(e)', like a ricocheting bullet; also, and more commonly heard, 3–4 short, sharp whistles, 'ki ki ki ki', during courtship. ♀ has loud, repeated purring 'brre-ah' (slightly down-slurred), often in flight.

Red-crested Pochard *Netta rufina*   V* / (r**B**5)

L 53–57 cm, WS 85–90 cm. Breeds on fairly large, reed-fringed eutrophic lowland lakes and sea-bays, also on larger lagoons and saline marshes. Migratory only in north. Regular vagrant, especially in autumn and winter, to Britain, 10–50 records annually, and a few breed. Habits much as for dabbling ducks, upending and dabbling for food in shallow, richly vegetated waters. Vegetarian.

IDENTIFICATION Rather *large*, with long, bulky body and *large, rounded head*. In all plumages, *very conspicuous, broad white wing-bars*. – Adult ♂ breeding: Rusty-orange head (palest on crown, but impression may vary depending on angle of light) and *striking coral-red bill*; breast, stern and centre of belly black, *flanks white*; swimming bird typically shows *white narrow patch at side of mantle*; upperparts plain brown. In flight, broad white wing-bars and peculiar *white oval flank patches lined with black* distinctive. Immediately told from ♂ Pochard by red bill and darker, brown back. – Adult ♀: General appearance like a large, pale ♀ Common Scoter, with off-white sides of head and foreneck and dark brown forehead, crown and hindneck; body plain brown, often diffusely patchy on flanks; *bill dark grey with pink near tip*. Eye brown. – Adult ♂ eclipse: Like adult ♀ but *bill all red*. – Juvenile: Like adult ♀ but *bill all dark*.

VOICE Rather silent. ♂ has a loud, often repeated 'baeht', and a 'stifled sneeze'. ♀ pursued by ♂ utters hard 'wrah-wrah-wrah-...' (likened to distant barking dog), and subdued 'rerr-rerr' during courtship on the water.

Ferruginous Duck *Aythya nyroca*   V*

L 38–42 cm, WS 60–67 cm. Breeds in same habitats as Pochard; less gregarious (and scarcer!); also appears more shy. Rare visitor in Britain autumn–spring.

IDENTIFICATION Medium-small, rather short-bodied, with *long bill* and neck, and characteristically high forehead and *peaked crown*. In flight, very striking, *broad, pure white wing-bars running to tip of wing*. On all but some juveniles, *white undertail* and (separated) *white belly patch* are obvious. Beware of some ♀♀ Tufted Ducks with white on undertail (though less pure and extensive). – Adult ♂ breeding: Whole plumage *deep chestnut* (tinged purplish), darkest on back and palest on flanks, except for *white undertail and belly*; narrow black neck-collar (rarely seen). *Eye white.* – Adult ♀: Dark brown with reddish tinge on head. *Pure white undertail. Eye dark.* Note *head/bill profile*, which separates from Tufted Duck, as does *longer and broader white wing-bars* in flight. – Adult ♂ eclipse: Like adult ♀, but tinged more obviously reddish, and retains *white eye*. – Juvenile: Resembles adult ♀ with dark eye, but main colour more dull brown, and both belly and undertail have less pure and extensive white patches (i.e. resembles Tufted Duck most).

VOICE Rather silent. ♀ has a purring flight-call with characteristic *dry ring*, snoring 'errr errr errr errr...'. ♂ has Tufted Duck-like '**wee-whew**' during courtship and also hard, nasal staccato-notes, 'chk-chk-chk-...'.

Pochard

Red-crested Pochard

Ferruginous Duck

POCHARD

ad. ♀

ad. w. ♂

greyish

sloping forehead

diffuse head markings

some lack pale on bill

pale grey band

more uniform brownish than ad. summer ♀

grey in winter (browner and more variegated in summer)

ad. s. ♀

ad. w. ♂

peaked

juv.

ad. wint. ♀

back as pale as flank

ad. ♂ eclipse

ad. wint. ♂

RED-CRESTED POCHARD

ad. ♀

ad. ♂

very conspicuous

at times, fiery-orange crown of ♂♂ appears almost to glow

rounded

resembles large, pale ♀ Common Scoter

all dark

white flank

black centre

pink spot

'coiffed' look

ad. ♀

coral-red

juv.

eye and bill red on eclipse ♂

back darker than flank

ad. ♂ eclipse

ad. ♂

FERRUGINOUS DUCK

ad. ♀

ad. ♂

very con-spicuous

white belly

peaked

beware: some ♀ Tufted Ducks have conspicuous white undertail

faint band

diffusely paler

juv.

rather long

ad. ♀

ad. ♀

ad. ♂

only nail black

duller white on juv.

conspicuous, *sharply de-fined* white patch at stern

white iris

very dark

ad. ♂ eclipse

ad. ♂

KM

(Greater) Scaup *Aythya marila* **W**4 / (r**B**5)

L 42–51 cm, WS 71–80 cm. Breeds in marine habitats (fjords, archipelagos) on salt or brackish water, or on freshwater lakes and pools in mountains (birch and willow zones) and tundra. Migratory. Wintering birds gregarious, diving for molluscs on open sea, or are seen along coasts and in bays.

IDENTIFICATION Medium-sized with rather large, rounded head, and sloping end of body when swimming. ♂ distinctive. Other plumages similar to Tufted Duck, with similar *white wing-bar*, but note: *no hint of crest* on hindcrown; *head shape more elongated and profile smoothly rounded*; body longer; and less black on tip of bill (mainly *just nail black*; Tufted has whole tip black). Beware of confusing hybrids (see p. 34), and the possibility of a rare vagrant Lesser Scaup (N America; p. 46). – Adult ♂ breeding: Head black with *green gloss*, eye yellow; breast and stern black; flanks white, *back greyish-white* (fine vermiculation appar-ent only at closer range). Bill pale grey with small, fan-shaped black nail patch. In flight, light back usually ob-vious, as are white wing-bars; upper fore-edge of wing vermiculated, appearing medium grey (darker than back but not black as on Tufted). – Adult ♀: Dull brown with pale brown-grey flanks and slightly darker back, latter with some greyish vermiculation (visible at close range only) and strik-ing *broad white band surrounding base of bill* (Tufted may have some white at bill-base though usually less, and only rarely extending clearly over culmen). In late spring/sum-mer, a *prominent pale patch on ear-coverts* on most. – Adult ♂ eclipse: Rather similar to breeding, but with brown cast on head, breast and back, and often a little white around bill-base. – Juvenile: Resembles adult ♀, but less white at bill-base, slightly darker bill with indistinct darker nail area, and paler flanks. Young ♂ attains most of adult plumage by end of 1st winter (Feb–Apr), but not fully until 2nd winter.

VOICE ♂ mostly silent. During display a chorus of low whistles, 'vü-vüp vü-vo vüpüvee...', recalling Tufted Duck but lower-pitched; now and then more typical falling 'püooh' (voice reminiscent of Eider). Growling call of ♀ like Tufted Duck's but is deeper, more drawn out, and voice raucous, 'krrah krrah krrah...'.

Tufted Duck *Aythya fuligula* r**B**4 / **W**2

L 40–47 cm, WS 65–72 cm. Versatile in choice of breeding habitat, therefore common: open, clear, oligotrophic lakes in forested areas; densely vegetated, eutrophic lowland lakes and marshes; along seashores; on tundra pools; slow-flowing rivers; reservoirs; park lakes; etc. Mainly migratory. Gregarious when not breeding, forming large, dense flocks.

IDENTIFICATION *Small*, short-bodied, with narrow neck and rather *large head with crest* (long and drooping on

breeding ♂, only a shorter tuft in other plumages) at hindcrown. Head 'unevenly rounded' with *high forehead and flattish crown*; *bill short but broad, pale blue-grey with most of tip 'dipped' in black*. In flight, *white wing-bars* prominent (as on Scaup). – Adult ♂ breeding: Swimming bird distinctive, has *black plumage with sharply defined rectangular white flanks* and *long drooping crest*. Eye yellow. – Adult ♀: Brownish with paler flanks diffusely blotched darker on most. Resembles Scaup, especially when showing white at bill-base, but told by *smaller size, hint of crest, different head shape, darker brown, especially above*, and *more black at tip of bill*. A few have some white on undertail, especially in autumn, recalling Ferruginous Duck, but told by different head shape, hint of crest, and *shorter and less prominent white wing-bars*. Eye usually deep yellow. – Adult ♂ eclipse: Crest short, flanks dull brown, black parts of breeding plumage tinged brown. – Juvenile: Like adult ♀, but a little paler brown on head with a little buff feathering at bill-base. Eye brown. Young ♂ attains most of adult plumage in 1st winter (Dec–Mar), but fully only in 2nd winter.

VOICE Display-call of ♂ a nervously quick, 'bubbly' or even giggly series of accelerating notes on slightly falling pitch (also in flight at night), 'vip vee-veevüvüp' ('Turkey in falsetto'). ♀ has growling call typical of *Aythya* ♀♀, 'krr krr krr...', slightly faster and higher-pitched, and each segment shorter, than in Pochard; higher-pitched than Scaup.

Ring-necked Duck *Aythya collaris* **V**∗

L 37–46 cm. American species. Certainly at times genuine vagrant in Europe (annually in Britain & Ireland), but many are likely escapes from wildfowl collections. Birds tend to reappear at same sites in several consecutive years.

IDENTIFICATION Small, short-bodied, rather large and of distinctive shape: *forehead very high and steep, crown peaked at hindcrown*, and hint of rounded crest at hindcrown creating *small indentation between crown and nape*. *Bill* of adults slate-grey with wide black tip and white subterminal band, thus *is three-coloured*; breeding ♂ also has narrow white band at base, and nostrils encircled white. *Tail* rather *longer* than on Tufted Duck, and often *conspicuously raised*. In flight, very similar to Tufted but told by *grey wing-bar*, not white. Beware of confusing hybrids (see p. 34), and moulting ♂ Tufted Ducks. – Adult ♂ breeding: Plumage like Tufted, but *flanks pure white only at front, forming vertical 'spur'*, and diffusely along upper edge, *otherwise grey*; also, *upper edge of flank panel more markedly S-curved* than on Tufted. Indistinct purplish-brown neck-collar (often hidden). – Adult ♀: Best told by head shape and bill pattern (though latter less clear than on ♂, esp. in early autumn, and thin white band along bill-base missing); plumage recalls ♀ Pochard more than Tufted, having *pale throat, diffuse pale patch bordering bill-base, pale eye-ring* and, often, a *pale line behind eye*. Eye dark. – Adult ♂ eclipse: Resembles adult ♀ with some pale feathering bordering bill, and lack of distinctive white band at bill-base, but *retains yellow eye*, and head *lacks pale eye-ring and line behind eye*; also, head and breast darker, blackish-brown. – Juvenile: Very similar to adult ♀, but possibly distinguished by *almost all-dark bill*, pale subterminal band developing during 1st autumn. Young ♂ attains adult plumage as early as (late) 1st winter.

Scaup

Tufted Duck

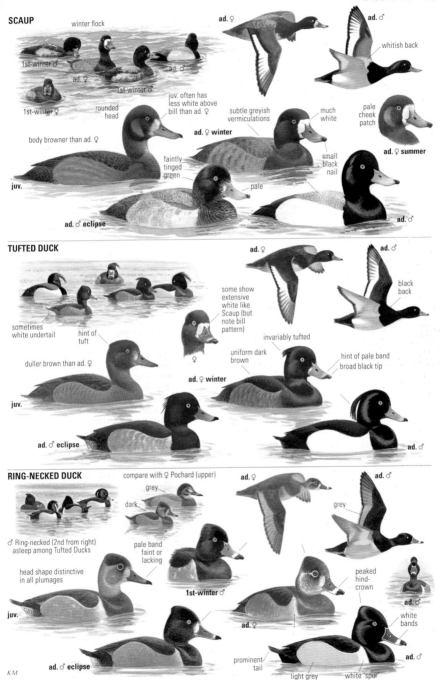

SCAUP

winter flock

ad. ♀

ad. ♂

whitish back

1st-winter ♂

ad. ♀

ad. ♂

1st-winter ♂

1st-winter ♀

rounded head

juv. often has less white above bill than ad. ♀

subtle greyish vermiculations

much white

pale cheek patch

body browner than ad. ♀

ad. ♀ winter

small black nail

ad. ♀ summer

faintly tinged green

pale

juv.

ad. ♂ eclipse

ad. ♂

TUFTED DUCK

ad. ♀

ad. ♂

black back

some show extensive white like Scaup (but note bill pattern)

sometimes white undertail

hint of tuft

♀

invariably tufted

hint of pale band

duller brown than ad. ♀

ad. ♀ winter

uniform dark brown

broad black tip

juv.

ad. ♂ eclipse

ad. ♂

RING-NECKED DUCK

compare with ♀ Pochard (upper)

ad. ♀

ad. ♂

grey

grey

dark

♂ Ring-necked (2nd from right) asleep among Tufted Ducks

pale band faint or lacking

peaked hind-crown

ad. ♂

head shape distinctive in all plumages

1st-winter ♂

white bands

juv.

ad. ♀

ad. ♂

prominent tail

ad. ♂ eclipse

light grey

white 'spur'

KM

WILDFOWL HYBRIDS

Although wild birds do not normally mate with birds other than representatives of their own species, exceptions do occur. Generally, a cross between two different species is most likely to occur in areas where one of the two is much less common than the other; interspecific barriers may break down if a bird in breeding season cannot easily find a mate of its own kind, and reasonably similar ones are at hand.

In our part of the world, hybrids appear to be more common among wildfowl and gamebirds than in other groups. There may be genetic reasons, as yet unexplored, for this situation; but it may be no more than an impression.

Firstly, birds of both groups are handled in thousands by hunters, who will report on oddities which they happen to bag. Secondly, the impression may be partly due to the widespread keeping of wildfowl in parks; in these artificial conditions, where birds of some species are singles or, at least, in a minority among congeners, and where perhaps the normal sexual activity during courtship cannot take place to a full extent owing to pinioned wings or lack of space, hybrids become more frequent. In many wildfowl parks, locally hatched birds are allowed to fly freely, which means that we are likely to see a few of these hybrids later on at natural sites. Another likely reason for the apparent frequency of wildfowl and gamebird hybrids is that they are fairly obvious and easy to observe. Hybrids among warblers and chats may be just as common, but who would notice it?

A rather puzzling fact is that, among wildfowl, hybrids seem to occur more commonly among geese and members of the 'pochards', genus *Aythya*, than among dabbling ducks. The reason for this is not clear.

Below, some of the most frequent and confusing hybrids of 'pochards' are described. Emphasis is put on males, not because they are more common but because they are easier to detect for the average birdwatcher. Hybrid females are notoriously difficult to spot and designate correctly, a task for the expert. In the majority of cases, a fair guess can be made as to which are the two parent species of a certain hybrid. To say which was the male and which was the female is usually much more difficult.

When studying a hybrid duck, always pay close attention to: *pattern*, *size and shape of bill*; *iris colour*; *shape of head seen in profile*; and presence or not of *fine vermiculation* on grey or grey-brown parts.

HYBRID

Pochard ♂ × Tufted Duck ♀

This hybrid has Pochard ♂ and Tufted Duck ♀ as parents. It can be deceptively similar to a Tufted Duck ♂ moulting out of eclipse plumage (which you would see in late autumn), but note *darkish eye*; *very short 'stand-off' crest* like some

Tufted Duck × Ring-necked Duck

This hybrid has some characters intermediate between the parent species but is often very similar to Ring-necked Duck and at a casual glance could be mistaken for one. Clues to its mixed parentage are usually the presence of a tiny crest (not just a peaked hindcrown), and both the white band at the

Scaup × Tufted Duck

This hybrid between Scaup and Tufted Duck looks superficially like a Scaup ♂, with head glossed predominantly green, but close attention to detail will reveal: *darker grey*

CONFUSION SPECIES

Tufted Duck ♂ eclipse

adult ♀ Tufted Ducks; and *not entirely black upperparts*, being *very finely vermiculated* on some (visible at closest range only); the *flanks are rather darker towards belly*, sometimes *finely vermiculated*.

Ring-necked Duck ♂

base of the bill and the white 'spur' at the fore-edge of the flank (characters typical of a Ring-necked Duck) being reduced in extent or lacking altogether. Note that this hybrid often has a wholly or partly white wing-bar, unlike a true Ring-necked Duck.

Scaup ♂

back and scapulars than genuine Scaup, *vermiculation often being finer and more diffuse*; also, *more black on tip of bill* (Scaup has only the nail black); many 'Scaup ♂ type' hybrids have a *firm, pointed crest* (like the one shown here, at bottom of p. 34), giving a characteristically peaked or angled crown never seen on Scaup (but those without any hint of crest can

Tufted Duck ♂ × Pochard ♀

A now well-known hybrid is that between Tufted Duck ♂ and Pochard ♀. Surprisingly, this cross frequently produces offspring which resemble a third species, the Lesser Scaup of N America (see p. 46), a rare vagrant to Europe. The head is tinged purplish-brown and the crown is peaked, rather like on that species. Important field marks which separate from Lesser Scaup are: *rather dark grey back with very fine, indis-*

Pochard × Ferruginous Duck?

This hybrid is more rarely encountered than the others and is liable to be mistaken for a Redhead, a North American species (very few records in Europe; see p. 46). One such hybrid was seen in Malmö in S Sweden in the 1960s, originally identified as the first record ever of Redhead in Europe (yellow eye, rather rounded head), but the record was later withdrawn as the bird was obviously just a hybrid. Certainly

Pochard × Ferruginous Duck

Another tricky but much more Ferruginous-like hybrid can result from the same crossing as above. It closely resembles Ferruginous Duck, but usually shows a combination of two or more of the following differences: *red or yellow eye* (instead of white as on ♂ Ferruginous Duck); *breast contrastingly*

Scaup × Tufted Duck

This ♀ hybrid between Scaup and Tufted Duck is most similar to a ♀ Scaup, but note the following distinguishing points: much white feathering at the base of the bill (indicating Scaup), yet *extensive black at the tip of the bill* (indicating Tufted Duck); no hint of a crest on the rounded head (indi-

look markedly Scaup-like, with rounded head); other hybrids have a *shorter bill with broader tip*, just like Tufted Duck.

Lesser Scaup ♂

tinct vermiculation (Lesser Scaup has white back with rather coarse dark vermiculation, still giving a fairly light back, and vermiculation readily visible even at some distance); *tip of bill much more extensively black*, and base of bill usually dusky grey (on Lesser Scaup whole bill pale grey, with only nail black). Check wing-bar, too, *white only on secondaries* on true Lesser Scaup.

Redhead ♂

Pochard is one of its parents; the other parent species is more uncertain, but Ferruginous Duck seems the most likely. Points to note which reveal that it is not a genuine Redhead are: *bill pattern of Pochard ♂*, with both much black at tip and dark grey base (Redhead has pale grey bill with tip 'dipped in black' and a subterminal whitish band); *medium-dark grey back* (Redhead light grey).

Ferruginous Duck ♂

darker than flanks (more uniformly chestnut on Ferruginous Duck); *fine vermiculation on back and flanks* visible at close range (never any vermiculation on Ferruginous Duck); *undertail not purely and extensively white; tip of bill with much black* (only the nail black on Ferruginous Duck).

Scaup ♀

KM

cating Scaup), yet *noticeably dark back, clearly darker than the flanks* (indicating Tufted Duck). Other ♀ hybrids between these two parent species can have a hint of a crest, a more Scaup-like paler back or reduced black on the bill, and some will be very difficult to recognize as hybrids.

(Common) **Eider** *Somateria mollissima* rB3

L 60–70 cm, WS 95–105 cm. Marine in habits. Breeds commonly along coasts and in archipelagos on salt and brackish waters. Visits fresh water occasionally. Large flocks of immatures and other non-breeders common sight along coasts. Dive for crustaceans and molluscs (primarily sea mussels) at moderate depth. Nests close to water, often openly, ♀ sits tight. Nest lined with (celebrated) down.

IDENTIFICATION Large, *heavily built*, with fairly *short neck*, large head with *long, wedge-shaped bill*. In flight gives heavy impression (possible to confuse with geese at horizon range), *wings broad* and comparatively short, wingbeats fairly slow, *head held rather low*. Flock formation often rather disorderly, loose clusters or irregular, long lines. – Adult ♂ breeding: *Largely white, with black belly*, sides and stern, white rounded 'thigh patch', white tertials; *head white* with black crown; sides of nape pale green. – Adult ♀: Brown, barred dark; dark speculum bordered white. – Adult ♂ eclipse (late Jun–Sep): *Very dark, unbarred; upperwing-coverts and long, curved tertials white* (if not moulted; generally in Aug); some white bases to feathers of mantle and upper scapulars sometimes show. – Juvenile: Like adult ♀, but *dark speculum usually not bordered white*, and *head and upperparts more uniform*; juvenile ♂ has *dark head with pale stripe over eye* (like 1st-summer eclipse ♂); *tertials rather short and not strongly curved*. – 1st-year ♂: Juvenile plumage moulted from late autumn, when some white feathers appear on breast, mantle and back, then scapulars and flanks. – 2nd-year ♂: Like adult ♂, but in breeding plumage *upperwing-coverts and tertials not pure white*, shape of latter less curved.

VOICE Vocal during breeding season. ♂ has characteristic cooing display-call during communal courtship, a far-carrying, deep 'a-**ooh**-e' (at long range may recall distant Eagle Owl). ♀ has incessant chuckling 'gak-ak-ak-ak-...', like distant throbbing of vessel engine.

King Eider *Somateria spectabilis* V*

L 55–63 cm, WS 87–100 cm. Breeds in Arctic on tundra pools and, less commonly, along sheltered, shallow seashores, wintering along sea coasts. Habits much as for Eider, and single birds often seen with Eiders. Rare winter visitor in Britain, occurring especially in N.

IDENTIFICATION Slightly *smaller and more compact* than Eider, body and bill shorter. – Adult ♂ breeding: Unmistakable; *prominent orange knob at base of bill*; *coral-red bill*; crown and (well-rounded) nape blue with purplish tinge; lower neck and breast salmon-pink; mantle white (as on Eider) but *back and scapulars black* (Eider: white), *scapulars with triangular erect 'sails'*. In flight, white patches on forewing. – Adult ♀: Distinguished from Eider by *shorter,*

darker bill, with *bulging feathering above culmen reaching farther down on bill than feathering at sides*; *dark marks on body plumage more open*, U-shaped and have dark central spot, especially on mantle and flanks; size slightly smaller; often rather pale cheeks and area around bill-base, dark gape contrasting (and has 'happy-looking' expression). – Adult ♂ eclipse: Resembles imm. ♂ (since knob shrinks) but keeps *white forewing patches*. Bill pale red. General impression is dark, but often shows traces of white on side of breast. – Juvenile: Like ♀, but more similar to Eider in colour (colder brown, not so rusty) and pattern (more simple barring, less U-shaped 'scallops'). Told by size, shape of bill, and feathering at base of bill. – 1st-summer ♂: Like Eider, but *bill shorter and pink* (not dull yellowish-grey), and *back never has any white* (only mantle).

VOICE Rather less vocal than Eider. ♂ has a deep, vibrating cooing in spring, 'hroo, roo roo-e', falling slightly in pitch at the end; flock calls can recall cooing Black Grouse. Clucking of ♀ similar to that of ♀ Eider.

Steller's Eider *Polysticta stelleri* V***

L 42–48 cm, WS 68–77 cm. Breeds on arctic tundra, winters along seashores in N. Prefers shallow coastal waters, especially with inflowing fresh water from creeks, swimming right up to shoreline, upending among boulders and seaweed; also good diver, but rarely seen far out from land.

IDENTIFICATION Medium-small. Differs from other diving ducks in having shape and some habits more like a dabbling duck; body elongated, tail long, *bill oblong, not so wedge-shaped*; shape of head characteristic, *crown flat, forehead and nape rather angled*; wings rather long and narrow, permitting easy take-off from sea. Flight rapid and light. – Adult ♂ breeding: *White* with *orange-buff sides*, darkening to chestnut on belly (and black on centre of lower belly); chin, eye-surround, collar and back of neck, mantle, back and stern black, *scapulars and tertials black and white* giving striped effect; rounded tuft on rear crown, and loral patch, green; a black spot on side of breast visible at waterline level on swimming bird; bill lead-grey; in flight, *forewing is white*. – Adult ♀: Dark brown; *speculum blue, distinctly bordered white* in Mallard fashion; tertials dark, almost invariably *with pale tips*, shape well curved. – Adult ♂ eclipse: Like ♀ but readily separated in flight by white forewing. – Ju-venile/1st-winter: Juvenile like adult ♀, but *speculum dull brown*, insignificantly bordered off-white; *tertials dull brown and short*, shape less curved. Often a rather obvious *pale eye-ring and pale line behind eye* (more so than on adult ♀). ♂ has *chin, forehead and nape patch dark*, darker than on ♀; from late winter, sex positively revealed by the first white feathers appearing on head. – 2nd-summer ♂: Similar to adult ♂, but is distinguished by a varying amount of dark feathers lingering on head and upperwing-coverts.

VOICE Rather silent. Raucous quacking 'gah gaah geeah'. Wing noise a fine whistling (slightly more obvious than Mallard's).

Eider   King Eider   Steller's Eider

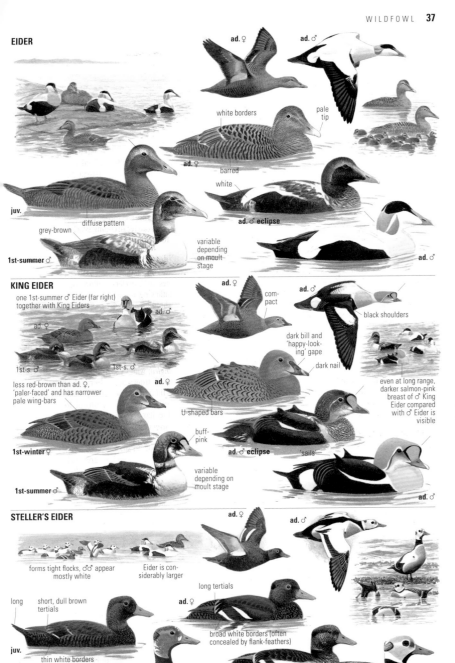

EIDER

ad. ♀

ad. ♂

white borders

pale tip

ad. ♀

barred

white

juv.

diffuse pattern

grey-brown

ad. ♂ eclipse

1st-summer ♂

variable depending on moult stage

ad. ♂

KING EIDER

one 1st-summer ♂ Eider (far right) together with King Eiders

ad. ♂

ad. ♀

com-pact

black shoulders

ad. ♀

1st-s. ♂   1st-s. ♂

dark bill and 'happy-look-ing' gape

dark nail

ad. ♀

less red-brown than ad. ♀, 'paler-faced' and has narrower pale wing-bars

U-shaped bars

even at long range, darker salmon-pink breast of ♂ King Eider compared with ♂ Eider is visible

buff-pink

1st-winter ♀

ad. ♂ eclipse

'sails'

variable depending on moult stage

1st-summer ♂

ad. ♂

STELLER'S EIDER

ad. ♀

ad. ♂

forms tight flocks, ♂♂ appear mostly white

Eider is con-siderably larger

long tertials

long tertials

ad. ♀

long

short, dull brown tertials

broad white borders (often concealed by flank-feathers)

juv.

thin white borders

1st-summer ♂

ad. ♂ eclipse

ad. ♂

KM

Common Scoter *Melanitta nigra* m(r)**B**5 / **W**3

L 44–54 cm, WS 70–84 cm. Breeds near lakes and rivers in boreal forests (upper coniferous, birch/willow) and close to tundra waters. Migratory, spring migration largely on broad front over land at night, autumn migration diurnal mainly along coasts and over sea. Most ♂♂ return S as early as late summer to S Denmark to moult. Gregarious, can form very large flocks.

IDENTIFICATION Medium-sized, plump-bodied. Bill fairly small, ♂ *with knob at base*. Longish tail often exposed when swimming. All-dark plumage *without white in wing*. Dives usually with small leap, *wings kept folded tight to body* (see p. 40). Wing-flapping punctuated with quick downward thrust of head. Resting *flocks usually very dense*, more so than with Velvet Scoter, and frequently larger. – Adult ♂: Whole plumage *black*. In flight *primaries are contrastingly paler*, thus wings two-toned, especially in strong light. *Centre of bill* (culmen) *yellow*, but *small bill knob black* seen from the side (cf. rare vagrant Black Scoter, p. 46, which has large orange-yellow 'bulge' on bill and lacks the black at the base). – Adult ♀: *Sooty-brown with paler cheeks* and dark brown forehead and crown; superficially recalling a swimming ♀ or juvenile Red-crested Pochard, though smaller and darker-bodied, invariably with all-dark bill, and habitat usually different. (A few birds have centre of cheek mottled brown, creating a hint of the 'two-pale-spots pattern' typical of Surf Scoter, which see.) – Juvenile: Like adult ♀ but *belly paler*, brownish-white instead of medium brown-grey; plumage somewhat browner, less grey. (Moulting 1st-winter ♂♂ can also show hint of two pale spots on cheeks just as some ♀♀.) Young ♂♂ attain adult plumage largely in 1st winter but retain pale belly (can look strikingly white in flight!), and by spring wings noticeably faded, brownish.

VOICE ♂ has soft piping 'pyu', regularly repeated about once per second; heard during display, but most often during spring nights (late Apr/early May) over land in N Europe from NE-bound flocks. Also drawn-out 'pyu-ih' and other short, repeated calls during courtship. ♀ has a repeated 'karrr', similar to many other ♀ diving ducks. Wingbeats produce fine whistling noise.

Velvet Scoter *Melanitta fusca* **W**4

L 51–58 cm, WS 79–97 cm. Breeds along coasts on salt or brackish waters (such as the Baltic Sea coasts), and on fresh waters of mountain and tundra; often found on large lakes and rivers in boreal coniferous forests. Breeds late, often a month later than Eider. Migratory, movements usually along coasts. Nests on ground, usually not far from water.

IDENTIFICATION Medium-large, stocky, rather long-bodied with *thick lower neck* (though upper neck can look thin

when neck is stretched). *Wedge-shaped bill* rather heavy. *All-dark plumage* with *white secondaries* makes it distinctive in flight even at long range; wings rather broad, but appearing narrower when seen against the sky owing to the white secondaries. White in wing often, but not always, visible as a narrow patch on swimming birds as well. Resting flocks rarely so large and dense as those of Common Scoter. Dives without leap, *wings semi-open* (see p. 40). Wing-flaps with head held up and bill raised, showing white secondaries. – Adult ♂: Whole plumage *black* except for the white secondaries, and tiny white mark under eye (difficult to see except at close range). *Bill has pale orange-yellow band along sides*, readily visible at a distance. Cf. rare vagrant White-winged Scoter (p. 46), which has larger white eye patch, curling up at rear, a black bill knob and more reddish bill colour. – Adult ♀: Sooty-brown with white secondaries; belly only slightly paler, brown-grey. Head pattern variable: on breeding birds usually a *rather large diffuse pale loral patch*, generally also *a smaller but rather distinct pale patch on ear-coverts*; a few look darker with very diffuse patches only; during later stages of breeding, wear and bleaching can cause sides of head to appear much paler. – Juvenile: Like adult ♀ but *belly much paler* (esp. juv. ♀), being off-white, and *facial patches* generally *more distinct*.

VOICE During courtship 'morning flights' by pair in spring, ♀ has slightly hoarse, vibrating 'braa-ah... braa-ah... braa-ah...'. Apart from this, rather silent. ♂ has an insignificant, low, nasal call during courtship, 'aah-er'.

Surf Scoter *Melanitta perspicillata* **V***

L 45–56 cm. American species, rare vagrant in Europe, mainly ♂♂ recorded (♀♀ probably easily overlooked) in autumn, winter or spring. In Britain & Ireland, a few recorded annually.

IDENTIFICATION Only a trifle larger than Common Scoter, appearing of same size; clearly somewhat smaller than Velvet Scoter. As two congeners, largely all-dark plumage; *no white in wing*. Bill *large and triangular*, making head look heavy. Feathering at base of culmen extends forward horizontally (cf. Common Scoter). *Crown in profile flatter* than on Common Scoter. Swimming birds often have *tail prominently cocked*. Dives with little leap forward and *wings half-opened* (p. 40). Wing-flaps with rigid (not drooping) neck. – Adult ♂: Black, with *large white patch at nape* (rarely partially lost in summer–early autumn) and *smaller white patch on forehead*. Bill, swollen at base, appearing *bright yellow-orange* at distance; if seen close, bill has white base and black rounded patch at base of bill-side (appearing as extension of feathering). *Eye white*. – Adult ♀: Sooty-brown, as ♀ Velvet Scoter with *two pale patches on side of head* (variable, can be indistinct or missing), but additionally often has *small white patch on nape*. Note heavy triangular bill with straight or even slightly convex culmen (Velvet has slightly concave culmen), and dark feathering extending horizontally over base of culmen. Eye usually dark, but sometimes paler (brownish-white). – Juvenile: Like adult ♀, but *belly whitish* (not pale brown) and cheeks and throat usually paler; never pale patch on nape; eye dark. Young ♂♂ acquire most of adult pattern in 1st winter, but white patch on forehead and full size and colours of bill not developed until 2nd winter.

Common Scoter

Velvet Scoter

COMMON SCOTER

typical offshore flock

paler primaries

typical flock formation

ad. ♀

juv.

all young scoters have a pale belly

variation with dark cheek-divide

♀♀ have strongly 'two-toned' head pattern

primaries flash 'silver' in sunlight

characteristically, wing-flapping is combined with a quick downward toss of the head

fades to whitish

1st-sum-mer ♂

imm. / ♀

yellow on bill can be very striking, even at long range

'knob'

yellow

juv.

wings fade to brownish by spring

young ♂♂ lack the swollen 'knob' of adult

long, pointed

slender neck

1st-summer ♂

ad. ♂

VELVET SCOTER

typically spread-out flock formation

at long range, looks narrow-winged in flight

ad. ♀

compact

ad. ♂

juv.

pale

compare with ♀ Scaup (right)

head pattern variable, a pale cheek patch usually visible in summer

head held raised throughout wing-flapping

juv.

slightly concave profile

ad. ♀

fresh

worn

white wing patch often hidden in wing pocket

white 'speculum' often looks like a thin line

stretched neck when worried

1st-w. ♂

ad. ♂

SURF SCOTER

♂ Surf Scoter among Common Scoters

uniform dark wings

ad. ♂

juv.

pale

ad. ♀

heavy head and bill

head held high, rather small wings beating vigorously

compare with ♀ Eider (right)

some ad. ♀♀ have pale nape patch

courtship

juv.

often held high

straight or slightly swollen profile

ad. ♀

1st-w. ♂

bill pattern develops through 1st winter

ad. ♂

K M

Long-tailed Duck *Clangula hyemalis* **W3**

L 39–47 cm (excl. elongated tail-feathers of ♂ 10–15), WS 65–82 cm. Breeds commonly in the Arctic on tundra pools and marshes, also along sea coasts (local in the Baltic) and on still-standing mountain waters. Gregarious. Winters at sea, often in large, dense flocks, mainly off coasts; then restless and active, taking off and flying agilely, low over the water, in short pursuit-flights, constantly changing direction, landing with splash; dives for molluscs and other animal food. Spring migration through Baltic spectacular; main exodus from Gulf of Finland to tundras in NE on late-May evenings with tail-wind, involving over 100,000 birds on peak days. Complex moult produces at least three different-looking plumages annually; most scapulars and feathers of sides of head and neck are moulted three times a year, breast, upper mantle and rest of head and neck twice, and rest of plumage once a year.

IDENTIFICATION Rather small, brown, black and white duck. Head rounded, *bill short* and stubby, *central tail-feathers of ♂ elongated to thin line* (Pintail ♂ only other duck with same feature). Wings rather narrow and pointed, all dark, *wingbeats fast and elastic, mainly below horizontal*, adding to peculiar, almost bat-like impression, and *wings slightly swept back*. Plumages and moults complex. – Adult ♂ spring/summer (from late Apr): Largely *brown-black with grey-white flanks, white stern* and *white patch at side of head*; scapulars blackish, edged rufous-yellow. Summer moult to full eclipse during later stages of breeding changes little in appearance (new scapulars shorter and duller). Bill has pink band across centre in spring, is often all black in summer. – Adult ♂ autumn (from Sep): Becomes much whiter; upper mantle, *most of scapulars, neck and head white*, sides of lower head/upper neck with dark patch. Bill acquires pink band again. – Adult ♂ winter (from late Oct): As in autumn, but forehead and sides of head pale grey-brown, not white,

and patch on side of head/upper neck black (with chestnut lower rim), not mottled brown-grey. – Adult ♀ summer (May–Aug): Like adult ♂ but *lacking elongated tail-feathers*, and colours *duller*; also, pale patch on side of head narrower and less sharply set off, and invariably a *light collar around neck*. – Adult ♀ autumn/winter (Aug–Feb): Main difference from summer is *head*, which is *off-white with blackish crown and patch at lower cheek/upper neck*; scapulars are longer and edged more brightly rufous-buff (or even partly whitish). – Juvenile: Like adult ♀ autumn, but with less dark and distinct patch at lower cheek/upper neck, and with shorter and blunter tertials and scapulars.

VOICE Vocal. ♂ has characteristic, far-carrying, nasal yodelling, 'ow **ow**-ow**de**lee, ow-owde**lee**', used both in display and during migration; chorus from flocks becomes pleasing song like distant bagpipes; also drawn-out wailing short version, 'a-**gleh**-ah', as if with falsetto 'skid' in the middle. Both sexes have nasal, low 'gak', often heard from migrating flocks.

Harlequin Duck *Histrionicus histrionicus* **V★★★**

L 38–45 cm, WS 63–70 cm. Breeds on fast-flowing streams in arctic tundra, within region only in Iceland, where sedentary; Icelandic population does not exceed 3000 pairs. Winters on rough waters along coasts. Accidentals outside range very rare, and not all believed to be genuine. In Britain fewer than ten records since 1950. Gregarious, though parties generally rather small. Swims energetically, even into strong currents. Dives from surface, from perch or directly from wing, not avoiding rough parts of streams.

IDENTIFICATION Rather small diving duck; *bill small, forehead high and head rounded, neck thick, tail pointed*. In all plumages *dark, with conspicuous white marks on sides of head*. Swims high, often with bobbing head. Flight rapid, *wings all dark*. – Adult ♂ breeding: *Dark blue-grey* (blue

A COMPARISON OF FEMALES AND DIVING ACTIONS OF SCOTERS
Full accounts of these species on pp. 38–39.

COMMON SCOTER
Dives with small leap and wings held tight against body.

VELVET SCOTER
Flip-dives without leap, wings partially opened.

SURF SCOTER
Dives with tiny leap, wings flicked open just before it disappears.

KM

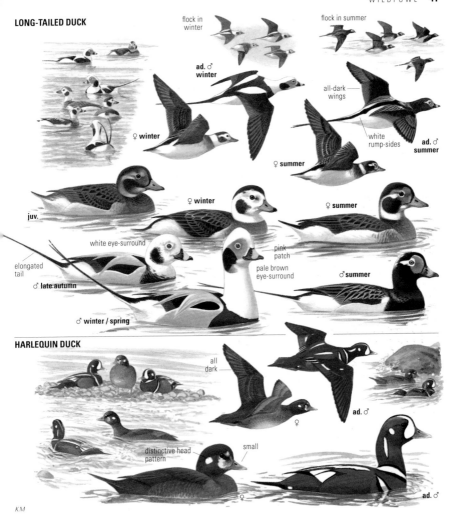

LONG-TAILED DUCK

flock in winter

flock in summer

ad. ♂ winter

all-dark wings

white rump-sides

ad. ♂ summer

♀ winter

♀ summer

juv.

white eye-surround

♀ winter

♀ summer

pink patch

elongated tail

pale brown eye-surround

♂ summer

♂ late autumn

♂ winter / spring

HARLEQUIN DUCK

all dark

ad. ♂
♀

distinctive head pattern

small

ad. ♂

♀

ad. ♂

KM

Long-tailed Duck

Harlequin Duck

colour surprisingly difficult to see at a distance, looking almost black) with *chestnut flanks* and blackish belly and stern, attractively *marked white* (white outlined black) *on*

head, neck, side of breast, and back. –Adult ♀: Head and neck dark grey; body sooty-brown except belly, which is dusky white mottled with grey; *large whitish patch between bill and ear-coverts, whitish spot above lores, and small white patch on ear-coverts* combine to form distinctive face pattern. –Adult ♂ eclipse: Dull, dark, rather similar to ♀, but with some traces of breeding pattern (some white on scapulars often visible, white line on side of breast indicated, a little chestnut remaining on flanks). – Juvenile: Very similar to adult ♀, only slightly browner, less greyish, and secondaries duller, lacking purplish-brown metallic gloss of adult ♀.

VOICE Rather silent. Fine nasal, piping 'vee-ah vee-ah...' and 'vee' heard from displaying ♂♂.

(Common) **Goldeneye** *Bucephala clangula* r**B**5 / **W**3
L 40–48 cm, WS 62–77 cm. Breeds in forested areas by lakes, mainly clear, oligotrophic, and by slow-flowing rivers and in archipelagos; less commonly at open coasts. Mainly migratory, hardy, returning early to breeding sites. Less gregarious than most other diving ducks, often seen in pairs or smaller groups. Shy. Food animal. Courtship display in early spring by ♂ includes tossing head back onto back and then stretching neck up with bill pointing upwards, at times also splashing water with feet. Nests in tree-hole (old Black Woodpecker nest) or box.

IDENTIFICATION Medium-sized, *compact* duck with *large, rounded head* like a knob on fairly short neck; *crown slightly peaked and shape of head triangular.* Bill rather small and triangular. *Speculum white,* rest of underwing appearing blackish; much *white on upper forewing* on adults. Flight rapid but appearing laborious, wings beating fast. For separation from similar Barrow's Goldeneye, see that species. – Adult ♂ breeding: Head black with green gloss; *large white rounded loral spot;* eye yellow; *breast and flanks gleaming white;* stern and much of upperparts black; *narrow black scapular lines* hanging down over white flanks. – Adult ♀: *Head brown; eye pale yellow; collar* and belly *white,* breast and flanks ash-grey; bill dark with a yellowish band across the outer part when breeding, usually all dark at other seasons. – Adult ♂ eclipse: Like adult ♀ but retains wing pattern, with extensive white on upper forewing. – Juvenile: Like adult ♀ but head duller and greyer brown, and lacks white collar. Juvenile ♂ has wing pattern similar to adult ♀, juvenile ♀ has less white (narrow white bar on greater coverts, no pale patch on medians).

VOICE Display-call of ♂♂ squeezed double-note, 'bee-**beeech**', usually accompanied by a low, dry, Garganey-like rattle,'drrudrrir'. ♀ has loud purring or grating call, often in flight,'brra, brra, ...', very like Tufted Duck but perhaps slightly softer and slower. Most often heard and characteristic sound is *loud, musical whistling produced by wing-beats,* especially strong from ♂♂ in winter and spring (but almost absent in case of juveniles).

Barrow's Goldeneye *Bucephala islandica* **V*****
L 42–53 cm, WS 67–82 cm. Breeds in Iceland, Greenland and N America on tundra pools and lakes. Resident or moving to open waters. Nests in tree-hole if available or on ground in scrub.

IDENTIFICATION Somewhat *larger* than Goldeneye, which it resembles. Apart from size, head profile differs slightly but clearly, *head appearing longer* and *less triangular, forehead being higher, crown flatter* (indistinct peak well to front) and nape well rounded. Breeding ♂ has purple gloss on head

(imm. ♂ Goldeneye can be confusingly dull dark brown-black), and *white crescent on lores reaches above eye* (never so on moulting imm. ♂ Goldeneyes, which can have a partly crescent-shaped loral spot); *black of upperparts more extensive* and *reaching far down on sides of breast; scapulars have white encircled as narrow 'windows'.* – Adult ♀ and juvenile very similar to Goldeneye, but note somewhat different shape of head, and generally *more extensive yellow on bill in breeding season.*

VOICE Display-call of ♂ weak, grunting, short notes in staccato, 'wa wa-wa...'. ♀ has similar call to Goldeneye but deeper and more coarse, 'krrah krrah...'. Wing noise of ♂ similar to Goldeneye, but possibly lower-pitched.

SAWBILLS

Diving ducks of variable size, from large to rather small. Bill has small hook at tip and tooth-like lamellae along cutting edges, enabling better grip of main prey, fish. Three species, of which larger two (p. 44) similar and have elongated body and rather long neck. The third, the Smew, is smaller and more compact. Expert divers, and pursue prey very agilely; at times fish communally. Frequent various habitats, both freshwater and marine. Nest usually in hole or cavity, sometimes on ground under cover.

Smew *Mergellus albellus* **W**4–5
L 38–44 cm, WS 56–69 cm. Breeds in northern boreal forests by clear lakes or calm rivers. Rather scarce. Migratory. Shy and restless, pairs or small parties on wintering grounds inclined to make aerial excursions, spreading out rather than keeping together (though small and medium-sized flocks can be seen at the most favoured localities), feeding close to reedbeds in shallow waters. Nests in tree-hole (old nest of Black Woodpecker) or box.

IDENTIFICATION Small, rather compact sawbill, more similar to the *Bucephala* species (and indeed occasionally interbreeds with Goldeneye). – Adult ♂ breeding: Distinctive, being *largely dazzlingly white,* with neat black patterning; head white with black loral 'mask' from bill backwards around eye; black line or patch at side of nape. In flight, appears surprisingly *pied black and white.* A *large oval white patch* is visible *on upper forewing.* – Adult ♀: Dull brown-grey with *white cheeks and dark chestnut-brown forehead and crown; lores* blackish, becoming more brownish during breeding. Wing pattern similar to that of ♂, but white more restricted. – Adult ♂ eclipse (Jul–early Nov): Like adult ♀ summer (with brownish lores), but retains more extensive white wing patch, and is blacker, not so greyish, on back. – Juvenile: Very similar to adult ♀, but belly a little less whitish, more mottled grey, and loral area brown at least in 1st autumn, and sometimes longer.

VOICE Mostly silent. Display-call of ♂ low and rarely heard, a deep, accelerating, frog-like croaking ending with a hiccup, 'gr-r-r-rrr-**chic**'. ♀ has a hoarse 'krrr'.

Goldeneye

Barrow's Goldeneye

Smew

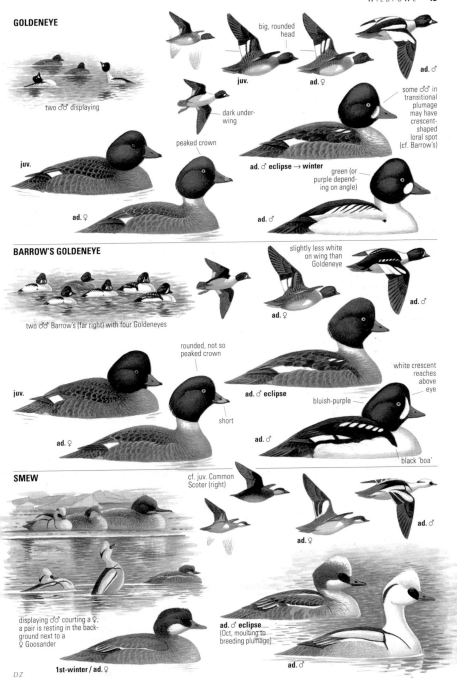

GOLDENEYE

two ♂♂ displaying

big, rounded head

juv.

ad. ♀

ad. ♂

dark under-wing

peaked crown

some ♂♂ in transitional plumage may have crescent-shaped loral spot (cf. Barrow's)

juv.

ad. ♂ eclipse → winter

green (or purple depending on angle)

ad. ♀

ad. ♂

BARROW'S GOLDENEYE

two ♂♂ Barrow's (far right) with four Goldeneyes

slightly less white on wing than Goldeneye

ad. ♂

ad. ♀

rounded, not so peaked crown

white crescent reaches above eye

juv.

ad. ♂ eclipse

bluish-purple

short

ad. ♀

ad. ♂

black 'boa'

SMEW

cf. juv. Common Scoter (right)

ad. ♂

ad. ♀

displaying ♂♂ courting a ♀; a pair is resting in the background next to a ♀ Goosander

ad. ♂ eclipse (Oct, moulting to breeding plumage)

1st-winter / ad. ♀

ad. ♂

DZ

Goosander *Mergus merganser* rB4 / (W4)

L 58–68 cm, WS 78–94 cm. Breeds on lakes and slow rivers in forested areas, requiring fairly deep, clear, fish-holding waters and mature trees with holes for nesting. Nests also in nestbox, under or in house, in crevice, etc. Gregarious except when breeding; can form very large flocks (tens of thousands recorded) when ♂♂ gather to moult in late summer, or for communal fishing on favourite lakes in late autumn. Expert diver. Hardy, in winter many retreat only from ice.

IDENTIFICATION Large, *long-bodied*, streamlined, with *long, narrow, red bill ending in small hook*. Head appears rather large owing to *full crest* at hindcrown and nape. Neck long, but swimming birds often retract most of it, looking short-necked. In flight, silhouette with long, *outstretched neck* and rather *shallow, fast wingbeats with straight wings* and direct flight at times recall loons and grebes more than other ducks. – Adult ♂ breeding: Largely *white* (tinged salmon-pink in winter–early spring), with *black head/upper neck glossed green* (green cast visible at closer range or in good light only); crest mane-like, 'combed with grease', giving *odd bulge shape to nape*. Back largely black. In flight, *very large white 'boxes' on inner wing*. – Adult ♀: *Head dark red-brown* except *clearly demarcated white throat patch*; crest looser than on ♂, 'hair combed dry', often forming two tufts, one at hindcrown, one at nape; *sharp division between brown upper neck and whitish lower foreneck*; body greyish. In flight, white squares formed by secondaries and greater coverts (on some, *indistinct, thin* dark line along tips of all or a few outer greater coverts). For distinction from similar ♀ Red-breasted Merganser, see latter. – Adult ♂ eclipse: Like adult ♀, but retains extensive white on upperwing. – Juvenile: Like adult ♀, but bill duller red, crest shorter, throat less clean white, has pale loral line, and iris paler. 1st-winter ♂ has more extensive white on inner wing than ♀.

VOICE Display-call of ♂ is a pleasant, deep, hard, muffled ringing 'krroo-krraa', not loud, yet rather far-carrying. ♀ has a hard 'pra, pra-pra...' when courted. In flight, a louder, repeated 'prrah, prrah, prrah...', and a conversational, fast chuckling 'chakerak-ak-ak-ak'.

Red-breasted Merganser *Mergus serrator* rB4 / W3

L 52–58 cm, WS 67–82 cm. Breeds along coasts, in archipelagos and at clear inland waters both in taiga and in mountains and on tundra. Gregarious, even in breeding season, but rarely forms very large flocks (as Goosander habitually does). Nests on ground among sheltering vegetation.

IDENTIFICATION Medium-large, slimmer than Goosander but otherwise very similar. *Bill is narrower, especially at base*, and crest is more brush-like and shaggy (on ♂ almost 'punk'), giving a slightly *small-headed* appearance; neck

also narrower. – Adult ♂ breeding: Head black, glossed green; *white collar* above *black-streaked rufous-brown breast*; *black area with large white spots at side of breast* is distinctive. Most of upperparts black, flanks grey. In flight, large white patch on inner wing, but not so extensive as on Goosander. – Adult ♀: Resembles ♀ Goosander. For separation, note the following: *lack of sharp division between brown head and off-white lower neck*; *pale throat patch diffusely set off*; *bill thin at base*; *crest shorter and 'spikier'*; *head more cinnamon-brown*, not so dark red; more *'striped' face*, with pale lores and dark lines above and below; *back darker tinged brown*, and *flanks less clean grey*; in flight, *head and neck appear slimmer*, *white square on inner wing invariably divided by obvious black line*; *upper forewing dark brown-grey*, darker and less clean grey than on Goosander. – Adult ♂ eclipse: Like adult ♀, but retains extensive white on wing. – Juvenile: Like adult ♀, but bill duller red, and crest shorter. – 1st-summer ♂: Attains most of adult ♂ plumage except for head and neck, head being tinged brownish, neck largely lacking the *white collar*; white on wing not so extensive as on adult ♂.

VOICE Display-call of ♂ weak and not often heard, rhythmic, one hiccup and a sneeze well spaced, 'chika ... pitchee'. Call of ♀ repeated hard, grating 'prrak prrak prrak...'.

White-headed Duck *Oxyura leucocephala* —

L 43–48 cm (incl. tail 8–10). Breeds on shallow, richly vegetated eutrophic lakes. Local and rare. Mainly resident. Reluctant to fly, prefers to swim for cover.

IDENTIFICATION Small, but somewhat larger than Ruddy Duck, with proportionately somewhat larger head, immediately told by *large bill with swollen culmen* basally. – ♂: In spring characteristic, with *huge, pale blue bill*, pale brown body and largely *pure white head*; narrow blackish crown-stripe, and variably black on nape, neck and around eye (black eye patch usually isolated on white side of head). *No white on undertail-coverts*. In winter duller. – ♀: Dull brown plumage, similar to Ruddy Duck, but note *distinctive bill shape*, larger size, and more *well-marked dark cheek-stripe*.

VOICE Mainly silent.

Ruddy Duck *Oxyura jamaicensis* rB4

L 35–43 cm (incl. tail 6–8). American species, introduced in Britain and spreading to W Europe. Has hybridized with White-headed Duck in Spain (this posing a serious conservation problem). Breeds on shallow, eutrophic lakes.

IDENTIFICATION Small and compact, with *large head, strong bill*, rounded body and *long, stiff tail*, often cocked. – ♂: In spring, *body largely deep chestnut*, crown (down to eye) and hindneck black, *cheeks*, chin *and undertail-coverts white*. Bill pale blue. In winter, bill is grey and body dull brown, but white cheeks retained. – ♀: Like ♂ winter, but head dull dark brown and off-white and has a *diffuse* (strength variable) *dark horizontal stripe across cheeks*.

VOICE Mainly silent. Displaying ♂ produces varying tapping sounds by beating its bill against the chest.

Goosander

Red-breasted Merganser

White-headed Duck

GOOSANDER

communal fishing flock in Nov

flanks diffusely pale grey

pale grey

all white

white

ad. ♀

pale 'face' markings

head shape (crest!) changes with mood

large white wing patch

thick base

ad. ♂ eclipse

juv.

full crest

white chin

sharp border

ad. ♀♀

ad. ♂

RED-BREASTED MERGANSER

courtship display

flanks rather dark grey

black line

rather dark

dark

ad. ♀

crest at times erect

thin base

large white wing patch

ad. ♂ eclipse

crest usually short, untidy

diffuse border

ad. ♂

ad. ♀♀

ad. ♂

WHITE-HEADED DUCK

ad. ♂

RUDDY DUCK

ad. ♂

swollen profile

concave profile

♀ / juv.

ad. ♀

juv.

ad. ♂ winter

1st-summer ♂
(dark-headed variant)

ad. ♂ winter

ad. ♂ summer

ad. ♂ summer

DZ

VAGRANT WILDFOWL

All species on this spread are scarce to very rare visitors to Europe, mainly from N America.

Green-winged Teal *Anas carolinensis* V*

L 34–38 cm. N American species, closely related to Eurasian Teal (p. 28). Many records annually in Britain, and found in most other W European countries. Several records presumably referrable to a number of birds now resident in Europe.

IDENTIFICATION *Small*. Adult ♂ breeding has a *vertical white bar across side of breast* (not a horizontal stripe above flanks), and lacks most of the yellow lines bordering the dark green patch on sides of head. Centre of breast somewhat brighter pinkish-buff, and in flight *fore upper wing-bar usually has more rufous-buff tinge*. Other plumages like Eurasian Teal, but crown and eye-stripe on average darker, and fore upper wing-bar generally has more obvious rusty tinge.

Lesser Scaup *Aythya affinis* V*

L 38–45 cm. N American species now regularly recorded in Britain and in many other W European countries. Records in the UK have increased lately (5–15 annually).

IDENTIFICATION Like Greater Scaup (p. 32) but *smaller* (smaller even than a Tufted Duck) and told by: *crown peaked at rear* almost giving hint of crest like in ♀ Tufted (Greater Scaup has smoothly rounded crown); *wing-bar* above white on 'arm' but *brownish-grey on primaries* (all white on Greater; odd Lesser have more diffuse difference between brown and white); *underwing not as all white* as on Tufted and Greater, has darker bases to flight-feathers; adult ♂ has *coarser dark vermiculation on back* and *more purple* than green *sheen* on head. ♀ like Greater Scaup, told on size, head shape and wing pattern. Compared with ♀ Tufted, body is somewhat paler grey-brown, making darker head contrast more.

Black Scoter *Melanitta americana* V**

L 44–54 cm. Breeds in E Siberia and N America. Rare vagrant to coastal Europe. Often feeds among surfs close to land.

IDENTIFICATION Resembles Common Scoter, but adult ♂ breeding instantly recognized (also in flight) by *large, bright yellowish-orange bulge on bill* lacking black near forehead. The bulge is usually yellow on top but orange at base. Note *in flight* all-dark wings, but *primaries appear quite pale* as in Common Scoter, especially below. ♀ like Common but has a trifle more pronounced hook on bill-tip and is more extensively dark on nape. A few adult ♀♀ have some dull yellow markings on otherwise dark bill, not seen in Common Scoter.

White-winged Scoter *Melanitta deglandi* —

L 50–57 cm. Breeds in E Siberia (ssp. *stejnegeri*) and in northern N America (*deglandi*). Very rare vagrant to Europe, recorded in Finland (*stejnegeri*) and Iceland (*deglandi*).

IDENTIFICATION Like Velvet Scoter, but adult ♂ breeding told by small but *obvious black knob on culmen of bill* (in Velvet: at the most a faint ridge), *bill-tip with more pink and orange-red than yellow* (Velvet: mainly bright yellow), and a *larger white eye patch which curls up at rear behind eye* (Velvet: small horizontal patch). The E Siberian ♂ has more sloping transition between crown and forehead, more angled bill knob, yellow lower edge of otherwise orange-red bill-tip, and

less black along edge of lower mandible. ♀ differs subtly from Velvet in shape of bill and head, and outline of feathering, but identification difficult and requires close range.

Redhead *Aythya americana* V***

L 44–51 cm. N American species. Very rare vagrant to Europe (Britain, Iceland, Ireland).

IDENTIFICATION Subtly larger than Pochard (p. 30), which it resembles: ♂ has *reddish head* and *black breast*. Recognized by: *rounder head shape* and *steeper forehead*; more *evenly thick bill* (Pochard: more triangular in profile); *bill mainly light grey but whole tip black* ('dipped in ink'); *yellow eye*. ♀ told by rounded head shape, bill shape and, again, the broad dark tip. Unlike the similar ♀ Pochard, breast is somewhat paler (and more warm rufous-brown), contrasting less or not at all with flanks. Note on the brown head a pale eye-ring and a pale stripe curling back from eye on cheeks. Swims more often than Pochard with *tail raised above surface*.

Canvasback *Aythya valisineria* V***

L 49–56 cm. N American species. Very rare vagrant to Europe (Britain, Iceland, Netherlands).

IDENTIFICATION Compared with Pochard, adult ♂ breeding can be picked out through its *larger size* and *paler body* (silky white; Pochard: pale grey), but the *all-dark and longer bill with attenuated, fine tip* is also striking. Chestnut brown head often swarthy around bill-base and back towards eye. ♀ best told by all-dark, long and pointed bill typical of the species, and by *slightly more uniform grey-brown head* with a little *darker loral area* than in Pochard.

Bufflehead *Bucephala albeola* V**

L 32–39 cm. Breeds in N America. Several of the records in Iceland and Britain & Ireland regarded as genuine, whereas a majority of other records in Europe are generally treated as referring to escapes from parks.

IDENTIFICATION The smallest relative of Goldeneye, smaller even than Tufted Duck, with typically *large head*. ♂ has black back and black head (green and purple gloss at close range) with *a large white patch at rear* (behind and above eye, around neck). White wing patches and shoulder straps obvious in flight, much as in Goldeneye. ♀ has a small white patch on the cheek, otherwise rather nondescript brown with darker back than breast and flanks. *Eye dark*, bill pale grey.

Hooded Merganser *Lophodytes cucullatus* V**

L 42–50 cm. Breeds in N America. A few records (e.g. many in Britain, Ireland and Iceland) are presumed to be genuine, whereas birds seen in several other European countries are usually regarded as escapes from wildfowl collections.

IDENTIFICATION ♂ has black head with *a large black-rimmed white crest*, which can be erected but which is commonly folded, e.g. in flight. *Flanks rufous-ochre*, and *chest* attractively patterned *black-and-white* with double black bars on each side. Eye mustard-yellow. ♀ has a slightly smaller *rusty-tinged crest*, is otherwise rather plainly grey-brown with paler breast. *Lower edge of bill yellow*. Both sexes have *black-and-white elongated tertials* creating a striped pattern, but note that these feathers can be moulted or worn and hence not so conspicuous.

GREEN-WINGED TEAL

one ♂ Green-winged among Teal

hybrid Teal × Green-winged

hybrids show 'mix' of characters

ad. ♀

lacks Teal's diagnostic white scapular stripe

indistinct buff borders to green eye-stripe

ad. ♂

diagnostic white bar

♀-type birds practically identical to Teal; strong rust colour of wing-bar may give clue, head pattern possibly more striking, but much overlap

LESSER SCAUP

approximately same size as Tufted Duck (left)

ad. ♀

greyish 'back'

less dark at tip than Tufted Duck

1st-w. ♂ (Jan–Feb) advanced birds may be very adult-like

dark at tip restricted to 'nail'

peak

coarse vermiculation

ad. ♂

light grey

ad. ♂

lighter 'back'

peak far ahead

peak at rear

ad. ♂♂

with Greater Scaup (left)

ad. ♂

only inner wing-bar white

BLACK SCOTER

beware of Commons with excessive yellow!

Black (front) has more prominent yellow on bill and thicker neck than Common

base mostly grey-black

ad. ♂ Common (see p. 38)

some have extensive yellow

ad. ♀

entire swollen base yellow

colour appears more intense in lower 'half'

ad. ♂

WHITE-WINGED SCOTER
deglandi

ad. ♀ *deglandi*

'broken-nosed' profile

ad ♂ VELVET SCOTER (bottom)

pinkish-red with yellow strip

ad. ♂ *stejnegeri* (Asia)

pinkish-red

ad. ♂ *deglandi* (N America)

REDHEAD

REDHEAD (winter)

POCHARD (winter ♀)

head shape and bill pattern distinctive

black tip, narrow subterminal band

REDHEAD

PO-CHARD

steep forehead

ad. ♀

ad. ♂

distinctive bill pattern

CANVASBACK

longer-necked than Pochard

CANVASBACK

very pale

longer bodied than Pochard

POCHARD (winter ♀)

light patch

pale grey band

in summer, ♀ Pochard may have an all-dark bill

ad. ♀

all black

ad. ♂

BUFFLEHEAD

check legs for rings!

ad. ♂

tiny little duck!

♀/imm.w.

around late winter, 1st-w ♂♂ develop dark mantle and white on back of head

ad. ♂

HOODED MERGANSER

ad. ♀

ad. ♂

'mobile' crest

ad. ♀

ad. ♂

crest lowered

ad. ♂

long tail; broadly white-edged tertials

KM

GROUSE *Tetraoninae*

Medium-sized to large, sturdy and thickset, non-migratory birds of mountain slopes and boreal forests. Spend much time on ground. Feathered nostrils, strong completely or partly feathered feet and lack of spurs are common features, as are rounded wings with stiff, downcurved primaries and, in flight, noisy, rapid bursts of wingbeats interrupted by glides. Mainly vegetarians, but take insects in summer. Nest on ground. Young precocial, already capable of short flights after 1–2 weeks; tended solely or primarily by ♀.

Willow Ptarmigan *Lagopus l. lagopus* —

(Alt. name: Willow Grouse.) L 35–43 cm. Sedentary in boreal forests, preferably birch (but also found in coniferous and willow), and on tundra with willows, dwarf birch, heather, etc. In winter often at lower altitudes and in more sheltered habitats, coniferous and denser birch forests.

IDENTIFICATION The two *Lagopus* species are very similar, best separated by calls, fine details in plumage, size of bill, and choice of habitat. With the exception of Red Grouse (the Willow Ptarmigan race of the British Isles; see below), they are told from other grouse by largely *white wings* all through the year. Rounded body, smallish head, feathered toes and short tail are other characteristics. In winter, almost completely white, in summer mainly rufous-, tawny- or greyish-brown. Basically, ♂ has three plumages, ♀ two. – Adult ♂ spring (*c.* Mar–early Jun): *Head, neck and upper breast deep rufous* or chestnut-red, *almost uniform* without much barring; *belly* and parts of flanks and back *white*, white gradually reduced, partly replaced by brown; upperparts black with brown barring. – Adult ♂ summer (Jun–Sep): Gradually acquires *barring on head, neck and upper breast*, and more *ochrous and tawny colours*, less rufous. – Adult ♀ breeding (–Sep): Similar to summer ♂, but plumage on average tawnier, less rufous, and more obviously barred buff and black. – Winter (Nov–Mar): From Sep, slowly attaining nearly *all-white plumage* (except for black tail); told from Rock Ptarmigan by invariably *white lores*, not black, and slightly larger size (esp. ♂♂); habitat also good (though not infallible) clue, Willow Ptarmigan preferring cover in birch and coniferous forests, rarely entering willow and alpine zones in winter. When seen close, note that the *bill is heavier* than Rock Ptarmigan's (and claws on average paler).

VOICE Main call (with function of song) from perched cock during spring nights is an accelerating series of loud, nasal, bouncing barks, ending in a trill, 'kau, kau kau-ka-ka-kakarrrrrr'. Alternative call, uttered partly in flight, is '**ke**-u, **ke**-kerrrrr-ke-kerr**ehe ehe che**', slowing down at end and often followed (or preceded) by a few '**kowah**' ('go back!'). ♀ call 'nyau'(clever mimic will attract ♂♂).

Red Grouse *Lagopus lagopus scotica* rB2

L 33–38 cm. A local race of Willow Ptarmigan. Sedentary on heather moors and in highlands of N and W Britain and in Ireland, largely avoiding forests.

IDENTIFICATION Shape and habits as Willow Ptarmigan, differing in summer in *dark, not white, wings* and, especially ♂♂, in darker plumage overall; striking difference in winter, when *no white plumage* is adopted. In *all plumages basically dark brown with reddish tinge,* ♂ always more uniform and darker and deeper rufous, ♀ paler and with ochrous edges or bars. Sexing possible when breeding pair seen together; often more difficult when single birds are encountered.

VOICE Identical to that of Willow Ptarmigan.

Rock Ptarmigan *Lagopus muta* rB3

L 31–35 cm. Breeds on mountainsides and tundra, barren and rocky terrain with only little vegetation. Sedentary, but in winter sometimes in more sheltered habitats, e.g. in willow scrub and open birch forests at upper tree-limit.

IDENTIFICATION Slightly *smaller* than Willow Ptarmigan (although Svalbard birds are much larger) but very similar in shape, with rounded body and smallish head; toes feathered. In winter, nearly *all white with short, black tail.* In summer, brown colours are more grey-tinged than rufous, making it reasonably distinctive if seen well. – Adult ♂ spring (May–Jun [Jul]): *Head, neck and upper breast* and parts of flanks and back *dark grey-brown and black, finely vermiculated* black and edged or barred buff and white, giving a mainly *dark grey general impression* in the field (Willow Ptarmigan deep red-brown and hardly barred at all); *lores blackish; belly, wings and parts of flanks white.* – Adult ♂ summer (Jul–Sep [Nov]): As in spring, but colour *paler grey,* less dark brown-grey; *sparse black patches on back* typical; lores brown, speckled white. Paler than Willow Ptarmigan at same time of year, lacking rufous tinge. – Adult ♀ spring–summer: Differs from ♂ in having more *coarsely barred plumage* with *more yellowish tinge,* dark brown and black, barred and spotted yellowish-buff and off-white. Lores rather pale. Compared with Willow Ptarmigan, *throat and sides of head are a little darker and more greyish* (not tinged buff and ochrous, *wings* white). – Winter: Plumage *all white,* except for black tail and *black loral stripe* on all ♂♂ and, thinly, on some ♀♀; most ♀♀ have white lores (and may therefore be difficult to distinguish from Willow Ptarmigan). When seen close, note that the *bill is smaller* than on Willow Ptarmigan (and claws are on average darker). Generally forms large flocks in winter, keeping on open mountainsides or in scrub and heather rather than in forests (cf. Willow Ptarmigan).

VOICE All calls rather similar, mostly low belching or dry, snoring sounds. Only one call typical, an almost Garganey-like creaking with characteristic rhythm, 'arr orr ka-**karrr**' ('here comes the bride'; sometimes only two or three syllables); appears to have song function (but at times given by ♀, too), mostly delivered from ground in spring, sometimes when flushed. A related shorter 'urr-errr' has been described as 'perch-song'. In song-flight in early spring, ♂ gives a belching sound while descending on stiff wings, ending in a cackle, 'ahrrrr-ka-ka-ka-ka-ka-ka'. Alarm or aggression shown by a 'kwa' and variations. Call of ♀ a soft 'kee-a'.

Red Grouse   Willow Ptarmigan   Rock Ptarmigan

WILLOW PTARMIGAN

juv.

dark, short central primary

ad. ♂ summer

displaying ♂ in Apr

heavy bill

ad. ♀ late summer

ad. ♂ late summer (Aug/Sep)

big red eyebrow

deep rufous

tawny-brown, barred

♀

♂

winter

ad. ♀ spring

ad. ♂ spring (May/Jun)

RED GROUSE

retains brown plumage all the year around

ad. ♂

on a Scottish moor

ad. ♀

ad. ♂

ROCK PTARMIGAN

ad. ♀ late summer

ad. ♂ late summer (Aug/Sep)

ad. ♂ summer

among lichen-covered rocks on bare mountain slope

♀

delicate bill

black lores

buff with grey admixed

greyish, spangled white

♂

winter

ad. ♀ spring

ad. ♂ spring (May/Jun)

DZ

(Western) Capercaillie *Tetrao urogallus* ⌐B4

L ♂ 74–90 cm (incl. tail *c.* 25), ♀ 54–63 cm. Sedentary in mature coniferous forests, preferring areas with old pines on rocky ground with abundance of berry-bearing shrubs and moss and with element of aspen and spruces, avoiding open ground. Spectacular communal display in late spring.

IDENTIFICATION *Very large*, ♂ being about a third larger than ♀. Takes off with *very loud wing noise*; longer flights fast and direct, with bursts of rapid wingbeats interspersed between glides. – ♂: Not likely to be confused with any other grouse through mere size; that aside, has *very long tail* (cocked and spread as a fan during display), *long neck* (erect in display, with raised feathers) and *dark colours*. Heavy, strongly curved *bill straw-yellow*. A conspicuous *rounded, white patch at base of forewing*. In flight characteristic, with *long, folded, straight tail* and *long, thick, outstretched neck* making wings look comparatively small. – ♀: Considerably smaller than ♂, still distinctly larger than ♀ Black Grouse. In spite of size difference can be confused with Black Grouse, which it resembles, the two being brown, barred dark, and found in similar habitat, but told by: *throat and parts of upper breast unmarked orange-brown*; *more obvious strings of white patches on scapulars* ('braces'); often slightly paler and more rufous plumage; slightly longer and more rounded tail, which is somewhat more rufous (though rufous tinge to some degree often found on ♀ Black Grouse, too).

VOICE At evening gathering of ♂♂, belching, bellowing 'ko-**krerk**-korohr' and variants. Call of ♀, mainly from perch in tree, a slowly repeated cackling 'grak', often at peak lek season (early May in N). Song, at dawn, first from tree perch, then from ground in group, consists of clicking double notes, accelerating into 'cork-pop' note, immediately followed by fine, grinding hissing, ground-displaying ♂♂ at times replacing last element with noisy wing-flutter. Whole song lasts 5–7 sec., audible at only 200–300 m.

Black Grouse *Lyrurus tetrix* ⌐B4

L ♂ 49–58 cm (incl. tail *c.*15), ♀ 40–45 cm. Sedentary in variety of habitats, ranging from moors, bogs, and clear-fellings in forested areas to heaths and barren islands. Communal display in *early* spring, *many* ♂♂ on ground in *open* area (incl. ice-covered lakes). Single ♂♂ may sing from treetops, even in summer. In winter often seen in birches, eating buds.

IDENTIFICATION Medium-large, small-headed grouse with small bill. Wing noise on take-off moderately loud. Flight as Capercaillie, but Black Grouse appears more compact. – ♂: Black (with purplish gloss) except for *white undertail-feathers* (prominently exposed during display), white under-wing, and a *white wing-bar*; comb-shaped *red eyebrows*. Most striking feature is *lyre-shaped tail*, cocked and spread during

display, but in flight folded and simply appearing *long*. – ♀: Black-barred greyish-brown all over, incl. throat and breast (cf. ♀ Capercaillie). Tail grey-brown, often showing slight tinge of rufous (less than on Capercaillie), square or slightly forked. In flight, shows *narrow whitish wing-bar*.

VOICE ♀ has a rapid, cackling call, often ending with a nasal, drawn-out note, 'kakakakakaka**keh**-ah'. Song of ♂ a far-carrying, bubbling, prolonged rookooing, 'rro-perre-**oo**-ohr rro-per**roo**...'. Another distinctive call, partly with song function, is a strong, harsh hissing 'choo-**iiish**'.

Caucasian Grouse *Lyrurus mlokosiewiczi* —

L ♂ 50–55 cm (incl. tail *c.* 18), ♀ 37–42 cm. Sedentary in Caucasus and NE Turkey on mountainsides (at 1500–3000 m) at tree-limit and on open slopes with low rhododendron scrub and herb meadows; somewhat lower in winter.

IDENTIFICATION ♂ resembles Black Grouse, differing in following points: *longer, straighter tail, deeply forked* with less obvious lyre-shape; *black undertail-coverts; no white wing-bar*. – ♀ is similar to Black Grouse, but (at close range) separated by: *greyer*, less rufous tinge; *longer tail*; barring of underparts finer and less regular; somewhat less extensive white visible on undertail-coverts; and paler supercilium and darker cheek patch.

VOICE ♂ almost mute. During display, flutter-jumps produce thin, whistling wing sound, audible at *c.*150 m. ♀ has cackling call reminiscent of Black Grouse.

Hazel Grouse *Tetrastes bonasia* —

L 34–39 cm. Sedentary in mixed coniferous (rarely broad-leaved), closed forests, preferring damp and densely under-grown areas with old spruces. Spends much time on the ground in shaded places, but may walk along the limbs of trees, and take a perch high up in trees. Difficult to see, cleverly keeps in cover; does not squat, flies early.

IDENTIFICATION Rather *small*, roughly Jackdaw-sized, has plump body, small head and bill. Intricate plumage pattern with greyish upperparts, brownish wings and whitish underparts, latter marked dark brown and rufous. – ♂: *Short crest*, which can be erected; *throat bib blackish*, bordered white; *tail greyish* with prominent black terminal band. – ♀: Like ♂, but smaller crest, and *throat brown, speckled white*, not black; colours duller, upperparts less clean grey.

VOICE Anxiety-call of ♀ a liquid 'piih-tettettettett'. Alarm repeated 'plit'. Song of ♂ a characteristic rhythmic phrase of *very fine notes* (comparable only to Yellow-browed Warbler, Goldcrest and Penduline Tit), 'tsiii-u-**iih** ti, ti-ti-ti-ti', the last series of notes, after a hint of a pause, rapid and falling slightly in pitch. Wing noise from flushed birds (often not seen in dense forest!) typical, 'burr, burr, burr,...'.

Capercaillie

Black Grouse

Caucasian Grouse

Hazel Grouse

CAPERCAILLIE

wing rather plain

♀

large, rounded rufous tail contrasts against greyish rump

a lek at dawn

orange-brown, unbarred

♀

♂

takes off with loud wing noise!

BLACK GROUSE

light 'braces'

Black Grouse ♀

coarsely barred

Capercaillie ♀

a lek on a bog

tawny-brown, barred

lyre-shaped

white

♀

♂

white wing-bar

white under-wing-coverts

♀

♂

CAUCASIAN GROUSE

displaying ♂♂ on a mountainside

♀

long

only moderately curved

black

♂

long

no wing-bars

♀

♂

HAZEL GROUSE

display posture

throat speckled white

black throat

♀

♂

dark terminal band

♂

DZ

PARTRIDGES and PHEASANTS *Phasianinae*

A very heterogeneous group, but all preferring open terrain rather than woodland. Found both on open plains and on mountain slopes above treeline. Even more ground-dwelling than the grouse; run quickly and athletically, flying only when need dictates. Nostrils unfeathered. Most have bare tarsi, and ♂♂ of several species are equipped with spurs. Food plant material and insects. Nest on ground.

Caucasian Snowcock *Tetraogallus caucasicus* —

L 50–60 cm. Very closely related to other snowcocks, including Caspian (see below). Endemic to Caucasus. Breeds on bare, rocky, boulder-strewn mountains at 2000–4000 m (rarely lower, mostly in winter), with only patchy low vegetation. Gregarious when not breeding, but flocks small. Shy and difficult to approach; soon takes wing and drops like a stone out of sight behind precipice edge or runs off even when several hundred metres away. Food plant matter.

IDENTIFICATION *Big and robust* with very large head, thick neck and long tail. Plumage intricately marked in grey, brown, white and black, but looks mostly *rather dark grey* at distance. *White patch on neck-side* and *rusty tinge on nape*. Flanks tinged reddish-brown, feathers heavily patterned black. *Breast and mantle finely vermiculated dark*. Large *white wing patches* (as on bustard; common to all snowcocks). Sexes similar, but ♀ slightly duller and with nape not so obviously rusty-brown. (For differences from Caspian Snowcock, see latter.) Agile and persistent mountain-climber, bounds and runs nimbly up slopes, often with tail raised enough to reveal white, *fluffed-up undertail-coverts*. Flies down at breathtakingly steep angle with long glides and short bursts of wingbeats. – Juvenile: Like ♀ but duller; upperparts lack red-brown elements and flanks lack contrasting markings.

VOICE Essentially as for other snowcocks. See Caspian Snowcock for a description. Song differs from latter's in that final notes drop in pitch.

Caspian Snowcock *Tetraogallus caspius* —

L 56–63 cm. Close relative of Caucasian Snowcock. Breeds at scattered sites in mountain regions, including in E Turkey (and so is the species most easily seen by W Europeans) and Armenia. Habits and habitat much as Caucasian but less of a high-alpine bird, mostly found lower down (1800–3000 m).

IDENTIFICATION *Big and hefty* like Caucasian Snowcock, or even a trifle bigger. Plumage similar, but overall appearance is *slightly paler*. If seen through telescope at moderate range, some minor differences are revealed: *nape is mainly grey*, lacking rusty-red; *breast light grey* finely spotted dark at centre (not vermiculated), *paler than flanks*; mantle and

breast-sides uniform grey, not vermiculated; *a small yellow-brown spot behind eye*; *flanks more grey-brown* than red-brown. Sexes similar, but slightly smaller ♀ is duller.

VOICE Flight-call a series of loud cackles, 'chok-ok-ok-ok-ok-...'. Rather muffled, chuckling conversational notes 'buk-buk-buk...', which now and then (when agitated) turn into drawn-out, bubbling trills (recalling ♀ Cuckoo) terminating in a whining 'loop', 'buk-buk-bu-bubububrrrrrr-rrrrrreyah'. Song, given with head thrown back and, near end, bill wide-open, a desolate, very far-carrying and echoing whistle ascending the scale in clear stages, 'sooo-luuu-dlee-**iiih**'; at distance somewhat like Curlew's call. Song heard almost solely at daybreak and in early morning.

Black Francolin *Francolinus francolinus* —

L 33–36 cm. Breeds on dry plains with access to water, especially cultivated fields but also tall-grass fields, untilled areas, e.g. riversides, with tamarisks and other shrubbery. Sedentary. Terrestrial. Shy and unobtrusive.

IDENTIFICATION Barely the size of a Grey Partridge but often with fairly *upright posture*, thus looks rather large. Somewhat slimmer build than Grey Partridge. – Adult ♂: Distinctive, with *black head and breast* separated by *broad reddish-brown neck-band*. White cheek patch, white spots on mantle and breast-sides. Belly black with white U-shaped spots. – ♀: More anonymous-looking: yellowish-brown with heavy dark barring and spotting. Told by *red-brown nape* and *black outer tail-feathers*.

VOICE Song highly characteristic and normally the first sign of the bird's presence, a constantly repeated, seven-syllable call with distinctive rhythm and cracked, mechanical voice, 'kyok, **kiiik** ki-**kii**-ko ki-**kiik**'; heard mostly at dawn.

Double-spurred Francolin *Pternistis bicalcaratus* —

L 30–33 cm. Local and rare breeder in Morocco (main range S of Sahara) in open woodland, wooded pasture, palm groves, cultivated fields, along river shores with thickets and also in vegetated wadis. Sedentary. Habits as Black Francolin. Shy, difficult to see, slips away smartly.

IDENTIFICATION Size of Barbary Partridge (which occurs in same area) but has *red-brown crown and nape*, *all-brown tail* without red-brown outer feathers, *heavily black-streaked underparts* (not barred flanks), black forehead and narrow white supercilium. Bill and legs yellowish-grey or greenish. Sexes alike, but ♂ has small spurs.

VOICE Utters an irregularly repeated, disyllabic 'i-teck'. Best located by its song, delivered mostly at dawn from an elevated perch (tree, rock, mound, post), a repeated, low-pitched 'kuarr kuarr'. Also gives loud hacking and jarring calls, 'krrrrr krrrrr...'.

Caucasian Snowcock

Caspian Snowcock

Black Francolin

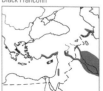

Double-spurred Francolin

CAUCASIAN SNOWCOCK

display posture,
calling ♂

greyish-tinged
rusty-red, often
some cinnamon

dark

dark,
heavily
barred
and ver-
miculated

heavy
flank-stripes

birds in more arid areas
paler and brighter, some
with cinnamon 'moustache'

CASPIAN SNOWCOCK

no overlap
in distribution
so confusion
unlikely

dark
grey

pale

plenty
of white
in wing
(both species)

pale
grey,
sparsely
spotted

weak
flank-stripes

some birds in Armenia have
plain, buffier breast, hence are
closer to nominate *caspius* of Iran

'tauricus'
(Turkey)

BLACK FRANCOLIN

black tail in
both sexes

♂

♀

red-brown

white
cheek
patch

black

♀

♂

DOUBLE-SPURRED FRANCOLIN

sometimes
perches in
trees

rufous
cap

pale
super-
cilium

red-brown

heavy
striping

grey
rump

BARBARY
PARTRIDGE (to compare)

DZ

Red-legged Partridge *Alectoris rufa*   rB2

L 32–35 cm. Breeds mainly in lowlands, selecting varied habitats, including cultivated farmland with pasture, heathland, unworked sandy or stony terrain with low vegetation, coastal meadows; sometimes found also in mountain areas above treeline. Introduced in Britain. Resident. Gregarious, often in flocks. Wary. Runs away quickly when disturbed. Flies only if pressed, flight then low with rapid beats and stiff-winged glides. Most often located by its calls.

IDENTIFICATION Virtually no range overlap exists between the four similar *Alectoris* species. Nevertheless, best to check: *small white bib* framed by coarse, *broad black border* which at lower edge breaks up into *heavy black streaking on breast*; *brown hindneck* (greyer on other species). Like Grey Partridge has *reddish-brown tail-corners*, but is *grey on lower back*, *rump and central tail*, this well visible when flushed.

VOICE Rhythmic series of hoarse notes. Territorial call, often heard morning and evening (but midday, too), begins with a few clucks only to turn into a three-syllable phrase at cantering pace, the last two notes raucous, 'chu chu chu chu ka-**cheh-cheh**, ka-**cheh-cheh**, ka-**cheh-cheh**,...'; the final phrases are sometimes five-syllabled, 'ku-**kak** kaka-**cheh**,...'.

Rock Partridge *Alectoris graeca*   —

L 33–36 cm. Breeds on rocky alpine slopes generally above treeline (but also in clearings and among scattered trees), mostly at 1000–3000 m, often with grass, scrub and low bushes; avoids N-facing slopes. Has declined. Resident. Gregarious. Behaviour as for Red-legged Partridge.

IDENTIFICATION Like Chukar; differs in: *pure white bib*; *sharply demarcated black border to bib* (varying in width in breast-centre: broad in Alps, narrow or even broken in Sicily); *narrow* white supercilium which often *also crosses forehead*; a little *black bordering base of upper mandible*.

VOICE Territorial call usually irregular series of short, throaty, very hard notes, now and then interrupted by a rapid choking series, 'chi chek pe-ti-**chek** chi-chek chik chi-chek chi-**cheay-cheay-cheay**, pe-**te**-ti-ti**chek** chik te...' and so on, thus often without the repetitive pattern of its relatives.

Barbary Partridge *Alectoris barbara*   —

L 32–35 cm. Breeds in a wide variety of habitats and at different altitudes, from bare, rocky mountain slopes (up to 3000 m) among boulders and shrubs, through open woodland and clearings at lower levels, down to dry, open terrain in lowlands. Occurs spontaneously in Canary Islands; introduced in Madeira. Sedentary. Gregarious. Behaviour as for Red-legged Partridge.

IDENTIFICATION Has *broad, dark reddish-brown neck-collar bestrewn with white spots*; dark *red-brown central crown-band*;

no white bib, is instead *light grey on throat and head-sides* with diffuse light brown band backwards from cheek; *shoulders attractively patterned in pale blue-grey and rusty-brown*.

VOICE Territorial call often typical: series of short, impure notes (sometimes hint of double syllable) with an occasional hoarse, drawn-out interjection, 'tre tre tre tre tre tre **cheehch**e tre tre tre tre tre...' (trotting rhythm with a stumble). But at times very like Red-legged Partridge, with series of trisyllabic 'tra-**che-che**'. Also an upslurred 'tshuuih'.

Chukar *Alectoris chukar*   [V***]

L 32–35 cm. Breeds in similar habitats to closely related Rock Partridge, but also descends lower and is found around cultivations, pasture on barren terrain etc.; also in mountain deserts. Has declined in Europe. Said to be shyer and more difficult to see than Rock Partridge.

IDENTIFICATION Like Rock Partridge; differs in: *yellow-tinged* white bib; sometimes some *black spots at bottom of bib* in breast-centre; *broad, diffuse pale supercilium* not reaching forehead; *no black bordering base of upper mandible*; *brown streak behind eye*; browner upperparts.

VOICE Territorial call variable, often rhythmic series at fast galloping pace and with hoarse voice, introduced by several short, monosyllabic, shrill notes, 'ga ga ga ga cha-**chakera-chakera-chakera**-...'; structure at times simpler, e.g. 'chak chak chak che-chak **truc**-ku **truc**-ku **truc**-ku **truc**-ku...'.

Sand Partridge *Ammoperdix heyi*   —

L 22–25 cm. Breeds in mountain deserts, on rocky slopes, in wadis and similar but with access to water. Often not shy, and easy to approach. Behaviour otherwise as *Alectoris*.

IDENTIFICATION Like a *small pale* Grey Partridge, and has similar *red-brown tail-corners*. – ♂: Sandy-brown with *blue-grey head* and *white loral/cheek patch*. Bill orange-yellow. Flanks with wavy striping of white/black/brown. – ♀: Much plainer than ♂; lacks white on head. Bill dull yellow.

VOICE Song a usually rapidly repeated short call (1½–2 per sec.), slightly downslurred and ending sharply, 'kiwa kiwa kiwa...', or (at distance) slightly softer 'ua ua ua...'.

See-see Partridge *Ammoperdix griseogularis*   —

L 22–25 cm. Breeds in similar terrain to Sand Partridge, but on average on more level terrain and sometimes also at somewhat higher altitude (to 2000 m).

IDENTIFICATION Like Sand Partridge, but ♂ has *tapering black band from forehead to nape*, bordering white cheek patch; *grey neck-sides are speckled white*. ♀ duller.

VOICE Song a repeated upslurred whistle (a good 1 per sec.), at distance recalling Spotted Crake (of all species out in middle of the desert!), 'ho**it**, ho**it**, ho**it**,...'.

Red-legged P.   Chukar   Rock Partridge

Barbary Partridge   Sand Partridge

See-See Partridge

RED-LEGGED PARTRIDGE

brown back

reddish-brown

broad, black-streaked 'neck-shawl'

ROCK PARTRIDGE

black rim along entire base of upper mandible; narrow light supercilium

white

BARBARY PARTRIDGE

dark median crown-stripe

light grey

red-dish-brown

CHUKAR

black rim along base of upper mandible only on forehead; broad, diffuse light supercilium

cream

SAND PARTRIDGE

♂

white patches on cheek and lore

like a washed-out version of ♂

plain

♀

♂

SEE-SEE PARTRIDGE

♂

black supercilium and forehead

very similar to ♀ Sand Partridge

speckled white

♀

♂

Grey Partridge *Perdix perdix* rB3

L 28–32 cm. Breeds in open farmland with some hedges and other shelter, preferably in or beside cultivated fields; also meadowland. Has declined greatly in Britain. Resident. Terrestrial. Gregarious, flocks keep tightly together. Nervous and wary. Freezes or runs away in good time. Food seeds and leaves. Nests on ground; clutch large, 10–20 eggs.

IDENTIFICATION Compact, *rounded body* and small, rounded head. Usually encountered as an entire flock is flushed, all at once, and with excited calls and noisy, rapid series of wingbeats and short stiff-winged glides flies a short way low over the ground; the image caught is of grey-brown birds with *rusty-red on the tail*. Seen closer, this is a handsome bird, with *orange-brown 'face' and throat*, chestnut-brown flank-bars, delicately vermiculated *ash-grey breast* and a large *horseshoe-shaped blackish-brown belly patch*. Sexes similar (♂ 'cleaner', with better-marked pattern). ♀'s belly patch smaller, less clear. – Juvenile: Yellow-brown and grey-brown; lacks orange throat, grey breast and dark belly patch. (More like a chick Pheasant or adult Quail than a Grey Partridge.) After 2–3½ months assumes adult-like plumage.

VOICE On rising, gives sharp, excited, short 'prri prri prri...' or 'rick-rick-rick...'. Song (by both sexes) a hoarse and abruptly clipped 'kierr-ik', repeated a few times at moderate intervals, often heard at dusk or at night.

(Common) Quail *Coturnix coturnix* mB4–5

L 16–18 cm. Breeds on open farmland, drawn to vast plains with clover pasture and young corn fields. Warmth-loving; winters in Africa. Heavily hunted on passage in Mediterranean region. In warm springs quite a number of night-flying migrants 'overshoot' their destination and end up farther N, when far more than usual heard calling in May/Jun. Nests on ground, often in crops; clutch 8–13.

IDENTIFICATION Often heard but rarely seen; keeps well hidden, reluctant to fly by day. *Very small*, like a rather small Pheasant chick, which it somewhat resembles in plumage: *buff-brown* with dark brown markings above and on breast and head, and with narrow whitish streaks above and on flanks. ♂ is black on centre of throat (extent variable, often just a narrow band, thus best judged in front view), ♀ is dirty white. If you do flush a Quail, or see a raptor do so, it looks unexpectedly *long-winged*, different from other gallinaceous birds; *wingbeats fast, flightpath low and direct*; quickly drops back into cover.

VOICE When flushed, a wader-like soft, rolling 'wrree'. Song, mainly at dusk and daybreak (but also at other times of day), a rhythmic, trisyllabic phrase rapidly (1 per sec.) repeated 3–8 times: a sharp whistle, far-carrying, with stress on first and last syllables and with the final two syllables close together, '**büt** bül-**üt**' ('wet-my-lips'); close to, a throaty mechanical secondary note (roughly as that of Spotted Crake) is audible; sometimes also a few muffled, nasal, creaky 'mau-wau' introductory notes.

Small Buttonquail *Turnix sylvaticus* —

(Old evocative name: Andalusian Hemipode.) L 15–17 cm. One of the area's least-known species, now extinct in Europe and very rare in Morocco (but widely distributed in sub-Saharan Africa). Related to the waders but kept here due to superficial similarity with Quail. Lives mostly on dry heaths with low dwarf palm, asphodel, etc. Also in overgrown pumpkin fields. Resident. Roles of sexes reversed (as in phalaropes). Nests on ground. Long breeding season, Apr–Sep.

IDENTIFICATION Rarely seen, presence noted mostly by its voice. Hard to flush more than once. *Very small*. Short-tailed. Brown above and whitish below, with *orange patch on breast* and *coarse dark spotting on breast-sides*. ♀ more brightly coloured and somewhat bigger than ♂. Rises with loud *wing noise*. Is quite *short- and round-winged*, unlike Quail. On landing, stands momentarily in characteristic upright manner with wings held out. *Upperwing-coverts* yellowish-brown, contrastingly *paler than dark primaries*.

VOICE ♀'s territorial call a low-pitched, straight, second-long growling hoot, repeated a few times at 2–3-sec. intervals, 'hoooo... hoooo... hoooo...', each hoot initially increasing slightly in strength. Heard most at dusk and dawn.

Corncrake *Crex crex* mB4

L 22–25 cm. Belongs to the rails and crakes (pp. 124–127), but is not so tied to wet marshland as its relatives. Breeds on damp meadows by marshy lowland lakes, but also on lush meadowland and hay fields (where usually wiped out by silage-cutting, haymaking, etc.) with access to wetter spot with taller vegetation. Has declined. Now very rare and local in Britain & Ireland. Summer visitor, winters E Africa.

IDENTIFICATION Half the size of a Grey Partridge and much slimmer, like a Water Rail in shape (quite long neck, rounded body) but with *short, stubby bill*. Plumage *greyish yellow-brown* with *blue-grey band over eye and grey on breast-sides* (grey colour less obvious on ♀); *heavily dark-spotted above*. In flight, which appears 'flappy' and unsure, shows *red-brown wing patch* and prominent, *dangling legs*. Hard to flush, sneaks away cleverly. Most often seen at dusk and dawn, when courting ♂'s head sticks up above the vegetation.

VOICE Display-call is heard from dusk to morning (at times briefly in daytime, too), a persistently repeated hoarse and mechanical, sharp rasping 'ehrp-ehrp' (or, if you like, 'crex-crex'), once per second and for hours on end with only brief rests. Otherwise rather silent.

Grey Partridge

Quail

Small Buttonquail

Corncrake

GREY PARTRIDGE

hardy, will endure much snow and cold weather

orange-brown

compared with Quail, bigger and chunkier; can also be mistaken for a Pheasant chick (p. 58)

dark brown belly patch

♀

juv.

ad. ♂

(♀ is similar, only somewhat duller and has diffuse supercilium and smaller, or no, belly patch)

QUAIL

long-winged

pale throat

black throat-centre

difficult to flush, steals discreetly away in cover

♀

♂

SMALL BUTTONQUAIL

pale iris

orange-brown in ♀ Capercaillie fashion

short-tailed, chicken-like look

ad. ♀
(♂ has slightly duller colours)

pale coverts

hard to flush, squats tightly

flight action recalls Jack Snipe rather than Quail!

CORNCRAKE

legs dangling before alighting

red-brown wings

long legs

usually keeps in lush cover when displaying

(Common) **Pheasant** *Phasianus colchicus* ［r**B**1

L ♂ 70–90 cm (incl. tail *c*. 35–45), ♀ 55–70 cm (tail *c*. 20–25). Introduced in Europe (in Britain probably from 11th/12th century); wild populations occur E Black Sea and in Caucasus, eastwards to China. Breeds in farmland areas with cover such as copses with dense undergrowth, scrubby thickets, conifer plantations, also in large gardens and parks. Resident. Roosts in trees, sometimes in reeds. Food plant material, insects etc. Nests on ground.

IDENTIFICATION Adult ♂ : Unmistakable with *long, barred brown tail*, the dark *head glossed greenish-black and violet* and with bare, warty, *red head-sides*. Many in Britain have narrow *white neck-collar* and grey rump (ssp. *torquatus* from China), but ♂♂ lacking white collar and with brown rump (typical of nominate *colchicus* from SW Asia, incl. Caucasus) not uncommon; most, however, a mixture of these (and other, also introduced) races. – ♀: Buff-brown with dark angular spots and feather centres. Rather nondescript, but long tail identifies it as a pheasant. See also under the other pheasant species.

VOICE Alarm, often when flushed, a hacking series of disyllabic, slightly hoarse calls with stress on first syllable, '**ku**-tuk **ku**-tuk **ku**-tuk...', often decreasing in strength. ♂'s song far-carrying, abrupt, an allied hacking call of two syllables, not unlike clearing one's throat, usually immediately followed by a short, noisy wing-flutter, '**ko-ohrk**-tuk (burrrr)'; the cock repeats this at long intervals (1–15 min.) from a well visible perch on the ground, log pile, fence post or similar site. A range of other soft calls.

Golden Pheasant *Chrysolophus pictus* ［r**B**4?]

L ♂ 90–105 cm (incl. tail *c*. 60–70), ♀ 60–80 cm (tail *c*. 30–35). Indigenous in mountains of C China. Introduced in Britain since late 1800s; several self-sustaining populations (main ones SW Scotland, S & E England, Anglesey), inhabiting younger dense, dark, rather bare-floored pine and larch forest, rarely also mixed forest with richer understorey. Closely related to Lady Amherst's Pheasant (see below).

IDENTIFICATION Adult ♂: Unmistakable. Despite its *bright red colours on underbody and tail-sides*, and the yellow areas on its crown and back, the cock often avoids detection by slipping stealthily through the vegetation. *Orange-yellow nuchal cape* can be raised like a *fan* when displaying. – ♀: Resembles ♀ Common Pheasant, though proportionally longer tail and legs impart more elegant impression. Both this and ♀ Lady Amherst's differ from ♀ Pheasant in overall slightly darker, more *densely barred* plumage and subtly discrete 'cloak' of feathers over nape. ♀ Golden has somewhat *paler* yellow-brown *forehead*/forecrown, comparatively *unobvious yellowish-pink orbital skin, yellow-tinged legs,*

barred belly and a little *less boldly barred tail* than ♀ Lady Amherst's.

VOICE ♂'s song a one- or two-syllable, gruff, piercing'ehk' or 'eh-aik' (with second syllable one note higher), usually heard in spring; higher in pitch than Common Pheasant's territorial call. Otherwise rather silent.

Lady Amherst's Pheasant *Chrysolophus amherstiae* [r**B**5]

L ♂ 105–120 cm (incl. tail *c*. 75–90), ♀ 60–80 cm (tail *c*. 25–30). Breeds in mountain forests of SW East Asia. Introduced in Britain in 1900s; a few small feral populations survive very locally in S England (esp. Bedfordshire), where they live in woods and thickets with rich, dense understorey (e.g. bramble, rhododendron). Close relative of Golden Pheasant, and the two occasionally hybridize in England.

IDENTIFICATION Adult ♂ unmistakable, pyrotechnical display of feathered splendour! The broad white feathers with black (blue-glossed at certain angles) tips which adorn the nape and normally droop like a *silver-white cape* can be raised to form a *fan* when courting the ♀. The herringbone-patterned *grey-white tail is extremely long*. – ♀: Basically like Common Pheasant, but is more *finely and evenly barred above and on neck*, not so coarsely spotted as latter. Very like ♀ Golden Pheasant; differs in *darker red-brown forehead, crown and throat*, more conspicuous *pale greyish orbital skin, greyish legs*, unbarred belly and more *boldly barred tail*.

VOICE Song often heard at nightfall, an 'aahk aik-aik', thus as Golden Pheasant's but with *double-note* ending.

Helmeted Guineafowl *Numida meleagris* —

L 58–68 cm. Now probably extinct in Morocco, the last place where a once healthy NW African population remained. Introduced and domesticated in e.g. S France. Breeds in drier, savanna-type country with bushes and scattered trees, also in farmed areas with some clumps of trees and bushy cover; shuns deserts and marshes. Gregarious, seen mostly in small or rather large (often of *c*. 25 birds) flocks. Roosts in trees, like pheasants; also takes 'siesta' in shade. Resident. Food mostly plant material, some insects.

IDENTIFICATION Fully as *big as a chicken*, with heavy, *compact body*, narrow neck, small head, broad, rounded wings, rather long tail and fairly long legs (thus somewhat turkey-like). Plumage violet-tinged *grey, sprinkled with white spots*. Head striking: a brownish *knob* on crown, and *bare blue-white and red skin on sides and on forehead*. Greyish-black upper neck and sparse, 'unkempt' hair on nape and hindneck. Sexes alike. When disturbed, the flock runs quickly into cover, all birds with neck extended. – Juvenile: grey-brown with pale spots. Head markings much as those of a ♀ pheasant.

VOICE All calls raucous, unmusical, with several loud. Wide variation, and many calls merge into each other depending on mood etc. ♀'s include a disyllabic'ka-bak'. Other calls are hard staccato 'tek-tek-tek-tek-...', at times mixed with hard rolling calls, 'tk trrrrrrrrrrr tk tk tk...', also a plaintive 'cher-chih, cher-chih,...' and so on.

Pheasant   Golden Pheasant   Lady Amherst's Pheasant

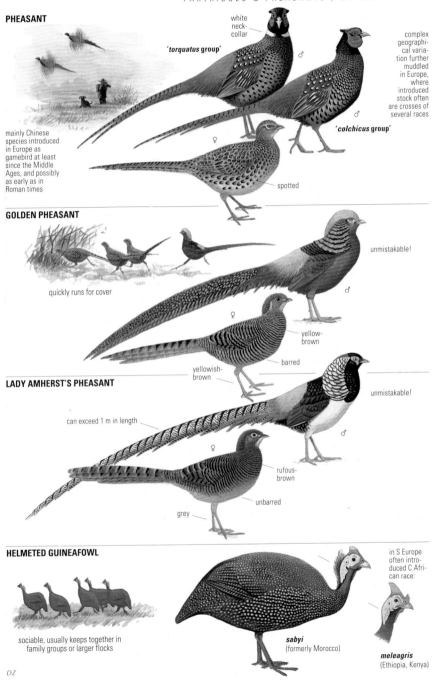

PHEASANT

white neck-collar

'torquatus group'

♂

complex geographical variation further muddled in Europe, where introduced stock often are crosses of several races

'colchicus group'

♂

mainly Chinese species introduced in Europe as gamebird at least since the Middle Ages, and possibly as early as in Roman times

♀

spotted

GOLDEN PHEASANT

unmistakable!

♂

quickly runs for cover

♀

yellow-brown

barred

yellowish-brown

LADY AMHERST'S PHEASANT

unmistakable!

♂

can exceed 1 m in length

♀

rufous-brown

unbarred

grey

HELMETED GUINEAFOWL

in S Europe often introduced C African race:

sociable, usually keeps together in family groups or larger flocks

sabyi
(formerly Morocco)

meleagris
(Ethiopia, Kenya)

DZ

LOONS (DIVERS) *Gaviidae*

Four regular species, all rather large. Loons, also known as divers, usually swim low (like cormorants, unlike ducks), but float higher when resting or preening. Long body and neck. Feet placed far back, making movements on land ungainly. Differ from cormorants in silhouette, shape of bill and in being 'tailless', from grebes in larger size and thicker neck, from mergansers in bill shape and longer neck.

To catch food (mainly fish), they dive with smooth, neat 'bow'(not leaping, as cormorants often do), often remaining submerged for a minute or more, covering sometimes long distances under water. Dives often preceded by recon-naissance from surface with head submerged for long pe-riods ('snorkelling').

Breed mainly at pools and lakes in woodland or on tundra. Nest a shallow cup close to the water. Require stable water-level by the nest. Winter along coasts or in bays, sometimes on lakes, at times in larger, loosely scattered gatherings, more often seen singly or in small parties. Migrate (Apr–Jun and Sep–Nov) mostly over sea, singly or in small parties (rarely larger flocks), birds within flocks widely spaced, never in dense formation (like ducks or cormorants). Often fly high when wind is light, or in tail-winds, with continu-ous beats of long narrow wings, and on direct course.

GREAT CRESTED GREBE — winter — 'flickering' quick wingbeats — grey-buff — winter — white — *ad. winter*

RED-THROATED LOON — ad. winter

GREAT NORTHERN LOON — light — ad. winter — feathers often raised — dark — *ad. winter* — sloping — long tail, floats on surface or submerged

CORMORANT — young — young

K M

IDENTIFICATION OF LOONS

Identification easy when in distinctive and strikingly beau-tiful adult summer plumages. Immature and adult winter plumages less distinctive, but swimming birds can general-ly be identified, given reasonably close views. Look careful-ly for the full range of differences before making a firm diagnosis: these mainly involve size; bill shape and bill colour; precise pattern and extent of light and dark areas on head and neck (often easier to discern at long range than size and bill shape); and pattern of fine markings—if any—on upperparts.

Identification of flying birds much more tricky: requires much practice in making rather difficult assessments of size; structure (esp. thickness of neck, 'hunchbackedness', and foot projection); and depth and speed of wingbeats (variable according to wind speed and relativewind direction). At closer ranges, look also for the diagnostic differences in colour, shape and size of bill, and in the pattern and extent of light and dark on head and neck. Odd flying loons at a distance must be left unidentified even by experienced ob-servers. Flying loons are illustrated on p. 62.

Red-throated Loon *Gavia stellata*   r**B**4 / **W**3

L 55–67 cm (excl. feet *c.* 7), WS 91–110 cm. Commonest loon in most regions. Breeds on often small and fishless pools on tundra or on forest bogs, commuting to larger lakes or coast for food.

IDENTIFICATION Smallest loon, but only slightly smaller than Black-throated. Bill uptilted with straight culmen, usually held *pointing slightly upwards*. Forehead flat, and inclination to have angled hindcrown. Usually noticeably flat-chested (useful esp. at long range; Black-throated has prominent chest). In flight, look for *slimmer neck*; also, more 'sagging' neck giving hunchbacked impression; bill often pointing slightly upwards in flight; usually *mod-est foot projection*, making wings set behind centre (but a few have feet projecting more, as on Black-throated); generally somewhat *faster and deeper wingbeats* than Black-throated in comparable winds; and rather common *habit of moving head* as if 'calibrating vision'(Black-throated does this only infrequently). – Adult summer: Rufous neck patch is dark, and can look black at a distance, so (pale) grey throat and plain upperparts best long-range marks. – Adult winter: In profile, *more than half of neck is white* (50/50 on Black-throated), visible in flight too, and, seen from behind, some white visible on sides of neck (Black-throated usually en-tirely dark-necked from rear), and a little white surrounds eye (esp. in front); whole upperparts finely speckled with white. Flanks show as a complete pale band (mottled) above waterline (if any pale visible), usually not as a conspicuous white patch at rear. – Juvenile (often retained to midwinter, partly to first spring): As adult winter, except white areas on head and *neck have fine, dusky streaking* (often extensive), not pure white, and upperparts have duller, greyer speck-ling. – 1st-summer: Summer plumage only partial. Often twin dark stripes on foreneck in transition.

VOICE Flight-call loud and monotonous (somewhat goose-like) cackling, 'kah kah...'. Song is far-carrying duet of ♂'s loud, 'grinding' 'oo **rroo**-uh, oo **rroo**-uh, oo **rroo**-uh,...' and ♀'s stronger, higher-pitched '**arro**-arr o-arro-...'. Also a drawn-out, wailing 'eeaaooh'. Some calls resemble barking of a fox. Silent in winter.

RED-THROATED LOON

juv. has less white on head and neck than ad. winter

ad. sommar

looks black at distance

white sides

ad. winter

white in front of eye

low, sloping forehead

finely striped

uniform, grey

extensive white

tilted upwards

plain

narrow rusty strip

juv.

finely speckled white

ad. winter

flat-chested

ad. summer

BLACK-THROATED LOON

juv. is paler and more brown-grey than ad. winter

ad. summer, displaying

no white sides

ad. winter

often darker edge

discreet scaling

lacks white eye-ring

often peaked forehead

velvety, light grey

blackish

wide black patch

juv.

riding high (relaxed), shows large white flank panel

blackish, almost uniform

white 'blocking'

level

darker border

'full' chest

ad. summer

ad. winter

distinct white flash

KM

Black-throated Loon (Arctic Diver)

Gavia arctica   r**B**4 / **W**4

L 63–75 cm (excl. feet *c.*10), WS 100–122 (127) cm. Unlike Red-throated, breeds on clear, fish-holding freshwater lakes (or sea-bays without tide), builds nest on small islet near waterline. Often gregarious in winter.

IDENTIFICATION Appears somewhat larger than Mallard, though shape slimmer. (A few large individuals same size as smallest Great Northern.) Dagger-shaped bill with slightly curved culmen, slimmer than Great Northern's; *carried horizontally* (or nearly; Red-throated's more clearly upwards). Forehead often steep, hindcrown smoothly rounded. *Thick neck* (same-thickness head and neck), and has prominent chest at waterline (Red-throated more flat-chested). In flight, compared with Red-throated, *neck slightly thicker and carried straighter* (less 'sagging'), as is bill; feet always large and obvious, *projecting prominently*, making wings set close to centre; slightly slower and shallower wingbeats on average than Red-throated in comparable winds (but dependent on individual size; small birds more like Red-throated, large more like Great Northern). – Adult summer: Velvety pale grey crown and hindneck, black foreneck and throat, and a pattern of *distinct white patches on upperbody*. – Adult winter: In profile, *at least half of neck dark*, visible in flight too, and border between grey hindneck and white foreneck usually *highlighted by swarthy line.*

Generally no white around eye (but beware juv. Red-throated, which can show reduced white). Swimming bird often *shows distinctive, isolated white patch on rearmost flanks*. Upperparts plain dark grey. Bill (pale) grey with blackish tip. – Juvenile (often retained to midwinter, partly to first spring): As adult winter, except pale feather fringes on upperparts form neat, fine scaly pattern; upperparts slightly paler and browner; tip of bill less blackish. – 1st-summer: Summer plumage only partial.

VOICE Hard, croaking 'knarr-knorr' and drawn-out gull-like wailing '**aaah**-oh' are heard from territory, often at night. Song is loud, desolate but evocative, rhythmic whistling 'cloo**ee**-co-cloo**ee**-co-cloo**ee**-co-cloo**ee**', each 'cloo**ee**' a strongly rising whistle. When display-diving, sometimes a fierce, abruptly cut-off 'co**eet**!'. Generally silent in flight (only rarely subdued 'karr-arr-arr'). Silent in winter.

Red-throated Loon

Black-throated Loon

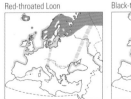

Great Northern Loon *Gavia immer*   **P / W**4

L 73–88 cm (excl. feet *c.* 13), WS 122–148 cm. Breeds on large, deep lakes and bays in woodland, and on tundra. About 300 pairs in Iceland. Main range in North America.

IDENTIFICATION Large and strongly built, much as a Cormorant. (With size variation, smallest match large Black-throated Loons, largest may equal Yellow-billeds.) When swimming, note *heavy, horizontally held bill*, large head, and on most rather thick neck; *forehead steep*, and can be raised to create small *crest-like 'bump'* making *crown appear concave*. When floating high, pale (mottled) flanks often show as complete light band above waterline (as on Red-throated). In flight: thick neck held straight out; *large well-protruding feet*, often with spread toes adding to large size; comparatively slow wingbeats with 'elastic' outer wing sometimes characteristic (but still not a great difference from the flight of large individuals of Black-throated); and *greater bulk and bill size* at times obvious. – Adult summer: *All-black head and white-chequered upperparts* and white-striped patch on side of neck. – Adult winter: Bill, carried near *horizontally*, is bluish *grey-white with dark tip and culmen*. Check pattern of dark and light on head and neck: prominent *pale eye-ring*, rather *distinct demarcation between dark hindneck and white foreneck*, and *blackish half-collar on lower side of neck* emphasized by *white indentation* above it. Contrast between darker hindneck and paler upperbody. Lacks juvenile's scaly upperparts, and retains some inconspicuous white-spotted lesser wing-coverts from summer. – Juvenile: As adult winter, except that rather broad, pale fringes on upperparts form prominent, *neat scaly pattern* (at long range giving paler appearance to upperparts than on juv. Black-throated); wing-coverts lack fine white spots; and pale eye-ring less obvious. – 1st-summer: Remains in fully juvenile-like plumage. – 2nd-winter: As adult winter, but upperparts often slightly darker, and has paler, not so blackish bill-tip; lacks white-spotted lesser wing-coverts.

VOICE Frequently heard (incl. as 'atmosphere' in films) is anxiety/contact-call, a repeated high-pitched eerie laughter which often opens with a lower note, 'ho-yeyeyeyeya'. The song is a wailing with structure reminiscent of Black-throated Loon, 'aaoooh... **wee** we-a **wee** we-a **wee** we-a'. Also a monosyllabic, moaning, drawn-out call.

Yellow-billed Loon (White-billed Diver)
Gavia adamsii   **V**∗∗

L 77–90 cm (excl. feet *c.* 14), WS 135–150 cm. Breeds on arctic coasts of Russia, wintering in small numbers along coasts of N Norway and rarely in North Sea and Baltic.

IDENTIFICATION Largest loon, on average slightly larger than Great Northern. Bill usually held pointing clearly upwards, and practically straight culmen (exceptions insignificant) and marked upward angle of gonys add to *distinctive uptilted look*. *Bill-tip and outer part of culmen invariably pale*. Neck thick (same-thickness head and neck). Against dark background, pale colour of bill makes it look large, but at long range or against pale background not always so massive-looking. In flight, slightly larger size than Great Northern makes 'weight' and slower wingbeats more obvious in comparison with Black-throated. – Adult summer: *Bill yellowish-white* or ivory-coloured (all black on Great Northern, but can gleam pale in strong light, or rarely have extensive whitish tip in summer). White markings on neck and upperparts larger and fewer than on Great Northern. – Adult winter: Bill less yellowish and often with dark shade at base (esp. along culmen, at times to 2/3 of length). *Head and hindneck paler grey-brown*, and light throat/foreneck more tinged brownish than on Great Northern (which is more contrasting greyish-black and white); also *more pale around eye and on side of hindneck*, behind ear. (In late winter, a few much-faded Great Northerns may approach, but never quite match, Yellow-billed's paleness.) The more extensively retained summer wing-coverts have larger and

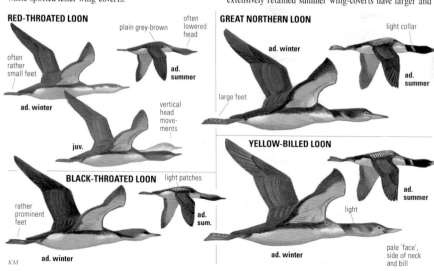

RED-THROATED LOON   often lowered head   plain grey-brown

often rather small feet

ad. summer

ad. winter

vertical head movements

juv.

BLACK-THROATED LOON   light patches

rather prominent feet

ad. sum.

ad. winter

GREAT NORTHERN LOON   light collar

ad. winter

large feet

ad. summer

YELLOW-BILLED LOON

ad. summer

light

ad. winter

pale 'face', side of neck and bill

KM

GREAT NORTHERN LOON

dark

neat scaly pattern

juv.

dark cul-
men and
tip of bill

'snorkelling'

often very
faded

usually steep fore-
head and flat crown

greyish
white

1st-summer (worn)

level, grey-white

dark half-collar, and
white indentation
above

ad. spring
(at start of moult)

dark, plain

ad. winter

ad. summer

ad. summer

YELLOW-BILLED LOON

juv. → **1st-winter**

pale
neck-
side

bold scaly pattern

juv.

pale tip and
outer culmen

grey, becoming
buff-yellow
toward tip

appearance of bill colour
and bill shape vary according to background

barring a little more
pronounced than in
Great Northern

pointing upwards,
yellowish-white

half-collar often more
extensive than in Great
Northern

bill strikingly
pale, often
visible at
great distance

ad. winter

rather large

broad
white
stripes

ad. summer

ad. summer

KM

therefore more easily seen white spots than those of Great
Northern Loon. – Juvenile: Upperparts have *neat, scaly
pattern*. Pale overall impression. Less typical bill shape than
in later stages. – 1st-summer: Remains in fully juvenile-like
plumage. – 2nd-winter: Upperparts appear all dark. No
white-spotted wing-coverts. – 2nd-summer: Mixture of
adult and juvenile. Scattered 'sugar lumps' on upperparts,
and head and neck (incl. foreneck) largely uniformly sooty-
grey (lacking white markings of adult).
  VOICE Similar to that of Great Northern Loon.

Great Northern Loon

Yellow-billed Loon

GREBES *Podicipedidae*

Aquatic birds, highly specialized and accomplished divers (counterpart to the loons in shallow, well-vegetated lakes and shallow sea-bays). Streamlined body with feet placed far back to facilitate fast diving. Move awkwardly on land. Toes lobed (feet thus lacking the complete webbing of loons and wildfowl). Flight almost invariably close over water, appears laborious, wings small. Grebes in flight shown on p. 66. Food fish, aquatic insects, plants, etc. Build a floating nest. Five breeding species and an American vagrant.

Slavonian Grebe (Horned Grebe)
Podiceps auritus   m**B**5 / **W**4

L 31–38 cm, WS 46–55 cm. Breeds on shallow, well-vegetated lakes; accepts small ponds with little open water so long as they do not dry up in summer; also on clear, open ponds in uplands in N Fenno-Scandia. Winters on coastal waters, lakes, reservoirs, mainly at W European coasts.

IDENTIFICATION Rather small, about as large as a Teal. Longish neck, *flat crown peaking at rear*, flat-topped head when viewed head-on, and longish body give shape like miniature Great Crested rather than Little Grebe. 'Powder puff' rear end not usually so prominent as on Black-necked. *Bill straight and rather short*, often with *pale tip*. In flight, shows *small white shoulder patch* and *white speculum confined to secondaries*. Flight somewhat auk-like, path veering; migrants have habit of raising head high (at times feet too) in flight for a brief moment or longer. – Adult summer: Unmistakable, but in transition to or from summer plumage dark on ear-coverts can give pattern like winter Black-necked; at distance, reddish foreneck can look black. – Adult winter: Sharply contrasting black-and-white pattern on head; neat black cap sharply demarcated from white cheeks, with rather *straight division from bill through eye* (no prominent dusky downward bulge on ear-coverts) and only *slight if any 'hook-back' at nape*; head-on, cap appears as *forward-tilted 'black beret'*; from rear, *very thin dark line down upper hindneck* between 'wrap-around' white cheeks; dusky band across upper foreneck sometimes extensive (as on Black-necked); at close range, *whitish spot on lores*. – Juvenile: Like adult winter, but with dusky band across cheek, slightly browner above, and with more extensive pale on bill-base. Caution: some individuals less well marked, and distance or bright light can give Black-necked more contrasting black-and-white head pattern than usual.

VOICE Commonest call a rather feeble but far-carrying, trembling or rattling 'hii-arrr', falling in pitch. Display-call is a squealing, *pulsating* trill, each wave of rapid, giggling notes turning into whinnying ones which descend and end with a nasal drawn-out note. Mainly silent in winter.

Black-necked Grebe *Podiceps nigricollis*   m**B**5 / **W**4

L 28–34 cm. Breeds, usually in small colonies, on shallow ponds and lakes with much emergent vegetation, frequently associated with Black-headed Gulls or Black Terns; prefers access to rather large patches of open water. Winters on coastal waters, lakes, reservoirs in W and S Europe.

IDENTIFICATION Slightly smaller than Slavonian Grebe, and more elegant with *thinner neck*. Dumpy body and usually prominent 'powder-puff' rear end give shape recalling Little Grebe; *steep forehead*, with *crown rounded* or *peaking over eye*; *bill rather tiny, uptilted* to sharp point (shape obvious only at close range). In flight, white on secondaries *extends to inner primaries*, and *lacks white shoulder patch*. – Adult summer: Note especially *drooping, thin fan of yellowish-white cheek feathers*. – Adult winter: Head pattern *less sharply contrasting black and white* than Slavonian (but beware distance or bright light) because of *grey downward bulge on rear ear-coverts*; *prominent whitish 'hook-back' up* sides of nape. Caution: juvenile Little Grebe can show similar head pattern, but coloration dark brown/buff (not black/grey/white); also eye black, and different bill shape.

VOICE Territorial call easiest to recognize, a rising, somewhat strained, repeated 'pü-**iii**(ch)'. More unobtrusive, short whistling 'wit'. In display, and territorial encounters, a fast, vibrant, rather low, hard trill. Silent in winter.

Little Grebe *Tachybaptus ruficollis*   r**B**3

L 23–29 cm. Breeds in vegetated margins of often small inland waters (ponds, even ditches). Shy, keeps in cover of reeds and sedges for long periods. Winters on often less vegetated lakes, reservoirs, sheltered coasts of W Europe.

IDENTIFICATION *Smallest grebe* (can be mistaken for duckling!), with shortish neck, *very dumpy body*, and *prominent 'powder-puff' rear end*; tiny straight bill. *Lacks white on wing*. – Adult summer: *Chestnut cheeks and foreneck*, with *prominent pale yellow fleshy gape*. – Adult winter: General coloration *brown and buff*, lacking black, grey and white contrasts of Slavonian and Black-necked; foreneck and cheek sandy-buff. – Juvenile: Like adult winter, but with short dark stripes behind and below eye, and bill-base yellowish-flesh; plumage very faded sandy-buff by first winter.

VOICE Varying high-pitched calls, e.g. 'bee-eep' and 'bit, bit bit', accelerating into a far-carrying, rattling, high-pitched trill, often drawn-out and in chorus, 'bibibibi-bibibi...'; somewhat like call of ♀ Cuckoo. Silent in winter.

Pied-billed Grebe *Podilymbus podiceps*   **V**∗∗

L 31–38 cm. Rare vagrant from America; often stays long.

IDENTIFICATION Shape and plumage like a large, stocky, heavily built adult winter Little Grebe (and could thus easily be overlooked at distance or if size not obvious), but *bill very thick and heavy*. – Adult summer: Bill is whitish *with a thick black vertical band*; *thin white eye-ring* and *black chin*. – Winter: Bill whitish or yellowish, at most with hint of band; no black chin.

VOICE Silent in winter.

Slavonian Grebe

Black-necked Grebe

Little Grebe

SLAVONIAN GREBE

'shallow' cap

compare with Great Crested Grebe (centre) in winter

flat crown

flattish crown

yellow feathers *above* eye

ad. s.

typical head shape visible at longer range than neck colour

winter

spring (moulting to breeding plumage)

weak stripe

peak

sloping

juv.

white

pale tip (often visible at con- siderable range)

reddish

light grey collar

ad. summer

ad. winter

downy young

BLACK-NECKED GREBE

'full' cap

more 'black and white' than Little Grebe

rounded crown

yellow feathers *below* eye

ad. s.

yellow ear-tufts sometimes difficult to see at a distance

winter

sometimes tinged reddish!

buffish

dark bulge on light cheek

peak

spring (moulting to breeding plumage)

juv.

steep

pointed, rather fine

black

ad. summer

dark grey collar

downy young

ad. winter

LITTLE GREBE

more brown and buff than winter Slavonian/Black- necked Grebes

rounded, dumpy shape

gape spot visible at considerable distance

ad. s.

winter

rounded

chestnut

juv.

buff- brown

spring (moulting to breeding plumage)

reddish

pale spot

yellow gape

ad. summer

downy young

ad. winter

PIED-BILLED GREBE

PIED-BILLED   LITTLE

summer bill pattern and black chin develop quite early

narrow light eye-ring and greyish-white bill

plain

pale head- sides

black

winter

long, when alert

spring (moulting to breeding plumage)

black

thick, heavy

winter

white

ad. summer

KM

Great Crested Grebe *Podiceps cristatus* r**B**4 / **W**3

L 46–51 cm, WS 59–73 cm. Breeds commonly on reeded larger waters. In winter, offshore or on lakes, reservoirs, mostly in W Europe. Not shy, spends much time openly on unvegetated waters. Nest a large mound of reed stems.

IDENTIFICATION *Largest* and most familiar grebe in most of Europe. Shape distinctive: long low body and *long slender neck* either held erect or (when inactive) lowered with head resting in mid-back; *bill* long, slim and *more extensively pale* than other grebes. In flight, exceptionally *long skinny outline* with flickering wingbeats, slightly stern-heavy, and blackish upcurled feet projecting; *white secondaries and shoulder patch prominent.* – Adult summer: Unmistakable; *head plumes* held compressed when alert, or raised, fanned and vigorously shaken in head-to-head courtship displays which climax in 'penguin dance' in which pair-members raise whole body upright, breast to breast. Bill reddish-pink. –Winter: Head, foreneck and flanks more extensively white than on other grebes, with *white above eye* and *black loral line from eye to bill.* – Juvenile: Striped cheeks; bill pale pink. –1st-summer: Head plumes are incomplete or lacking.

VOICE Highly vocal, giving *loud, far-carrying calls.* During display (incl. on spring nights) has very strong, *rolling* 'crrra-ahrr', slowly repeated; hard, moaning, nasal sounds; and shorter 'krro'; also series of hard 'vrek-vrek-vrek-...' during head-shaking phase of display. Young beg with persistent, clear, whistling 'pli(e), pli(e), pli(e),...', somewhat like distant Oystercatcher.

Red-necked Grebe *Podiceps grisegena* **W**4

L 40–46 cm. Breeds on reeded waters, often smaller and more vegetated than for Great Crested. Winters on coasts, occasionally lakes, reservoirs, in W and S Europe. Mostly shy and retiring habits, reluctant to expose itself much.

IDENTIFICATION Slightly smaller than Great Crested, with *shorter, stockier neck* and often *small 'powder-puff' rear end* (shape thus between Great Crested and Slavonian); bill slightly shorter, more wedge-shaped and held slightly down, often giving *subtle drooping effect* unlike Great Crested. In flight, less skinny than Great Crested, with shorter, dark neck and less prominent white on shoulder. Unlike Great Crested (but like smaller grebes), *dives usually with pronounced jump clear of water.* –Adult summer: Unmistakable; courtship display as Great Crested. Small *clear-cut yellow patch at base of bill.* –Winter: *Head pattern without sharp black and white contrast* as on Great Crested and Slavonian; pale throat and light 'hook-back' behind ear-coverts contrast diffusely against *dusky ear-coverts* and *extensively grey upper foreneck*; breast often whitest part. A hint of the yellow patch at bill-base visible when close. – Juvenile: Like winter, but cheeks and ear-coverts whitish with thin dark stripes, and *upper foreneck and neck-sides reddish-brown*; diffuse *yellow bill-base* more extensive than on adult.

VOICE Noisy when breeding. Display-calls include loud, repeated, grating 'cherk, cherk, cherk...' and drawn-out, wailing and mournful 'aaoouuh', which can turn into pig-like squeals (recalling Water Rail but much louder and longer, and more baleful in tone). Silent in winter.

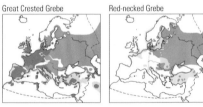

Great Crested Grebe   Red-necked Grebe

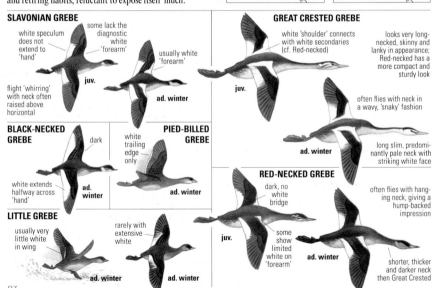

SLAVONIAN GREBE
some lack the diagnostic white 'forearm'
white speculum does not extend to 'hand'
usually white 'forearm'
flight 'whirring' with neck often raised above horizontal
juv.
ad. winter

GREAT CRESTED GREBE
white 'shoulder' connects with white secondaries (cf. Red-necked)
looks very long-necked, skinny and lanky in appearance; Red-necked has a more compact and sturdy look
juv.
often flies with neck in a wavy, 'snaky' fashion
ad. winter
long slim, predominantly pale neck with striking white face

BLACK-NECKED GREBE
dark
white trailing edge only
PIED-BILLED GREBE
ad. winter
ad. winter
white extends halfway across 'hand'

RED-NECKED GREBE
dark, no white bridge
juv.
often flies with hanging neck, giving a hump-backed impression
some show limited white on 'forearm'
ad. winter
shorter, thicker and darker neck then Great Crested

LITTLE GREBE
usually very little white in wing
rarely with extensive white
ad. winter
ad. winter

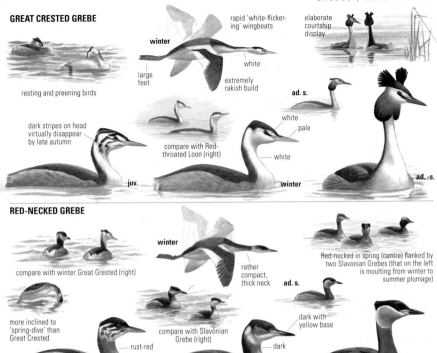

GREAT CRESTED GREBE

resting and preening birds

rapid 'white-flicker-ing' wingbeats

elaborate courtship display

winter

white

large feet

extremely rakish build

ad. s.

dark stripes on head virtually disappear by late autumn

white
pale

compare with Red-throated Loon (right)

white

juv.

winter

ad. -s.

RED-NECKED GREBE

compare with winter Great Crested (right)

winter

rather compact, thick neck **ad. s.**

Red-necked in spring (centre) flanked by two Slavonian Grebes (that on the left is moulting from winter to summer plumage)

more inclined to 'spring-dive' than Great Crested

rust-red

compare with Slavonian Grebe (right)

dark with yellow base

dark

juv.

winter

ad. s.

K M

WATCHING SEABIRDS

Commercial ferry routes can provide sometimes excellent opportunities for seeing seabirds. On a large boat, a *sheltered* spot on an open *upper* deck near the *front* is usually the best position. Special 'pelagic' boat trips for birdwatchers, which go far out into the ocean, provide the best chances for seeing shearwaters and petrels in their offshore feeding areas. Otherwise, these species come close inshore only when driven by severe storms, and they come in to their nesting sites only at night.

For most birdwatchers, seawatching from a coastal headland is the most usual and most convenient way to look for seabirds. It involves often long periods of looking out to sea. Fresh or strong onshore winds are usually required to bring the birds in close.

Seawatching is a somewhat acquired taste. This is because

in overcast weather

in afternoon sun

Note how different light can create dramatically different impressions of the same species (above, Manx Shearwater).

the birds are often distant, just 'dots on the horizon', with identification based on silhouette and flight action, requiring a good deal of experience and identification skill. For the beginner, seawatching can thus be a daunting experience.

Giving accurate directions is important so that others can quickly locate any potentially interesting passing seabird. Agreeing particular reference points (buoys, ships, offshore rocks, etc.) and deciding what is 'straight out' is a good idea at the beginning of any seawatch. Directions to a particular bird should always include *direction from the observers* (the usual method is to use the observers as the centre of an imaginary clockface, with '12 o'clock' being straight out), whether it is *flying left or right*, whether it is *flying above or below the horizon*, an approximation of *distance*, and *when it is passing one of the previously agreed reference points*: 'Skua at 2 o'clock, flying left, just below horizon, middle distance... approaching the red buoy and passing it 3...2...1... Now!' is obviously better than the frustratingly vague directions so often heard on seawatches! Using a binocular field of view, although varying somewhat according to makes, as an approximate unit of distance can also be useful, e.g.: 'Half a field above the horizon' or 'Two fields left of the yacht'. Once the basics have been learned, a good seawatch can provide very exciting birdwatching, and a convenient chance of seeing seabirds from remote places.

SHEARWATERS, PETRELS et al. *Procellariiformes*
A group of highly pelagic birds: albatrosses (*Diomedeidae*; see p. 76), fulmars, shearwaters and petrels (*Procellariidae*), and storm petrels (*Hydrobatidae*). The larger species have long, narrow wings and can glide for hours over the sea with hardly a wingbeat. The smallest species, storm petrels, are shorter-winged and use more wing action in flight. Straight bill with hooked tip and tube-shaped nostrils; nasal glands secrete the seawater's salt. Take food from the sea (fish, crustaceans, plankton, molluscs). Visit nests (on cliff-ledge or in burrow) mostly at night.

(Northern) Fulmar *Fulmarus glacialis* rB2

L 43–52 cm, WS 101–117 cm. Breeds in loose colonies on ledges on steep coastal cliffs or in burrows on inaccessible slopes, but locally also in exposed sites on buildings along N Atlantic coasts. Diurnal at nest site (unlike relatives). Chick spits oily gastric juice at intruders. Versatile forager (takes crustaceans, fish offal, whale flesh), and large gatherings may be seen far from colonies.

IDENTIFICATION Looks like a small, compact Herring Gull, but flight and shape differ clearly, even at long range when plumage details not visible. *Glides* over sea or back and forth along breeding precipice *on stiff, straight wings*, now and then giving a series of *stiff, shallow, quite rapid wingbeats*. Floats high when swimming, and needs a pattering run-up in order to take off. Has *short, thick neck and large head*. Bill short and deep, lead-grey and yellowish. Wings are more bluntly rounded at tip than on large gulls. Commonest plumages are *pale medium grey above* (*base of 'hand' with diffuse pale patch*) and *white below*; head white with small dark spot at corner of eye. Occurs in various morphs (not well differentiated), from light grey above and white below to fairly dark grey above and somewhat paler grey below; dark birds ('Blue Fulmar') commonest in far north.

VOICE Series of throaty, guttural cackling notes, varying in speed, often accelerating and with increasing power in duet between the sexes; heard mostly at colonies.

Cory's Shearwater *Calonectris diomedea borealis* P4

L 50–56 cm, WS 118–126 cm. Closely related to 'Scopoli's Shearwater', the two here treated as races of the same species (though separation into two species favoured by others). Breeds colonially in burrows or caves on rocky coasts of E Atlantic islands S of 40°, locally in small numbers also in Mediterranean in SE Iberia (e.g. near Almeria). Winters in S Atlantic, perhaps mainly off South America. Straggles in summer–autumn (mainly Aug–Sep) rather regularly to North Sea, some even to Sweden. Small loose flocks often seen foraging, visible from land.

IDENTIFICATION Roughly the size of a Lesser Black-backed Gull, and characterized by *long, flexible wings* which at all times are held slightly or distinctly *bowed*. Flies with *long glides* close above the water (even in calm conditions!), relieved by 3–4 (rarely up to 6–7) *relaxed, flexible beats* of the bowed wings; wing-tips look somewhat blunt, and '*hand' is angled backwards*. Brownish-grey above ('hand' somewhat darker), with diffusely paler uppertail-coverts (normally seen only at close range), whitish below (no dark marks on axillaries or belly) with characteristic *dirty grey side of breast, neck and head. Bill pale pinkish-yellow* with dark near tip; in good light, pale bill can be seen at long range. Differences from Great Shearwater are described under that species.

VOICE After dusk noisy at colonies, giving raw, crow-like sounds and a strained, nasal series of three similar notes (voice a little like Kittiwake's) ending with a different one, roughly 'ga**oo**ha ga**oo**ha-ga**oo**ha-wääh'.

'Scopoli's Shearwater' *Calonectris d. diomedea* V***

L 45–52 cm, WS 112–122 cm. Closely related to Cory's Shearwater, here treated as a race of it. Breeds on rocky islands and precipitous coasts mainly in Mediterranean but locally and rarely also on Iberian Atlantic coast (Portugal, Biscay). Most thought to winter in S Atlantic off S Africa, but at least some remain in C Mediterranean (e.g. S of Sicily). Very rare vagrant in North Sea, with odd records in Britain and Sweden. Habits as for Cory's.

IDENTIFICATION Frequently more difficult to reliably separate from Cory's than many imagine: the two are often inseparable due to individual or age-related variation, or to field conditions (light, distance). When photographed or seen close in favourable, soft light, focus on following subtle differences: *inner webs of much of primaries on underwing whitish*, leaving less dark on wing-tip and rear 'hand' (Cory's: nearly all of visible primaries dark on underwing), but beware of variation and effects of light; *slightly smaller general size and bill*; subtly paler or *duller yellow bill*, not quite as vividly yellow as in Cory's (average difference only); subtly paler brown-grey upperparts (much overlap).

VOICE Calls very similar to those of Cory's Shearwater but said to differ in being slightly less coarse and guttural.

Great Shearwater *Puffinus gravis* P4

L 43–51 cm, WS 105–122 cm. Breeds on islands in S Atlantic, wintering (during European summer) at sea east of N America. On southward migration takes a more easterly route and found regularly in W Europe, mainly Aug–Oct; rarer in North Sea. Often attracted to vessels.

IDENTIFICATION Despite its name, a shade smaller than Cory's Shearwater, which it somewhat resembles; has e.g. *white uppertail-coverts*. Distinguished by *stiffer and more rapid wingbeats* (reminiscent more of Fulmar and Manx Shearwater); clearly *contrasting dark* (sooty-brown) *cap*, accentuated by *conspicuous whitish neck-side*; *discrete dark mark at side of breast; dark bars and spots in 'armpit' and on inner part of underwing*; diffuse *dark patch on central belly* ('oil stain'; but often hard to see) and dusky (not white) undertail-coverts; *black bill*; and somewhat more pointed wing-tip.

Fulmar

'Scopoli's Shearwater'

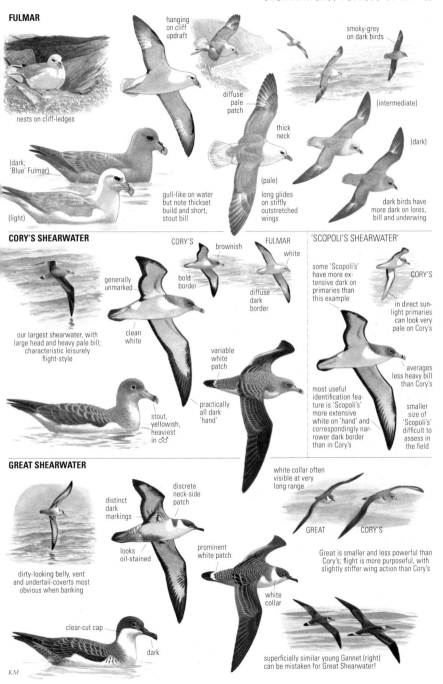

FULMAR

nests on cliff-ledges

hanging on cliff updraft

diffuse pale patch

smoky-grey on dark birds

(intermediate)

thick neck

(dark)

(dark; 'Blue' Fulmar)

(light)

gull-like on water but note thickset build and short, stout bill

(pale)

long glides on stiffly outstretched wings

dark birds have more dark on lores, bill and underwing

CORY'S SHEARWATER

our largest shearwater, with large head and heavy pale bill; characteristic leisurely flight-style

generally unmarked

CORY'S

brownish

bold border

clean white

FULMAR

white

diffuse dark border

variable white patch

practically all dark 'hand'

stout, yellowish, heaviest in ♂♂

'SCOPOLI'S SHEARWATER'

some 'Scopoli's' have more extensive dark on primaries than this example

CORY'S

in direct sunlight primaries can look very pale on Cory's

averages less heavy bill than Cory's

most useful identification feature is 'Scopoli's' more extensive white on 'hand' and correspondingly narrower dark border than in Cory's

smaller size of 'Scopoli's' difficult to assess in the field

GREAT SHEARWATER

dirty-looking belly, vent and undertail-coverts most obvious when banking

distinct dark markings

discrete neck-side patch

looks oil-stained

prominent white patch

white collar often visible at very long range

GREAT   CORY'S

Great is smaller and less powerful than Cory's; flight is more purposeful, with slightly stiffer wing action than Cory's

clear-cut cap

dark

white collar

superficially similar young Gannet (right) can be mistaken for Great Shearwater!

KM

Macaronesian **Shearwater** *Puffinus baroli* V***

L 25–30 cm, WS 58–67 cm. Breeds in rather small colonies in spring on N Atlantic islands (e.g. Azores, Madeira, Selvagens, Canaries), often under boulders. Only rarely seen N of breeding areas.

IDENTIFICATION A small version of the Manx Shearwater (latter common in N Atlantic), and similarly very dark above and white below. Besides *smaller* size, differs as follows: *shorter, broader wings* with more rounded tip; often appears to have *narrower dark border to underwing* (although there seems to be some variation as to this, a few birds having all-dark flight-feathers below); proportionately *more slender bill* and *more rounded head*, which is often *held high* and sometimes raised in flight; upperwing more two-toned with a *diffuse pale area on inner wing* (greater coverts often white-tipped); *white of cheeks extends upwards to enclose dark eye* and accentuates dark of neck-sides. In comparable winds, flight is more flapping with *quicker wingbeats* in *longer series* than in Manx (but comparison with auk still far-fetched), and glides are often short. Occasionally hangs in wind above sea surface with wings raised and feet dangling, snatching food.

VOICE On dark, moonless nights, usually after nightfall, high-pitched laughing, rhythmic series with emphasis on second note heard from colonies, 'ka-**ki**-kukukur-kaa'.

Manx **Shearwater** *Puffinus puffinus* m**B**2

L 30–35 cm, WS 71–83 cm. Breeds colonially in burrows above coastal cliffs on islands in N Britain; huge numbers in Britain & Ireland. Gregarious. Nest visited at night. Numerous on passage, especially in autumn. Does not normally follow ships.

IDENTIFICATION Medium-large shearwater, *uniformly sooty-black above* and *white below*. Head dark to below eye, dark of upperparts extends to breast-sides, and between these often *hint of paler semi-collar* (white 'notch'). Note good *contrast between black and white on underwing*, invariably *white undertail-coverts*, and not (or only very slightly) protruding feet. In strong light bleached summer birds can look more brownish above. Sexes and ages similar in plumage. Typical shearwater flight, in calmer weather low over water with series of rapid, shallow *beats of stiff, straight wings*, relieved by long *glides with slightly down-turned, rigid wings*; in strong wind almost exclusively gliding flight, one moment shearing over the sea surface in the wave troughs and the next sweeping up (sometimes to many metres) and veering down again. Cf. Yelkouan and Balearic Shearwaters.

VOICE At colonies on dark nights, raucous, coughing calls, four-syllable 'ak-ka **chich**-ach' and the like, the latter notes hoarser, more drawn-out and deeper.

Yelkouan **Shearwater** *Puffinus yelkouan* —

L 30–35 cm, WS 70–84 cm. Breeds in burrows above coastal cliffs in C & E Mediterranean. Locally numerous; large flocks regularly seen passing through Bosporus, presumably en route between feeding waters in Black Sea and breeding or roosting sites in Mediterranean. Due to similarity with Manx and some Balearic Shearwaters its normal movements are not fully established, but it appears that the whole population stays the winter in the Mediterranean.

IDENTIFICATION Very similar to Manx Shearwater, being similarly large and of same shape. Plumage similar, too, with dark upperparts and whitish underparts, and the two are not safely separable outside their ranges in normal encounters. Under ideal close observations, and on photographs, note the following: *upperparts brown-grey*, not as blackish as in Manx (though may look blackish in overcast weather); underwing not so contrasting, *flight-feathers* being *grey* rather than blackish; *feet protrude* on average *more* (only a very little or not at all in Manx); lacks pale semi-collar behind dark cheeks, has more *straight division on head and neck between dark and pale*; sometimes a hint of a *paler eye-ring*. Some have dark undertail-coverts not seen in Manx, but many are paler and alike in this respect.

VOICE Similar to Balearic and Manx Shearwaters, but calls perhaps slightly higher-pitched, 'a-ha **ga-iih**-ah'.

Balearic **Shearwater** *Puffinus mauretanicus* P4

L 34–39 cm, WS 78–90 cm. Breeds colonially in caves in coastal cliffs in Balearic Islands (mainly Formentera). Winters in W Mediterranean, but usually moults flight-feathers in Bay of Biscay. Regularly seen in summer and autumn north to Britain & Ireland. Vulnerable after recent marked decline, apparently now < 2000 pairs remaining.

IDENTIFICATION Slightly larger than Manx and Yelkouan Shearwaters, and often appearing pot-bellied and heavy. Short tail makes *feet protrude* a little. *Greyish-brown above* and variably *dirty brownish-white below, lacking contrast between dark and pale* on sides of head and breast. Underwing lacks contrasts, and *underwing-coverts have some darker markings*, are never all-white. *Undertail-coverts, sides of rump and 'armpit' invariably dusky*. Darkest birds invite confusion with Sooty Shearwater but Balearic is smaller, has slightly broader, shorter and less pointed wings, is more *pot-bellied*, whereas Sooty has stronger chest, and flight of Sooty is more fast and dashing. Any pale seen on underbody eliminates Sooty, which has silvery-white centres to underwing only but invariably all-dark body.

VOICE At colonies on dark nights, utters repeated drawling calls with a drawn-out, stressed falsetto note, 'ah-**ii**-ah eech'. Harsher and deeper-voiced than Manx and Yelkouan.

Macaronesian Shearwater

Manx Shearwater

Yelkouan Shearwater

Balearic Shearwater

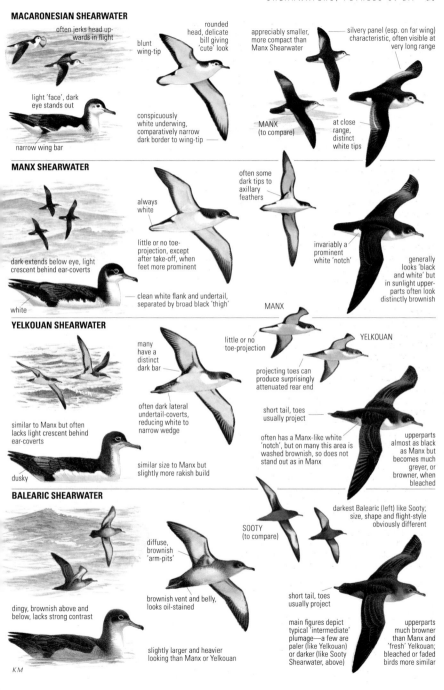

MACARONESIAN SHEARWATER

often jerks head up-wards in flight

light 'face', dark eye stands out

narrow wing bar

rounded head, delicate bill giving 'cute' look

blunt wing-tip

conspicuously white underwing, comparatively narrow dark border to wing-tip

appreciably smaller, more compact than Manx Shearwater

silvery panel (esp. on far wing) characteristic, often visible at very long range

MANX (to compare)

at close range, distinct white tips

at close range, distinct white tips

MANX SHEARWATER

dark extends below eye, light crescent behind ear-coverts

white

always white

little or no toe-projection, except after take-off, when feet more prominent

clean white flank and undertail, separated by broad black 'thigh'

often some dark tips to axillary feathers

MANX

invariably a prominent white 'notch'

generally looks 'black and white' but in sunlight upper-parts often look distinctly brownish

YELKOUAN SHEARWATER

similar to Manx but often lacks light crescent behind ear-coverts

dusky

many have a distinct dark bar

often dark lateral undertail-coverts, reducing white to narrow wedge

similar size to Manx but slightly more rakish build

little or no toe-projection

YELKOUAN

projecting toes can produce surprisingly attenuated rear end

short tail, toes usually project

often has a Manx-like white 'notch', but on many this area is washed brownish, so does not stand out as in Manx

upperparts almost as black as Manx but becomes much greyer, or browner, when bleached

BALEARIC SHEARWATER

dingy, brownish above and below, lacks strong contrast

diffuse, brownish 'arm-pits'

brownish vent and belly, looks oil-stained

slightly larger and heavier looking than Manx or Yelkouan

SOOTY (to compare)

darkest Balearic (left) like Sooty; size, shape and flight-style obviously different

short tail, toes usually project

main figures depict typical 'intermediate' plumage—a few are paler (like Yelkouan) or darker (like Sooty Shearwater, above)

upperparts much browner than Manx and 'fresh' Yelkouan; bleached or faded birds more similar

KM

Sooty Shearwater *Puffinus griseus* P3

L 40–50 cm, WS 93–106 cm. Breeds in southern hemisphere (mainly New Zealand, E Australia, southernmost South America) but spends 'winter' in northern; when southward passage begins many are in W European waters and are then seen in association with storms, mainly Aug–Oct; regular (incl. flocks) off British & Irish coasts. Feeds on surface or dives for it, at times in groups, down to 60 m, food being mainly squid, crustaceans and small fish.

IDENTIFICATION A medium-sized, dark shearwater with long pointed wings. Flight fast, powerful and direct, the *pointed, narrow wings slightly bowed and with the 'hand' somewhat backswept* in strong wind. As soon as the wind is sufficiently strong, Sooty will outpace all other seabirds, and frequently you lose the bird you follow in your telescope and have to search anew and much further ahead over the sea than you thought! In poor light Sooty looks *all black*, including on underwing, but in good conditions it looks more chocolate-brown with diffusely outlined, *paler, almost silvery-looking centre of underwing* (of somewhat variable prominence). Can be confused with darkest Balearic Shearwater, the only other shearwater likely to turn up in our waters that does not have white belly, but that species is smaller and shorter-winged, often appears more pot-bellied and is not so fast and athletic a flyer as Sooty. Dark-morph skua employing shearwater flight, and juvenile Gannet, can also recall Sooty Shearwater, but this would be only at very long range. Cf. Bulwer's Petrel, too.

Bulwer's Petrel *Bulweria bulwerii* —

L 25–29 cm, WS 67–73 cm. Breeds colonially on islands in E Atlantic, in our region mainly in Madeira but also a few in Canaries (e.g. off Lanzarote), Jun–Oct. Nests among boulders or in cliff cavity, often near sea. Visits colonies at night. Forages alone; does not flock, but loose small parties can be seen when feeding or after chumming.

IDENTIFICATION A moderately big, largely all-dark, graceful petrel with *long, narrow wings* and *long tail*. The only feature in the otherwise *uniformly dark brown plumage* is *paler diagonal wing-covert panels* above. Dark plumage may suggest Sooty Shearwater when size difference not obvious, but Bulwer's has longer tail, *all-dark underwing*, and flight action is quite different, with obviously *looser, deeper wingbeats*. Often glides on downturned, slightly bowed wings. The long and pointed tail is rather surprisingly of limited use as a field mark since it frequently merges with the dark background of the sea (and Bulwer's Petrel keeps low, close to the water). More useful, often, is the comparatively *slim body with short neck and small head in relation to the narrow and long wings*. Bill is quite strong.

VOICE Silent at sea and when returning to colonies at dusk. At night, moderate-paced series of hollow, hoarse dog-like yapping sounds heard from nest burrows, 'hroo hroo hroo...' (recalling the distant sound of a slow steam-engine).

Fea's Petrel *Pterodroma feae* V**

L 33–36 cm, WS 86–94 cm. Breeds in late summer and *autumn* in small colonies in burrows in Madeira (Bugio, Desertas) and Cape Verdes, possibly also in Azores. Small population (*c.* 1250 pairs), vulnerable. Visits nest at night, forages by day at sea. There are now a fair number of records in W Europe thought to refer to this species, and in a few cases the evidence is deemed to be sufficient for a positive identification (incl. in Britain).

IDENTIFICATION First impression may be of 'a small Cory's or Great Shearwater', but *underwing is almost entirely blackish*, leaving only *a whitish triangle at the foremost base*; primary bases, too, can show a small pale patch in certain lights and angles. Upperparts *grey with dark W-pattern* (obviousness depending on light and range). *Uppertail uniformly pale grey*. Sides of breast washed grey (but lacks complete dark breast-band of similar-looking rare vagrant Soft-plumaged Petrel; p. 418). Fast flight, in strong winds careening high with angled wings. Fea's and Zino's Petrels are generally near-inseparable in the field, at least when only one bird is seen at a time; Fea's is *larger* than Zino's and has *slightly heavier, more bulbous bill*.

VOICE Drawn-out mournful and deeply howling sounds, 'aaoooooh', sometimes terminating with a hiccup, at night at colonies. The sound seems ideal as background noise in horror movies!

Zino's Petrel *Pterodroma madeira* —

L 32–34 cm, WS 80–86 cm. Nests *late spring*–summer in Madeira (main island), in inaccessible rock crevices near top of inland mountain (1650 m). Highly threatened (only *c.* 75 pairs known). Visits nest at night, forages by day at sea.

IDENTIFICATION Very like Fea's Petrel (which see). Differs on *smaller general size* and *smaller bill*. Claimed to have on average a *more rounded wing-tip* and perhaps a *slightly shorter 'hand'* than Fea's, proportionately. Another possible difference is the apparent absence of any hint of a pale supercilium (although this needs to be further tested). Upperwing might have a less well-marked dark 'W', and underwing a more prominent whitish patch on primary-bases.

VOICE The calls resemble Fea's Petrel, but are slightly more high-pitched and feeble, as would be expected from a slightly smaller bird, a drawn-out wailing sound which might recall a distant Tawny Owl due to the often slightly vibrant tremolo, 'oooeeh... oh-ho-o-o-ooeh' and similar variations.

Sooty Shearwater

Bulwer's Petrel

Madeira

Canaries

Fea's Petrel

Madeira

Canaries

Zino's Petrel

Madeira

Canaries

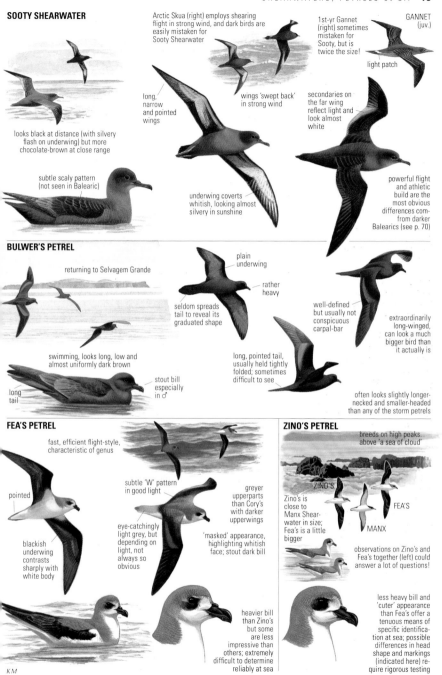

SOOTY SHEARWATER

looks black at distance (with silvery flash on underwing) but more chocolate-brown at close range

subtle scaly pattern (not seen in Balearic)

Arctic Skua (right) employs shearing flight in strong wind, and dark birds are easily mistaken for Sooty Shearwater

long, narrow and pointed wings

wings 'swept back' in strong wind

underwing coverts whitish, looking almost silvery in sunshine

1st-yr Gannet (right) sometimes mistaken for Sooty, but is twice the size!

GANNET (juv.)

light patch

secondaries on the far wing reflect light and look almost white

powerful flight and athletic build are the most obvious differences com- from darker Balearics (see p. 70)

BULWER'S PETREL

returning to Selvagem Grande

plain underwing

rather heavy

seldom spreads tail to reveal its graduated shape

swimming, looks long, low and almost uniformly dark brown

stout bill especially in ♂

long tail

long, pointed tail, usually held tightly folded; sometimes difficult to see

well-defined but usually not conspicuous carpal-bar

extraordinarily long-winged, can look a much bigger bird than it actually is

often looks slightly longer-necked and smaller-headed than any of the storm petrels

FEA'S PETREL

fast, efficient flight-style, characteristic of genus

pointed

subtle 'W' pattern in good light

greyer upperparts than Cory's with darker upperwings

eye-catchingly light grey, but depending on light, not always so obvious

'masked' appearance, highlighting whitish face; stout dark bill

blackish underwing contrasts sharply with white body

heavier bill than Zino's but some are less impressive than others; extremely difficult to determine reliably at sea

ZINO'S PETREL

breeds on high peaks, above 'a sea of cloud'

ZINO'S

FEA'S

MANX

Zino's is close to Manx Shear-water in size; Fea's is a little bigger

observations on Zino's and Fea's together (left) could answer a lot of questions!

less heavy bill and 'cuter' appearance than Fea's offer a tenuous means of specific identifica-tion at sea; possible differences in head shape and markings (indicated here) re-quire rigorous testing

KM

(European) **Storm Petrel** *Hydrobates pelagicus* m**B**3 / P
L 15–16 cm, WS 37–41 cm. Breeds on inaccessible rocky islands and coasts in N Atlantic and W Mediterranean; especially abundant W Ireland, NW Scotland and Faroes. Nests quite near water in burrow or among rocks. Passage migrants sometimes more numerous in autumn storms.

IDENTIFICATION The *smallest* storm petrel in European waters, barely the size of a House Martin (which, with its dark upperparts and snow-white rump, it can momentarily recall), and with *more fluttering and 'busier' flight* than its relatives. Commonest confusion is with Leach's Storm Petrel, but flight action of that species usually clearly indicates its larger size; *lacks obvious covert band on upperwing* (though a narrow whitish bar is shown by juveniles) but has *diagnostic broad white band on underwing; broader and uniformly white rump patch extends far onto sides*.

VOICE From nest burrow at night gives a purring sound with interposed grunts (like stomach-rumbling).

Wilson's Storm Petrel *Oceanites oceanicus* V**

L 16–18½ cm, WS 38–42 cm. Southern-hemisphere breeder, visiting N Atlantic Jun–Oct (Nov) (e.g. Biscay, waters SW of Britain & Ireland, off NW Africa); seldom seen from land. Like Storm Petrel, often follows ships.

IDENTIFICATION Closest to Storm Petrel, being *all dark with broad white rump* and *square-cut tail*. Differs in being somewhat bigger, with more *obvious pale covert band on upperwing, plain underwings*, and *longer legs* (project slightly beyond tail in flight). *Broader*-winged than Storm Petrel, *wings mostly held straight*, with less obvious carpal. Capable of extraordinarily sustained glides on almost flat wings.

Leach's Storm Petrel *Oceanodroma leucorhoa* m**B**3 / P

L 18–21 cm, WS 43–48 cm. Breeds northernmost N Atlantic; nests in cavities on rocky coasts. Winters S Atlantic. Sometimes large numbers on autumn passage in association with autumn storms. Does not follow ships.

IDENTIFICATION A fairly large and long-winged storm petrel, considerably larger than Storm Petrel. Note *rather long, somewhat V-shaped white rump patch with hint of central divide* (visible only at very close range), *prominent pale covert band* on upperwing (often more obvious than white rump!), *dark underwings* and *more pointed and proportionately longer wings*. Forked shape of tail visible only from certain angles, but sides often raised to give rather ragged shape in profile. In strong wind flight quite powerful: glides on slightly bowed, *angled wings* (carpals pressed forward to give 'headless' silhouette), gives a few *powerful wingbeats*, veers off, pulls up, hangs in air, etc.

VOICE From burrow at night gives strange rattling cooing, now and then interrupted by a falsetto 'wuee-cha' and ending with spirited 'chu-chatte**rich**a chu-chitte**ri**!'

Madeiran Storm Petrel *Oceanodroma castro* V***

L 19–21 cm, WS 43–46 cm. Breeds on islands on rocky coasts, incl. off Portugal, on Madeira and Canaries. Elusive; pelagic by day, not seen in flocks off (usually small) colonies; does not follow ships. Nests both late summer and late autumn, in burrows or rock crevices.

IDENTIFICATION *Dark with white rump* and thus difficult to separate from several relatives. In almost all respects much more like Leach's Storm Petrel than Wilson's or European, but following differences most useful: *less prominent covert band on upperwing; negligible fork in tail; white rump patch seems more wide than long* (usually the reverse on Leach's) and on many the white extends prominently onto the sides of rump. At close range *bill noticeably heavy*. Thorough familiarity with storm petrels a prerequisite to judging subtle wing-shape and flight-action differences: Madeiran's wing often seems to have less pronounced angle at carpal (but much more 'arm' than Wilson's), and typical flight incorporates tight twists and short glides on direct switchback course.

VOICE From burrow at night gives cooing 'kr'r'r'r'r' and squeaky 'chiwee' ('rubbing finger on windowpane').

Swinhoe's Storm Petrel *Oceanodroma monorhis* V***

L 18–21 cm, WS 45–48 cm. This species, with main breeding distribution in E Asia, has, surprisingly, been recorded several times in N Atlantic and trapped, e.g., at Selvagens N of Canaries and on rocky coast of NE England, in Norway, France, Spain, Portugal, Italy and Israel. 1st N Atlantic record in 1983, but species is probably overlooked owing to its nocturnal habits and similarity to Leach's Storm Petrel (same size, same wing pattern, forked tail). Differs from Leach's in having *rump dark*, like back, and *primary shafts pale at base*.

White-faced Storm Petrel *Pelagodroma marina* —

L 19–21 cm, WS 41–44 cm. Breeds on Madeira and Selvagens N of Canaries, visiting burrows at night. Rarely seen at sea. Follows whale and dolphin schools.

IDENTIFICATION The region's only storm petrel with *white underparts*. Shape most similar to Wilson's Storm Petrel, but looks bigger, with even longer legs, longer bill and *large butterfly-like wings*. Dark 'mask' surrounded by white, and grey of mantle extends prominently onto side of breast. Sails low over sea on stiffly outstretched wings (recalling flying fish!) and uses its long legs to 'kick off' from the surface. Possibly confusable with winter-plumaged phalaropes (which are pelagic), but this would require great imagination.

VOICE Stifled slowly repeated 'koo, koo, koo, ...', sometimes with squeaky falsetto 'kyi' notes interspersed, heard from nest.

Storm Petrel

Leach's Storm Petrel

Madeiran Storm Petrel

STORM PETREL

smallest storm petrel

broad white band

square or round-ended **juv.**

rather angled wings

plain, faded brownish wing-coverts in summer

white extensive on side (cf. Leach's) (worn)

typical flight views

thin wing-bar on fresh juv. (Oct)

WILSON'S STORM PETREL

shorter 'arm' than Storm Petrel and Leach's

bronzy sheen, lacks white band

square or round-ended

smoothly curved leading, edge, no angle

short pale panel

toes project

primary moult almost complete (late summer)

white nearly encircles tail

typical flight views

long glides on flat wing

from some angles tail appears notched!

projecting toes difficult to discern

long legs

LEACH'S STORM PETREL

strong contrast on upperwing

forked

long pale panel

hint of central divide

'face' often greyish

reduced white on side

typical flight views

long, arched wings, forked tail

in direct comparison with Storm Petrel, Leach's appears much bigger

MADEIRAN STORM PETREL

somewhat less contrast on upperwing than Leach's

slightly indented

inconspicuous pale panel

heavy bill

white more extensive on side than in Leach's

slightly shorter wings than Leach's

square or slightly notched

width of rump patch greater than length

LEACH'S STORM PETREL (to compare)

forked

length of rump patch greater than width

SWINHOE'S STORM PETREL

beware: some Leach's show very little white on rump

forked

uniform with upperparts, entirely lacking white

wing shape as Leach's

light shafts (difficult to see in the field)

BULWER'S PETREL (to compare)

long, narrow

WHITE-FACED STORM PETREL

sails on stiffly outstreched, flat wings (cf. phalaropes, p. 162)

long legs, prominent feet

long-winged, wings broadcentred

typical flight views

appears to 'bounce' from side to side over surface

paddle-shaped wings

KM

GANNETS *Sulidae*

Large seabirds with long, pointed wings and long, pointed tail, plus straight, pointed bill. Dive for fish, often spectacularly, from a height. Sexes alike. Two species.

(Northern) **Gannet** *Morus bassanus* r+m**B**2

L 85–97 cm, WS 170–192 cm. Breeds colonially along steep rocky coasts and on inaccessible rocky islands in N Atlantic, with largest concentration in W Scotland (St Kilda). Pelagic and mobile outside breeding season: some reach W Africa, others W Mediterranean, many remain in breeding area. Numerous along W European coast in autumn. Nest of seaweed, on cliff-ledge or steep slope.

IDENTIFICATION With its *considerable size*, its *long, narrow wings* and its characteristic flight in *quite fast, shallow and uniform wingbeats interspersed with short glides*, generally easy to identify even at distance. In very windy conditions inclined to bank and shear like a shearwater and, especially in the case of immatures, may be mistaken for one of the large shearwater species, but Gannet is much bigger and has *more projecting, wedge-shaped tail* and *longer head/neck*. Makes stunning steep, diagonal dives for fish from height of 10–40 m, wings thrown back just prior to striking surface. – Adult: *White*, with head tinged yellow-buff, and *black wing-tips*. – 4th-winter: As adult, but central tail-feathers and odd scattered secondaries are usually black. – 3rd-winter: Most tail-feathers and secondaries black, but scattered white ones intermixed. Head and body largely as adult. – 2nd-winter: Underparts and head largely white, as are uppertail-coverts and usually some lesser upperwing-coverts. Often some dark on crown and partial 'collar' (at distance may recall Great Shearwater). – Juvenile: Plumage wholly grey-brown but for whitish uppertail-coverts. At close range, entire plumage seen to be finely speckled white, giving beautiful silvery sheen to head and back in good light. – Note that individual variation leads to occasional deviations from the plumage development described above.

VOICE Loud, grating sounds at colonies.

Brown Booby *Sula leucogaster* —

L 65–75 cm, WS 135–150 cm. Breeds in tropical coastal waters, including islands in Red Sea; seen regularly in e.g. Gulf of Aqaba (Eilat). *Smaller* than Gannet, with somewhat *more rounded tail*. *All dark brown above*. Below, brown with *white secondary-coverts* and *sharply defined white belly*. (Immature has pale brown-grey belly.) *Pale bill* and *pale facial skin* contrast with otherwise dark head. (Bill-base bluish in breeding season.) Dives steeply like Gannet, but also at oddly shallow angles. Often rests on buoys.

Black-browed Albatross
Thalassarche melanophris **V****

L 80–95 cm, WS 200–235 cm. S Atlantic species. Very rare summer and autumn visitor. Single birds have oversummered in Gannet colonies in Scotland. Wingspan as Mute Swan's, thus a huge bird. *Underwing white with broad black frame* (broadest at front). *Dark, short eye-stripe. Bill yellow* with orange at tip. Immature has grey bill and *narrow grey neck-collar*, as well as dusky underwing.

PELICANS *Pelecanidae*

Very big waterbirds with huge bill equipped with elastic skin pouch on lower mandible, with which fish are captured. Skilled soaring birds which, unlike storks and large raptors, *circle in orderly flocks with large parts of the flock in synchronized motion*, not a swarm of individualists. Sexes alike. Two species; both have declined greatly and are in need of protection. Breed at sheltered, shallow, fish-rich waters. Nest a pile of plant material on ground.

White Pelican *Pelecanus onocrotalus* [V]

L 140–175 cm (neck extended), WS 245–295 cm. Rare and local breeder at coastal swamps and shallow inland lakes. Migrates overland to winter in NE Africa.

IDENTIFICATION Adult has *white plumage* with, in breeding season, *faint yellow-pink tinge*; usually looks pure white at distance. *Bill-pouch greyish-yellow* (brighter yellow when breeding). *Feet yellow-pink* (more reddish when breeding). *Iris dark* (red), *framed by flesh-coloured bare skin* (orange when breeding). In flight, *black flight-feathers below* contrast strongly with white wing-coverts (thus same pattern as White Stork, but different soaring behaviour; see intro above). – Juvenile: Rather *dark brown-grey on upperparts*; dirty white below. *Bill-pouch* yellowish, *orbital skin and legs pink*. In flight resembles Dalmatian Pelican, but has *much darker remiges below* and *greater contrast between remiges and paler central panel*.

VOICE Colonies produce a buzzing hum of stifled mumbling and various unarticulated grunting sounds.

Dalmatian Pelican *Pelecanus crispus* —

L 160–180 cm (neck extended), WS 270–320 cm. Very rare and local breeder at shallow lakes and swamps; also accepts smaller inland waters. Short-distance migrant.

IDENTIFICATION Like White Pelican but averages *slightly larger*. Standing adult differs in *curly nape-feathers, grey feet, grey-tinged* white plumage (but as White Pelican with yellowish breast patch), *pale iris very small area of pale surrounding skin*, plus *reddish Bill-pouch* when breeding. In flight nearly *all-pale underwing* (grey-white, *whiter central band*), *with only wing-tips dark*. – Juvenile: Fairly pale brown-grey above and dirty white below, *slightly paler and greyer* than juv. White, but best told by tufted nape feathers. Has *grey feet* and *greyish-white orbital skin*.

VOICE Poorly known. As White Pelican's but higher.

Gannet

White Pelican

Dalmatian Pelican

GANNET

juv.

4th-cal.-yr (summer)

4th-cal.-yr (variation)

2nd-cal.-yr (May–Oct)

ad. summer

3rd-cal.-yr (Aug–Sep)

juv.

ad. winter

BROWN BOOBY

juvs.

turns yellowish

GANNET **1st-yr**

white

pale-chested

brownish

greyish at first …

dark brown

leucogaster (Atlantic)

ad.

ochre/pinkish in breeding ♀; blue orbital skin indicative of ♂

imm.

ad. ♂

can be mistaken for imm. Cormorant (left)

BLACK-BROWED ALBATROSS

imm. (1–3 years old)

greyish-pink, darker tip

dark tail

ad.

wide black border

darker than in adult

stout, pale bill

ad.

white

GREAT BL.-B. GULL

'jizz' recalls giant Fulmar more than Gannet

long pointed

GANNET, imm.

black 'brow'

short

sits high on water

exceptionally long, narrow wings!

WHITE PELICAN

black

1st-yr

comical look

ad.

sub-ad.

1st-yr

flesh-pink/ yellowish

pink 'mask', 'beady' eye

flowing crest

bill-pouch yellow-ochre in *all* ages

ad. ♂ spring

DALMATIAN PELICAN

imm.

ad.

limited black

1st-yr

ad.

not so 'friendly' a face

bill-pouch is pinkish or yellowish in immatures, orange in adults

unkempt look

pale iris

1st-yr

greyish

sub-ad.

tinged greyish

intensely red-orange

ad. ♂ spring

KM

CORMORANTS *Phalacrocoracidae*

Medium-sized to large aquatic, mainly dark-plumaged birds. Body elongated, neck and tail long, latter much rounded. Bill strong, straight with hooked tip. All four toes webbed. Swim with body low and bill raised. Dive, often with leap from surface, for food, mainly fish. Frequently fly far between fishing waters and breeding sites or roosts. After dives, often perch on rocks or elevated sites with wings spread to dry in 'prehistoric' fashion. Breed colonially.

(Great) **Cormorant** *Phalacrocorax carbo* r**B**4 / **P** / **W**

L 77–94 cm, WS 121–149 cm. Breeds on cliff-ledges along sea coasts (ssp. *carbo*, Atlantic; *maroccanus*, Morocco) or in trees at lakes or coasts, at times in reedbeds or on ground (*sinensis*; E and S Europe, presumably all breeders in Baltic). Northern birds migratory, British & Irish disperse locally. Roosts on sandbanks, rocks or take-net poles etc., often in large, dense flocks. Nest of seaweed, reed, twigs. Tree-nesting birds eventually kill their nest-trees by their droppings. Numbers have increased recently.

IDENTIFICATION Large with long, thick neck. *Head profile rather wedge-shaped and angular*, bill strong (though at times confusingly thin on some juvs.). In flight superficially goose-like (similar size and wingbeat rate, neck outstretched, often flying at some height, incl. over land, generally in formation when in flock), but told by incidence of *brief glides breaking active flight*, by slightly shallower wingbeats, *longer tail*, *kinked neck*, and less orderly flock formation; also bill and tail held rather high. Single birds often fly low over the water (cf. Shag). Swims with body low and bill pointing up, and at long range could be confused with Yellow-billed Loon, but note *angular hindcrown, more erect neck* and *more strongly inclined bill*, and habit of frequently *leaping into dive*, and of usually diving without prior surveillance with head submerged in water, so typical of loons. – Adult: Black with bluish and some green gloss (amount of green not sufficient for picking out ssp. *sinensis*), wings tinged bronze and scaled black. *Bare skin at base of lower mandible yellow, surrounded by white* area. A white thigh patch is worn in early breeding season only (often lost as early as Jun). Crown and hindneck with varying amount of white feathers, on average more on old birds and in ssp. *sinensis* (but overlap considerable); much of these white feathers lost during summer. In winter less glossy, and white on cheeks and throat becomes duller and less clear-cut. – Juvenile: Dark brown above, underparts with varying amount of white, generally on centre of throat, breast and belly. 1st-year birds can look very white beneath, or have *patchy breast and belly with contrast between pure white and dark streaks*. – Immature: Gradually darker and more glossy

over two-year period until adult plumage attained. – Variation: In Morocco, adults have some white on foreneck and upper breast (ssp. *maroccanus*), or have all-white breast and upper belly contrasting with solidly dark lower belly (southern race *lucidus* or extreme variation of *maroccanus*?).

VOICE Various deep, guttural calls at colony.

(European) **Shag** *Phalacrocorax aristotelis* r**B**3

L 68–78 cm, WS 95–110 cm. Breeds in loose colonies on coastal cliffs. Nest, in crevice, small cave or under large boulder, a heap of vegetation.

IDENTIFICATION Somewhat smaller and slimmer than Cormorant, with *thinner neck, smaller, more rounded head* (*forehead steep, crown rounded*) and *thinner bill*. Differs in flight in *neck being straighter, outer wing slightly more blunt, wings appearing attached 'astern of midships'*, flight path lower (usually close to water), wingbeats somewhat more elastic and fast without interspersed brief glides. Dives with more pronounced leap clear of water. – Adult breeding: Black with green gloss, wings tinged purplish, scaled black. *Gape prominently yellow, contrasting with dark surround* including bill (Atlantic race; Mediterranean ssp. *desmarestii* has paler, more yellowish bill, reducing contrast). During early breeding has *upcurved, black crest on forecrown*, subsequently lost. – Adult non-breeding: Slightly duller, less glossy. Chin pale. *Bill yellowish*. – Juvenile: Dark brown above, pale brown below, chin whitish (ssp. *desmarestii* usually entirely or extensively off-white below). Small and thin-billed Cormorants in paler plumage told by: patchy dark and white breast, not so *uniform and pale brown* as Shag; bill horn-coloured with prominent yellow patch at base, not *all pale without yellow patch at base* as on Shag; feet dark, not *pale* (webs flesh-coloured) as Shag; also, note *different head shape and face pattern*. Pale tips to upperwing-coverts form *pale wing-panel* in flight (Cormorant usually all dark, but pale imm. may have a hint). – Immature: Gradually darker and more glossy over two years until adult plumage attained.

VOICE Various clicking and grunting calls at colony.

Pygmy Cormorant *Phalacrocorax pygmeus* —

L 45–55 cm, WS 75–90 cm. Breeds colonially at freshwater lakes and coastal deltas with rich vegetation. Swims in dense parties, perches on branches or climbs on reed stems. Nest of twigs and grass, in low, dense trees, or in reedbeds.

IDENTIFICATION Rather *small*, about Coot-sized. *Tail very long; bill proportionately short and thick* ('baby-faced'); *neck* appears short too, but *can be stretched when swimming*. Blackish with dark green and bronzy gloss. – Adult breeding: Head, neck and underparts have small white feather tufts, subsequently lost in summer. – Adult non-breeding: Whitish chin and paler, brown-tinged chest. Has no white spots on head, neck or body. – Juvenile: Duller and browner; slightly streaked pattern on head and breast.

VOICE Little known. Silent away from nesting sites. Croaking and grunting calls at colony.

Cormorant   Shag   Pygmy Cormorant

CORMORANT

drying wings

1st-year

ad. courtship

thick, kinked

Cormorants and Shags often sit together on rocks

long tail

dives with half-leap

swallowing large fish

all except right-most figure depict Atlantic ssp. *carbo*

light greyish

often white

always dark

white patch

lucidus / maroccanus (NW Africa)

juv.

1st-year

ad. non-breeding

ad. courtship

ad.

SHAG

juv.

CORMORANT, juv.

prominent yellow patch

often pale panel

1st-year

rather short, rounded

narrow, straight **ad.**

yellowish

pot-bellied

ad.

crown peaked at rear

thick

light greyish (never yellow)

CORMORANT, juv.

crown peaked at front

slender

dull yellowish

juv.

yellowish

discrete pale throat

uniform brown-ish

pale

pale tips

juv.

1st-year

crest

dark

glossy green

ad. non-breeding

ad. courtship

much whiter than Atlantic ssp. (though exceptionally these too can be very white)

desmarestii (Mediterranean)

1st-year

PYGMY CORMORANT

short

ad. non-breeding

panting in heat

long

short, thick

ad. courtship

swimming birds can at times stretch neck and look surprisingly long-necked

some juvs are extensively whitish on belly

juv.

ad. courtship

juv.

KM

HERONS, STORKS and IBISES *Ciconiiformes*

A group of large or medium-sized wading birds with long legs, neck and bill. Toes not webbed. Included are bitterns, egrets and herons (*Ardeidae*), which retract their long necks in flight, storks (*Ciconiidae*), which hold their necks outstretched in flight and which often soar high up, and ibises and spoonbills (*Threskiornithidae*), which have characteristic bill shapes but otherwise most resemble the storks.

(Great) **Bittern** *Botaurus stellaris*   r**B**5 / **W**4

L 69–81 cm, WS 100–130 cm. Breeds only in extensive *Phragmites* reedbeds. Retreats from ice, but is hardy, returns early. Food mainly fish, frogs, insects. Polygamous. Nest a platform of dead reeds at water-level in reedbed.

IDENTIFICATION *Large buff-brown heron*; slightly smaller than Grey Heron, but shape much more *compact and stocky*, with *thicker neck* (usually hunched into shoulders), loose throat feathering, and shorter legs. Flies rather infrequently, mainly in spring and early autumn. In flight, note *broad wings, quicker wingbeats* than Grey Heron, *'thick' front end* and trailing feet; *at distance looks all ginger-brown* (broad wings and brown colour recalling buzzard or owl) with only *slightly paler wing-coverts than darker* (barred) *flight-feathers*. Rarely seen on ground (stays hidden in reedbed), and presence mostly revealed by *distinctive voice*; when alarmed in reedbed, adopts camouflage posture with bill pointing up ('bittering posture'). –Juvenile: Like adult, but crown and shorter moustachial stripe browner (not black).

VOICE Flight-call a single or repeated, deep and hoarse 'graoh', somewhat recalling fox or large gull. Song of ♂ (so-called 'booming') is peculiar, far-carrying (up to 5 km on calm nights), very low, foghorn-like, exhaled, breathy 'whump', usually repeated 3–5 times with 2½-sec. intervals; when close, quieter inhalation also audible, 'uh-**whump**', and brief series of subdued, short opening notes.

American Bittern *Botaurus lentiginosus*   **V*****

L 59–70 cm, WS 95–115 cm. Breeds in N America; very rare vagrant in W Europe, mostly in Oct–Nov. Smaller than Bittern, with thinner bill; *size and flight action may recall Night Heron*. Plumage Bittern-like except *flight-feathers plain blackish* (not barred), and *upperwing-coverts much paler*; *crown brown* (not blackish); *black moustachial stripe thicker* (though narrow and less black on juv.); and loose feathering of *foreneck more prominently striped with brown*.

Little Bittern *Ixobrychus minutus*   **V****

L 33–38 cm, WS 49–58 cm. Breeds in extensive or small reedbeds, overgrown reedy ponds, ditches. Summer visitor (mostly Apr–Oct); winters in Africa. Rare vagrant in Britain

& Ireland. Food fish, frogs, insects. Nest a platform of reeds often raised above water-level in reedbeds or bushes.

IDENTIFICATION Very small heron, *smaller than a Moorhen*. Secretive, but not infrequently comes to feed at edges of reedbeds etc.; often retreats by climbing and running like a crake rather than flying. In flight, quick jerky wingbeats distinctive, and *creamy wing patches unmistakable* on adults, less clear-cut on juveniles. –Adult ♂: Crown and upperparts *black*; wing patch buff-white, neat, contrasting. –Adult ♀: Black parts of ♂ are *brown*; wing patch buff-brown, unstreaked but less contrasting. – Juvenile: Like dull ♀ with *streaked* neck, upperparts and wing patch.

VOICE Flight-call an abrupt, nasal 'kwekwekwe'; also shorter 'kwer' or 'kerack'. Song, mostly at night but at all hours, monotonous, continuous series of a low and muffled croaking (still far-carrying) 'hogh', repeated every 2½ sec.

(Black-crowned) **Night Heron**
Nycticorax nycticorax   **V***

L 58–65 cm, WS 90–100 cm. Breeds colonially at marshes, ponds and riversides. Summer visitor (mainly Mar–Oct); winters in Africa. Rare visitor to mainly S Britain & Ireland, (esp. Apr–Jun). Food fish, frogs, insects. Active mainly at night, roosting communally in trees during daylight. Nests in trees, sometimes reedbeds; nest is platform of sticks.

IDENTIFICATION Short-billed, very *stocky*, medium-sized heron. In flight, has *compact, stumpy-ended outline* (feet not much noticeable) with *slightly raised body and drooping bill*, and stiff wingbeats. Most often seen at dawn or dusk, flying to and from roosts, but not always strictly nocturnal, and daytime sightings not infrequent. – Adult: Unmistakable in good views, but in flight often looks *all grey* (black back often difficult to see). Legs dull pinkish-yellow, reddish when breeding. – Juvenile/1st-winter: *Brown and buff* with *prominent white spotting* (gradually lost). – 2nd-winter: Like adult but crown and back faded grey-brown, not black.

VOICE Noisy at colonies; various raucous croaking notes. Flight-call a coarse, nasal, rather frog-like croak, 'quark'.

Striated Heron *Butorides striata*   **V*****

L 40–47 cm, WS 60–73 cm. In W Palearctic, breeds only in S Sinai in mangroves; rare visitor N to Eilat, S Israel. A few records in Europe (Britain, Iceland, Ireland), and on Azores, were of N American close relative and similar-looking Green-backed Heron (see p. 410). In non-breeding season, often found in marinas and on harbour piers.

IDENTIFICATION A small and dark, *long-billed* heron, slim-necked when alert, *thickset* when relaxed. – Adult: *Dark slate-grey, silvery-grey or brownish-grey above* (often just looks 'dark' at distance), with *wing-feathers neatly fringed buff-white* or rufous. *Sides of head, neck and breast varying from buff-grey to deep rufous*. – Juvenile: Brownish, with white spots/streaks on wings and bold stripes on neck. Cf. juv. Night Heron (which is larger, and shorter-billed).

VOICE A sharp 'skyek' call when flushed.

Bittern

Little Bittern

Night Heron

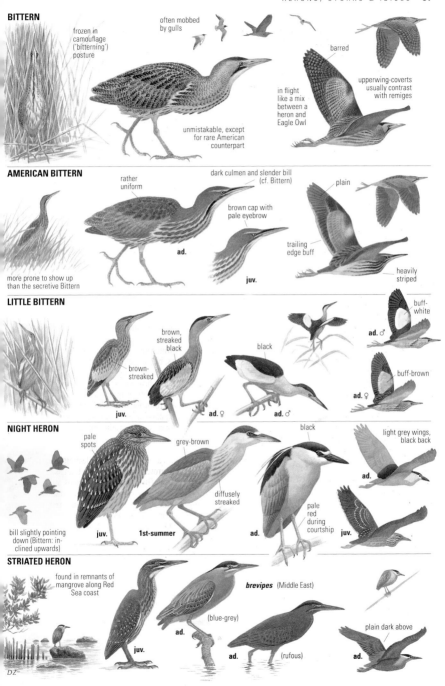

BITTERN

frozen in camouflage ('bitterning') posture

often mobbed by gulls

barred

in flight like a mix between a heron and Eagle Owl

upperwing-coverts usually contrast with remiges

unmistakable, except for rare American counterpart

AMERICAN BITTERN

rather uniform

dark culmen and slender bill (cf. Bittern)

plain

brown cap with pale eyebrow

trailing edge buff

heavily striped

ad.

juv.

more prone to show up than the secretive Bittern

LITTLE BITTERN

brown, streaked black

black

buff-white

ad. ♂

brown-streaked

buff-brown

ad. ♀

juv.

ad. ♀

ad. ♂

NIGHT HERON

pale spots

grey-brown

black

light grey wings, black back

ad.

diffusely streaked

pale red during courtship

bill slightly pointing down (Bittern: in-clined upwards)

juv.

1st-summer

ad.

juv.

STRIATED HERON

found in remnants of mangrove along Red Sea coast

brevipes (Middle East)

plain dark above

(blue-grey)

ad.

juv.

ad.

(rufous)

ad.

DZ

Cattle Egret *Bubulcus ibis* **V**∗∗

L 45–52 cm, WS 82–95 cm. Nests colonially in trees and bushes by lakes and rivers, but forages in *often dry habitats*, among grazing cattle, on fields and grassland, but certainly also in wet marshes. Short-distance migrant. Range expanding. Eats insects; grasshopper specialist.

IDENTIFICATION Quite *small, white* heron with *compact body, short bill* and fairly short neck often held retracted. *Rounded head with well-feathered chin* ('undershot jaw'). Bill and legs generally greyish-yellow; pinky-orange when breeding, when also has *orange tone to crown, breast and mantle*. Active and mobile, makes short dashes. Flies with fast wingbeats in small disorderly flocks and lines.

VOICE In flight may give monosyllabic, quite soft croaks, 'kre' and 'ehg'. Commonest call otherwise a gruff 'rick-**reck**'. Hoarse buzz heard from colonies.

Squacco Heron *Ardeola ralloides* **V**∗∗

L 40–49 cm, WS 71–86 cm. Breeds in small colonies at shallow marshy lakes, ponds and rivers with reeds, bushes and lines of trees. Outside breeding season mostly solitary or in small groups. Summer visitor, winters Africa. Rare vagrant Britain. Food insects, small fish, amphibians.

IDENTIFICATION Quite small, *buffy-brown heron with white wings*; startling 'quick-change number' when light brown bird takes off and in flight looks almost all white! – Adult summer: Elongated nape-feathers. Back pale ochre with faint violet shimmer. Head and neck-sides almost un-streaked yellowish buff-white. Bill bluish with black tip, legs yellowish-pink (briefly coral-red during courtship). – Adult winter/immature: Lacks ornate nape-feathers, head- and neck-sides extensively but finely streaked, upper-parts duller, darker grey-brown. Bill yellowish with darker culmen/tip, legs greenish.

VOICE Often gives a hoarse, croaking 'kaahk', quite like ♀ Mallard. Pounding, nasal 'kak kak kak...' and more thick 'ah ah ah...' heard from colonies.

Little Egret *Egretta garzetta* r**B**4 / **W**4

L 55–65 cm, WS 88–106 cm. Nests colonially in dense trees and bushes at shallow marshy lakes, rivers and coastal lagoons. Has recently colonized S England. A resident or short-moving species in W Europe, but a migrant in E. Takes fish, frogs, insects, snails etc. from shallow lakes, fish ponds, flooded fields and so on, often in small groups.

IDENTIFICATION Medium-sized white heron, closer in size to Cattle Egret than to Great Egret, but *slim and elegant*. *Black legs with sharply contrasting bright yellow toes*, unlike any other W Palearctic white heron (but see similar Reef Egret, and rare vagrant Snowy Egret, p. 410). In

flight, *legs moderately projecting*, wings look near centrally placed (Great Egret very long-legged, wings 'fixed well forward'). *Bill black*; lores blue-grey for greater part of year, reddish during courtship period. In nuptial plumage delicate plumes formed by *two elongated nape-feathers*.

VOICE Mostly silent away from colonies. On rising, sometimes a Rook-like, hoarse yell, 'aaah'. Greeting call a loud 'da-**wah**'. At colonies several hoarse hard calls and a typically gargling 'gulla-gulla-gulla-...'.

(Western) Reef Egret *Egretta gularis* —

L 55–68 cm, WS 88–112 cm. In region treated here breeds only in S Sinai, otherwise locally along Red Sea coasts (race *schistacea*); regular in small numbers at Eilat and Suez. Very rare vagrant Morocco from breeding sites mainly in Senegal (race *gularis*). Coastal habits; rests on shores, buoys, jetties, fishing poles, etc. Fish-eater.

IDENTIFICATION Two morphs, *white* and *dark grey*. Close relative of Little Egret, and white morph very like latter; same size and shape, except that *bill is usually a touch longer*, often deeper-based but through rest of its length narrower and almost always *slightly curved* (hint of sabre shape). Bill colour variable, usually *yellowish* with darker culmen. *Legs greenish grey-black to halfway or more down tarsus, rest of tarsus and toes dull yellow*; legs can be more extensively greyish-yellow. Immature white morph often has scattered dark remiges.

Great Egret *Casmerodius albus* **V**∗∗

L 85–100 cm, WS 145–170 cm. Breeds in colonies at large, shallow swampy lakes, preferably in reeds with some low bushes and trees. Partial migrant, winters in Mediterranean region or Africa. Rare but annual vagrant in Britain. Feeds (on fish, aquatic insects) in flood meadows and along rivers, but also in somewhat drier terrain.

IDENTIFICATION A *very big all-white heron*. Almost the size of Grey Heron but a shade more elegant, with longer legs and neck. Needs to be told primarily from Little Egret and white-morph Reef Egret. (Cattle Egret also white, but is smaller and podgier and often has distinct ochre elements in plumage, so hardly confusable.) Differs from Little Egret in: *slower, more dignified wingbeats*; proportionately *longer legs, which project a very long way past tail in flight*; *yellowish tibia* ('thigh') *and in part tarsus*; *yellowish bill except in breeding season*; *lacks nape plumes*; larger size. Told from white-morph Reef Egret by e.g. larger size, longer legs, plus straight dagger-shaped bill (not long and narrow with hint of sabre shape).

VOICE Silent except at colonies. Calls include a thick, rolling almost ricocheting 'kr'r'rah', dry and wooden.

Cattle Egret

Squacco Heron

Little Egret

Great Egret

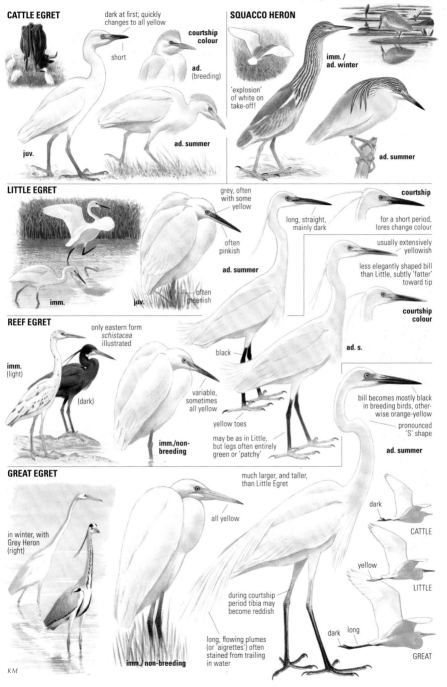

CATTLE EGRET

dark at first; quickly changes to all yellow

courtship colour

short

ad. (breeding)

juv.

ad. summer

SQUACCO HERON

imm. / ad. winter

'explosion' of white on take-off!

ad. summer

LITTLE EGRET

grey, often with some yellow

long, straight, mainly dark

courtship

for a short period, lores change colour

often pinkish

ad. summer

usually extensively yellowish

less elegantly shaped bill than Little, subtly 'fatter' toward tip

imm.

juv.

often greenish

courtship colour

ad. s.

REEF EGRET

only eastern form *schistacea* illustrated

imm. (light)

(dark)

black

bill becomes mostly black in breeding birds, other-wise orange-yellow

pronounced 'S' shape

variable, sometimes all yellow

yellow toes

ad. summer

imm./non-breeding

may be as in Little, but legs often entirely green or 'patchy'

GREAT EGRET

much larger, and taller, than Little Egret

in winter, with Grey Heron (right)

all yellow

dark

CATTLE

yellow

LITTLE

during courtship period tibia may become reddish

long, flowing plumes (or 'aigrettes') often stained from trailing in water

imm. / non-breeding

dark   long

GREAT

KM

Grey Heron *Ardea cinerea* rB3

L 84–102 cm (neck extended), WS 155–175 cm. Breeds in colonies, or sometimes solitarily, in woodland with tall trees beside lakes and brackish sea-bays. Waits patiently, stock-still, for prey (mostly fish) on lakeshores and riversides; rests on one leg in shallow water, often at edge of reeds. Hardy, just retreats from ice in N, but some migrate to W Europe. Nest a flat basket of sticks in tree crown.

IDENTIFICATION Very *big*, strongly built heron, mostly *medium grey above* and greyish-white below. Distinguished from Crane by *retracted neck* in flight, and often when standing. *Bill straight, powerful*, greyish-yellow (orangey when breeding), legs greyish-yellow or grey. Flies with *slow, somewhat irregular beats*, all the time with *wings strongly bowed*, often high up. Upperwings bicoloured, grey with black remiges and primary-coverts; also *two paler patches at carpal*, clearly visible in front view. (For differences from Purple Heron, see latter.) – Adult: *Forehead, crown-centre and head-sides white; crown-sides and nape black; long, narrow black nape plume* (seldom visible); *neck-sides pale greyish-white* with black-streaked white central band. – Juvenile/1st-winter: *Forehead and crown grey*; nape greyish-black with short plume; *head-sides and neck-sides medium grey*, as back; neck-centre buffish.

VOICE Commonest call, often heard at dusk from birds flying to roost, a loud, harsh and croaking 'kah-**ahrk**', which often has a sing-song, echoing quality. At colonies gives knocking and croaking series.

Purple Heron *Ardea purpurea* V*

L 70–90 cm (neck extended), WS 120–138 cm. Breeds in colonies at extensive, shallow swampy lakes with reedbeds. Forages in shallow water in gaps among reeds and along rivers, or more in open in wet meadows. Summer visitor (mostly Apr–Oct), wintering in tropical Africa. Feeds on fish, frogs, insects. Nests in reeds (often Europe) or trees.

IDENTIFICATION Big, *lanky* heron, usually *quite dark-looking* in the field. Main confusion risk is Grey Heron. Note the following differences: *longer, more uniformly narrow bill* ('bayonet' rather than 'dagger'); *narrower head* (merges more into bill); *wings* in front view *slightly bent at carpal*, and *hint of jerkiness in wingbeats*; somewhat *narrower neck*, which in flight often forms *more angular* 'neck keel'; *longer toes*, sometimes held *more splayed and disarranged*; slightly smaller size. – Adult: Head-sides and *neck-sides reddish-brown*, neck with distinct, thin black border. *Back dark grey*, wing-coverts uniform dark grey with *purple-brown cast* (most marked on ♂). – Juvenile/1st-winter: Head- and neck-sides, *back and upperwings ochre-brown*, upperparts mottled (feather centres dark, fringes

ochre), dark margins along neck-side indistinct. – 1st-summer: Neck and back almost as adult, but *wings still variegated brown* (juv. feathers, now worn).

VOICE In flight a gruff, monosyllabic 'krrek', like Grey Heron's but shorter, straighter (not faintly disyllabic as Grey Heron's) and 'deader', less resonant.

White Stork *Ciconia ciconia* P5

L 95–110 cm, WS 180–218 cm. Breeds in open farmland with access to swampy riversides, marshes and floodlands. Stronghold in E Europe, declining in NW (overhead cables and draining probably main causes). Nests, often in small colonies, in stick nest (readily accepts specially erected carriage wheels and platforms) on house roofs, church towers, telephone poles; thus not shy, is approachable. Summer visitor, winters in tropical Africa. Notable passage at Bosporus and Gibraltar end Aug, returns in Apr. Food frogs, insects, also snakes, young birds etc.

IDENTIFICATION *Very big*, with *long neck* and *long legs*. *Black and white* with *red bill* and *red legs* (imm. has duller red bill with slightly darker tip; close views needed). Unmistakable when seen on the ground (slow, dignified walk); in flight at distance could be confused with White Pelican, but *holds neck straight out*, has *projecting legs* and *soars in much more disordered fashion*, not with contingents well synchronized like the pelicans. (Cranes can also look almost black and white in strong light, but their legs project further and the neck is a bit longer.)

VOICE Virtually silent, but compensates for this by loud bill-clappering, especially when partners meet at nest.

Black Stork *Ciconia nigra* V**

L 90–105 cm, WS 173–205 cm. Breeds in vast, swampy forests, mostly in aged mixed coniferous forest interspersed with rivers and marshes. Nest of sticks high in tree crown. Shy and withdrawn at nest site. Summer visitor, winters in Africa; migrates a month later than White Stork, returns in May. Food mainly amphibians, insects.

IDENTIFICATION Barely smaller than White Stork, which it resembles in shape. At close range, easily told from White by *black head, neck, breast and back with metallic green or violet gloss*. In flight at distance the differences can be more difficult to see than expected; note that underwing has only *white triangles on axillaries*. – Adult: Bill and legs red. – Juvenile: Bill and legs *grey-green*.

VOICE Seldom heard owing to discreet nesting habits and shy behaviour. At nest gives rasping series, 'shi-luu shi-luu shi-luu...', with first syllable hoarse, second clear. Said also to have a buzzard-like mewing 'piu'. Bill-clappering quiet and rarely used.

Grey Heron

Purple Heron

White Stork

Black Stork

GREY HERON

harpoon fishing
in shallow water

toes
usually
kept folded

ad.

'neck
keel'
rounded

dark crown

strong,
dagger-
shaped

white
forehead

juv. **2nd-winter** **ad.**

PURPLE HERON

'jerky' flight on rakish wings

toes
long,
often
spread

ad.

'neck
keel'
often
angular

bill often pointing
upwards

juv.

long
and
narrow

narrow

striped
'face'

juv. **ad.**

WHITE STORK

breeds near
humans, often
nesting on
buildings

white

dark
tip

juv. **ad.**

BLACK STORK

in strong light,
upper-parts
can appear
as pale
as on
White
Stork

black

shy, nests in
deep forests

violet and green
metallic gloss

grey-
green

dull greenish-black

grey-green **juv.** **ad.**

Glossy Ibis *Plegadis falcinellus* **V***

L 55–65 cm, WS 88–105 cm. Breeds colonially at shallow, well-vegetated marshes; nests in trees (often with herons) or reeds. Most migrate in winter to Africa, returning in Apr. Gregarious; feeds (on insects, frogs) in small groups in wet marsh, wading in shallow water.

IDENTIFICATION *Dark. Legs and neck quite long,* curlew-like *downcurved bill*. Plumage *dark purple-brown, with green gloss* on wings. Superficially like Bald Ibis, but has longer legs and neck, and in flight, *when feet project past tail-tip,* shows *longer, narrower neck,* and *curlew-like wing-beats* are more mechanical, with *frequently interspersed short glides* (so distant head-on silhouette like Pygmy Cormorant, another dark bird that may be seen in flocks in same habitat). Often flies in lines, flock's *flightpath undulating.* – Adult breeding: Head and neck dark. Lores blue with narrow white border. – Adult summer/winter: As breeding, but head/neck somewhat duller brown and dotted with white. Lores all dark. – Juvenile/ 1st-winter: As adult non-breeding, but largely *dull greenish* lacking obvious tinge of purple or brown.

VOICE Not vocal. Hoarse grunting 'grru' sometimes heard. Grunting and croaking sounds at breeding site.

(Northern) **Bald Ibis** *Geronticus eremita* —

L 70–80 cm, WS 120–135 cm. One of the region's most threatened species; extinct since 1989 in Turkey (only free-flying captive-bred birds left) and merely remnants left in Syria and Morocco. Breeds colonially in rocky semi-desert, with proximity to running water. Nests on ledge or in cavity in cliff face. Syrian birds migrants, wintering in S Arabia and E Africa. Food insects and small animals, captured on dry ground, fields and at edges of rivers.

IDENTIFICATION Large, *black,* rather *short-legged* but *long-winged* and long-tailed; *downcurved bill*. Black plumage has metallic green and purple-brown gloss. *Bill red. Head 'bald', skin red. Nape-feathers elongated, droop* like unkempt mane. Wing shape characterized by rather short 'arm' and *long, quite narrow 'hand'*. *Feet do not project past tail* in flight (cf. Glossy Ibis). *Wingbeats shallow* but notably *powerful and flexible,* and not necessarily relieved by short glides as in Glossy Ibis on longer flights (though gliding not uncommon; wings then held slightly bowed).

VOICE Silent away from colonies. At latter, calls include short, guttural 'hrump' and hoarse, high 'hyoh'.

(Eurasian) **Spoonbill** *Platalea leucorodia* **P**5

L 80–93 cm, WS 120–135 cm. Breeds in colonies (sparse, local) in large reed swamps with some bushes and trees. Nests in trees or reeds, normally not mixed with other

species. Requires access to sheltered shallow open water; also salt ponds. Food molluscs, crustaceans, small fish.

IDENTIFICATION Unmistakable at close range, with *long bill with spatula-shaped tip*. At distance can be confused only with one of the white egrets owing to its size and *all-white plumage*. Note *bushy nuchal crest* ('Indian chief' style) and typical foraging behaviour: bill is held lowered in the water and *head is swung from side to side* while the bird wades forwards. *Flies with neck extended* (like storks, unlike herons), *wingbeats quite fast* (quicker than those of the large egrets) interspersed with glides. – Adult breeding: Bushy crest (longer on ♂), ochre breast patch (as on pelicans), all-white primaries, black bill with yellow tip. – Adult summer/winter: As breeding, but lacks crest and ochre breast patch. – Juvenile/1st-winter: As adult non-breeding, but *black tips to outer primaries* and with initially pinkish legs and bill which gradually darken.

VOICE Silent.

FLAMINGOS *Phoenicopteriformes*

Big wading birds inhabiting shallow salt or brackish water. Extremely long neck and long legs. Hefty downward-bent bill with lamellae used for filtering food from water. Colonial; do not breed every year. Food small aquatic animals, plankton. Nest a pile of mud on sheltered island or inaccesible shore.

(Greater) **Flamingo** *Phoenicopterus roseus* [**V****]

L 120–145 cm, WS 140–170 cm. Breeds in a few but large colonies on low islands and banks on extensive, open, low, muddy beaches or at salt lakes, sea-bays, etc. Sensitive to disturbance. S European population *c.* 35,000 pairs.

IDENTIFICATION White with pinky tone, red wing-coverts and black flight-feathers. Note that bill is mostly pink with *only extreme tip black,* that *legs are entirely pink*, and that plumage is *more white than pink*. (See also related Chilean Flamingo and American Flamingo, p. 424, two occasional escapes from European bird collections.) Often seen in large, tightly packed flocks (appearing like whitish-pink strip on the horizon). In flight the *red on the wings is conspicuous,* and neck and legs are held fully extended (neck almost 'ridiculously' long, drooping a little basally). Wingbeats continuous, fairly rapid. – Adults similar but ♂ averages larger and often has slightly stronger pink-red colours. – Immature: Brownish juv. plumage quickly bleaches and is moulted to near-white with darker shaft-streaks and dark-tipped wing-coverts. Bill pale grey, eye dark. Pink attained slowly from 2nd-year. – Juvenile: Grey-brown, white-bellied.

VOICE Loud cackles, somewhat recalling Greylag and Bean Goose. Large flocks give a continuouoes grunting murmur.

Glossy Ibis  Bald Ibis  Spoonbill  Flamingo

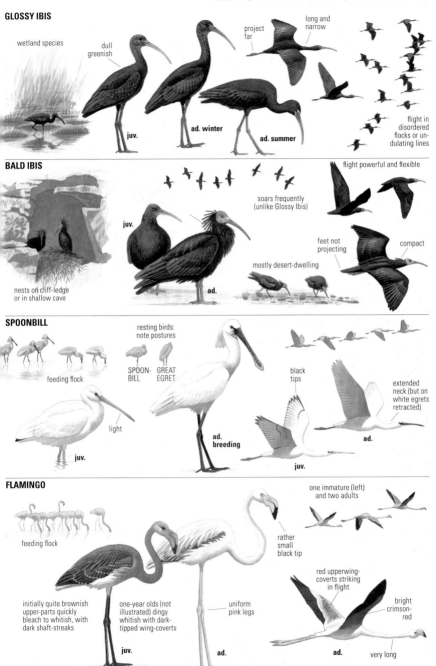

GLOSSY IBIS

wetland species

dull greenish

project far

long and narrow

juv.

ad. winter

ad. summer

flight in disordered flocks or un-dulating lines

BALD IBIS

flight powerful and flexible

soars frequently (unlike Glossy Ibis)

juv.

feet not projecting

compact

mostly desert-dwelling

ad.

nests on cliff-ledge or in shallow cave

SPOONBILL

resting birds: note postures

SPOON-BILL GREAT EGRET

feeding flock

black tips

extended neck (but on white egrets retracted)

light

ad. breeding

ad.

juv.

juv.

FLAMINGO

one immature (left) and two adults

feeding flock

rather small black tip

red upperwing-coverts striking in flight

bright crimson-red

initially quite brownish upper-parts quickly bleach to whitish, with dark shaft-streaks

one-year olds (not illustrated) dingy whitish with dark-tipped wing-coverts

uniform pink legs

juv.

ad.

ad.

very long

DZ

BIRDS OF PREY *Accipitriformes*

47 diurnal species of varying size and shape, belonging to three different families. All are mainly carnivorous, with hooked bill and strong feet. Most of the species catch and kill their prey, except for the vultures.

Accipitridae: hawks, buzzards, eagles, kites, vultures and harriers. Largest family, 34 species. Broad,'fingered' wings suitable for soaring and gliding.

Pandionidae: only one species, the Osprey. Specializes in catching fish after hovering over water and diving in feet-first. Feet and talons very strong.

Falconidae: falcons. 12 species of fast, skilful flyers with pointed wings. Capable of catching their prey in the air, at times after long, dashing dive.

IDENTIFICATION OF BIRDS OF PREY

This group is very difficult to master in the field owing to generally long-range or brief sightings, variation in plumages, similarities in flight silhouette between related species, infrequency of calls, and the rareness of most species offering few opportunities to practise. Do not hope for or pretend reliable identification of all birds of prey in the field—*ever*. But with sensible practice you can come a long way.

Visit any major *raptor migration site* in the area (Eilat, Bosporus, Gibraltar, Falsterbo, etc.) and start out on a trip with an *experienced leader*. Memorize the appearance of *key species* like Common Buzzard, Sparrowhawk, Marsh Harrier and Black Kite at various angles and in various lights.

Follow all the birds which have come quite close, and which have been positively identified, in your *telescope* (essential for raptor studies) as they disappear, for *as far as you can possibly see them*. This working 'backwards from the answer' is the best way to learn on your own how the various birds of prey look when they are far away and are seen under less favourable conditions—unfortunately the most common situation in raptor studies. *Observe actively* by asking yourself questions while you look at the bird.

The telescope is an essential piece of equipment, as already mentioned. But just as important is to fix it on a *stable tripod* (lightweight or budget tripods do not provide a steady view, as they shake when you move the telescope or in a wind) equipped with a high-quality *movable head-joint* (designed for film-cameras; a ball-and-socket joint is less suitable). This enables you to study *flying* raptors, which is what it is all about for 95% of the time. And, for raptor studies, it is

LESSER SPOTTED EAGLE

The whiteness of clouds 'burns out' colours and details of the plumage. Wait until the bird comes out against the blue sky, or a very dark cloud, which provides a better view.

generally more comfortable to have an *angled* eye-piece than a conventional straight telescope, although this is a matter of taste.

Optical and *size illusions* frequently occur when observing birds of prey, especially at long range. It is easy to misjudge the size of a soaring eagle: its slow-motion-like action and dark colour against the bright sky tend to make it appear larger than it is. And experience shows that the size of raptors is misjudged even when other birds are present for comparison. Few birdwatchers realize how *extensive* the *normal size variation* is within one single species. Study the wingspan ranges in this book; they are carefully calculated from skins, photographs and live birds, and demonstrate the large overlap between various species. Still, they give a useful hint on the relative sizes, which cannot be disregarded.

Mode of flight provides a clue to identification, but it is essential to realize that to some extent it *varies* within a species *in relation to size*: the largest birds have slower wing-beats and movements than the smallest.

Plumage variation is also extensive, especially in buzzards and some eagles, but it pays to learn the full range of these variations. Unlike size and proportions, a positively seen plumage detail is far less susceptible to influence from illusions or the observer's subjectivity.

The whiteness of clouds 'burns out' much of the colours of birds flying in front of them. Hang on to a confusing bird until it comes out *against the blue sky, or against a dark cloud*, which will reveal more of its plumage.

An *understanding of moult*, finally, is essential for the correct identification, and especially ageing, of raptors. See pp. 11–12 for some general guidelines.

juvenile   adult

SPOTTED EAGLE

A slight difference in general shape between adults and juveniles can be observed in many raptor species in flight. Adults have more evenly broad wings with broader wing-tip and shorter tail, whereas juveniles have wings with more S-curved rear edge, narrower 'hand' and slightly longer tail.

1st centre

2nd centre

WHITE-TAILED EAGLE

Active moult of flight-feathers gives a clue to ageing (and is often important for reliable identification). This White-tailed Eagle in summer may look like a juvenile, but the moult pattern conclusively shows it to be at least in its 2nd summer, with two active moult centres among the primaries.

LAMMERGEIER

patrols mountainsides with wings slightly arched down

shadow sometimes easier to spot than the bird itself against light mountainside

apart from flesh, readily eats bone marrow by swallowing and digesting bones

soaring young have untypically compact, almost eagle-like silhouette

juv.

long and pointed wing

very dark

long, wedge-shaped

dark

juv.

ad.

pale mantle

some are slightly whiter

juv.

juv.

ad.

ad.

VULTURES *Gypaetus, Neophron, Gyps* et al.

Six species. Mostly very large and broad-winged raptors, scavengers which soar and glide to find carrion or prey remnants. Wing muscles rather weak in relation to body weight and wing-surface area, thus dependent on thermals, and generally not active in the morning hours. Often seen in parties. Many species have an unfeathered head, and some have a long neck which is withdrawn in flight in heron fashion. Claws comparatively short.

Lammergeier *Gypaetus barbatus* —

L 105–125 cm, WS 235–275 cm. Very rare and local resident in high mountains; declining, only *c.* 500 pairs left within region. Confined to most inaccessible, steep cliffs but patrols slopes and valleys too. Food meat from freshly killed animals and secondarily carrion, incl. bone marrow; has habit of dropping bones from height onto rock to break them into pieces, which are swallowed. Takes tortoises in talons and drops them from height onto rocks to open shell. Pair-bonds last for life. Huge nest in a cave or deep crevice.
IDENTIFICATION Huge and *long-winged*. Flight silhouette characteristic, with uniquely *narrow and pointed wings* (for such a large raptor) and *long, wedge-shaped tail, clearly longer than width of wing* (silhouette approached only by Egyptian Vulture, which is half the size, shorter-tailed and blunter-winged). Soars and glides on generally flattish

wings, only slightly arched down, but at times lowers primaries more when gliding. Patrols mountainsides endlessly on motionless wings, flight giving slow-motion impression (owing to large size and generally long range). In strong light, when upperparts look pale and merge with mountainside, often best spotted by dark shadow moving over the ground. – Adult: Underbody light, whitish with a varying degree of buffish-yellow or often rather deep rufous-buff tinge (acquired through sand-bathing!), contrasting with dark underwing. In good light lesser and median underwing-coverts are darkest, being jet-black. Upperparts lead-grey with pale feather shafts. – Juvenile: Body dull grey with contrasting dark grey head, neck and upper breast ('Hooded Crow pattern'). Upperparts not uniformly dark with lighter shafts (as on adult) but variegated; mantle, rump and some wing-coverts light. Also, silhouette is different, with *shorter tail, broader wing and more blunt wing-tip.* – Immature: Adult pattern is attained in *c.* 5 years, subadults keeping the dark head rather long.

VOICE Usually silent; during aerial display at breeding site, utters shrill, loud whistling notes or a trill.

Lammergeier

(Eurasian) Griffon Vulture *Gyps fulvus* V***

L 95–110 cm, WS 230–265 cm. Resident in mountains of Mediterranean area, Turkey, Caucasus. Accidental in N Europe. Declining, but still *c.* 20,000 pairs, of which 90% in Spain. Soars and glides frequently, often appearing in loose flocks, keeping to ridges and peaks of mountains. Nests on cliff-ledge, often in loose colonies of 10–20 pairs.

IDENTIFICATION Huge, clearly bigger than most large eagles. At a distance, size apparent by *slow-motion-like movements* in the air. *Broad wings* with *very long 'fingers'*; a tendency to have *bulging secondaries* and *indented inner primaries* (Black Vulture has more evenly broad wings). *Tail short*, generally well rounded (a hint of wedge-shape on many). Head appears small in flight. *Tips of secondaries rounded*, do not give distinct saw-tooth appearance. Flight heavy, wingbeats very slow and rather deep, a few at a time relieved by gliding. *Wings held in shallow V when soaring*, like Golden Eagle. When gliding, adopts flatter or more arched wing posture. When soaring, now and then takes a single, deep, 'embracing' wingbeat, peculiar to the large vultures. Basically *two-coloured*, but beware birds seen against strong light, which can look all dark. – Adult: Broad *pale buff tips to upperwing-coverts*; rather dark, *medium brown lesser underwing-coverts traversed by one or two narrow light bands*; medium brown underbody; whitish ruff, visible at closer range. Bill yellowish, face often swarthy. – Juvenile: *Upper greater coverts lack pale tips*; underwing-coverts cream-coloured, somewhat *lighter than on adult and more uniform*; under-body lighter than on adult; ruff buff-brown. Bill grey, face often pale. – Immature: Adult plumage attained in 5–6 years.

VOICE Somewhat more vocal than other vultures. A variety of hissing and hoarse grunting notes, mainly at gatherings by carrion or at roosts.

Rüppell's Vulture *Gyps rueppellii* —

L 90–105 cm, WS 220–255 cm. Breeds in sub-Saharan Africa. Rarely strays to N Egypt and Spain, and singles have stayed in Griffon colonies in S Spain since the 1990s.

IDENTIFICATION Somewhat *smaller* than Griffon (*c.* 10% difference, obvious when the two are seen together). Adult is *dark with dense pale spotting*. Immatures (those usually seen in Europe) are rather *all dark above and below* with a *bold white band inside leading edge on underwing*. Adult develops more and narrower white bands on forewing below.

(Eurasian) Black Vulture *Aegypius monachus* —

L 100–115 cm, WS 250–285 cm. Very rare, until recently having declined markedly (in Europe 1000+ pairs remain, most of them in Spain). Breeds both in arid, wild mountains and in vast lowland forests with hills or rocky outcrops.

Food mainly carrion. Nest almost invariably in tree, a huge eyrie of branches and twigs.

IDENTIFICATION Huge. Size of distant soaring bird apparent mostly by *slow-motion-like movements*. *Wings broad* with *very long 'fingers'*. Flight silhouette more eagle-like than in Griffon owing to more *evenly broad wings*, more apparent head and *all-dark plumage*. Tips of secondaries pointed, giving *saw-toothed trailing edge*. Confusion with an eagle prevented by the outer wings being more *deeply 'fingered'*, and the outermost two primaries flex upwards more prominently than on eagles. *Tail short* as on Griffon Vulture. Head-on silhouette when soaring significantly different from Griffon in that *inner wings* invariably are *held flat, with primaries slightly lowered*. Unlike other large, broad-winged vultures, glides in to land *without dangling legs*; instead, often holds tail up. Colour of feet variable: bluish grey-white, pale pinkish, pale yellowish. – Adult: *Underwing-coverts not uniformly dark* but show some paler grey-brown on lesser and median (generally forming one or two diffuse bands). Also *ruff and head are pale brownish*, not blackish as on juvenile. – Juvenile: *Underwing-coverts uniformly black*, darker than the flight-feathers. Head and ruff, too, are blackish, no lighter brown as on adult. Mature in *c.* 6 years.

Lappet-faced Vulture *Torgos tracheliotus* —

L 98–112 cm, WS 250–280 cm. Used to breed in S Israel, now extinct. Extremely rare straggler within region. Breeds in mountains or deserts. Food mainly carrion. Nest in tree or, less commonly, on cliff-ledge, a huge eyrie of branches.

IDENTIFICATION *Huge*. Flight silhouette most like Black Vulture, with *saw-toothed trailing edge*, although *secondaries bulge a little*, almost as on Griffon, and *tail is proportionately shorter* than on both. Head-on silhouette when soaring flattish. Best separated by plumage characters: *head is light, ruff is dark*, *dark breast* of adult is *striped and mottled whitish*, with some white also along flanks. Upperwing-coverts and back brownish, contrasting somewhat against darker flight and tail-feathers (though less than on Griffon). – Adult: Underbody striped dark and pale. *Underwing-coverts black with one pale band near leading edge.* – Juvenile: Underbody and underwing-coverts rather uniform dark. – Immature: Adult plumage acquired in 6–7 years.

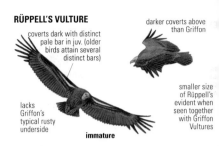

RÜPPELL'S VULTURE

coverts dark with distinct pale bar in juv. (older birds attain several distinct bars)

darker coverts above than Griffon

lacks Griffon's typical rusty underside

smaller size of Rüppell's evident when seen together with Griffon Vultures

immature

Griffon Vulture

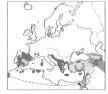

Black Vulture

Lappet-faced Vulture

GRIFFON VULTURE

soars and glides on slightly raised wings (form shallow V)

gliding

long 'fingers'

wings appear pointed from certain angles

juv.

often broad pale lines

ad.

coverts paler than flight-feathers

bill yellowish

outer primaries flex upwards prominently

insignificant pale tips on greater coverts

bill grey

pale brown

white

buff-brown

juv.

ad.

juv.

buff-brown

bulging 'arm'

narrower 'hand'

buff-brown

ad.

greater coverts prominently tipped pale

BLACK VULTURE

soars and glides on flat wings, often with outer wing slightly lowered

gliding

pale feet

ad.

coverts darker than flight-feathers

black

black

black

pale brown

pale

juv.

gives very dark impression

brown-black

black

black

black

juv.

ad.

Black Vulture often holds tail up before landing

Griffon often holds legs dangling before landing

relatively evenly broad, rear edge saw-toothed

ad.

LAPPET-FACED VULTURE

soars and glides on flat wings, often with outer wing slightly lowered

gliding

pale vent

ad.

narrow pale line

pale

juv.

juv.

striped black

ad.

broad wings, bulging 'arm', rear edge saw-toothed

perched in acacia

ad.

Egyptian Vulture *Neophron percnopterus*   **V**∗∗∗

L 55–65 cm, WS 155–170 cm. Rare summer visitor mainly to mountains in Mediterranean area, more common only in Spain/NW Africa and Turkey; winters in Africa. Declining. Food carrion, offal at refuse dumps, etc. Nests on cliffs.

IDENTIFICATION Medium-large. Flight silhouette characteristic, with *wedge-shaped tail* (slightly shorter than width of wing) and *small head with narrow, long bill*; wings comparatively broad and well 'fingered'. Head-on silhouette fairly flat when soaring; more arched when gliding, with primaries lowered. Flight with rather slow and deep wing-beats. Peculiar white centres of flight-feathers on upper-parts unique. – Adult: Body, head, tail, lesser and median wing-coverts white; head, neck, breast and mantle with a varying degree of yellowish- or grey-brown tinge. Strong contrast on underside between white parts and black flight-feathers. Yellow cere and bare skin on head show up at some distance. – Juvenile: *Dark brown, with strongly contrasting ochrous-buff* (bleaching to whitish) *broad tips* to tail-feathers, upperwing-coverts and feathers of mantle, rump, scapulars and lower belly/vent. With wear, this plumage becomes more dull brown, lacking in contrast. There is some variation, too: some juveniles have less prominent light tips and look more uniform brown than others. – Immature: Plumage less distinctly patterned, more dull brown, with pale or whitish feathers first appearing in 2nd summer on upper mantle, rump/lower back, belly and wing-coverts. Adult pattern gradually acquired over *c.* 5 years.

VOICE Usually silent.

White-tailed Eagle *Haliaeetus albicilla*   r**B**5

L 76–92 cm, WS 190–240 cm. Breeds along sea coasts and by larger lakes or rivers. Rare (except locally in Norway). Reintroduced in Scotland. Adults mainly resident except in far N, juveniles more migratory. Food fish, also waterbirds, carrion, offal. Nest huge, in crown of old tree or (along Atlantic coasts) on cliff-ledge; re-used if pair is undisturbed.

IDENTIFICATION Very large, with *long, broad wings*, outer wings well 'fingered'. *Fairly short, bluntly wedge-shaped tail* and *long neck* and *heavy bill* combine to give characteristic flight silhouette. Head-on, wings flat/somewhat arched ('arm' raised, 'hand' lowered). Active flight with *long* series of relaxed, rather shallow wingbeats, now and then relieved by short glide (cf. Golden Eagle); characteristic is sudden slight rise or descent of flight path. Can soar for long periods at supreme height. – Adult: Brown, with *paler, yellowish-brown head and neck*. Some upperwing-coverts and feathers of mantle and scapulars irregularly tipped pale yellowish, creating 'untidy', scalloped effect. *Bill yellow. Tail pure white*, or white with brown mottling at base and tip. – Juvenile:

Whole plumage uniformly fresh/worn. Compared with adult, outer secondaries longer, creating bulging outline, and tail longer and less graduated. Head, neck and body dark brown, breast, belly, back and upperwing rufous-brown with extensive blackish tips, lesser coverts being darkest. (Darkness variable, some appearing blackish, others rather brown.) Underwing dark, with diffuse *light patch on axillaries* and *narrow light band on median coverts*. Tail-feathers largely dark with light centres, can appear all dark when folded (even in flight!) but quite pale when spread in good light. Bill dark. *Lores whitish*, forming a rather conspicuous light patch. – 1st-immature (1st–2nd summer): Plumage not uniformly fresh/worn (owing to protracted moult). Rufous-brown colours bleached, and some white-based feathers on breast and mantle. 'Trousers' all dark. Bill slightly paler grey. – 2nd-immature (2nd–3rd summer): Variable, but consistently more uniformly dark. Breast, mantle and wings become more brown. Some still blotched white on back and breast. Bill greyish-yellow. – 3rd-immature (3rd–5th summer): Markedly different from preceding immatures in greater similarity to adult, though still showing scattered whitish feathers, and head/neck still rather grey-brown. Bill pale yellow, eye pale brown.

VOICE Generally silent outside breeding season. Main call, especially in vicinity of nest, a loud series of shrill, cackling yaps, reminiscent of spring call of Black Woodpecker, ♂ higher-pitched than ♀, 'klee klee klee klee ...'. Alarm lower-pitched, hard knocking 'klek', a few slowly repeated.

Osprey *Pandion haliaetus*   m**B**4

L 52–60 cm, WS 152–167 cm. Summer visitor (Mar–)Apr–Sep; winters in Africa. In Britain, very rare breeder (*c.* 150 pairs, in Scotland). Breeds on clear freshwater lakes, also on coasts at brackish water, in Mediterranean at salt. Food fish, caught after dive. Nests in very top of pine tree.

IDENTIFICATION Medium-large, long-winged, ventrally pale raptor with unique flight silhouette, *wings being narrow with long 'hand'*, having only four 'fingers', *tail short and square-cut*. *Wings angled, carpals held forward when gliding*, wings in head-on silhouette distinctly bowed and at times recalls Great Black-backed Gull. Hovers (wingbeats heavy) over water and dives feet-first for fish. Upperparts grey-brown, *underparts white*, underwing with blackish marks. – Adult: Upperparts uniformly grey-brown, no white feather tips. Crown white. Greater coverts below all blackish; secondaries dark, barring usually indistinct. Eye yellow. ♂ has on average less prominent brown breast-band than ♀. – Juvenile: Feathers of upperparts tipped whitish. Crown white with dark streaks. Greater coverts below white, barred dark; secondaries pale but coarsely barred dark. Eye orange.

VOICE During aerial display (undulating flight high up, feet dangling), whistles mournfully 'u-eelp u-eelp u-eelp...'. The alarm, often given in flight, a hoarse but still sharp 'kew-kew-kew-kew-...'. Contact-call is a short, loud and 'sudden' whistle, 'pyep!'.

Egyptian Vulture

White-tailed Eagle

Osprey

EGYPTIAN VULTURE

at times feeding on refuse dumps

juv.

variation

juv.

subad.

ad.

wedge-shaped

narrow bill

typical silhouette with wedge-shaped tail

juv.

subad.

ad.

juv.

ad.

WHITE-TAILED EAGLE

attended by Hooded Crows on spring ice

pale bill-base

head often darkest

long, heavy, dark

long 'fingers'

varia-tion

2nd-winter

pursued by Common Gulls

juv.

juv.

ad.

narrow light band

light 'armpit'

light 'armpit' retained several years

yellow bill

pale mantle

2nd-winter

yellowish-brown

juv.

predomi-nantly dark but can appear light against the sky

white, wedge-shaped

ad.

juv.

subad.

ad.

OSPREY

light-tipped

uniform

juv.

ad.

hovers and dives for fish

finely cross-barred

darker trailing band

juv.

greater coverts cross-barred

dark band

ad.

juv.

ad.

head-on silhouette

EAGLES *Aquila, Circaetus, Haliaeetus* et al.

Common name for large, broad-winged, powerful diurnal birds of prey with strong bill and sharp talons; however, the group is not clearly defined and includes species of different size which are not all closely related. Sexes alike in plumage, but the female is generally larger. Adult plumage attained only after several years. (One species, the White-tailed Eagle, is treated on the preceding spread.)

Golden Eagle *Aquila chrysaetos* ⌐B4

L 80–93 cm, WS 190–225 cm. Breeds in mountains and vast upland forests, sometimes also in more restricted lowland forests. In southern part of range resident; in N more migratory, especially young birds. Food mammals (hare, rabbit, squirrel, rodents, even young fox), gamebirds, carrion. Versatile hunter, soaring high up looking for prey, or flies low attempting to surprise game in Goshawk fashion. Will also sit in treetops for long periods on the lookout. Nest huge, in old tree or on cliff-ledge; re-used if pair is undisturbed.

IDENTIFICATION Very large, long-winged eagle with typically *long tail*, about as long as width of wing. Silhouette typical also in wings being slightly narrower at base and inner 'hand', producing *S-curved rear edge*, most pronounced on young birds. Flight powerful, often 6–7 rather deep, slow wingbeats relieved by glide of 1–2 sec., then further short series of beats, etc.; flight path usually straight. Head-on silhouette when soaring, and often when gliding, typical, with *wings raised in shallow V*; at times also glides on flatter or somewhat arched wings. Common to all plumages is rather dark brown colour with yellowish-brown or light rufous-brown (due to variation, not age criterion) nape-shawl. – Adult: Flight- and tail-feathers basally grey with 3–5 broad, coarse, dark cross-bars, widely tipped blackish. Upper median and inner greater wing-coverts bleached and worn, forming irregular pale panel (varying in prominence). Often paler rufous breast patch and median underwing-coverts. – Juvenile: *Large pure white areas on central wing* (primaries and outer secondaries white-based) *and inner tail*. Rarely, amount of white in wing restricted to small patch (and thus no age criterion). Upperwing-coverts uniformly dark brown, *no pale panel*. Rear edge of wing S-curved, all wing-feathers uniformly fresh/worn. – 1st-immature (1st–2nd summer): Similar to juvenile but has pale panel on upperwing, and plumage not uniformly fresh/worn. – 2nd-immature (3rd–5th summer): Like 1st-immature, but a varying number of adult-type flight-feathers, grey and barred, among juvenile-type. White on base of tail kept longest, sometimes still when wings appear adult.

VOICE Rather silent. A thin, fluty whistle, 'klüh...', sometimes in flight. Eaglets and ♀ beg with disyllabic 'pee-chulp'.

(Eastern) Imperial Eagle *Aquila heliaca* —

L 70–83 cm, WS 175–205 cm. Rare and local breeder in ▶ and SE Europe in forests on steppe or open plains, also in upland forests. Food mammals, birds. Nests in tree.

IDENTIFICATION Very large. Most similar to Golden Eagle but *tail shorter*, about ¾ or ⁴/₅ of width of wing, and wings on average more evenly broad. Head-on silhouette often differs, too, *wings held flattish*, but can also be slightly raised (almost as on Golden). Often soars with tail folded. *Tail i◀ more square and has sharper corners* than on Steppe Eagle – Adult: Very dark, brown-black, with pale golden nape shawl, paler than on Golden Eagle. Generally *no pale pane◀ on wing-coverts above*, but at times an indistinct one. Flight feathers grey, diffusely barred, *underwing-coverts contras◀ ingly black. Inner tail paler* and more densely barred than o◀ Golden, outer with *broad black band at tip* (at long range two-coloured tail can recall pattern of young Golden▶ *White patches on shoulders* (sometimes small and difficult t◀ see). – Juvenile: Body and wing-coverts sand-coloured◀ breast, mantle and coverts coarsely streaked cold brown◀ typically has *contrast between streaked* (at distance: dark◀ *breast and mantle, and unmarked pale belly and lower back* respectively. Flight- and tail-feathers and upper greater cov◀ erts blackish, widely tipped whitish; *inner three primarie◀ typically contrastingly paler*, visible on both surfaces. *Pal◀ primary patch insignificant*, giving slightly different upper◀ wing pattern compared with immature Steppe Eagle or th◀ two spotted eagles. (Exceptionally shows hint of light mid◀ wing-band below, recalling young Steppe Eagle, but rest of◀ characters should be conclusive.) – 1st- and 2nd-immature◀ Similar to juvenile but, owing to moult, plumage is not uni◀ formly worn, with one and two active primary moult centre◀ respectively. Pale trailing edges of wings and tail less con◀ spicuous. – 3rd- and 4th-immatures: Gradually, dark feath◀ ers grow, commencing on throat, upperbody and lesser cov◀ erts; can look quite *pied* or mottled in intermediate stages◀ Adult plumage attained from *c.* 6 years of age.

VOICE Fairly vocal. Calls deeper than Golden Eagle's◀ Commonly, a quick series of harsh barking notes, 'owk-owk◀ owk-...', often during dives in aerial display.

Spanish Imperial Eagle *Aquila adalberti* —

L 72–85 cm, WS 180–210 cm. Closely related to Imperia◀ Eagle (treated as conspecific by some). Very rare breeder in C◀ and SW Spain (*c.* 150 pairs). Nests in tree.

IDENTIFICATION Very similar to similarly sized Imperia◀ Eagle, differing in following ways: Adult has pure *white lead◀ ing edge of wing above and below*, and usually larger whit◀ shoulder patches (located further forward, often appearing◀ as extensions of the white upper leading edge of wing▶ *Flight-feathers* on averag◀ *darker, less barred. Inner ta◀ is paler* grey, appearing un◀ barred. – Juvenile is mor◀ *rufous-brown* (not sandy-buff◀ and *unstreaked* (or very finel◀ marked on breast at the most◀ The bill is slightly heavier.

VOICE Calls very similar t◀ those of Imperial Eagle.

Golden Eagle

Imperial Eagle

Spanish Imperial Eagle

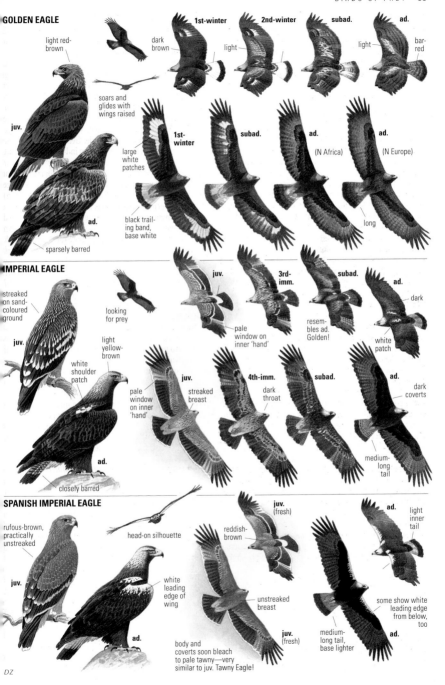

GOLDEN EAGLE

light red-brown

dark brown

light — **1st-winter**

2nd-winter

subad.

light

ad.

bar-red

juv.

soars and glides with wings raised

large white patches

light

1st-winter

subad.

ad.
(N Africa)

ad.
(N Europe)

ad.

sparsely barred

black trailing band, base white

long

IMPERIAL EAGLE

streaked on sand-coloured ground

juv.

looking for prey

juv.

3rd-imm.

pale window on inner 'hand'

subad.
resembles ad. Golden!

ad.

dark

white patch

light yellow-brown

white shoulder patch

pale window on inner 'hand'

juv.
streaked breast

4th-imm.
dark throat

subad.

ad.

dark coverts

medium-long tail

ad.

closely barred

SPANISH IMPERIAL EAGLE

rufous-brown, practically unstreaked

head-on silhouette

juv.
(fresh)

reddish-brown

ad.
light inner tail

juv.

white leading edge of wing

unstreaked breast

juv.
(fresh)

some show white leading edge from below, too

ad.

body and coverts soon bleach to pale tawny—very similar to juv. Tawny Eagle!

medium-long tail, base lighter

ad.

DZ

Lesser Spotted Eagle *Aquila pomarina* —

L 55–65 cm, WS 143–168 cm. Summer visitor Apr–Sep (Oct), winters in Africa; locally rather numerous migrant through Turkey and Middle East. Breeds in forests on open or wooded plains. Food small mammals, amphibians, some birds and insects. Nests in tree.

IDENTIFICATION Medium-sized, dark and compact eagle. Wings evenly broad, tips well 'fingered', although not quite so prominently as on the other, larger *Aquila* eagles; 'hand' less ample, 7th primary 'finger' minute (juv.) or short (ad.). Tail rounded, tail length ⅔ to ¾ of width of wing. In head-on profile, *wings are slightly angled at carpal joints and primaries lowered*, especially when gliding, somewhat also when soaring (lowered primaries when gliding occur among other species too, notably Spotted and Steppe Eagles). Wingbeats rather quick. Head and neck buzzard-like, *bill small* for an eagle. Nostrils round (characteristic of spotted eagles). Best identified by plumage: *dark brown with contrastingly paler head/neck and upperwing-coverts*; a *small whitish primary patch above* is characteristic; a little white on uppertail-coverts on most; underwing usually has *wing-coverts paler than flight-feathers* (milk-chocolate brown against dark grey; when seen well in good light, very few ambiguous). Flight-feathers generally densely barred dark (at distance usually invisible), dark bars at least as wide as paler bars in between; barring prominent right out to tip; some birds are completely unbarred. – Adult: Lesser and median upperwing-coverts *pale yellowish-brown*, forming well-defined area on forewing; upper head pale; back dark (most often) or somewhat paler, depending on moult and wear. White primary patch small and well defined. No obvious white tips to greater coverts, secondaries or tail-feathers. Eye yellow-brown (Spotted: dark). – Juvenile: Wing-coverts (lesser and median, both sides) darker than on adult, medium brown, still usually slightly paler than greater coverts and flight-feathers (though some have confusingly dark underwing-coverts). Head brown, with *rufous-golden* (white when bleached) *patch on nape*. White primary patch is rather extensive, often 'spreading' across inner primaries. Greater coverts (both sides), trailing edge of wings and tail narrowly tipped white; outer upper median coverts also tipped white, but not always visible in the field. Tendency to have slightly bulging secondaries (unlike ad.). – Immature: Adult plumage acquired over a period of *c.* 5 years, but ageing in the field is difficult except for adults and juveniles; however, moult of flight-feathers can give a clue—only inner primaries renewed in 1st summer, both some inner and some outer in following summers.

VOICE Vocal in breeding area. High-pitched bark, 'k-yeep', sometimes likened to yapping of small dog. ♂ has drawn-out whistle, 'wiiiik', during aerial display.

(Greater) **Spotted Eagle** *Aquila clanga* **V∗∗∗**

L 59–69 cm, WS 153–177 cm. Breeds in forests, often interspersed with rivers and marshes. Short-range migrant, wintering in southern part of range and in Middle East. Rare. Food mammals, carrion. Nests in tree.

IDENTIFICATION Large, dark, compact eagle. Flight silhouette very similar to Lesser Spotted Eagle (which see); differs very slightly only in having more ample 'hand' with longer 'fingers'. Bill medium-large. Nostrils round. Best identified by plumage: dark brown, with usually *darker coverts than flight-feathers below* (the opposite of Lesser Spotted) and *lacking well-defined pale upperwing-coverts*. White primary patch above less distinct, formed mainly by basally white shafts of primaries, and more 'spread out', *reaching almost to leading edge of wing*. Flight-feathers below generally densely barred dark, dark bars narrow and becoming fainter towards tip; some birds are unbarred. – Adult, typical: *Dark brown* with slightly and diffusely paler brown head, back and upperwing-coverts, dark plumage relieved only by diffuse, whitish primary patches and, on many birds, some white on uppertail-coverts. Under forewing often darker than greater coverts and flight-feathers, but many look uniformly dark below. A *prominent single white crescent or patch below at the base of the outer primaries* is a good mark, but some have two narrow crescents, one at the primary base and a fainter second at the base of the primary-coverts, just as on Lesser Spotted. – Adult, pale: Some are paler brown on head and wing-coverts, recalling Lesser Spotted Eagle, but lack dark back and sharp division between pale covert area above and dark rest of wing, are only *diffusely* paler brown on forewing and mantle. – Juvenile, typical: Purplish *brown-black* with buff/white spots and feather tips of varying number and prominence. Birds with large spots very typical, *spots forming prominent rows along all wing-coverts*; those with smaller spots approach most prominently spotted juvenile Lesser Spotted Eagle. Exceptionally, pale spots on head merge to form patch on nape as on Lesser Spotted. *Head very dark with contrasting yellow gape-flange.* Under forewing brown-black, clearly darker than medium grey greater coverts and flight-feathers. Secondaries long and 'bulging' in flight silhouette, plumage evenly fresh/worn. – Juvenile, pale: Variable; less dark ground colour, partly or extensively rufous-brown, often heavily spotted or striped yellowish, can be confused with Lesser Spotted at long range, but generally distinguished by dark brown feathers or pattern on head, neck and lesser upperwing- and underwing-coverts, thus less homogeneously brown; washed-out primary patch above also good clue. – Palest form (var. *fulvescens*; very rare) has all body-feathers pale golden (wear to whitish in 1st summer) and pale tips so extensive that it looks more like pale morph N African Tawny Eagle or bleached juv. Imperial Eagle; distinguished—with difficulty—by size, proportions, size of bill and sometimes head-on silhouette and plumage details. Palest form probably occurs among immatures, too. – Immature: Spotted plumage of juvenile repeated in at least one more plumage (spots only slightly smaller and less distinct), thus adult plumage is not developed until after 5–6 years.

VOICE Calls similar to Lesser Spotted Eagle's but lower-pitched. Main call barking 'kyak', singly or repeated.

Lesser Spotted Eagle   Spotted Eagle

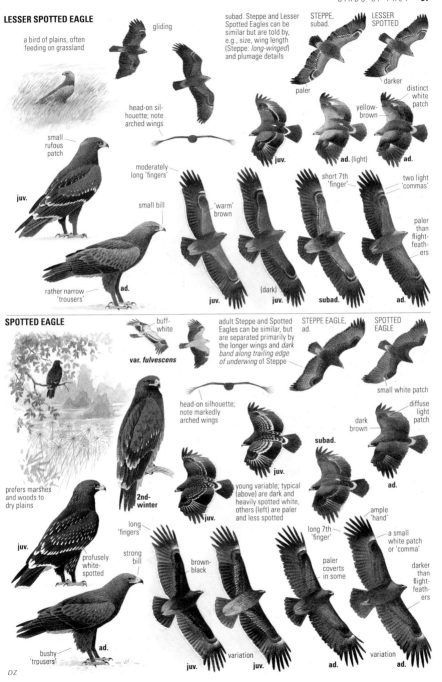

LESSER SPOTTED EAGLE

a bird of plains, often feeding on grassland

gliding

subad. Steppe and Lesser Spotted Eagles can be similar but are told by, e.g., size, wing length (Steppe: *long-winged*) and plumage details

STEPPE, subad.

LESSER SPOTTED

paler

darker

distinct white patch

head-on silhouette; note arched wings

small rufous patch

yellow-brown

juv.

ad. (light)

ad.

moderately long 'fingers'

'warm' brown

short 7th 'finger'

two light 'commas'

small bill

paler than flight-feathers

juv.

rather narrow 'trousers'

ad.

(dark)

juv.

juv.

subad.

ad.

SPOTTED EAGLE

buff-white

adult Steppe and Spotted Eagles can be similar, but are separated primarily by the longer wings and *dark band along trailing edge of underwing* of Steppe

STEPPE EAGLE, ad.

SPOTTED EAGLE

var. fulvescens

small white patch

head-on silhouette; note markedly arched wings

diffuse light patch

dark brown

subad.

ad.

2nd-winter

young variable; typical (above) are dark and heavily spotted white, others (left) are paler and less spotted

juv.

prefers marshes and woods to dry plains

juv.

long 'fingers'

strong bill

profusely white-spotted

brown-black

paler coverts in some

long 7th 'finger'

ample 'hand'

a small white patch or 'comma'

darker than flight-feathers

bushy 'trousers'

ad.

variation

juv.

juv.

variation

ad.

ad.

DZ

Steppe Eagle *Aquila nipalensis* —

L 62–74 cm, WS 165–190 cm. Closely related to Tawny Eagle. Breeds on open, dry plains and foothills, in Europe on the vast plains north of Caspian Sea; summer visitor (Mar–Oct/Nov), wintering in Africa; locally numerous migrant in Middle East. Food rodents and other mammals, bird nestlings, insects (in winter especially termites), carrion, etc. Nests in tree or on mound on ground.

IDENTIFICATION *Large* to very large. Wings long and broad, tips prominently 'fingered'. *Head and neck* somewhat more *prominent* than on the spotted eagles, and 'arm' proportionately longer. Other clues are heavy bill and long, conspicuous, yellow gape-flange (reaching to rear edge of eye). With experience, the more ample 'hand', with more prominent, longer 'fingers' with long 7th primary, separates from Lesser Spotted Eagle. Size helpful only for largest birds and in direct comparison. Nostrils oval (round on spotted eagles). Useful plumage details: *Flight- and tail-feathers coarsely barred dark*, *bars well spaced and always prominent* at close range, covering central 'hand' ('palm'), too (spotted eagles unbarred there); barring not so obvious at a distance, but often shows on distal parts of inner primaries; *terminal band wider on adult, forming dark trailing edge of wing*, visible on most birds at moderate range. Chin and throat pale in all plumages. Only tips of primaries ('fingers') black below, *'palm' paler grey*. On perched birds, 'trousers' are wider and more prominent than on especially Lesser Spotted Eagle (which has tight 'stockings' only). Many have a small white patch on the centre of the back, but this can be seen on both spotted eagles as well. – Adult: Dark or mid brown with coarsely barred flight- and tail-feathers, dark trailing edge of wings unbroken; nape often diffusely paler brown. Dark birds recall Spotted Eagle, but have *dark trailing edge to wings and tail*, and have slightly longer wings. Pale birds recall Lesser Spotted Eagle, but, apart from structural differences (cf. above), often (many exceptions!) have back and upperwing-coverts uniformly light brown (back always darker than wing-coverts on Lesser Spotted); underbody darker than underwing; rather prominent darkish carpal area below. – Juvenile: Characteristic, mid brown with *broad white band along centre of underwing* (exceptionally vestigial) and *broad white trailing edge to wings and tail*. On some, white midwing-band expands to large cream-white patch at carpals. Upper greater coverts broadly tipped white, merging with usually prominent whitish primary patch. Uppertail-coverts with much white. Plumage uniformly fresh/worn. – Immature: Variable, but has complete or partial white midwing-band, and traces of white trailing edge to wings and tail, especially on inner primaries. Plumage unevenly worn or in moult. – Subadult: Similar to adult, but

a few flight-feathers remaining without a prominent dark tip, and traces of the white midwing-band generally visible (remaining for longest on inner primary-coverts).

VOICE Rather silent. Calls basically similar to those of the two spotted eagles, but somewhat lower-pitched.

Tawny Eagle *Aquila rapax* —

L 62–72 cm, WS 165–185 cm. Closely related to Steppe Eagle. N African race (ssp. *belisarius*) resident in mountains and forests, rare. Food as for Steppe Eagle; piracy often recorded. Nests in tree.

IDENTIFICATION Compared with Steppe Eagle a smaller, somewhat more chunky bird, only marginally larger than Spotted Eagle. Plumages differ from Steppe Eagle in being more variable and *often quite pale*. Flight- and tail-feathers *densely but insignificantly barred*, or unbarred, some adults with a faint dark trailing edge. *Lower back to uppertail-coverts almost invariably very pale* (typical of young Imperial Eagle, too, but rare for Steppe). Pale birds usually have *markedly pale inner three primaries* (and sometimes base to rest of primaries below), recalling pattern of young Imperial Eagle. Yellow gape-flange prominent but reaching only to centre of eye (rear edge on Steppe; tiny difference, still useful when seen well). *Throat not markedly paler than rest of underbody* (Steppe Eagle has pale throat). *Young birds lack striking white band on underwing* of young Steppe Eagle. – Adult: Rather variable, dark, medium, light or rufous-brown with darker flight- and tail-feathers. Upperwing-coverts and many body-feathers dark-centred, giving streaked overall impression, and through gradual moult a ragged, less uniform appearance shown by many (except for the darkest, but these are very rare within region). Pale patch at base of inner primaries insignificant. Iris light (dark in Steppe). – Juvenile: Usually pale rufous-brown, bleaching to creamy-white; a few are darker mid brown. Underwing-coverts are pale, the greaters dark-centred. Plumage uniformly fresh/worn.

Verreaux's Eagle *Aquila verreauxii* —

L 78–88 cm, WS 190–210 cm. Extremely rare breeder in SE corner of region (Sinai). Found in mountains in desert. Favourite food hyrax. Nests on cliff-ledge.

IDENTIFICATION Almost size of Golden Eagle (may look larger owing to dark plumage). Adult striking, with *black plumage, large whitish 'windows' on primaries* above and below, and *white slashes to mantle and white rump/uppertail-coverts*, together forming a large U (in flight) or V (at rest) on upperbody. An approaching bird at long range can be taken for a dark-morph Long-legged Buzzard before silhouette and upperparts are revealed. Flight silhouette very characteristic, with *bulging secondaries, narrow wing-base and narrow inner 'hand'*. At distance, wing-tips often appear rather pointed. *Soars with wings raised in distinct V*. – Juvenile: Vaguely reminiscent of adult Golden Eagle, being rather dark with pale rufous-golden nape-shawl, differing in having *distinctly light-based primaries, different wing shape, whitish uppertail-coverts*, slightly shorter tail, proportionately smaller head and longer, slimmer black neck, longer legs and larger feet. Sides of head and throat black. Upperwing-coverts dark, tipped pale buff, appearing 'scaly'.

Steppe Eagle

Tawny Eagle

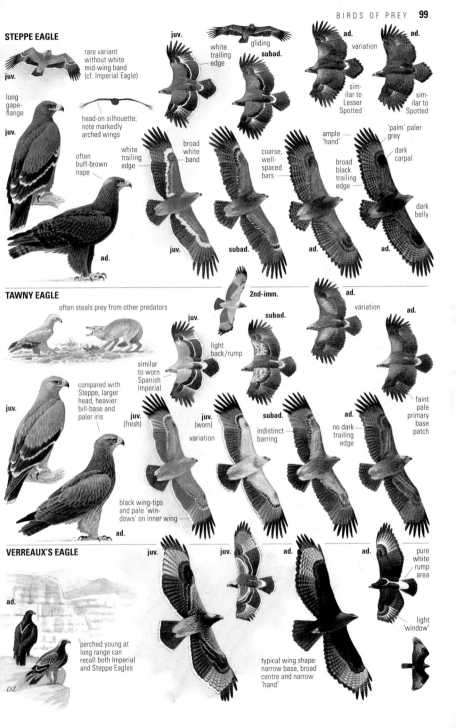

STEPPE EAGLE

juv.

rare variant without white mid-wing band (cf. Imperial Eagle)

long gape-flange

juv.

head-on silhouette; note markedly arched wings

often buff-brown nape

white trailing edge

ad.

white trailing edge

juv.

gliding

subad.

broad white band

broad white band

coarse, well-spaced bars

juv.

subad.

ad. variation

similar to Lesser Spotted

ad.

similar to Spotted

ample 'hand'

'palm' paler grey

dark carpal

broad black trailing edge

dark belly

ad.

ad.

TAWNY EAGLE

often steals prey from other predators

compared with Steppe, larger head, heavier bill-base and paler iris

juv.

juv.

light back/rump

similar to worn Spanish Imperial

2nd-imm.

ad.

variation

subad.

ad.

juv. (fresh)

juv. (worn)

variation

subad.

indistinct barring

ad.

no dark trailing edge

faint pale primary base patch

black wing-tips and pale 'windows' on inner wing

ad.

VERREAUX'S EAGLE

ad.

perched young at long range can recall both Imperial and Steppe Eagles

juv.

juv.

ad.

ad.

pure white rump area

light 'window'

typical wing shape: narrow base, broad centre and narrow 'hand'

DZ

Short-toed Eagle *Circaetus gallicus*   V***

L 62–69 cm, WS 162–178 cm. Summer visitor (late Mar–Oct), winters in Africa. Scarce, *c.* 10 000 pairs within region, mostly in southwest. Breeds in mainly open, arid country with mountains and scattered woods, in north also in river valleys, forests. Food reptiles. Nests in tree.

IDENTIFICATION Large, pale eagle with both *long and broad wings*. Tail length about ⁴/₅ of or equal to width of wing, *tail narrow at base and square-cut* when folded, corners sharp. *Neck short and head broad*, almost owl-like on perched birds (but not always apparent on flying). Flight silhouette when gliding characteristic, *carpals held well forward* and *wings bowed* or arched when seen head-on. Outer secondaries and inner primaries long, giving rather *straight rear edge of wing even when carpals are pressed forward*. Soars on flattish wings. *Hovers*, or hangs motionless in the wind by making small wing adjustments. *Very pale underparts* separate it from most other species; silvery-white with *sharply set-off dark grey head/breast* is typical, rows of dark spots across belly and wing-coverts and loosely barred flight- and tail-feathers. *Tail has three evenly spaced dark bars*, with a hint of a fourth on some, bars visible above as well. Distinguished from Osprey and pale-morph buzzards by *lack of dark carpal patches*, and by tail pattern. Some are darker-patterned, others so pale that only a hint of the tail-bands and some faint barring are visible. Tips of outer primaries shaded grey, darkest along edges, unlike buzzards (which have solidly blackish tips). – Adult/Juvenile: Ageing difficult. Juvenile has more bulging secondaries and narrower 'hand', and all flight-feathers are evenly fresh/worn, finely tipped white.

VOICE Rather vocal in breeding season. One call likened to 'kyo' call of Golden Oriole, given singly or repeated. Also a plaintive 'mee-ok'.

Booted Eagle *Aquila pennata*   V***

L 42–51 cm, WS 110–135 cm. Summer visitor (late Mar–Oct), winters in Africa (a few in SW Spain). Breeds in forests with mixture of open ground and hills or mountains. Food mammals, reptiles, birds, mostly caught on or near ground after spectacular stoop with wings folded. Nests in tree or (rarely) on cliff.

IDENTIFICATION Small eagle, of similar size and shape to a buzzard, though 'hand' is more ample with one more 'finger' (6 instead of 5), making the wing more evenly broad than on buzzards. Flight is also more eagle-like, with straighter course, often faster speed and longer glides. Tail ⁴/₅ of or about equal to width of wing, *square-cut, sides straight* (or even slightly concave near tip). Two morphs: pale has *white underbody* (with some brown or dusky shades on breast and around eyes) and *white underwing-coverts* (with scattered

dark spots) *strongly contrasting against black flight-feathers* (*inner three primaries typically paler* and barred); dark morph has dark brown underbody (rufous tinge, streaked) and underwing-coverts (often somewhat paler towards leading edge), and flight-feathers similar to pale morph although on average not so black, more brown and barred; most striking feature is *paler three inner primaries*. Undertail on both morphs grey with *darker tip and centre*. (There is some variation in both morphs: pale morph with rufous tinge on underbody; and—more common—dark morph being mid brown with paler rufous leading edge of under-wing, sometimes called 'rufous morph'.) About 75% of all show very characteristic *pure white small patch on base of leading edge of wing*, visible when seen head-on ('landing lights'); exceptionally, something similar can be seen on Honey Buzzard (though not such pure white and well-marked patches). Upperparts on both morphs have *pale ochrous-buff panel across inner upperwing*, *whitish uppertail-coverts* and some pale feathers on scapulars.

VOICE Highly vocal in breeding season. Main call shrill 'kli kli kli'. Also buzzard-like 'hiyaah' and long series of 'gü'.

Bonelli's Eagle *Aquila fasciata*   —

L 55–65 cm, WS 145–165 cm. Breeds in forests or mountains. Rare, *c.* 800 pairs within region, mainly in SW. Food usually medium-sized mammals or birds. Nests in cave or on ledge on steep, inaccessible cliff, sometimes in tall tree.

IDENTIFICATION Medium-large, powerful eagle with broad wings and *straight, rather broad, square-cut tail* of about the same length as width of wing, or slightly less. *Wing-tips only moderately 'fingered'* for such a large raptor. *Carpals pressed forward when gliding*, and combination of this and *fairly small head* and length of tail recalls Honey Buzzard. Soars and glides on flattish wings, primaries at times slightly lowered. – Adult: At distance in flight, combination of *whitish body and dark wings and tail* unique. When seen closer from below, *leading edge of wings is white* (extent variable, involving lesser and sometimes some median coverts), outer primaries are basally diffusely pale, and greater coverts are darkest, being blackish; *tail is pale grey with broad dark terminal band*. White areas narrowly streaked dark. Upperparts dark grey with characteristic *whitish patch on back* (of varying size; can be missing). – Juvenile: Underbody and underwing-coverts *pale rufous-buff* (finely streaked dark), *wing-tips blackish*, flight- and tail-feathers pale grey and narrowly barred dark, no broad terminal band on tail. Greater coverts can be rufous-buff without any dark, or have some dark bases forming a faint band on outer and on primary-coverts. – Immature: At first, underbody is pale brown, *prominently streaked dark*. Gradually, greater coverts become darker and body whiter. Dark terminal tail-band develops from 3rd winter, adult plumage from 5th.

VOICE Apparently rather silent. Various barks and shrill notes have been recorded at breeding sites.

Short-toed Eagle

Booted Eagle

Bonelli's Eagle

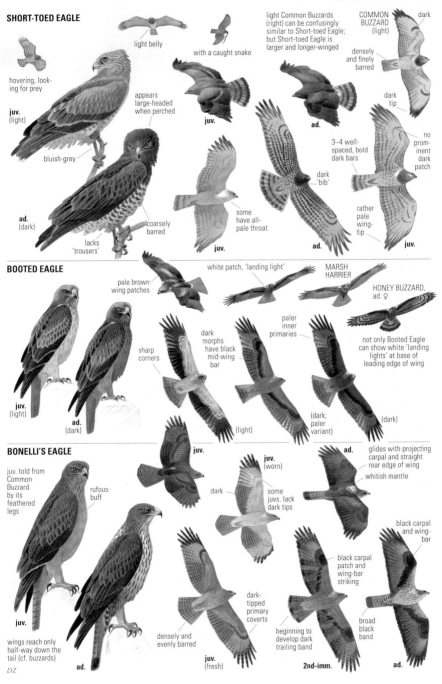

SHORT-TOED EAGLE

light belly

with a caught snake

light Common Buzzards (right) can be confusingly similar to Short-toed Eagle; but Short-toed Eagle is larger and longer-winged

COMMON BUZZARD (light)

dark

densely and finely barred

hovering, looking for prey

juv. (light)

appears large-headed when perched

juv.

dark tip

bluish-grey

no prominent dark patch

3–4 well-spaced, bold dark bars

ad.

dark 'bib'

rather pale wing-tip

ad. (dark)

coarsely barred

lacks 'trousers'

some have all-pale throat

juv.

ad.

juv.

BOOTED EAGLE

pale brown wing patches

white patch, 'landing light'

MARSH HARRIER

HONEY BUZZARD, ad. ♀

paler inner primaries

sharp corners

dark morphs have black mid-wing bar

not only Booted Eagle can show white 'landing lights' at base of leading edge of wing

juv. (light)

ad. (dark)

(light)

(dark; paler variant)

(dark)

BONELLI'S EAGLE

juv. told from Common Buzzard by its feathered legs

rufous-buff

juv.

juv. (worn)

dark

some juvs. lack dark tips

ad.

glides with projecting carpal and straight rear edge of wing

whitish mantle

black carpal and wing-bar

black carpal patch and wing-bar striking

dark-tipped primary coverts

black carpal patch and wing-bar striking

broad black band

juv.

wings reach only half-way down the tail (cf. buzzards)

DZ

ad.

densely and evenly barred

juv. (fresh)

beginning to develop dark trailing band

2nd-imm.

ad.

KITES *Milvus* et al.

Medium-sized to large, often long-winged and long-tailed, raptors. Experts at soaring and gliding, twisting their long and in many species forked tails when manoeuvring. Feed on a variety of smaller animals, often carrion or traffic kills, visit refuse dumps and patrol lakesides or rivers looking for dead fish. Often food-parasites, taking prey from corvids or other raptors. Nest in twig nest high up in tall, mature tree. One species treated on p. 114.

Red Kite *Milvus milvus*   r**B**4

L 61–72 cm, WS 140–165 cm. Breeds in forests near lakes, interspersed with open fields. Mainly migratory in northern part of range. In Britain, smaller population resident in Wales; reintroduced at several sites in England and Scotland, now totalling *c.* 1000 pairs. Food fish, offal, refuse, insects. Often steals food from other birds. Nests in tall tree.

IDENTIFICATION Medium-large, long-winged, with characteristic *long, deeply forked tail*. Wings rather narrow and evenly broad, held somewhat arched and flexed at the carpals, *long 'hand'* lowered. In profile, *looks stooped, with tail and head hanging down*. Flight buoyant and leisurely, almost crow-like, with constant *twisting of tail*. White 'windows' below on inner 'hand'* typical. *Tail rufous above,* paler below. *Pale yellowish-brown panel diagonally across dark inner wing,* paler and more prominent than on Black Kite, and visible in head-on profile as pale wing-bend. Head pale. – Adult: No light tips to greater upperwing-coverts. Breast and belly deep rufous, narrowly streaked black. Undertail-coverts darkish. – Juvenile: Narrow white tips to upper greater coverts. Breast and belly rufous with yellowish-buff central streaks, giving a paler appearance than dark underwing. Undertail-coverts pale.

VOICE Rather silent. Thin, piping, Common Buzzard-like mewing followed by drawn-out notes rising and falling, 'weee-ooh, ee oo ee oo ee oo'. Also 'wee-oh' repeated.

Black Kite *Milvus migrans*   **V****

L 48–58 cm, WS 130–155 cm. Summer visitor, winters in Africa. Breeds in forests near lakes, rivers or wetlands, also close to human settlements. Annual vagrant to Britain & Ireland. Food fish, offal, refuse, sometimes picked up in harbours or by motorways. Social habits, and where numerous fairly large flocks can be seen. Nests in tree.

IDENTIFICATION Medium-sized, with slightly forked tail. Distinguished from Red Kite by dark plumage, slightly smaller size, proportionately shorter wings and tail, *broader wing-tip* and *less forked tail*, which is *grey-brown above,* never rufous. Dark brown with paler bases to primaries below (European adults only slightly paler; juveniles, and many

eastern migrants in Middle East, much whiter, approaching Red Kite). Upperparts dark brown with *pale mid brown panel across inner wing* (same pattern as on Red Kite but darker and duller). If tail-fork not seen, possible to confuse with dark-morph Booted Eagle, but latter has light upper-tail-coverts and usually white patch at base of forewing. Marsh Harrier separated by much less 'fingered' wing-tips, lack of pale primary bases below, and head-on silhouette with raised 'arm' and flatter 'hand', not arched wings with lowered primaries. – Adult/Juvenile: Ageing much as for Red Kite. – Variation: 'YELLOW-BILLED KITE' (ssp. *aegyptius*) of Egypt and S Arabia differs on all-yellow bill and more uniform and pale brown plumage with finer barring of flight-feathers.

VOICE Main call like Red Kite but is faster, a whinnying 'pee-e-e-e-e'. Also a Common Buzzard-like mewing call.

HARRIERS *Circus*

Medium-sized long-winged and long-tailed raptors. Often seen patrolling in leisurely flight low over reeds, marshes, meadows, heaths, steppes or deserts, beating their wings few times, then gliding with slightly raised wings, looking for prey. Food includes small birds, voles, insects and lizards. Nest on ground in cover of reeds or other low vegetation.

(Western) **Marsh Harrier** *Circus aeruginosus*   m**B**4 / **P**5

L 43–55 cm, WS 115–140 cm. Breeds on shallow freshwater lakes or rivers with lining of fairly extensive reedbeds. Except in south and west mainly migratory, wintering in Africa. Small population in E Britain, now about 350 pairs. Food small mammals, birds, insects. Nests in tall reedbed.

IDENTIFICATION Largest harrier, somewhat bigger than Common Buzzard but with slimmer body, narrower wings and longer tail. Soars *with wings raised in shallow V,* glides *with 'arm' raised and 'hand' more level.* At distance possible to confuse with Black Kite and dark-morph Booted Eagle, but has *less 'fingered' wing-tips, rounded tail* and different head-on profile. – Adult ♂: *Distinctly set-off black wing-tips, uniformly pale blue-grey tail,* head and breast pale yellowish-white, upperwing three- or four-coloured. Some are very pale, at distance recalling ♂ Hen Harrier, but have chestnut belly and chestnut on upperwing-coverts. – Adult ♀: Dark brown with creamy-white crown, throat and forewing (variable amount). Often *pale patch on breast.* ♂-like pale birds distinguished by *lack of well-marked black wing-tips* and *by brown tail.* – Juvenile: Blackish-brown with golden crown and throat, sometimes also forewing; a rare morph is all dark. Differs from adult ♀ in narrow golden line along greater upperwing-coverts, very dark underwing-coverts, darker tail, narrower 'hand' and more bulging 'arm'.

VOICE ♀ has thin, piping whistle, 'psee-ee' when receiving food from ♂. Alarm a rapid 'quek-ek-ek-ek-ek'. In breeding season during aerial display high up in undulating flight, ♂ has whining, nasal 'way-e', repeated twice with 2-sec. pause, in tone recalling Lapwing alarm.

Red Kite

Black Kite

Marsh Harrier

RED KITE

juv.

juv.

juv.

bright rufous

ad.

pale-tipped coverts on young birds

deep rufous

ad.

short legs and drawn-out rear parts typical of kites

paler vent easiest way to pick out young bird at some distance

long and deeply forked tail

prominent light 'window'

BLACK KITE

grey-brown

ad.

'YELLOW-BILLED KITE'

aegyptius (Egypt)

pale-tipped coverts in juv.

tail-fork disappears when tail is spread

juv.

juv.

ad.

quite deep fork

yellow bill in ad.

dark 'mask' and eye

juv.

streaked belly and dark mask typical for young individuals

dark brown body

slightly forked

ad.

ad.

greyish head and pale eye in ad.

ad.

diffuse 'window'

more rufous underpart then nominate

ad.

easterly variant (Central Asia)

distinct 'window'

MARSH HARRIER

glides and soars on lifted wings

♂ HEN HARRIER

long wings

♂ MARSH HARRIER

unusually pale bird; still, note the dark lower belly

like mixture of ♂ and ♀

coverts dark

2nd-s. ♂

warm brown

characteristic tricoloured upperparts

ad. ♀

juv.

ad. ♂

well-marked black wing-tip

dark coverts

pale brown coverts

pale under-wing and rufous body

ad. ♀

♂ features obvious on underwing

ad. ♀

ad. ♂

2nd-s. ♂

♀♀ highly variable; typical individual shown; many as dark as juv.

ad. ♂

ad. ♀

ad. ♂

DZ

Hen Harrier *Circus cyaneus* r(m)**B**4

L 45–55 cm, WS 97–118 cm. Breeds on bogs in open taiga, marshes or shallow lakes with much vegetation, also on moors and heathland in flat country or hills. Open country in winter. Food small mammals, birds. Nests on ground.

IDENTIFICATION Recognized as a harrier by *long wings and tail*, and *low flight with wings raised in shallow V when gliding*. Separated from Marsh Harrier by plumage, lighter build and flight, from the two narrow-winged species by *shorter and more ample 'hand' with 5th primary long*, and *broader 'arm'*. – Adult ♂: Wing-tips black, *underparts white with blue-grey head and breast sharply set off*, darkish trailing edge to underwing; upperparts blue-grey with *white uppertail-coverts*. – Adult ♀: Upperparts brown with white uppertail-coverts, inner wing with variable yellowish panel; underparts buffish-white, streaked brown. – Juvenile: Distinguished from adult ♀ by more *rufous-yellow underparts* with fewer streaks, especially on belly, and *more prominent and more rufous panel above on inner wing*, not so yellowish and indistinct; *pale tips to greater upperwing-coverts*.

VOICE ♀ has thin whistling 'piih-eh' when receiving food from ♂. Alarm of ♀ a twittering 'chit-it-it-it-et-it-et-it-et...' with varying pitch towards end; ♂ has a more display-like, straighter 'chek-ek-ek-ek'. In breeding season during aerial display, ♂ gives rapid, bouncing series of notes, 'chuk-uk-uk-uk-uk', somewhat recalling alarm of Little Gull.

Montagu's Harrier *Circus pygargus* m**B**5

L 39–50 cm, WS 96–116 cm. Summer visitor (Apr–Sep), winters in Africa. Breeds on open plains, bogs, wasteland among arable fields, heathland with low bushes. In Britain, sporadic breeder only, fewer than 10 pairs. Food small birds or mammals, lizards, insects. Nests on ground.

IDENTIFICATION Very long and narrow wings (5th primary short, unlike on broader-winged Hen Harrier), *slim body* and narrow, long tail. *Flight buoyant and tern-like*, especially ♂. ♀ very similar to ♀ Pallid Harrier. Melanistic birds occur rarely in W Europe. – Adult ♂: Resembles ♂ Hen Harrier, but differs in shape, flight and in having *black bands across secondaries* (two below, one above), more extensive black on wing-tips, *darker grey back/inner wing* (upperparts appearing *tricoloured*, with paler grey-white outer wing inside black tip). Chestnut streaks on belly detectable at close range. (Very old ♂♂ have extensive blue-grey breast/upper belly and lose much, occasionally all, of the chestnut streaking on belly.) – Adult ♀: Best separated from Hen Harrier by shape. From Pallid by slightly narrower wing-base; *obvious dark band across brown secondaries above*, just outside coverts; widely spaced prominent dark bands across pale buff secondaries below, *pale colour not darkening towards body* as on

Pallid; *dark trailing edge on 'hand' below*; and larger underwing-coverts and *axillaries distinctly cross-barred*. – Juvenile: Darker and more rufous than adult ♀, *underparts uniformly chestnut or golden-rufous*, except for some dark streaks on sides of breast on many (very rare on Pallid). *Secondaries* dark, unbarred above and *almost unbarred dark grey below*. Narrow pale band along greater coverts. Best distinguished from similar juvenile Pallid by *lack of prominent unstreaked buff-white neck-collar* or uniform dark brown sides of neck—there is *no dark 'boa'*. ♂♂ have more irregular barring below on primaries. – 1st-summer ♂: Variable amount of dark grey feathers on head, neck and upper breast, often creating (potentially misleading) effect of pale collar. Central tail-feathers often moulted to greyish from Apr. Flight-feathers still juvenile (brown, barred).

VOICE ♀ and fledged young have thin whistle, 'pee-ee' (not falling in pitch), when receiving food. Alarm similar for both sexes, a rapid bouncing 'chit-er chit-er chit-it-it-it-it', slightly recalling Turnstone. In summer during aerial display, ♂ has a rather nasal, Jackdaw-like 'kyeh kyeh kyeh'.

Pallid Harrier *Circus macrourus* **V*****

L 40–50 cm, WS 97–118 cm. Summer visitor (Mar–Sep), winters in Africa. Vagrant to Britain, about one every three years. Habitat, food and nest much as for Montagu's Harrier.

IDENTIFICATION Shape similar to Montagu's Harrier, but adult ♀ differs on average in having *slightly broader wing-base* and shorter and more ample 'hand', thus is *a trifle more compact* and buzzard-like, and ♂ has shorter 'hand'. Hunting flight of narrow-winged ♂ often with quick Kestrel-like wingbeats, different from Montagu's. Reliable identification, however, can be made only by using plumage characters, too. – Adult ♂: Very pale; *pearl-grey above* (*without prominent white uppertail-coverts* of ♂ Hen Harrier), white below with *only faintly darker head and upper breast*, wing-tips with *narrow black wedges*. – Adult ♀: Very similar to ♀ Montagu's but told by: almost *uniformly dark brown secondaries above*, lacking a blackish band (but can have a hint); overall *darker secondaries below with denser dark bands*, and *pale bands darkening towards body*; trailing edge of 'hand' paler, and barring of primaries less strong distally; larger *underwing-coverts and axillaries dark*, only finely spotted pale, 'armpits' same or diffusely barred rufous. Told from very similar ♀ Hen by different wing shape. – Juvenile: Very similar to juvenile Montagu's but separated by *prominent, unspotted pale buff-white neck-collar*, enhanced by *uniform dark brown sides of neck* ('boa'). Primary bases below often unbarred, leaving *pale 'boomerang' outside dark tips of coverts*. – 1st-summer: Usually like juvenile (but paler and worn), though a few ♂♂ moult early and show pale belly, a few thin rufous streaks on breast, and greyish central tail-feathers even in Mar.

VOICE ♀ has thin, whining, monosyllabic whistle, 'piih', when receiving food from ♂. Alarm similar to Montagu's but on average shorter. Display-call of ♂ is a high-pitched whinnying trill, 'dir-r-r-r-r'.

Hen Harrier

Montagu's Harrier

Pallid Harrier

HEN HARRIER

boldly streaked

juv.

rather compact

juv.

ad. ♀

broad black tip

ad. ♂

juv.

wingtip rather blunt

five visible 'fingers'

breast tinged rusty-yellow, streaked

dark

breast streaked brown on white ground

dark trailing edge

2nd-w. ♂ (Sep)

broad 'arm'

juv.

ad. ♀

ad. ♂

ad. ♀

ad. ♂

MONTAGU'S HARRIER

distinct

juv.

juv.

ad. ♀

ad. ♂

may have light collar

plain rufous

pointed

dark frame to hand

four visible 'fingers'

dark band just outside coverts

black band

juv.

ad. ♀

extensive black on wing-tip

coverts distinctly barred

broad light band

2nd-w. ♂ (Sep)

juv.

ad. ♀

ad. ♂

PALLID HARRIER

no dark

juv.

juv.

ad. ♀

ad. ♂

light collar

plain rufous

pointed

rather pale edge

four visible 'fingers'

dark wedge

almost uniformly grey-brown

dark 'boa' and light collar

dark secondary-coverts

black wedge

very white

juv.

juv.

inner secondaries dusky

2nd-w. ♂ (Sep)

ad. ♀

light

ad. ♂

ad. ♀

ad. ♂

DZ

BUZZARDS *Buteo*

Medium-sized, broad-winged raptors. They frequently soar high up using rising warm air. Hover or sit motionless, often on telegraph poles or tree watching for prey, mainly small rodents but also insects, lizards and earthworms, which are taken on ground. *Buteo* buzzards are not closely related to honey buzzards (*Pernis*, p. 110) but appear very similar at a distance, and in flight require care to be separated.

Long-legged Buzzard *Buteo rufinus* —

L 50–58 cm, WS 130–155 cm (ssp. *rufinus*; SE Europe, Asia); L 43–50 cm, WS 105–125 cm (ssp. *cirtensis*; N Africa). Breeds in arid steppe or semi-desert, also in mountains (Africa). Food small mammals, reptiles, insects. Nests often on cliff-ledge. Winters mainly in Middle East and N Africa.

IDENTIFICATION Ssp. *rufinus*: *Large and long-winged* buzzard, similar in shape to Rough-legged Buzzard. *Silhouette and flight as for Rough-legged Buzzard* (incl. hovering), but at times glides and *soars on slightly more raised wings*. Often best distinguished by plumage, but, with some very similarly plumaged 'Steppe Buzzards' (cf. under that and below), size and flight must be considered, too. At least three different colour morphs: pale (commonest within treated range), rufous and dark, but difference between first two far from clear-cut. Common to all are *extensive dark carpal patches* (slightly on upperwing as well, which is rare for 'Steppe Buzzard'), *dark trailing edge of wings, very white primary bases* below, and for adults *unbarred pale rufous tail* (there may be a hint of a dark end-band); in strong light and at a distance, the *adult tail can look whitish* above, especially when bleached. All pale and many rufous morphs have *pale head and breast and progressively darker belly*. Underwing-coverts uniform or streaked, not barred. Upperwing has pale greyish-white primary bases (with often visible barring at close range) and often *pale rufous or yellowish-brown forewing, paler than dark grey flight-feathers*. Easily confused with rufous morph of 'Steppe Buzzard', and best recognized by larger size, proportionally longer wings, slower wingbeats and unhurried soaring in wider turns, by dark belly and paler head in pale morph and by usually paler upperwing-coverts. Note that tail of juvenile is greyish-buff and finely barred, very similar to some 'Steppe Buzzards'. *Dark morph* (easternmost parts of range) has body and wing-coverts blackish, tail- and flight-feathers white with distinct barring; very similar to dark-morph Rough-legged Buzzard (but hardly occurring in same area) and dark-morph 'Steppe Buzzard', but is larger and longer-winged than latter. – Adult: Tail unbarred pale rufous above, slightly paler and greyer below; broad black band along trailing edge of wing. – Juvenile: Tail greyish-buff, finely barred grey; indistinct

dark band along trailing edge of wing. – Ssp. *cirtensis*: Smaller and a little more compact, similar to Common Buzzard in size and shape. Head often pale and belly rufous; dark carpal patches below can be missing. Cere and feet of juveniles often pale brownish blue-grey rather than yellow. No dark morph.

VOICE Rather silent. Calls resemble those of Rough-legged Buzzard, but are shorter and more mellow in tone.

'Steppe Buzzard' *Buteo buteo vulpinus* —

L 40–48 cm, WS 100–125 cm. A northeasterly subspecies of Common Buzzard (p. 108), here treated separately due to its slightly different appearance, its migratory habits and its at times great similarity to Long-legged Buzzard. Breeds in NE and E Europe in coniferous or mixed forests, often near glades, bogs or rivers. Migrates east of Mediterranean to winter in E and S Africa. Most abundant raptor seen on migration both at Bosporus and through Israel. Food and nesting habits similar to Common Buzzard.

IDENTIFICATION Shape similar to that of Common Buzzard, but has on average *slightly narrower and more pointed wings* and longer tail, differences which are subtle and often difficult to ascertain in the field. The *slightly smaller size* and *quicker wingbeats* can be more useful to the trained eye. Plumage characters include generally *warmer, more rusty colours below and on uppertail* (but full scale of plumage variation occurs as in most other buzzards, except that whitish morph is missing), rather *whiter flight-feathers below* with finer but more distinct barring, and more *well-marked black wing-tip and trailing edge to wings* in adults. *Upperwing* nearly always *has pale patch on bases of outermost few primaries*, and *rufous-tinged uppertail is progressively paler towards base. Rufous and darker morphs* as a rule lack the pale band across lower breast so typical of W European Common Buzzard, *are uniformly dark on underbody*. Told from often very similarly plumaged Long-legged Buzzard mainly on smaller size and more agile movements, on *soaring on more level, not raised wings* (similarly, glides on slightly lowered outer wings, whereas Long-legged holds wings level). Also, on 'Steppe' pale primary patches above small and ill-defined (Long-legged usually has large pale brown-grey or even whitish 'palms') and *upperwing-coverts* only rarely prominently paler, usually *rather uniformly dark* (Long-legged often has strikingly paler yellowish or rusty-buff coverts). Also useful to note that *median underwing-coverts are somewhat lighter* than coverts, and *rufous-tinged uppertail is progressively paler towards* on most 'Steppe' but not on Long-leggeds, which have rather uniform buff-white or rufous lesser and median coverts and often darker greater coverts forming a dark line between coverts and flight-feathers.

Long-legged Buzzard

'Steppe Buzzard'

Common Buzzard   (See p. 108.)

Rough-legged Buzzard   (See p. 108.)

LONG-LEGGED BUZZARD

rufinus
(SE Europe)

often perches on ground

large pale area

grey-buff, finely barred

soars with lifted wings in Golden Eagle-like fashion

cirtensis
(N Africa, Sinai)

pale birds might cause confusion

Common Buzzard-like

juv.

Rough-legged Buzzard-like

juv.

juv.

dark band

ad.
(dark)

rufous with white base

ad.
(pale)

ad.

juv.

dark birds very like dark 'Steppe Buzzards' (below)

ad.
(dark)

ad.
(rufous)

ad.
(pale)

prominent black carpal patches typical

rufous belly

ad.

juv.

ad.

heavy bill

long neck evident when hovering

smaller than *rufinus*; like 'Steppe Buzzard' in size

dark 'trousers' (cf Rough-Legged)

dark brown belly and 'trousers'

active flight heavy and eagle-like

rufous belly and 'trousers'

'STEPPE BUZZARD'
vulpinus

juv.
(dark)

juv.
(fox-red)

juv. lacks dark trailing band

heavily streaked

fairly quick wingbeats

pale eyebrow

soaring together with Long-legged Buzzard (centre) on migration

extremely variable, some birds identical to nominate Common Buzzard

similar to Long-legged

distinct breast-band

ad.
(dark)

dark end-band excludes Long-legged

ad.
(fox-red)

ad.
(dark)

ad.
(fox-red)

ad.
(dark rufous)

agile flight usually reveals the smaller size

DZ

Common Buzzard *Buteo buteo* rB3

L 48–56 cm, WS 110-130 cm. Fairly common, and along with Sparrowhawk most numerous European raptor. Breeds in forests or small woods with access to open land, farmland, meadows or marshes. Food mainly voles, also birds, rabbits, reptiles, amphibians, insects, earthworms. Nests in tree. Resident except in much of Fenno-Scandia, from which rather large exodus takes place Sep–Oct. Replaced in NE Europe by 'Steppe Buzzard'. (Map on p. 106.)

IDENTIFICATION Medium-sized, *broad-winged*, rather compact raptor with *broad, short neck* and medium-length tail. Often seen perched on fence posts or telegraph poles watching for prey. Also soars (moderate height), showing *fan-shaped, spread tail*, and wings held pressed forward and *lifted in shallow V*. Glides on flatter wings, generally with slightly raised inner wing and slightly lowered outer wing, *showing more angled carpal bend than Honey Buzzard*. In fast glide carpals are pushed forward and primaries flexed back. As with most *Buteo* buzzards, *tail is slightly or obviously shorter than width of wing*, has straight sides and sharp corners (Honey Buzzard usually has slightly longer tail, equalling width of wing, slightly convex sides and round corners). Long-distant migrant 'Steppe Buzzard' ssp. *vulpinus* (see p. 106) on average slightly smaller, more narrow-winged and appearing longer-tailed than short-distance migrant or resident W European ssp. *buteo*. *Active flight with rather quick, stiff wingbeats*, with experience appreciably faster pace than Honey and Rough-legged Buzzards. Plumage variable, from very dark to very pale; still, several characters useful. Off-white *tail densely barred grey*; regardless of colour morph *a pale band across lower breast*, separating dark upper breast from (sides of) belly; from below, trailing edge of wing and wing-tips blackish, involving whole 'fingers'; rest of flight-feathers whitish, barred dark (5–7 bars; outer primaries have only faint bars). Pale morph vaguely similar to pale-morph Booted Eagle, but base of primaries extensively white on underside, and *dark carpal patch*, often in shape of a bold comma, diagnostic; upperparts frequently with some degree of *pure white on forewing* (lesser and median coverts) and rump. Differences from Long-legged Buzzard treated under that species (p. 106). Note that rarely light immatures may resemble light-morph immature Long-legged, differing on smaller size, more broken-up dark carpal patch, often presence of some barred underwing-coverts, and hint of a pale band along mid-wing formed by median coverts. – Adult: *Terminal tail-band blackish, distinct and much broader than inner bars*. Light feathers of breast, lower belly/ 'trousers' and greater underwing-coverts *finely cross-barred*, not blotched or streaked. Wings rather evenly broad with more ample 'hand', and tail slightly shorter (esp. on ♂) than juvenile. – Juvenile: *No distinctly broader terminal tail-band*. Underparts coarsely *streaked*, especially on breast; some feathers blotched or with indication of cross-bars, but streaking predominates. 'Hand' narrow, tail often slightly longer than on adult. Eye paler than adult's.

VOICE Highly vocal for a bird of prey, especially in spring. Main call a loud, mewing 'piiiyay', falling in pitch (and skilfully mimicked by Jay!); given mainly in flight. Begging call of fledged young similar but drawn out and with tremolo, and more whining in tone.

Rough-legged Buzzard *Buteo lagopus* P / W5

L 49–59 cm, WS 125–148 cm. Breeds on fells, in mountain valleys or upland tundra, sometimes in lowland forests with minor hills or crags. In Britain & Ireland, rare passage and winter visitor. Food mainly small mammals. Nests on cliff-ledge or steep slope, at times quite accessible, or in tree. Short-distance migrant, wintering in S Fenno-Scandia and NC Europe. (Map on p. 106.)

IDENTIFICATION Typical buzzard, with broad wings and medium-length tail, but compared with Common Buzzard larger and *longer-winged*. Glides on *wings clearly bent at carpal joints* (bend slightly more marked than on Common), inner wing raised and primaries flat. *Active flight with slower wingbeats* than Common Buzzard. *Hovers frequently*, much more commonly than Common Buzzard. Best recognized by *white inner tail above and below* (pale-morph Common can have whitish inner tail, too, but not so clearly contrasting against dark upperwing/rump and distal tail, more diffusely set off, and coinciding with white upperwing-coverts and rump not found on Rough-legged Buzzard). Underside whitish with bold pattern, always incl. *blackish carpal patches* and *dark belly*. Plumage variation largely referable to differences between sexes and ages. (Very rarely a *dark morph* has been recorded within region, blackish on whole body and wing-coverts, with white tail- and flight-feathers heavily barred black; very difficult to distinguish from dark Long-legged Buzzard, and except for size from dark 'Steppe Buzzard'.) – Adult ♂: Ample 'hand', evenly broad wings and rather short tail; trailing edge often somewhat uneven through wear and moult. *2–4 black cross-bars inside broad black end-band above on outer tail*, 1–3 of these visible from below; flight-feathers narrowly but distinctly cross-barred; typically has *head/breast darker than belly*, latter profusely broken up by light barring; 'trousers' cross-barred; *underwing-coverts dark*, often giving Common Buzzard-like impression, not seen on adult ♀; upperparts usually all dark with small, diffuse paler patch at base of outermost primaries, and leading edge of 'arm' a little paler. – Adult ♀: Differs from adult ♂ in having *only one narrow black bar* (very rarely two bars) *inside broad black end-band on tail*, visible above and below; *dark belly patch usually prominent and almost invariably darker than head/breast*; underwing-coverts variable but never so dark as on darkest adult ♂. – Juvenile: *Narrower 'hand' and slightly bulging secondaries*; all flight-feathers uniformly fresh (or worn). Ground colour yellowish-white, *breast and underwing-coverts typically loosely streaked*; *belly solidly black without light barring*; tail has *diffusely defined dark grey broad end-band, often rather pale from below*, on ♂ frequently with hint of several cross-bars through it; flight-feathers only diffusely barred, end-band diffuse; *upperwing typically with large pale primary patches*. – Subadult ♂: Often intermediate between adult ♂ and adult ♀, making sexing in the field more difficult except for typical extremes. Underwing-coverts and head/breast not so dark as on typical adult ♂, and tail with fewer bands.

VOICE Vocal in breeding season. Main call a wailing miaow, recalling Common Buzzard, but drop in pitch perhaps more pronounced, and tone often sounds more whining and piercing. Also used as alarm, at times vibrant (esp. when taking off from perch).

COMMON BUZZARD

extremely pale variants (mainly N Germany–S Sweden) are very odd-looking; typically show mainly white head, underside, upperwing-coverts and tail-base

juv. (light)

HONEY BUZZARD: hanging wings in head-on view, slim wingbase and rounded, long tail

white base only

often perched on fence posts along roadsides

juv. (light)

black 'comma'

ad.

COMMON BUZZARD: level or slightly raised wings and compact shape

dark birds rather non-descript above

compact shape

streaked breast in juv.

barred breast in ad.

ad.

juv. (light)

pale mid-wing band typical in all ages

dark trailing edge in ad.

pale breast generally typical

juv.

juv.

ad.

ad.

ROUGH-LEGGED BUZZARD

hovering

ROUGH-LEGGED BUZZARD

GOLDEN EAGLE

head-on view with clearly raised wings

hunting voles in winter

diffuse trailing band in juv.

buff

juv.

single diffuse bar

ad. ♀

1–2 bars

shape and flight action at times reminiscent of Marsh Harrier

ad. ♂

3–4 bars

streaked

diffuse trailing band

juv.

large pale patch in juv.

no pale patch in adults

feathered legs

juv.

ad. ♂

barred

dark carpals typical in most plumages

light

distinct trailing bands

two or more bands

dark wing-coverts and hood in ♂♂

some dark individuals confusable with Common when tail-pattern hidden; note blotched upperpart and contrasting 'face' with pale 'blaze'

large black belly patch in ♀

ad. ♀

ad. ♂

variation

ad. ♂

DZ

HONEY BUZZARDS *Pernis*

Medium-sized, broad-winged raptors, although not closely related to the *Buteo* buzzards (pp. 106–109), still at times very similar to them, in particular in flight. The two groups share similar habits of frequently soaring on thermals, therefore have developed similar wing-shape. Honey buzzards have longer, thinner neck and smaller head. They feed their young mainly on nests of wasps, which they locate by intently watching the adult insects returning to their nests, then dig out the wasp nests with their claws. Feathers around bill and eyes are scale-like as protection from wasp stings. Diet includes numerous other small animals, especially in rainy summers with fewer wasps. Long-distance migrants, wintering in tropical Africa or S Asia.

(European) **Honey Buzzard** *Pernis apivorus*   m**B**5

L 52–59 cm, WS 113–135 cm. Summer visitor (late Apr/May–late Aug/Sep), wintering in tropical Africa. 2nd-calendar year birds remain in Africa (exceedingly rare with confirmed records north of Africa), returning to breed in 3rd-calendar year. Breeds in forests with clearings, glades, small wetlands, fields. In Britain, rare away from few breeding sites (only *c.* 40 in all). Food mainly larvae and nests of wasps, also reptiles, amphibians, nestlings of small birds, worms, etc. Nests in tall tree, lined with fresh leaves during breeding.

IDENTIFICATION Slightly larger and more *long-winged* than Common Buzzard, but at distance easily confused with it. At closer range, several finer points of distinction: *neck slim and head held forward* in flight in Cuckoo fashion; *tail* rather long, *about as long as width of wing, sides slightly convex* and *corners rounded*; head-on silhouette in glide with *smoothly downcurved wings*, unlike Common Buzzard in lacking obvious bend at carpal joints; soars on flattish wings; *active flight with slower, more elastic wingbeats*. Plumage variable, incl. dark, medium, pale and rufous morphs both as adults and as juveniles. Adults usually distinctive, but juveniles much more similar to Common Buzzard. Common to both age categories is presence of rectangular or oval-shaped dark carpal patch. – Adult ♂: Ample 'hand' and rather long tail. *Only tips of longest primaries black, sharply set off*; prominent black trailing edge to wing; *long step to next dark bar across pale grey flight-feathers*; tail similar, *prominent dark end-band and long step to inner one or two narrower dark bars at base*; head largely blue-grey; *upperparts brown-grey, with black bars of underside visible above as well*; cere grey, eye yellow. – Adult ♀: Differs from adult ♂ in having slightly *more extensive dark on 'fingers'*, diffusely set off; *shorter step between dark bars at end and base of tail- and flight-feathers, showing more bars* (usually 2 basally, rarely hint of 3rd);

Honey Buzzard   Crested Honey Buzzard

secondaries often dusky; no or rather little blue-grey on head; *upperparts dark brown, hardly showing any bars.* – Juvenile: Slightly shorter inner primaries and sometimes shorter tail, giving *more Common Buzzard-like silhouette*; *wing-tips more extensively dark*, as on Common Buzzard; *flight-feathers more densely barred than on adult* (4–5 bars instead of 2–3), recalling juvenile Common Buzzard (but latter has *c.* 6); *secondaries and tail often darkish, with 4 or 5 evenly spaced bars*; cere yellow, eye dark. Dark morph (fairly common) uniformly dark brown on underbody, *lacking pale breast-band* of most Common Buzzards (ssp. *buteo*); lightest part on inner underwing is greater coverts (bases of secondaries on Common Buzzard). Pale and medium morphs often coarsely *streaked on breast/belly*, not cross-barred as most adults.

VOICE Silent except in breeding season. Main call a plaintive whistling '**peee**-lu' or 'glü-**i**-yü', at times recalling Grey Plover in general structure. Near nest an odd, mechanical, rapidly ticking call as from a distant mowing-machine.

Crested Honey Buzzard *Pernis ptilorhyncus*   —

(Alt. names: Oriental or Siberian Honey Buzzard.) L 55–65 cm, WS 130–155 cm. Breeds in the taiga of SC and E Siberia (ssp. *orientalis*), often near bogs, rivers or clearings, wintering in tropical Asia, from Arabia in the west (scarce!) to SE Asia. Near-annual occurrence of odd migrants in Eilat, S Israel. Food mainly larvae and nests of wasps, as in European Honey Buzzard. Nests high up in tree.

IDENTIFICATION Slightly larger and more eagle-like and *broader-winged* than European Honey Buzzard, with *'fuller'* wing-tips (6th primary longer, constituting one more 'finger'); the broad wings make head look small and give the bird a rather 'heavy' impression in flight. As its name implies, has elongated crown-feathers forming small crest, but this is rarely erected or seen. Recognition first alerted by size and shape, then confirmed by plumage characters. – Adult ♂: Readily told by characteristic *tail pattern with only two and very broad and prominent dark bands, one at tip, the other near base* (♂ of European Honey Buzzard has two narrow bands basally apart from broad terminal one). – Adult ♀: Unlike ♂, much more similar to ♀ European Honey Buzzard, with two similar narrower basal bands on tail, but possible to separate on: *lack of dark carpal patches*; *more bars on secondaries* (3–4 inside terminal bar rather than 2) *which are more evenly spread*, lacking wide step between terminal and penultimate typical of European Honey Buzzard; *narrow and poorly marked darker trailing edge to wing* compared to that of European Honey Buzzard; *barring of primaries denser and more irregular* ('messy') than in European Honey Buzzard; often an obvious *dark brown or black marking on sides of throat* and across lower throat. – Juvenile: Variable in plumage, most very similar to corresponding morphs of juvenile European Honey Buzzard, only invariably *lacks dark carpal patches* (whereas Honey Buzzard only exceptionally lacks carpal patches). Generally noticed due to *large size* and *broad wings* making *head look particularly small*, after which identification is confirmed by absence of carpal patches, details in wing shape, plumage, etc. Usually, *secondaries are less dark* than in European Honey Buzzard and therefore *show more obvious dense barring* (usually 5–6 bars visible as against commonly 4 in European Honey Buzzard).

HONEY BUZZARD

ad. ♂

typical shape with protruding head and long, rounded tail

smoothly downcurved wings in head-on silhouette

yellow cere and dark eye in juv.

long rear and short legs—kite-like jizz!

juv.

very complex plumage variation, especially in juvenile. Only some samples shown

yellow eye

ad. ♂

ad. ♀

juv.

ad. ♀
pale 'hand'

ad. ♂
distinct bars

distinctive tail pattern in both sexes

typical shapes

ad. ♀

pale mid-wing

lacks pale breast-band (cf Common Buzzard)

dusky wing-tip and denser barring in ♀♀

juv.

juv.

juv.

ad. ♀

ad. ♀

translucent 'hand'

wide gap between dark bands

dark carpal

ad. ♂

ad. ♂

ad. ♂

CRESTED HONEY BUZZARD

associating with Honey Buzzards on migration through the Middle East

juv.

ad. ♂

unique pattern

heavy, eagle-like impression

6th 'finger'

gorget may be lacking

usually 5–6 bars (4 in Honey)

6th 'finger'

tail pattern as ♂ Honey B.

irregular speckling/ dense barring in ♀♀

ad. ♀

ad. ♀

unmarked carpal

juv.
Short-toed Eagle-like

juv.
Bonelli's Eagle-like

juv.
Spotted Eagle-like

dark eye

defined 'gorget' diagnostic

ad. ♂

pale eye

ad. ♀

broad and full 'hand'

more evenly-spaced dark bands (cf. ♂ Honey B.)

lack of dark carpal patch typical!

comparatively short tail

gorget

unique pattern

ad. ♂

ad. ♂

DZ

HAWKS *Accipiter*

Small to medium-sized, broad-winged and long-tailed raptors. Hunt by surprise attack or fast pursuit in the air, relying on speed and ambush strategy, but hunts preceded either by stalk from perch or from soaring surveillance in flight, at times high up. Rounded wing-tips enable quick turns and hunts through dense forests. Food mainly birds but also some smaller mammals. Birds caught on ground or in air, then plucked and eaten on the ground under some cover. Nest in twig nest high up in dense tree, often a conifer.

(Eurasian) **Sparrowhawk** *Accipiter nisus* ꜰ**B**3

L ♂ 29–34 cm, ♀ 35–41 cm, WS ♂ 58–65 cm, ♀ 67–80 cm. Breeds in forests, also near human settlements; sometimes in dense parts of large parks or gardens. Along with Common Buzzard most numerous raptor in Europe. Food small birds. Hunts with surprise attack and fast flight, often low among trees, bushes and buildings. Will even pursue small birds on foot under and through bushes. Nest in tree, often 6–12 m up in spruce, newly built each year. Secretive near nest, often evades notice. Resident in large parts of Europe, but Fenno-Scandian breeders largely migrate to Continental Europe, young ♂♂ moving furthest and adult ♀♀ shortest distance.

IDENTIFICATION A small raptor with rather short, broad, blunt-tipped wings and long tail. *Tail always longer than width of wing* and *has 4–5 (6) bars, also on central tail-feathers* (though often indistinct on ad. ♂). Flight characteristic, a few quick wingbeats relieved by a short glide, *pigeon-quick wingbeats* making the bird ascend, and brief glides descend, thus *flight path slightly undulating*, not so steady as that of Goshawk. ♀ could be confused with ♂ Goshawk (as size difference not always obvious) but, apart from quicker wingbeats and lighter, more undulating flight, the flight silhouette is slightly different, with *narrower tail-base, sharper corners to squarer tail, proportionately shorter inner wing and larger head, shorter neck and slimmer body*. Often flies low over ground making surprise attacks; does not shun urban areas and gardens (as Goshawk normally does). Perched birds (then adopting upright stance) identified by *comparatively small bill, thin legs* (rule of thumb: tarsus thinner than diameter of eye, the opposite in Goshawk) *and slim lower body*. – Adult ♂: Small; slate-grey upperparts, often tinged bluish; rufous cheeks; barring on breast and belly is rufous (rarely brown-grey like ♀). – Adult ♀: Larger than ♂, with steadier flight and on average slightly more poised wingbeats, still as a rule perceptibly quicker than in Goshawk; slate-grey upperparts; barring below is brown-grey (rarely faintly tinged rufous, and then similar to the least rufous ♂♂). – Juvenile: *Dark brown upperparts*, feathers of forewing edged rufous, visible at close range. Barring

of underparts often coarse, broken up and irregular on breast, most noticeable on ♂♂.

VOICE Mainly silent outside breeding season, and even near nest often very quiet. Main call a rapid chattering or cackling 'kewkewkewkewkew...' or 'kekekekekeke', with minor variations used for contact, display and alarm. Food-call of ♀ a thin wailing '**pii**-ih'.

(Northern) **Goshawk** *Accipiter gentilis* ꜰ**B**4

L ♂ 49–56 cm, ♀ 58–64 cm, WS ♂ 90–105 cm, ♀ 108–120 cm. Breeds in forests, generally in mature conifer stands but also found in younger forests (but then requiring at least some denser and taller parts for cover and nesting). British population at least 400 pairs. Although normally a shy bird which stays clear of humans, in some areas it can adapt and even perch on roofs and enter towns in search of prey. Food birds and mammals, up to size of Pheasant and hare, but more commonly takes pigeons, corvids, smaller grouse, thrushes, squirrels and rabbits. Hunting technique similar to Sparrowhawk's, but more often soars at high altitude when looking for prey. Nest in tree, bulky, about 10–16 m up, often re-used. Twigs with green leaves are added as lining on the nest during breeding. Largely resident, but young birds, and populations in NE Europe, are usually short-distance migrants.

IDENTIFICATION Medium-large, strong raptor, broad-winged and long-tailed, by and large very similar to Sparrowhawk, only one full size bigger. Marked sexual size difference, wingspan of ♂ similar to Carrion Crow, of ♀ to Common Buzzard. Usually seen flying at treetop level or slightly higher, a few *relaxed wingbeats* (slower pace than Jackdaw) relieved by short *straight glide* (unlike Sparrowhawk does not lose height when gliding). Prey species like Woodpigeons and corvids show great respect for an approaching Goshawk, taking off in massive flushes. A hunting Goshawk initially often soars or sits watching, body erect, showing *strong lower body* ('hip-heavy'), and sometimes *extensive, 'fluffy' white undertail-coverts*. Note that a Goshawk has very *powerful tarsi and toes*, the legs being about three times as thick as the 'knitting needles' of a Sparrowhawk (a clear indication of prey choices). Adult ♂ confusable with ♀ Sparrowhawk despite larger size; note somewhat slower wingbeats and more steady, confident flight, *proportionately longer and more pointed wings, rounded tail-corners, broader tail-base*, and *slightly more protruding head* due to longer neck. – Adult: Bluish-grey (♂) or slate-grey (♀) above; whitish below, finely barred grey. Supercilium white, contrasting with near-black crown and ear-coverts. – Juvenile: Brown above; *buff-white below, coarsely streaked brown* ('armpits' blotched, can look 'chequered' below in flight). – Variation: Breeders in Russia and W Siberia (ssp. *buteoides*), rare stragglers to Europe in winter, are larger and paler, adults whiter below due to finer barring, juveniles being paler cream-white below with finer brown streaking, and upperwing-coverts and tail-base often partly marbled or speckled white.

VOICE Mainly silent outside breeding season. Calls, heard mostly at dawn, much as Sparrowhawk's but distinctly louder, lower-pitched and more fierce, and cackling slower, 'kya-kya-kya-kya-...'. Begging-call of ♀ and young a wild, melancholy 'piii-lih'. Both calls cleverly mimicked by Jay!

Sparrowhawk · · · · · · · · · · Goshawk

SPARROWHAWK

pale-fringed coverts

SPARROWHAWK   KESTREL

slim

juv.

ad. ♂

some ♀♀ are as rufous as ♂♂

usually pale supercilium in ♀

juv.

juv.

long tail, slim base

ad. ♀

ad. ♂

ad. ♀

ad. ♂

rufous cheek and breast

ad. ♂

coarsely barred

slim legs

sharp

slim hip   'no' neck

GOSHAWK

juv.

mottled pattern above

rounded   long wing

shape differs from Sparrow-hawk (above)

broad hip   long neck

2nd plum-age ♀

coarse barring

striped belly

ad. ♀

dark crown and cheeks

juv. ♂

juv.

fine bars

black hood

heavy hip

short tail

streaked below, flanks variable (chevron-, drop- or heart-shaped markings)

ad. ♂

strong legs

ad. ♂

DZ

Levant Sparrowhawk *Accipiter brevipes* —

L 30–37 cm, WS 63–76 cm. Summer visitor (Apr–Sep) to SE Europe, winters in Africa. Typically gregarious during migration, and migration peaks rather concentrated (tens of thousands birds recorded in a day at Bosporus and in Israel, often around 25 Apr in spring and 20 Sep in autumn). Breeds in forests mainly at lower altitudes, also on plains with undisturbed copses, orchards, riparian forests. Food lizards, large insects, small birds and small mammals. Hunting technique much as Sparrowhawk. Nest in tree, often 5–10 m up, lined with green leaves.

IDENTIFICATION Resembles Sparrowhawk, but sexes more similar in size, *wings narrower and more pointed, tail slightly shorter*, giving almost falcon-like silhouette when gliding, but action still hawk-like. *Dark wing-tips above and below* characteristic (esp. on ♂, intermediate on ♀, less so on juv.), as are *unbarred central tail-feathers* (Sparrowhawk has invariably barred central tail-feathers, although bars can be subdued on some ad. ♂♂). *Tail has thinner and more (6–8) bars* (Sparrowhawk: usually 4–5). *Thin dark central throat-stripe* (also called gular or mesial stripe) of adult ♀ and juvenile characteristic (when seen). At close range, *dark eye and uniformly grey sides of head* of adults (thus no white supercilium) are useful points, as are *prominent yellow cere* (much more vividly yellow and obvious than on Sparrowhawk) and stronger legs with shorter toes. *Underwing appears whitish* on adults (but is actually finely barred rufous when seen at close range). Habit of *migrating in dense, at times large flocks* separates it from Sparrowhawk (100 or more can occur, but even ten birds soaring close together would signal Levant Sparrowhawk). – Adult ♂: Blue-grey above, with well-marked blackish wing-tips. – Adult ♀: Slate-grey above, with slightly darker wing-tips; somewhat larger than ♂; has often a faint central throat-stripe. – Juvenile: Dark brown above with somewhat darker wing-tips discernible in strong light. Underside streaked or spotted on breast and blotched or barred dark on flanks and underwing, recalling more juvenile Goshawk than Sparrowhawk on plumage pattern. Central throat-stripe well marked.

VOICE A characteristic, shrill, clear 'kewick', often repeated a few times, 'kewick-kewick-kewick', therefore surprisingly recalling both Turnstone and Collared Pratincole in voice.

Dark Chanting Goshawk *Melierax metabates* —

L 39–47 cm, WS 85–105 cm. Dwindling resident local population in S Morocco (Sous valley) now perhaps extinct or at least increasingly difficult to find (main distribution south of Sahara). Breeds in open parkland, semi-deserts and scrub. Nests in crown of dense tree.

IDENTIFICATION Like a mixture of ♂ Goshawk and ♂ Hen Harrier, *broad-winged* with fairly long tail, *blue-grey* (belly and underwing finely barred white), with *dark iris, red feet* and *red cere; wing-tips and central uppertail black*. Whereas the general plumage pattern thus could recall ♂ Hen Harrier with extensive black wing-tips, silvery grey wings, darker grey chest and forewings and whitish rump, flight mode differs of course clearly, and wings are much broader and uppertail is much darker. Hunting flight often low and fast, mixing series of wingbeats with shorter glides much like an *Accipiter* hawk. Will also soar at higher altitude with *slightly raised wings*. Perches upright on exposed perch, usually showing long red legs well. – Adult (sexes alike): Lead-grey above, finely vermiculated grey and white on belly. Cere and legs deep red. – Juvenile: Brown above, breast streaked, belly coarsely barred dark; on flying bird, note *pale rump*. Iris, feet and cere yellow.

VOICE Vocal during breeding season. Series of loud, clear whistling notes, 'pee-pee-pee-...' or a series of slower whistles with descending end, 'pee-a pee-a pee-a...'. Alarm a rapid cackling 'klew-klew-klew-...'.

Black-winged Kite *Elanus caeruleus* —

(Alt. name:'Black-shouldered Kite', although this generally refers to an Australian species.) L 31–36 cm, WS 76–88 cm. Breeds in savanna, semi-deserts, open plains, forest fringes by rivers, in drier agricultural landscape with woods and copses interspersed (one might say in rather insignificant habitats). Has fairly recently colonized SW Europe from Morocco. Rare still in Europe (but more common in large parts of S Africa and S Asia). Hunts from exposed perch or in flight. Food mainly insects and lizards but also small mammals and birds. Nests in tree, building a new nest nearly every year. (Two more kite species are treated on p. 102.)

IDENTIFICATION Unmistakable: the *size of a Hobby*, but has bulkier body, a proportionately *large head* and *short neck* giving the bird a thick-set look, also has *pointed* but *broad wings* and *short, square-cut tail*. *Pale blue-grey above with black forewing*. Seen from below in flight, white with blackish primaries and slightly shaded secondaries; head white with *black 'face-mask'*. The *large eye is amber red* and the short legs are yellow. *Hovers* like a Kestrel, *soars and glides with raised wings* like a harrier, when hunting recalls Short-eared Owl due to rather deliberate, deep wingbeats with sometimes hint of quicker upbeat, and in active flight can even vaguely resemble a large Black Tern. – Juvenile: Back and shoulders dark grey, feathers tipped white; greater upperwing-coverts tipped white. Crown brownish. Breast tinged rufous.

VOICE Mainly silent. Sharp, shrill 'chee-ark', vaguely reminiscent of Grey Partridge display, can be heard from both sexes. A high-pitched, piping whistle is used as alarm.

Levant Sparrowhawk

Dark Chanting Goshawk

Black-winged Kite

LEVANT SPARROWHAWK

SPARROWHAWK
(to compare)

migrates in large flocks

falcon-like

black tip

3–4 (5) bars

plain centre

ad. ♂

blunt, no dark tip

ad. ♀

dark cheek

central throat-stripe

grey cheek, dark eye

dark eye

coarse rufous barring

longer and more pointed wings and shorter tail than Sparrowhawk

dark tip

ad. ♀

boldly streaked, flanks barred

barred flank

6–8 bars

plain

note short toes, adaptation to different prey than Sparrowhawk— takes lizards and insects

streaked body

juv.

very pale with black wingtip

ad. ♂

juv.

ad. ♂

DARK CHANTING GOSHAWK

pale primaries

tricoloured upperparts in ad.

pale rump

juv.

soars on lifted wings

active flight re- calls Goshawk

pale rump

ad.

yellow cere and eye in juv.

dark eye

red cere

juv.

black

ad.

watches from ex- posed perch (unlike *Accipiter* hawks)

juv.

ad.

black wingtip and grey hood of adults re- miniscent of Hen Harrier

boldly barred

long red legs

darker second- aries, paler primaries

note typical shape with evenly broad wings and rounded tail

BLACK-WINGED KITE

unmistakable!

ad.

ad.

hovering

glides on lifted wings

juv.

black coverts

scaly back in juv.

juv.

ad.

juv. may show ochre-tinged breast

short

black 'hand'

DZ

FALCONS *Falconidae*
Small to medium-sized raptors, wings rather pointed, flight agile and quick, most take their prey in air. Bill has blunt 'tooth' on cutting edges of upper mandible.

(Common) Kestrel *Falco tinnunculus*   r(m)**B**3

L 31–37 cm, WS 68–78 cm. Found in open country, on plains, by airfields, motorways, arable fields, heaths and marshes interspersed with woods or copses, also lower fells up to birch and willow zones. Food voles, insects. Nests in tree, often in old nest of corvid; or building, in hole or niche.

IDENTIFICATION Medium-sized falcon with *long wings and tail*, wings rather narrow at base and slightly blunt at tips when spread. *Hovers frequently* with hanging tail spread like a fan. Active flight with *rather looser* (relaxed, mechanical) *wingbeats* than other falcons. General shape at times recalls Sparrowhawk, but confusion prevented by continuous flight *with much less gliding*, by narrower wings which are held more straight out (Sparrowhawk has somewhat flexed wings at carpals, with projecting wing-bend), and plumage: *back and upperwing-coverts reddish-brown, contrasting with darker flight-feathers*. – Adult ♂: Rump and uppertail *unbarred blue-grey*, tail with wide dark terminal band; head greyish, finely streaked; back and upperwing-coverts *deep chestnut with small black spots*. – 1st-summer ♂: Some are distinguished by obvious dark barring of rump and uppertail. – Adult ♀: Rump and uppertail *brown, finely barred dark*, terminal tail-band somewhat broader; head brownish, distinctly streaked; upperparts warm brown, *less reddish than* ♂, barred dark rather than spotted. (Rarely, head, rump and inner tail are greyish, but tail virtually always barred.) – Juvenile: Similar to adult ♀ but more yellowish red-brown above, and breast more boldly and diffusely streaked; *a hint of a pale band along upper primary-coverts* often visible. A few ♂♂ have (barred) blue-grey rump and uppertail.

VOICE Main call a fast series of short, sharp notes,'kee-kee-kee-kee-...', rather shorter than Hobby and Merlin calls. Young and ♀ beg with whining, vibrant trills,'keerrrl...', repeated a few times.

Lesser Kestrel *Falco naumanni*   V***

L 27–33 cm, WS 63–72 cm. Mainly summer visitor (Mar–Sep). Breeds *colonially* in towns, steep cliffs, ruins, etc., usually at low level. Social also on migration. Rare vagrant to Britain & Ireland. Food insects, caught on ground (frequently walks briefly on ground between dives, like Redfooted Falcon) or in the air. Nests in niche or hole in building, on cliff-ledge; scrape, no material.

IDENTIFICATION Small, elegant falcon, very similar to Kestrel but separated by *diagnostic* call (see below) and *slightly*

shorter wings and tail, which when fully spread are broader. On a windy day, still looks elegant, with pointed wings, narrow tail and *quick wingbeats*. Slightly elongated central tail-feathers on some (only exceptionally so on Kestrel). *Claws pale* (Kestrel: dark).– Adult ♂: Similar to Kestrel but *lacks black spots on upperparts*, and has a narrow *bluish wing-panel above* between chestnut forewing and dark flight-feathers. Head blue, with *no dark moustachial stripe* as on Kestrel (but shadow can create this effect). *Underwing more whitish* with *darker wing-tip* than on Kestrel (but some are more similar to Kestrel). Spots on underparts smaller and rounder than on Kestrel, and *breast often darker rufous-buff* (tinged vinous). – 1st-summer ♂: Very Kestrel-like, as *blue-grey on upperwing is missing*, and *dark spots are present on chestnut upperparts* (on coverts and tertials). Often only central tail-feathers of adult type, rest barred. No real dark moustaches. – Adult ♀/Juvenile: Similar to Kestrel, and often inseparable by plumage alone. On average paler underside with finer streaks, and finer moustachial stripe, and *lacks dark streak from eye and back*, has slightly paler cheek patch than on Kestrel giving 'kinder' expression. (Blue-grey rump/uppertail is not a clue, can be seen on some Kestrels, too.)

VOICE Diagnostic contact-call trisyllabic, hoarse, rather high-pitched, rasping 'chay-chay-chay' (tone recalls Grey Partridge), often in chorus in colonies. Also Kestrel-like series of short notes, though faster and more chattering. Begging-call is very like Kestrel's, a vibrant 'keerrrl...'.

Red-footed Falcon *Falco vespertinus*   V*

L 28–34 cm, WS 65–76 cm. Summer visitor (Apr–Sep), winters in Africa. Vagrant to Britain & Ireland, usually about 5–10 annually. Breeds colonially, rarely singly, in open country with stands of trees, on steppe, by meadows and in open river valleys. Food mainly insects. Nests in old nest of corvid, often in rookery; may breed among Rooks.

IDENTIFICATION Medium-small falcon, *silhouette recalling both Kestrel and Hobby* depending on how tail length is perceived. *Active flight* often Kestrel-like *with rather loose wingbeats*, but habit of *catching insects in flight* strongly recalls Hobby. *Often hovers, alights on ground, walks, hovers again*, etc. – Adult ♂: Unique combination of *dark blue-grey body and coverts, dull red 'trousers' and undertail*, and *pale silvery-grey flight-feathers*. Cere and feet are deep orange. – 1st-summer ♂: Variable amount of blue-grey in plumage, but invariably *underwing is barred* as on juvenile, and *underbody at least partly rufous*. – Adult ♀: *Virtually unstreaked* pale rufous-buff *underparts*; *back slate-grey, barred dark*; head yellowish-white; feet and cere orange. – Juvenile: Underparts rich buff, *streaked dark*; short, dark moustachial stripe and mark through eye; *underwing has prominent dark trailing edge*; cere and feet yellow.

VOICE Vocal in breeding season and at roosts. Chattering calls,'kekekeke...', ♂ generally faster and higher-pitched than ♀. Flight-call of ♂ resembles similar call of Hobby, a whining 'kew kew kew kew...'.

Kestrel

Lesser Kestrel

Red-footed Falcon

KESTREL

dark moustachial stripe

bold, dense spotting

ad. ♀

ad. ♂

black spots

watching for voles while hovering, or from perch

ad. ♀

ad. ♂

black claws

at times confusingly rounded wing-tips

ad. ♀

ad. ♂

LESSER KESTREL

plain grey cheek

ad. ♀

1st-s. ♂

ad. ♂

blue-grey panel

unmarked red-brown

small, sparse, rounded spots

ad. ♀

pale

ad. ♀

pale claws

longer wings

deep rufous-buff

ad. ♀

ad. ♂

RED-FOOTED FALCON

often briefly alights on ground while chasing insects

brown with a tinge of grey

juv.

streaked

ad. ♂

red

rufous-buff

ad. ♀

blue-grey, barred

primaries dark (unlike in ad.)

silvery-grey primaries above

rufous-buff

ad. ♂

ad. ♀

barred (cf. Hooby)

dark trailing edge

juv.

1st-summer ♂

mixture of old and new flight-feathers

plain grey

juv.

barred, retained juv. primaries

1st-summer ♂

2nd-winter ♂ (Sep)

ad. ♀

ad. ♂

DZ

(Eurasian) Hobby *Falco subbuteo*   m**B**4

L 29–35 cm, WS 70–84 cm. Summer visitor (Apr/May–Sep), winters in Africa. Breeds in variety of habitats: arable land with lakes and woods; bogs and rivers in taiga; mountain forests. In Britain scarce, *c.* 2000 pairs. Food insects and birds, caught in the air by supreme velocity in chase or stoop. Nests in tree, using abandoned nest of other bird, often crow.

IDENTIFICATION Medium-sized, elegant falcon with *long, pointed wings* and *medium-length, square-cut tail. Dashing flight*, and ability to catch birds in the air (incl. swallows and even swifts) generally causes alarm when Hobby appears; often seen in low, fast flight over reedbeds or marshes, *wing-beats powerful, somewhat spaced, clipping*. When catching dragonflies at the height of summer flight is more relaxed, with short accelerations and turns; prey then eaten in the air during brief glide. Silhouette often typical, with quite pointed wings when *long primaries* are flexed back, but at times wing-tips appear more blunt, recalling Red-footed Falcon (which has only a trifle longer tail and shorter wings, and thus might be mistaken for a Hobby at a distance). At normal range, adult looks *dark grey above* and *darkish below with white throat*; at closer range *rusty-red 'trousers' and vent* show, as does bold streaking of breast and belly, and black moustachial stripe. In the Mediterranean area, Eleonora's Falcon is a potential confusion risk, too, but is slightly larger and has proportionately somewhat longer tail. Also, Hobby has more uniform wings and tail below. – Juvenile: Differs from adult in lack of red 'trousers' and vent (still lacking in 1st summer), having pale tips to upper primary and greater coverts, paler forehead, browner cast on upperparts, and yellowish-buff (instead of whitish) underparts.

VOICE Main call a scolding 'kew-kew-kew-kew-...', not unlike song of Wryneck but with less whining, pleading ring; considerable variations in tempo. Also, repeated, agitated, sharp 'kit-chic' from both sexes in air; single sharp 'kit'. Begging-call drawn-out, urging '**peee**-eh', recalling Merlin (but different from vibrant trill of Kestrel).

Eleonora's Falcon *Falco eleonorae*   **V***

L 36–42 cm, WS 87–104 cm. Summer visitor (late Apr/May–Oct), winters in Africa. In Britain, vagrant twice. Breeds colonially on mainly Mediterranean islands and coastal cliffs. Breeding postponed until late summer–autumn to coincide with autumn migration of small passerines, which are caught over sea. Food small birds. Nests on cliff-ledge or in rock crevice.

IDENTIFICATION Medium-large, with *long, comparatively narrow wings, long tail* and *rather slim body*. Like a large Hobby or slim ♂ Peregrine, but separated from both by longer tail and from the latter by proportionately longer and

narrower wings. However, the difference does not always appear significant, e.g. at long range or from tricky angles, and it is unwise invariably to expect immediate recognition by silhouette alone; a prolonged and more all-embracing observation, or closer range, will be conclusive, though. *Flight relaxed, with soft but forceful wingbeats* (can recall a skua *Stercorarius*!), but very quick in pursuit-flight. Spends much time gliding on flattish wings, looking out for prey over the sea. – Adult, light morph: In plumage resembles juvenile Hobby, but separated by *contrast between very dark underwing-coverts and paler grey bases to flight-feathers*; also, *underbody is deeper rufous-buff* or rusty-brown (not buff-white). – Adult, dark morph: Entirely dark brownish-black but, just as light morph, shows contrast between blackish underwing-coverts and somewhat paler bases to flight-feathers. – Juvenile: Rather similar to juvenile Hobby, but distinguished by silhouette, size, and paler *bases to barred flight-feathers below, contrasting with dark wing-tip and wing-coverts*. Rich buff below, finely streaked dark. Throat buffish, not so white as on Hobby.

VOICE Main call a nasal, somewhat grating 'kyeh kyeh kyeh kyah'. Various sharp notes as alarm in vicinity of nest, e.g. 'kak kak' and 'kekekeke...'.

Sooty Falcon *Falco concolor*   —

L 32–37 cm, WS 78–90 cm. Summer visitor to extreme SE corner of range (Israel, Egypt, Red Sea) in (Apr) May–Nov, winters in E Africa. Breeds in lower mountains or on cliffs in deserts and on coral islands and coastal cliffs. Habits much as those of Eleonora's Falcon. Food small birds, in winter insects, too. Nests on cliff-ledge or in rock crevice.

IDENTIFICATION Medium-sized, *slightly larger than Hobby*, which it otherwise resembles in silhouette, and is somewhat smaller than Eleonora's Falcon. *Head proportionately larger than Eleonora's, and tail shorter.* Yellow cere generally more prominent than on adult Eleonora's. *Uniform bluish-grey plumage of adult* is unique—dark-morph Eleonora's has blackish underwing-coverts contrasting with paler bases to flight-feathers (Sooty uniform); and adult ♂ Red-footed Falcon has silvery-grey primaries, paler than blue-grey body, and has dull dark red 'trousers' and vent (lacking on Sooty). Darkness of adults varies from rather pale blue-grey (paler than any Eleonora's) to dark, blackish lead-grey, extremes bridged by intermediates. ♂♂ apparently more frequently pale, but plumage-wear and age might also play a part. – Juvenile: Resembles both juvenile Hobby and juvenile Eleonora's (also juv. Red-footed Falcon). Distinguished, apart from size and silhouette, by *broad dark trailing edge to undertail and underwing* (unlike Hobby) and lack of contrastingly dark underwing-coverts (unlike Eleonora's); *dark and unbarred central tail-feathers* (visible from below as well) which are somewhat longer than rest, giving *slightly wedge-shaped tail*; and tendency towards *dark breast-band*.

VOICE Little known. Some calls resemble Kestrel's but are slower and coarser.

Hobby

Eleonora's Falcon

Sooty Falcon

HOBBY

feeding on
dragonflies on
calm summer day

juv.

light-
tipped
coverts

juv.

ad.

plain

ad.

red-
brown

buff
vent

evenly
barred

red 'trousers'

ad.

ELEONORA'S FALCON

juv.

darker
trailing
edge

ad. (dark)

ad. (light)

dark
cov-
erts

rusty-
brown

ad.
(dark)

long tail
and wings

juv.

rounded
white cheek
(cf. Hobby)

serrated
pattern

dark
subtermi-
nal band

ad.

ad.
(light)

juv.

SOOTY FALCON

broad dark
trailing edge

juv.

dark 'hand' con-
trasts with dove-
grey upperparts

ad.

ad.

ad.

plain
grey

broad band

continuous
individual
variation in
greyness
of ad.

short tail
and very
long and
slender
wings

juv.

wings project
beyond tail

ad.

no barring
visible
above

juv.

Peregrine Falcon *Falco peregrinus*   r**B**4

L ♂ 38–45 cm, ♀ 46–51 cm, WS ♂ 89–100 cm, ♀ 104–113 cm. Circumpolar. Drastic decline owing to biocides and persecution in 1950s and 1960s, since when some recovery. Breeds on steep coastal cliffs or in mountains, also on cliffs in lowlands and on ground on open bogs in taiga. Migrant in north and east. Food small or medium-sized birds, caught in the air after quick horizontal pursuit or spectacular stoop from height with closed wings. Nests on cliff-ledge, rarely in old nest in tree or (esp. in northeastern taiga) on ground.

IDENTIFICATION Medium-sized to large; pronounced sexual size difference, ♀ larger. Strongest of falcons in relation to size, *bulky body*, *'chest-heavy'*. *Wings pointed, 'arm' rather broad, tail medium-length*. In flight, wings are usually slightly flexed, *carpal bends showing well*. Active flight with fairly quick, rather shallow wingbeats, speed moderate. When prey spotted, wing action more determined, and speed suddenly increases. – Adult: *Slate-grey above, lower back/rump/upper-tail-coverts paler blue-grey*. Underparts white, lower breast/belly finely barred, leaving cheeks/throat/upper breast plain white (most extensive white on ♂) and contrasting strongly with *black hood and broad, distinct moustachial stripe*. – Variation: Mediterranean ssp. *brookei* is slightly smaller, darker above, tinged rufous below, and has some rufous on nape (cf. Barbary Falcon). – Juvenile: Best recognized by prominently *streaked* (not cross-barred) *breast and belly*; further, *upperparts are brownish*, with thin pale fringes along tips of primary and greater coverts; cere and feet dull yellowish-green or bluish-green. Confusion is possible with dark juvenile Gyrfalcon, but, apart from silhouette and, for ♂♂, size, the *moustachial stripe is more distinctly broad and black, and does not merge with streaked cheek; underwing uniformly greyish* (Gyr has contrastingly dark coverts); undertail-coverts barred or marked with wedge-shaped blotches (not streaked as on Gyr); colour of cere and feet differ on average, Gyrfalcon often having purer blue in 1st winter.

VOICE Mostly silent away from breeding site. Main call harsh cackling at medium pace, scolding and persistent, sometimes repeated at length, 'rehk rehk rehk rehk...', used as alarm and contact-call. Both sexes have a disyllabic, sharp 'ee-chip' during display, often in air. Begging-call a vibrant 'kyi-i-i-ih'.

Barbary Falcon *Falco pelegrinoides*   —

L ♂ 33–37 cm, ♀ 36–39 cm, WS ♂ 76–86 cm, ♀ 89–98 cm. Closely related to Peregrine Falcon (by some regarded as conspecific). Breeds in semi-desert with cliffs and smaller mountains, further E also in foothills and mountains. Food as that of Peregrine Falcon. Nests on cliff-ledge.

IDENTIFICATION Very much a smaller version of Peregrine, and at times difficult to identify reliably. In size, ♂ only marginally larger than Hobby (but bulkier and 'chest-heavy' like Peregrine). In flight silhouette, 'arm' is fractionally narrower than on Peregrine. Owing to size, wingbeats on average slightly quicker in comparable situations. Identification based primarily on plumage. – Adult: Compared with Peregrine, *upperparts are paler blue-grey; flight-feathers from below paler with darker wing-tips and rear edge of outer wing; underbody and underwing-coverts tinged rufous (fresh) or creamy* (bleached), but note that Peregrine of ssp. *brookei* has faint rufous tinge, too. Other criteria for Barbary are *barring of underparts less extensive* (least barring on ♂); tendency to show *dark terminal band on undertail; hint of dark 'comma' below on tips of primary-coverts; moustachial stripe narrower; nape rusty-red* (requires close range, and not that Peregrine ssp. *brookei* has rufous tinge on nape, too). – Juvenile: Resembles juvenile Peregrine, but *head is not so dark*, forehead, nape and supercilium being rufous-buff, *moustachial stripe narrower*; tendency to show *dark end-band on undertail*; and *streaking on underbody narrow* and concentrated into central area, *throat/upper breast and 'trousers' lower belly almost unmarked buff-white*.

VOICE Like Peregrine's, but slightly more high-pitched.

Merlin *Falco columbarius*   r+m**B**4 / **W**

L 26–33 cm, WS 55–69 cm. Breeds mainly in birch and willow zones in fells, also on bogs in taiga and in drier coastal areas and heathland. Migrant in north. Food mainly small birds, caught in the air after dashing horizontal chase or prolonged aerial pursuits. Nests in tree (in abandoned nest of crow or raptor), in scrub or on ground.

IDENTIFICATION Smallest falcon. Flight silhouette not unlike Peregrine, rather 'chest-heavy' with pointed 'hand', fairly broad 'arm' and medium-length tail, and confusion for a brief moment actually possible when size not apparent. However, active flight with *much quicker wingbeats*. Hunts in low flight, at times with *thrush-like camouflage-flight* (gently bouncing, tendency for series of wingbeats relieved by fleeting close-winged glides), or in quick aerial pursuit. Very agile on wing. – Adult ♂: *Upperparts blue-grey, tail with broad black terminal band, primaries dark*. Breast *rusty yellow, finely streaked* dark, throat white. Rather indistinct narrow moustachial stripe. – Adult ♀: *Upperparts brown-grey, tail heavily barred dark* (c. 5 dense, broad bars, terminal only slightly wider). *Breast buff-white, boldly streaked dark*. Moustachial stripe rather narrow, not so distinct. – Juvenile: Very similar to adult ♀, and generally not separable in the field. On average richer brown above, without the greyish cast on back and tail apparent on many adult ♀♀. Primary coverts barred pale rufous (on ad. ♀ more uniform).

VOICE Silent except in nest area. Alarm quick accelerating series of sharp notes 'kikikiki...', slightly higher pitched and more rapid in ♂ than in ♀. Begging-call resembles Hobby's (even Curlew's), a plaintive 'pee-eh'.

Peregrine Falcon   Barbary Falcon   Merlin

PEREGRINE FALCON

juv.

PEREGRINE   GYRFALCON

ad.

note darker wing-coverts and
broader wings of Gyrfalcon

paler
tail-base

juv.

ad.

variation

juv.

juv.

streaked

broad
mous-
tache

barred

ad.

tip dark

pointed

dark
'hand'

evenly barred

darker to-
wards tip

BARBARY FALCON

buff
hue

LANNER
erlangeri

← compare →

rufous

ad.   ad.

juv.

lightly
streaked

not so
broad
mous-
tache

finely
barred

ad.

juv.

dark
tip

ad.

MERLIN

SPARROWHAWK (to compare)

dark marking on
cheek appears
'double
moustache'

wing-
and tail-
feathers
boldly
barred

chasing passerines

dark outer
wing

♀

black termi-
nal band

blue-
grey

juv.

orange
hue

ad. ♂

♀

ad. ♂

ad. ♂

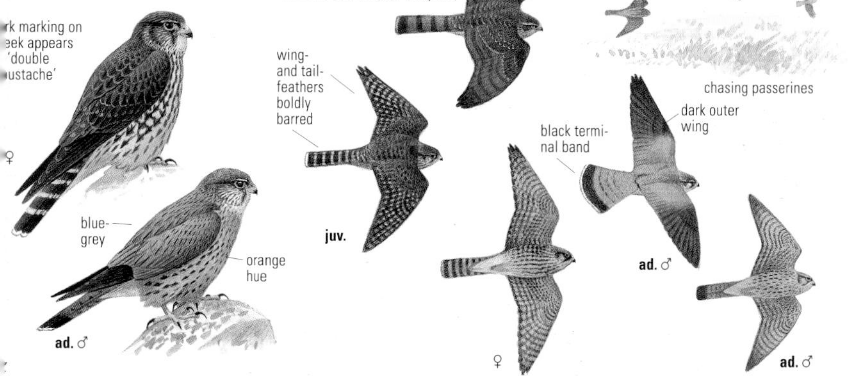

Gyrfalcon *Falco rusticolus* **V**∗∗

L 53–63 cm, WS 109–134 cm. Resident in mountains and tundra, preferring canyons in birch zone, or tundra with crags and patches of trees and scrub. Vagrant to Britain & Ireland, about 3–4 records annually. Food taken on ground or in air, birds (mainly Rock Ptarmigan), lemmings and other small mammals. Nest on cliff-ledge, often revealed by abundance of yellowish-green lichens growing beneath on rock.

IDENTIFICATION Largest falcon, *wingspan as Common Buzzard*. *Wings broad, wing-tips rather rounded* for a falcon. *Tail slightly longer than on Peregrine Falcon*, the other large falcon within same range. Body weighty, as Peregrine's, but longer, and *base of tail broad* (undertail-coverts bulky, as on Goshawk). Flight similar to Peregrine, but owing to larger size and *slightly longer 'arm'* more stable, with *somewhat slower and more shallow, elastic wingbeats*. Aerial hunts mainly by horizontal pursuit, often low over ground; ♂ and ♀ can co-operate to exhaust prey; only rarely stoops from height like Peregrine; prey often taken on ground. Plumage variation basically geographical, but overlap and variation substantial: attractive white morph (predominating Greenland, also occurring Siberia) is *white with dark wing-tips* and some dark spotting; light grey morph (Iceland) is pale grey with whitish barring; dark grey morph (Fenno-Scandia, Russia) is medium grey with pale grey barring. Last two have *narrow and indistinct moustachial stripe, pale supercilium*, and *finely streaked or dusky cheeks*. – Adult, grey morphs: Cere and feet are yellow; upperparts greyish (or greyish-brown), barred or spotted pale ash-grey; underparts rather finely spotted and streaked dark, some cross-barring on flanks and 'trousers' on most. –Juvenile: *Cere and feet bluish* (at least through 1st winter); upperparts are dark brown-grey, with very little or no pale barring or spotting except on tail; *underparts boldly streaked dark, no barring.*

VOICE Has a loud, hoarse rasping or harsh cackling,'wray (-eh) wray(-eh) wray(-eh)...', pace slower, voice gruffer and individual notes more clearly disyllabic than Peregrine's.

Saker Falcon *Falco cherrug* [**V**∗∗∗]

L 47–55 cm, WS 105–129 cm. Breeds in wooded steppe, by grassland close to foothills or forests, in gallery forests along rivers in open country. Partial migrant. Food rodents (susliks, voles), also birds. Nests on cliff-ledge or in old stick nest in tree, often in heronry. Rare (only *c*. 600 pairs in Europe).

IDENTIFICATION *Large*, almost as Gyrfalcon, and resembles that in flight silhouette, with *rather broad wings with blunt tips*. Separated with practice by slightly narrower wings and slimmer base of tail and 'hips', but as a rule necessary to consider plumage characters as well. – Adult, normal: Most birds distinguished from similar Lanner by *tawny-brown back*

and inner wing, contrasting with darker grey flight-feather almost as on ♀ Kestrel (Lanner usually slate-grey with da ker flight-feathers, like ♀ Red-footed). Head usually rath pale brown, with whitish supercilium (absent on Lanner, rufous-buff and indistinct), *crown finely streaked* (lackin dark forecrown and rufous nape of Lanner), *moustachi stripe narrow*, indistinct on many, more prominent on som some look almost white-headed at distance. *Uppertail ofte looks uniform brown* when folded, but variable amount c buff bars present on outer or all tail-feathers (Lanner c average more prominently and more completely barred, b much overlap makes this character unreliable). (Cautio young Lanners in N Africa and Middle East are cinnamo brown above, often with very pale head, and unbarred u pertail!) Underparts usually prominently *streaked from low throat and down*, often boldly *on flanks*, and often *'trouser are all dark* (Lanner is more finely spotted on breast, and h cross-bars on flanks and 'trousers'). Underwing similar Lanner, with more or less contrasting dark underwin coverts forming a *dark fore-wing* (unlike Peregrine ar Barbary). – Adult, grey morph ('*saceroides*'; very rare Europe): Upperparts greyish and barred, much resemblir Lanner, possibly told by size, shape, and head patter – Juvenile: Very similar to adult; *upperparts tawny eart brown, underparts boldly streaked*; head on average dark than on adult but with more distinct white superciliur Difficult to distinguish by plumage from juvenile Lanner, b *'trousers' often all dark* (almost like Rough-legged Buzzard and *upperparts slightly paler with a tawny tinge* compare with European juvenile Lanners. Cere and feet dull gre green, or grey tinged bluish-green.

VOICE Rather similar to Peregrine Falcon's, though som what slower and more whining, less hoarse.

Lanner Falcon *Falco biarmicus*

L 43–50 cm, WS 95–105 cm. Resident in desert, arid stepp lower mountains. Food mainly small and medium-size birds, caught in the air by fast stoop or chase; ♂ and ♀ oft co-operate when hunting. Nests on cliff-ledge, rarely in tre using old stick nest. Rare (only *c*. 300 pairs in Europe).

IDENTIFICATION Slightly smaller than Saker, with som what narrower wings and tail (like a short-tailed Kestrel silhouette). – Adult: *Upperparts slate-grey or brownish-gre barred dark* (European ssp. *feldeggii*),or ground colour pal bluish-grey (N African ssp. *erlangeri*), not cinnamon-brov as most Sakers. Underparts *less streaked* than Saker, mo finely *spotted*, and *are cross-barred on flanks/'trousers'*. Ten ency towards dark forecrown-band. Nape and supercilium present) *pale chestnut*. Blackish moustachial stripe pror inent but narrow. (For further distinctions, cf. Saker.) – Juv nile: Very similar to bo adult and juvenile Saker, b the *'trousers' are streake never all dark*. *Upperpa either very dark* on man (Europe), or paler, warm brown (N Africa, Mid East) like many Sakers.

VOICE Resembles that Gyrfalcon, but is less loud

Gyrfalcon   Saker Falcon

Lanner Falcon

YRFALCON

dusky cheek and broad, diffuse moustache

blue-grey cere

'hooded'

juv.

blunt tip

juv.

uniform tail and rump pattern (cf. Peregrine)

ad.

dark mid-wing panel

coverts darker than remiges (cf. Peregrine)

bluish-grey feet

ings orter an tail

dark

PEREGRINE, juv. (to compare)

uniform

juv.

heavy, bulky

ad.

(white morph; Greenland)

ad.

more 'athletic' build

AKER FALCON

v. usually has dark ousers' (never en on nner)

juv.

pale head, narrow moustache

juv.

dark vent (due to dark 'trousers')

ad.

tawny-brown coverts

ad.

streaked

long tail

ad.

usually dark 'trousers'

juv.

broad dark covert band

ANNER FALCON

juv. Lanner very similar to juv. Saker – note size and proportions

Saker-like, but note dark 'tiara'

erlangeri, **juv.**

juv.

pale vent

feldeggii **ad.**

evenly barred tail (cf. Peregrine)

ad.

rufous

feldeggii, **juv.**

warm buff

barred

slate-grey, barred brown

blue-grey, scaly

light

narrow dark covert band

barred 'trousers'

erlangeri (N Africa), **ad.**

feldeggii (S Europe), **ad.**

juv.

DZ

RAILS and CRAKES *Rallidae*

Small to medium-sized, aquatic birds living in marsh vegetation, usually concealed. Podgy body, in some species laterally flattened, and rounded wings. Several species have loud calls, heard mostly at night. Migrate at night. Food plant material, small animals. Nest in dense vegetation, generally in marshes. (One species, Corncrake, is treated on p. 56.)

Water Rail *Rallus aquaticus* rB4 / W3

L 23–26 cm (incl. bill 3–4½). Breeds at thick-reeded shallow lakes and sedge marshes. Northern breeders winter in W and SW Europe. Usually hidden away on reedbed floor or among dense sedge, but sometimes seen scurrying across a muddy gap in reeds; in winter, may come out onto open ice.

IDENTIFICATION Rather small with rounded rear body, flat breast and quite long neck, and *long, narrow, slightly decurved bill*. At distance looks *rather dark*; just *undertail-coverts gleaming light* (white with buff admixed). Moves cautiously one moment, nimbly the next; flies into cover with rapid wingbeats and dangling legs. – Adult: Black-spotted olive-brown above, 'face' and breast dark blue-grey, flanks and belly barred black and white. *Bill red* with darker culmen and tip, legs dirty red. – Juvenile: *Throat and centre of breast white*, sides of head and breast brownish and diffusely barred; often suggestion of *darker mask and narrow pale line over lores*. Reduced red on bill.

VOICE Rich repertoire: a discontented piglet-like squeal, soon dying away (cf. Red-necked Grebe), 'grüiit grroit grui gru'; a weary, 'all-in', choking moan, 'ooouuuh'; short 'kip' notes when disturbed. ♂'s song, often at night, a persistent pounding 'küpp küpp küpp küpp...' in long, slightly accelerating series; ♀♀ (unmated?) at times utter a high, characteristic 'püp püp pü-errrrr' (structure as call of ♀ Little Crake, but higher-pitched and with Water Rail voice); the terminal trill, 'pü-errrrr', is heard in spring from flying rails (probably also ♀♀) at night.

Spotted Crake *Porzana porzana* mB5 / P5

L 19–22½ cm. Breeds mostly in soggy sedge bogs and damp tussocky meadows (avoids reedbeds). Hard to flush, slips away unseen even in low sedge. Rare and local in Britain.

IDENTIFICATION A bit smaller than Water Rail, which it resembles in general shape and habits. Told by *short, straight bill*, usually buff undertail-coverts, *greenish legs* and, in fluttering flight, *white leading edge to wing*. Close to, shows *fine white spotting on neck, breast and upperparts, yellow-grey bill* with spot of red at base. – Juvenile: Lacks lead-grey colour on supercilium, throat and breast; throat often whitish.

VOICE Song highly distinctive, a high, gently upslurred, sharp 'wolf-whistle' (first part only), rhythmically repeated

(fully 1 per sec.) at night, 'huitt, huitt, huitt,...'; audible at 1–2 km on still nights, at distance can recall dripping water close to, a cracked, grunting secondary note heard.

Sora *Porzana carolina* V**

L 18–21 cm. Very rare vagrant from N America. Like Spotted Crake in appearance and habitat choice. In all plumages differs in *unstreaked dark median crown-strip* (Spotted is streaked), *lack of white spots on head, neck and breast*, and dark *tertials with pale edges* (not wavy bars as on Spotted). Calls include a Little Grebe-like trill which rapidly drops in pitch and slows down; song a slowly repeated low-pitched, drawling whistle rising at end, 'ku-vü' (like prelude to Bar-tailed Godwit's display-call).

Little Crake *Porzana parva* V**

L 17–19 cm. Breeds in reeds in somewhat deeper water. Most at home in natural reedbeds with some bulrush, open areas and channels. Mostly E European species, rare vagrant in Britain & Ireland. Located almost solely by loud calls.

IDENTIFICATION *Small*. Like Baillon's Crake; differs in *long primary projection*, rather *long tail*, in adult *a touch of red at base of bill*, *almost no white markings above* (though younger birds can have some white *streaks or spots* (but not loops) along with *restricted barring on rear flanks*. Sexes differ: ♂ blue-grey below, ♀ buffy grey-white with a hint of blue-grey on head. Immature like ♀ but has strikingly *whiter face and underparts*, with more *extensive barring on body-sides*; told from young Baillon's by longer primary projection.

VOICE ♂'s song is heard at night, a far-carrying (often audible at 1–2 km), nasal yapping 'kua', repeated slowly at first, then faster and faster, finally dropping in pitch and becoming a stammer; many renditions left unfinished. ♀ has a short phrase with the same nasal, high-pitched tone, 'kua-kua-kvarrrr' (cf. Water Rail).

Baillon's Crake *Porzana pusilla* V**

L 16–18 cm. Breeds in sodden sedge bogs where water is no more than knee-deep, sometimes with a few reeds. Rare in Europe, more numerous in E; a rare vagrant in Britain.

IDENTIFICATION *Small*. Like Little Crake; told by *short primary projection* (primary tips reaching only a little beyond outside tertials), rather *short tail, no red on bill-base*, *white loops and squiggles on reddish-brown upperparts*, and also, below, *more distinct and more extensive dark and white barring on rear body*. Sexes similar (♂ cleaner and darker blue-grey); immature dirty brown below, with no blue-grey.

VOICE Song a soft, dry rattle varying in volume, reminiscent of edible frog (and to lesser extent of Garganey), 2–3 sec. long, not audible beyond 300 m (often only 75 m!).

Water Rail

Spotted Crake

Little Crake

Baillon's Crake

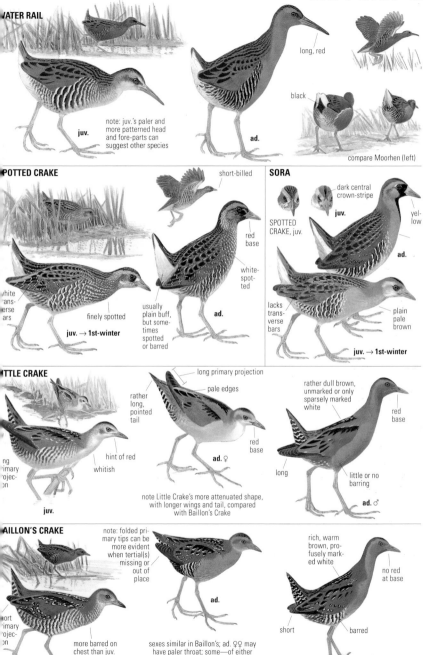

WATER RAIL

long, red

black

note: juv.'s paler and more patterned head and fore-parts can suggest other species

juv.

ad.

compare Moorhen (left)

SPOTTED CRAKE

short-billed

red base

white-spotted

white transverse bars

finely spotted

usually plain buff, but sometimes spotted or barred

juv. → 1st-winter

ad.

SORA

dark central crown-stripe

juv.

SPOTTED CRAKE, juv.

yellow

ad.

lacks transverse bars

plain pale brown

juv. → 1st-winter

LITTLE CRAKE

long primary projection

pale edges

rather long, pointed tail

red base

ad. ♀

rather dull brown, unmarked or only sparsely marked white

red base

long

little or no barring

ad. ♂

long primary projection

hint of red

whitish

note Little Crake's more attenuated shape, with longer wings and tail, compared with Baillon's Crake

juv.

BAILLON'S CRAKE

note: folded primary tips can be more evident when tertial(s) missing or out of place

ad.

rich, warm brown, profusely marked white

no red at base

short

barred

ad.

short primary projection

more barred on chest than juv. Little Crake

juv.

sexes similar in Baillon's; ad. ♀♀ may have paler throat; some—of either sex—have brown cheek spot

M

(Common) **Moorhen** *Gallinula chloropus* r**B**2 / **W**2

L 27–31 cm. Common breeder on smaller lakes, ponds, pools and rivers with dense vegetation cover. Usually secretive but in some places quite bold, and may be seen strolling in open on grass and waterside meadows. Northern breeders migrate to W and S Europe, returning in Mar–Apr. Nest a roofed basket well hidden in dense vegetation.

IDENTIFICATION A pigeon-sized, *dark* bird with rather long tail and *long green legs* and *long toes*. *Bill red with yellow tip*. Slate-grey plumage (upperparts brown-tinged) with a *white line along side of body* and *white sides to undertail-coverts* (form inverted white V in rear view). Carries tail high and *jerks it* when walking, also jerks head when swimming. – Juvenile: Grey-brown with dirty white chin and throat; identified by shape and actions, also by pale line along side and white undertail-coverts. Downy young is like that of Coot, black with red bill, but *lacks yellow-brown collar*.

VOICE Wide repertoire, and heard more often than seen. Most typical is a short, explosive, bubbling or gargling call, 'kyorrrl!', which reveals its presence within reeds. Other calls are a sharp 'ki-keck' (annoyance) and a trisyllabic, fast, cracked 'kreck-kreck-kreck', which may be repeated for long periods and is also given at night in flight.

(Eurasian) **Coot** *Fulica atra* r**B**3 / **W**2

L 36–42 cm. Breeds on lakes and slow rivers with ample vegetation but also open-water areas. Many winter in northern harbours, others migrate to W Europe; large winter flocks on some British & Irish reservoirs. Nest a pile of dead reeds at reedbed edge, often fairly visible. Defends territory fiercely, swims menacingly at intruders and charges them.

IDENTIFICATION *Plump*, broad body, short tail and small, rounded head. *Sooty-grey body* and *black head* with *white bill and white frontal plate*. In flight shows narrow *white trailing edge to secondaries*. Legs strong, toes long and lobed (like grebes; not webbed as on ducks). Swims with gently *nodding head*. Dives with a small leap, soon resurfacing (floats like a cork). When taking off from water, runs along surface with splashing feet. For differences from similar Red-knobbed Coot, see latter. – Juvenile: Greyish-white on head-sides, foreneck and breast, otherwise grey-brown. Note that undertail is dark (pale on young Moorhen). Downy young is black with *red and blue on head* and with *yellow-brown collar*.

VOICE Repertoire rich and varied; livens up lowland lakes at night. Loudest call is a cracked, monosyllabic 'kowk' or 'kruke', often repeated. Frequently utters an explosive, sharp 'pitts!' (like striking a stone with a metal bar). On spring nights, a nasal trumpeting 'pe pe-**eu**', somewhat jolting in rhythm, is often heard from Coots flying around territory. Young beg with pitiful, whimpering, lisping '**üh**-lif' notes.

Red-knobbed **Coot** *Fulica cristata* –

L 39–44 cm. Breeds in N Morocco, and locally (rare) in SW Spain. Habitat the same as for Coot, and the two can coexist. Perhaps a trifle shyer than Coot, and seems to be more reclusive than that species. Resident.

IDENTIFICATION Very like Coot, and safely identified only at reasonable range and with some experience of variation in latter. During breeding season *small dark red nodules on forehead* diagnostic, but these rapidly shrink after breeding and are often surprisingly hard to see in the field at a distance (look simply dark). Best characters: *feathering on side of upper mandible is gently rounded* (not forming pointed wedge as on Coot); bill white with *tinge of blue-grey* (not creamy-white or pinky-white as on ad. Coot); frontal shield extends a bit more towards crown and is *squared off at top; neck is narrower*; in flight *wing all dark* (lacks Coot's white trailing edge to secondaries). Often swims with weight at front and 'stern' raised, and head looks conical (but that shape and posture at times very like Coot). Bends neck at centre when picking food from water's surface ('neck given way in middle'; Coot bends entire neck).

VOICE Very different from Coot's. Often gives a disyllabic very fast 'ker**re**', nasal and shrill (can recall Little Crake), also a hollow, nasal, slightly hoarse 'ka-hah'. Occasionally utters long series of rolling 'krre' notes.

Purple Swamphen *Porphyrio porphyrio* –

L 45–50 cm. Breeds at marshes and smaller lakes with lush but rather low growth of reeds, bulrushes and sedges, often with plenty of broken and fallen stems for clambering about, at times climbs high up in reed clumps, unlike its relatives.

IDENTIFICATION Almost *size of Black Grouse*, with small shimmering dark blue and purple plumage, big red bill (with large frontal plate) and powerful, prawnshell-red legs with long toes. *Undertail-coverts white*. Bill a little duller in winter, but frontal plate usually remains deep red. Immatures have somewhat duller plumage with greyer head, a few greyish-white feathers on central underparts, and initially more greyish-pink on bill and legs. Flies with fast-cranking wingbeats and long, dangling legs. Swims if need be. – Variation: Egyptian population (ssp. *mada- gascariensis*) has green colour on back, scapulars and tertials; Caspian and Turkish breeders (*poliocephalus*) have paler, greyish head.

VOICE Very loud and startling (almost as if someone is 'pulling your leg'!). Rich repertoire. Often emits (esp. at night) drawn-out tooting calls with nasal, bleating tone and alternating in pitch, repeated in long series. Also hard, straight, mechanical trills with cracked voice, 'prrih prrri prrih...', varying in intensity, at times very shrill and 'indignant'. Contact-call a clicking 'chuck'.

Moorhen

Coot

Red-knobbed Coot

Purple Swamphen

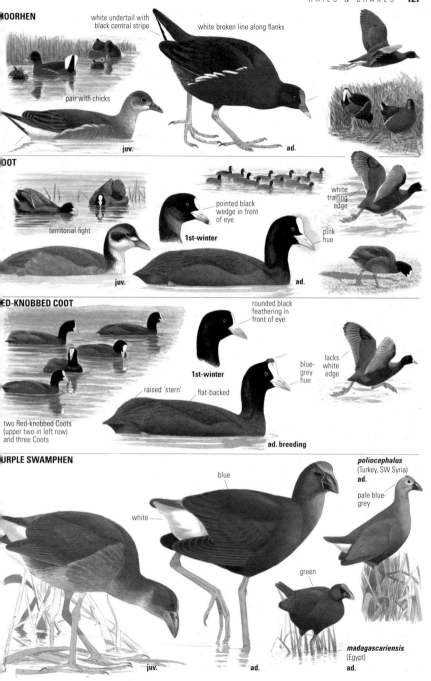

MOORHEN

white undertail with black central stripe

white broken line along flanks

pair with chicks

juv.

ad.

COOT

territorial fight

pointed black wedge in front of eye

1st-winter

white trailing edge

pink hue

juv.

ad.

RED-KNOBBED COOT

rounded black feathering in front of eye

1st-winter

raised 'stern'

flat-backed

blue-grey hue

lacks white edge

two Red-knobbed Coots (upper two in left row) and three Coots

ad. breeding

PURPLE SWAMPHEN

blue

white

green

poliocephalus (Turkey, SW Syria) **ad.**

pale blue-grey

madagascariensis (Egypt) **ad.**

juv.

ad.

ad.

CRANES *Gruidae*

Very large with long legs and long neck. Bill straight, medium-long, dagger-shaped. Toes unwebbed. Tertials elongated and bulky, seem to form the tail (true tail short, hidden underneath). Experts in soaring flight, flying with neck extended like storks; form V or oblique line on migration. Note outstretched legs, slow pace, much gliding, incidence of soaring on thermals and high flight altitude. Feeding action can be described as 'near-sighted', involving petty scrutiny and flashing pick-up of food with bill.

(Common) Crane *Grus grus*   P4–5 / V*

L 96–119 cm, WS 180–222 cm. Breeds sparsely on bogs in boreal forests, in reedbeds in lakes or along rivers in forested areas. Long-distance migrant (arriving late Mar–1st half Apr, leaving Sep–Oct). Spectacular dancing display mainly in spring by pair, or involving hundreds of birds (mostly imm.) at favoured sites. Food plant material, grain, old potatoes, insects. Nest mound of vegetation.

   IDENTIFICATION *Huge*, with *very long legs* and *long, narrow neck*. Plumage mainly *pale bluish-grey*, but most breeding birds acquire *rusty-brown back* from staining with (ferruginous and muddy) bog water during incubation. *Head and upper neck black and white*; bare skin of *hindcrown red* (size of red varying). *Elongated tertials extremely bulky* and disorderly, giving impression of almost Ostrich-like, bushy 'tail'. Flight-feathers grey-black, contrastingly darker than coverts and body. Sexes similar, but ♂ on average slightly larger when pair seen together. When duetting (neck erect, bill pointing up, tertials and breast-feathers fluffed up), wings of ♂ half-open, of ♀ not. *Wings long and rectangular*, tips deeply 'fingered'. Majestic flight with slow wingbeats and rather flat wings, *neck and legs outstretched* as on a stork (though White Stork has proportionately slightly shorter legs and neck, at least the shorter legs often discernible in flight; Grey Heron differs in having neck withdrawn, wings deeply arched on downstroke, wing-tips not deeply 'fingered'). Expert in soaring flight but will also use active flight on migration, then forming V-shaped flocks or oblique line, flight altitude often supremely high, speed moderate. Gait measured and stately, but can be quite quick when chasing a fox away from young (performed confidently!). – Juvenile: Immediately distinguished by rather *pale rufous-brown head and upper neck*, lacking the black, white and red pattern of adult, and by less bulky tertials.

   VOICE Vocal, calls loud and far-carrying. Contact-call (mainly in flight) deep, resounding, trilling or jarring 'krro' and 'karr'. Young in autumn have peculiar piping 'cheerp', mixing with deep adult calls. Pair sounds the reveille (duet-call) at breeding site in early dawn (or during light summer

nights in N), ♂ uttering loud trumpeting 'krrroo' and (immediately after) lower-pitched 'kraw' and sometimes als hard, knocking 'ka-ka-ka'. Other duets may run 'krruu-n kraw, **krru**-kraw, ...'.

Demoiselle Crane *Grus virgo*

L 85–100 cm, WS 155–180 cm. Breeds on dry natural stepp sometimes at high altitudes. Long-distance migrant, wes ern populations wintering in Africa, arriving in spring breeding sites at same time as Crane, but leaving one mon earlier (mainly late Aug); passage through Middle Ea largely unnoticed (seen regularly only in Jiddah, so may tak easterly route; however, also often in Cyprus in Aug). Grega ious; often forms larger flocks than Crane on migration.

   IDENTIFICATION *Very large*, similar to Crane, but has sligh ly larger, more rounded head, and somewhat shorter bill ar shorter and thinner neck. Although almost 20% smaller tha Crane can look similar-sized in the field, and plumage diffe ence not obvious at long range, especially in flight. As wi Crane, *body and wing-coverts are pale ash-grey* with contras ingly *darker flight-feathers*; unlike Crane, breeding birds c not acquire rusty-brown back in summer (no bog wat available; breed in drier habitats), but stay pale grey. C ground, at long range, differently shaped *tertials* best clu being *long, pointed, straight* and *hanging down in orde fashion*, lacking 'bushy' look of Crane. At closer range, di ferent pattern of head and neck evident: *black on neck reach further down*, to breast, and consists of *elongated, loose blac feathers*. No red on crown as on Crane. Sexes similar, thoug ♂ slightly larger on average. In flight, Demoiselle Crane very similar to Crane, but note *black breast*, and *greyer pr mary-coverts* creating *less distinct contrast with black prim ries*. The proportionately *slightly larger, more rounded hea is sometimes noticeable. Any moult-gap in the wing (Jur Oct) indicates Demoiselle Crane (Crane moults all primari simultaneously, like wildfowl). – Juvenile: Generally dulle than adult, having *very pale head* (appearing whitish a distance) with vestigial tufts; foreneck and breast slate-gre (not black) in centre, feathers only slightly elongated.

   VOICE In general recalling Crane, but slightly more hig pitched and drier, not so resounding. Contact-call fro migrating birds, 'grro', at distance similar to Crane. Cour ship-calls said to be more guttural and hoarse than in Cran

Siberian Crane *Grus leucogeranus*

L 105–145 cm, WS 205–245 cm. Extremely rare breeder W Siberia (Ob basin, now probably less than 10 pairs) c taiga bogs, and on migration through region seen near-a nually in the Volga delta late Mar–mid Apr, and fewer autumn, on their way to and from winter grounds in N Ira A larger population exists in E Siberia (Yakutia, *c.* 150 pairs), wintering in E China. Favours wet habitats.

   IDENTIFICATION *Huge*, even somewhat larger than Cran *Pure white* with *black wing-tips* and *red 'face'* (unfeathere skin). *Legs reddish*. Sexes similar, but ♂ on average slight larger with longer bill. – Juvenile: Slightly smaller. Pa plumage extensively 'stained' by rust-brown tinge and spo Wholly feathered head, lacking red 'face'.

   VOICE Common call drier than Crane, display-calls on th other hand more clear and bugling and drawn out.

Crane   Demoiselle Crane

CRANE

GREY HERON

front-heavy with withdrawn neck

'landing lights'

ad.

pair on breeding site at bog

brownish head

back grey outside breeding season

juv.

short 'rear'

dancing display

ad. summer

bushy, dishevelled 'tail' formed by the tertials

DEMOISELLE CRANE

note typical elegant shape

less contrast

'landing-lights' are lacking

characters separating from Crane surprisingly subdued at distance

long white plumes

ad.

CRANE

DEMOI-SELLE CRANE

slightly broader base

shorter neck

ad.

black

short 'rear'

pattern reminiscent of that of ad.'s

juv.

long, black, hanging breast-feathers

long and narrow, neat 'rear'

SIBERIAN CRANE

much black

WHITE STORK

juv.

ad.

juv.

migrants at Volga delta

feeding in aquatic habitat

ad.

red 'face'

BUSTARDS *Otididae*

Large, heavily built terrestrial birds with long, strong feet. Wings long and broad with 'fingered' tips and striking white patterns visible in flight. Birds of large plains, semi-cultivated or natural steppe or desert with scrub, also grassy fields and open slopes. Gait slow and deliberate if undisturbed. Feeding action much like in cranes. Wary, will often withdraw by running in vegetation to avoid close encounter.

Great Bustard *Otis tarda*   V***

L ♂ 90–105 cm, ♀ 75–85 cm, WS ♂ 210–240 cm, ♀ 170–190 cm; weight ♂ 8–16 kg, ♀ 3½–5 kg. Breeds on open plains, preferably natural steppe. Marked decline in numbers. Gregarious, having spectacular communal displays, when ♂♂ turn much of plumage 'inside-out', showing, and shaking, loose white feathers all over ('foam bath'), with head and neck completely withdrawn onto back.

IDENTIFICATION *Huge*; largest ♂♂ heaviest birds in region, much larger than ♀♀. Strongly built, with *heavy chest*. Wings long and deeply 'fingered', flight action with uninterrupted, measured wingbeats, majestic. On ground, recognized by *rufous tinge on breast* and *plain greyish head*. Wing has much white above, broad panel on wing-coverts contrasts with blackish secondaries and ochrous-brown or tawny forewing, white invading *washed-out outer wing*. – Adult ♂: *Huge*; breast and sides of lower neck rufous (less in nonbreeding); *head and upper neck blue-grey*; *neck very thick*, even when relaxed; *long bristle-like white feathers at sides of chin* (erected in display, lost in non-breeding); often some white visible on wing, even when folded. Displaying ♂ immediately recognized by *much white exposed*. – ♀: Smaller than ♂. Plumage differs from breeding ♂ in *less blue tinge to head*, being pale ash-grey with yellow tinge; *often lacking rufous on lower neck* (being pale blue-grey) *and breast* (being more pale ochrous-brown); and having less white on upperwing, *white only rarely showing on folded wing*; still, some ♀♀ are very similar to subadult ♂♂, especially in winter.

VOICE Mostly silent. Alarm-call is a short, nasal bark, 'ongh'. Young have a fine, plaintive whistle, 'cheeoo'.

Houbara Bustard *Chlamydotis undulata*   V***

L 45–65 cm, WS 115–150 cm. Breeds in deserts and semi-deserts with sufficient scrub and bushes for cover. Sensitive to disturbance. Has declined markedly.

IDENTIFICATION *Large*, with long, narrow neck and rather *elongated body and long tail*. Wings long and evenly broad, well 'fingered'. Flight impressive like eagle, though wingbeats deeper, and hint of jerky upstroke as with *Asio* owls. Main feature *black band along side of neck*. Upperparts creamy sand-brown with dark vermiculation and cross-bars.

In flight, *white primary patch* on upperwing striking, *contrasting with black secondaries and rest of primaries*. Sexes similar, but ♀ smaller, and has greyer tinge above, less pale forehead, and less clean-cut black neck-band. Very similar to Macqueen's Bustard but *central crown all white* and is *erected to short 'crest' during display*; 'tufted' *feathers on side of neck black; breast white, finely vermiculated;* and *undertail largely white*. – Variation: Local and rare race on Fuerteventura and Lanzarote (*fuertaventurae*) is darker above and smaller.

VOICE Rarely heard. Has display call of 3–5 very deep booming notes, like mechanical noise from distant factory, one per 2 sec., hardly audible over 50 m.

Macqueen's Bustard *Chlamydotis macqueenii*   –

L 55–65 cm, WS 130–150 cm. Breeds in semi-deserts, or dry steppe and in wadis with low bushes. Sensitive to disturbance. Has declined, much due to cultivation and falconry.

IDENTIFICATION Very similar to Houbara Bustard, differing in the following ways: elongated *crown-feathers white with some black admixed, folded over bill during display;* 'tufted' *feathers on side of neck black with white central band; breast uniformly grey;* and *undertail* partly *finely spotted and barred*. Upperparts more pinkish pale brown or ochrous tinged with *bold cross-bars* or 'arrows', thus a bolder pattern on more rufous-tinged ground than in Houbara.

VOICE Rarely heard. Display call (only audible short distance) is a series of deep knocking notes, rather machine-like, consisting of *c.* 25–40 notes in 12–18 sec.

Little Bustard *Tetrax tetrax*   V***

L 40–45 cm, WS 83–91 cm. Breeds in open terrain with vegetation tall enough to give cover. Display, apart from call, includes foot-stamping and sometimes wing-flash with short leap in the air. Monogamous, or one ♂ can have 2–4 ♀♀. Food mainly plant material and invertebrates.

IDENTIFICATION Smallest bustard within the region, still roughly of ♀ Pheasant size. *Flight grouse-like*, with noisy take-off, and quick, shallow wingbeats, interrupted by short glides on medium-long wings with rounded tips and somewhat arched primaries. *Much white on wing, including secondaries*, contrasting with *black tips to outer primaries and primary-coverts*. – Adult ♂ breeding: Striking pattern on head and neck, *neck being black with two white collars*; also neck-feathers erected in display, making neck look inflated (even in flight!); *head and throat lead-grey*. Body dark-marked sandy-brown above, whitish below. 4th outermost primary short and peculiarly emarginated to produce high-pitched whistling wing sound. – ♀ and non-breeding ♂: Like breeding ♂ but lacking black, white and grey pattern of head and neck, being sandy-brown with fine dark marks.

VOICE Rather silent. Song slowly repeated (every 10 sec.), dry snort, 'prrt', soft but still audible at 500 m. Also, ♂ has whistling wingbeat sound. When flushed, ♀ may utter low staccato chuckle. Young have a fine, plaintive call, 'cheeoo', very similar to that of Great Bustard.

Great Bustard

Houbara B.   Macqueen's B.

Little Bustard

GREAT BUSTARD

heavy, like an eagle in flight

slim neck

♀

trace of chestnut

imm. ♂

ad. ♂

sturdy neck

much white in wing

heavy angular chest in flight

bright chestnut in adult ♂♂

early stages of the display

full display ('foam bath')

HOUBARA BUSTARD

white patch

flight action like a crane

crouching from passing Bedouin party

white crest

MACQUEEN'S BUSTARD

male in display

black and white crest

juvenile lacks the black neck-stripe

juv.

ad.

plain grey

fuertaventurae
(Fuerteventura, Canary Islands)

extensively barred and verniculated above

ad.

heavily barred

ad.

grey, freckled (seen at close range)

undulata
(NW Africa)

LITTLE BUSTARD

perform air-leaps during peak of display

ad. ♂

black neck aquired in 3rd cal.-yr birds

ad. ♂

extensive white in wing and grouse-size render the species unmistakable

two ♂♂ in display showing erected neck-feathers

♀

WADERS *Charadriiformes*

Waders comprise a large group of long-legged birds, usually living near water along shores or on bogs and marshes; a few have adapted to drier habitats. On migration and in winter often seen in dense flocks on tidal mudflats and on seaweed along the shoreline. They include several families, two of which are large: plovers and lapwings *Charadriidae*, and sandpipers and allies *Scolopacidae*; see below. Waders are closely related to skuas, gulls, terns and auks, all being members of the same large order, *Charadriiformes*. Recently it has also been established that the buttonquails *Turnicidae* should be included in this order, not among the cranes as earlier believed. The family has only one representative within the treated range, Small Buttonquail (p. 56), which for pedagogic reasons is presented together with Quail and Corncrake.

Waders feed on insects, worms, molluscs and crustaceans; sometimes also on plant material, small fish, etc.

The nest is usually a simple scrape on the ground, and the incubating parent relies either on camouflaging plumage or on cryptic egg pattern. Young are down-covered on hatching and immediately leave the nest, those of most species being selffeeding. They are tended by both parents (commonest), by the female alone (Ruff, Jack Snipe, Great Snipe, Woodcock) or by the male alone (e.g. Painted Snipe, Dotterel, Purple Sandpiper, Spotted Redshank, Wood Sandpiper, phalaropes). Among those species where both parents tend the young it is common for the female to leave the young before they are fledged, whereas the male stays longer.

The plovers (pp.138–145) form a distinctive group of 14 species. They are all short-billed, and when relaxed have a compact, rather rounded, horizontal and short-necked shape. When approached, they head-bob nervously, and when feeding have an especially characteristic 'run-stop-peck' action. *Charadrius* plovers have particularly striking head patterns and breast-bands. The region's five lapwings (pp. 146–147) are closely related to the plovers, but are usually larger and have broader, more blunt-tipped wings.

The family *Scolopacidae* (pp. 148–173) includes 51 of the region's waders. It contains such well-known groups as the *Calidris* sandpipers, the curlews and godwits, the snipes, the *Tringa* sandpipers ('shanks') and the phalaropes. Most breed in northern Europe and Siberia, nesting in open, sparsely vegetated upland or arctic habitats. In the rest of Europe they are seen mainly during winter or on migration, forming flocks on marshes and coasts. Sandpipers feed by restless picking from the surface or by probing in the mud or seaweed with their usually long and slender bills.

IDENTIFICATION OF WADERS

Beginners usually find waders difficult. This is because waders have different winter, summer and immature plumages, as well as transitional stages as they moult from one plumage to the next. Also, there is a confusing array of vagrants, most of which are very similar to their regular European counterparts. Then there is the practical problem that they are often seen at a distance, making it difficult to make out sufficient detail.

The first steps

The following suggestions should remove most of the initial confusion and uncertainty about wader identification.

Most of the main plumages displayed during the course of a year are described and illustrated. Bear in mind that they are not all seen at the same time of year, so the situation is not so complex as a first look at the plates might suggest. Only *those plumages relevant to the season need to be considered.*

If you find a 'different' wader in a flock, the overwhelming likelihood is that it will be another common species (or one in a different plumage from others of the same species around it) rather than a rarity. With this in mind, we have *grouped the regular European sandpipers together* on pp. 148–163, with similar species compared on the same or an adjacent page. All of these species are fairly distinctive and, given reasonable views (admittedly not always easy), it should be possible to identify the species fairly quickly. Until the common species can be identified with confidence, it is best to ignore the remote chance that you will find one of the vagrant species, which are treated on pp. 162–173 (plus two on pp. 152–153). This approach, and the arrangement of the plates, should simplify things. (For plovers and lapwings the number of possible vagrants is smaller, and regular species and vagrants are therefore treated together.)

Although you will quickly be able to identify distinctive waders such as Curlew, Redshank or an adult summer Dunlin even at a distance, most species require close views. The ideal situation is a close flock, such as from a hide at a bird reserve. A *telescope* greatly increases the potential for identification, and is essential for any serious wader-watching.

Waders are identified by a combination of features, especially *size* (though beware of pitfalls; see p.13), *bill shape and length*, *leg length*, *plumage patterns*, *colour*, *voice* and *behaviour*. Note that shape can vary: for example, resting birds hunch their neck into the shoulders, whereas it is extended when feeding; in cold weather waders fluff up their body plumage, looking fatter and shorter-legged.

With experience, waders can be identified at increasing distances, mainly on the basis of silhouette, behaviour, and especially their distinctive flight-calls.

Advanced techniques

Wader identification can be taken much further. It may be possible to tell the sex, or what subspecies it belongs to; the various plumages can be recognized, including first-winter and first-summer, which are rarely illustrated in the plates. These abilities make identification easier. To identify difficult vagrants, it is essential to make comparisons with similar species in the same plumage. All this requires detailed observation, and knowledge of topography and moult.

A typical plover, the Ringed Plover. It has a short bill suitable for picking up.

A typical sandpiper, the Dunlin. Long bill suitable for probing.

Autumn is the best time for studying waders, because juveniles are present and it is the main moult period. Try spending an hour or two on a flock of close waders and practise ageing techniques, locating the various feather tracts, or studying their moult, Practice of this kind, repeated whenever the opportunity arises, provides the only real way of mastering wader identification and ageing.

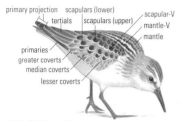

A Little Stint in juvenile plumage.

Topography. Practise locating the features labelled on the chart above when you look at the plates, or when studying magazine and book photographs, and birds in the field.

Moult. *Adults* have two moults each year, the 'spring' moult (mainly Jan–Apr) which produces summer plumage, and the 'autumn' moult (mainly Aug–Nov) which produces winter plumage. The spring moult is partial (head, body and variable amount of inner wing-coverts), and the autumn moult is complete (including wings and tail). Waders thus change their body plumage twice each year, but the wing- and tail-feathers only once.

Typical moult cycle of wader.

Unlike the adult, juveniles have only a partial moult in autumn, the 'post-juvenile moult' (mainly Sep–Nov), which produces first-winter plumage. Subsequent moults are as for adult. All moults take place in or near the wintering areas. The post-juvenile moult starts soon after arrival, and the spring moult is completed before the northward migration.

Ageing. Typical juvenile, adult winter and adult summer plumages, as well as any sexual difference in summer plumage, are described and illustrated in the plates. Note the following points about these plumages, including the period in which each occurs:

Juvenile: Jul–Oct (Nov). Juveniles do not usually move far from the breeding areas until mid Aug; later they usually outnumber adults in migrant flocks. Juvenile plumage is all fresh and neatly patterned, unlike the worn, variegated appearance of adults in autumn. The difference is sometimes difficult to see, and is best looked for on the tertials and greater coverts (unworn and neatly patterned on juveniles, much worn and plainer on adults).

Adult summer: (Mar) Apr–Sep (Oct). In spring, the fresh summer plumage of most *Calidris* and some other species has whitish fringes giving a distinctive 'frosted' appearance. These not very resistant broad fringes disappear quickly to reveal full summer colour, usually by the time the birds arrive in the breeding areas. By late summer, the plumage is generally highly worn and darker above. In autumn, any wing moult indicates adult; this is useful for ageing adults in flight, on which the tatty, moulting wings are obvious.

Adult winter: Sep–Mar. In most species, winter plumage is plainer, less colourful, and quite unlike juvenile and adult summer plumages.

The following plumages are usually not described:

First-winter: (Sep) Oct–Mar. Because the wings are not included in the post-juvenile moult, first-winters can be told from adult winter by the pattern of the retained juvenile wing-feathers. The difference shows best on the tertials or greater coverts. As shown in the plates, the pattern on these feathers is, on most species, quite different from the usually plainer adult winter ones. Wear reduces the difference, however, and it is usually difficult to discern after midwinter.

First-summer: May–Jul. First-summers often stay in the wintering areas all summer, and do not return north with the adults. In most species, the first spring moult produces only partial summer plumage, or a plumage little different from winter. In some species, adult summer females are less bright than males, so only extreme winter-like birds can safely be aged as first-summer. First-summers often start their moult early, so any individuals in full winter plumage in Jul–Aug are probably second-winter. Before May and after Jul, it is not safe to distinguish between first-summers and late- or early-moulting adults.

Transitional plumages: Moulting birds show mixed characters of adjacent plumages, and are referred to as e.g. 'adult winter moulting to summer' or 'juvenile moulting to first-winter'. Transitional plumages are especially noticeable in autumn, with the usually plainer and less colourful winter feathers producing a patchy appearance against the summer or juvenile plumage.

Juvenile Dunlin moulting to 1st-winter plumage.

1st-winter Dunlin: note diagnostic retained juvenile wing-feathers. Cf. adult winter.

Adult winter Dunlin: note plain wing-feathers. Cf. 1st-winter.

Adult summer Dunlin moulting to winter plumage.

(Eurasian) **Oystercatcher**

Haematopus ostralegus r+m**B**2

L 39–44 cm (incl. bill 6½–9), WS 72–83 cm. Breeds on open, flat coasts. Commonly feeds on tidal flats or fields and other open ground (generally no more than *c.* 20 km from sea, but locally much further inland); roosts communally in winter. Food mainly cockles, mussels, worms; also crustaceans, insects. Vertically flattened, blade-like blunt-tipped bill specially adapted for hammering and opening cockles, mussels, etc. Nest is bare scrape on ground, preferring pebbly patches, sand or rocky ground to grassy meadows.

IDENTIFICATION Large, compact and deep-chested. Striking combination of *long and straight orange-red bill* and *black and white plumage*. Broad *white wing-bar* conspicuous in flight. Flight direct with quick wingbeats, somewhat recalling a duck. – Adult summer: Orbital ring orange-red, eye deep red. ♂ has on average shorter, thicker bill than ♀; sometimes ♂ also has brighter red base of bill, but sexing of single birds usually not possible. Legs pink. – Adult winter: Like adult summer, but has obvious white 'chinstrap'. – Juvenile: Dull brownish-black above, not black; extensive black tip to pale orange bill; legs greyish. – 1st-year: Distinguished from adult by narrower black tail-band; extensively dark-tipped bill; brownish upperparts; dull flesh legs; white chinstrap usually retained in 1st summer; wings (retained from juv.) much faded, brownish by 1st summer; orbital ring dull yellow-orange and eye brownish-red. Some older birds may retain dark bill-tip, so 1st-year safely aged only by taking notes of full range of features.

VOICE Noisy. Common flight-call (also when perched) is distinctive far-carrying, shrill 'peep' or 'k-peep!'. So-called piping call (usually from ground, open bill pointing down, often maintained by pair or dense little group) is of similar shrill quality; typical phrases, often run together, are accelerating 'kip kip kip-kip-kip-...' running into a fast 'kli-klikli...', and a distinctive bubbling trill, 'prrrrrrrr...'. Flight-song (by ♂, or ♂+♀, in wide-circling display-flight, typically with slow-motion, deep wingbeats) contains slower 'plee-ah plee-ah plee-ah...' in time with wingbeats.

(Pied) **Avocet**

Recurvirostra avosetta r+m**B**4

L 42–46 cm (incl. bill *c.* 8; excl. leg projection *c.* 10), WS 67–77 cm. Breeds on flat, open seashores or shallow lagoons with brackish or salt water, with bare margins and sandy shoals; in winter, also frequents estuaries, mudflats. Migratory in N and E, wintering in Mediterranean and Africa. In Britain & Ireland, uncommon away from well-known breeding colonies and wintering sites in E and S England. Food mainly insects and small crustaceans, typically obtained by distinctive feeding action in which slightly opened bill-tip is swept from side to side through water or soft mud, detecting prey by touch. Also feeds by pecking at surface of mud; occasionally swims, upending to feed like a duck. Nest is sparsely or fully lined scrape on bare ground near water, occasionally more substantial collection of vegetation forming raised mound in shallow water. Wary, giving full alarm in flight when disturbed during breeding.

IDENTIFICATION In quick or distant view, can be overlooked as a gull because of general white appearance. Otherwise *unmistakable pattern of black and white*, slender *upcurved black bill* and *long pale blue legs*. In flight, pied pattern distinctive, though at long range can be taken for ♂ Goosander or Shelduck. – Adult: ♂ told by longer bill which appears less sharply curved, and head markings always black and neat; ♀ has shorter, sharply curved bill, and head markings are sometimes less clear-cut and tinged brownish. – Juvenile: Markings dull brownish, not black; white areas sullied brownish or barred; legs dull grey. – 1st-year: Distinguishable from adult so long as juvenile features retained; primaries much faded and brownish by 1st summer.

VOICE Common call is loud, piping 'klüp klüp klüp...' with characteristic ring; used also for alarm.

Black-winged Stilt *Himantopus himantopus* **B**5? / V

L 33–36 cm (incl. bill *c.* 6½; excl. leg projection 14–17) Frequents areas of shallow fresh, brackish or salt water, e.g lagoons, saltpans, estuaries. Migratory, generally wintering in Africa. Vagrant to Britain & Ireland. Food mainly insects long legs enable feeding in deeper water than used by other waders. Nest is lined scrape on islet or near water, occasionally a raised mound of vegetation in shallow water.

IDENTIFICATION Extraordinarily elegant wader; unmistakable, with *needle-fine, straight bill*, amazingly *long reddish-pink legs*, and *black and white plumage*. In flight, head and legs extended, giving bizarrely elongated outline; *all black underwing* and flicking wing action with quite pointed wings add to distinctive appearance. – Adult: ♂ readily told by whole upperparts green-glossed black; ♀ has gloss brown mantle and scapulars forming brown saddle that contrasts with black wings. ♂ averages more black on head than ♀, but much overlap and head and neck pattern not safe difference between sexes. – Juvenile: Dull brown with scaly pattern of pale feather fringes, and pale-tipped secondaries and white-tipped inner primaries; legs dull flesh. – 1st-year Told from adult so long as dull legs and juvenile flight feathers retained; many indistinguishable by 1st summer.

VOICE Very noisy during breeding season, otherwise rather silent. Common call is shrill, squeaky, quickly repeated 'kyik kyik kyik...', recalling Avocet (similar fluty, ringing tone but higher-pitched and harder, less liquid). Potential predators near colony chased with loud and grating 'kreet kreet kreet...', rather like Spur-winged Lapwing. A rolling 'krre' can sound rather like Black-headed Gull. Juvenile has different call, 'gip!', recalling Wood Sandpiper, which can be heard still in 2nd cal.-year spring.

Oystercatcher   Avocet   Black-winged Stilt

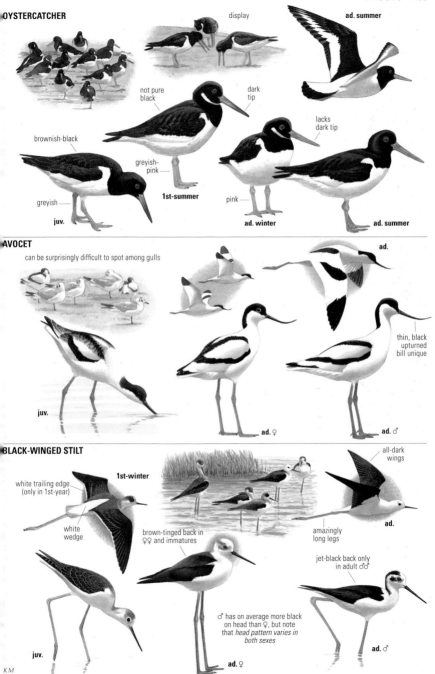

OYSTERCATCHER

display

ad. summer

not pure black

dark tip

lacks dark tip

brownish-black

greyish-pink

1st-summer

greyish

juv.

pink

ad. winter

ad. summer

AVOCET

can be surprisingly difficult to spot among gulls

ad.

thin, black upturned bill unique

juv.

ad. ♀

ad. ♂

BLACK-WINGED STILT

1st-winter

all-dark wings

white trailing edge (only in 1st-year)

white wedge

brown-tinged back in ♀♀ and immatures

amazingly long legs

ad.

jet-black back only in adult ♂♂

juv.

♂ has on average more black on head than ♀, but note that *head pattern varies in both sexes*

ad. ♀

ad. ♂

KM

Stone Curlew *Burhinus oedicnemus*   m**B**4

L 38–45 cm, WS 76–88 cm. Breeds on bare or sparsely vegetated open ground (heathland, dry, stony pastures, dried mud, steppe margins, etc.). Summer visitor (Mar–Oct), winters in SW Europe and Africa. In Britain & Ireland, rare away from few breeding areas in S and E England. Most active dusk to dawn, but daytime activity and calling not unusual. Food mainly insects. Nest is bare scrape.

IDENTIFICATION General shape and actions like oversized plover. Walks stealthily or runs with long body held horizontal and head often hunched into shoulders. Well camouflaged when stationary; *large yellow eye* and *black-tipped bill* are often most eye-catching features. Distinctive in flight, with *strong black-and-white wing pattern*, *hunchbacked profile*, and *shallow beats of long, stiffly bowed wings*. Will sometimes squat on ground to avoid detection. – Adult: Standing ♂ has horizontal white covert bar more strongly bordered with black than ♀, difference obvious when pair seen together. – Juvenile: Lacks prominent whitish supercilium; coverts plain, sandy (lacking adult's obvious blackish and whitish bars); greater coverts broadly fringed white.

VOICE Noisy during breeding season. Some calls rather reminiscent of Curlew, like rising, whistling 'küü-**liie**' or 'kur-li-**lii**' and alarm, a rapid 'kü-vü-**vü**', repeated in long series. Also heard is a thin piping 'kiiiie'. Display-call often heard at night, a series of loud, wailing, mournful notes with first syllable coarsely trilled and second clear, 'klii-ürr-**lee**, klürr-**lee**, klürr-**leee**, klürr-**leee**, klürr-**leee**,...'.

Senegal Thick-knee *Burhinus senegalensis*   —

L 35–39 cm. Resident Egypt, mainly Nile delta and Cairo region. Inhabits open ground like Stone Curlew, but showing marked preference for wet habitats and water margins.

IDENTIFICATION Like Stone Curlew, but *bill longer, stouter*, with relatively *small area of yellow at base*, and broad, *uniform pale grey midwing-panel*, lacking adult Stone Curlew's dark double wing-bars with a narrow white base between. Legs on average duller greyish-yellow, not so bright yellow.

VOICE Long series of mournful, ringing whistles, first risig and accelerating, then falling and fading, 'püpi pi-pi-pi-**pii**-**pii**-**pii**-pii-pi pi...', vaguely recalling Oystercatcher.

Cream-coloured Courser *Cursorius cursor*   **V*****

L 24–27 cm. Requires bare, flat terrain to accommodate ground-running behaviour, especially sandy semi-desert. Uncommon, but sometimes in small groups. Vagrant to Europe, including Britain & Ireland. Food mainly insects.

IDENTIFICATION Unmistakable through its *all sandy-buff plumage* and *pale legs*. Smaller than Golden Plover. Rather plover-like in feeding action and structure, but *longer-*

legged, and *bill curved*. Flight action flicking, rather dove-like, interspersed with short glides. Distant flight pattern very distinctive: *black outer wing* and 'flashing' *black underwing* contrast sharply with pale sandy-buff remainder. Juvenile has plainer head and faint scaly pattern above.

VOICE Usual flight-call is a subdued, ventriloquial 'quett'. Display-flight high up giving well-spaced 'quett' calls, now and then relieved by a more subdued 'cheah'.

Collared Pratincole *Glareola pratincola*   **V****

L 24–28 cm, WS 60–70 cm. Breeds in loose colonies in extensive flat, dry terrain with low vegetation, on wet meadows, by reedbeds, saltpans, etc. Summer visitor (Apr–Oct), winters in Africa. Vagrant to N Europe; annual in Britain & Ireland. Catches insects in flight, interspersing graceful Black Tern-like action with erratic swooping manoeuvres.

IDENTIFICATION Long wings and forked tail giving *tern-like shape* but with longer legs and stubby bill. At distance, look for distinctive combination of *agile feeding flight* (see above) and *square white rump and white belly contrasting with all-dark remainder*. Collared difficult to tell from Black-winged: at close range, *dull red underwing-coverts* and *white trailing edge of inner wing* are diagnostic of Collared, but even then both may be surprisingly difficult to discern. Also generally paler above than Black-winged, and adult has more red at base of bill. (Cf. also rare vagrant Oriental Pratincole, p. 412.) At very close range, adult summer ♂ has black lores (brownish on ♀) and slightly sharper head pattern than ♀. Neat black throat-surround lost in winter. – Juvenile: Dense dark spotting forming broad breast-band. Head plain; scaly above; all-black bill.

VOICE Calls nasal but piercing. Common flight-call is a high-pitched, sharp 'kit' or 'kittik' recalling Little Tern. From flocks in spring are heard excited, rhythmic five-syllable calls ('song'), '**kür**re-kek, kit**tik**'.

Black-winged Pratincole *Glareola nordmanni*   **V****

L 24–28 cm. Breeding habitat as for Collared Pratincole but more often found at wet meadows, reedy marshes and in vegetated areas. Summer visitor (Apr–Oct), winters mainly in S Africa. Vagrant to W Europe.

IDENTIFICATION Very similar to Collared Pratincole, but *darker and more uniform upperwing* (invariably lacking light trailing edge to inner wing), *black underwing* (Collared, too, can look black in poor light!), at rest *slightly longer legs* and on average *shorter tail*. Adult has *less red at base of bill* than Collared and often slightly *darker lores and forehead*.

VOICE Similar to Collared Pratincole's but slightly lower pitched and harder and drier ('knocking'), often 'kettek' 'ke-ti-tik' or 'kett'. Alarm is dry, hard, rolling 'tr-r-rt'.

Stone Curlew   Cream-coloured Courser   Collared Pratincole   Black-winged Pratincole

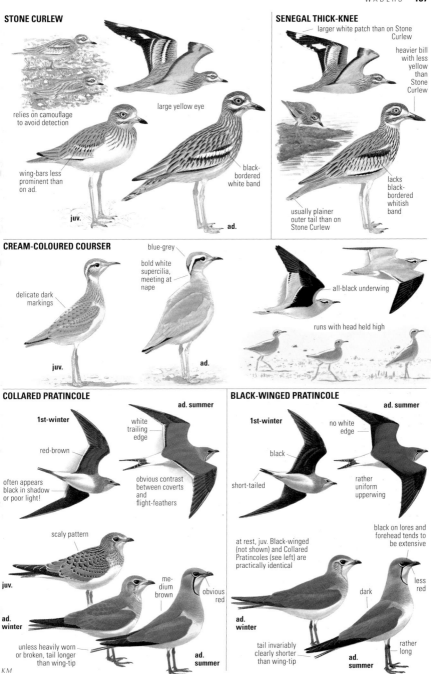

STONE CURLEW

relies on camouflage to avoid detection

large yellow eye

wing-bars less prominent than on ad.

black-bordered white band

juv.

ad.

SENEGAL THICK-KNEE

larger white patch than on Stone Curlew

heavier bill with less yellow than Stone Curlew

lacks black-bordered whitish band

usually plainer outer tail than on Stone Curlew

CREAM-COLOURED COURSER

blue-grey

bold white supercilia, meeting at nape

delicate dark markings

all-black underwing

runs with head held high

juv.

ad.

COLLARED PRATINCOLE

1st-winter

ad. summer

white trailing edge

red-brown

often appears black in shadow or poor light!

obvious contrast between coverts and flight-feathers

scaly pattern

medium brown

obvious red

juv.

ad. winter

ad. summer

unless heavily worn or broken, tail longer than wing-tip

BLACK-WINGED PRATINCOLE

1st-winter

ad. summer

no white edge

black

short-tailed

rather uniform upperwing

at rest, juv. Black-winged (not shown) and Collared Pratincoles (see left) are practically identical

black on lores and forehead tends to be extensive

less red

dark

ad. winter

tail invariably clearly shorter than wing-tip

rather long

ad. summer

KM

Little Ringed Plover *Charadrius dubius* mB4

L 15½–18 cm, WS 32–35 cm. Breeds on bare, usually sandy or gravelly terrain near fresh water, e.g. river or lake edges or islands, often at man-made sites such as gravel-pits, reservoirs or saltpans. Summer visitor to Europe (mostly Apr–Sep/Oct), winters in Africa. Rather scarce in most areas, but sometimes seen in small groups on migration. Generally prefers freshwater margins and estuaries to coastal mudflats. Food mainly insects. Nest is shallow scrape on bare ground.

IDENTIFICATION Slightly smaller than Ringed Plover, with slimmer, longer bill and more slender rear end. Often appears slightly *longer-legged* owing to full' chest and belly. *No light wing-bar* or only a very faint one; *dark earcovert patch often rather pointed at lower rear corner*, not rounded. Long tertials cover most of primaries (thus shorter primary projection than on Ringed Plover). – Adult: Distinctive combination of *obvious yellow orbital ring*, all-black bill, and pale brownish- or greyish-pink (not orange) legs. ♀ has much brown in black head markings and breast-band. In winter, black on head and breast becomes brown, forehead and supercilium buffish; yellow orbital ring usually still visible. – Juvenile: Told from Ringed by *smaller and more diffuse pale area on forehead, not extending back behind eye*, thus lacks an obvious supercilium (is tinged buff above lores and eye); has *pale yellow orbital ring* visible at close range.

VOICE Commonest call, often heard from lone bird in flight, a slightly downslurred, short, piping '**piu**' (or 'pew'). Alarm-call a similar but straighter and more strident and hard 'pree'. On territory, sometimes an almost Sand Martin-like rasping 'rererere...'. Song, either in daytime in slow-winged, bat-like display circuit or heard at night, consists of quick, hard 'pri-pri-pri-...' and rather slow, rhythmic, coarse, buzzy '**crree**-a, **crree**-a, **crree**-a, ...'.

(Common) Ringed Plover
Charadrius hiaticula r(m)**B**4 / **P**+**W**

L 17–19½ cm, WS 35–41 cm. Breeds on open shores by sea or lakes, preferring gravelly or sandy patches among short grass; also above treeline on fells and tundra. Outside breeding season fairly common on inland water margins, estuaries and tidal flats; may form sizable groups on migration and in winter. Food is wide variety of freshwater and marine invertebrates. Nest is scrape on sand, shingle or other bare ground near inland water or on the coast.

IDENTIFICATION Compared with Little Ringed Plover, more compact and full-chested with slightly shorter tertials and longer primary projection. Flight rapid with rather loosely 'clipped' wingbeats; compared with e.g. Dunlin, is longer-winged with less protruding head. Adult's *orange legs and bill-base*, and *prominent white wing-bar* striking; ♀

Little Ringed Plover   Ringed Plover

has often much brown in black head markings and breast-band, giving less clear-cut and smart appearance than ♂. In winter, legs are sometimes a little duller orange, and bill sometimes is all dark. – Juvenile: *Dark brown* instead of black *on head and breast; breast-band reduced* or broken in centre; upperparts finely pale-fringed; bill all dark or with a little yellowish at the base, *legs* dull orange to *yellowish*. Black on head and breast usually acquired by Dec / Jan, after which 1st-winter rarely distinguishable from adult. – Variation: Very slight; northerly birds of Fenno-Scandian mountains and arctic coasts (ssp. *tundrae*) on average slightly smaller and slimmer than in rest of Europe (*hiaticula*) and a tinge darker above; they also moult flight-feathers in winter quarters, instead of near breeding grounds as *hiaticula*.

VOICE Call very characteristic: a soft, distinctly disyllabic, rising '**poo**-eep'. Alarm is a monosyllabic, piping 'peep'. Song usually alternates between a quick, rhythmic, mellow 'tee-**too**-e tee-**too**-e tee-**too**-e tee-**too**-e...' and a hurried, more simple 't'weea-t'weea-t'weea-t'weea-...', usually in slow-winged, bat-like display circuit low over the ground.

Semipalmated Plover *Charadrius semipalmatus* V***

L 16–17½ cm. Vagrant from N America; only two British records, and one Spanish, but probably overlooked.

IDENTIFICATION Very similar to Ringed Plover, except is slightly smaller, and all *front toes are obviously webbed* (Ringed has tiny web only between middle and outer toes), *bill averages shorter and thicker-based* (but some have same bill size), and slightly smaller and more slim-bodied shape recalling Little Ringed Plover (but beware smaller northern Ringed Plover ssp. *tundrae*). In all plumages has *thin pale orbital ring* (lacking on Ringed except some ad. ♂♂, and note that juv. Ringed often shows a suggestion of a pale eye-ring or crescent in front of or below the eye). – Adult: Dark breast-band averages narrower, and on most lacks obvious breast-side bulges, *often strikingly thin and of even width*; ♂ *usually lacks white supercilium behind eye* (almost always present on Ringed). – Juvenile: At close range, look for more or less large wedge of white extending from side of throat above corner of gape (juv. Ringed practically always dark from lores down to gape). Also, upperwing-coverts often more contrastingly pale-fringed, and breast-band seldom if ever broken in centre.

VOICE Differs markedly from Ringed's: either a quick, hoarse, rising whistle, 'che**wee**', not so distinctly disyllabic as Ringed's, and with stress on second syllable, recalling a distant Spotted Redshank; or a more drawn-out 'che-**weee**'.

Killdeer *Charadrius vociferus* V**

L 23½–26 cm. Vagrant from N America, annual in Britain & Ireland. Mostly recorded in winter. Prefers meadows and short-grass fields, but can also be seen on sandy seashores.

IDENTIFICATION Obviously larger than Ringed Plover and with longer neck, long all-dark bill, and much *longer tail* producing strikingly elongated rear end. Double dark breast-band and *orange-brown rump* diagnostic in all plumages. 1st-year distinguishable only by faint pale fringes on retained juvenile upperpart feathers.

VOICE Thin, far-carrying, high-pitched, rising 'klüee', sometimes breaking into fast, whinnying 'klüee-i-i-i-...'.

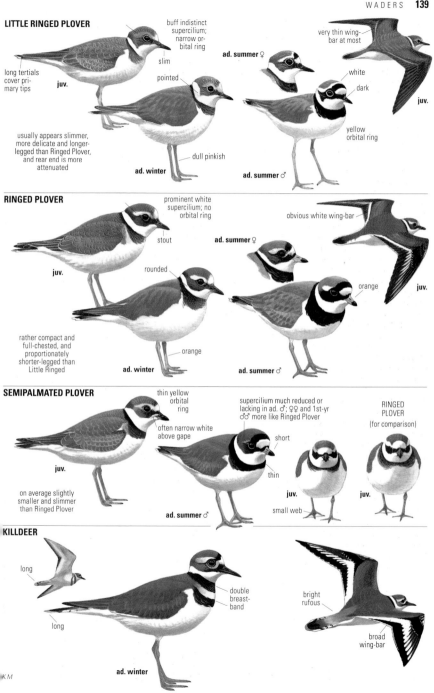

LITTLE RINGED PLOVER

buff indistinct supercilium; narrow orbital ring

slim

pointed

ad. summer ♀

long tertials cover primary tips

juv.

very thin wing-bar at most

white

dark

juv.

yellow orbital ring

usually appears slimmer, more delicate and longer-legged than Ringed Plover, and rear end is more attenuated

dull pinkish

ad. winter

ad. summer ♂

RINGED PLOVER

prominent white supercilium; no orbital ring

stout

ad. summer ♀

obvious white wing-bar

juv.

rounded

orange

juv.

rather compact and full-chested, and proportionately shorter-legged than Little Ringed

orange

ad. winter

ad. summer ♂

SEMIPALMATED PLOVER

thin yellow orbital ring

often narrow white above gape

supercilium much reduced or lacking in ad. ♂; ♀♀ and 1st-yr ♂♂ more like Ringed Plover

RINGED PLOVER (for comparison)

short

juv.

thin

juv.

on average slightly smaller and slimmer than Ringed Plover

small web

ad. summer ♂

KILLDEER

long

long

double breast-band

bright rufous

broad wing-bar

ad. winter

KM

Kentish Plover *Charadrius alexandrinus* **P**5

L 15–17 cm. Breeds on mudflats or sparsely vegetated ground near coasts. Summer visitor to N Europe (mostly Mar–Oct). Locally common, but has declined in many areas. Food insects, crustaceans, worms. Nest on bare ground.

IDENTIFICATION Slightly smaller, longer-legged and *thinner-billed* than Ringed Plover, with *short 'stern' and flat-crowned, broad, large head. Paler than Ringed*. In all plumages has *small dark breast-side patches* rather than complete or centrally broken breast-band; bill black; *legs blackish* (or sometimes paler grey-brown); in flight, shows prominent wing-bar and *all-white tail-sides*. – Adult: ♂ has neat black markings on head and breast-sides, and *crown and nape with* variable amount of *rusty colour*; pattern and colour less well marked in winter. ♀ has grey-brown (not black) markings on head and breast (rarely, may have complete brown breast-band), and no or very little rusty on crown. In non-breeding plumage, note *complete white neck-collar*. – Juvenile: Like adult ♀, but head and breast-side markings paler and has fine pale fringes on upperparts. – 1st-summer ♂: Lacks fully developed black markings and rusty crown.

VOICE Calls on ground, often in anxiety, are a quiet, rolling 'dr'r'rp' and an ascending, soft whistle, 'bew-it' (almost like Spotted Crake). Flight-call a short 'bip' or 'bipip'. The song is rarely heard, a mixture of rhythmical '**che**-ke **che**-ke **che**-ke...' and somewhat Dunlin-like hard rolling, guttural 'drru**urree-rre-rre**', usually in butterfly-like display-flight.

Kittlitz's Plover *Charadrius pecuarius* —

L 14–16 cm. Found on mudflats, shores, near ponds, etc. Locally common in Egypt. Mainly sedentary; vagrant to Israel. Nest is scrape on bare ground, often near water.

IDENTIFICATION Adult summer *head pattern and orange-washed underparts unmistakable*. Winter adult and juvenile more difficult: resemble Kentish Plover but slightly smaller (appear to have a shorter body), *longer-billed*, and *longer-legged*. In all plumages (esp. breeding), has rather *patterned, 'faded-looking' upperparts* caused by diffuse pale feather fringes, not uniform upperparts as on Kentish Plover; dark lesser coverts sometimes show as dark 'shoulder'; and *obvious buff or pinkish tone to supercilium, hindneck-collar, breast-side patches and across breast*, not whitish as on Kentish Plover. Legs are grey, sometimes faintly tinged green or brown. Toes project well beyond tail in flight.

VOICE Rather silent. Short 'chik' in flight, and a dry 'drrr'.

Lesser Sand Plover *Charadrius mongolus* **V**∗∗∗

L 17–19 cm. Vagrant to Europe from Central or E Asia; scarce migrant in Middle East. Often seen on shorelines, mudflats and fields. Winters E Africa and S Asia.

IDENTIFICATION Only slightly larger than Ringed Plover but somewhat longer-legged and longer-billed. All plumages very similar to Greater Sand Plover, esp. non-breeding, so size and structure important: is slightly *smaller*; structure recalls Greater Sand Plover owing to proportionately *rather large head*, similar leg length *without noticeably long tibiae*, and *small bill*, length of which is equal to or less than distance from bill-base to rear of eye, and *bill-tip is blunter* (Greater often has *noticeably long tibiae*, and usually *heavier bill*, length of which is usually greater than bill-base to rear of eye, and *bill-tip is more pointed*); *legs dark* grey-green (often slightly paler on Greater); and *even-width, full-length wing-bar* (on Greater, inner wing-bar faint, outer wing-bar broad). Tends to take fewer steps (1–4) between shorter pauses when feeding (Greater often takes 3–10 steps and pauses longer, but much overlap). – Adult: Summer ♂ has on average broader rufous breast-band and more black on head than often somewhat duller ♀, but sometimes sexes are more similar. – 1st-summer: Only partly acquires adult plumage. Winter adult and immature plumages recall Kentish, from which told by larger size, absence of white collar and lack of all-white tail-sides. – Variation: In Central Asia (ssp. *pamirensis* of *atrifrons* group), legs and bill slightly longer, wings shorter, forehead on summer ♂ black with only tiny white loral spots, and rufous breast-band lacks black upper border; in E Asia (*mongolus* group), legs and bill shorter, wings longer, forehead on summer ♂ with large white patch, and rufous breast-band has thin black upper border.

VOICE Hard, short 'trik' or 'tirrik'; also has soft, more rolling 'trrrp', very similar to call of Greater Sand Plover.

Greater Sand Plover *Charadrius leschenaultii* **V**∗∗

L 19–22 cm. Scarce breeder in C Turkey and Middle East on dry, sparsely vegetated steppe. Uncommon migrant Middle East; vagrant to Europe. On migration and in winter frequents beaches or water margins near coast. Food insects, small crustaceans and marine worms.

IDENTIFICATION Proportions recall Grey Plover rather than Kentish, but smaller, slimmer, much longer-legged (esp. tibiae), and bill longer. In all plumages resembles Lesser Sand Plover, which see for detailed comparison. Note particularly *slightly larger size, longer legs and bill*, latter with somewhat *stronger base and more pointed, attenuated tip*, also often noticeably *longer body* and *slightly longer and thinner neck*. *Legs* on average *paler* greenish- or yellowish-grey than on Lesser. Sexes appear to differ clearly, ♀ being much duller and greyer. – Variation: Breeders in Turkey and W Middle East (ssp. *columbinus*) have slightly shorter bill, and summer ♂ much rufous on breast and often on flanks and back, too. Birds E of Caspian Sea (*crassirostris*) similar but with longer, stronger bill. In E Asia (*leschenaultii*), rufous breast-band is narrow.

VOICE Call a trilling 'trrr', often quickly doubled or trebled; quality recalls Curlew Sandpiper or Turnstone. Difficult to tell from some Lesser Sand Plover calls without much practice.

Kentish Plover · Kittlitz's Plover · Greater Sand Plover

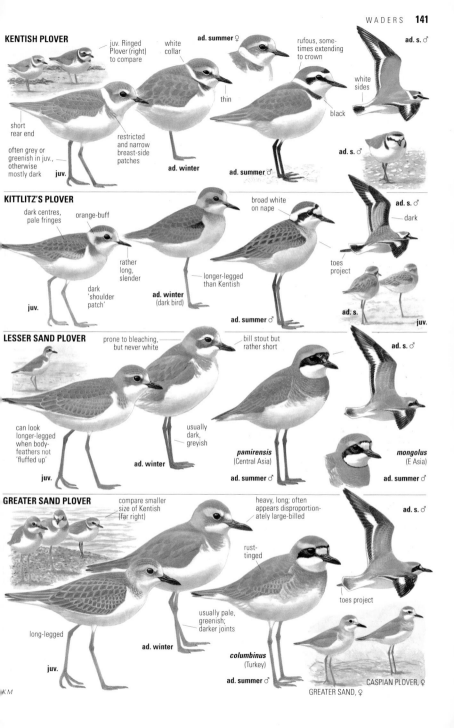

KENTISH PLOVER

juv. Ringed Plover (right) to compare

white collar

ad. summer ♀

rufous, sometimes extending to crown

ad. s. ♂

white sides

thin

ad. s. ♂

black

short rear end

restricted and narrow breast-side patches

often grey or greenish in juv., otherwise mostly dark **juv.**

ad. winter

ad. summer ♂

KITTLITZ'S PLOVER

dark centres, pale fringes

orange-buff

broad white on nape

ad. s. ♂

dark

rather long, slender

toes project

dark 'shoulder patch'

longer-legged than Kentish

ad. winter (dark bird)

juv.

ad. summer ♂

ad. s.

juv.

LESSER SAND PLOVER

prone to bleaching, but never white

bill stout but rather short

ad. s. ♂

can look longer-legged when body-feathers not 'fluffed up'

usually dark, greyish

pamirensis (Central Asia)

mongolus (E Asia)

juv.

ad. winter

ad. summer ♂

ad. summer ♂

GREATER SAND PLOVER

compare smaller size of Kentish (far right)

heavy, long; often appears disproportionately large-billed

ad. s. ♂

rust-tinged

long-legged

usually pale, greenish; darker joints

toes project

juv.

ad. winter

columbinus (Turkey)

ad. summer ♂

CASPIAN PLOVER, ♀

KM

GREATER SAND, ♀

Grey Plover *Pluvialis squatarola* P / W3

L 26–29 cm, WS 56–63 cm. Breeds in high Arctic on tundra, where adults spend Jun–Jul, then migrates to W Europe, where some stop to moult, and even stay through winter (mostly ♂♂), whereas a majority pass through to winter in W Africa (mostly ♀♀ in tropical Africa); juveniles migrate *c.* 1½ months later (late Aug–Sep) than adults in autumn. On autumn migration seen on coastline, often singles or a few birds only, loosely spread out along shore or on shingle bank, behaviour poised (even sluggish). In winter frequents tidal flats and adjacent freshwater pools, can be seen in large gatherings at high tide. In spring crosses N Europe in one or two long legs in May, usually passing unnoticed at great height in fair weather, but in cold, adverse conditions large, dense flocks can sometimes be seen resting. Food mainly marine worms, molluscs and crustaceans.

IDENTIFICATION Distinctively big plover with *bulky body*, *large head* and *heavy bill*; posture more hunched and feeding action more ponderous than with smaller plovers. In flight, *black axillaries* diagnostic, *bold white wing-bar* and *whitish rump* characteristic. – Adult summer: ♂ has solid black below, ♀ has white intermixed. – Adult winter: Underparts whitish, and grey upperpart feathers and wing-coverts diffusely fringed or barred whitish. – Juvenile: Greyish-black upperparts and wing-coverts neatly spotted and notched *yellowish-buff*; underparts pale yellowish-buff with fine grey streaking; *black 'armpits'* distinctive in flight; general coloration often yellowish-brown, recalling juvenile Golden Plover, and since it has a well-marked pale supercilium and often a hint of a dark ear patch it may even resemble an American Golden Plover (p.144). – 1st-summer: Like adult winter or with very little black below. Many stay through summer in W Europe and W Africa.

VOICE Vocal. Call is a distinctive, mournful, trisyllabic whistle with middle syllable lower-pitched and sometimes stressed, 'peee-**uu**-ee'; at times, often with juveniles, voice is shrill and hoarse. Display-song, often in butterfly-like slow-motion flight, is a different trisyllabic whistle with first and last syllables lower and stressed, '**plu**-ee-**uu**'; other calls heard on breeding grounds, usually in anxiety, are a straight 'plüüh' and a Curlew-like 'pluu-ee'.

(Eurasian) Dotterel *Charadrius morinellus* mB4

L 20½–24 cm. Breeds high in mountain areas with low cover and scree, often preferring partly flat areas; also on open tundra. Occurs on northerly breeding sites mostly mid Jun–mid Aug. Rare on migration, often singles or in small groups on arable or short-grass fields. Food insects. Nest is bare scrape. ♂ does most of incubation and care of young. Some ♂♂ are extremely tame just before eggs hatch.

IDENTIFICATION Somewhat smaller than Golden Plover, and is slightly more compact, with shorter neck and wings and proportionally larger head with smaller bill. In all plumages, has *long white or ochrous-buff supercilium* and *plain upperwing* with narrow pale leading edge (white shaft to outermost primary). – Adult summer: Combination of *orange-rusty breast*, *black belly*, narrow *black and white breast-band* and well-marked *white supercilium* contrasting with *dark hindcrown* makes it unmistakable; as with phalaropes, ♀ is brighter-coloured, smarter than ♂. – Adult winter: Greyish breast and flanks tinged buff, with *faint whitish breast-band*, and upperpart feathers and wing-coverts dark (slaty-) grey with thin rufous-buff fringes. – Juvenile: Like adult winter, but upperparts and larger wing-coverts blackish-brown with neat *creamy-buff fringes broken at tip by black central streak*; underparts have on average more obvious dark mottling on breast and upper flanks. – 1st-winter: Told by retained juvenile wing-coverts.

VOICE Flight-call on migration (often on take-off) is a soft, rolling, somewhat descending 'pyürr'. Contact-call also a repeated song-like 'pwit'. Song by ♀ is a simple, repeated whistling, 'pwit pwit pwit...', *c.* 2 calls per sec., delivered in flight over territory with shallow, shivering wingbeats.

Caspian Plover *Charadrius asiaticus* V***

L 19–21 cm. Breeds in Central Asia, mainly N and E of Caspian Sea, on dry plains, often lowland and saline habitats near water; often forms loose breeding colony, nests *c.* 50 m apart. Summer visitor (mostly late Apr–early Aug), winters Africa. Rare in Middle East on migration, vagrant Europe (recent British records on Scilly Is and in Scotland). Nest is sparsely lined scrape on bare or thinly vegetated ground.

IDENTIFICATION Resembles the sand plovers, but *shape and carriage much more elegant*, *with slimmer, more elongated rear end*, has longer, slimmer bill than Lesser and *long legs* like Greater. Slightly larger than Ringed Plover, about size of Lesser Sand Plover, but slender, elegant shape makes it appear larger than it actually is. In all plumages, has *much bolder pale rear supercilium* than sand plovers, lacks obvious white on tail-sides (white confined to terminal fringe), and has only a *faint wing-bar*. – Summer ♂: Pure white forehead, *lores and cheeks*, dark ear-coverts, and *broad chestnut breast-band* distinctly *bordered black at lower edge*; 1st-summer can be less well marked. – Summer ♀: Supercilium and cheeks *off-white*, not pure white; ear-coverts very *pale brown*; *breast-band grey-brown with just a little chestnut*, and *no black lower border*; border between off-white throat and grey-brown breast-band diffuse. – Winter: Adult plumage resembles that of a sand plover (cf. p. 140), but has *an extensive grey-brown unbroken breast-band*, not defined breast-side patches, and *more grey underwing*. Juvenile similar to adult winter, but has more obvious pale fringes on upperparts.

VOICE Call on ground a short 'chüp', often quickly repeated twice, but usually given singly in flight. Song is a repeated ringing trisyllabic call.

Grey Plover

Dotterel

Caspian Plover

GREY PLOVER

two ad. summers and one 1st-summer

juv.

diagnostic black 'armpit' patches

ad. summer ♂

bold white wing-bar

white

neatly spotted and notched yellow-buff

black axillaries

heavy

juv.

ad. winter

ad. summer ♀

ad. summer ♂

DOTTEREL

white shaft

juv.

plain above

ad. s. ♀

lacks wing-bar

dark with creamy-buff fringes; some scapulars renewed

broad supercilium, buff at rear

more evenly grey than ♂

bright rufous

pale breast-band

juv. → 1st-winter

ad. winter

greenish-yellow

ad. summer ♀

ad. s. ♂

CASPIAN PLOVER

juv.

ad. s. ♂

short white wing-bar

toes project

♂ and ♀

less distinct fringes than juv. Dotterel (often looking rather plain)

broad creamy or buffish

much white

rufous

long, thin

black border

long tibia

grey-green

juv.

ad. winter

ad. summer ♀

ad. summer ♂

KM

(European) **Golden Plover** *Pluvialis apricaria* ʀ**B**3 / **W**2
L 25–28 cm, WS 53–59 cm. Breeds on moors, bogs, upland pastures, mountainsides above treeline, tundra; on British breeding sites mostly late Mar–mid Jul, but on northern tundras Jun–Aug. In winter on lowland fields or pastures, often with Lapwings and locally in large flocks. Migrating flocks dense, flight rapid; roosting movements at dawn and dusk involve much manoeuvring (flock flicking from yellow-brown to white as birds turn). Rarely seen wading, prefers drier ground. Rather shy and wary. Food insects, worms, berries and seeds. Nest is scrape in heather or grass.

IDENTIFICATION Slightly smaller and daintier than Grey Plover, and has somewhat *narrower and more pointed wings* and faster wingbeats. At distance looks brown above, but seen close has dark grey upperparts with feather edges densely notched greenish- or ochrous-yellow and white. In flight, shows *diffuse, rather narrow whitish wing-bar*, dark rump and flashing *white axillaries and inner underwing*. For separation from very similar Pacific and American Golden Plovers, see under those. – Adult: Summer ♂ shows more black below than♀, but because of individual variation only very dark adult summer ♂♂ in N Europe, or breeding pair seen together, safely sexable. In winter, underparts lack black. – Juvenile: Very similar to adult winter but more uniformly fresh and neatly patterned, and *much of belly and flanks finely barred grey-brown* (ad. not so neatly patterned, and has *whiter belly*). – 1st-summer: Often like adult winter or with very little black below. – Variation: Birds from S of breeding range ('*apricaria*' type) on average acquire less striking summer plumage than N birds ('*altifrons*'), which average more black below, but much variation in all parts of breeding range, so subspecific status not fully justified.

VOICE Plaintive flat whistle,'püü'or'tüü(u)'(only slightly downslurred), in flight or persistently repeated on ground when alarmed. Plaintive, rhythmic song given in butterfly-like display-flight,'pü-**peee**-oo, pü-**peee**-oo, pü-**peee**-oo...', often followed by a quicker, subdued 'per**purr**lya-per-**purr**lya-per**purr**lya-...' when alighting, or on ground; latter call can be heard outside breeding season, too.

Pacific Golden Plover *Pluvialis fulva*   **V**∗∗
L 21–25 cm. Closely related to American Golden Plover and formerly treated as conspecific with it. Breeds in N Siberia and W Alaska. Vagrant to W Europe, mostly in late summer (ads.) and autumn (juvs.). Often seen on coastal lagoons and mudflats, but sometimes on fields and pastures.

IDENTIFICATION Very similar to Golden Plover, size and structure differing in following ways: slightly *smaller* and *slimmer-bodied*, with proportionately slightly *larger head* and longer neck, *longer legs*, and in flight *slimmer wings* and *toes projecting somewhat beyond tail-tip*. On average, the bill is slightly longer and more evenly thick (Golden has bill a little shorter with finer tip and deeper base, but subtle difference and overlap). In all plumages, *axillaries and inner underwing are grey-brown* (same as on

Golden Plover

American, but diagnostic compared with Golden, which is pure white). Even more difficult to separate from American Golden Plover, but note this: a trifle *smaller* size; wings *shorter* but *tertials longer*, resulting in moderate primary projection with *3 primary tips showing beyond tertials* (note that longest two primaries are close together), instead of 4 (or even 5) as on American; some juvenile Pacific have shorter tertials and a 4th primary tip barely showing; beware also of moult or wear altering this feature. Further, Pacific has *slightly longer legs* than American. – Adult summer: Like Golden Plover but: has *more white on forehead* and *along sides of neck and upper breast* (forming hint of white 'bulge') but less along flanks; *narrow white rim along flanks* often *boldly marked black* or even vestigial; undertail-coverts variably pied or predominantly black (rarely appearing all black in the field); *upperparts more coarsely patterned*. – Juvenile: Very similar to juvenile Golden, best told by size, structure, underwing colour, and voice. On average *slightly more yellow-buff tinge on head, neck and breast*, with finer and more distinct spots, but some are practically alike in the field. See American for distinctions from that.

VOICE Flight-call is whistled, disyllabic 'chu-it!', remarkably similar to Spotted Redshank. Straight 'peee', as from Golden Plover, downslurred 'pluu-e', or trisyllabic 'tuu-ee-uh', have been noted from flocks on migration.

American Golden Plover *Pluvialis dominica*   **V**∗∗
L 24–27 cm. Closely related to Pacific Golden Plover and formerly treated as conspecific with it. Breeds in N America. Vagrant to Europe. Seen in same habitats as Pacific.

IDENTIFICATION Very similar to Pacific Golden Plover, sharing with it *grey-brown axillaries and inner underwing* (white on Golden), but on average *slightly larger* and *slimmer-bodied, longer-winged* and *shorter-legged* (toes usual*l*y *do not project beyond tail-tip* in flight). Crucial to establish long primary projection on resting bird: *4* (sometimes even 5) *primary tips visible beyond tertials*, and *wing-tips usually project clearly beyond tail-tip* (same length as tail, or only slightly beyond, on most Pacific). Told from Golden by colour of underwing, slightly *smaller size* and *slimmer build*, with *wings appearing longer and thinner* in flight. Several plumage characters are important, too (see below). – Adult summer ♂: *All-black underparts* including under tail-coverts; *clear-cut white shawl* along sides of neck and upper breast ending in *prominent white 'bulge'*; *no white along flanks*. – Adult summer ♀/1st-summer: Plumage can be very similar to Pacific, best told by structure and proportions. – Juvenile: Compared with Golden, and especially Pacific, a *more greyish* bird, lacking yellowish-buff tinge on head, neck and breast; *supercilium* on average *more whitish and distinct*, and *ear-spot* and *'loral smudge' darker; dark cap* and *dark mantle* but paler greyish nape; *golden spangling on upperparts* often *confined to fore-mantle, upper scapulars and tertials*, remainder of upperparts and wing-coverts spangled whitish or buff (Pacific and Golden have whole upperparts and wing-coverts more uniformly gold-spangled).

VOICE Common call is a '**clu**-ee', unlike Pacific (and Spotted Redshank) with more stress on first syllable, and second syllable sometimes barely audible, 'cluu(e)'. Also trisyllabic'**dlu**-ee-uh', second note higher and trembling.

GOLDEN PLOVER

feeding winter flock

ad. summer ♀
'Southern'

ad. s. ♂
'Southern'
or
ad. s. ♀
'Northern'

ad. summer ♂

white axillaries

finely
notched

juv.

juv. and winter
plumages very
similar

winter

ad. summer ♂
'Northern'

spring flock

PACIFIC GOLDEN PLOVER

PACIFIC
GOLDEN,
juv.

GOLDEN,
juv.

ad. s. → w.
(Aug)

rather short

can, at times, look
remarkably dumpy
and not noticeably
long-legged

more coarsely
patterned than
Golden

coarsely
notched

ad. s. ♂

toes
project

grey-brown

longer tertials than
American Golden,
and only moderate
primary projection

very
long

some are
as black as
American
Golden

juv.

smaller, more delicate than
Golden, with proportionately
longer legs and bill

ad. summer ♂

AMERICAN GOLDEN PLOVER

GREY, juv.

AMERICAN
GOLDEN,
juv.

ad. s. → w.
(Sep)

subtly longer body
and smaller bill than
Pacific Golden

ad. s. ♂

bold
pattern

coarsely
notched

flanks
black

grey-brown

attenuated

prominent supercilium

usually greyer, less
yellowish-brown than
Golden and Pacific Golden;
stronger head pattern and
more attenuated rear end

rather long
primary
projection

long-
legged

moulting ad. (centre) in
autumn with Golden Plovers

juv.

ad. summer ♂

KM

(Northern) Lapwing *Vanellus vanellus* r(m)**B**2 / **W**1

L 28–31 cm, WS 67–72 cm. A common breeding bird of a variety of inland and coastal open country, usually on arable fields, pastureland or seashore or lakeside meadows; in winter forms sometimes large flocks on farmland, marshes. Food worms and insects. Nest is lined scrape on ground.

IDENTIFICATION Unmistakable *black and white*, pigeon-sized, stocky plover. Has unique *long, thin, wispy crest*, and at close range beautiful *green and purple iridescence on dark upperparts*. In flight, has extraordinary *strongly rounded wing-tip* ('frying-pan wings') and deliberate, flappy beats, wings alternately showing white below and dark above, so that the typically rather closely packed flocks give a distinctive 'flickering' effect even at great distance. – Adult summer: ♂ told by longer crest and solid black on foreneck and face; also, in flight, broader, rounder 'hand'. ♀ has shorter crest, and black on foreneck is less extensive and speckled white, giving less smart head pattern; also narrower, more pointed 'hand'. – Adult winter: Chin and foreneck become white, and upperparts and coverts fringed with buff at tips, giving scaly pattern. – Juvenile: Crest stumpy; resembles adult winter, but pale fringes on upperparts, coverts and tertials complete (not confined to tips) and finely scalloped or spotted. – 1st-winter: Told from adult winter by pattern of retained juvenile coverts and tertials, but difference lost quickly through wear of feather fringes.

VOICE Highly vocal, and often heard at night. Flight-call when fully alarmed is a heartbreakingly shrill, breathy 'pwaay-**eech**' or 'wa**a**y-ach'. Highly aerobatic rolling and tumbling display-flight of ♂ low over field in early spring accompanied by bubbling, wheezy 'song' and (at close range) vibrant throbbing produced by beating wings; song goes 'chae-widdle**wip**, i-**wip** i-**wip**... cheee-o-**wip**'.

Spur-winged Lapwing *Vanellus spinosus* —

L 25–28 cm. Scarce breeder in SE Europe; locally numerous in Middle East. Vagrant to W Europe. Frequents bare or sparsely vegetated open ground, usually near water. Food mainly insects. Nest is lined scrape on bare ground.

IDENTIFICATION Slightly smaller than Lapwing, and a little slimmer and *longer-legged*. At all ages, distinctive *large white cheek patch* on otherwise black head and has *black breast, flanks* and *tail*, black bill and dark grey legs. In flight at distance, best told from Sociable Lapwing by *black secondaries*. – Adult: Upperparts uniform pale brown. – Juvenile: Like adult, but has some pale scaly feather fringes above (e.g. on forehead) and black parts are tinged brownish-grey.

VOICE Vocal and noisy. Squeaky single or quickly repeated 'p(v)ik', somewhat recalling Black-winged Stilt. Song a series of rapidly repeated high-pitched '**titi-te**rit'.

Red-wattled Lapwing *Vanellus indicus* —

L 32–35 cm. Breeds across southern Asia; within treated region, found regularly only on Tigris river around Cizre in SE Turkey. Accidental outside breeding area in Middle East.

IDENTIFICATION Structure, actions and general coloration as Spur-winged Lapwing, but somewhat *larger*. Has diagnostic combination of *long yellowish legs, black throat and central breast* but *white flanks*, (in flight) *white tail crossed by even-width, fairly narrow black band*, and (at close range) *red bill-base, lores and orbital ring*. Juvenile separable by white forehead and chin, and white-mottled black throat patch.

VOICE Strident, grating 'cree', 'crik' and other variations, repeated or extended into tern-like chattering, e.g. 'cree-crik cree-ki-koo-it cree-ki-koo-it...'.

Sociable Lapwing *Vanellus gregarius* **V****

L 27–30 cm. Breeds on wide, dry steppes of Central Asia and SW Siberia, often rather far from water; winters in NE Africa, Syria, Iraq and India, sometimes also in S Israel (Negev). Scarce passage migrant in SE Europe and Middle East; vagrant elsewhere, usually appearing with Lapwings, which have similar habitat and food requirements.

IDENTIFICATION Behaviour and structure as Lapwing, but slightly smaller and *longer-legged*. At all ages, unmistakable combination of *bold white supercilium, blackish legs*, and *three-coloured upperwing pattern* (recalling Sabine's Gull). Recalls White-tailed Lapwing, but not so slender or long-legged and has prominent supercilium, black legs and *black-banded tail*. – Adult summer: Crown and eye-stripe black, breast pinkish grey-buff, *belly black* with deep chestnut rear part. – Adult winter: Breast paler, *belly white*. – Juvenile: Diffusely streaked breast and prominent scaly pattern above. – 1st-winter: Told by retained juvenile coverts and tertials.

VOICE Utters single harsh 'kretch' or repeated, chattering 'kretch-etch-etch...' in flight.

White-tailed Lapwing *Vanellus leucurus* **V*****

L 26–29 cm. Breeds Central Asia W to Iraq, winters E Africa and India. Scarce migrant SE Europe and Middle East; very rare vagrant to W Europe. Often seen at well-vegetated ponds, lush river edges, along canals and ditches with good cover, rather than on open mudflats. Behaviour often rather skulking, taking cover and being flushed rather close.

IDENTIFICATION Slightly smaller than Spur-winged Lapwing and with much more elegant shape and *very long legs*. At all ages, rather uniformly *pale grey-brown head and body* without prominent supercilium, and *very long yellow legs* diagnostic, even before sight of striking flight pattern with *all-white tail*. Juvenile has obvious scaly pattern above.

VOICE Migrants usually silent.

Lapwing

Spur-winged Lapwing

Red-wattled Lapwing

Sociable Lapwing

LAPWING

short

pale-scalloped

juv.

pale fringes

forms large flocks in winter

ad. s. ♀

ad. winter

ad. summer ♂

even-width wing

ad. s. ♀

ad. s. ♂

very broad 'hand'

SPUR-WINGED LAPWING

ad.

(small spur)

ad.

wide black band

black belly

subdued scaly pattern

juv.

ad.

RED-WATTLED LAPWING

ad.

ad.

broad white terminal band

red

yellow

ad.

ad.

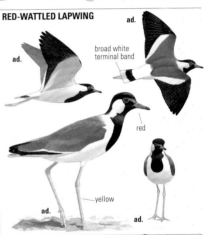

SOCIABLE LAPWING

white secondaries

Dotterel-like head markings

juv.

black band

broad super-cilium

dark

dark

ad. winter

ad. summer

WHITE-TAILED LAPWING

ad.

ad.

all white

rather plain

long

heavily spotted

yellow

very long

juv.

ad.

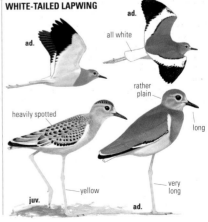

KM

(Red) **Knot** *Calidris canutus*   P / W1–2

L 23–26 cm, WS 47–53 cm. Breeds in high Arctic. Winters on wide tidal flats, locally in huge flocks; smaller numbers occasionally on freshwater pools near coast. Concentrated spring migration late May–early Jun largely unnoticed.

IDENTIFICATION Rather *short-legged, plump wader*, larger and stockier than Dunlin. Bill usually *straight and rather thick and short* (about same length as head), but some have surprisingly long and slightly decurved bill. In flight, has long-winged, short-bodied outline with thin white wing-bar and *pale grey* (actually finely barred) *rump patch*. In winter and juvenile plumages, best told by *shape, general greyness*, prominent light *supercilium* and rather short *grey-ish-green legs*. (Also, cf. similar but very rare vagrant Great Knot, p. 412.) – Adult summer: ♂ averages more uniform and richer *rufous-orange below* than ♀. – Adult winter: Plain grey above with thin white fringes to scapulars, white supercilium, grey-streaked breast, and grey bars and chevrons on flanks and undertail-coverts. – Juvenile: Like adult winter at distance, but usually with *orange-buff wash on underparts* (occasionally so strong as to suggest adult summer) and breast and flanks more finely spotted and barred with grey; at close range, upperparts and especially wing-coverts *neatly pale-fringed, with dark subterminal crescents*.

VOICE Flight-call a short, nasal, rather soft 'whet-whet'.

Sanderling *Calidris alba*   P / W3

L 18–21 cm. Breeds in high Arctic; seen on passage and in winter in sometimes large flocks on sandy beaches, mudflats or pools near coast; rare inland. Feeds with *distinctive dashing running action ahead of breaking waves*, but also more methodically on calmer, shallow shorelines or grassy areas.

IDENTIFICATION Dunlin-sized, but more compact. Could be confused with stints owing to lively action, especially lone bird away from typical shore habitat, but note *larger size, stouter bill, much broader black-bordered white wing-bar*, and diagnostic lack of hind toe. In winter, appears *strikingly pale* among other shore waders and shows *distinctive blackish wing-bend* (when this is not covered by breast-side feathers). – Adult summer: In spring, head and *clear-cut dark-spotted breast-band* grey/rufous, and upperparts blackish and rufous with broad pale fringes; in summer, head, breast and upperparts become strongly rufous. – Adult winter: Generally plain, very pale grey above with blackish lesser coverts; wing-coverts and tertials diffusely white-fringed. – Juvenile: General pattern like juvenile stint, but whole plumage more contrastingly black and white, *mantle and scapulars more spangled or spotted with white* (not neat scaly pattern) and lacks white mantle- and scapular-Vs.

VOICE Flight-call is an emphatic, slightly liquid 'plit'.

Purple Sandpiper *Calidris maritima*   r(m)B5 / P+W1–2

L 19–22 cm, WS 37–42 cm. Breeds on coasts, rocky tundra and in marshes on barren mountainsides up to snowline; in winter and on passage found on coastal rocks (look for it on extensive areas of wave-washed, seaweed-covered rocks) usually in flocks, often with Turnstones, occasionally also at pools near coast; rare inland. Nest is scrape on bare ground.

IDENTIFICATION A little larger than Dunlin and with similar bill shape, but *fatter body* and *shorter legs* give distinctive *dumpy, rather shuffling look*, and *tail longish*, projecting beyond wing-tips. In winter, told by habitat and distinctive combination of shape, *uniform brown-grey head* (no supercilium), *upperparts and breast*, and *mustard-yellow* (or sometimes pale reddish-brown) *legs and bill-base*. Dark general coloration contrasting with whitish belly, 'armpits' and rump-sides, and *narrow, poorly marked wing-bar* distinctive in flight. Lone bird feeding in manner of other waders, in untypical habitat (e.g. muddy pool edge), can be temporarily very confusing, especially juvenile in neat scaly plumage. – Adult summer: Breast and flanks are mottled blackish; *mantle/scapulars edged whitish and rufous*. – Juvenile: Neat whitish fringes on scapulars and wings form *strong scaly pattern*; rusty on crown and mantle; actually rather like a juvenile Dunlin, but *legs and bill-base paler* and *tail longer*.

VOICE Usual flight-calls sharp 'quit' or 'quit-it'. Alarm is a loud, rapid laughing 'pehehehehehe...'. In display-flights, utters Dunlin-like buzzing 'prruee-prruee-prruee-...' and hard twittering 'kewick-kewick-wick-wick-wick-...'.

(Ruddy) **Turnstone** *Arenaria interpres*   P / W3

L 21–24 cm, WS 43–49 cm. Breeds on mainly stony or rocky coasts and in treeless archipelagos; wary, keeping watch from top of shore or boulder. Winters in a variety of coastal habitats. Food highly varied; uses bill to turn over stones or seaweed to obtain prey beneath. Nests on ground.

IDENTIFICATION Medium-sized, *stocky* wader with distinctive *wedge-shaped, pointed bill, orange legs* and *pied wing pattern*. – Adult summer: ♂ has extensive bright orange-brown on scapulars and wing-coverts, and clear black-and-white head pattern; ♀ (and 1st-summer ♂) has less clear orange (often lacking on wing-coverts) and less smart head pattern. – Adult winter: Distinctive pale-centred breast-side patches; rather dark grey-brown above, often with some dull orange on scapulars; wing-coverts and tertials *plain* grey-brown or *diffusely* edged paler; legs bright orange. – Juvenile: Like adult winter, but with neat ginger-brown or whitish fringes giving *distinct scaly pattern* above.

VOICE Flight-calls chuckling 'tuk-a-tuk-tuk' and short, yelping 'k(l)ew'. Both alarm and display-song an emphatic, accelerating rattle, 'chuvee-chuvee-vitvitvitvitvitvit'.

Knot

Sanderling

Purple Sandpiper

Turnstone

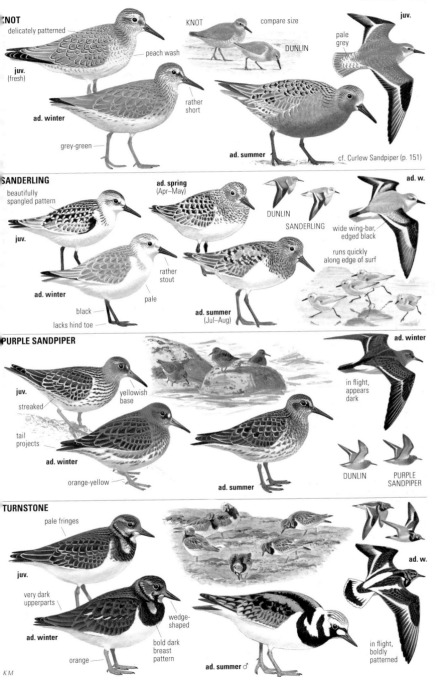

KNOT
- delicately patterned
- peach wash
- juv. (fresh)
- ad. winter
- grey-green

KNOT
compare size
DUNLIN
rather short

juv.
pale grey

ad. summer
cf. Curlew Sandpiper (p. 151)

SANDERLING
- beautifully spangled pattern
- juv.
- ad. winter
- black
- lacks hind toe
- rather stout
- pale

ad. spring (Apr–May)

DUNLIN
SANDERLING

ad. w.
wide wing-bar, edged black
runs quickly along edge of surf

ad. summer (Jul–Aug)

PURPLE SANDPIPER
- juv.
- streaked
- tail projects
- ad. winter
- orange-yellow
- yellowish base
- ad. summer

ad. winter
in flight, appears dark

DUNLIN
PURPLE SANDPIPER

TURNSTONE
- pale fringes
- juv.
- very dark upperparts
- ad. winter
- orange
- wedge-shaped
- bold dark breast pattern
- ad. summer ♂

ad. w.
in flight, boldly patterned

KM

Dunlin *Calidris alpina* r(m)**B**4 / **P**+**W**2

L 17–21 cm, WS 32–36 cm. Breeds on low or high ground, in wet short-grass or tundra habitats. The region's commonest small wader on migration (in autumn: ads. in late Jul–Aug, juvs. in late Aug–Oct) and in winter, then found in a variety of marshy or coastal habitats, but most numerous on tidal flats or on banks of seaweed on shallow shores.

IDENTIFICATION Familiarity with different Dunlin plumages will provide yardstick against other species. It will also help in initially picking out other species as 'something different'. – Just under Starling-sized wader with variable bill length (see Variation), the shortest and straightest almost stint-like, the longest and most decurved approaching Curlew Sandpiper. – Adult summer: *Black belly patch* diagnostic; variable *rufous on mantle/scapulars*, becoming more obvious through wear; ♂ brighter than ♀, with contrasting pale grey nape (brownish on ♀). – Adult winter: Drab and rather featureless plumage is in itself rather distinctive, combining plain brownish-grey upperparts (fine whitish fringes, esp. on wing-coverts); rather clear-cut pale grey breast-band contrasting with white belly and finely streaked or clear white flanks; *rather plain head* with indistinct short supercilium; and usually fairly long decurved black bill (but beware long and short extremes). – Juvenile: Dark brown above, neatly fringed pale rufous or rich rufous, usually with white scapular- and mantle-Vs; pattern of underparts distinctive, with neat breast-band of diffuse streaks and *band of rather bold blackish spots on belly-sides* leaving broad, *usually unmarked white flank-band*; whole *head, nape and upper breast washed ginger*, without obvious supercilium: even at long range, therefore, shows distinctive combination of rather plain gingery head, spotted belly-sides and white flank-band. Note variability of belly-side spotting, from often very dense (easily confused with moulting adult summer) to very rarely totally lacking. – Variation: Three subspecies occur within treated region, differing in average bill length and brightness of rufous on upperparts: *arctica* (Greenland; winters mainly W Africa) shortest-billed, least rufous; *schinzii* (Iceland, W Europe; winters mainly W Africa) intermediate bill length and rufousness; and *alpina* (Arctic; winters W Europe, Mediterranean) longest-billed, richest rufous. Only extreme individuals safely identifiable: ♀♀ average longer-billed than ♂♂, so shortest-billed individuals probably ♂ *arctica/schinzii*, and longest-billed probably ♀ *alpina*.

VOICE Flight-call a buzzing or harshly rolling 'chrrreet'. Conversational calls from flocks short, rippling 'plip-ip-ip'. In display-flight, utters ascending, strained '**rrü**ee-**rrü**ee-rrüee-...', which turns into a hard, descending, slightly slowing trill, 'rü**rrüü**rürürürü-rü-ru-ru ru'.

Curlew Sandpiper *Calidris ferruginea* **P**◀

L 19–21½ cm. Breeds in arctic Siberia; winters mainly in Africa, scarcely in Mediterranean region, and rarely W Europe. On passage, usually rather scarce, but much commoner in some autumns; frequents marshes and coast, often in mixed flocks with Dunlin. Food mainly invertebrates.

IDENTIFICATION Slightly larger than Dunlin, but with *longer, finer-tipped and more evenly decurved bill*; longer neck and legs (esp. tibiae) give *more elegant outline*; *white rump* obvious in flight. – Adult summer: ♂ averages more uniform (less white-flecked) and brighter *brick-red below* than ♀. – Adult winter (rarely seen in Europe): Resembles Dunlin (esp. long-billed *alpina*) except for more elegant shape, paler-grey upperparts, *prominent white fore-supercilium* and white rump. – Juvenile: Greyer than Dunlin, with *uniform scaly pattern* (lacking Vs), more *prominent and white supercilium, streaking on underparts confined to breast-sides* and usually obvious *orange-buff wash across breast*.

VOICE Flight-call a short trilling or jingling 'chürrip' or 'kürürip', a little like call of Temminck's Stint but lower-pitched and coarser, quite different from buzzing call of Dunlin (but note that giggly conversational calls from flocks of feeding Dunlins can sound rather similar).

Broad-billed Sandpiper *Limicola falcinellus* **V**∗∗

L 15–18 cm. Breeds in wettest parts of upland or taiga bogs, summer visitor (mostly late May–Aug), winters E Africa, S Asia on or near coasts. Scarce passage migrant anywhere in Europe but more common in E than W; in Britain & Ireland 2–3 records annually. During migration seen resting both on inland marshes and on coasts. Nest is lined cup in tussock.

IDENTIFICATION Easily overlooked as a young Dunlin, but is *slightly smaller*, enhanced by somewhat *shorter legs*, and usually slower feeding action giving more stocky and furtive impression; bill Dunlin-like but rather long and straighter in profile, with subtle *downward kink only near tip*; even at long range juveniles often picked out by size, slow movements, *dark upperparts* and *white belly*; while in Europe, has *boldly striped head pattern* (obvious dark eye-stripe, strong pale supercilium, and *thin pale lateral stripes on dark crown*). – Adult summer: Distinctive combination of heavily streaked flanks (often with arrowheads) and *complete breast-band* with *dark upperparts* with thin white mantle- and scapular-Vs; in fresh plumage whitish fringes give frosted look, but when worn *upperparts and breast become very dark* and Vs reduced or lacking. – Adult winter: Dunlin-like, but a hint of distinctive head pattern retained. – Juvenile: Upperparts like strongly marked Dunlin, but underpart pattern is very different, with well-marked, brown-washed breast-band streaked at sides, contrasting with *all-white belly and flanks*.

VOICE Flight-call is a dry, high-pitched, buzzing trill, a little upward-inflected, 'brrre-eet'; also short 'chep'. In display-flight has rhythmic, buzzing 'brre, brre, brre **ber**re **ber**re brebrebrebre...', recognized by the call-note voice (Sand Martin-like).

Dunlin

Curlew Sandpiper

Broad-billed Sandpiper

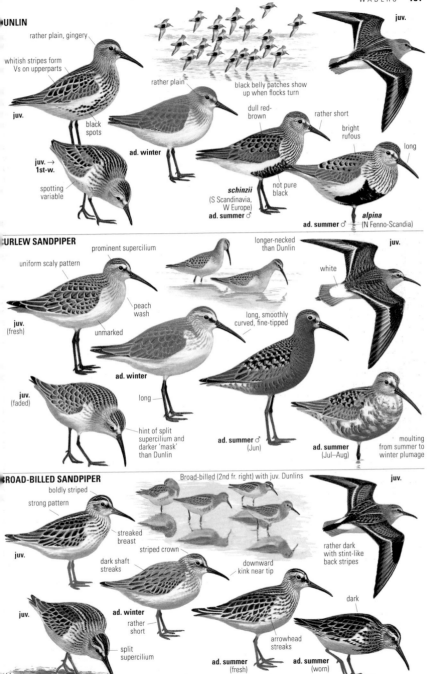

DUNLIN

rather plain, gingery

whitish stripes form Vs on upperparts

juv.

black spots

juv. → 1st-w.

spotting variable

rather plain

black belly patches show up when flocks turn

ad. winter

dull red-brown

rather short

bright rufous

long

juv.

schinzii (S Scandinavia, W Europe) **ad. summer ♂**

not pure black

ad. summer ♂

alpina (N Fenno-Scandia)

CURLEW SANDPIPER

prominent supercilium

uniform scaly pattern

juv. (fresh)

peach wash

unmarked

longer-necked than Dunlin

white

juv.

long, smoothly curved, fine-tipped

ad. winter

long

juv. (faded)

hint of split supercilium and darker 'mask' than Dunlin

ad. summer ♂ (Jun)

ad. summer (Jul–Aug)

moulting from summer to winter plumage

BROAD-BILLED SANDPIPER

boldly striped

strong pattern

Broad-billed (2nd fr. right) with juv. Dunlins

juv.

juv.

streaked breast

striped crown

dark shaft streaks

downward kink near tip

rather dark with stint-like back stripes

juv.

ad. winter

rather short

split supercilium

arrowhead streaks

dark

ad. summer (fresh)

ad. summer (worn)

KM

Temminck's Stint *Calidris temminckii* m**B**5 / **P**5

L 13½–15 cm. Breeds on bogs, marshes or river deltas in upland taiga or on tundra or on fells; also in coastal habitats. Winters mainly Africa. On passage (mostly May and Aug) frequents pools and marshes, lakesides or rivers; in Britain & Ireland, much less common than Little Stint, but just as likely to occur in spring as in autumn. Nest is lined scrape in open or in low vegetation, often willow scrub, in damp area.

IDENTIFICATION Same size as Little Stint, but slightly *longer-bodied* (tail projects slightly beyond wing-tips) and *shorter-legged*, with finer, *slightly decurved bill*. Unobtrusive, unlike Little Stint typically choosing thinly vegetated rather than open sections of mudflats; also, usually *feeds with distinctive creeping action* on flexed legs and with slower pecking rate. Unlike Little Stint, *flies off high on erratic course*, continuously giving *distinctive trilling call*. Outer tail-feathers all white (grey on Little). *Legs pale* (yellowish, greenish or brownish). In winter and juvenile plumages *recalls a miniature Common Sandpiper* because of shape, rather plain brownish upperparts, and well-marked breast. Beware that winter Little Stint frequently misidentified as Temminck's, especially one with sleek plumage, thus looking slimmer and longer-bodied than usual, with mud-coated (thus pale-looking) legs, and in bright sun (when outer tail-feathers can look white in flight); Temminck's, however, has *brownish* upperparts (pale grey with dark feather centres on Little), rather *plain brownish head* (strong supercilium and whitish face on Little), *clear-cut brownish breast* (pale grey on Little) and *shorter legs*. – Adult summer: Variable number of black-centred, rufous-edged feathers on upperparts, mainly among scapulars and on mantle, at distance giving distinctive spotted effect. – Adult winter: Plain, dull grey-brown upperparts and breast contrasting with pure white underparts. – Juvenile: At close range, *pale fringes and dark subterminal line on scapulars and wing-coverts*; upper scapulars contrastingly dark-centred.

VOICE Flight-call when flushed distinctive, a loud, dry trilling, usually repeated 'tirrr-tirr-tirr...'. In display-flight, ♂ hovers giving continuous cricket-like reel, 'titititititi...'.

Little Stint *Calidris minuta* **P**4 / **W**5

L 14–15½ cm, WS 27–30 cm. Breeds on tundra, usually near coast. On passage in W Europe, less common in spring than autumn (due to more easterly route in spring), and in autumn numbers highly variable depending on breeding success. Frequents muddy edges of pools, coastal flats etc., often in sizable flocks, and often mixing with Dunlins.

IDENTIFICATION A small wader, about ²⁄₃ length of Dunlin. The standard stint of region; familiarity with all its plumages essential for picking out Temminck's or vagrant stints

(p. 164, and below). *Small size*; 'runs around feet' of othe small waders, e.g. Dunlin, but even when alone sma size indicated by usually *quicker actions and pecking rat* Fine-tipped, *straight, short bill* (shorter than shortest-bille Dunlin) gives distinctive *spiky-billed look* at distance; mo *extensive white on head and breast* gives whiter front end tha other small waders except Sanderling. *Legs black*, but loo brownish against dark background, or pale when mu coated. –Adult summer: Rusty-orange tones on head, breas and upperparts variable (strongest on ♂), becoming mo obvious during summer as whitish fringes of fresh plumag disappear. *Split supercilium* and *yellowish mantle-V* usual obvious. – Adult winter: Grey above, with dark feathe centres giving mottled effect; grey breast-sides (or breas band: then easily confused with Temminck's). – Juvenil *Blackish-centred upperpart feathers* and wing-coverts wi neat *pale or rich rufous fringes* and usually prominen mantle- and scapular-Vs; underparts white except for cluste of fine streaks over rusty wash at breast-sides; *strong whi split supercilium*, and *dark 'ridge' down centre of crown*.

VOICE Flight-call is single or repeated sharp, high-pitche 'stit'. Song (in flight or from ground) a weak 'swee-swee swee-swee-...', now and then relieved by silvery 'svirrr-r-r'.

Long-toed Stint *Calidris subminuta* **V****

L 14–15½ cm. Breeds in Siberia. Rare vagrant to Europe.

IDENTIFICATION Like Least Sandpiper (pale legs, no pri mary projection), but slightly larger, bill with usually ob vious pale base to lower mandible, and longer neck and leg (esp. tibiae) giving general outline often recalling miniatur Wood Sandpiper rather than stint; toes noticeably long with middle toe slightly longer than tarsus (a little shorte on Least); toes project somewhat beyond tail in flight; fore head often dark down to bill (supercilia usually join acros forehead on Least); supercilia whiter and broader, ofte with split over eye; lores rather pale (usually darker on Least and often rufous-tinged 'cap'. –Adult summer: *Broad rufou fringes on upperparts and tertials*. –Adult winter: Clear-cu dark centres to grey-brown scapulars. – Juvenile: At leas *median and lesser coverts fringed whitish*, contrasting wit rufous-fringed scapulars/tertials (juv. Least has wing coverts also fringed rufous, thus no contrast).

VOICE Call a trilled 'chrip', like weak Pectoral Sandpipe

Least Sandpiper *Calidris minutilla* **V***

L 13–14½ cm. N American species. Vagrant to W Europe mostly Jun–Oct. Averages one annually Britain & Ireland.

IDENTIFICATION Similar to Long-toed Stint, which see fo differences. Told from other stints by *pale* (usually yellowis or greenish) *legs* (except otherwise dissimilar Temminck's) and *short or nonexistent primary projection*; more extensiv *breast streaking* (often complete breast-band), *duller hea pattern*, and *browner coloration* jointly give *less white, mor drab general appearance*. –Adult summer: Thin rufou fringes on upperparts and tertials, often lacking when worn – Adult winter: Mousy grey-brown above, with diffuse dar feather centres giving scaly impression at distance. – Juve nile: Often bright rufous fringes on upperparts and wings and has thin white mantle- and scapular-Vs.

VOICE Call a soft, high-pitched, vibrant, rising 'trre-eep'

Temminck's Stint   Little Stint

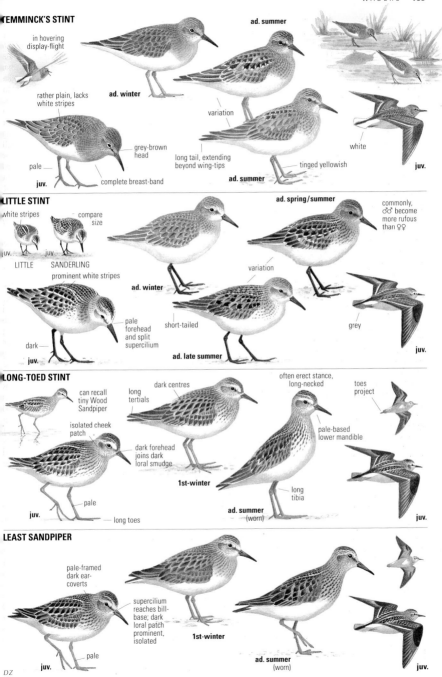

TEMMINCK'S STINT

in hovering display-flight

rather plain, lacks white stripes

ad. winter

ad. summer

grey-brown head

pale

variation

long tail, extending beyond wing-tips

tinged yellowish

white

juv.

complete breast-band

ad. summer

juv.

LITTLE STINT

ad. spring/summer

commonly, ♂♂ become more rufous than ♀♀

white stripes

compare size

juv. **juv.**

LITTLE SANDERLING

prominent white stripes

pale forehead and split supercilium

ad. winter

short-tailed

grey

dark

variation

juv.

juv.

ad. late summer

LONG-TOED STINT

can recall tiny Wood Sandpiper

dark centres

long tertials

often erect stance, long-necked

toes project

isolated cheek patch

dark forehead joins dark loral smudge

pale-based lower mandible

1st-winter

long tibia

juv.

pale

long toes

ad. summer (worn)

juv.

LEAST SANDPIPER

pale-framed dark ear-coverts

supercilium reaches bill-base; dark loral patch prominent, isolated

1st-winter

pale

juv.

ad. summer (worn)

juv.

DZ

Wood Sandpiper *Tringa glareola*   m**B**5 / **P**4

L 18½–21 cm, WS 35–39 cm. Breeds on bogs and marshes in taiga. On passage, frequents inland or coastal marshes, shallow pools, often in large flocks (except in NW); winters mainly Africa. Nest is usually in dense vegetation on ground.

IDENTIFICATION General shape similar to Redshank, but ⅓ smaller in size and bill length. Similar to Green Sandpiper, differing in *paler brown* colour; *larger pale markings on upperparts*; *longer legs*; *fine barring of outer tail*; breast and flanks loosely streaked and barred, *markings not forming a clear-cut breast-band*; light *supercilium extends behind eye*; and *yellowish-green legs*. Bobs rear body like Common Sandpiper when agitated. In flight, unlike Green Sandpiper, shows *pale underwing, whole of toes projecting beyond tail*, and no strong black/white contrast. – Adult summer: Upperparts coarsely and irregulary mottled whitish and buff. –Adult winter: Plumage greyer and breast rather plain. – Juvenile: Upperparts *densely*, neatly and *strongly spotted buff-white*. *Breast finely streaked*, not nearly uniform.

VOICE Flight-call distinctive, a quick, high-pitched whistling 'chiff-if-if', all syllables on same note. Alarm is persistent sharp 'gip gip gip...'. Song (in display-flight) a rapid burst of fluty, yodelling 'tü-**lüll**ti**lüll**ti**lüll**ti**lüll**ti...'.

Green Sandpiper *Tringa ochropus*   m**B**5 / **P**+**W**4

L 20–24 cm, WS 39–44 cm. Breeds in waterlogged wooded areas, on bogs and marshes. Summer visitor (mostly Mar/Apr–Aug); winters in S Europe and Africa. On passage, shows preference for ditches, pond edges, lakesides, etc., seldom open mudflats. Occurs singly or in small groups. Nest is in tree in old nest of other bird, usually thrush.

IDENTIFICATION Size and structure as Wood Sandpiper but *more robust*, with *slightly shorter legs*. Typically sighted when flushed unexpectedly from some unlikely ditch or puddle, flying off with distinctive calls, showing *blackish wings* (even below) and *white rump*, pattern like oversized House Martin; *toes barely project beyond tail*. Outer tail has just *a few very broad, dense black bars*, looking *all dark*. On ground, differs from Wood Sandpiper in *much finer speckling* and darker colour which make *upperparts look uniform blackish*; *dark breast clear-cut against white underparts*; eye-ring white; *white supercilium does not extend beyond eye*; and *greenish-grey legs*. – Adult winter: Breast and upperparts almost uniform dark. – Juvenile: Upperparts *neatly and finely spotted* with buff-white (thus more like Wood).

VOICE Flight-call a clear, ringing whistle, '**tlu**eet-wit-wit', first note with rising inflection, the next two higher-pitched. Alarm-call a clipping 'tlip-tlip-tlip-...'. Song (in display-flight) more complex than Wood Sandpiper's, a rapid, rhythmical phrase, '**tluu**i-**tü**i **tluu**i-**tü**i **tluu**i-**tü**i **tluu**i-**tü**i...'.

Common Sandpiper *Actitis hypoleucos*   m**B**3 / **P**4 / **W**

L 18–20½ cm, WS 32–35 cm. Breeds near water in forested areas, preferring stony or gravelly shores, most common at lakes and rivers in taiga, also at coasts, in archipelagos. Summer visitor (mostly Apr/May–Aug), winters in Africa, rarely in S Europe. Migrates singly or in small parties frequently by night, when revealed by its calls. Nest is placed in vegetation, usually in forest, close to water.

IDENTIFICATION Told by behaviour and shape long before plumage is seen: a medium-small wader with horizontal semi-crouched carriage, short-necked, *long-tailed*, rather short-legged, which habitually *bobs rear body*, especially after every quick movement or on landing (the other sandpipers on this page share this habit, but in less exaggerated form). Flight action highly distinctive, with *flightpath close over water* and quick bursts of *shallow, pulsating beats interrupted by fleeting glides on stiffly down-bowed wings*. Brown above, with *distinctive white divide between carpal area and clear-cut breast-side patch*; legs greenish, brownish or dull yellowish-grey. – Adult: Plainer than juvenile, especially in winter. – Juvenile: More obvious barring above, wing coverts and tertials finely tipped and edged pale buff.

VOICE Vocal. Flight-call a rapid series of clear, high-pitched whistling notes falling in pitch slightly, 'swee-swee-swee-swee-swüü', often heard on dark August nights. Alarm-call is a drawn-out whistling note, 'heeeep'. Song, in pulsating song-flight, is rhythmic repetition of finely twittering phrase, 'swidi**dii**-dide-swidi**dii**-dide-swidi**dii**-dide-...'.

Terek Sandpiper *Xenus cinereus*   **V**∗∗

L 22–25 cm. Breeds in lowland boreal taiga zone by rivers and oligotrophic lakes; also in harbours or at sawmill factories with floating logs. Winters in Africa, Asia and Arabia. On passage, seen mainly on coastal mudflats and pools, rare in W Europe. Nest is lined scrape in short vegetation.

IDENTIFICATION Shape rather like large Common Sandpiper, but with *very long upcurved bill*. Occasionally wags rear body, and habitually picks insects from surface with *quick, dashing actions*. In flight, *broad white trailing edge to* wing recalling Redshank's, but narrower and less contrasting; *rump and tail grey*. Upperparts and breast *pale grey*, with *dark carpal area* and (on adult summer only) prominent *irregular dark stripes on upper scapulars*. Bill black (adult summer) or with dull yellow or orange base; short legs dull or bright yellow, or (on adult) often orange.

VOICE Flight-call a rapid ringing series of 2–5 short, whistling notes, a high-pitched ringing 'vüvüvü', at times recalling Whimbrel, though softer. Song a repetition of a usually trisyllabic unit, e.g. 'ka klee-**rreee**, ka klee-**rreee**, ka klee-**rreee**, ...', somewhat recalling song of Stone Curlew.

Wood Sandpiper   Green Sandpiper

Common Sandpiper   Terek Sandpiper

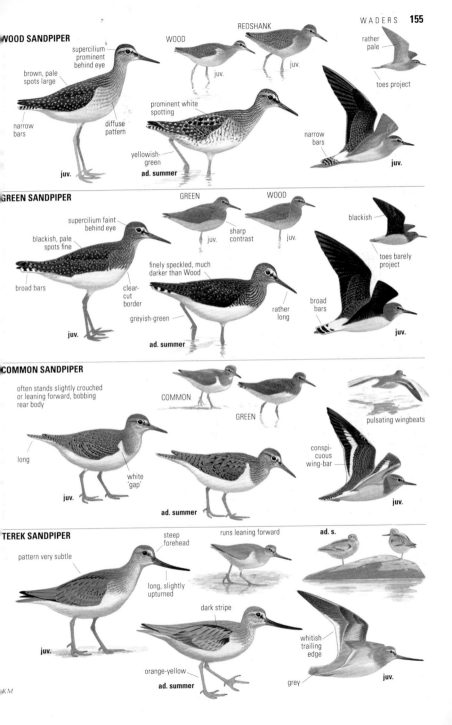

WOOD SANDPIPER

WOOD

REDSHANK

rather pale

supercilium prominent behind eye

brown, pale spots large

juv.

juv.

toes project

narrow bars

diffuse pattern

prominent white spotting

narrow bars

yellowish-green

juv.

juv.

ad. summer

GREEN SANDPIPER

GREEN

WOOD

blackish

supercilium faint behind eye

blackish, pale spots fine

sharp contrast

juv.

juv.

toes barely project

broad bars

clear-cut border

finely speckled, much darker than Wood

greyish-green

rather long

broad bars

juv.

ad. summer

juv.

COMMON SANDPIPER

often stands slightly crouched or leaning forward, bobbing rear body

COMMON

pulsating wingbeats

GREEN

long

conspi-cuous wing-bar

white 'gap'

juv.

ad. summer

juv.

TEREK SANDPIPER

steep forehead

runs leaning forward

ad. s.

pattern very subtle

long, slightly upturned

dark stripe

juv.

whitish trailing edge

orange-yellow

grey

juv.

ad. summer

KM

(Common) **Redshank** *Tringa totanus* r(m)**B**3 / **P**+**W**2

L 24–27 cm, WS 47–53 cm. Breeds on inland and coastal marshes, wet meadows and moorland. On passage and in winter, singly or in flocks mainly on or near coast; less commonly far inland. Wary and noisy. Nests on ground.

IDENTIFICATION Commonest medium-sized wader in most of region. *Red legs and bill-base* (shared only by Spotted Redshank, but beware that some Ruffs can be similar, having orange-red legs and bill-base). In all plumages, brownish above, with streaked breast and flanks, and whitish eye-ring and fore-supercilium. In flight, pointed white rump and diagnostic *white hind-wing* obvious. – Adult summer: Upperparts and underparts with irregular, coarse dark markings; often, whitish eye-ring only prominent feature on head. – Adult winter: Rather *plain grey-brown above and on breast*; flanks and undertail-coverts *only sparsely barred or streaked*. – Juvenile: Feather fringes on *upperparts neatly notched with buff*; neck and breast distinctly streaked; flanks and belly finely but *evenly patterned*; *legs yellow-orange* (less red than adult).

VOICE Call a distinctive, musical 'teu' or melancholy, downslurred 'teu-hu' or 'teu-huhu'. Alarm-call a persistent 'kyip-kyip-kyip...'. Song (often in flight) a loud, musical 'tül-tül-tül-tül-tul**iiu**-tul**iiu**-tul**iiu**-...' and 'l**ee**o-l**ee**o-l**ee**o-...'.

Spotted Redshank *Tringa erythropus* **P**+**W**4

L 29–33 cm. Breeds in open, arctic taiga near bogs and marshes, and on tundra. On passage, singly or (mainly in spring) in small groups, on inland and coastal marshes, lagoons, less commonly on coasts; ♀♀ move S as early as Jun.

IDENTIFICATION Like Redshank, but *outline slimmer* and more elegant, with slightly *longer legs* and *longer, finer bill* which has *distinctive but very subtle downward droop near tip*; also, *red on bill restricted to base of lower mandible* (not whole bill-base as Redshank); winter and juvenile plumages show much more *clear-cut white fore-supercilium*. When feeding, typically wades more deeply than Redshank, often swims and upends. Flight dashingly fast and direct; wings plain, with *white rump extending in thin 'slit' far up back*. – Adult summer: ♂ *all black* (with fine white speckling above); legs blackish. ♀ similar, but with extensive white flecking on ♀ flanks, belly, undertail-coverts. – Adult winter: Pale grey above, with *clear white underparts* and whitish head/neck (Redshank obviously browner above and on breast); legs red. – Juvenile: *Underparts rather uniformly barred* (lacks Redshank's breast-band); feather edges on greater coverts and tertials more finely barred than adult winter. Legs red.

VOICE Call is very distinctive, quick, emphatic, disyllabic whistle, 'chu-it!'. Song (in flight) is a rhythmic, 'grinding' or rolling whistle, 'crru**eee**-a crru**eee**-a crru**eee**-a ...'.

(Common) **Greenshank** *Tringa nebularia* r+m**B**4 / **P**.

L 30–34 cm, WS 55–62 cm. Breeds mainly on dry ground in northern mature pine forests near bogs and water. On migration to Africa, usually travels alone or in small, loose parties only. Food insects and worms, but a very agile bird and also habitually dashes after small fish in shallow water.

IDENTIFICATION Slightly larger and more heavily built than Redshank, with slightly *longer and broader-based bill with slight or distinct upcurve*; bill-base and *legs grey* or *greenish grey*. At distance, adult winter and juvenile look grey above with *whitish head/neck*. In fast and powerful flight, wings dark, with *white rump extending in point up back*, and very *pale tail*. – Adult summer: Variable number of scapulars dark-centred, contrasting with otherwise brownish-grey upperparts; head, breast and flanks coarsely streaked and barred. – Adult winter: Rather *pale grey above* with fine scaly pattern; *foreneck and centre of breast white*. – Juvenile: Fairly *dark grey-brown above*, with neat pale fringes rather striped pattern; *breast neatly and uniformly streaked*.

VOICE Vocal. Call is a powerful, trisyllabic whistling 'tyew-tyew-tyew', all on one pitch (juv. often with shrill voice). Alarm a persistent 'kyu!'. Song (in flight, high up) is sustained grinding 'clu-**wee** clu-**wee** clu-**wee** clu-**wee** ...'.

Marsh Sandpiper *Tringa stagnatilis* **V**★★

L 22–25 cm. Breeds in grassy lowland marshes, including on steppe, or on open taiga bogs. Migrates through E Europe to winter in Africa and S Middle East (also in India), usually resting at inland marshes, ponds and mudflats.

IDENTIFICATION Size between Wood Sandpiper and Redshank, but *long, straight* (very subtly upturned on some) *needle-fine bill*, slim body and neck, and *very long, rather spindly legs* (esp. long tibiae) give extreme slenderness and elegance unlike any other *Tringa*, at times even reminiscent of Black-winged Stilt. In flight, dark wings, *long white 'slit' up back* (extending almost to nape!), and *long leg projection beyond tail-tip*. Diffusely white supercilium and feathering around base of bill prominent in all plumages. Legs dull yellowish (rarely orange on ad. summer) or greenish-grey. – Adult summer: Brown-grey above, *prominently spotted and barred black*. – Adult winter: Grey above, white below, and whitish head/neck, giving impression of small, 'delicate' Greenshank. – Juvenile: Upperparts dark with neat pale feather fringes and finely notched greater coverts and tertials, rather like juv. Greenshank but whiter centre of breast.

VOICE Flight-call 'kyew', with diphthong, or 'kyu-kyu-kyu', like Greenshank but quicker and higher-pitched. Song a melodious, rhythmic 'tu-**lee**-a tu-**lee**-a tu-**lee**-a...', recalling Redshank but more drawling and woeful in tone, and rapid twittering series, 'chip! chip! chip-ipepepepepepep'.

Redshank

Spotted Redshank

Greenshank

Marsh Sandpiper

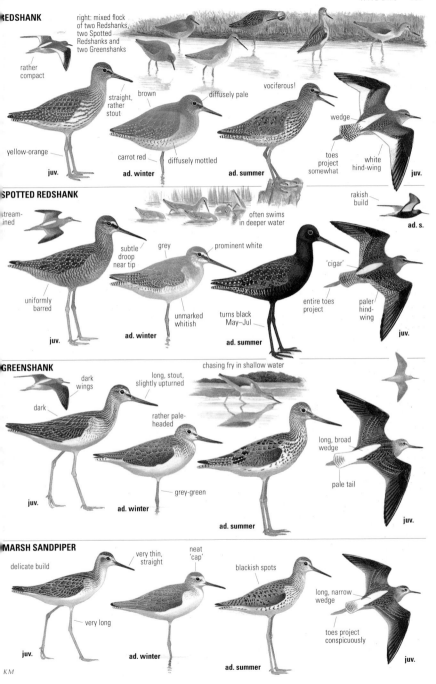

REDSHANK

right: mixed flock of two Redshanks, two Spotted Redshanks and two Greenshanks

rather compact

straight, rather stout

brown

diffusely pale

vociferous!

wedge

yellow-orange

carrot red

diffusely mottled

toes project somewhat

white hind-wing

juv.

ad. winter

ad. summer

juv.

SPOTTED REDSHANK

stream-ined

often swims in deeper water

rakish build

ad. s.

subtle droop near tip

grey

prominent white

'cigar'

uniformly barred

unmarked whitish

turns black May–Jul

entire toes project

paler hind-wing

juv.

ad. winter

ad. summer

juv.

GREENSHANK

dark wings

chasing fry in shallow water

dark

long, stout, slightly upturned

rather pale-headed

long, broad wedge

pale tail

juv.

grey-green

ad. winter

ad. summer

juv.

MARSH SANDPIPER

delicate build

very thin, straight

neat 'cap'

blackish spots

long, narrow wedge

very long

toes project conspicuously

juv.

ad. winter

ad. summer

juv.

KM

Black-tailed Godwit *Limosa limosa* r(m)**B**5 / **P**+**W**4

L 37–42 cm (incl. bill 8–11), WS 63–74 cm. Breeds on extensive wet meadows, grassy marshes, boggy moorland. On passage and in winter, frequents mainly estuaries, saltings, coastal mudflats and lagoons; also inland marshes.

IDENTIFICATION About the size of Whimbrel, but outline slimmer. When plumage differences not discernible, told from Bar-tailed Godwit on ground by *longer tibiae*, slightly *longer and straighter bill* (straight or very slightly upcurved, more obviously curved on Bar-tailed). Unmistakable in flight: *broad white wing-bar, black tail-band, square white rump*, and equally long projection of bill/head and tail/legs at each end of body giving *distinctive elongated look*. Bill in summer yellowish or orange-flesh with dark tip, pink-based in winter. – Adult summer: ♂ acquires much more extensive and brighter orange-rufous summer plumage than ♀; *belly and undertail-coverts whitish*, and *flanks barred* (ad. summer ♂ Bar-tailed entirely darker rufous below). – Adult winter: *Plain* brown-grey breast and upperparts (Bar-tailed streaked). – Juvenile: *Neck and breast tinged cinnamon-buff* (of variable strength); *wing-coverts pale-fringed with black subterminal spots* (Bar-tailed more streaked pattern). – Variation: Icelandic birds (ssp. *islandica*) have shorter bill and legs, and deeper rufous summer plumage below extending further down on the breast, than in rest of Europe (*limosa*).

VOICE Flight-call a quick, reedy, slightly nasal 'vi-vi-ve!' and a Lapwing-like 'vääh-it'. Noisy when breeding: song (partly in complex display-flight) opens with nasal, drawling 'kehi-**e**-itt kehi-**e**-itt kehi-**e**-itt …', turning into rapid, rhythmic 'weddy-**whit**-o weddy-**whit**-o weddy-**whit**-o …'.

Bar-tailed Godwit *Limosa lapponica* **P** / **W**3

L 33–41cm (incl. bill 7–11), WS 62–72 cm. Breeds on open tundra and taiga bogs. Migrates (late May and late Jul–Aug) mainly along coasts and over sea to and from winter grounds in W Europe and W Africa. Large flocks on some tidal flats.

IDENTIFICATION About equal in size to Black-tailed Godwit, from which told by *shorter legs* (esp. tibiae) and slightly shorter, usually *clearly upcurved bill* (beware that bill is longer on ♀ than on ♂ and hence more like Black-tailed). The two godwits best told by plumage: *barred tail*; white rump extending in *point up back* (as on Curlew); plain wings *lacking wing-bar*. Leg projection shorter than on Black-tailed. Bill dark with pink base. – Adult summer: ♂ has *whole underparts dark rufous* (darker than Black-tailed), whereas much paler underparts of ♀ (off-white, washed pale apricot) usually *entirely lack rufous*, and breast is streaked. Bill black with a little pale at base. – Adult winter: Upperparts pale brownish-grey, with dark feather centres giving *streaked effect* (Black-tailed plain); off-white below without

buff; faintly streaked on breast-sides only. Bill pink-based. – Juvenile: Like adult winter but darker above, with *neat buff-spotted edges to scapulars and* (most obvious) *tertials*; breast all streaked; buff wash on foreneck/breast (cf. ad. ♀).

VOICE Call is nasal 'cew**ee**-cew**ee**' (sharper than Knot). Song (in display-flight) is persistent, nasal 'kuwe-kuwe-kuwe-…', often preceded by series of rapid 'witwitwet'.

(Eurasian) Curlew *Numenius arquata* r(m)**B**2 / **P**+**W**

L 48–57 cm (incl. bill 9–15), WS 89–106 cm. Breeds in taiga on open bogs or arable fields along rivers, on wide coastal (usually wet) meadows, wide pastures, arable plain with patches of wet meadows, moorland, etc. On passage (mostly Apr and Jun–Sep) and in winter, usually in flocks on coasts or pasture. Often wary during breeding.

IDENTIFICATION *Largest wader, with unmistakable very long, evenly decurved bill* (longest on ♀, shorter on ♂ and juv.). Easily confused only with usually less common Whimbrel. Uniformly streaked and barred grey-brown, with no outstanding plumage features. Flight rather slow, gull-like. Note *pointed white rump* and darker outer primaries.

VOICE Call a far-carrying, rising, fluty, melancholy whistle, 'cour-**lii**'; on migration often a 'cue-cue-cew'. Alarm a fierce 'vi-vi-vü!'. Song (in flight with shivering wings, then descending glide) starts with drawling notes, merging into distinctive rhythmic, rippling trill, 'oo-ot, oo-ot oo-eet trru-ee trrru-eel trrrru-eel trrrru-eel trrru-uhl'.

Whimbrel *Numenius phaeopus* m**B**4 / **P**4

L 37–45 cm (incl. bill 6–9), WS 78–88 cm. Breeds on taiga bogs, on mountainsides above tree-limit (e.g. cranberry scrub), tundra. On passage (mostly late Apr–May and Jul–Aug) seen singly or in small parties; frequents mainly pasture and coasts; winters on African coasts, rarely in Europe.

IDENTIFICATION Like Curlew but slightly *smaller*, and has proportionally somewhat *shorter bill* (beware juv. Curlew, especially ♂, also with rather short bill), *dark crown-sides* and *rather dark eye-stripe* giving *strong supercilium* and (when viewed head-on) neat *light median crown-stripe* (beware faint crown-stripe often shown by Curlew, too). Wingbeats faster, and underwing looks on average slightly darker than Curlew's (wing-coverts are more patterned). – Variation: American birds (ssp. *hudsonicus*; possibly a separate species), rarely recorded in W Europe, lack white rump.

VOICE Call a loud, fast, rippling whistle, 'püpüpÿpÿpü-püpü', all on same note, vaguely recalling ♀ Cuckoo or Little Grebe. Alarm at breeding site similar to call but longer and not so even-pitched. Song begins like Curlew's but breaks into bubbling, long-drawn-out, straight trill, *lacking Curlew's pulsating rhythm*, and has more woeful ring.

Black-tailed Godwit

Bar-tailed Godwit

Curlew

Whimbrel

LACK-TAILED GODWIT

long, nearly straight

orange-buff

coarsely spotted

ven pale inges

plain

islandica
(Iceland)

ad. summer ♂

ad. summer ♂

ad. winter

toes project conspicuously

orange base

white square

longer than on Bar-tailed

black and white tail

white

barred

white wing-bar

juv.
islandica
(Iceland)

short super-cilium

ad. winter

spring flock

AR-TAILED GODWIT

somewhat shorter than on Black-tailed, and more upcurved

streak-ed

ad. summer ♀

ad. winter

toes pro-ject slightly

ale otch-ng

barred tail

dark

white wedge

lacks wing-bar

streaked

rufous

unbarred

juv.

ad. winter

longer super-cilium

ad. summer ♂

in winter

URLEW

rather plain head

sometimes hint of pale crown-stripe

moulting primaries (Sep–Oct)

autumn

rather short

♂

♀

often noticeably longer bill

juv. is finely streaked on breast-sides, not so 'spot-barred' as ad.

juv.

ad.

ad. ♀

VHIMBREL

prominent pale median crown-stripe

more distinct dark eye-stripe than on Curlew

no white

hudsonicus
(N America)

in autumn, juv. is neat and fresh, whereas ad. is worn and duller

juv.

ad. summer (worn)

migrants

ad.

M

(Eurasian) **Woodcock** *Scolopax rusticola*   r(m)**B**3–4 / **P**3

L 33–38 cm (incl. bill 6–8), WS 55–65 cm. Breeds in moist woodland (deciduous or mixed), interspersed with glades, rides or fields, with wet soil, shade and at least some undergrowth. On passage and in winter, also sometimes drier scrub or bushy terrain; usually solitary. Crepuscular habits; rarely active in daytime unless accidentally flushed. Best chance of seeing Woodcock is *dusk* visit to panoramic viewpoint in known breeding area, especially Apr–Jun when 'roding' ♂ patrols large area in level, direct flight over treetops. Nest is lined cup in sparse cover in shaded wood.

IDENTIFICATION Pigeon-sized, *thick-bodied* 'snipe of the woods'. If flushed from day roost, flies off with whirring, rather loud *wing noise*, twisting through trees, showing much *rusty-brown on rump and tail*. In roding flight, note unmistakable *fat-bodied, long-billed silhouette* with *stiff, flickering action* (actually rapidly doubled wingbeats) of *broad, blunt-ended wings*, slightly tail-heavy carriage and with bill pointing obliquely down. All plumages similar.

VOICE Roding ♂ repeatedly gives 3–4 grunted or growling notes followed by high-pitched, short explosive sound, 'wart wart wart-wart **pissp**!'; moderately far-carrying (300 m). When encountering rival in air, close pursuit accompanied by frenetic 'plip-plip pissp psi-plip...', etc.

Great Snipe *Gallinago media*   V∗∗

L 26–30 cm (incl. bill 6–7), WS 43–50 cm. Breeds on open wet meadows in lowland or (in N Europe) on mountainside, where ♂♂ meet on traditional leks. On passage frequents wet meadows, lush pastures and drier fields. Scarce; only a vagrant to British Isles (often Sep–Oct). Winters in Africa.

IDENTIFICATION Slightly *larger, stockier* and shorter-billed than Snipe, with flight steadier and not so high or lengthy, but beware slow-flying (tired or injured) and thus large-looking Snipe. Flushes (at *c*. 5 m) with *audible wing sound*; on upperwing, best clues are *white-bordered, dark midwing-panel* on greater coverts and *white-tipped primary-coverts* (not white trailing edge as on Snipe); *underparts and underwing coarsely barred* (not extensive white belly and striped underwing as on Snipe); and *much white on outer tail-feathers* (not visible in normal flight: best looked for when tail spread on landing or take-off). On ground, *white tips form obvious lines on wing-coverts*, and *underparts wholly barred* (only slightly less complete on centre of belly). All plumages similar; juvenile has less white on tail than adult.

VOICE Quiet; gruff croak occasionally given when flushed. Song from ♂♂ gathered at display 'arena' lasts 4–6 sec., in three merging parts: rapid rising then falling twittering ('bibbling'); accelerating clicking (like bouncing table-tennis ball); then whining 'whizzing' (audible *c*. 300 m).

(Common) **Snipe** *Gallinago gallinago*   r(m)**B**3 / **P**+**W**

L 23–28 cm (incl. bill *c*. 7), WS 39–45 cm. Breeds i marshes, bogs and damp meadows with short, dense vege tation; on passage and in winter, also small muddy patche pool margins, ditches, seashores, pastures, etc., often i small groups. Nest usually well concealed in vegetation.

IDENTIFICATION Medium-sized wader; prefers cover, bu when seen in open shows distinctive combination of *dispro portionately long, straight bill*, rather *dumpy shape, shor legged, crouching posture*, and *striped head and body*. Whe flushed (usually within 10–15 m), *zigzags fast*, giving *dis tinctive call* and showing *thin white trailing edge to wing*, the *towers quickly* on *twisting course*. Plumage basically brow with boldly striped head and upperparts. Flanks barre dark, *belly white* (which from some angles is difficult to se falsely 'signalling' Great Snipe). All plumages similar.

VOICE Flight-call on take-off is abrupt, scraping 'catch (like a rubber boot being pulled out of soft mud); tends to b disyllabic, 'ca-**atch**!' (like a muffled sneeze). Vocal son (from post or other perch) often long-sustained '**chip**-pe **chip**-per **chip**-per...'; but as main display has distinctiv 'drumming' sound, a throbbing, bleating 'hühühühühühü hühü...', produced by air vibrating through spread oute tail-feathers in intermittent short, steep dives during wide circling display-flight, mostly at dusk.

Jack Snipe *Lymnocryptes minimus*   P / W2–

L 18–20 cm (incl. bill 4), WS 33–36 cm. Breeds in extensiv waterlogged bogs in N Europe. On passage (mostly Apr early May and Sep–Oct) and in winter, muddy pool margir etc. Uncommon or local, but probably overlooked owing t skulking habits in often inaccessible patches.

IDENTIFICATION About ²⁄₃ size of Snipe, with *much shorte bill* (about 1½ times length of head; Snipe's bill at least twic head length), but which is proportionately *deeper-base* Skulking: *flushes only when almost trodden on*, flies u *silently*, showing more *rounded wing-tips, pointed tail* an less erratic flight than Snipe, *usually landing not far awa* When seen on ground, head shows strong pattern, with *spl supercilium, dark crescent below eye*, always strong dar loral patch, and *lacks median crown-stripe; breast and flank strongly streaked* (not barred); *bold yellowish back-stripe* green sheen on mantle and scapulars; and unique, ofte prolonged *bouncing action* when feeding, as if whole bod on springs ('sewing machine'). All plumages similar.

VOICE Rarely, gives quiet, harsh 'gatch' when flushed. I long, fast, shallow descent during display-flight, hollo 'ogogok-ogogok-ogogok-...', like distant cantering hors lasting *c*. 8 sec., followed by a series of muffled, high-pitche fizzing notes; sound distinctive, but hard to locate.

Woodcock

Great Snipe

Snipe

Jack Snipe

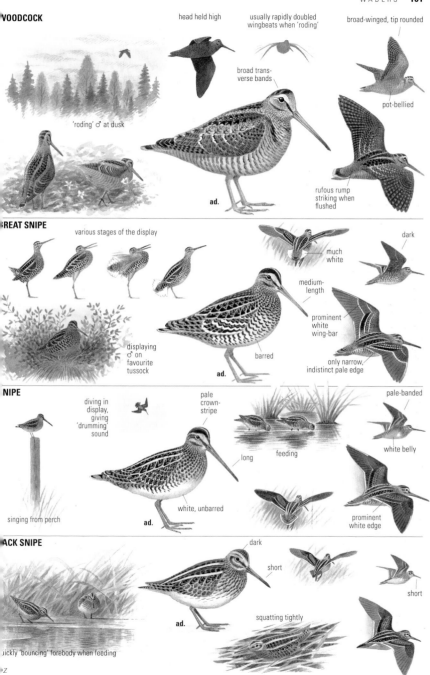

WOODCOCK

head held high

usually rapidly doubled
wingbeats when 'roding'

broad-winged, tip rounded

broad trans-
verse bands

pot-bellied

'roding' ♂ at dusk

rufous rump
striking when
flushed

ad.

GREAT SNIPE

various stages of the display

dark

much
white

medium-
length

prominent
white
wing-bar

displaying
♂ on
favourite
tussock

barred

only narrow,
indistinct pale edge

ad.

SNIPE

diving in
display,
giving
'drumming'
sound

pale
crown-
stripe

pale-banded

white belly

long

feeding

singing from perch

white, unbarred

prominent
white edge

ad.

JACK SNIPE

dark

short

short

quickly 'bouncing' forebody when feeding

ad.

squatting tightly

Z

Red Phalarope (Grey Phalarope)
Phalaropus fulicarius   **P**4–5

L 20–22 cm, WS 36–41 cm. Breeds in Arctic on mainly coastal wet tundra, sometimes in drier area, but always near fresh or brackish water. Winters in Atlantic off S and W Africa, via migration routes well out in N Atlantic. In W Europe, very rare in late spring; more regular Sep–Oct (esp. W coast Britain & Ireland), when gales bring singles or small groups (at times large numbers) close inshore, or odd birds to coastal pools. Nest in tussock near shallow water. Common sex roles reversed; more attractively plumaged ♀♀ gather in flocks to compete for ♂♂, which tend eggs/young.

IDENTIFICATION Dunlin-sized, usually very tame wader. Outside breeding areas, in black, pale grey and white plumage, often seen swimming (when shape and colour recall miniature Little Gull) or in flight over sea, when could be overlooked as Sanderling (but *white wing-bar pronounced only on the 'arm'*, thinner on primaries, dark 'mask' often discernible at long range, and looks *distinctly strong-chested*) until it *plops down and swims*; also, flying parties over sea usually looser, not so cohesive as Sanderling (and other *Calidris*). When swimming, pecks quickly all around for food on surface. In all plumages, told from Red-necked by *slightly thicker and* (when viewed head-on) *flattened bill*; identification difficult at distance if bill shape not discernible, especially if in moult (e.g. juvenile to 1st-winter), when plumage differences obscured. – Adult summer: Rarely seen away from breeding areas; unmistakable. ♂ has underparts often white-flecked and slightly less bright, and head pattern less clear-cut than ♀. Bill yellow with dark tip. In flight, distinctive *contrast between dark body and white underwing-coverts*. – Adult winter: *Almost plain pale blue-grey upperparts* and wings (Red-necked has whitish mantle-V and scapular fringes); *dark-bordered white crown* (ad. Red-necked white crown, juv. solidly black); bill dark, usually with hint of *yellow-brown base*. – Juvenile: Resembles juvenile Red-necked Phalarope, but *ochre-yellow mantle-V thinner and scapular-V usually lacking*, and rather quickly acquires its first pale grey 1st-winter scapulars (juv. Red-necked moults somewhat later). Has *light apricot-buff wash on neck/breast* (more pinkish-tinged grey at first on Red-necked).

VOICE Flight-call sharp, high-pitched, metallic 'pit' (can recall Coot). Song (by ♀ in slow-winged, circling display-flight) buzzing, far-carrying 'brrreep', vaguely recalling Broad-billed Sandpiper. Excited 'bip bip bip...' on water.

Red-necked Phalarope *Phalaropus lobatus*   m**B**5 / **P**4

L 17–19 cm, WS 30–34 cm. Breeds on wet marshes and pools and osier delta lands on mountainsides above tree limit or on tundra; extends farther inland and to higher altitude than Red Phalarope. Migrates in late Jul–Aug/early Se(earlier than Red), mainly non-stop towards SE across Eu rope to winter in Arabian Sea, where it leads pelagic life fa from land. Scarce on autumn passage, feeding at coasts inland marshes and pools. In W Europe rare in spring, to but in N Baltic and further east may occur in very large flock in late May. Nest in vegetation near water. Sex roles reverse as with Red Phalarope.

IDENTIFICATION General size, shape and behaviour much a described for Red, but a little *smaller and daintier*, wit slightly *shorter body* (less gull-like when swimming), an *bill very thin* (not laterally flattened) and always all black – Adult summer: Crown, cheeks, back of neck, breast-side and mantle lead-grey, *throat and small spot above eye whit* Variable amount of *red or rufous-ochre on sides of neck* neater and more intense on ♀, less colourful and with les clear-cut pattern on ♂. – Adult winter (rarely seen in Eu rope): Medium grey above, with *distinct whitish mantle-and scapular fringes* (Red Phalarope paler and almost plain – Juvenile: Unlike Red Phalarope, has prominent *ochrous buff* (or whitish when faded) *Vs above* on both mantle an scapulars, a *solid dark cap*, and a faint pinkish-grey hue c neck/breast-sides (which, however, quickly fades to white)

VOICE Common call is a short, nasal and slightly throat 'chep' or longer (almost disyllabic) 'cherre'. The throaty cal and sharper variants, 'chik', are used in chorus in display.

VAGRANT WADERS

Most of the waders on the following pages are vagrants Europe from N America and Asia (as are a few species o pp. 138–139, 144–145 and 152–153). Accordingly, the are unlikely to be seen, and will therefore require carefu observation and full description before they can be reporte confirmed, and details published. Some vagrants are unmis takable, but most are similar to regular European specie and identification requires much caution and thorough fam iliarity with commoner counterparts. Read the genera advice on wader identification on pp. 132–133, too.

Wilson's Phalarope *Phalaropus tricolor*   **V**

L 22–24 cm. Breeds in N America. Vagrant to Europe. A fe records annually in Britain & Ireland, mostly 1st-winte birds in Aug–Oct, seen at coastal pools, marshes.

IDENTIFICATION Slightly larger than the other phalarope and oddly proportioned, with *longer neck*, *longer, needle fine bill*, and *longer legs*. Swims, but also *commonly walks* shallow water or on muddy edges, snapping at insects *stealthy, crouched posture, rear end pointing up, breast almo touching ground*, or pecks with distinctive rocking-hors action as if whole body on pivot. *Legs* black on adult sum mer, otherwise *yellow*. In flight, *plain wings and squar white rump*. – Adult summe Unmistakable; ♀ brighter than ♂. – Adult winter: Unifor *very pale grey upperparts and wing-coverts*, white belov looks strikingly white at distance; long white superciliu diffuse greyish stripe on side of neck; *lacks dark head mark ings*. – Juvenile/1st-winter: Commonest plumage seen Europe: as adult winter, but brown, pale-fringed juvenil wing-coverts/tertials and often some scapulars retained.

VOICE Flight-call a short, nasal 'vit'.

Red Phalarope

Red-necked Phalarope

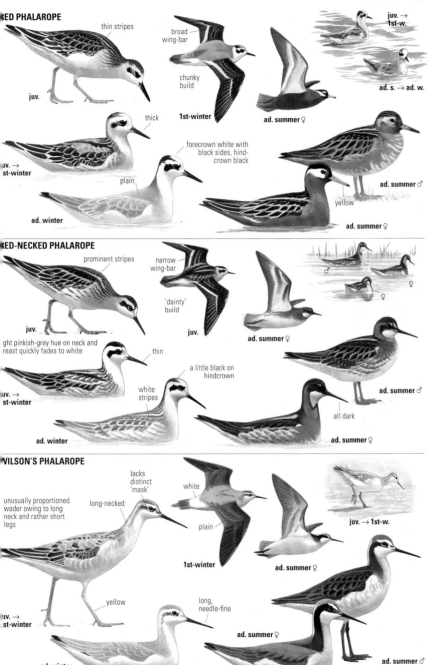

⌐ED PHALAROPE

thin stripes

broad wing-bar

chunky build

juv.

1st-winter

ad. summer ♀

juv. → 1st-w.

ad. s. → ad. w.

thick

forecrown white with black sides, hind-crown black

⌐uv. → ⌐st-winter

plain

ad. summer ♂

yellow

ad. winter

ad. summer ♀

⌐ED-NECKED PHALAROPE

prominent stripes

narrow wing-bar

'dainty' build

juv.

juv.

ad. summer ♀

♂

♀

♀

ght pinkish-grey hue on neck and ⌐reast quickly fades to white

thin

a little black on hindcrown

⌐uv. → ⌐st-winter

white stripes

ad. winter

all dark

ad. summer ♂

ad. summer ♀

⌐VILSON'S PHALAROPE

lacks distinct 'mask'

white

unusually proportioned wader owing to long neck and rather short legs

long-necked

plain

juv. → 1st-w.

1st-winter

ad. summer ♀

yellow

long, needle-fine

⌐uv. → ⌐st-winter

ad. summer ♀

ad. winter

ad. summer ♂

⌐M

Baird's Sandpiper *Calidris bairdii* V*

L 14–17 cm. Breeds in N America, NE Siberia. Vagrant to Europe. Often on dry or short-grass terrain but also beaches, especially with banks of storm-deposited seaweed.

IDENTIFICATION Slightly smaller than Dunlin, with delicate build like a stint. Unusual 'flattened oval' shape to body when viewed from in front or behind. *Wing-tips project beyond tail*, and *long-bodied, short-legged outline* often even more striking than on White-rumped Sandpiper, from which it differs in all plumages in *lack of white rump*, less prominent supercilium, *finer-based bill with straighter culmen*, and all-black bill. – Adult summer: Grey-buff upperparts with variable amount of dark feather centres; *buffy wash across breast, streaked brown at sides*. Beware of Sanderling in confusing, somewhat similar transitional plumages (Sanderling has chunky build, longer legs, stouter bill). – Adult winter: As adult summer, but upperparts plainer grey-brown with pale feather fringes. – Juvenile: General coloration on head and breast *usually strikingly buff*; upperparts a shade greyer, with broad whitish tips forming a *uniform scaly pattern* (similar to juv. Curlew Sandpiper), lacking obvious white Vs or rufous fringes; throat and indistinct supercilium pale buff; *usually complete gorget of brown breast streaking* on otherwise white underparts.

VOICE Flight-call short purring trill, 'prrreet'.

White-rumped Sandpiper *Calidris fuscicollis* V*

L 16–18 cm. Breeds in N America. Vagrant to Europe.

IDENTIFICATION A little smaller than Dunlin. With Baird's Sandpiper shares distinctive *long-bodied, rather short-legged outline*, unlike other small waders; *wing-tips project beyond tail*. In all plumages, white uppertail-coverts forming *U-shaped white patch*, prominent *whitish supercilium*, and medium-length, rather thick-based *bill slightly decurved at tip*; at close range, *brownish base to lower mandible*. – Adult summer: When fresh, greyish above with neat black feather centres; breast and *flanks streaked black*. Narrow white *mantle*, *stripes* and, when worn, *conspicuously pale bases to second lowest row of scapulars*. – Adult winter: Closely resembles winter Dunlin, best told by: small size; slim, attenuated shape; shorter bill. – Juvenile: Attenuated shape, *whiteness of underparts* and *prominence of supercilium* usually most striking differences from Dunlin. Pattern of upperpart feathers rather like juvenile Baird's, but usually *obvious rufous edges to mantle, upper rows of scapulars and tertials*. Also, has distinct, stint-like *white mantle-and scapular-Vs*, crown tinged rufous, and flanks finely streaked grey.

VOICE High, squeaky, mouse-orbat-like 'tzeet'.

Semipalmated Sandpiper *Calidris pusilla* V**

L 13–15 cm. Breeds in N America. Vagrant to Europe.

IDENTIFICATION Compare Little Stint (p. 152) and Western Sandpiper. Typically stockier than Little Stint, with *short, straight, thick-tipped bill* (on average thicker and straighter than Little's), but beware some longer-billed birds with slight decurve. *Toes half-webbed* (eliminates all stints except Western). – Adult summer: *No obvious rufous colour or light Vs on upperparts*, thus rather *drab grey-brown with black scapular centres*; rather *well-marked breast-band* with a few streaks extending to upper flanks. – Adult winter: Best told

by palmations and call; from Western by more *diffuse breast side streaks* and very *slightly longer primary projection* – Juvenile: Usually noticeably less bright and overall *mor uniformly patterned* than Little Stint. May show faint, diffuse pale mantle-V (often more obvious at long range) an slightly darker upper-scapular tract, but generally *lack contrasting rufous zones or distinct light Vs*. Ear-coverts an lores usually a shade darker, and crown more evenly streaked contrasting with rather clear-cut supercilium. Breast-side more diffusely streaked and *primary projection shorter tha* on Little Stint. A few are rather bright above, being ver difficult to distinguish from juvenile Western, but still ar more uniformly coloured, with upper rows of scapulars sam shade as crown (tend to be brighter rufous on Western).

VOICE Most distinctive flight-call a short, harsh 'tchrp' but also a more Little Stint-like 'tüpp'.

Western Sandpiper *Calidris maura* V***

L 14–17 cm. Breeds in N America. The rarest of the Ameri can stints in Europe.

IDENTIFICATION Compare Little Stint (p. 152) and Semi palmated Sandpiper. *Bill usually fine-tipped, slightly de curved and rather long*: this, and flat-backed, longish-legge shape, *typically recall miniature Dunlin rather than stint* Crown-sides and ear-coverts typically rather pale; whit supercilium prominent in front of eye. *Toes half-webbe* (eliminates all stints except Semipalmated). – Adult sum mer: Extensive rufous on black-centred scapulars, contrast with *rather plain grey wing-coverts*; usually *rufous on crown sides and ear-coverts*; breast streaking breaks into *obviou arrowhead markings on upper flanks*. – Adult winter: Ver difficult to distinguish from winter Semipalmated. Sharp fine streaks *often extend across breast*. – Juvenile: *Prominen rich rufous on upper scapulars* contrasts with basically *gre lower scapulars and wing-coverts*; lower scapulars typicall have *pointed* blackish centres and tips (broader, mor anchor-shaped black markings on Semipalmated); fain mantle- and scapular-Vs; neat breast-side streaks.

VOICE Flight-call a high, vibrant 'jeet'.

Red-necked Stint *Calidris ruficollis* V**

L 13–16 cm. Breeds in NE Siberia. Rare vagrant to Europe.

IDENTIFICATION Compare Little Stint (p. 152), and juve nile and adult winter Semipalmated Sandpiper. Usuall subtly but *distinctly shorter-legged/longer-bodied shape an* shorter-billed than Little Stint. – Adult summer: Throat upper breast *plain, bright orange, bordered below with black streaks or arrowhead marks*; orange tones extend to supercil ium and nape; scapulars black-centred with rufous fringes contrasting with plain or dark-centred *grey wing-coverts an tertials*; sometimes prominent *whitish mantle-V*. Orange o breast/head and rufous above often largely obscured by whitish fringes in fresh plumage. – Adult winter: Told fron Semipalmated by shape and *unwebbed toes*; probably no safely told from Little Stint, but possibly by shape (of classi individual) and call. – Juvenile: Like Little Stint, but lowe scapulars and wing-coverts greyer, thus *contrast* with black centred, rufous-fringed upper scapulars and mantle; man tle-V and split supercilium less prominent or lacking.

VOICE Flight-call a high-pitched, slightly hoarse 'chriit'

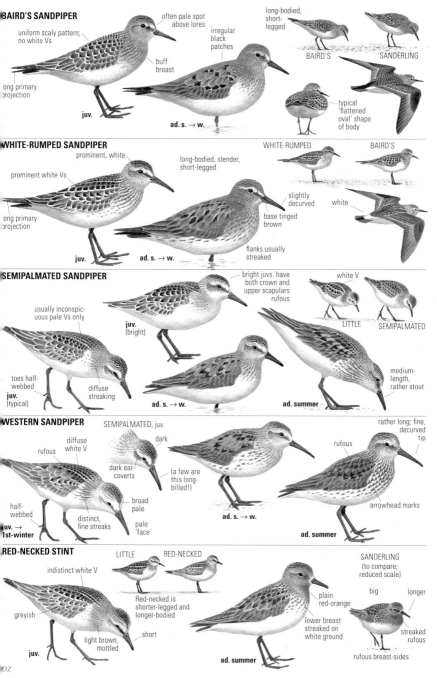

BAIRD'S SANDPIPER

uniform scaly pattern; no white Vs

often pale spot above lores

irregular black patches

long-bodied, short-legged

buff breast

long primary projection

juv.

ad. s. → w.

BAIRD'S

SANDERLING

typical 'flattened oval' shape of body

WHITE-RUMPED SANDPIPER

prominent, white

prominent white Vs

long-bodied, slender, short-legged

WHITE-RUMPED

BAIRD'S

slightly decurved

base tinged brown

white

long primary projection

flanks usually streaked

juv.

ad. s. → w.

SEMIPALMATED SANDPIPER

bright juvs. have both crown and upper scapulars rufous

white V

usually inconspic-uous pale Vs only

juv. (bright)

LITTLE

SEMIPALMATED

toes half-webbed

juv. (typical)

diffuse streaking

medium-length, rather stout

ad. s. → w.

ad. summer

WESTERN SANDPIPER

SEMIPALMATED, juv.

dark

diffuse white V

rufous

rather long; fine, decurved tip

rufous

dark ear-coverts

(a few are this long-billed!)

half-webbed

distinct, fine streaks

broad pale

pale 'face'

arrowhead marks

juv. → 1st-winter

ad. s. → w.

ad. summer

RED-NECKED STINT

LITTLE

RED-NECKED

indistinct white V

SANDERLING (to compare; reduced scale)

greyish

Red-necked is shorter-legged and longer-bodied

plain red-orange

big

longer

light brown, mottled

short

lower breast streaked on white ground

streaked rufous

juv.

ad. summer

rufous breast-sides

DZ

Ruff *Philomachus pugnax* mB5 / P+W4

♂ L 29–32, WS 54–60 cm, ♀ L 22–26 cm, WS 46–49 cm.
Not a rare vagrant but a common European breeder, treated here to facilitate comparison with some similar-looking species. Breeds on taiga bogs, in marshes and pools on mountainsides and tundra, and in wet grass meadows at lakesides or seashores. Summer visitor, winters mainly in Africa; on passage and in winter frequents marshes, shallow pools, estuaries, ploughed and stubble fields. Usually seen as a few or in smaller parties, but in spring at favoured sites can occur in very large flocks. ♂♂ display in communal lek on traditional arenas, which ♀♀ visit for mating. Arena, commonly used by *c.* 5–20 ♂♂, open grassy patch, often with bare soil on favourite spots, *c.* 1 m apart. Display silent, involves wing-flutter, short jumps, crouching with ruff erected, brief fluttering attacks on rivals with bill, legs or wings, sudden erect stance, or forward-bent stance with ruff raised and drooped wings and tail.

IDENTIFICATION Marked sexual size difference: ♂♂ (Ruffs) considerably larger than ♀♀ (Reeves). Largest ♂♂ slightly bigger and bulkier-bodied than Redshank; smallest ♀♀ only slightly larger-bodied than Dunlin (but with longer neck and legs). Small-headed/large-bodied look, *rather long neck* (when not hunched), and *medium-length bill* which, unlike some confusion species, is *slightly curved*, give rather distinctive shape which is especially useful for identifying some immatures or ♀ plumages that are otherwise confusingly variable in pattern, bare-part colours and size. Frequent habit (shared only with Black-tailed Godwit) of slightly raising mantle-feathers when feeding, appearing as *loose, pointed 'crest' at top of back*. Wingbeats subtly deeper and slower than other small or medium-sized waders, giving rather distinctive *lazier flight action*; often long glides before alighting. Thin, indistinct light wing-bar but *bold white oval patch on each side of uppertail*. Bill and leg colour variable: some non-breeding ♂♂ with bright orange bill-base and legs easily confused with Redshank (p. 156), but shape different, and *upperparts largely scaly* (with some coarse bars at rear), not largely plain (with some fine barring and spotting). – Adult summer: Display-plumage of ♂ has *erectable crest and ruff* in variable combinations of plain or coarsely barred *black, deep rufous, orange or white*; worn only in May–Jun and rarely seen away from breeding areas; bill, warty skin on face, and legs usually orange. ♀ brown, variably mottled or boldly spotted with black; typically, bill all dark but legs orange or dull red (rarely brown or greyish-green). – Adult winter: Sexes similar except for size. Bare parts usually duller and darker. ♂ has no crest, ruff or bare skin on face (but *pale feathering at bill-base*). Rather *plain pale grey-brown*, but some are more whitish on head and neck; tertials and large scapulars diffusely pale-fringed or *coarsely barred dark*. – Juvenile: Like adult winter, but often *stronger buff tinge below*, and *upperparts have neat scaly pattern with solidly dark feather centres*.

VOICE Nearly mute, has only low grunting sounds.

Ruff

Buff-breasted Sandpiper *Tryngites subruficollis* V

L 18–20 cm. Breeds in N America. Occurs annually in Britain & Ireland, often appearing in small groups. Often seen on short-grass fields (pasture, airfields, golf courses) but also on pools and mudflats on seashores.

IDENTIFICATION Dunlin-sized wader *preferring dry or short-grass terrain*. Like small juvenile ♀ Ruff, but *bill shorter, finer, straighter and black* (Ruff's usually pale-based); head and underparts more uniform sandy-buff, with *neat dark-streaked cap* and *black eye prominent on plain face*; *breast-sides neatly spotted*; and *legs mustard-yellowish*. In flight resembles miniature Golden Plover, *shows no white above* and has *white underwing with dark crescent* on primary-coverts. Adult has broadly pale-edged scapulars and plainer wing-coverts, whereas juvenile has neat and uniform scaly upperparts with intricately patterned wing-coverts.

VOICE Usually silent; low, muffled, monosyllabic 'gerf'.

Pectoral Sandpiper *Calidris melanotos* V

L 19–23 cm, WS 38–44 cm. Breeds in N America, NE Siberia. Commonest American vagrant to Britain & Ireland, most records Sep–Oct; in rest of Europe often in May.

IDENTIFICATION Slightly larger than Dunlin, with rather short, slightly decurved *pale-based bill*; general shape and flight pattern recall small Ruff, particularly when alert and adopts *erect stance with neck stretched*. In all plumages, has *sharply demarcated gorget of streaks* coming to point in centre of breast (belly thus unmarked), and *pale legs tinged yellowish* (at times more grey-green). – Adult summer: As juvenile, but rusty fringes and white Vs less prominent or lacking, less neat scaly pattern especially on worn wing-coverts, and supercilium less prominent. ♂ has coarsely mottled, not neatly streaked, breast-band. – Adult winter (extremely rare in Europe): Grey-brown above, with dark feather centres more diffuse than on juvenile. – Juvenile: Upperpart feathers blackish-centred with rufous fringes, forming neat scaly pattern; white mantle- and scapular-Vs usually prominent; creamy, finely streaked supercilium.

VOICE Flight-call a short, trilling, slightly throaty 'krrrt', a little like that of Curlew Sandpiper (or Baird's) but deeper.

Sharp-tailed Sandpiper *Calidris acuminata* V**

L 17–21 cm. Breeds in E Siberia. Rare vagrant to Britain & Ireland. European records in May and Jul–Sep.

IDENTIFICATION Resembles Pectoral Sandpiper, but in all plumages *lacks its distinctive breast pattern*, has *whiter supercilium* (esp. behind eye) against *clear-cut rustier cap* and more *prominent white eye-ring*. – Adult summer: Streaked upper breast, with variable but usually *extensive dark chevrons* (or arrowheads) *on breast-sides, flanks and undertail-coverts*. – Adult winter: Streaked breast, *extending as streaks or chevrons* (arrowheads) *onto flanks and undertail-coverts* (where streaks, if any, are very few and fine on Pectoral). – Juvenile: *Usually strong orange-buff wash on breast*, with *streaking confined to thin necklace across upper breast; cap brighter, more fiery-rufous* than on Pectoral, with *clear white supercilium very prominent*, especially behind eye.

VOICE Flight-call a very distinctive mellow, plaintive, rather subdued 'wheep' or 'pleep', often doubled or repeated several times; can recall Barn Swallow.

RUFF

juv. ♂

juv. ♀

juv. ♂

juv. ♀

ad. ♂

♂

♀

flock in spring

white sides

prominent feet

often raised, appear 'loosely attached'

autumn flock showing variation in size and plumage

plain

communal lek in spring

short primary projection

juv. ♀

ad. ♂ winter

pale bill-surround

ad. ♀ spring

ad. ♂ breeding

BUFF-BREASTED SANDPIPER

pale eye-ring

recalls small plover

neat scaly pattern

spotted sides

patchy

lacks obvious wing-bar

conspicuous dark crescent

juv.

long primary projection

juv.

mustard-coloured

ad. s. → w.

no white

PECTORAL SANDPIPER

SHARP-TAILED SANDPIPER

DUNLIN, juv.

short

juv.

point

inconspicuous wing-bar

'full' chest

ad. summer

PECTORAL, juv.

conspicuous white stripes

stronger head pattern than Pectoral

streaked necklace

breast streaks end abruptly

sharp 'cut-off'

plain orange-buff, streaking restricted to sides

juv.

longer primary projection than Ruff or Dunlin

white, unmarked

juv.

V-shaped markings

rufous

hint of 'spot-bars'

ad. s. → w.

ad. s. → w.

KM

Lesser Yellowlegs *Tringa flavipes* V*

L 23–25 cm. Breeds in N America. Fairly regular vagrant to Europe, mainly in late summer and autumn, and about 5–10 records annually in Britain & Ireland. Frequents short-grass marshes, muddy coastal pools, or inland marshy lakesides.

IDENTIFICATION Resembles Redshank but smaller (about same size as Marsh Sandpiper), with noticeably rakish build, *long*, *slender neck*, *long wings* and *long legs*. Unusual shape and proportions as likely to attract attention as any plumage feature or colour of legs. *Bill* virtually *straight*, *fine* and *practically all dark*; *long legs* always *bright yellow*. Distinguished from Marsh Sandpiper and Greenshank in flight by having *square white rump patch* (without pointed extension up back). Can be confused with Wood Sandpiper (which has similar white rump patch and rarely has dull yellowish rather than greenish legs) but, apart from differences in structure (note esp. Lesser Yellowlegs' *longer primary projection* and *extension of the wing-tips beyond tail*), Lesser Yellowlegs has *shorter supercilium* (not continuing behind eye) and lacks the distinct dark cap and two-tone bill of Wood Sandpiper. In flight, has much less compact look than Wood Sandpiper owing to longer neck, longer legs and longer wings. When size is not apparent can easily be confused with Greater Yellowlegs, but note: *slender build*; *narrow* and slightly *shorter bill* (only slightly longer than head; clearly longer on Greater); *straighter bill* (straight or only faintly upturned on Greater); and darker bill (all dark or with only hint of yellow-brown base; always paler-based on non-breeding Greater). Also, secondaries and primaries are uniformly dark (secondaries and inner primaries at least partly finely spotted light on Greater). – Adult summer: Unlike Greater Yellowlegs, has *flanks sparsely barred or unmarked*, and belly usually completely unmarked white. – Adult winter: Plumage much as Greater Yellowlegs. – Juvenile: Differs from adult winter in having darker mantle, wing-coverts and tertials with more contrasting light spots. Plumage similar to juvenile Greater Yellowlegs, but *breast more diffusely and finely streaked*.

VOICE Call rather high-pitched, clear 'tew', often uttered singly, sometimes 2–4 notes in quick succession at same pitch; voice a mixture of Redshank and Marsh Sandpiper.

Greater Yellowlegs *Tringa melanoleuca* V**

L 29–33 cm. Breeds in N America. Rare vagrant to Europe. Almost annual in Britain & Ireland.

IDENTIFICATION Greenshank-like, but in all plumages has *bright yellow legs* (sometimes slightly orange-yellow), *generally browner* (less grey) coloration, upperpart feathers more prominently notched with white, and *white rump not extending in V up back*. Can be confused with Lesser Yellowlegs, which see for detailed comparison, but *size and shape like Greenshank*, especially *stouter*, slightly *longer bill* (considerably more than head length), which is *obviously upturned* and *grey-based*. Caution: some Greenshanks have dull greyish-yellow legs, and some juvenile Redshanks have pale orange legs. – Adult summer: Much *more and coarser barring on flanks* than Lesser Yellowlegs, sometimes *also on belly*. – Adult winter: Largely grey above and white below. Like juvenile, but paler, greyer and less strongly patterned above. – Juvenile: Dark brownish above with neat pattern of

notches and spots on feather fringes; *breast is more strongly streaked* than on Lesser Yellowlegs.

VOICE Flight-call three-syllable whistle, very like Greenshank, but sometimes seems 'livelier', and slightly higher-pitched and almost invariably with lower-pitched third syllable, 'peu-peu-pew'. Typically somewhat louder and clearer than call of Lesser Yellowlegs (though some calls close), and three-syllabic pattern often good clue.

Solitary Sandpiper *Tringa solitaria* V***

L 18–21 cm. Breeds in N America. Vagrant to Europe, usually in summer–autumn.

IDENTIFICATION Very similar to Green Sandpiper except *lacks white rump* (rump is dark, with barred tail-sides) and strength of white spotting above intermediate between Green and Wood Sandpipers. In flight, *rather dark underwings* recall Green Sandpiper, but Solitary is noticeably slighter in build, with longer-looking, narrower wings. When seen well on the ground, *bill and legs average slightly longer*, and *bill usually slightly decurved* (unlike Green and Wood Sandpipers); *white eye-ring usually bolder than on Green Sandpiper*. – Adult summer: Head, neck and breast more coarsely streaked with white than Green Sandpiper; usually prominent white spotting on mantle and scapulars, with contrastingly darker-looking wing-coverts. – Juvenile: Compared with juvenile Green Sandpiper, a shade lighter brown above, with longer primary projection and extension of wing-tips beyond tail. Both have short supercilium and white eye-ring, but eye-ring often more conspicuous on Solitary Sandpiper.

VOICE Call like Green Sandpiper's in quality, but softer, 'tewit-weet'.

Spotted Sandpiper *Actitis macularius* V*

L 18–20 cm. Breeds in N America. Fairly regular vagrant to Europe, with annual records in Britain & Ireland, including one breeding attempt in Scotland; usually in late summer and autumn, but a few spring and overwintering records.

IDENTIFICATION Very similar to Common Sandpiper; in all plumages generally *greyer above* (less brown), *tail projection short*, and *short wing-bar* not reaching to body. – Adult summer: *Black spots below* diagnostic, but sometimes sparse and difficult to see. – Adult winter: Plain brown-grey above, no spots below. Very similar to Common Sandpiper; at very close range tertial edges unmarked (faintly barred on Common), but identification best confirmed by tail length, extent of wing-bar, and call. – Juvenile: *Tertials and greater coverts plain, with barring restricted to tips, leaving strongly barred area confined to median and lesser coverts* (tertial fringes and coverts fully barred on Common); legs pale, yellowish (usually dull brownish or greenish on Common). Juvenile also differs from Common Sandpiper in more uniform ('smoother') and greyer breast-side patches; bill often rather pale pinkish-horn with darker tip. In flight, shows less white on outer tail than Common Sandpiper (and seems less inclined than Common to spread tail).

VOICE Variable. Single vagrants often frustratingly silent! Can at times sound very like Common Sandpiper, but more often utters a short, whistled 'peet!', or doubled 'peet-weet' with tone reminiscent of Green Sandpiper.

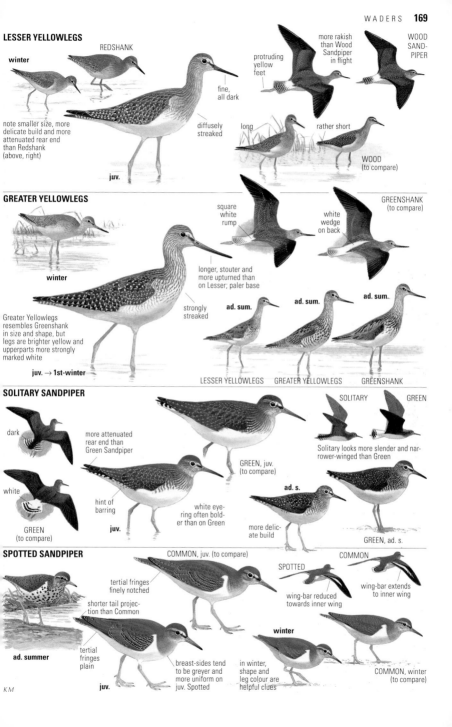

LESSER YELLOWLEGS

winter

REDSHANK

more rakish than Wood Sandpiper in flight

WOOD SAND-PIPER

protruding yellow feet

fine, all dark

note smaller size, more delicate build and more attenuated rear end than Redshank (above, right)

diffusely streaked

long

rather short

juv.

WOOD (to compare)

GREATER YELLOWLEGS

square white rump

white wedge on back

GREENSHANK (to compare)

winter

longer, stouter and more upturned than on Lesser; paler base

Greater Yellowlegs resembles Greenshank in size and shape, but legs are brighter yellow and upperparts more strongly marked white

strongly streaked

ad. sum.

ad. sum.

ad. sum..

juv. → 1st-winter

LESSER YELLOWLEGS   GREATER YELLOWLEGS   GREENSHANK

SOLITARY SANDPIPER

dark

more attenuated rear end than Green Sandpiper

SOLITARY   GREEN

Solitary looks more slender and narrower-winged than Green

white

hint of barring

GREEN, juv. (to compare)

white eye-ring often bolder than on Green

ad. s.

GREEN (to compare)

juv.

more delicate build

GREEN, ad. s.

SPOTTED SANDPIPER

COMMON, juv. (to compare)

COMMON

SPOTTED

tertial fringes finely notched

wing-bar extends to inner wing

shorter tail projection than Common

wing-bar reduced towards inner wing

winter

tertial fringes plain

ad. summer

breast-sides tend to be greyer and more uniform on juv. Spotted

in winter, shape and leg colour are helpful clues

COMMON, winter (to compare)

juv.

KM

Slender-billed Curlew *Numenius tenuirostris* **V*****(?)

L 36–41 cm (incl. bill *c.* 8), WS 77–88 cm. Now extremely rare or even extinct, the most recent confirmed or convincing sightings were in 1999 (Greece and Oman). Used to breed in W Siberia on remote taiga bogs, and possibly also at lakes in N Central Asia, and winter around Mediterranean and near coasts from NW Africa to Middle East, then frequenting meadows or saline steppe near marshes.

IDENTIFICATION Whimbrel-sized (p. 158), but slightly more delicate build, and *bill thinner* and *tapering to finer tip* (making it sometimes appear longer than it is). Ground colour on *sides of head, throat and breast whiter*, and black streaks more sharply contrasting than on both Whimbrel and Curlew; breast streaking gradually breaks up into diagnostic *distinct rounded or drop-shaped spots on lower breast and flanks*, contrasting sharply against white ground. Head pattern intermediate between Curlew and Whimbrel, has *rather dark loral stripe and crown-side* creating impression of *light supercilium* (though nothing like on Whimbrel, and lacks distinct pale median crown-stripe of that species). *Underwing-coverts all white*; rump white and *uppertail very pale* (looking whitish, paler than on Curlew and Whimbrel); secondaries and inner primaries distinctly barred light, giving obvious contrast with darker outer wing (but hardly a diagnostic difference unless directly compared with the other two species). – Juvenile: Flanks streaked, not spotted (some spots acquired in 1st winter). Caution: Curlew can have proportionately as short a bill, white underwing, and contrasting flank markings, so crucial to be sure of small size, different bill shape, and (on ads.) shape of flank spots.

VOICE Call is a Curlew-like 'cour-lee' but *sweeter, higher-pitched*, repeated *faster* and can turn into *a giggling trill*. Alarm in flight a sharper and shorter 'cu-ee'.

Little Curlew *Numenius minutus* **V*****

L 29–32 cm (incl. bill *c.* 4½), WS 57–63 cm. Breeds in N and NE Siberia, winters mainly in N Australia. Extremely rare vagrant to Europe (e.g. in Sweden, Norway) in late summer–autumn.

IDENTIFICATION Obvious curlew, but *tiny* (slightly smaller than Golden Plover) and has *small head*, rather *short neck* and a proportionately *much shorter and less decurved bill* than its congeners. In all plumages, *Whimbrel-like head pattern* (dark crown-sides, narrow light median crown-stripe), wholly *grey-brown underwing-coverts*, rather finely mottled *buffish underparts*, and *lacks white rump* (caution: Whimbrel of the N American ssp. *hudsonicus*, p. 158, lacks white rump, too; see also Upland Sandpiper). Forehead and lores often pale, and this together with prominent light buff supercilium gives *pale-faced impression*. Primaries and secondaries are rather uniformly dark, contrasting with *pale upperwing-coverts* on inner wing. Base of lower mandible pinkish. – Very similar also to exceedingly rare (possibly now extinct) N American relative Eskimo Curlew *Numenius borealis* (p. 419), but this has shorter legs, slightly longer, more downcurved bill, longer wings (wing-tips projecting beyond tail when folded), and darker-patterned underparts. At long range the short bill, buffish colours and general shape can recall a Ruff more than a curlew. Vagrants to W Europe sometimes associate with Golden Plovers on stubble fields,

then often spotted by steady and methodical gait when feeding (Golden Plover stands still, briskly walks a few steps, then stops again, etc.).

VOICE Flight-call is a rapid laughing 'kui-kui-kui-kui-kui-...', distinctly higher-pitched and softer than Whimbrel. In anxiety, may utter a slightly sharper 'quit-quit'.

Upland Sandpiper *Bartramia longicauda* **V****

L 28–32 cm, WS 50–55 cm. Breeds in N America. Vagrant to Europe, annual in Britain & Ireland. Prefers grass fields (pastures, airfields, golf courses) but will feed on stubble fields and low-growing crops, too.

IDENTIFICATION Redshank-sized wader with small head and shape like miniature Curlew or Whimbrel, thus not unlike Little Curlew, but with *short, straight bill* (with pale yellow-brown lower mandible), *very long tail extending well past wing-tips*, and long, *very thin neck* when alert; *no obvious white above in flight*. In all plumages has *dark cap* with *thin light median crown-stripe, prominent dark eye on plain face*, and dark arrowhead marks on breast and flanks. Legs are dull yellowish. Wing-feathers strongly barred. Wanders about, leisurely bobbing rear body when stopping. – Adult: Dark-barred scapulars, tertials and wing-coverts. – Juvenile: Pale-fringed feathers and pale-notched tertials, giving neater, stronger scaly pattern than adult.

VOICE Clear, whistled, bubbly 'quip-ip-ip-ip' in flight, strongly recalling ♀ Cuckoo or Little Grebe.

Stilt Sandpiper *Calidris himantopus* **V*****

L 18–23 cm, WS 37–42 cm. Breeds in N America. Vagrant to Britain & Ireland, mostly adults in late summer. Generally found with flocks of small sandpipers on coastal mudflats, but also mingles with *Tringa* species (e.g. Redshank) and Ruff, and then will also wade on deeper water.

IDENTIFICATION Slightly larger than Dunlin, but longer-winged, *longer-necked* and *much longer-legged*. Curved bill, general shape and *white rump* recall Curlew Sandpiper, but *legs noticeably longer*, giving stilt-like impression (and longer foot projection in flight), *legs yellowish to dull green* (not black), and *bill more even in thickness, less fine-tipped*, giving rather tubular impression. Distinctive body shape, remarkably *angular* in profile and peculiarly *laterally compressed* when viewed from in front or behind (not so rounded as e.g. Redshank). Wings rather uniformly dark above, *lacking obvious wing-bars*, but trailing edge rather pale. Often wades in deep water, and extraordinary length of legs may then not be obvious. – Adult summer: Partially or completely *barred underparts*; *rusty ear-coverts* contrasting with white supercilium. Indistinct short pale median crown-stripe towards rear of 'cap'. At long range looks plain and rather dark, prompting thoughts of dowitcher or even juvenile Spotted Redshank. – Adult winter: Plain grey-brown above; breast more extensively streaked than Curlew Sandpiper, extending to flanks and undertail-coverts. – Juvenile: Blackish feather centres and whitish fringes give less grey, more contrasting scaly pattern on upperparts than Curlew Sandpiper; prominent supercilium contrasts with dark cap and dark mask across ear-coverts.

VOICE Flight-call is a soft trilled 'trrrp' recalling that of Curlew Sandpiper.

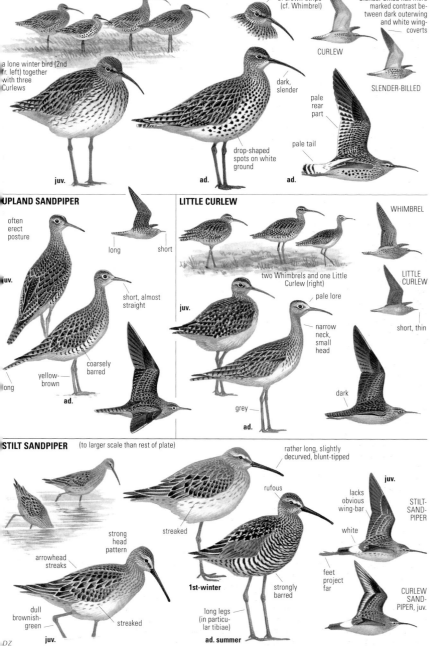

SLENDER-BILLED CURLEW

some have hint of crown-stripe (cf. Whimbrel)

compare with Curlew, Slender-billed has more marked contrast between dark outerwing and white wing-coverts

CURLEW

SLENDER-BILLED

a lone winter bird (2nd fr. left) together with three Curlews

dark, slender

pale rear part

drop-shaped spots on white ground

pale tail

juv.

ad.

ad.

UPLAND SANDPIPER

often erect posture

long   short

juv.

short, almost straight

long

coarsely barred

yellow-brown

ad.

LITTLE CURLEW

two Whimbrels and one Little Curlew (right)

juv.

pale lore

narrow neck, small head

grey

dark

WHIMBREL

LITTLE CURLEW

short, thin

ad.

STILT SANDPIPER   (to larger scale than rest of plate)

rather long, slightly decurved, blunt-tipped

rufous

streaked

juv.

lacks obvious wing-bar

white

feet project far

STILT-SAND-PIPER

strong head pattern

arrowhead streaks

1st-winter

strongly barred

dull brownish-green

streaked

juv.

long legs (in particular tibiae)

ad. summer

CURLEW SAND-PIPER, juv.

DZ

Long-billed Dowitcher *Limnodromus scolopaceus* **V***
L 27–30 cm (incl. bill 5½–7½), WS 42–49 cm. Breeds in N
America and E Siberia. Vagrant to Europe; several records
annually in Britain & Ireland.
IDENTIFICATION Dowitchers are Snipe-sized, with shape,
actions and plumage suggesting cross between Snipe and
Bar-tailed Godwit; very *long bill*, prominent supercilium,
greenish moderate-length *legs* and (in flight) *thin white 'slit'
up back* and *prominent whitish trailing edge to wing*. Typical
of dowitchers is that *blunt-tipped bill is a fraction decurved
distally. Given good views, identification of juveniles straight-
forward, but adult summer and adult winter difficult, often
requiring diagnostic voice* to confirm. Bill lengths overlap
extensively, and useful only for extremes: *twice head length or
more indicates* Long-billed, about 1½ times head length
indicates Short-billed; on Long-billed *black bars on tail-
feathers broader than light bars*, on Short-billed black bars
narrower than light bars (those with equal-width bars are
indeterminate). – Adult summer: Underparts (often except
undertail-coverts) rusty-orange, *including belly*, with *dense
spotting on foreneck and upper breast*, and *strong barring* or
transverse blotching *on breast-sides, flanks and undertail-
coverts.* –Adult winter: Flank barring and *whole breast* grey,
with *rather abrupt border against white belly.* – Juvenile:
Tertials, greater coverts and scapulars *solidly dark-centred*,
with neat (sometimes finely scalloped) whitish or rufous
fringes (or sometimes small internal markings near tips).
VOICE Single shrill, short, sharp 'yip' or 'kyip', quality
recalling Wood Sandpiper or distant Oystercatcher; often
calls singly every 2–3 sec. in flight, or quickly doubled or
trebled ('kyip-ip' or 'kyipipip'), especially when alarmed.

Short-billed Dowitcher *Limnodromus griseus* **V*****
L 25–29 cm (incl. bill 5–6½). Breeds in N America. Very
rare vagrant to Europe, much rarer than Long-billed Dowitcher. Document any presumed sighting well!
IDENTIFICATION See Long-billed for general characters of
dowitchers, and for bill-length and tail-pattern differences.
– Adult summer: Base colour of underparts usually washed-
out orange (paler than Long-billed), either with *faintly
spotted breast* (looking plain orange at distance), *sparsely*
barred flanks and *spotted* undertail-coverts (ssp. *hendersoni*)
or with *white belly* and at times stronger markings on breast,
flanks and undertail-coverts as Long-billed (ssp. *griseus*),
although *spots on sides of breast are on average more rounded
blotches*, with less transverse barring. – Adult winter: Near-
identical to Long-billed. Flank barring and upper breast
grey, breaking up into *very fine mottling and speckling on
lower breast* before white belly. – Juvenile: Breast and upperparts usually tinged stronger rufous than on Long-billed;

tertials, greater coverts and scapulars have *obvious internal
rufous cross-bars, spots or stripes.*
VOICE Flight-call is a fast, clear double or triple note,'tüdü
or 'tüdlü', with rattling quality like Turnstone's flight-call.

Pin-tailed Snipe *Gallinago stenura* —
L 25–27 cm (incl. bill *c.* 6). Breeds in N Ural range and E
through N Siberia, in open tundra and flat river valleys,
winters in S Asia; rare vagrant to E Europe and Middle East.
IDENTIFICATION Similar to Snipe (p. 160), but note: *very
thin and diffuse pale trailing edge to wing* (not prominent
white as on Snipe); median and lesser coverts form *paler
panel on upperwing*; bill slightly *shorter* and rather stout,
especially at base; rather *short-tailed*, and *toes often project
beyond tail-tip* in flight; *wings rather blunt-tipped*, and primaries more brownish (darker on Snipe); *primary-coverts
narrowly tipped pale* (on average more so than on Snipe);
underwing darker and uniformly barred (lacking white bands
of Snipe); *creamy stripes on upperparts not so prominent* as
on Snipe, and *different voice*. When seen on ground, main
colour of *back is brown with much barring and spotting* (not
so unmarked brown-black as on Snipe), median and lesser
coverts are rather *barred* brown (not predominantly spotted
and streaked as on Snipe), and lower scapulars are fringed
whitish *around tips* (not broadly fringed white only on outer
webs and ochrous-brown on inner webs as on Snipe). *Thin,
pin-like outer tail-feathers* unique.
VOICE Call when flushed is short, sharp, harsh and throaty,
monosyllabic 'chree!', slightly lower-pitched and coarser
than Snipe's. The 'song', delivered in dives during display-
flight, mostly in evenings, is a peculiar scratchy, sharpening
sound culminating in vibrating and whizzing notes, shorter,
more high-pitched and faster than in Swinhoe's Snipe.

Painted Snipe *Rostratula benghalensis* —
L 23–26 cm. Breeds in Egypt in overgrown swamps, pools,
ditches; vagrant in Israel. Rather secretive, and rarely seen
unless flushed. As with phalaropes, sex roles reversed; ♀ may
lay several clutches, incubated by different ♂♂.
IDENTIFICATION Roughly Snipe-sized, but looks distinctly
heavier, more lethargic when it takes off on *broad round-
tipped wings*, with long legs obvious, *entire toes projecting
beyond tail-tip*. Shortish, rather *drooping bill which is fairly
thick and pale* (dull pinkish); *thick light eye-ring continuing
in stripe behind eye*; and *yellow-buff breast/shoulder-stripe
joining mantle-V* striking (especially on ♀). Narrow buff
median crown-stripe. Underwing white, broadly bordered
dark. – Adult ♂: Dull *grey-brown on throat* (upper throat
mottled white), *neck and upper breast; narrow grey-black
breast-band*. Indistinct buff-white eye-stripe. Upperwing
and scapulars brownish-green, *wing-coverts broadly barred
buff.* –Adult ♀: Brighter than ♂; *rufous on throat, neck and
upper breast*, bordered with a *wide black band across breast*.
Distinct white eye-stripe bordered blackish. *Upperwing and
scapulars rather uniform greenish*, finely vermiculated dark.
– Juvenile: Resembles adult ♂ but breast more diffusely
mottled, lacking distinct border towards belly, and upperparts have more buff barring.
VOICE ♀ gives low hooting 'koot' in Woodcock-like roding
display-flight at dusk; otherwise mainly silent.

Pin-tailed Snipe   Painted Snipe

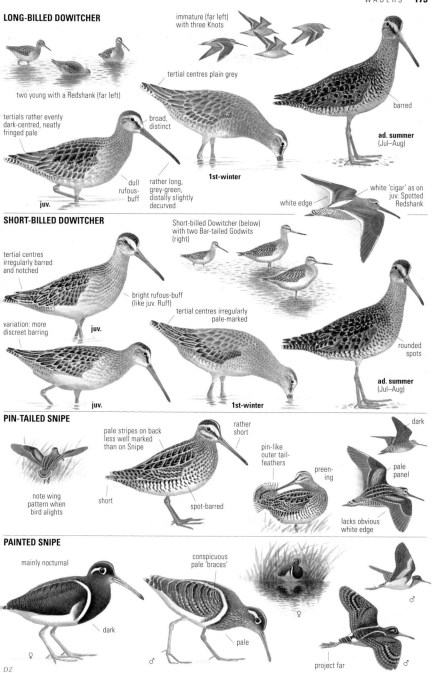

LONG-BILLED DOWITCHER

immature (far left) with three Knots

two young with a Redshank (far left)

tertial centres plain grey

barred

ad. summer (Jul–Aug)

tertials rather evenly dark-centred, neatly fringed pale

broad, distinct

dull rufous-buff

rather long, grey-green, distally slightly decurved

juv.

1st-winter

white 'cigar' as on juv. Spotted Redshank

white edge

SHORT-BILLED DOWITCHER

tertial centres irregularly barred and notched

Short-billed Dowitcher (below) with two Bar-tailed Godwits (right)

bright rufous-buff (like juv. Ruff)

variation: more discreet barring

juv.

tertial centres irregularly pale-marked

rounded spots

ad. summer (Jul–Aug)

juv.

1st-winter

PIN-TAILED SNIPE

pale stripes on back less well marked than on Snipe

rather short

dark

pin-like outer tail-feathers

preen-ing

pale panel

note wing pattern when bird alights

short

spot-barred

lacks obvious white edge

PAINTED SNIPE

mainly nocturnal

conspicuous pale 'braces'

♀

dark

♀

pale

♂

♂

project far

♂

DZ

SKUAS *Stercorariidae*

Medium-large, darkish, gull-like birds, mostly maritime. Breed in N Europe on tundra, coastal moors, remote islands, etc. Migrate along coasts and at sea, and to a lesser extent overland, to winter in N Indian Ocean and off W Africa, S America and southern N America, where they lead a pelagic life. Food during breeding lemmings, eggs and young of other birds, fish, insects and berries, during rest of year mainly fish; also other small animals, carrion and offal. Fish generally obtained through piracy from other birds.

Predatory habits reflected by hooked tip of bill and sharp, curved claws. Piratical chases of other birds performed with impressively rapid and acrobatic flight; victim, usually gull or tern, pursued incessantly and very closely, the skua demonstrating fabulously quick reactions, until victim disgorges the contents of its crop, which are caught in the air by the pursuing skua. Auks and other fish-eating birds are also attacked. Gulls and terns generally fly up when a skua approaches, as for a bird of prey.

Skuas are mainly grey-brown and off-white with a light pattern at the primary bases. Dark, light and intermediate morphs occur in some species. Sexes alike. Adults of the smaller species have long, projecting tail-feathers of characteristic shape. Often rest on sea, but alight 'reluctantly' after long glide and short hovering.

FIELD IDENTIFICATION OF SKUAS

The successful identification of skuas in the field requires a lot of practice. Below, some advice is offered, addressed especially to the beginner.

Most skuas look dark in the field, and this generally holds true for young and light-morph birds as well. They normally *appear even darker than most juvenile gulls*, a common alternative at long range (>1 km) or at tricky angles. As a rule, the lighter head and back of young gulls are visible even at a distance, as are the more conspicuous light rump and, on most, paler inner primaries. Apart from plumage distinctions, skuas are recognized by *more pointed 'hand'* (not so valid for the broad-winged Great Skua, of course), faster flight, more elegant movements, in relaxed flight *more shallow, even and powerful wingbeats*; in short, more 'purposeful' flight in comparison with the rather deliberate, almost lazy (in strong winds somewhat hesitant) wingbeats of gulls, which also are deeper in corresponding relaxed flight.

An Arctic Skua is chasing a tern to force it to disgorge the fish in its crop.

The remote risk of confusion with a shearwater remains to be eliminated, especially in strong winds and at long range. In gales, the skuas, too, practise the shearwater mode of banking flight: a fast low glide before the wind along the waves and a steep climb obliquely into the wind, banking to one side, again followed by low glide before the wind, etc. As soon as the bird beats its wings (usually immediately before the climb) the experienced eye can tell them apart: the flight of skuas is falcon-like, buoyant and 'athletic' with rather slower and shallow wingbeats; the *shearwaters give quick bursts of beats* (though somewhat slower for the larger species) with their narrower and *straighter wings*.

The most difficult part of skua identification is to separate non-adults of Arctic, Pomarine and Long-tailed Skuas. The crucial points to note are given in the species accounts. Some general advice: *Size* is, as always, difficult to be certain of when watching a bird through a telescope far out over the sea, but an attempt should be made to establish whether the bird is decidedly large or small. Note that the dark plumage of skuas gives a misleading impression when comparing with the whiteness of gulls: dark birds seen against a bright sky appear larger than white ones; and, conversely, dark birds against a dark sea often appear smaller than white ones, which appear big, especially so if seen in strong light.

Then pay attention to *structure and proportions* (such as width of the 'arm') and to *mode of flight*. Check whether the bird has a central tail projection of any kind, and whether there are any prominent light areas in the plumage. For birds approaching closer, it is useful to note prominence of primary patches, size and coloration of bill, and shape of short tail projections (if any). And do not expect to be able to identify all skuas in flight at the beginning. It takes a great deal of experience before the more difficult individuals can be sorted out. Even experts (at least sensible experts!) leave some birds unidentified.

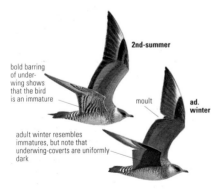

2nd-summer

bold barring of under-wing shows that the bird is an immature

moult

ad. winter

adult winter resembles immatures, but note that underwing-coverts are uniformly dark

Ageing of skuas (Arctic Skua shown above) in autumn and winter is made easier if you pay attention to the underwing and moult.

Great Skua

Pomarine Skua

GREAT SKUA

juv.

ad.

juv.
(dull)

on average
smaller
wing
'flashes'
than ad.

imm. / ad. (dark)

short
tail

extent of pale streaking
and blotching very
variable, probably
age-related; sequence
of immature plumages
poorly understood

wing 'flashes'
conspicuous at
long range

juv.
(bright)

head
darker
than
body

barrel-shaped

plain

ad. in threat/
display posture

very
heavy

ad.
(pale)

variable pale
markings

conspicuous white
wing 'flash'

Great Skua attacking Gannet

juv.

plain (unbarred)
underbody and
undertail-coverts

dark

ad.

KM

Great Skua *Stercorarius skua*   mB4 / P3

L 50–58 cm, WS 125–140 cm. Breeds on rocky islands and damp, elevated coastal moors, usually in loose colony; winters in N and S Atlantic. In Britain & Ireland, breeds in Scotland and seen irregularly along coasts, especially during or after autumn storms. Food mainly fish, taken from surface of sea or from other birds, or behind trawlers. Nest a shallow depression, defended with fierce dives.

IDENTIFICATION Large and heavy, roughly of Herring Gull size. In flight, *all dark with large white primary patches above and below*. Bill heavy. *Head and neck powerful*. Central tail-feathers broad, tips rounded, not or only slightly projecting. Soars more frequently than congeners, can give Buzzard-like impression. Floats high on sea. – Adult summer: Coarsely streaked yellow-brown on nape, neck and upperparts, ♀ on average more so. – 1st-year: More uniformly brown than adult and *tinged reddish-brown*, especially below; the white 'flashes' on upper 'hand' often less extensive and prominent than on adult. – Main risks of confusion with young large gull (esp. an oiled one) and with juvenile Pomarine Skua or, at distance, adult dark-morph Pomarine. Told from young gull by *flight being more steady and purposeful, wings beating more flexibly* (wingbeat pace otherwise similar); proportionately *heavier body* ('flying barrel'); *broader wing-bases* but often more pointed 'hand'; and—most reliably—*white wing patches* (see above). From juvenile Pomarine by *heavier body and flight; lack of light cross-barring* (thus never has contrasting pale rump, vent or underwing-coverts); often *slightly darker head than rest of body* (Pomarine usually uniformly dark, or head a shade paler than chest); broader and proportionately somewhat shorter wings; and generally clearly larger and purer white wing patches reaching right in to arm. (However, some juv. Great Skuas have confusingly restricted patches on upperwing.)

From adult dark-morph Pomarine by silhouette and flight, and certainly by lack of broad, long tail projections (though can be absent through moult on Pomarine).

VOICE Silent away from breeding grounds. There, single or series of short, slightly nasal 'gok' can be heard; in display utters squeezed, rolling 'chirr'.

Pomarine Skua *Stercorarius pomarinus*   P4

L 42–50 cm (excl. ad. summer tail projections 5½–11), WS 115–125 cm. Breeds on arctic tundra; passes N and W Europe on migration, mostly in May and Oct–Nov, to and from winter areas e.g. off W Africa. Rare but regular on passage in Britain (esp. E and S coasts) & Ireland. Usually seen singly or in small groups, sometimes very large parties (50+). Juveniles can be quite fearless.

IDENTIFICATION In flight, appears definitely larger than Common Gull and *slightly smaller than Herring Gull*. On the ground surprisingly small, smallest birds appear hardly as big as Kittiwakes but always look heavy-chested. Long-winged; *'arm'* rather broad, appears somewhat wider than distance from rear edge of wing to tip of tail (projections disregarded); 'hand' pointed, much as on Arctic Skua. *Flight relaxed and steady*, wingbeats measured and path more direct, not light and jerky with energetic wingbeats as sometimes with Arctic. In all plumages, *light primary patch above and below* roughly as on Arctic and considerably less than on normal Great. – Adult summer: *Long and broad tail projections*, look blunt-ended in profile (central tail-feathers twisted 90°; 'carries a spoon'); moulted twice annually, Nov–Dec and Mar–Apr (and at other times very rarely broken off). Two morphs: light (commoner) has dark cap (reaches below gape), dark and *coarse* breast-band (on ♂ fairly often broken on centre of breast, and can be missing), insignificant or prominent barring of flanks, and

POMARINE SKUA

juv.

2nd-s.

ad. s.

some ♂♂ lack breast-band

immatures show barred underwing

ad. s. (dark)

often pale-based under primary-coverts

square-ended

broad 'arm'

heavy bill, pale-based

juv. (intermed.)

juv. (dark)

ad. s. (pale)

pale, barred

juv. (pale)

subtly spotted

large head

spoon-shaped

dark bill-surround

dark

juv.

juv. (pale/intermed.)

ad.

usually barred

light grey

ad. summer (pale)

juv. (dark/intermed.)

KM

extensive dark vent distinctly demarcated (element of light barring on many); dark morph is all dark (except for wing patches). – Juvenile: Varying from mid brown with yellowish-buff bars, especially on rump and vent, to uniform dark blackish-brown, but *almost always looks dark in the field*. Told from similar juvenile Arctic Skua by *calmer, more steady flight; broader 'arm'; fuller belly; heavier bill* being more noticeably *pale blue-grey* (readily visible at 300 m); *tipped dark*; light morph has pale 'double patch' on underwing, i.e. *both primaries and primary-coverts basally light* on most (hard to judge beyond 300 m); and *blunt-ended, broad central tail-feathers which project only insignificantly, if at all*. Juvenile practically never has strongly contrasting pale head or belly (cf. Arctic and Long-tailed). Darkest dark morphs unbarred like dark adult. – 2nd cal.-year, light morph: Neck and belly paler, central tail-feathers shortish or half-length and not twisted. – 3rd cal.-year, light morph: Neck and belly pale, dark cap beginning to show, but underwing-coverts still barred, not all dark as on adult. – Adult winter, light morph: Similar to 3rd cal.-year, tail projections being half-length, but told by all-dark, unbarred underwing-coverts.

Arctic Skua *Stercorarius parasiticus*   m**B**4 / **P**3
L 37–44 cm (excl. ad. summer tail projections 5–8½), WS 108–118 cm. Breeds on tundra, coastal moors and barren islands, locally in loose colony; winters mainly S of the Equator. In Britain fairly common locally in N Scotland. Nest a shallow depression. Often keeps watch from top of island or a moorland ridge.

IDENTIFICATION Like a *dark* gull with pointed wings and *fast, flexible, almost falcon-like flight*. *light primary patches* and on adult *half-length, pointed tail projections*. In flight, appears to be of *Kittiwake size*. Wings long and fairly narrow, 'hand' pointed; width of 'arm' appears

equal to or slightly less than distance from rear edge of wing to tip of tail (projections disregarded). Flight lighter than Pomarine's, but beware, may at times show some weightiness; in strong winds Arctic recalls a Kittiwake in lightness, quick wingbeats and sudden lunges, whereas Pomarine is decidedly heavier and more steady, like a large gull. *Bill a little finer* than on Pomarine. Plumages much as for Pomarine—thus large variation—but note the following distinctions: dark cap of adult less black, does not solidly surround gape, and always excludes a *small light patch above base of bill*; if breast-band present, usually a dark ill-defined shade, not coarsely patterned; palest juveniles lighter than Pomarine, have *contrasting light head and neck* (finely streaked; discernible at close range) and *often belly, too*. Darkest juveniles extremely similar to darkest Pomarine, most reliably identified by bill size, tail shape and overall proportions; note also pointed central tail-feathers projecting 1–3 cm, forming tiny *'double point'* at rear of tail, primary-coverts below lacking pale base (exceptions: 1 out of 20) and *bill generally looking dark*, not strikingly pale-based (tendency to paler base hardly visible beyond 150 m).

VOICE Most often heard is a nasal, mewing 'eh-glow', repeated a few times in display. Alarm a short 'pjew!'.

Long-tailed Skua *Stercorarius longicaudus*   **P**4
L 35–41 cm (excl. ad. summer tail-streamers 12–24), WS 105–112 cm. Breeds on heathy fells above tree-limit and on drier tundra; winters mainly in S Atlantic. Feeds largely on lemmings in summer, and accordingly fluctuates in numbers. Scarce passage migrant in W Europe in May and Aug–Sep (Oct, rarely later). Occasionally, juveniles rest on inland fields, appearing fearless.

IDENTIFICATION *Small and slim*, body like Black-headed Gull, but *long, slender wings* give a larger impression in

ARCTIC SKUA

juv. (inter-med.)

usually conspicuous white 'flash'

pointed

juv. (inter-med.)

2nd-s.

imm. has barred underwing

ad. s. (dark)

long, pointed

juv. (dark)

blackish

ad. s.

sharp division

ad. s. (pale)

juv. (pale)

rufous fringes

pale tips

streaked

rather slender, only tip darker

pale and dark morphs often pair

pale bill-surround

juv. (intermed.)

generally warmer, more rusty-tinged than juv
Long-tailed

little or no barring

juv. (dark)

ad. summer (pale)

LONG-TAILED SKUA

usually more slender-winged and long-tailed than Arctic (but recently fledged can look fairly dumpy)

juv. (intermed.)

pale breast patch

blunt tip

juv. (dark)

lacks extensive wing 'flash'

very long, flexible

ad.

brown-grey

ad.

dark grey

no sharp division

ad.

lacks prominent wing 'flashes'

juv. (intermed.)

juv. (pale)

no pale tips

fairly short, rather thick, about half black

rather uniform

juv. (intermed.)

generally more greyish-brown than Arctic, not rufous

distinctly barred

juv. (dark)

juv. and ad.

ad. summer

KM

flight. *Flight light*, almost tern-like. Inclined to stall and settle on the sea, even on migration. Often hovers on breeding grounds. – Adult summer: Distinct black cap, *contrast above between dark remiges and paler brown-grey wing-coverts and back*, *no light primary patch below*, and *very long and pointed tail-streamers*. – Juvenile: Variation as pronounced as in Arctic Skua; told from that by silhouette (slim body, narrow wings, longer-looking rear end) and flight, and in light and median morphs *greyer, less brown general coloration* (but dark morph very like Arctic), by similarly (1–3 cm) protruding central tail-feathers being rather rounded, and by *bill* being *short* with *pale base*. A few have longer stubby 'points' at rear of tail than any juvenile Arctic.

VOICE Higher-pitched than that of congeners, repeated 'kew' and 'kriep' and quarrelling 'kre-krep'. In display wailing '**glee**-ah', reminiscent of Common Gull.

Arctic Skua

Long-tailed Skua

GULLS *Laridae*

Some of the region's 19 breeding species are very common, and frequently occur near humans, with the result that gulls are a familiar group of birds. For breeding, they require habitats near coasts, marshes or inland waters, often forming large colonies. Nest is a collection of twigs and plant material sited on bare ground or among low vegetation.

Especially outside the breeding season, some species are common inland, and they also then form large nocturnal roosts on sheltered coastal waters, or on lakes or reservoirs. The distance between roosting and feeding areas can be as much as 40 km: morning and evening flights are often high, with flocks frequently in lines or V-formations. The smaller species are generally rather dainty feeders, picking from the surface (sometimes plunging more deeply for fish, etc.), and in summer large numbers often gather to hawk flying ants high in the sky. The larger species are scavengers or take fish, and are often predatory on eggs and young of seabirds. Gulls are attracted to fishing ports, trawlers, rubbish tips and sewage outfalls, and some species form large flocks, mainly in search of worms, on arable fields and pasture, especially when ploughing is in progress.

IDENTIFICATION OF GULLS

Adult gulls are usually fairly easy to identify, because each species has a diagnostic combination of size, upperpart colour, wing-tip pattern, and bill and leg colour. The immatures are generally more difficult, having comparatively nondescript brown or barred plumage, and bill and leg colours which have yet to acquire adult coloration. The great variety of immature plumages can at first seem very complex and daunting, but the picture will be greatly clarified if the two following areas of basic information about gull plumages and moult are fully understood:

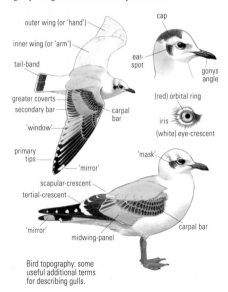

Bird topography: some useful additional terms for describing gulls.

(1) Length of immaturity

Each species belongs to one or other of three categories:

Two age-groups. This category contains most of the small species. Black-headed Gull is a common example of a species with two age-groups. Adult plumage is acquired when the bird is just over one year old. Thus, for most of the year, only two different age-groups are recognizable, i.e. 1st-year (juv., 1st-winter, or 1st-summer) and adult (ad. winter or ad. summer). Note that during Jul–early Sep, however, three age-groups can be recognized: juvenile (recently fledged); 1st-summer (just over one year old); and adult summer (just over two or more years old).

Three age-groups. This category contains the medium-sized species and Little Gull. Common Gull is a familiar example of this category. Adult plumage is acquired at just over two years of age. Thus, for most of the year, three age-groups are recognizable, i.e. 1st-year (juv., 1st-winter, or 1st-summer); 2nd-year (2nd-winter or 2nd-summer); and adult (ad. winter or ad. summer). During Jul–early Sep, however, four age-groups can be recognized: juvenile (recently fledged); 1st-summer (one year old); 2nd-summer (two years old); and adult summer (three or more years old).

Four age-groups. This category contains all the large species. Herring Gull is a common example of a gull with four age-groups. Adult plumage is acquired at just over three years of age. Thus, for most of the year, four age-groups are recognizable, i.e. 1st-year (juv., 1st-winter, or 1st-summer); 2nd-year (2nd-winter or 2nd-summer); 3rd-year (3rd-winter or 3rd-summer); and adult (ad. winter or ad. summer). During Jul–early Sep, however, five age-groups can be recognized: juvenile (recently fledged); 1st-summer (one year old); 2nd-summer (two years old); 3rd-summer (three years old); and adult summer (four or more years old).

(2) Moult

The change from juvenile to adult plumage is achieved by regular moults, in which old feathers are replaced by new ones. The first moult of all gulls is the *post-juvenile moult*, which for most species takes place shortly after the bird first flies: it is a partial moult, replacing only the juvenile head-and body-feathers, and results in 1st-winter plumage. In every year following the one in which it hatched, every gull has a *spring moult* (a partial moult, in which only the head-and body-feathers are renewed) and an *autumn moult* (in which the whole plumage is renewed).

In all moults, the feathers are replaced a few at a time. The partial moults take 1 or 2 months, and complete autumn moult takes 3 or 4 months. In the case of an immature gull, each moult brings its appearance gradually closer to that of the adult. In adults, summer plumage is acquired by the spring moult, and winter plumage by the autumn moult. The diagram (opposite) shows the complete sequence of moults (and the resulting appearance from juvenile to adult) in gulls with two, three, and four age-groups. Note that the wing-and tail-feathers are retained for a whole year, and by the summer can appear as a very worn and faded version of their original pattern, especially on first-summer birds.

Like the plumage, the colour of the bill and legs of immatures changes gradually until the adult colour is acquired, and thus provides further clues to the age of a young gull.

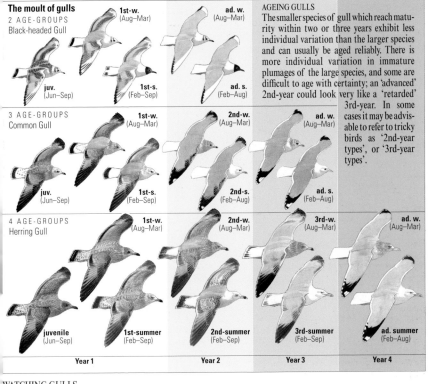

The moult of gulls

2 AGE-GROUPS
Black-headed Gull

juv.
(Jun–Sep)

1st-w.
(Aug–Mar)

1st-s.
(Feb–Sep)

ad. w.
(Aug–Mar)

ad. s.
(Feb–Aug)

AGEING GULLS
The smaller species of gull which reach maturity within two or three years exhibit less individual variation than the larger species and can usually be aged reliably. There is more individual variation in immature plumages of the large species, and some are difficult to age with certainty; an 'advanced' 2nd-year could look very like a 'retarded' 3rd-year. In some cases it may be advisable to refer to tricky birds as '2nd-year types', or '3rd-year types'.

3 AGE-GROUPS
Common Gull

juv.
(Jun–Sep)

1st-w.
(Aug–Mar)

1st-s.
(Feb–Sep)

2nd-w.
(Aug–Mar)

2nd-s.
(Feb–Aug)

ad. w.
(Aug–Mar)

ad. s.
(Feb–Aug)

4 AGE-GROUPS
Herring Gull

juvenile
(Jun–Sep)

1st-w.
(Aug–Mar)

1st-summer
(Feb–Sep)

2nd-w.
(Aug–Mar)

2nd-summer
(Feb–Sep)

3rd-w.
(Aug–Mar)

3rd-summer
(Feb–Sep)

ad. w.
(Aug–Mar)

ad. summer
(Feb–Aug)

Year 1	Year 2	Year 3	Year 4

WATCHING GULLS

Perhaps the best time to study gull plumages is in winter, when their patterns are relatively fresh after the complete autumn moult. Find a close flock to start with, and begin to identify the two age-groups of Black-headed Gull (1st-winter and ad. winter), the three age-groups of Common Gull (1st-winter, 2nd-winter and ad. winter), and the four age-groups of Herring Gull (1st-winter, 2nd-winter, 3rd-winter and ad. winter). In spring, note the progress of the spring moult of adults (Black-headed acquiring its hood, or Herring Gull losing its head streaks and acquiring an all-white head). In summer and autumn, notice how the wing pattern of immatures becomes sometimes much worn and faded,

and also how gulls look very ragged in flight as their wing- and tail-feathers are renewed during the autumn moult.

Such studies will increase your skill at identifying and ageing common species, and thus also increase the chance of noticing some rare gull among them. Note that a few gulls, especially immatures of species with four age-groups, can be very tricky to identify or age, even for the expert.

Because they are common, large and approachable, gulls provide valuable exercises in studying moult and the progress of immature plumages. This knowledge can also be applied to many birds other than gulls, so time spent studying gulls is worthwhile not only for its own interest.

Two adult summer Lesser Black-backed Gulls *Larus fuscus graellsii*. Note that tone of grey changes to some extent with angle of view (or light), important to bear in mind when looking at gulls.

juv.

1st-summer

Note that retained juvenile wing- and tail-feathers can become much worn and faded by 1st-summer plumage, when they are one year old, as shown on these Kittiwakes.

Black-headed Gull

Chroicocephalus ridibundus   r(m)**B**2 / **W**1

L 35–39 cm, WS 86–99 cm. Common colonial breeder at lakes in vast reedbeds or marshy areas, also on ponds near coasts. Migratory in far north, retreating in winter from ice. Abundant feeder on ploughed fields and in towns. Not shy.

IDENTIFICATION Two age-groups (see p. 178). In flight, told instantly from other common gulls by *white leading edge to outer wing*; this is its best field mark, often being visible at extremely long range. Similar wing pattern shared only by Slender-billed Gull and vagrant Bonaparte's Gull; see also Grey-headed Gull (p. 413). Also note: *obvious blackish area bordering white leading edge of underwing; pointed* (rather tern-like) *wings*; and proportionately *smaller head, longer neck* and *shorter tail*: these all add to distinctive appearance compared with other common gulls. When standing, also told by combination of small size, *head pattern* (hood or dark ear-spot) and (on adult) *reddish bill and legs*. – Adult summer: *Dark brown hood* (often looks black) and *dark, dull red bill and legs*. – Adult winter: *Bold dark ear-spot* and *red or brown-red legs and bill*, latter with dark tip. – Juvenile (Jun–Sep): Looks strikingly 'different', having extensive ginger-brown upperparts and head markings, brown wing markings, black tail-band, and yellowish-flesh legs and bill, latter with a dark tip. –1st-winter: Juvenile wings, tail, and bill and leg colour retained, but head and body like adult winter. –1st-summer: Like 1st-winter, but many acquire partial hood (at times full hood like ad. summer); brown on wings often faded and much reduced in extent (look for brown-centred tertials); and bill-base and legs more orange-red.

VOICE Noisy at colonies and when feeding in flock; the noise from large colonies in spring can be earsplitting. Calls include strident, downslurred, single or repeated 'krreearr', with many variations, and short, sharp 'kek' or 'kekekek'.

Slender-billed Gull *Chroicocephalus genei*   **V*****

L 37–42, WS 90–102 cm. Habitat and habits much as Black-headed Gull, but more coastal outside breeding season. Rather scarce even in Mediterranean breeding areas, where much rarer than Black-headed. Vagrant to S Britain.

IDENTIFICATION Mainly two age-groups (see p. 178). A little larger than Black-headed Gull, which it closely resembles in wing pattern and general appearance, but *head white* (lacks hood) in summer; has only *faint ear-spot, if any*, in winter; and has usually obvious *yellowish or whitish iris* (however, can look dark-eyed depending on light; Black-headed Gull always dark-eyed). Also very important for identification is its *peculiar head-and-bill shape*, produced jointly by *longer* (but actually not more slender) *bill*, more *elongated forehead*, and very long neck when fully extended; beware

Black-headed Gull   Slender-billed Gull

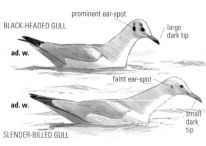

prominent ear-spot

BLACK-HEADED GULL

large dark tip

ad. w.

faint ear-spot

ad. w.

small dark tip

SLENDER-BILLED GULL

Slender-billed Gull has longer bill, forehead and, when alert longer neck than Black-headed Gull.

that Black-headed can give a rather similar impression (but never quite so exaggerated), so important to check other differences carefully to avoid misidentification. Legs comparatively long. Adult summer differs further in: usually *strong pink wash on underparts*; and darker red bill (often looking blackish) and legs. Adult winter has less or no pink, has a pale grey ear-spot (if any), and less dark red bill and legs. In addition to the structural differences from Black-headed, 1st-year has usually paler brown wing markings, paler ear-spot (if any); *paler yellowish-brown or orange-flesh bill and legs*, and *bill with dark tip much smaller or lacking*. – Some 2nd-year birds told by dark-centred tertials.

VOICE Vaguely recalling Black-headed's, but the rolling call is a lower, drier and harder 'krerrr' with a somewhat strained voice. At breeding site some low, gruff, nasal calls.

Bonaparte's Gull *Chroicocephalus philadelphia*   **V****

L 31–34, WS 79–84 cm. Vagrant from N America; a few records annually in Britain & Ireland.

IDENTIFICATION Two age-groups (see p. 178). Small version of Black-headed, which it closely resembles in upperwing pattern and general appearance. Small size and quicker wingbeats may suggest Little Gull. Lacks Black-headed's dark grey on outer underwing, thus *from below primaries are translucent white* or pale grey *with distinct black trailing edge*. Beware of 1st-summer Black-headed with faded and translucent primaries, and which are often slightly smaller than adults! Other all-plumage differences from Black-headed are its *proportionately smaller bill*, which is *black* or mainly black; proportionately *shorter legs*; and slightly darker grey upperparts which accentuate the white leading edge on the outer upperwing. Adult summer differs further in *blackish hood* (not brown), usually pink-flushed underparts (unusual on summer Black-headed), and more orange-red or pinkish legs. Adult winter usually has *pale flesh-pink legs* (not red). 1st-year differs from Black-headed in blacker (less brown) upperwing markings; *different pattern on primary-coverts*, inner being whitish, outer marked dark (1st-year Black-headed rather dark inner, white central, and one or two dark-marked outer); larger ear-spot; and grey often extending strongly onto hindneck and breast-sides (head / neck pattern thus can recall ad. winter Kittiwake).

VOICE Common feeding call a high, nasal, tern-like 'chirp', quite unlike Black-headed Gull's.

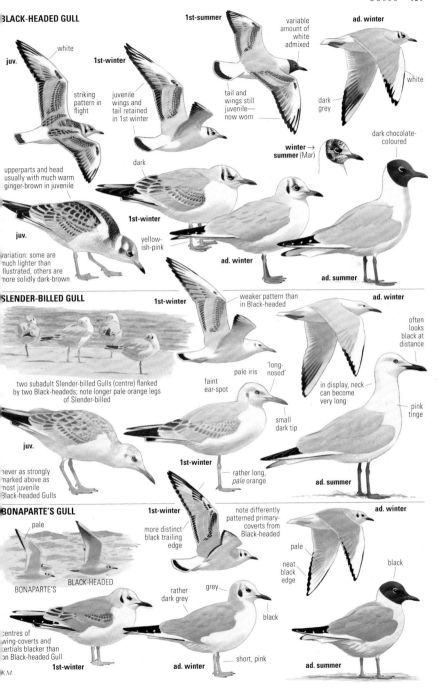

BLACK-HEADED GULL

juv.

white

striking pattern in flight

1st-winter

juvenile wings and tail retained in 1st winter

dark

1st-summer

variable amount of white admixed

tail and wings still juvenile— now worn

ad. winter

white

dark grey

dark chocolate-coloured

upperparts and head usually with much warm ginger-brown in juvenile

juv.

variation: some are much lighter than illustrated, others are more solidly dark-brown

1st-winter

yellow-ish-pink

**winter →
summer** (Mar)

ad. winter

ad. summer

SLENDER-BILLED GULL

1st-winter

weaker pattern than in Black-headed

ad. winter

often looks black at distance

two subadult Slender-billed Gulls (centre) flanked by two Black-headeds; note longer pale orange legs of Slender-billed

faint ear-spot

pale iris

'long-nosed'

small dark tip

in display, neck can become very long

pink tinge

juv.

never as strongly marked above as most juvenile Black-headed Gulls

1st-winter

rather long, *pale* orange

ad. summer

BONAPARTE'S GULL

pale

BLACK-HEADED

BONAPARTE'S

1st-winter

more distinct black trailing edge

note differently patterned primary-coverts from Black-headed

ad. winter

pale

neat black edge

black

centres of wing-coverts and tertials blacker than on Black-headed Gull

rather dark grey

grey

black

short, pink

1st-winter

ad. winter

black

ad. summer

KM

Common Gull (Mew Gull) *Larus canus* r**B**3 / **W**2

L 40–46 cm, WS 100–115 cm. Breeds colonially or singly along coasts, on islands, in marshes, along rivers or at inland lakes. Lined nest usually near water on ground, preferably somewhat elevated, on boulders, poles in harbours, occasionally low trees, roofs of buildings, etc.

IDENTIFICATION Three age-groups (see p. 178). Somewhat larger than Black-headed Gull. *Much larger* Herring Gull (p. 184) is the only similar common species; when size difficult to judge, look for *smaller, narrower bill*, more *rounded head*, much daintier general shape and quicker actions, less heavy, more active flight with often *deeper and more energetic wingbeats*, and *narrower wings*; adult also often shows *more obvious tertial-crescent* and *larger 'mirrors'*; 1st-year has *plain* (not barred) *midwing-panel* giving neater upperwing pattern, and broad black, *clear-cut band on white tail*; also *more neatly patterned underwing* than immature Herring. Standing adult summer also resembles Kittiwake, but *legs long and pale* (not short and black), *forehead less steeply rounded*, and *primaries are white-tipped* and more extensively black. – Adult: Head white (streaked grey in winter), bill and legs greenish-yellow (bill duller in winter, with thin dark band). – Juvenile: Grey-brown head and breast, scaly brown upperparts. –1st-winter: Head and body similar to adult winter, but retains juvenile wings and tail; bill greyish with clear-cut black tip. –1st-summer: Like 1st-winter, but wings and tail often much faded, and bill-base yellowish or pinkish. –2nd-year: Like adult, but more black on wing-tip, lacks prominent white primary tips, has smaller 'mirrors', and legs and black-banded or black-tipped bill often bluish green-grey.

VOICE All calls much higher-pitched than Herring Gull's, including 'laughing' call, 'ke ke ke **kleeeh**-a... **kleeeh**-a... kay-a kay-a kay-a ke ke'. Call a distinctive, thinly yelping 'keea', often repeated. Persistent 'klee-**u** klee-**u**...' when alarmed (often good hint of bird of prey in vicinity!).

Ring-billed Gull *Larus delawarensis* **V***

L 41–49 cm, WS 112–124 cm. Vagrant from N America. Numerous records annually in Britain & Ireland.

IDENTIFICATION Three age-groups (see p. 178). Resembles Common Gull, but slightly larger, heavier-bodied, and has *thicker bill*. – Adult: *Clear-cut, broad black bill-band* (beware sometimes obvious band on ad. winter or 2nd-year Common); *yellow iris* (brown on Common); *paler grey upperparts* (similar to Black-headed); *faint tertial-crescent* (prominent on Common); and *smaller 'mirror'*. – 1st-year: Told from Common Gull by *paler grey upperparts and midwing-panel*, often *a few obvious dark crescents among scapulars* (unmarked or with only faint shaft streaks on Common), *narrower pale tips to tertials*, and *more variegated tail*

Common Gull

Mediterranean Gull

MEDITERRANEAN GULL, ad. summer

Note how posture can change hood shape of gulls.

pattern (broad band on white tail on Common). – 2nd-year: Like adult, but has more black on outer wing; white primary tips and 'mirror' (if any) smaller, and often greenish-yellow or greyish bill-base and legs; commonly has *variable tail-band* (usually lacking on Common). Beware that immature Herring Gull can provide easy pitfall if actually much heavier and longer bill and much larger size are not apparent (e.g. on lone bird); 2nd-year Herring resembles 1st-year Ring-billed, but tertials and coverts more mottled or barred (rather solidly dark-centred on Ring-billed), underwing rather uniformly dusky (neatly dark-flecked, and secondary bar showing through, on Ring-billed), and tail more extensively dark; 3rd-year Herring resembles 2nd-year Ring-billed, but wing and tail patterns less neat, and legs pink – Has habit of walking about a lot. Also, adults and 2nd-winter birds appear to have remarkably loose plumage.

VOICE Like Herring Gull's but higher-pitched and nasal.

Mediterranean Gull *Larus melanocephalus* m**B**5/**W**4

L 37–40 cm, WS 94–102 cm. Nests colonially. Breeding populations of W and S Europe small, usually among Black-headed colonies, including (rare) S and SE England. Scarce annual visitor to Britain & Ireland, mainly in S and E (mostly migrants from Black Sea).

IDENTIFICATION Three age-groups (see p. 178). A trifle larger than Black-headed Gull, with *shorter, thicker, more obtuse bill*; less pointed, *broader, slightly shorter wings*; and longer legs. Flight with *quicker and stiffer wingbeats* than Common Gull. – Adult: Unmistakable; *all-white flight-feathers*, very *pale grey upperparts* (silvery-white at distance), *scarlet bill and legs* (less bright in winter), and *large black hood* (dark 'mask' in winter). – 1st-year: Resembles Common Gull, but differs in *all-white underwing-coverts*, *paler grey midwing-panel*, slightly *thinner tail-band*, *reddish-based dark or wholly black bill*, *reddish or black legs*, and (from Sep) *paler grey upperparts, dark 'mask'* (or partial hood in 1st summer) and *white underparts*. – 2nd-year: Like adult, but with variable, small black markings on wing-tip.

VOICE Call is distinctive, rising-then-falling note, like enthusiastic, nasal 'yeah!', but with slightly whining tone.

HERRING GULL   RING-BILLED GULL

2nd-winter

tertials and coverts finely vermiculated or barred

rather plain

1st-winter

Beware of similarity between immature Herring and Ring-billed Gulls (see text).

COMMON GULL

large white 'mirror'

more black than on ad.

1st-winter

small 'mirror'

2nd-winter

ad. winter

dusky brown-grey

dark grey

coarsely marked

broad, clear-cut band

ad. w.

thick band

thin band

medium grey

neat

juv.

pale

white primary tips small

1st-winter

2nd-winter

broad white

ad. summer

yellow-green

RING-BILLED GULL

small 'mirror'

more black than on ad.

1st-winter

little or no 'mirror'

2nd-winter

ad. winter

pale grey

pale grey

variegated band

coarsely marked

often partial band or dark spots

ad. w.

pale grey, some dark crescents

stout

pale grey

pale iris on ad.

thin pale fringes

broad band

pale

white primary tips small

1st-winter

2nd-winter

faint white crescent

ad. summer

yellowish

MEDITERRANEAN GULL

black

white

when fully spread, small white spots are revealed

1st-winter

variable black markings

2nd-winter

ad. winter

pale grey

unmarked white

narrow band

white

ad. w.

dark 'mask'

very pale

black

more distinct white fringes than in Common Gull

rather plain

short, stout

blood-red

whiter below than juv. Common

long, dark

variable black markings

juv.

1st-winter

2nd-winter

white

ad. summer

red

KM

LARGE WHITE-HEADED GULLS

Much attention and interest have in recent years been focused on a group of large white-headed gulls including such widespread and well-known species as Herring Gull and Lesser Black-backed Gull. The close attention has yielded new insights not only concerning the finer variations and how these varying forms can be identified, but also on the best taxonomy of the group, resulting in the recent split of several taxa. It is clear that there has been a quick radiation of subtly different species, meaning also that reliable identification can at times be depressingly difficult. Working through every immature gull on a rubbish dump or harbour pier for hours is a type of birding that attracts only a few, but these 'gull aficionados' can become extremely skilful at what they do and detect rare gulls among the common. On the next few spreads, the large white-headed gulls are presented together with a few similar-looking species.

(European) Herring Gull *Larus argentatus* ᵣB2 / **W**2

L 54–60 cm, WS 123–148 cm. A widespread and common species of N Europe. Frequently abundant on and near coasts, but seen inland, too, feeding on fields, rubbish dumps, etc. Roosts on harbour piers or secluded islands; flights to these at height, often in formation. Frequently soars high up in high-pressure weather, often in large parties. Nests, mainly colonially but at times in single pairs, on coastal islands, cliffs or at lakes. Omnivorous, taking fish, crustaceans, earthworms, offal and roadkill, eggs and young of other birds, etc.

IDENTIFICATION Four age-groups (see p. 178). Told from all other common gulls by *large size* and *pale grey upperparts* (cf. smaller Common Gull), but see very similar Yellow-legged and Caspian Gulls for differences from these. Important to note shape also; compared to both Yellow-legged and Caspian, Herring has not quite as long and pointed wings, easiest to see on a standing bird which shows *only moderate primary projection* and a *rather compact* and 'bulky' look. Plumage development and ageing can be worked out from the plate, but keep in mind the individual variation. Thus, grey on upperparts develops only from 2nd winter onwards, so 'all-brown' 1st-years are more difficult to tell from Lesser and Great Black-backed Gulls (p. 190). *Head* of adult white in summer, *boldly streaked grey-brown in autumn* (Sep–Jan). Bill colour develops from dark with variable pale base on juvenile/1st-winter to yellow with orange-red spot on adult; iris from dark brown of juvenile to pale yellow (with yellow or orange-red orbital ring) of adult, *eye looking* invariably *pale* at a distance. *Legs pink* at all ages (but see below). – Variation: Birds of W Europe and Iceland (ssp. *argenteus*) are relatively small and round-headed, have back pale grey and wing-tip with relatively much black and more limited white. Breeders in Scandinavia and around the Baltic (*argentatus*), wintering partly in W Europe, average larger, slightly darker grey on back, extremes having little black and much white on wing-tip; birds of E Baltic

Herring Gull

and N Fenno-Scandia like *argentatus*, but locally some of most have legs yellowish or yellow (var. '*omissus*'), inviting confusion with Yellow-legged or Caspian Gulls. Still, '*omissus*' usually told by typical compact 'Herring Gull shape', pale eye, extensive white tips to exposed primaries, and in autumn–early winter densely streaked hood.

VOICE Calls include strident 'kyow', repeated and loud when used as alarm. In anxiety a distinctive 'gag-ag-ag'. Familiar exalted 'laughing' display-call is a loud, deep and clanging 'aau... kyy**aa**-kya-kya-kya-kya-kya kya...ky**au**'.

American Herring Gull *Larus smithsonianus* **V**∗

L 52–58 cm, WS 120–144 cm. Recently split from Herring Gull on account of distinct first-year plumage and slight genetic difference. Widespread in N America, straggles to W Europe (many records in e.g. Britain, France and Ireland).

IDENTIFICATION Four age-groups (see p. 178). Very similar to Herring Gull, and adults often inseparable. Most juveniles and 1st-winters, and several older immatures, can be told, though. Juvenile is *darker* on average than Herring with *nearly all-black tail* (fine pale barring visible at sides and base at the most, giving no white impression at all), and *rump and tail-coverts on both surfaces are densely dark-barred*, hardly contrasting with blackish tail. *Head and underbody are* not streaked or mottled as in Herring but are almost *uniformly dark brown*. A typical 1st-winter bird shows *good contrast between whitish head and still uniformly brown underbody* and has 'Glaucous Gull-patterned bill' (*pink with dark tip*), whereas others are very similar to Herring Gull. Later-stage immatures, and adults, increasingly more similar to Herring, often requiring specialist knowledge to be separated, and some must be left unidentified.

VOICE Very similar to Herring Gull, but long call perhaps slightly quicker and more high-pitched.

Comparison of typical juvenile American Herring Gull and Herring Gull alongside very dark juvenile Lesser Black-backed Gull.

AMERICAN HERRING GULL
'window' averages a little less obvious than in Herring Gull
often rather large birds, heavy-billed
heavily barred
dark, little or no white at base
densely barred
uniform

LESSER BLACK-BACKED GULL
(dark)
dark; lacks light 'window'
sometimes heavily barred
seldom if ever as densely barred as American Herring Gull
sometimes dark but rarely if ever uniform

HERRING GULL
typical juvenile (some are darker than this!)
light 'window'
white tail-base
streaked and mottled

ERRING GULL

wide individual variation

argenteus (W Europe)

more black, less white

ad. s.

brown

2nd-w.

usually grey

3rd-w.

argentatus (N Europe)

1st-w.

more white, less black

ad. s.

ESS. BLACK-B.

GT BLACK-B.

1st-w.

raellsii 1. w.

ad.

1st-w.

ixed gull flock in early winter; Herring, esser and Great Black-backed Gulls

argentatus ad. w.

ad. w.

3rd-w.

ad. w.

ad. w.

sually rather venly coloured, rtials and coverts stinctly 'notched'

juv.

2nd-w.

argenteus is a shade lighter grey above than *argentatus*

ad. w. *argenteus* (W Europe)

juv.

minority of birds, especially in N urope, have virtually unnotched rtials and coverts (above), ome looking like Lesser ack-backed Gull!

1st-w.

3rd-w. *argentatus*

argentatus averages slightly larger than *argenteus*

ad. w. *argentatus* (N Europe)

MERICAN HERRING GULL

averages more *solid* black

on average, more barred than Herring

2nd-w.

often well-defined black markings on secondaries

3rd-w.

primary pattern details may provide vital identification clues

ad. s.

st-years often eye-atchingly dark

tle or no hite in tail

densely barred

uniform, brown

1st-w.

often still heavily barred

extremely variable!

2nd-w.

a few adult-types show distinctive black tertial-spots

lighter grey than *argenteus*

streaking on head and breast often heavy, blotchy

ad. w.

M

Yellow-legged Gull *Larus michahellis* **P+W**4
L 52–58 cm, WS 120–140 cm. Habits similar to Herring Gull, to which it is closely related (though nowadays commonly treated as a separate species). Sedentary. Increasing in many areas, has expanded both along French Atlantic coast and into interiors of C Europe (where summer visitor only), and now frequently straggles in fair numbers in Jul–Oct (Dec) to NW, NC Europe and Baltic.

IDENTIFICATION Four age-groups (see p. 178). Very similar to Herring and Caspian Gulls in adult and subadult plumage. Immatures differ somewhat from Herring, are more like Lesser Black-backed Gull (and Caspian). Compared to Herring, adults consistently have *whiter head in autumn* (streaking mainly concentrated on eye-surround and nape, but some geographical variation, with palest birds in E Europe) and usually *bright yellow legs*, and often show all or many of the following characteristics: in flight *slightly longer wings*; heavy bill with strongly curved tip and well-marked gonys angle in most; on adults, *outer primaries have more black* and *smaller white spots; larger red spot on bill* often reaching onto upper mandible; on average *slightly darker grey back* with *less bluish cast*; more *nasal calls*. Young birds (juvs., 1st-winters) often spotted on combination of *pale head with suggestion of dark 'mask', all-black heavy bill, pale rump* and *neat black tail-band, pale underparts, no or only poorly suggested light 'window' on inner 'hand'*, rather *dark underwing-coverts*, and *long wings*. Again, cf. also Caspian Gull. – Variation: Birds of Atlantic Islands (*atlantis*), most typical birds on Azores, are even *darker grey on back*, almost as dark as British Lesser Black-backed (*graellsii*), have dense dark head streaking in winter plumage giving *dark-hooded* appearance, tend to be slightly smaller and shorter-winged than Mediterranean birds. Juveniles and immatures more like Lesser Black-backed and American Herring than Mediterranean Yellow-legged being quite dark. – W Iberian birds ('*lusitanius*') tend to be smaller and have less extensive black on wing-tip.

VOICE All calls more similar to Lesser Black-backed Gull's than Herring's owing to the nasal, rather deep voice.

Armenian Gull *Larus armenicus* —
L 50–56 cm, WS 115–135 cm. Closely related to Yellow-legged Gull genetically, but since it has an allopatric inland range and differs slightly structurally, now often, as here, treated as a separate species. Breeds at lakes in mountainous areas in Asia Minor, Transcaucasia and NW Iran. Winters mainly in Levant S to Sinai with largest concentrations in N Israel. Fairly common also in the Gulf but in rest of Arabia probably rare or accidental only.

IDENTIFICATION Four age-groups (see p. 178). Slightly smaller than Yellow-legged Gull with proportionately *shor er, stubbier-looking bill* and a trifle longer legs. *Head is mo rounded*, not so flat-crowned with angled rear crown as Yellow-legged. Adults frequently retain *black subtermin bar on bill-tip* even when breeding (thought not always), develop it more frequently and prominently in winter, tha Yellow-legged. *Iris on average darker* than in Yellow-legge Mantle subtly darker grey than Yellow-legged, and black c wing-tip slightly more extensive, with white 'mirror' usu ally only on outermost primary. Immatures very like Yellow legged, differing mainly in size and structure (small, roune headed, short-billed, long-legged).

VOICE Calls similar to those of Yellow-legged (and to som extent Lesser Black-backed Gull), voice being rather nas and deep, but are a trifle more high-pitched and hurrie sometimes vaguely recalling Common Gull.

Audouin's Gull *Larus audouinii* —
L 44–52 cm, WS 117–128 cm. Still a comparatively ra gull, but significant increase in W Mediterranean in rece years (huge population now in Ebro delta; c. 10,000 pairs Nests colonially or singly on rocky, small islands. Strict coastal or pelagic, frequenting wave-washed rocky coas and bays or sandy beaches. Rarely scavenges like other larg gulls, but feeds almost exclusively on fish, snatched in flig from near surface or in deeper plunge-dives. Biggest col nies usually dependent on presence of fishing industry.

IDENTIFICATION Four age-groups (see p. 178). Slight smaller than Herring Gull, with more elegant gener appearance; *shorter, stubbier bill; more elongated forehea* with feathering expanding towards nostrils; *longer neck* whe alert; *longer, narrower wings*; flies more gracefully, with mo gliding; bill often pointing slightly down when perched. breeding areas, Yellow-legged Gull is only similar adu gull, from which told by *dark red bill* (looks black at di tance), *very dark iris*, *dark grey or greenish legs* (often visibl in flight, too), *much paler grey* and slightly two-toned *uppe wing* lacking broad, clear-cut white trailing edge, and has *on tiny white 'mirror'* at wing-tip; *small white primary tips*, lik '*string of beads*'. – 1st-year: Juvenile similar to Yellow legged and Lesser Black-backed Gulls, but underwing mo strongly patterned with *dark-barred coverts and dark fligh feathers separated by light midwing-panel; rump darkish*, bu *white uppertail-coverts forming prominent U*; largely *blac tail-feathers broadly tipped white*; lower flanks usually dar *legs dark grey*; *bill deepest at tip, two-toned*, tip black an inner two-thirds grey. Develops adult-like upperbody earl gaining many pale grey feathers on scapulars and mant Feb–May. – 2nd-year: Told by *pale grey upperparts*; '*smar wing pattern* with *dark and neat secondary bar; narrow blac tail-band*; and *dark leg* – 3rd-year: Like adult, but little black on primary-co erts; 'mirror' lacking; an sometimes hint of tail-band

VOICE Calls when breedin include a nasal 'gleh-i-eh low 'ug-ug-uk' (alarm) an harsh, raucous 'argh' i endless series.

Yellow-legged Gull   Armenian Gull   Audouin's Gull

 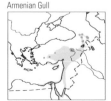 

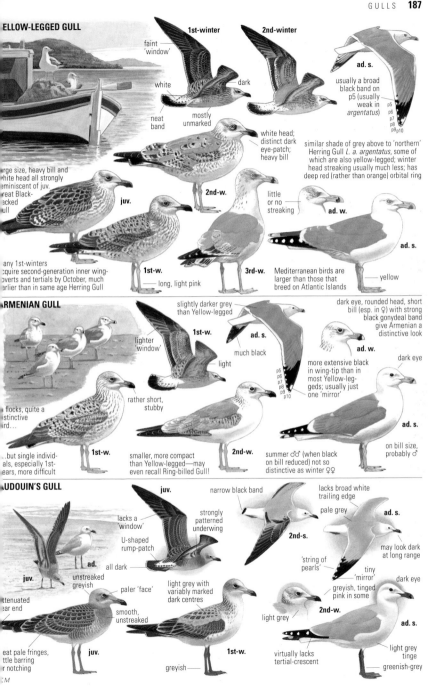

ELLOW-LEGGED GULL

1st-winter

2nd-winter

faint 'window'

white

dark

neat band

mostly unmarked

ad. s.

usually a broad black band on p5 (usually weak in *argentatus*)

p5
p6
p7
p8
p9 p10

rge size, heavy bill and hite head all strongly eminiscent of juv. reat Black- acked ull

white head; distinct dark eye-patch; heavy bill

similar shade of grey above to 'northern' Herring Gull *L. a. argentatus*, some of which are also yellow-legged; winter head streaking usually much less; has deep red (rather than orange) orbital ring

juv.

2nd-w.

little or no streaking

ad. w.

ad. s.

any 1st-winters cquire second-generation inner wing- overts and tertials by October, much arlier than in same age Herring Gull

1st-w.

long, light pink

3rd-w.

Mediterranean birds are larger than those that breed on Atlantic Islands

yellow

RMENIAN GULL

slightly darker grey than Yellow-legged

dark eye, rounded head, short bill (esp. in ♀) with strong black gonydeal band give Armenian a distinctive look

lighter 'window'

1st-w.

ad. s.

much black

light

ad. w.

dark eye

more extensive black in wing-tip than in most Yellow-leg- geds; usually just one 'mirror'

p5
p6
p7
p8
p9
p10

flocks, quite a stinctive ird...

rather short, stubby

ad. s.

..but single individ- als, especially 1st- ears, more difficult

1st-w.

smaller, more compact than Yellow-legged—may even recall Ring-billed Gull!

2nd-w.

summer ♂♂ (when black on bill reduced) not so distinctive as winter ♀♀

on bill size, probably ♂

UDOUIN'S GULL

juv.

narrow black band

lacks broad white trailing edge

pale grey

ad. s.

lacks a 'window'

strongly patterned underwing

2nd-s.

may look dark at long range

U-shaped rump-patch

ad.

juv.

unstreaked greyish

all dark

'string of pearls'

tiny 'mirror'

dark eye

ttenuated ear end

paler 'face'

light grey with variably marked dark centres

greyish, tinged pink in some

2nd-w.

smooth, unstreaked

light grey

eat pale fringes, ttle barring r notching

juv.

1st-w.

virtually lacks tertial-crescent

ad. s.

light grey tinge

greyish

greenish-grey

M

Caspian Gull *Larus cachinnans* V∗ / P+W4–5

L 55–60 cm, WS 138–147 cm. Closely related to Yellow-legged and Herring Gulls but recently commonly recognized as a separate species. Breeds on inland lakes and along rivers and at reservoirs in E Europe and Central Asia, has expanded westward in recent time to become a much more regular visitor to the Baltic and NW European coasts, mainly in summer–autumn. Limited hybridization occurs where Caspian meets both Herring and Yellow-legged Gulls (Poland), but breeding success of such mixed pairings still poorly known. Colonial breeder, preferring in W Black Sea area low islets and remote shallow coasts, as opposed to habit of Yellow-legged to select cliffs and roof-tops (but note that Yellow-legged in Mediterranean, e.g. Camargue area, also breeds on low islets in deltas, etc.)

IDENTIFICATION Four age-groups (see p. 178). About as large as Yellow-legged and Herring Gulls, but compared to latter slightly *slimmer-bodied, smaller-headed, longer-legged* and *longer-winged*, with *flatter forehead* and rather *angular rear crown*. Often adopts more *upright posture* than con-geners, either with neck stretched or with *protruded chest*, and *wings held rather low*. Bill long and evenly narrow with attenuated point, *lacking marked gonys angle*. Bill of adult outside breeding season *rather pale yellow-green*. Legs out-side breeding season usually pale buffish-pink, *paler than on Herring*. Red spot on bill restricted to lower mandible (cf. Yellow-legged). *Eye* on the other hand *often dark* (in about 75% of birds clearly darker than in both Yellow-legged and Herring). Compared to Yellow-legged and Armenian Gulls, adults have *paler grey mantle*, more like Fenno-Scandian Herring Gull. *Wing-tip has less extensive black* and *more white 'mirrors' and longer white 'tongues' penetrating into the black tips*. Adult winter often picked out on having *seem-ingly unstreaked white head* (very discreet streaking around eye and on crown can only be seen at closest range), the dark eye, and on the structural differences described above. – Ju-venile: *Whitish head* with *dark, grey-brown 'boa'* across lower hindneck, *all-black bill*, *whitish rump* and *neat black tail-band*. Underbody pale grey-brown, and *underwing-coverts on average lighter* than in Yellow-legged Gull of same age (but some come close). Inner primaries slightly paler brown than outer, forming indistinct paler 'window', but much less obvious than in Herring Gull. Tertials variable, from rather uniformly dark with broad white tips, to dark with nar-rower white tips, at times with faint subterminal pattern. Greater coverts usually finely patterned, sometimes with dark bases forming dark panel as in Yellow-legged, or ac-centuating suggestion of pale wing-bars. – 1st-winter: Like Yellow-legged Gull moults rather early (from Oct of 1st winter), progressively attaining paler grey feathers with dark spots or anchor-marks on mantle and scapulars, an developing whiter body, neck and head. Bill becom slightly two-coloured, showing somewhat paler base o most. – 2nd-winter: Very similar to 1st-winter but *bill mo markedly two-coloured*, inner two-thirds being pale pin brown. Towards end of winter moults body-feathers an wing-coverts to develop the first more adult-looking plu age emerging in 2nd summer. *Grey feathers of upperparts no have no or only faint dark marks*, and tertials become pr dominantly white-tipped pale grey with only variable amou of dark patches remaining. – 3rd-winter: Shows adult wi pattern, told on dark subterminal bar on bill and lingerin dark marks on primary-coverts, tail and tertials.

VOICE Long call of adult delivered with raised wings. Com pared to Yellow-legged Gull has higher-pitched more 'chi ish' voice and quicker pace, long call sometimes likened staccato utterance of sea-lion or ass.

Pallas's Gull (Great Black-headed Gull)
Larus ichthyaetus V∗

L 58–67 cm, WS 146–162 cm. Is obviously not a 'lar white-headed gull' but treated here for easy comparison d to its large size and because non-adults have mainly whitis head. Nests colonially on marshes, lake islands etc. in S Russia. Scarce but regular in Middle East in winter; vagra to Europe, with only one record in Britain (in 1859), no in Ireland, but scattered genuine later records in other and C European countries, north to Sweden. Food fis crustaceans, insects (also swarming in flight), small mar mals etc.

IDENTIFICATION Four age-groups (see p. 178), but pluma development initially quick as in a three-year gull. As lar as or larger than Herring Gull, but slightly slimmer-bodie and *longer-winged*, with *elongated forehead* which accent ates *length and heaviness of bill*. In flight, slightly slimme winged than other large gulls, with deep chest, long hea and large bill giving front-heavy look. *Very large size* ar *black hood* make adult summer unmistakable, but standin birds in other plumages could easily be overlooked as Ca pian Gulls, except for usually *obvious dark 'mask'* and *whi eye-crescents*. Also, important always to bear in mind pitfa of some other large gull with oil-stained head, thus esse tial to check other diagnostic features described below, esp cially the distinctive wing and tail patterns. – Adult: *Out wing mainly white, with black crescent across tip*; black-ba ded yellowish bill (with red towards tip in summer); ar yellowish legs. – Juvenile: Scaly brown above; brown breas sides or breast-band contrasting with *white rest of unde parts*; striking *pale grey midwing-panel*; *underwing-cover white*; *broad, clear-cut, black tail-band*; pale, dark-tippe bill; and greyish or brownish-flesh legs. – 1st-winter/1s summer: As juvenile, except adult-like *pale grey upperpart dark hindneck*; bill and legs often show some yellowish in 1 summer. – 2nd-year: Like adult, but much black on out wing, and thin black tail-band. – 3rd-year: Like adult, b more black on wing-tip, and sometimes trace of tail-band

VOICE Noisy at colonies. Flight-call is a deep, rather nas 'aagh', rather similar to the call of Lesser Black-back Gull. Mostly silent when not breeding.

Caspian Gull

Pallas's Gull

⊂ASPIAN GULL

1st-w.

2nd-w.

3rd-w.

pale
under-
wing

diffuse
wing-bars

ad. w.

note that black
extends strongly
to p5, and long,
pale 'tongues' on
outermost primaries

p5
p6
p7
p8
p9 p10

more extensive
streaking than
Caspian Gull

dult, with Herring Gull;
ong bill, long neck
nd long legs
reate distinctive
mpression

**HERRING
ad. w.**

compare extent head/neck streaking in autumn/early winter; colour
and shape of bill often
distinctive in Caspian

CASPIAN
1st-w.

compare with
Yellow-legged Gull
(right)

YELLOW-
LEGGED
1st-w.

argentatus
(N. Europe)
(yellow-legged
variety, *'omissus'*)

YEL.-LEG.
ad. w.

CASPIAN
ad. w.

michahellis
(Mediterranean
& Iberia)

eculiar shape and
upright' posture
ften the first thing
ɔ attract attention
ɔ Caspian Gull

long, slender bill
usually lacking
pronounced
gonys angle

juv.

2nd-w.

usually
unstreaked

ad. w. ♂
(Sep–Oct)

most (*c.* 75%)
adults have
darkish eyes

reater coverts finely
atterned, not coarsely
hequered

♂♂ have, on
average, larger
head and bill

extensive
white

ad. s.

usually whiter head and underparts than Herring Gull; diffuse
wing-bars, long legs and long,
slim bill create distinctive look

1st-w.

3rd-w.

leg colour varies, ranging
from dull greyish, through
flesh to bright yellow

characteristic
'upright' stance,
'full' chest and
long legs

⊂ALLAS'S GULL

spring gathering, with
two Caspian Gulls

variable
markings

2nd-w.

1st-yr

juv. has only
diffuse
'mask'

juv.

light
'window'

neat
band

white

ad. w.

outer wing
mostly white,
black crescent
across tip

some have a
breast-band

in all plumages, this is
one of the most easily
identified 'large' gulls

more black
than adult

strongly
bicoloured

1st-w.
(Jan)

ad. s.
(Feb–Jul)

plain, no
barring

2nd-w.

first-years acquire adult-like
grey upperparts very rapidly!

M

Great Black-backed Gull *Larus marinus* r(m)**B**3

L 61–74 cm, WS 144–166 cm. Breeds along coasts, singly or in small colonies, locally at larger lakes. Sedentary. Food fish, offal, birds (incl. fledged); kleptoparasitic, robbing sea ducks, cormorants, etc. Nest often high on rocky islands.

IDENTIFICATION Four age-groups (see p. 178). *Largest gull*, with bulky body, thick neck, and *very heavy bill. Majestic flight with slow wingbeats* perceived with practice. *Blackish upperparts* (develop in spring of 3rd calendar-year) and *large size* instantly indicate this species (or Lesser Black-backed Gull, which see). 1st-year and most 2nd-winter lack black on upperparts, thus also difficult to tell from Herring Gull. Bare parts develop gradually: bill from all black on juvenile/1st-winter to yellow with orange spot on adult; iris from brown to yellowish-grey (with red orbital ring). *Legs pale fleshy* at all ages. Given reasonable views, most immature large gulls can be aged by wing and tail patterns. – Juvenile: Whole plumage neat; scapulars and tertials mainly solidly dark-centred; head mottled. – 1st-winter: Resembles juvenile, but scapulars are more barred and head becomes slightly paler. – 2nd-winter: Resembles 1st-winter, but wing-coverts less strongly chequered, greater coverts rather plain or finely patterned, and usually acquires pale base of bill. Extensively blackish scapulars and some mantle-feathers usually acquired in spring moult of 3rd calendar-year (2nd summer). – 3rd-winter: Most wing-coverts blackish; a few barred. – 4th-winter: Many as adult, but a few retain some slight signs of immaturity on wing-coverts, tail, and bill-tip.

VOICE Like that of Lesser Black-backed and Herring Gulls, but *much deeper* and typically always *hoarse and gruff*; 'laughing' display-call is also slower and shorter.

Lesser Black-backed Gull *Larus fuscus* m+r**B**2 / P+**W**2

L 48–56 cm, WS 117-134 cm. Breeds colonially along coasts and at lakes. In N and E Europe long-distance migrant, crossing Europe along rivers to reach Africa, in W Europe shorter-distance migrant, and many winter in British Isles. Baltic and N Russian birds sometimes separated as different species, but here kept as races pending further research.

IDENTIFICATION Four age-groups (see p. 178), althoug Baltic birds often mature quicker. *Blackish upperparts* de velop from 2nd winter onwards, instantly narrowing choic to this species or Great Black-backed, from which told b *smaller size* (same as or slightly smaller than Herring), *slim mer body* and *longer, more pointed wings, and thinner, les massive bill.* In addition, adults have *yellow legs* (not pink *smaller white 'mirror'* at wing-tip and smaller white spots o primary tips, so *lack continuous white rear edge to 'hand';* i W Europe *obvious grey smudging on head Sep–Feb* (hea white or faintly streaked on Great Black-backed). 1st-yea more difficult to tell from Great Black-backed and, espe cially, Yellow-legged, Caspian and Herring Gulls. Bare par develop gradually: bill from all black on juvenile to yellow with red spot on adult; iris from dark to yellow (with re orbital ring); and legs from pink to yellow. See Great Black backed for general comment on ageing large gulls. –1s summer birds uncommon in breeding areas, at least in earl summer. – Variation: Birds of W Europe and Iceland (ss *graellsii*, 'BRITISH LESSER BLACK-BACKED GULL'), resident c wintering south to W Africa, have upperparts slate-grey contrasting with blacker wing-tips. Rather compact an bulky, much as Herring Gull. – Breeders in SW Scandinavi (*intermedius*), which winter commonly in W Europe an south to W Africa, have upperparts dark blackish-grey, ofte with slight contrast to jet-black wing-tips. – Birds of Balti (*fuscus*, 'BALTIC GULL'), wintering in Middle East and Africa, have upperparts all (brownish-)black, with little c no contrast with wing-tips, and head much less streaked i winter. On average a more delicate and long-winged bir with longer primary projection when standing (though be ware of variation: some ♂♂ are not that different in shap from *intermedius*, or even ♀ Herring Gull). – Birds of Russia (*heuglini*; 'HEUGLIN'S GULL') are slate-grey above lik *graellsii*, but are subtly longer-winged than that. A littl larger, bulkier and longer-legged than *fuscus*. Keeps sum mer plumage long (as *fuscus*), starts autumn moult late.

VOICE All calls similar to Herring Gull's, but noticeabl *deeper and more nasal*, thereby recalling Yellow-legged Gull

GREAT BLACK-BACKED GULL

2nd-winter (c. Mar)

broad-winged

ad. s.

1st-w.

2nd-w.

obvious pale 'window'

juv.

ad. w.

dark back attained late

broad white margin

stout

much white

tail-band narrow, broken up

broad pale tips

2nd-w.

heavy, deep

3rd-winter

large white spots

dull pink

ad. summe

K M **juv.**

ESSER BLACK-BACKED GULL
raellsii

2nd-winter

juv.

3rd-winter

colour of
inner primaries
can help age
atypical 2nd-w.
and 3rd-w.

**ad.
summer**

lacks
'window'

usually
brownish

prominent
white rump

neat
band

comparatively
light build

reeding adults

intermedius (right) averages
darker above than *graellsii*,
sometimes as black
as *fuscus*!

ad. s. → w.
(Oct)

ather dark; lacks
old notching on tertials,
ark outer greater coverts

juv.

2nd-winter

**3rd-
winter**

slate-grey

juv. primaries, tail and most
wing coverts retained up
to May of 2nd cal.-year

ad. summer

xtremely variable, can be
ery like Herring Gull (as here)

1st-w.

2nd cal.-yr

BALTIC GULL' *fuscus*

unlike *graellsii*, usually
develops adult-like
appearance in
3rd cal.-year

3rd cal.-yr
(Jun)

entirely
blackish,
very little
contrast

much
white

ad. winter

advanced moult

ad. summer
♀

averages smaller
and lighter than *graellsii*,
but much overlap

juv.

2nd cal.-yr (Jun)

♀♀ in particular are strikingly
elegant, with longer, narrower
wings than *graellsii*

HEUGLIN'S GULL' *heuglini*

in the Middle East, 'Heuglin's' and 'Steppe
Gulls' (*L. f. barabensis*) commonly occur
side by side; 'Heuglin's' is darker above
and moults later than 'Steppe Gull'

adult Steppe (left)
and Heuglin's Gulls,
(Oct.)

juv. primaries usually not replaced before
2nd cal.-yr spring, but moult of body,
tail and wing coverts averages
more extensive than *graellsii*

often
pale-
based

ad. summer

verages larger,
nore heavily built
nd longer-legged than
uscus*, but much overlap

juv.

2nd cal.-yr
(May–Jun)

very similar to *graellsii*, but
slight 'on-average' differences
in proportions and shape

KM

Great Black-backed Gull

Lesser Black-backed Gull

'Heuglin's Gull'

Ross's Gull *Rhodostethia rosea* V**

L 29–32 cm, WS 73–80 cm. Breeds in High Arctic of NE Siberia and N America. Vagrant to N Europe, recently with a few records annually in Britain & Ireland.

IDENTIFICATION Two age-groups (see p. 178). Resembles Little Gull in habits and size, but slightly larger, with *longer, more pointed wings* and *long tail* with *pointed tip*. At distance, likely to be overlooked as Little Gull. – Adult summer: Neat *black neck-ring*, pink-flushed underparts, white trailing edge to wing *broader than on Little Gull* and *not extending around wing-tip*. – Adult winter: Like adult summer, but neck-ring replaced by small dark ear-spot, dark area in front of eye, and pale grey on crown, neck and breast-sides; underparts often white. – 1st-year: Dark 'W-pattern' like Little Gull, but less black on primaries and lacks secondary bar, giving *broader white trailing edge*, accentuated below by *ashy-grey greater underwing-coverts*; lacks Little Gull's full tail-band and blackish cap. 1st-summer may have partial or full neck-ring.

VOICE Vagrants usually silent.

Little Gull *Hydrocoloeus minutus* P4 / W4–5

L 24–28 cm, WS 62–69 cm. Breeds in usually small colonies on freshwater marshes. Winters in W Europe and Mediterranean area, migrants (mainly Apr–May, Aug–Oct) seen along coasts, sometimes even at sea, or at inland waters. Picks small food items from water surface; hawks flying insects. Nest usually among vegetation in shallow water.

IDENTIFICATION Three age-groups (see p. 178). *Very small* (²⁄₃ size of Black-headed) and dainty, with quick wingbeats and more erratic, indirect flight path giving tern-like impression. Adult and 2nd-year flash alternately *blackish underwing* and pale grey upperwing, giving distinctive appearance even at long range; *distinct white border to entire wing* gives impression of *rounded wing-tip*. – Adult summer: *Black* hood, reddish-brown bill (looks black), scarlet legs, and pink-flushed underparts. – Adult winter: Black ear-spot and blackish cap; underparts white; bill black; legs dull red. – Juvenile: Extensive blackish-brown on head and upperparts, *dark 'W pattern' across wings*, faint secondary bar, black tail-band, all-white underwing (except for small dark patch where rear edge of wing meets body), black bill, and pink legs. – 1st-winter: Head and body like adult winter, remainder as juvenile. – 1st-summer: Like 1st-winter, with partial or full hood; by midsummer wings often worn, and 'W' faded pale brown; tail-band often broken in centre. – 2nd-year: As adult, but variable *black markings on upperwing-tip*, and *underwing-coverts paler or whitish*.

VOICE Short, hard, nasal 'keck', at times quickly repeated. Noisy display-call a rhythmic '**kay**-ke-**kay**-ke-**kay**-ke-...'.

(Black-legged) Kittiwake *Rissa tridactyla* m(r)B2 / P

L 37–42 cm, WS 93–105 cm. Rarely seen far from sea. Breeds colonially (at times several tens of thousands of pairs) on steep sea-cliffs, sometimes on buildings. Common near colonies and fishing ports, or boats at sea. Food inverte-brates and fish, including waste from commercial fishing.

IDENTIFICATION Three age-groups (see p. 178). A trifle larger than Black-headed, with slight notch in tail and short legs. Stiff, quick wingbeats and narrow outer wing give less 'lazy', more tern-like flight than other gulls. – Adult: *Small black triangle on wing-tip*; dark grey upperparts *shade to whitish before black wing-tip*, adding to distinctiveness of almost three-coloured wing pattern; bill yellowish; legs dark brown or blackish (rarely red, orange or flesh); head white in summer, with grey hindneck and crescentic blackish ear-spot in winter. – 1st-year: *Dark 'W-pattern' across wing* and black tail-band; juvenile and some 1st-winters have white head with black ear-spot and *black half-collar*; otherwise head as adult winter; 'W' often much faded by 1st summer, and bill becomes dull yellowish with dark marking at tip. Juvenile / 1st-winter similar to 1st-year Little Gull but with practice, recognized by different structure (Kittiwake is bigger-headed) and flight action. Kittiwake is darker grey above, has *cleaner white rear portion of wing*, and lacks Little Gull's dark cap and small black patch where undersecondaries meet body. – 2nd-year: Like adult, but told at close range by e.g. black fringes on outermost primaries, some black on bill, or winter head pattern in summer.

VOICE Noisy at or near colonies; commonest call is repeated, quick, nasal 'kitt-i-**waa**ke'. In flight, a short, nasal 'kya!'. Alarm short, knocking 'kt kt kt...'.

Sabine's Gull *Xema sabini* V* / P

L 30–36 cm, WS 80–87 cm. Breeds in High Arctic of N Siberia and N America. Numerous records annually off mainly W coasts of Britain & Ireland, mostly Aug–Oct, especially when mid-Atlantic southward migration route of N American population (which winters off SW Africa) is diverted by northwesterly gales.

IDENTIFICATION Two age-groups (see p. 178). Size between Little Gull and Black-headed Gull. Neat body but large wings. Notched tail. *Sharply contrasting upperwing pattern* distinctive at all ages, but beware that wing pattern of 1st-year Kittiwake can look very Sabine's-like even at fairly close range. Important, therefore, to note other confirmatory features before making firm identification: adult summer has *blackish-grey hood*, which is retained much later into autumn than that of other hooded gulls; adult winter and 2nd-year birds have variable, usually extensive blackish nape and hindneck; juvenile (Jul–Dec) has *pale-scaled mouse-grey upperparts*, much grey on head and breast-sides (*head looks mainly dark at distance*, not white like Kittiwake); also, *flight is somewhat lighter*. – 1st-summer: Like adult, but head as in winter, or only partial hood.

VOICE Juveniles have a piping 'prree' like a young tern.

Little Gull

Kittiwake

Sabine's Gull

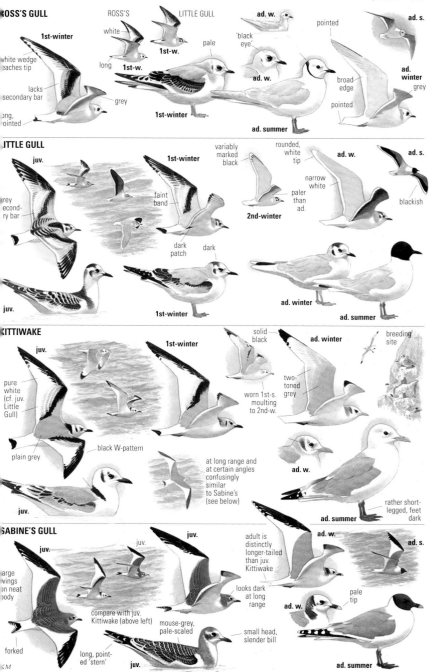

ROSS'S GULL

1st-winter

white wedge reaches tip

lacks secondary bar

long, pointed

ROSS'S

white

long

1st-w.

grey

1st-winter

LITTLE GULL

1st-w.

pale

'black eye'

ad. w.

pointed

ad. w.

broad edge

pointed

ad. s.

ad. winter

grey

ad. summer

LITTLE GULL

juv.

grey secondary bar

juv.

1st-winter

faint band

dark patch

dark

1st-winter

variably marked black

2nd-winter

rounded, white tip

narrow white

paler than ad.

ad. w.

ad. s.

blackish

ad. winter

ad. summer

KITTIWAKE

juv.

pure white (cf. juv. Little Gull)

plain grey

black W-pattern

juv.

1st-winter

solid black

worn 1st-s. moulting to 2nd-w.

at long range and at certain angles confusingly similar to Sabine's (see below)

ad. winter

breeding site

two-toned grey

ad. w.

ad. summer

rather short-legged, feet dark

SABINE'S GULL

juv.

large wings on neat body

forked

juv.

compare with juv. Kittiwake (above left)

long, pointed 'stern'

juv.

juv.

mouse-grey, pale-scaled

juv.

adult is distinctly longer-tailed than juv. Kittiwake

looks dark at long range

small head, slender bill

ad. w.

ad. w.

pale tip

ad. s.

ad. summer

KM

Glaucous Gull *Larus hyperboreus* P / W4

L 63–68 cm, WS 138–158 cm. Nests singly or in loose colonies on cliffs, islands, or other open ground near sea. South of breeding range, uncommon or rare (mostly Oct–Apr), mainly near coast among flocks of other large gulls. Food in summer mainly fish, eggs and young of other birds; in winter, mainly waste from fishing ports, rubbish tips, etc.

IDENTIFICATION Four age-groups (see p. 178). Glaucous and Iceland Gulls are the only *large* gulls with *whitish primaries*. Adults have paler grey upperparts than Herring Gulls; 1st-and 2nd-years in fresh plumage are rather *uniformly pale ochrous-brown* or café-au-lait-coloured, with fainter and finer barring on upperparts and wing-coverts than other large gulls; *unmarked* pale or white flight-feathers diagnostic at all ages, obvious in flight, especially from below, when they appear strikingly translucent (but beware of moulting Herring Gulls in Aug–Nov). Because they are similar in plumage at all ages, telling Glaucous from Iceland rests entirely on differences in size, structure, bill pattern (on 1st-years), and bill colour (on some adults and 3rd-years). Given good views, identification of most is easy: Glaucous is *obviously or slightly larger and longer-legged than Herring Gull* and more heavily built, with large bill, small eye, and sloping forehead and often angled hindcrown giving a mean, aggressive look; as a rule, *the bill is as long as or longer than the length of the* (short) *projection of primaries beyond the tail-tip*. Glaucous Gull is also correspondingly broader-winged, longer-necked and larger-headed, and more lumbering in flight than Iceland, but the structural and size differences are often more difficult to appreciate in flight than on the ground. Identification of 1st-years is made easier because of diagnostic difference in bill pattern: Glaucous invariably has *bill two-thirds pale* (pinkish) *with sharply demarcated black tip* (Iceland has more extensive black tip, shading into often only slightly paler base). Adult's orbital ring is yellow (reddish on Iceland). Occasional small or small-looking Glaucous or large or large-looking Iceland can be more difficult: in such cases, bill pattern (on 1st-years) and bill-length / primary-projection difference are the most useful features. – Caution is necessary over all-white albinistic individuals of other species, paler-than-normal (so-called leucistic) immature Herring Gulls, and Glaucous × Herring Gull hybrids. All-white albinistic gulls are rare, however, and in any case no Glaucous or Iceland is ever entirely pure white (although some much-faded 1st- and 2nd-years may appear so at long range). Leucism and hybridity show on immature gulls through the outer primaries, secondary bar or tail-band being at least slightly darker than the general colour of the rest of the plumage, and adult hybrids will show dark markings on the wing-tip; pure

Glaucous and nominate Iceland never show such features.
VOICE Resembles Herring Gull's, but some notes coarser.

Iceland Gull *Larus glaucoides* P / W4–5

L 52–60 cm, WS 123–139 cm. Breeds in Greenland and in N Canada (but not in Iceland!); rare winter visitor to NW Europe, usually much rarer than Glaucous. Habitat and food much as for Glaucous Gull.

IDENTIFICATION Four age-groups (see p. 178). Very similar to Glaucous Gull, having *whitish primaries*, adults with pale grey back and yellow bill with red spot, immatures with finely brown-barred plumage. However, Iceland Gull is smaller, even *slightly* orobviously *smaller and shorter-legged than Herring Gull* and usually more lightly built, with 'normal-sized' bill, proportionately larger eye and rather rounded head giving a comparatively gentle appearance; in flight it often gives a short-necked and small-headed impression. As a rule, the *bill is obviously shorter than the length of the* (long) *projection of primaries beyond the tail-tip*. 1st-years are usually told by bill pattern: *extensive dark tip* covering about half of bill, *shading into only slightly paler* (dull flesh or grey) *base*. Some Iceland retain diagnostic bill pattern well into 2nd year, but others subsequently acquire Glaucous-like pattern. Some adult and 3rd-year Iceland have *slightly greenish-yellow bill colour* (always yellowish on Glaucous, without greenish tone). Adult's orbital ring reddish. – Variation: Breeders in arctic NE Canada ('KUMLIEN'S GULL', ssp. *kumlieni*; vagrant to NW Europe, several records annually) like nominate *glaucoides* except often slightly larger and heavier-billed, and adults have diagnostic wingtip pattern (frosty-grey when fresh (though darkness variable) or grey-brown when faded, and often dark iris; 3rd- and 2nd-years have dark areas on outer webs of outer 3–4 primaries (reflecting eventual adult pattern), and often a darker tail-band; 1st-year indistinguishable from *glaucoides*. Breeders in arctic NW Canada ('THAYER'S GULL', *thayeri*; recorded several times in Europe) overall darker, with darker and more extensive markings on wingtips, approaching very pale Herring Gulls; on present knowledge difficult to identify reliably.

VOICE Like Herring Gull's, but slightly higher-pitched.

Ivory Gull *Pagophila eburnea* V**

L 41–47 cm, WS 100–113 cm. Rare vagrant south of High Arctic (just about annually recorded in Britain & Ireland), usually near fishing ports or at carrion on beach; often attracted to seal or cetacean corpses.

IDENTIFICATION Two age-groups (see p. 178). Size of Common Gull, but has *broader wings*, *shorter neck*, plumper body, longer tail and *shorter legs*. Flight light, wingbeats rather quick. – Adult: *All white*, but essential to note peculiar shape, *dark-based*, *yellow-tipped bill* and black legs to rule out chance of an all-white albinistic other gull. – 1st-year: Distinctive *finely black-spotted white plumage*; 'swarthy face'.

VOICE Tern-like 'krreeo'.

Glaucous Gull

Iceland Gull

Ivory Gull

GLAUCOUS GULL

buff-white tip

GLAUCOUS

translu-cent

dark

HERRING, 1st-w.

**juv. /
1st-w.**

less evenly patterned than 1st-w.

2nd-winter

3rd-winter

ad. w.

long 'weightier' neck, 'front-heavy'

long, pinkish, neat dark tip

pale iris

pale tip

broad wing, blunt tip

little or no tail-band

juv. / 1st-winter

1st-winter
(faded, spring)

2nd-winter

'blotchier' than 1st-w.

short wing-tip

ad. s.

ad. winter

ICELAND GULL

HERRING

GLAUCOUS

ICELAND

2nd-winter

3rd-winter

ad. w.

'short-necked'

**juv. /
1st-w.**

at all ages, shape (compared with Herring Gull) is safest means of identi-fication; all three above are 1st-w.

shortish, mostly dark

often acquires pale base in 1st winter

often surprisingly 'chunky' in flight; best to concentrate on head and bill

rather pointed

kumlieni

tinged greenish-yellow

ad. s.

attenuated 'stern'

juv. / 1st-winter

1st-winter
(faded, spring)

long wing-tip

2nd-winter

ad. w.

grey to blackish

ad. w.

kumlieni
(Arctic Canada)

IVORY GULL

bill colours diagnostic

unmarked white

dusky 'face'

feeding on seal carcase

black spots

**juv. /
1st-w.**

juv. / 1st-winter

ad.

short, black

ad.

ad.

KM

Sooty Gull *Larus hemprichii* —

L 42–45 cm, WS 105–113 cm. Breeds Jul–Sep in S Red Sea to Persian Gulf; in winter, occasionally wanders as far north as Suez and Eilat. Strictly coastal. Omnivorous; commonly scavenges along shores.

IDENTIFICATION Three age-groups (see p. 178). The size of Common Gull, but *dark colour* and *long wings* can give larger impression at distance; dark colour, especially dark underwing, can also give skua-like impression. Darkness and long-winged flying and perched silhouette, and *exceptionally long bill*, are characters also shared by slightly smaller White-eyed Gull, from which best told at all ages by *thick bill* (equally long but thinner on White-eyed), two-coloured and *sharply dark-tipped bill* (all dark-looking bill on White-eyed), and heavier build. –Adult summer: *Hood dark brown* (black on White-eyed); *white crescent* usually *above eye only* (thicker white crescents above and below eye on White-eyed); *breast and upperparts sooty-brown* (purer greyish and slightly paler on White-eyed); and *bill greenish with dark end* and red tip (dark red with black tip on White-eyed). – Adult winter: Hood less clear, and white half-collar obscured (on both species). – Juvenile: *Plain, pale greyish-brown head and neck* (darker brown on White-eyed); inconspicuous *thin pale crescent above eye* (prominent white crescents both above and below eye on White-eyed); and *sharply two-toned bill* (all dark on White-eyed). – 2nd-year: Like adult winter but head pattern less developed, and has variable black tail-band.

VOICE Resembles that of Kittiwake, especially falsetto laughing display-call 'veee**aah** ve ve **vah**, veee**aah**...'. Some calls are both shriller and harsher than White-eyed Gull's.

White-eyed Gull *Larus leucophthalmus* —

L 39–43 cm, WS 100–109 cm. Breeds in Red Sea and Gulf of Aden; small numbers straggle regularly north to Eilat. Mainly coastal. Food mainly fish.

IDENTIFICATION Three age-groups (see p. 178). Sooty Gull is only likely confusion species, sharing with White-eyed *dark colour, dark underwing*, slim shape and *long bill*; at all ages told from Sooty by following characters: slightly *smaller* size and lighter build; *thinner bill*, appearing *all dark* at distance. –Adult: *Black hood, thick white crescents above and below eye*, and *slightly paler and purer grey breast and upperparts*; bill at close range seen to be *dark red with black tip* (lacking greenish tinge of Sooty). – Juvenile: *Darker overall* than Sooty owing to duller pale tips to wing-coverts, and *darker and browner head and neck*; *prominent white crescents above and below eye*; and *all-dark bill*. – 2nd-year: Like adult winter, but has blackish 'mask' rather than hood, variable black tail-band, and all-dark bill.

VOICE Like Sooty Gull's, but usually slightly lower-pitched and less harsh.

Laughing Gull *Larus atricilla* V*

L 36–41 cm, WS 98–110 cm. Breeds in N America. Vagrant to Europe, a few records annually in Britain & Ireland.

IDENTIFICATION Three age-groups (see p. 178). Size between Black-headed and Common Gulls. Franklin's Gull is only likely confusion species, but also beware 2nd-or 3rd-winter Lesser Black-backed Gull, which could look similar if its actually much larger size not apparent (e.g. distant or brief views of lone bird). Laughing and Franklin's are much *darker grey above* than other small European gulls (nearly as dark as W European Lesser Black-backed *L. f. graellsii*). Laughing is slightly larger than Franklin's, with *longer bill*, legs and wings, latter giving *elongated rear end* when perched (Franklin's is more compact, Little Gull-like in shape). Adult and 2nd-year Laughing differ further from 1st-summer or older Franklin's in *all-black wing-tip* with *white primary tips tiny or lacking* (bold black and white pattern on Franklin's); tail all white (pale grey centre on Franklin's, although this not always obvious); *thinner white crescents* above and below eye (thick on Franklin's, and joined at rear); *less extensive dark on head* in winter (Franklin's has distinctive 'half-hooded' pattern in 1st-summer and adult winter plumages); and *dull red* bill and legs in adult summer (when Franklin's bright red). – 1st-winter: Note *grey on breast, flanks and underwing-coverts* (these areas white on Franklin's); less grey on head ('half-hood' on Franklin's); and *greyish tail with full, broad black band* (white with smaller band on Franklin's). – 1st-summer: As 1st-winter but upperwing-coverts often faded, pale brown (upperwing of Franklin's adult-like in 1st summer). – 2nd-year: As adult, but possible to age by black of wing-tip extending more or less strongly onto primary-coverts, and (on most) variable black band on tail.

VOICE Flight-call Common Gull-like, yelping 'kee-agh'.

Franklin's Gull *Larus pipixcan* V***

L 32–36 cm, WS 81–93 cm. Breeds in N America. Rare vagrant to Europe, often in summer but in other seasons, too.

IDENTIFICATION Two age-groups (see p. 178). Size of Black-headed Gull or slightly smaller. Only likely confusion species is Laughing Gull, which see for detailed comparison (but beware adult winter Kittiwake could look similar in brief or distant view). Main differences from Laughing: *slightly smaller size* and *more compact shape*; *shorter bill*; in adult plumage *bold black and white pattern at wing-tip*; *thicker white crescents above and below eye*, joining at rear; and *bright red bill and legs*. – 1st-summer: After complete moult in 1st spring, resembles adult winter, but with *white division* between grey of upperwing and black of wing-tip *narrow or lacking*; white primary tips smaller (but still obvious); and never acquires full hood (thus useful difference from adult summer); quite different appearance from 1st-summer Laughing Gull, which retains juvenile wings and tail. –2nd-winter: Like adult, but sometimes possible to age by less extensive white on wing-tip.

VOICE Soft 'krruk' or shrill, repeated 'guk' when feeding.

Sooty Gull   White-eyed Gull

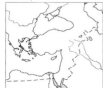

SOOTY GULL

juv.

2nd-winter

ad. summer

ad. s.

white

pale

lacks well-developed light neck-sides, solid dark hood and bill colours of adult

dark brown

broad pale fringes

heavy; pale base

brownish

green-ish, red tip

note: bill base initially blue-grey, colour gradually becomes more pink-ish by early winter

juv.

plain brown

ad. summer

2nd-winter

ad. w.

WHITE-EYED GULL

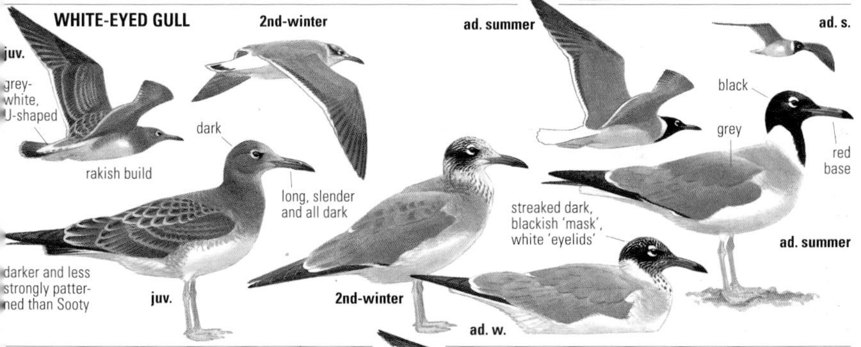

2nd-winter

ad. summer

ad. s.

juv.

grey-white, U-shaped

dark

black

grey

red base

rakish build

long, slender and all dark

streaked dark, blackish 'mask', white 'eyelids'

ad. summer

darker and less strongly patter-ned than Sooty

juv.

2nd-winter

ad. w.

LAUGHING GULL

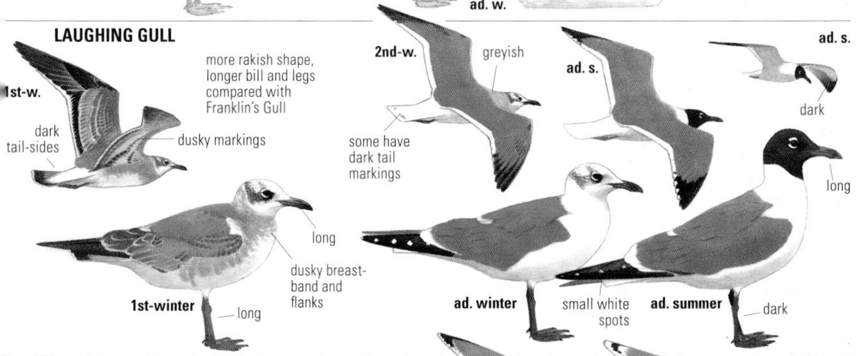

1st-w.

more rakish shape, longer bill and legs compared with Franklin's Gull

2nd-w.

greyish

ad. s.

ad. s.

dark tail-sides

dusky markings

dark

some have dark tail markings

long

long

dusky breast-band and flanks

1st-winter

long

ad. winter

small white spots

ad. summer

dark

FRANKLIN'S GULL

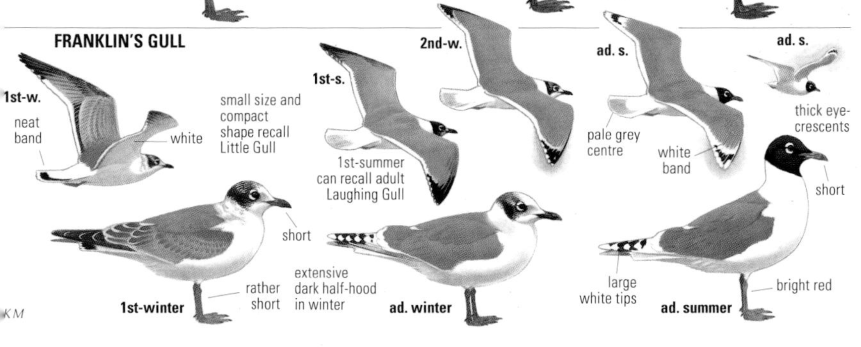

1st-w.

small size and compact shape recall Little Gull

2nd-w.

1st-s.

ad. s.

ad. s.

neat band

white

1st-summer can recall adult Laughing Gull

pale grey centre

white band

thick eye-crescents

short

short

rather short

extensive dark half-hood in winter

ad. winter

large white tips

ad. summer

bright red

KM

TERNS *Sternidae*

Relatively small to medium-sized, slender seabirds with long, narrow and pointed wings, usually long and deeply forked tail, and long and pointed bill. Masterly fliers, some migrating longer distance than any other birds. Flight light and elegant, at times buoyant, with elastic wingbeats, wings often kept bent at the carpal joint. Frequently hover before plunging into water for fish, but do not glide in the air, and only rarely swim like gulls. Nest on ground, often colonially.

Little Tern *Sternula albifrons* mB4

L 21–25 cm, WS 41–47 cm. Breeds in loose colonies along sandy, shallow coasts or islands with shingle and shells and low grass; also at lakes and wide rivers. Summer visitor (mid Apr–Oct), winters in Africa. Food mainly small fish.

IDENTIFICATION *Tiny* tern, appearing only about half the size of Common Tern, with *noticeably narrow wings* and *very fast wingbeats*; *hovers low over surface*, often prolonged, then plunges to catch fish; typically *repeats dives more quickly* than other terns. – Adult summer: *White forehead* and *black lores*; bill *yellow with small black tip*; legs orange-yellow; 2–3 outer primaries blackish. – Adult winter: Like adult summer, but from Aug lores white and forecrown paler. – Juvenile: Bill dark (with dirty yellow base), mantle/scapulars with dark scaly pattern, and upper forewing dusky; best identified by *flight action* and *small size*.

VOICE Distinctive sharp, rasping, often repeated 'kriet'; rapid chattering in display and alarm, '**ker**re-kiet **ker**re-kiet **ker**re-kiet...'. Young beg for food with a light, ringing 'plee' or 'plee-we'.

Sandwich Tern *Sterna sandvicensis* mB3 / P3–4

L 37–43 cm (incl. tail-streamers 6–9 on ad.), WS 85–97 cm. Breeds colonially on sandy beaches or low islands in salt or brackish water. Summer visitor to N Europe (mainly late Mar–late Sep), winters in S Europe and Africa. Scarce inland, unlike Caspian Tern strictly following Atlantic coasts on migration. Food mainly fish. Nest a scrape on ground.

IDENTIFICATION *Large size* (about same size as Black-headed Gull, but certainly slimmer) and *long, 'pencil' bill* distinctive. Rapid fishing flight with *measured, forceful wingbeats*, *wings slightly flexed, long and evenly narrow*; *also, appears short-tailed* (front-heavy) in flight. When feeding, often patrols at greater height than Common or Arctic Tern, with bill pointing down, making sudden and bold dive to catch fish. In stronger winds, typically feeds where waves break, sometimes in large and noisy gatherings. – Adult summer: Pale grey upperparts, white underparts (actually lightly washed creamy-pink) and broad white trailing edge give *very whitish impression* compared with Common or Arctic Tern; outer

4–5 primaries darken during summer to form blackish wedge or tip; *legs and bill black*, latter with *small yellow tip*. – Adult winter: *Forehead and crown white* (from Jun onwards), leaving dark 'mask', and (Nov–Mar) upperwing uniform frosty-grey. – Juvenile: *Bold scaly pattern above*, especially on scapulars; *complete dusky cap*; bill usually dark, shorter than adult's. – 1st-year: Head and body like adult winter, but retains juvenile dark markings on coverts, secondaries and tail; outer primaries and tail-feathers worn, blackish by spring. Yellow bill-tip tiny or lacking.

VOICE Noisy at colonies, and when feeding during summer–early autumn. Flight-call distinctive, a loud, grating 'ker**rick**' (like pressing amalgam into tooth). Begging-call of juvenile / 1st-winter thin, high-pitched, ringing 'sree-sri'.

Gull-billed Tern *Gelochelidon nilotica* V**

L 35–42 cm, WS 76–86 cm. Breeds colonially in open, flat country at lakes, marshes (fresh, brackish or saline), sheltered coastal waters, fields on irrigated plains, and mountain lakes (to *c.* 2000 m) with grassy meadows and sandy shores. Summer visitor (Apr–Sep), winters in Africa; in Britain & Ireland a very few records annually. Food mainly insects, also frogs, small mammals, etc.

IDENTIFICATION Distinguished from Sandwich Tern at all ages by *short, thick, all-black bill* only half length of cap (about same length as cap on Sandwich, but beware of not fully grown, all-black bill of immature Sandwich); slightly *broader wing-bases* (though wings still both long and pointed); shorter neck (the tail is shorter, too, but appears equal in practice); and *longer legs* when perched. Feeding behaviour and *inland habitats* also distinctive: *does not normally plunge-dive* like Sandwich, but with leisurely flight *hawks insects in air*, often over dry meadows, at times even in open forests, or swoops to pluck prey from water surface or vegetation. – Adult summer: *Uniform frosty-grey* upperparts, including *rump and* (white-sided) *tail*; primaries often uniform (though, on most, outer 4–5 wear to darker grey during summer, at times creating dark wedge on 'hand' much as on Sandwich); from *below, dark trailing edge on outer primaries* longer and more clear-cut than on Sandwich, often also showing from above. – Adult winter: *Isolated patch or streak behind eye*, not continuing across nape. – Juvenile: Bill slimmer than adult's; most have much *less patterned upperparts and wings* than juvenile Sandwich, and all have *less black on head*. – 1st-summer: Head and body as adult winter; retained juvenile wings and tail have less obvious dark markings than Sandwich. Slightly more Common Gull-like wing shape (on adult 'arm' is slightly broader, making outer wing look shorter and more pointed).

VOICE Noisy at colonies. Flight-call is a deep, nasal, rising 'gur-**wick**' (slightly reminiscent of Bar-tailed Godwit display); alarm is a quick, laughing 'gwic-gwic-gwic'. Also, from breeders a bubbly or rattling 'br-r-r-r-...'. Juveniles beg with a piping, high-pitched 'pe-eep' (lacking the ringing tone of juv. Sandwich).

Little Tern

Sandwich Tern

Gull-billed Tern

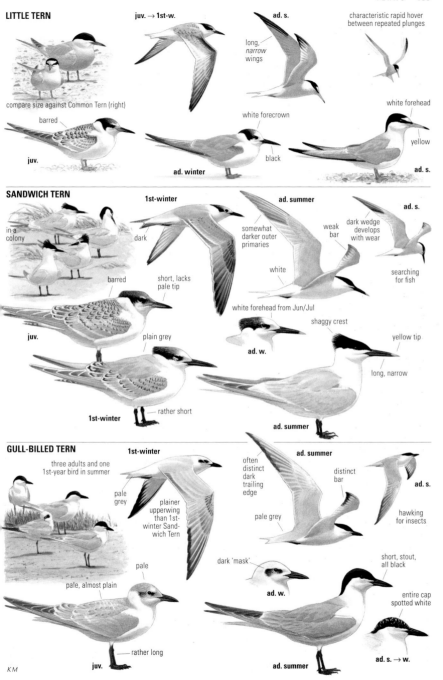

LITTLE TERN

juv. → 1st-w.

ad. s.

long, *narrow* wings

characteristic rapid hover between repeated plunges

compare size against Common Tern (right)

white forehead

white forecrown

yellow

barred

juv.

black

ad. winter

ad. s.

SANDWICH TERN

1st-winter

ad. summer

ad. s.

in a colony

dark

somewhat darker outer primaries

weak bar

dark wedge develops with wear

barred

short, lacks pale tip

white

searching for fish

juv.

plain grey

white forehead from Jun/Jul

shaggy crest

yellow tip

1st-winter

rather short

ad. w.

long, narrow

ad. summer

GULL-BILLED TERN

1st-winter

ad. summer

three adults and one 1st-year bird in summer

pale grey

plainer upperwing than 1st-winter Sandwich Tern

often distinct dark trailing edge

distinct bar

ad. s.

pale grey

hawking for insects

pale

dark 'mask'

short, stout, all black

pale, almost plain

ad. w.

entire cap spotted white

rather long

juv.

ad. summer

ad. s. → w.

KM

Common Tern *Sterna hirundo* mB3 / P2

L 34–37 cm (incl. tail-streamers 5–8 on ad.), WS 70–80 cm. Breeds colonially or singly both at coasts and at inland waters (lakes, rivers, etc.). Summer visitor (Apr–Oct), winters in W and S Africa. Food fish. Nest is scrape on ground.

IDENTIFICATION Very similar to Arctic Tern, but, with practice, separation nearly always possible. Apart from being slightly larger, different structure when perched useful: *longer bill* (often looking slightly decurved) and head, and *longer legs* (belly well clear of ground). Different structure also in flight: *longer neck/head and slightly broader wings*, which appear to be more or less central on body. Hovers, and dives directly for fish. – Adult summer: *Bill orange-red with black tip*, but note that black tip can be very small, especially by midsummer; underparts pale grey, usually without obvious contrast with white cheek; *short tail-streamers not extending beyond wing-tip*; outer 4–6 primaries darker than rest, forming *dark wedge on upperwing*, but note that wedge often is very faint in spring (only a dark 'notch' at trailing edge), becoming obvious by midsummer through wear; from below, *primaries white with broad, rather diffuse dark trailing edge*; translucency strongest on inner primaries. – Adult winter (some start moult in Jul; Arctic not until reaching winter range): White forehead and underparts, dark carpal bar, and all-black bill. – Juvenile: Upperwing has *pale square-shaped midwing-panel* between *narrow grey secondary bar* and *obvious dark carpal bar*; other differences from Arctic include: a little *orange at base of bill* (bill all black on Arctic from late summer); *forehead and upperparts obviously ginger-brown in tone* when recently fledged; extensive *carpal bar usually visible when perched* (smaller on Arctic, usually concealed). –1st-summer (usually stays in winter range; rare in Europe): White forehead and underparts; blackish bill; wholly or partly retained juvenile wings and tail much worn, giving striking variegated pattern. – 2nd-summer (many stay in winter range, scarce in Europe): Like adult summer, but has immature features such as white on forehead, patchy grey underparts, and blackish bill.

VOICE Noisy at colonies, including short, sharp 'kit', rapid series of quarrelling 'kt-kt-kt-kt-...' and typical '**kier**ri-**kier**ri-**kier**ri-...' with variations. Alarm is a disyllabic, drawn-out, downslurred 'kreee-arrr' or a sharp 'chip!'.

Arctic Tern *Sterna paradisaea* mB3 / P2

L 33–39 cm (incl. tail-streamers 7–11½ on ad.), WS 66–77cm. Breeds in colonies (in N often very large and dense) or singly along coasts, on islands in sea-bays, locally at ponds in taiga or barren mountains, and on tundra near water. Summer visitor (end Apr–Oct), winters off S Africa and Antarctica. Probably has the longest migration of all birds.

IDENTIFICATION Very similar to Common Tern (which see); Arctic told by: slightly smaller size; *shorter bill, head and neck* and *longer tail*, making wings look ahead of centre of body; *very short legs*; slightly *narrower wings*. Flight often more elastic and gracefully bouncing than Common Tern's, but display-flight of latter is just as elegant. Often dives with 'stepped hover', dropping short distance and hovering again before final plunge; at times snatches prey from surface in Black Tern fashion. – Adult summer: *Bill dark red* (blood-red) *without black tip*; *lower throat, breast and belly washed grey*, creating subtle contrast with white upper throat and cheeks, stronger than on Common; *long tail-streamers* extending beyond wing-tip when perched. *Upper-wing uniformly pale grey* (lacking dark wedge or 'notch' of Common), and *all flight-feathers near-white and translucent from below*, outer primaries with neat black trailing edge (narrower than on Common). – Juvenile: Carpal bar often fainter than on normal Common Tern, and *secondaries are whitish*, not shaded grey as on Common Tern; often a *white triangular area on hindwing*. Dark bill first red-based, from Aug/Sep all black. Forehead white (more clearly demarcated than on Common). – 1st-summer (scarce in Europe): White forehead, dark bill, a faint dark carpal bar (thus resembles adult winter).

VOICE Recalls Common Tern; includes piping, clear 'pi-pi-pi-pi-...', 'pyu pyupyu', and ringing '**prree**-eh', and quarrelling, hard rattling 'kt-kt-kt-krrr-kt-...'; alarm disyllabic 'krri-errrrr' (variable, like Common's or harder, drier).

Roseate Tern *Sterna dougallii* mB4

L 33–36 cm (incl. tail-streamers 8–11 on ad.), WS 67–76 cm. Breeds colonially at maritime coasts, on beach or offshore island. Summer visitor (mainly mid May–Aug), winters in W Africa. Scarce and local; less than 1000 pairs in Europe, and 100+ in Britain. Nest often in hollow.

IDENTIFICATION Compared with Common or Arctic, Roseate Tern has shorter wings and *faster, shallower wingbeats*, giving faster-looking more direct flight; hovers less and uses distinctive *angled power-dive, as if 'flying into water'* to catch fish. – Adult summer: Extreme paleness of grey upperparts and white-looking underparts give *outstanding whitish look* among greyer Common and Arctic; *bill all black or with red at base* (red becomes more extensive during summer); legs rather long, bright red; *underparts flushed very pale pink* (lacking any grey); *very long, flexible tail-streamers*; only outer 2–4 primaries darker than rest, forming *thin dark leading edge*; at rest, *whitish inner primaries* leave 'small dark wing-tip' (all primary tips rather dark on Common/Arctic). – Adult winter: White forehead and underparts, dark carpal bar, black bill; tail-streamers moulted Jul–Aug. – Juvenile: *Like miniature juvenile Sandwich Tern*, because of bold *scaly pattern on upperparts, complete blackish cap* (but forehead gradually turns white from Sep), all-black bill, and *blackish legs*.

VOICE Calls distinctive, a quick, disyllabic '**chiv**vick', slightly recalling Spotted Redshank, and a straight, deep, raucous 'krraahk'.

Common Tern

Arctic Tern

Roseate Tern

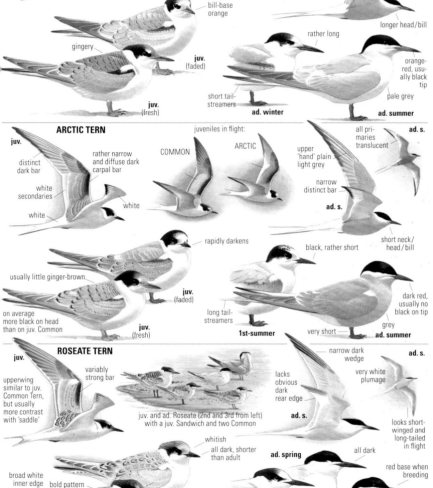

COMMON TERN

juv.

diffusely darker rear edge

dark carpal bar

dark secondary bar

pale grey

1st-summers in flight:

worn outer (juv.) primaries dark

all primaries fresh

COMMON

ARCTIC

dark wedge, faint in spring, more distinct in summer

only inner primaries translucent

ad. s.

broad diffuse bar

ad. s.

longer head/bill

bill-base orange

gingery

juv. (faded)

juv. (fresh)

rather long

oranged-red, usually black tip

pale grey

short tail-streamers

ad. winter

ad. summer

ARCTIC TERN

juv.

distinct dark bar

white secondaries

white

rather narrow and diffuse dark carpal bar

white

juveniles in flight:

COMMON

ARCTIC

upper 'hand' plain light grey

narrow distinct bar

all primaries translucent

ad. s.

ad. s.

short neck/head/bill

usually little ginger-brown

juv. (faded)

on average more black on head than on juv. Common

juv. (fresh)

rapidly darkens

long tail-streamers

1st-summer

black, rather short

very short

grey

ad. summer

dark red, usually no black on tip

ROSEATE TERN

juv.

upperwing similar to juv. Common Tern, but usually more contrast with 'saddle'

variably strong bar

juv. and ad. Roseate (2nd and 3rd from left) with a juv. Sandwich and two Common

lacks obvious dark rear edge

narrow dark wedge

very white plumage

ad. s.

ad. s.

looks short-winged and long-tailed in flight

broad white inner edge

bold pattern

whitish

all dark, shorter than adult

ad. spring

all dark

red base when breeding

juv. (pale)

majority have dusky forehead

dark

juv. (typical)

very pale

ad. winter

inner primaries pale with broad white inner edge

light rosy flush

ad. summer

KM

Caspian Tern *Hydroprogne caspia* V*

L 48–55 cm, WS 96–111 cm. Breeds colonially or singly, mostly on more remote offshore, low islands. Summer visitor (Apr–Oct; in Britain & Ireland about 5 records annually), winters in W Africa, occasionally Mediterranean Sea. Migration partly over land, following rivers over European continent, and crossing W Sahara in Mar and Nov. Food mainly fish, which during breeding can be caught far from colony (30–60 km, or more), often in fresh water. Nest is bare scrape on ground (sand or rocks).

IDENTIFICATION Largest tern; *wingspan larger than Common Gull*. Slow wingbeats and flight more ponderous than other terns, is more gull-like; also, looks decidedly *front-heavy* in flight. *Very large red bill* and *extensive dark* (blackish) *under primaries* at all ages. – Adult: *Bill bright red*, with small dark subterminal mark on most; legs black. In winter, *black on head extensive*, not thin-masked pattern of most other terns. – Juvenile: Full dark cap, extending further down on lores and cheeks than on adult; dusky-tipped orange bill; *pale legs*; rather faint dark markings on upperparts and tail; and inner wings rather uniformly pale, lacking obvious dark carpal bars of other large terns. – 1st-winter: Orange-red bill (only slightly paler than ad.winter) and dark legs develop quickly during first autumn. Forehead usually becomes slightly paler, too (still streaked dark), much as on adult winter. – 1st-summer: As adult winter, except wings and tail much worn or with obvious variegated pattern through moult.

VOICE Flight-call deep and fierce 'krre-ahk!' with hoarse, heron-like quality; the noise in colonies can be earsplitting. Begging-call of juvenile/1st-winter a repeated squeaky, whistled 'wee-**vi**'.

Royal Tern *Sterna maxima* V***

L 42–49 cm, WS 86–92 cm. Primarily an American species. In E Atlantic, breeds only in Mauritania, wandering north to NW African coast (exceptionally Spain). Rare vagrant to rest of W Europe, including Britain & Ireland, probably from N American population.

IDENTIFICATION Size between Sandwich and Caspian Terns. Distinguished from Caspian Tern at all ages by *smaller size*, *less stout bill*, slimmer wings, *more attenuated rear end*, *more forked tail*, and *dark below primaries confined to narrow trailing edge*, thus invariably lacking extensive blackish area of Caspian. Also similar to Lesser Crested Tern, which see for comparison (below). Told from Crested Tern (p. 414; a very rare vagrant to treated area) of Middle East race *velox* by distinctly paler grey back and upperwing (Crested is almost lead-grey) and orange bill (Crested yellow or duller greenish-yellow) which is straight and dagger-shaped (Crested

slightly slimmer and more downcurved). Also, in breeding plumage note white rump and tail-centre (Crested greyish), and black forehead (Crested narrowly white). Could also be confused with Elegant Tern (p. 414), extremely rare vagrant from C America, but is somewhat larger, has stouter, straighter and slightly more reddish-orange bill (Elegant slimmer, longer, slightly decurved bill which is orange with yellowish tip), and often more extensive white forehead and eye-surround in non-breeding plumage (Elegant invariably solidly black eye-surround). – Adult: Told from Caspian by *unmarked orange or orange-yellow bill* (red only exceptionally), which is slightly slimmer and more pointed; *shaggier crest*; and fact that cap is full only briefly in spring, then quickly reduced to *thin dark 'mask' in winter plumage*, developing progressively from as early as May–Jun. – 1st-year: Told from Caspian Tern by *yellowish or pale orange bill* (not orange-red); prominent dark carpal bar, greater-covert and secondary bars; and pale head pattern (like ad. winter). – Variation: Breeders in W Africa (ssp. *albididorsalis*), frequently (mainly in Jun–Oct) seen along Moroccan Atlantic coast, differ from American (*maxima*) in having on average longer, narrower and more yellowish-orange (less reddish-orange) bill, and in being slightly smaller and slimmer, but still much larger than Sandwich Tern. – NB: The identification of a vagrant orange-billed tern often depends on the assessment of a number of subjective criteria. Observe critically and carefully judge size, bill shape and length, mantle shade, colour of rump, etc.

VOICE Flight-call throaty 'kerriup' or 'krree-it', recalling Sandwich Tern but coarser and lower-pitched.

Lesser Crested Tern *Sterna bengalensis* V*** / (mB5)

L 33–40 cm (incl. tail-streamers *c.* 7on ad.), WS 76–82cm. Libyan breeding population winters along NW African coast; scarce in Gulfs of Suez and Aqaba (from Red Sea population). Very rare vagrant to Europe. Has bred a few times (Italy, Spain), in England some years in mixed pair with Sandwich Tern (apparently involving same bird which was present 1984–97). Habitat, food and breeding much as for Sandwich Tern.

IDENTIFICATION Size and plumage as Sandwich Tern, except that *bill is pale orange* (1st-year) *to bright orange* (adult), often shading to yellowish tip, and averaging slightly thicker-based and shorter than on Sandwich; compared with Sandwich Tern, *upperparts more uniform and slightly darker grey, including rump and tail-centre*; on juvenile, legs at first are dull brown-grey but quickly become blackish like adult. Distinguished from Royal Tern at all ages by: distinctly *smaller size*; *slimmer and more slender-tipped bill*; proportionately longer wings forming more upcurved rear end when perched; lighter flight action; more deeply forked tail; *upperparts darker grey* and (on adults) more uniform grey of about similar tone to Common Tern (Royal same tone as or paler grey than Sandwich Tern); and *rump and tail-centre grey* (white on Royal, but rump-centre grey on 1st-year Royal). – Juvenile/1st-winter: Like Sandwich Tern, but note yellowish-orange bill. (Very rarely, however, juvenile Sandwich Tern has a mostly pale orange or yellowish bill with just a dark culmen and tip.)

VOICE Flight-call a grating 'kerrick', like Sandwich Tern's.

Caspian Tern

Lesser Crested Tern

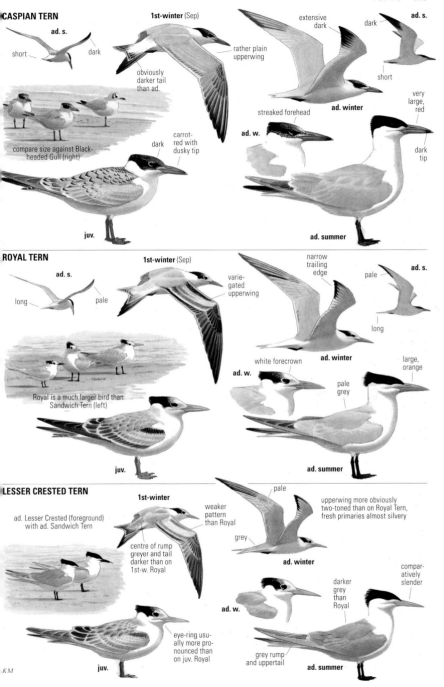

CASPIAN TERN

ad. s.

short

dark

compare size against Black-headed Gull (right)

1st-winter (Sep)

obviously darker tail than ad.

rather plain upperwing

dark

carrot-red with dusky tip

juv.

extensive dark

dark

ad. s.

short

streaked forehead

ad. winter

very large, red

ad. w.

dark tip

streaked forehead

ad. summer

ROYAL TERN

ad. s.

long

pale

Royal is a much larger bird than Sandwich Tern (left)

1st-winter (Sep)

varie-gated upperwing

narrow trailing edge

pale

ad. s.

long

white forecrown

ad. winter

large, orange

ad. w.

pale grey

juv.

ad. summer

LESSER CRESTED TERN

ad. Lesser Crested (foreground) with ad. Sandwich Tern

1st-winter

weaker pattern than Royal

centre of rump greyer and tail darker than on 1st-w. Royal

pale

upperwing more obviously two-toned than on Royal Tern, fresh primaries almost silvery

grey

ad. winter

compar-atively slender

ad. w.

darker grey than Royal

eye-ring usu-ally more pro-nounced than on juv. Royal

juv.

grey rump and uppertail

ad. summer

KM

Sooty Tern *Onychoprion fuscata* **V**∗∗∗

L 42–45 cm (incl. tail-streamers 7–10 on ad.), WS 72–80 cm. Breeds on islands in southern oceans and tropical waters, e.g. S Red Sea, Persian Gulf, and West Indies. Outside breeding, leads a partly pelagic life, rather like a skua. Rare vagrant to treated region, most often to W Europe and in summer.

IDENTIFICATION *Black-and-white* plumage, *dark rump*, *long*, narrow and pointed wings and long, deeply forked tail narrow choice down to this species and Bridled Tern (which see for comparison). Sooty Tern is *larger* (almost of Sandwich Tern size), with a proportionately *somewhat heavier head* and *longer bill*. Wings are slightly broader, and flight more powerful and steady (Bridled more narrow-winged with lighter flight). Primaries dark below (paler with dark tips on Bridled). – Adult: *Upperparts black* or very nearly black (clearly darker than Bridled, which is brown-grey, though beware of light effects making Sooty appear paler grey or Bridled look dark, too); *white forehead patch broad*, and *black loral stripe narrow at gape*. – Juvenile: Plumage largely sooty-grey, only lower belly whitish, and upperparts and upperwing-coverts having *white tips forming bars*. In flight recalls oversized, long-winged Black Tern, but note whitish lower belly (dark grey on Black) and dusky 'armpits' (light on Black). – Variation: Breeders of West Indies and S Atlantic (ssp. *fuscata*) have pure white underparts and underwing-coverts; breeders in Indian and Pacific Oceans (*nubilosa*) have a light grey wash below in fresh plumage.

VOICE Vagrants mainly silent, but occasionally utters very characteristic 'ker-**wac**ki-wack'.

Bridled Tern *Onychoprion anaethetus* **V**∗∗∗

L 37–42 cm (incl. tail-streamers 8–10 on ad.), WS 65–72 cm. Breeding areas include Caribbean, Mauritania, Red Sea and Persian Gulf. Habits similar to Sooty Tern's. Scarce vagrant to treated region, most often in summer.

IDENTIFICATION *Dark upperparts*—including rump and *deeply forked tail*—distinctive, shared only by similar Sooty Tern, but at long range White-cheeked Tern can be similar. Compared with Sooty, size and structure more like Common Tern, with narrow wings, long tail and rather small head, and *flight* often appears more *elegant and light*. – Adult: *White forehead patch narrow*, extending behind eye in a *sharp point*, and *bold black loral stripe*; pale collar on some (but can be absent; invariably lacking on Sooty); *upperparts paler*, *brown-grey*, with contrasting darker flight-feathers (at long range recalling adult Long-tailed Skua) and white outer tail-feathers (on Sooty, upperparts and upperwing uniform blackish, but beware effects of light making Sooty appear paler greyish, or Bridled look very dark). – Juvenile: Basic

pattern similar to adult, including whitish throat and breast, and whitish underwing with thin dark trailing edge, but head pattern less clear-cut, tail lacks obvious white outer feathers, and upperparts are either fringed buff, giving obvious scaly pattern, or darker and almost plain (1st-year Sooty quite different: sooty-grey on throat and breast, and sooty-black above, distinctly barred or spotted white).

VOICE Vagrants mainly silent.

White-cheeked Tern *Sterna repressa* —

L 32–34 cm, WS 58–63 cm. Breeds in Red Sea, S Sinai and in N Indian Ocean. Summer visitor (mainly Apr–Oct). Occasional in Gulf of Aqaba (Eilat).

IDENTIFICATION Rather similar to Common Tern, except slightly *longer bill*, narrower wings, and often more elegant, bouncing flight. – Adult summer: Like Common Tern, having black-tipped red bill, but: *darker grey above*; *grey rump and tail* concolorous with back; *darker grey underbody* with a clear contrast with *white cheeks* (recalling Whiskered Tern); whole *underwing has broad*, *diffuse*, *grey trailing edge*, contrasting with greyish-white underwing-coverts (often lightest along midwing-panel). – 1st-year: Like Common Tern, but juvenile usually more strongly marked dark above, and *rump/uppertail grey as back*; *plainer, more uniform grey upperwing* than on Common, with less obvious secondary bar (above) and, from most angles, a rather pale silvery-grey sheen on the primaries; bill soon becomes all black. Told from rather similar young Whiskered Tern by *longer*, *strongly forked tail*, and *longer bill* and longer, narrower wings.

VOICE Similar to that of Common Tern.

Forster's Tern *Sterna forsteri* **V**∗∗

L 33–36 cm (incl. tail-streamers *c.* 7½ on ad.), WS 64–70 cm. North American species. Vagrant to W Europe, mainly in winter; in Britain & Ireland near-annual.

IDENTIFICATION Size and shape much as Common Tern, except *slightly larger head*, *more wedge-shaped*, *conical bill*, *longer legs*, and *longer tail-streamers* extending well beyond wing-tips on perched adults. – Adult summer: Much as Common Tern, with orange-red black-tipped bill, but note: *tail centred pale grey*, leaving only rump pure white; *all primaries frosty-white above* (lacking dark outer wedge of Common); and *white underbody*. – Adult winter: *White crown*, leaving only a *prominent black 'mask'* around eye and on ear-coverts, and black bill, recalling Gull-billed Tern. Upperwing-coverts all grey, lacking dark carpal bar of winter Common Tern. (European records mainly in winter, so worth careful check of any tern in this season. Beware pitfall of distant adult winter Sandwich Tern, which has frosty-grey primaries Nov–Mar.) – 1st-year: Compared with young Common Tern, less contrasting dark wing and tail markings (lacks obvious carpal bar), and outer primaries only slightly darker than on adult, especially at tip; when perched, *dark-centred tertials* often obvious. As with adult, *heavy-based, long and pointed bill* useful clue. Overall whiteness and front-heavy, short-tailed look in flight recall Sandwich Tern, but Forster's has more distinct dark tips to outer primaries, bolder, more discrete 'mask', and reddish legs.

VOICE Call a rolling, slightly harsh 'kreerr', somewhat recalling Black-headed Gull.

Bridled Tern   White-cheeked Tern

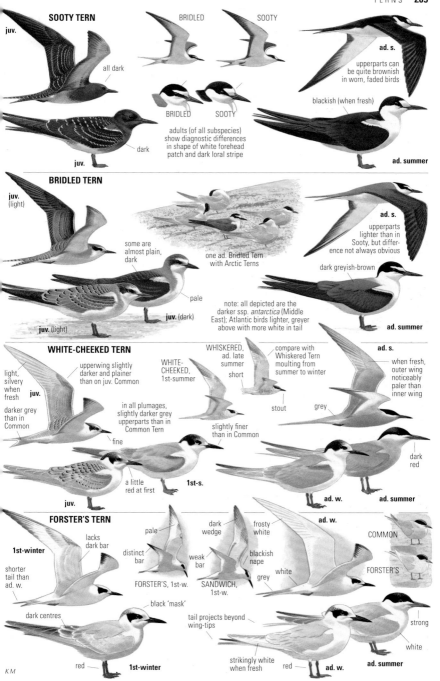

SOOTY TERN

juv.

all dark

dark

juv.

BRIDLED

SOOTY

BRIDLED

SOOTY

adults (of all subspecies)
show diagnostic differences
in shape of white forehead
patch and dark loral stripe

ad. s.

upperparts can
be quite brownish
in worn, faded birds

blackish (when fresh)

ad. summer

BRIDLED TERN

juv.
(light)

some are
almost plain,
dark

one ad. Bridled Tern
with Arctic Terns

pale

juv. (light)

juv. (dark)

note: all depicted are the
darker ssp. antarctica (Middle
East); Atlantic birds lighter, greyer
above with more white in tail

ad. s.

upperparts
lighter than in
Sooty, but differ-
ence not always obvious

dark greyish-brown

ad. summer

WHITE-CHEEKED TERN

WHISKERED,
ad. late
summer

compare with
Whiskered Tern
moulting from
summer to winter

ad. s.

WHITE-
CHEEKED,
1st-summer

short

when fresh,
outer wing
noticeably
paler than
inner wing

light,
silvery
when
fresh

upperwing slightly
darker and plainer
than on juv. Common

juv.

stout

grey

darker grey
than in
Common

in all plumages,
slightly darker grey
upperparts than in
Common Tern

slightly finer
than in Common

dark
red

fine

a little
red at first

1st-s.

juv.

ad. w.

ad. summer

FORSTER'S TERN

pale

dark
wedge

frosty
white

ad. w.

COMMON

1st-winter

lacks
dark bar

distinct
bar

weak
bar

blackish
nape

white

FORSTER'S

shorter
tail than
ad. w.

FORSTER'S, 1st-w.

SANDWICH,
1st-w.

grey

dark centres

black 'mask'

tail projects beyond
wing-tips

strong

red

1st-winter

strikingly white
when fresh

red

ad. w.

white

ad. summer

K M

MARSH TERNS *Chlidonias*

The three 'marsh terns' *Chlidonias* differ from 'sea terns' *Sterna* et al. in their lazier flight with more banking from side to side (sea terns have generally firmer wingbeats and steadier, more direct flight). Marsh terns also differ in having slightly shorter and broader wings and less forked tail than the sea terns. The marsh terns feed by dipping down to the water surface to take insects (only occasionally plunging more deeply after fish, etc.), or by catching insects in the air with agile, hawking flight. They do not hover-and-dive after fish like sea terns, but note that sea terns—especially immatures—will also dip-feed or hawk insects in the manner of marsh terns. Marsh terns breed in loose colonies in shallow freshwater or marshy habitats. On migration they also frequent coasts or large, open inland waters.

Black Tern *Chlidonias niger*   **P**4

L 22–26 cm, WS 56–62 cm. Commonest marsh tern, breeding in freshwater marshes with patches of low, floating vegetation, often among Black-headed Gulls for protection. Summer visitor (mainly late Apr–Sep), winters in Africa. In Britain & Ireland, occasional large migrant gatherings or coastal passage, but seldom numerous.

IDENTIFICATION Adult summer: All-black head and body, rather *uniform dark grey upperparts*, white undertail, and dark legs and bill; head and underparts mottled or patched with white on moulting adults in spring (before May) and summer/autumn (white feathers on head already appearing in Jun). – Adult winter (head and body plumage fully acquired Sep onwards): Much as juvenile, but *upperparts plain grey*, lacking scaliness, and *outer 5–7 primaries darker than rest*. – Juvenile: *Extensive black cap* and *obvious dark breast-side patches*; scaly pattern on darkish saddle and wing-coverts; darker carpal and secondary bars; and outer primaries not contrastingly darker than inner. Could be mistaken for juvenile Common / Arctic Tern, but note smaller size, *shorter, less forked tail*, slightly *shorter wings*, and dark breast-side patches; mode of flight (see above under 'Marsh terns') also good clue. – 1st-summer (usually stays in winter range): Variable; like adult winter but usually differing in having darker, more worn outer primaries and darker secondaries, and on some a few scattered dark feathers on underbody and head (these safely told only in May–Jun from moulting ad.).

VOICE Flight-call a harsh, almost multisyllabic 'k'shlet' (slightly recalling Little Tern), and short, nasal 'klit' and '**klee**-a'. Alarm-call energetic 'ki-ki-ki...', or a shrill 'kyeh'.

White-winged Tern *Chlidonias leucopterus*   **V***

L 20–24 cm, WS 50–56 cm. A rarer and more easterly relative of Black Tern, breeding in similar habitat. Summer visitor (May–Sep), winters in Africa. Vagrant to Britain & Ireland, 10–40 records annually.

IDENTIFICATION Structure subtly different from Black Tern: bill shorter, legs longer, wings broader, and tail only faintly notched. – Adult summer: *White forewing* (diffusely demarcated); *white rump*; contrasting *black saddle and underwing-coverts*; and *bright red legs*; moulting adults have white-mottled head and underparts as Black Tern, but are easily told by their retained *black underwing-coverts*. – Adult winter (head and body plumage fully acquired Sep onwards): Told from Black Tern by *whiter crown* and *lack of breast-side patches*; *whitish-grey rump and tail-sides*; and (at least while still in Europe) *narrower dark leading edge to outer wing* (only 1–4 outer primaries darker than rest), and often *some black retained on underwing-coverts*. – Juvenile/1st-winter: Told from Black Tern by *lack of dark breast-side patches*, and from late summer by *more white on forecrown*; *brownish-black saddle contrasting strongly* with *pale grey inner wing* and *white rump* (though beware of some Black Terns with darker saddle and very pale grey rump); tertials and rear scapulars rather plain (tipped pale on Black); *tail pale grey with white sides*. At first glance can look surprisingly like young Little Gull before different shape and upperpart plumage become apparent. – 1st-summer: As for Black Tern.

VOICE Lower-pitched and often harsher than Black Tern's: conversational, soft 'kek'; and typical rasping 'chr-re', recalling Grey Partridge (and Little Tern) in tone.

Whiskered Tern *Chlidonias hybrida*   **V****

L 24–28 cm, WS 57–63 cm. Breeds at lakes, rivers and marshes. Summer visitor (late Apr–Sep), winters in Africa, locally Mediterranean Sea. Vagrant to Britain & Ireland.

IDENTIFICATION Larger, longer-legged and often *stronger-billed* than other marsh terns, and flight action less 'nimble'; plumage can recall Common / Arctic Tern, so *shallow tail-fork* and *greyish rump and tail* important to note. – Adult summer: *Breast and belly dark grey*, contrasting strongly with *white cheek and underwing-coverts*; bill dark red; legs red. Can look quite pale at long range. – Adult winter: Rear crown finely streaked black, and a *solid black patch behind eye*; told from Black Tern by *dark smudge at breast-sides being faint or lacking*, from White-winged Tern by *plainer upperwing* lacking any contrastingly darker outer primaries. – Juvenile: Could be overlooked as White-winged Tern, but note: *broader wings*; often *stronger bill*; *pale grey rump and tail* (latter usually with thin dark end-band); *faint or no carpal bar*; *more coarse, black-spotted and scaly pattern on saddle*. From juvenile Common / Arctic Terns by: *grey rump and tail*; *paler wings*; and *contrasting saddle*. From juvenile Arctic also by *darkish secondaries*. – 1st-winter (head and body plumage moulted early Jul–mid Sep): Much as adult winter except juvenile wings (with dark secondaries) and tail retained. – 1st-summer: As for Black Tern.

VOICE Call is a squeaky, sharp 'zrik!' or a rasping 'cherk' (or 'krche'), almost recalling Corncrake in tone.

Black Tern

White-winged Tern

Whiskered Tern

BLACK TERN

juv.
(typical)

grey

dark breast patch

juv.
(variant with darker back and paler rump – cf. White-winged!)

ad. w.

grey

dark patch

ad. s.

pale

ad. s.

grey

ad. s. → w.

extensive black cap

conspicuous pale tips

dark breast patch

juv.

solidly black

comparatively long

ad. w.

plain grey

ad. s.

WHITE-WINGED TERN

white

juv.

white breast-side

juv. → 1st-w.

pale wings and pale rump contrast with very dark back ('saddle')

narrow pale tips

longer legs than Black Tern

white breast-side

juv.

ad. w.

pale grey

whitish

lacks dark patch

streaked

dark ear-spot

short

ad. w.

ad. s.

black

ad. s.

white

white

ad. s. → w.

note black underwing-coverts

greyish-black back

white coverts

ad. s.

WHISKERED TERN

gingery tinge

juv.

usually lacks dark carpal bar

juv. → 1st-w.

coarsely patterned

juv.

ad. w.

very pale grey

hint of breast-side patch

lower edge of cap rather straight

rather strong

ad. w.

ad. s.

ad. s.

tail shorter and rump greyer than Arctic Tern

pale silvery

ad. s. → w.

dark grey

ad. s.

KM

AUKS *Alcidae*

Black and white, medium-large or rather small marine birds, which come to land only to breed. Body elongated, with webbed feet placed far back; wings rather small and narrow. The 'penguins of the Northern Hemisphere', they dive skilfully to catch fish and crustaceans. Dive from surface with a flick of wings and 'fly' under water, using feet for steering. They practically always fly close to the water, with rapid, whirring wing-beats, in small 'trains' of several birds in closely spaced line. Winter at sea, more commonly seen at coasts during and after gales. Nest on cliff-ledge or in crevice.

An auk opens its wings and dives with typical flick.

'Train' of auks flying close to the water with rapid wingbeats. Here a Razorbill (last in line) and three Guillemots.

Little Auk *Alle alle*   W4

L 19–21 cm, WS 34–38 cm. Arctic species, breeding in huge colonies on coastal mountainsides, in places far from sea; winters south to North Sea. In N Britain, regular winter visitor, rare on other coasts, usually after gales. Often migrates and dives closer to land than its relatives. Food primarily planktonic. Nests in a crevice or under boulders in scree.

IDENTIFICATION *Very small*, appears only half size of Puffin, almost like small wader. *Bill very short* and stubby. Scapulars edged white. Flight light, more prone to veer than its relatives. Usually floats very low between dives, wings dragging; otherwise swims higher, with wings tucked in; may stretch neck. Takes off freely and without splash. Most obvious risk of confusion is with 1st-year Puffin because of latter's rather small, dark bill and fairly small size and similar *dark underwing*: best distinguished by *c*. 50% smaller size; *moth-quick, whirring wingbeats* with individual beats difficult to discern; somewhat narrower and *more backswept wings*, which appear central on body; white trailing edge to secondaries; smaller head; and lack of complete dark breast-band. Confusion with the larger auks possible, too, if observation is only fleeting and angle of view

poor; usually the whirring, quick wingbeats are obvious, but note also the grey (all-blackish looking) underwing.

VOICE Noisy at colonies, mostly silent elsewhere. Twittering, chattering notes turning into a laughter, 'krrii-ek ek ak ak ak', merge into a buzzing chorus. Alarm given on wing, a whinnying 'huhuhuhuhu...'.

(Atlantic) Puffin *Fratercula arctica*   m(r)B2

L 28–34 cm, WS 50–60 cm. Nests colonially in burrows on grassy, steep slopes by coastal cliffs. Common near colonies; winters further out to sea than other auks.

IDENTIFICATION Adult summer: *Parrot-like, huge, colourful bill*. At close range, unmistakable combination of bill, *large, pale, rounded 'cheek patch'* and dark unbroken breast-band. Floats high. Appears 60% the size of Guillemot. At distance in flight, told from that by *shorter, thicker body*; *larger, rounder head, pale in summer*; rather *dark underwing* (some greyish-white visible at close range); lack of white trailing edge to 'arm'; rump lacks white sides; 'stern' short, giving impression of *wings placed far back on body*; and orange feet often surprisingly obvious. – Adult winter: Bill smaller, eye-surround dark greyish. – Juvenile/1st-winter: Resembles adult winter, but bill much smaller and darker: can be confused with Little Auk (which see).

VOICE Deep, grunting 'arrr-uh' etc., mostly uttered from burrow.

Black Guillemot *Cepphus grylle*   r(m)B3

L 32–38 cm, WS 49–58 cm. Nests among boulders and in rock crevices, singly or in small, loose colonies. In Ireland and NW Britain fairly common; mainly sedentary.

IDENTIFICATION Clearly smaller than Guillemot, slightly larger than Puffin. Bill pointed, black. Head rather small, body plump and pot-bellied, appears pear-shaped. Flight close above water with *quick, whirring wingbeats*; *stern-heavy*. – Adult summer: *All black with large oval upperwing patches*; underwing largely pure white. Feet red. Heavy 'stern' together with rapid wingbeats immediately rules out drake Velvet Scoter (p. 38), even when long range makes different position of white wing patches difficult to see. – Adult winter: Very different, with *white underparts* and variable *white barring above*. In flight, superficially resembles a winter grebe, but told by shape, mode of flight, still obvious oval white upperwing patches, and extensively white underwing and rump area. – Juvenile/1st-winter: Light wing patch profusely barred dark; head, breast and rump usually duskier than on adult. – 1st-summer: Black body like adult, but barring on wing patch retained, on some birds more extensively and making wing look dark. (A rare all-black adult variety, *'motzfeldi'*, occurs in far N.)

VOICE Displaying ♂♂ call with a series of very fine pipit-like notes, 'sipp-sipp-sipp-...'. Sometimes a similar but scraping version is heard, 'sipp-sü-sipp-sü-sipp-sü...'. Also to be heard are very fine, drawn-out piping 'seeeeeeuu' calls.

Little Auk

Puffin

Black Guillemot

LITTLE AUK

PUFFIN, juv.   LITTLE AUK, winter

juvenile Puffins look small, but
Little Auks look tiny!

dusky
underwing

white
trailing
edge

white streaks

winter

white

ad. s.

beware of confusion
with 'baby' Razorbill!

ad. summer

small

white streaks
diagnostic

winter

winter

wings dragging between dives

PUFFIN

dusky
underwing

no white

RAZORBILL, winter   PUFFIN, winter

grey

dark 'thigh'

ad. winter

ad. s.

compared with Razorbill, Puffin is more
compact, has dark underwing (white wing-
coverts of Razorbill flicker white) and is
darker at both head and tail ends; also
has complete breast-band

note
small
bill

dusky grey

ad. winter

**ad.
summer**

juv.

BLACK GUILLEMOT

very pale

white

adult winter
at long range

ad. winter

point-
ing up

ad. s.

'vent-heavy'

compare with Velvet
Scoter (left)

some look all dark
at long range

dark spots

1st-winter

1st-s. (dark)

unmarked

ad. winter

ad. summer

juv.

KM

(Common) **Guillemot** *Uria aalge* r+m**B**1

L 38–46 cm (excl. exposed feet *c.* 4), WS 61–73 cm. Commonest auk in British Isles. Nests on bare, narrow cliff-ledge on steep coastal cliffs, in often large colonies. Single egg laid directly on ledge, pear-shaped to prevent rolling off, incubated by bird standing up. Parents recognize egg by pattern and young by voice. Young jumps from cliff when three weeks old, still not fledged, tended at sea by ♂ alone.

IDENTIFICATION Head and upperparts brownish, underparts white. Percentage of distinctive 'bridled' variety (see plate) increases towards north. At close range *slender, pointed bill* obvious. Greatest risk of confusion with Brünnich's Guillemot (which see), but much more common problem is to distinguish Guillemot from Razorbill at long range, when bill shape becomes surprisingly difficult to see. Guillemot has *paler greyish-brown upperparts* (Razorbill appearing almost black); difference obvious in W Europe (esp. small race *albionis* of S Britain is pale), less obvious but still present in N Norway. Also, Guillemot is *slightly larger* than Razorbill—in mixed 'trains' Guillemots are usually longer-bodied (but beware of variation, and in winter larger birds from the north may in some areas mix with smaller in the south). Variable amount of *dark streaking on flanks*, in 'armpits' and on underwing-coverts, but some populations (e.g. S Britain and Baltic) average less streaked and some unstreaked (Razorbill always pure white). Further, flying Guillemot usually told by *wide, dark brown rump area, generally with uneven border to narrow white rump-sides* (Razorbill has narrower black central rump band, broadly and evenly bordered pure white in Long-tailed Duck fashion); feet projecting beyond tip of short tail, giving *'untidy' end to stern* (feet tucked neatly beneath tail on Razorbill); slightly *hunchbacked outline* (Razorbill appears straighter-backed); and wings set a trifle ahead of centre (wings appear central on Razorbill). – Adult winter: Amount of white on side of head varies, just as on Razorbill, but Guillemot generally appears *more white-headed*. Acquires summer plumage early, (Nov) Dec–Feb, and can then be distinguished from 1st-winters, which moult later. – 1st-summer: Like adult, except variable amount of white on throat, and retained juvenile flight-feathers much faded, brownish.

VOICE Vocal in colony, e.g. 'stomach-rumbling' 'mmm...'; hard and nasal notes repeated in staccato, 'ha ha ha ha...', turning into prolonged bellowing '...ha-aahr', etc. Young give high-pitched, disyllabic 'plee-ü' after leaving nest.

Brünnich's Guillemot *Uria lomvia* V**

L 40–44 cm (excl. exposed feet *c.* 4), WS 64–75 cm. Arctic species. Nest site and breeding much as Guillemot (though selects narrower ledges on average), often in mixed colonies.

Winters south to C Norway; rare vagrant to British Isles.

IDENTIFICATION Told from Guillemot by *shorter, stouter bill* with pronounced gonydeal angle (but beware some variation in Guillemot); usually obvious *whitish line along upper edge of gape* (but note that exceptional Guillemots have hint of pale gape streak); *lack of obvious dark streaking on flanks* (a few have faint streaks low down); in summer, *white comes to sharp point on foreneck* (not rounded as on Guillemot). At longer range, flying birds can generally be separated from Guillemots by *slightly shorter body and more tucked-in neck* (valid only at breeding sites; in winter, comparison with smaller W European Guillemot *albionis* may be necessary, giving reversed size difference); *darker, almost blackish upperparts* with *broader white sides to black rump*, more *round-backed and pot-bellied outline*, and largely white 'armpits'. At distance in flight, Razorbill-like impression because of blackish upperparts and similar rump pattern, but Brünnich's told by slightly larger size, *shorter tail* and projecting feet, *round-backed outline, and pointed bill held inclined downwards*.– Adult winter: Wholly dark ear-coverts with no white behind eye. Greatest identification pitfall juvenile/1st-winter Razorbill, dark birds of which can have head pattern close to that of Brünnich's and a considerably weaker and more pointed bill than adult Razorbill; they are smaller, however, and usually have at least some white visible around ear-coverts and behind eye. – 1st-summer: Same differences as for Guillemot.

VOICE Calls are similar to Guillemot's, but tone is harder, 'meaner' and more crow-like.

Razorbill *Alca torda* m(r)**B**2

L 38–43 cm, WS 60–69 cm. Breeds on steep coastal cliffs, preferring broader, sheltered ledges; also among boulders or in burrows. Will accept lower rocks and islets if sufficiently remote.

IDENTIFICATION Distinctive combination of *deep, blunt bill*, flattened laterally, *black with white lines*; *blackish upperparts* (always blacker than Guillemot, but difference subtle in far N); *powerful neck*; *long, pointed tail*; and consistently *smaller size* than Guillemot at breeding sites (difference slightest in England and France). Bill shape surprisingly difficult to use when watching flying birds at some range. Concentrate on blacker upperparts, *straighter back, whiter underwing* and 'armpit', and *more neatly defined white rump-sides*. Long tail covers feet, whereas these generally visible at rear end on guillemots. See also Guillemot and Brünnich's Guillemot for detailed comparisons. – Adult summer: White line across tip of bill; white loral line. – Adult winter: Generally retains white bill line but lacks loral. – 1st-winter: Lacks white bill line (but often has hint of loral) and has *considerably weaker and more pointed bill*. At least has a small white smudge behind eye on ear-coverts. – 1st-summer: Same differences as for Guillemot.

VOICE Rather silent. Very deep, creaking 'urrr' heard from breeding birds.

Guillemot

Brünnich's Guillemot

Razorbill

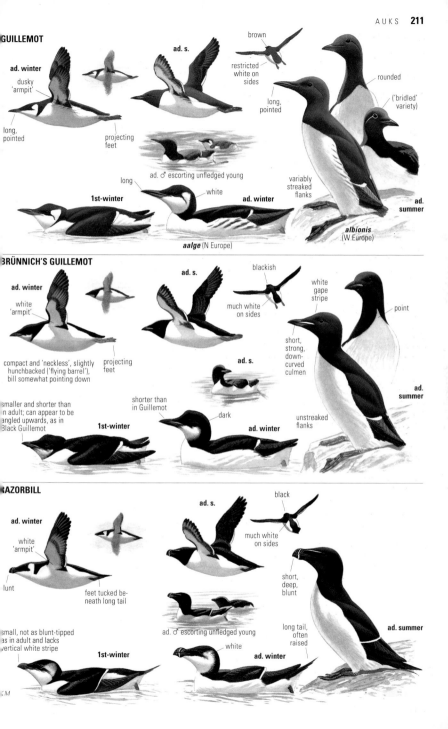

GUILLEMOT

ad. winter

dusky 'armpit'

long, pointed

projecting feet

ad. s.

brown

restricted white on sides

long, pointed

rounded

('bridled' variety)

variably streaked flanks

ad. ♂ escorting unfledged young

long

white

1st-winter

ad. winter

aalge (N Europe)

albionis (W Europe)

ad. summer

BRÜNNICH'S GUILLEMOT

ad. winter

white 'armpit'

compact and 'neckless', slightly hunchbacked ('flying barrel'), bill somewhat pointing down

projecting feet

ad. s.

blackish

much white on sides

white gape stripe

point

short, strong, down-curved culmen

ad. s.

smaller and shorter than in adult; can appear to be angled upwards, as in Black Guillemot

shorter than in Guillemot

dark

unstreaked flanks

1st-winter

ad. winter

ad. summer

RAZORBILL

ad. winter

white 'armpit'

blunt

feet tucked beneath long tail

ad. s.

black

much white on sides

short, deep, blunt

small, not as blunt-tipped as in adult and lacks vertical white stripe

ad. ♂ escorting unfledged young

white

long tail, often raised

ad. summer

1st-winter

ad. winter

KM

SANDGROUSE *Pteroclididae*

Medium-sized birds with small head, short neck and sturdy, compact body. Closely related to pigeons and waders, but also have partridge-like traits. Fast, strong flyers with long, pointed wings. Some species have pointed tail projections. Inhabit mostly deserts or arid plains. Flocks seen visiting freshwater sources in morning or evening, flying long distances to reach these. Partly nomadic in dry season, but details poorly known. Lay 2–3 eggs directly on ground.

Black-bellied Sandgrouse *Pterocles orientalis* —

L 30–35 cm. Breeds on poorly vegetated level plains; ascends to higher levels, is not so dependent on warm conditions as most congeners. Visits waterholes mostly in morning, sometimes flying very long. Wary and often shy.

IDENTIFICATION Big and *sturdy*, with *rather broad wings* and *distended silhouette*. Immediately told by *large black belly patch* on both sexes (but at long range can therefore recall Golden Plover, also suggested by golden-brown upperwing and rapid, straight flight on long, pointed wings). Strong contrast below between black remiges and white coverts. – ♂: *Unmarked greyish breast patch*. Rusty-yellow throat with black centre. Upperparts densely covered with large rusty-yellow spots. – ♀: *Yellow-brown breast patch finely spotted black*. Upperparts finely vermiculated.

VOICE Flight-call a rolling or bubbling '**chürrr´r**´re-ka´, slowing down at end (final syllable lower), at close range with a musical overtone, at distance roughly like a snorting horse (and then blurred into 'cheurrr´r´). On rising sometimes a high'chiiu', almost like Little Owl's call.

Pin-tailed Sandgrouse *Pterocles alchata* —

L 28–32 cm (plus tail projection 2–10), WS 55–63 cm. Breeds on open, dry lowland plains, sandy uncultivable tracts or dried-out lake beds and river mouths. Visits waterholes in morning. May be seen in flocks of thousands.

IDENTIFICATION Smaller and slimmer than Black-bellied Sandgrouse. *Long, narrow tail projection* (longest on adult ♂). *Belly and underwing-coverts white*, contrasting strongly with black remiges. Both sexes have *black loral stripe* and black-framed breast patch. – ♂: *Wing-coverts golden-green* with narrow black cross-bars; shoulders and back have rounded yellow-green spots. *Narrow black bib*. Crown, nape and neck-band plain olive-green. *Breast patch reddish-brown*. – ♀: Like ♂ but has *less greenish and more black-vermiculated wing-coverts*, densely black-vermiculated crown and nape, *whitish chin, two black bands between throat and breast* (♂ only one), and paler brown breast patch.

VOICE In flight a hard grating, descending'rreh-a'(nasal; at distance vaguely recalls Mediterranean Gull).

Spotted Sandgrouse *Pterocles senegallus* —

L 29–33 cm (plus tail projection 3–6). Like Crowned Sandgrouse, breeds in flat, desert-like, stony terrain at lower levels. Visits waterholes in morning, in often large flocks.

IDENTIFICATION Medium-sized, pale yellow-buff sandgrouse with protracted, attenuated rear body and *narrow tail projection*. In flight, *secondaries are contrastingly black below* (primaries, however, diffusely paler). A narrow black patch down belly-centre. – ♂: Crown, nape and breast light blue-grey, *throat, cheeks and neck-sides bright orange*. – ♀: Rather like ♂ Crowned, but has *tail projection*, more *buff ground colour* and *sparser spotting above*. Finely *dark streaked across breast*, also on crown and hindneck.

VOICE Flight-call in chorus from flocks visiting water holes a disyllabic, sharp 'ku**itt**-o' (akin to Lichtenstein's) sometimes preceded by a monosyllabic 'kvi'.

Crowned Sandgrouse *Pterocles coronatus* —

L 25–29 cm. Breeds at lower levels in deserts and semi-deserts and on dry, stony plains with sparse vegetation. Visits waterholes normally only in morning. Not over-shy.

IDENTIFICATION Between Lichtenstein's and Spotted Sandgrouse in size and plumage. In flight, *blackish remiges* contrast with paler coverts both above and below. – ♂: *Plain pinky-grey below*. Head- and neck-sides pale orange. A white-edged *black vertical mark on chin and forehead*, with white around eye. Crown-sides blue-grey and crown-centre reddish-brown. Upperwings have *dull yellow drop-shaped spots*. – ♀: Densely *spot-barred with black* (thus like Lichtenstein's), but has almost *unmarked buff-white throat*.

VOICE Flight-call a (four- or) five-syllable, rather hard, somewhat nasal 'chu-ku chu-ku-kurr' with jolting rhythm.

Lichtenstein's Sandgrouse *Pterocles lichtensteinii* —

L 22–25 cm. Within the region breeds in S Morocco, Sinai and S Israel in wadis and flatter mountain tracts in barren or desert-like, stony terrain. Visits waterholes at late dusk, flocks usually small (<50). Vigilant but not really shy.

IDENTIFICATION Compact, with *dark, densely vermiculated plumage*. Rather tall, and with rear end not so protracted and pointed as on congeners, and walks with tail somewhat more raised. Yellow bare skin around eye. – ♂: *Broad black and white bars on forecrown*. *Rusty-buff and black band across lower breast* and greater wing-coverts. Bill orange. – ♀: *Finely vermiculated all over*, without ♂'s head and breast markings. Bill grey.

VOICE Evening gatherings at waterholes give evocative full 'ku**it**(a)l' in chorus; the call has a metallic secondary note. Alarm-call a croaky, noisy, dry and excited'krre-krre krre-krre-...', also uttered on rising.

Black-bellied Sandgrouse   Pin-tailed Sandgrouse

Spotted Sandgrouse   Crowned Sandgrouse

BLACK-BELLIED SANDGROUSE

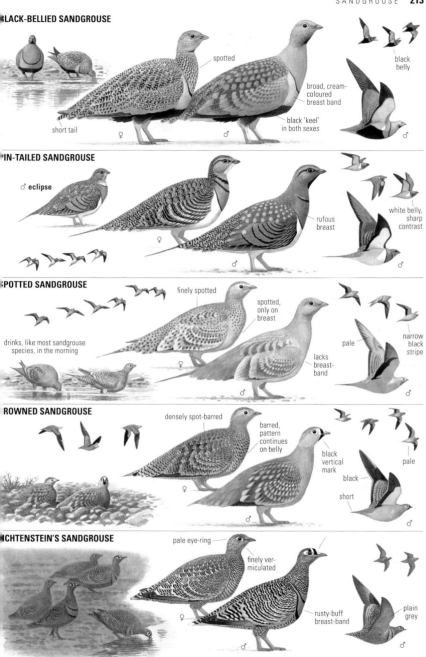

spotted

broad, cream-coloured breast band

black 'keel' in both sexes

black belly

short tail

♀

♂

♂

PIN-TAILED SANDGROUSE

♂ eclipse

rufous breast

white belly, sharp contrast

♀

♂

♂

SPOTTED SANDGROUSE

finely spotted

spotted, only on breast

drinks, like most sandgrouse species, in the morning

pale

narrow black stripe

lacks breast-band

♀

♂

CROWNED SANDGROUSE

densely spot-barred

barred, pattern continues on belly

black vertical mark

black

pale

black

short

♀

♂

♂

LICHTENSTEIN'S SANDGROUSE

pale eye-ring

finely ver-miculated

rusty-buff breast-band

plain grey

♀

♂

♂

drinks late at dusk

Z

PIGEONS and DOVES *Columbidae*

Medium-sized, compact birds with full, rounded breast (due to strong wing muscles!) and small head. Legs short, covered by scales. Fly rapidly in straight path with clipped wingbeats. Walk with bobbing head movements. Food mainly plant material (grain, seeds, shoots, fresh leaves) mostly taken on ground. Can drink with bill lowered in water (can suck up water, unlike other birds), and possibly due to this have completely or partly covered nostrils. Nest in holes in trees or rocks or build (casual) twig nest. The young are fed on 'pigeon-milk' from the crop.

Rock Dove *Columba livia* r**B**4

L 30–35 cm, WS 62–68 cm. Breeds in caves and on steep cliffs, mostly on sea coasts but also in mountains. Resident. Formerly widespread in Europe, now more local, mainly confined to Mediterranean area and coastal cliffs in W Europe. Small flocks are often seen on purposeful flight at high altitude between breeding and foraging sites. Is the ancestral form of the Feral Pigeon, and often reddish or pied birds, in all likelihood Feral Pigeons, join flocks of pure Rock Doves in wilderness, far from nearest town.

IDENTIFICATION Rather like Stock Dove in plumage and shape, but is *paler grey on back and upperwings*, making head, neck and breast stick out as contrastingly darker (Stock Dove is more uniform) and has *white underwings* (Stock Dove: grey). Has *snow-white rump* (light grey in Asia) and *two complete black wing-bars*. Eye reddish, bill dark. Flight fast and strong. In display, glides with wings held in V, just like Feral Pigeons practise in towns.

VOICE Coos like Feral Pigeon, a moaning 'drru**oo**-u', repeated several times, often with faint increase in volume.

Feral Pigeon *Columba livia* (domest.) r**B**2

L 29–35 cm, WS 60–68 cm. City pigeons (gone wild from dovecotes and pigeon-houses) originate from the wild Rock Dove. The species was also once used as the carrier pigeon, and is still domesticated in the form of the racing pigeon. Widespread and abundant in towns and cities. Resident.

IDENTIFICATION Identical to Rock Dove in shape, and some of the numerous plumage variants are also exactly the same as the wild form; other variants include variegated white and grey, all dark grey, dark pink-buff or dark piebald. Or they look just like a Rock Dove but without the white rump, or may be revealed by the presence of odd white remiges.

VOICE The cooing seems identical to that of Rock Dove.

Stock Dove *Columba oenas* r**B**2

L 28–32 cm, WS 60–66 cm. Breeds in wooded areas, forest edge and larger, undisturbed parks with mature oaks and other deciduous trees which provide nest holes (commones nest choice), locally also on rocky coasts or on buildings or ledges or in holes. Rather shy. Presence perhaps most often noted due to song.

IDENTIFICATION Rather like a Feral Pigeon in size and shape, thus smaller and *shorter-tailed than Wood Pigeon*, this obvious in mixed flocks in flight. Somewhat more vigorous flight with *faster flicking beats* than Wood Pigeon (though difference not great). Plumage blue-grey with shimmering green and vinous neck patch. In flight, show *grey underwings* (Rock Dove and Feral Pigeon largely white) and *paler blue-grey back and wing-covert panel above*. Small black wing patches near body but lacks full dark wing-bars (cf. Rock Dove) or white ones (cf. Wood Pigeon). Sexes and ages similar. Flight display with slow deep beats followed by glide on slightly raised wings and at times wing-claps above its back.

VOICE Song a rather monotonous series of disyllabic, hollow moaning '**ooo**-ue' notes, weaker at start of series, one note per sec. Wing-claps loudly on take-off, and has a fine whistling wing sound in normal flight.

(Common) Wood Pigeon *Columba palumbus* r**B**1/**W**+**P**

L 38–43 cm, WS 68–77 cm. Breeds in woods (esp. in arable farmland areas), parks and gardens, even in city centres. Forages on grass lawns and fields. Has increased greatly, and also become less shy, in Britain and N Europe. Resident in Britain & Ireland, where large numbers also visit in winter from N and E Europe, with migrating flocks 1000-strong seen in October, and as far south as the Pyrenees, large flocks can be seen in late autumn. Nest a simple (not to say careless) platform of sticks in tree, close to trunk, sometimes so thin as to make eggs visible from below through the nest! Prolonged breeding, can have recently fledged young May–Oct.

IDENTIFICATION A large pigeon, *clearly bigger than Feral Pigeon* and with proportionately somewhat longer tail and smaller head, and is also fuller-breasted but nonetheless *more elongated*. Best identified by *large white patch on neck side* and in flight a *white transverse band on upperwing*. In flight, differs from Stock Dove also in somewhat *slower wingbeats*, larger size and marginally *longer tail*. Sexes alike. Typical flight display with short steep climb, a few loud wing-claps and downward-sloping glide on stiff wings with spread tail. – Juvenile: Lacks white neck patches (acquired Aug–Dec, depending on date of hatching).

VOICE Clattering wing noise on take-off serves as alarm call. During breeding season often utters a hollow, hoarse, stifled, growling 'sigh', 'hooh-hrooo...'. Song a five-syllable hollow cooing with characteristic rhythm (some individual variation but nearly always first syllable emphasized, slightly longer pause before short fifth), '**dooh**-doo, daaw daaw... do'; the phrase is repeated 3–5 times without pause (which results in the short final syllable appearing to begin the next series).

Rock Dove

Stock Dove

Wood Pigeon

ROCK DOVE

narrow black border

white

reddish-brown eye

rather narrow band

dark bill

neat wing-bar

much lighter grey than head

white on back often concealed

white patch

bold black wing-bars

dark 'collar'

light grey

ad.

FERAL PIGEON

Feral Pigeons, as they are known, come in a wide variety of plumages, including a type that is identical to the 'pure' ancestral Rock Dove

STOCK DOVE

small, more compact than Wood Pigeon

more contrast than in Wood Pigeon

diffuse, light grey mid-wing panel

short-tailed

dark eye gives 'kind' look

lacks colour on neck

initially, bill is dark

same tone of grey as head

light-tipped reddish bill (cf Rock Dove)

completely lacks white markings

juv.

ad.

ad.

WOOD PIGEON

large, heavy, lots of wing noise on take-off

plain, low contrast

broad white band

rather long-tailed

ad.

initially, eyes and bill are greyish

pale yellow eye

prominent white patch

juvenile lacks white neck patches, but has the diagnostic white band on the wings

juv.

ad.

KM

Laurel Pigeon *Columba junoniae*   —

L 38–41 cm, WS 62–67 cm. Endemic resident breeder in Canaries (Palma, Gomera, Tenerife, Hierro; <2000 pairs?) in or immediately below the tree-heath and laurel zone on rocky mountain slopes. Usually only seen singly. Wary, but not really shy. Prolonged breeding season, often two clutches within Feb–Sep. Nests in rock crevice or on ground in steep slope under cover of trees and bushes.

IDENTIFICATION *Uniform dark grey-brown* with purple sheen, and has metallic gloss over much of head and neck. Lacks *contrast between coverts and remiges*, but has *paler bluish-grey rump/uppertail* and a diffusely marked broad (buff-tinged) *greyish-white terminal tail-band*. Eye red, bill yellow-white with pink base. *Long-necked* and *short-winged* in flight, which is *rather unsteady*, with *slow wingbeats*. Display-flight with slow wingbeats and horizontal glides in wide circles and with tail fully spread provides best view of pale tail-tip. – Juvenile: Resembles adult but colours are duller, plumage is more warm brown-tinged, has less purple tinge and metallic gloss. Scapulars and wing-coverts finely tipped paler.

VOICE Song opens with high-pitched drawn-out growl, then turns into a repeated trisyllabic cooing with second note in falsetto and last extended, 'hrrrüh... ru-pve**yuuh**... ru-pve**yuuh**... ru-pve**yuuh**... ', etc. Wing-claps at take-off.

Bolle's Pigeon *Columba bollii*   —

L 35–38 cm, WS 60–65 cm. Endemic resident breeder in Canary Islands (Palma, Gomera, Tenerife, Hierro; *c.* 3000 pairs) in laurel and tree-heath forests on mountain slopes. Shy (presumably due to ongoing illegal hunting), hard to observe when perched, but sometimes spotted by its frequent fluttering when rather clumsily feeding in canopy. Usually seen singly or in pairs, but sometimes in small flocks. Generally breeds in spring, but season is extended (Oct–Jul). Nests in dense canopy of tree. One egg only.

IDENTIFICATION Somewhat resembles a Stock Dove but is slightly larger and more *long-tailed*. Plumage is greyish-blue with *glossy vinous-pink and green patch on side of neck*. Remiges and uppertail are contrastingly darker, the *tail with hint of paler central cross-bar*. Rather short, rounded wings. Rapid flight, on take-off with loud clattering sound, in normal flight emitting a rhythmic whistle. Tail often kept folded in flight (looks square-ended). Bill red, eye yellow. – Juvenile: Very similar to adult but with duller and more brown-tinged plumage, and less dark terminal tail-band.

VOICE Song a subdued, four-syllable, hoarse, deep cooing, with third note slightly extended, and which is audible only at 100 m under favourable conditions, 'hwo hwo hwooh hwo', repeated 5–6 times in fairly quick succession. The first series usually extended to five syllables.

Trocaz Pigeon *Columba trocaz*   —

L 40–45 cm, WS 68–74 cm. Endemic resident breeder in Madeira (4000–5000 pairs) in caves and on rocky mountain slopes in the tree-heath and laurel zone, frequently preferring lower levels. Is entirely linked to these natural mountain forests, in particular to north-facing slopes, and the survival of the species is dependent on the preservation of the remains of the habitat. Shy (presumably due to ongoing illegal persecution). Feeds both in canopy and on ground. Small flocks of 10–30 birds sometimes aggregate where food is plentiful. Very long or scattered breeding season, probably linked to supply of laurel berry crops. Nests in tree, clutch generally only one egg.

IDENTIFICATION Appears like a big, short-winged Wood Pigeon without white markings on neck or wings but with *silver-grey neck patch* (which can appear both bluish and pink-red depending on angle). Plumage is *somewhat darker bluish-grey* than Wood Pigeon, lacking its paler grey back and rump is *nearly uniformly dark above. On uppertail characteristic light band runs across*, just behind centre, as wide as the dark end-band. *Underwing is dark grey* (cf. Rock Dove). Eye pale yellow-white. – Juvenile: Resembles adult but plumage is duller and more brown-tinged, not so pure bluish-grey, and has fine light tips to scapulars and wing-coverts.

VOICE Song recalls that of Wood Pigeon but is considerably hoarser and deeper, and also weaker, a six-syllable cooing with the middle two notes stressed and extended, 'uh-uh **hrooh-hrooh** ho-ho', this being repeated a few times.

Namaqua Dove *Oena capensis*   —

L 22–26 cm (incl. tail 10–12), WS 28–33 cm. Widespread in Africa south of Sahara and in Arabia, but breeds within the region only in S Israel. Found by cultivations in semi-desert with acacias and bushes. Often seen on ground. Rather shy and appears nervous in its actions. Nests rather low in tree.

IDENTIFICATION *Very small* (body only a trifle larger than a sparrow's) *and slim. Long and narrow, black tail*, which proves to be *wedge-shaped* when spread, e.g. at take-off and landing. Often flies low with fast, clipped wingbeats. Large rusty-brown primary patch (seen in flight). – Adult ♂: *Black forehead, throat and breast* diffusely bordered white. Bill white. *Bill yellow and red.* – Adult ♀: Lacks black on head and breast. Bill dark grey with reddish base. – Juvenile: Upperwing-coverts and shoulders coarsely marked with white, buff and dark grey.

VOICE A rather quiet species. Song consists of a low-pitched drawn-out, mourning, cooing note which increases in strength and faintly in pitch, preceded and ended with brief note, 'hu-huuoo**ooo**-hu' (ventriloquial voice).

Laurel Pigeon

Bolle's Pigeon

Trocaz Pigeon

Namaqua Dove

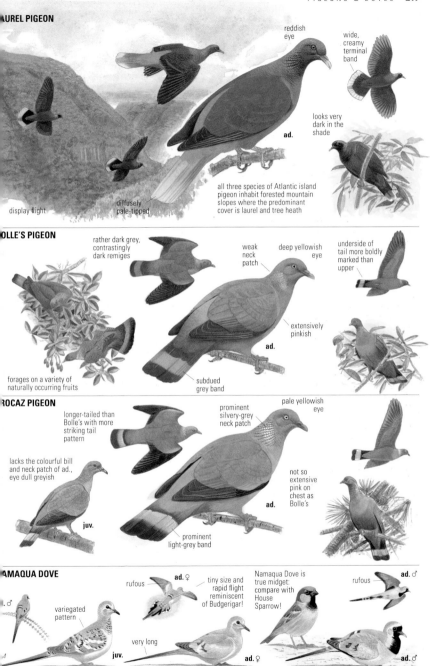

LAUREL PIGEON

reddish eye

wide, creamy terminal band

looks very dark in the shade

ad.

display flight

diffusely pale-tipped

all three species of Atlantic island pigeon inhabit forested mountain slopes where the predominant cover is laurel and tree heath

BOLLE'S PIGEON

rather dark grey, contrastingly dark remiges

weak neck patch

deep yellowish eye

underside of tail more boldly marked than upper

extensively pinkish

ad.

forages on a variety of naturally occurring fruits

subdued grey band

TROCAZ PIGEON

longer-tailed than Bolle's with more striking tail pattern

prominent silvery-grey neck patch

pale yellowish eye

lacks the colourful bill and neck patch of ad., eye dull greyish

not so extensive pink on chest as Bolle's

ad.

juv.

prominent light-grey band

NAMAQUA DOVE

ad. ♂

rufous

ad. ♀

tiny size and rapid flight reminiscent of Budgerigar!

Namaqua Dove is true midget: compare with House Sparrow!

rufous

ad. ♂

variegated pattern

very long

juv.

ad. ♀

ad. ♂

(Eurasian) Collared Dove *Streptopelia decaocto* r**B**2

L 29–33 cm, WS 48–53 cm. Breeds in lowlands around farmyards and in towns, in parks and gardens where there are dense trees for nesting; avoids open, unbroken country. Invaded Europe from SE in recent times. Resident. Food mostly vegetable matter; often spilled grain, frequents silos and cultivated fields, in winter often in flocks. Confiding.

IDENTIFICATION A medium-sized, *elongated*, rather 'elegant' dove with *long tail* and pale plumage. Best identified by narrow (white-edged) *black bar across neck-side*. Rather *uniformly pale buff-grey* with somewhat browner tone on back and blue-grey tinge to greater wing-coverts. Dark eye contrasts with the uniformly pale head. *Outer tail-feathers broadly tipped white* (obvious in flight, when tail is spread). Closed *tail looks almost all white from below*, contrasting with *ash-grey undertail-coverts*. Sexes similar, but ♂ on average purer pink with faint bluish sheen on head, neck and breast than ♀, which is more buffish-grey. – Juvenile: No black neck mark, has somewhat duller, browner colours.

VOICE In conflicts and on landing a noisy 'krreair', not unlike Black-headed Gull. Song a trisyllabic coo, repeated a few times, 2nd syllable drawn out and stressed and 3rd lower, 'doo-**doooo**-do', somewhat Cuckoo-like at distance.

(European) Turtle Dove *Streptopelia turtur* m**B**3

L 25–28 cm, WS 45–50 cm. Breeds in open lowland deciduous woods and copses with rich undergrowth, mainly in agricultural areas. Summer visitor (mostly May–Aug), winters S of Sahara. Rather shy and hard to observe closely, keeps well hidden in tree canopy but may be seen at distance on telephone wires and foraging on ground. Heavily hunted on passage through Mediterranean.

IDENTIFICATION Rather small and slim dove with fast, somewhat fitful or slightly pitching flight, e.g. when landing. In practice most like Laughing Dove, but differs in: *bright orange-brown feather edges* and *clearly demarcated black feather centres* above on scapulars and lesser wing-coverts (Laughing darker reddish-brown and almost uniform); *more restricted grey-blue panel in centre of wing*; *longer and more pointed wings* and *somewhat shorter tail*; *narrower white tail-sides* with *complete black inner border*; at close range shows *neck patch of black and white stripes* (the white at times blue-tinged), noticeable *black tail-base below*, and orange eye surrounded by rather distinct, reddish bare skin (cf. also Oriental Turtle Dove). – Juvenile: More uniformly dirty brown and buffy greyish-white, and lacks neck patch. Can be confused with juvenile Collared Dove, but note tail markings. Extremely similar to juvenile Oriental Turtle Dove (which see).

VOICE Wing-clatter on take-off. Song a deep, hard purring, 'turrrrrr turrrrrr turrrrrr', repeated several times.

Oriental Turtle Dove *Streptopelia orientalis* **V**⁕

L 30–35 cm, WS 54–62 cm. Breeds in C Siberian tai (ssp. *orientalis*), also in SW Siberian forest-steppe and low mountains of Central Asia (*meena*), these two perhaps bei separate species considering differences in plumage a song. Both races are vagrants to N and W Europe.

IDENTIFICATION Ssp. *orientalis*: Like Turtle Dove in p mage, but clearly *bigger* and *heavier-built*; in flight can rec a Stock Dove more than a Turtle Dove, with more *loose flicking beats* and *steadier course*. Adult differs from Tur Dove in: *tail-tip blue-grey*, *belly/undertail light grey* (Tur pure white); upperwing feathers have *bigger and more d fusely defined dark grey* (not jet-black) *centres* and *narrov and more rufous-brown* (not ochre) *fringes* as long as th are fresh (though will bleach to more buff); tends to sho two *narrow pale wing-bars* along tips of greater and medi coverts; *blue-grey panel on wing darker* and more restrict breast darker brownish-pink (Turtle: paler pink with fai bluish sheen); *back and rump forms uniform dark blue-gr area* (Turtle: nearly always some grey-brown admixe mantle darker, often with black blotches; *less conspicuo bare facial skin*. – Variation: Ssp. *meena*: Somewhat small the smallest almost as Turtle Dove. Main difference fr *orientalis* is *white colour of tail-tip and belly*, further inviti confusion with Turtle Dove. Upperparts and breast, how ver, much as in *orientalis*. – Juvenile: Very like juven Turtle Dove, but has darker mantle and breast and crud proportions. Note also slightly shorter primary projecti and on average less pale tips to primary-coverts.

VOICE Song of *orientalis* hollow and rather throaty: tv hoarse notes followed by two somewhat clearer ones, th two hoarse and so on, 'hru-hru oo-oo hru-hru oo-oo. Song of *meena* has three initial muffled, hoarse notes a a forth clearer and lower, '**hruuh**... hru-**hru**... woo'.

Laughing Dove *Streptopelia senegalensis* [**V**⁕⁕

L 23–26 cm, WS 40–45 cm. Breeds in small woods, plan tions (e.g. of date palms) and gardens, often close to hous also in urban areas so long as vegetation (e.g. lines of tre exists; typical bird of desert oases (where partly int duced). Resident. Bold. Nests in thick bushes and trees.

IDENTIFICATION A *small and slim*, rather long-tailed a short-winged dove. In the field appears mainly *reddis brown* at moderate range; close to, shows *blue-grey panel wings* between dark grey remiges and blotchy red-brov upperparts (blue more extensive than on Turtle Dove), a *all-white undertail* (and belly). *Black and ochre pattern crop* is diagnostic. In flight rather like Turtle Dove (e.g. si ilarly patterned uppertail and dark underwing), but told *more uniformly brown back* (without black feather centr and usually visibly *sligh longer tail and shorter win*

VOICE Song a hollow c of six short syllables at fa slightly jolting pace, 3rd a 4th a little higher-pitch and more stressed, 'do-d **du-du**-do-do'; several ca ing in unison sounds alm like Black Grouse cooing.

Collared Dove   Turtle Dove

Laughing Dove

OLLARED DOVE

lacks strong contrast in upper-tail pattern

long-tailed

off-white

ad.

light grey-buff

sexes very similar

folded tail largely white below

long

ad.

ad.

TURTLE DOVE

striking tail pattern

slight build

white

ad.

ash-grey

grey

striped patch

pink with blue cast

ad.

acquires bright, adult-type plumage in first autumn

juv. primary-coverts prominently tipped buff

juv.

juv. → 1st-w.

ad.

ORIENTAL TURTLE DOVE

extraordinarily long-necked when alarmed!

slightly shorter primary projection than Turtle Dove

pinkish-brown

heavy build and broad wings giving Stock Dove-like shape

pinkish-brown

blotched blackish

dark blue-grey

wing-bars

dark, with usually very narrow light tips

meena juv. → 1st-w.

meena (Central Asisa)

ad. meena (Central Asia)

juv. → 1st-w.

worn juv. → 1st-w. (February)

white in meena

darker overall than juv. Turtle Dove, with darker feather centres; however, rapid fading can complicate matters!

ad.

both races darker, more rufous than Turtle Dove

dark, duller than meena

white tips in meena (like Turtle Dove)

tips usually greyish in orientalis

juv.

clearly a larger, heavier and darker bird than Turtle Dove

ad. orientalis (Siberia)

LAUGHING DOVE

LAUGHING

TURTLE

much white, but less bold pattern than Turtle Dove

darker

all white

black base

plain

Laughing Dove has longer tail and shorter, blunter wings than Turtle

ad. ♀

also known 'Palm Dove'

often a little less colourful than ♂

delicate gorget

ad. ♂

CUCKOOS *Cuculidae*

Four breeding species, plus two vagrants from N America. Medium-sized, long-bodied birds with long tail. Short legs. Three of the breeding species are nest parasites.

(Common) Cuckoo *Cuculus canorus* **mB**3

L 32–36 cm, WS 54–60 cm. Seen in all kinds of woodland, farmland, alpine terrain, coastal heath—versatile in habitat choice. Summer visitor (Apr–Sep), winters in tropical Africa. Easy to observe in breeding season, perches in open, call very familiar. Parasitizes various hosts, each ♀ Cuckoo specializing on a certain species, laying mimicking eggs.

IDENTIFICATION Medium-sized, *slim*, with *long, rounded tail* and *pointed wings*. Flies with regular wingbeats (glides uncommon and short), wings mostly *beaten below horizontal plane* (unlike Kestrel and Sparrowhawk); moderate speed; 'straight-backed' posture with *forward-pointing bill*. Often settles right in open on telephone wire or fence post, landing a little clumsily. Perched bird often droops its wings and cocks tail a little. – Adult ♂: Plain blue-grey above and on head and breast; sharp border below against finely dark-barred white belly. Iris, orbital ring, bill-base and feet yellow. – Adult ♀: Two morphs: grey like ♂, but at close range told by rusty-buff tinge and some dark barring on breast; brown morph rusty-brown above and on breast and often has wholly dark-barred plumage. – Juvenile: Slate-grey with ± obvious element of rusty-brown. Best told by *white patch on nape* and *narrow white feather fringes above*.

VOICE Song the familiar disyllabic call with emphasis on first syllable, **'goo**-ko', when excited sometimes trisyllabic. In pursuit-flight often a guttural hoarse '**goch**-che-che'. ♀ has a loud, rapid bubbling trill, somewhat recalling Little Grebe, 'pühühühühühü...'. Young has a stubbornly repeated, squeaky, uninflected, slightly throaty call, 'psrih'.

Oriental Cuckoo *Cuculus optatus* **V**∗∗∗

L 30–34 cm. Breeds in taiga and scrubby forest. Arrives at its European breeding grounds mid May–Jun; winters in SE Asia. Insectivorous. Nest parasite.

IDENTIFICATION Very like Cuckoo, and usually hard to separate by appearance alone; close, lengthy observation always essential. Slightly *smaller* size with *proportionately slightly heavier bill*. Grey morph of both sexes is a shade darker on back and upperwings than Cuckoo (hard to judge in field). Dark barring on white belly basically the same as on Cuckoo, but some have slightly thicker and sparser bars. *Undertail-coverts often rusty-buff and almost unmarked* (some Cuckoos similar). Under primary-coverts often unmarked or with reduced barring (also shown by some Cuckoos, and is difficult to use in field!). Brown morph (only ♀; common)

always entirely and heavily barred, usually with *broad black bars above than red-brown ones* (normally the rever on ♀ Cuckoo), especially on tail.

VOICE ♂'s song differs clearly from Cuckoo's, is safest wa to separate the species: a *fast* series of *equally stressed* pair notes with *Hoopoe-like tone*, usually introduced (e.g. o landing) by stuttering series of 4–8 hoots, '(pupupupup pu-pu pu-pu pu-pu pu-pu...'. ♀ has a trill like ♀ Cuckoo's bu slower, harder and more rattling, almost hawk-like.

Great Spotted Cuckoo *Clamator glandarius* **V**∗∗

L 35–39 cm. Breeds in savanna-like heathland, often wi cork oak or stone pine, also in olive groves and the like. Ne parasite on crows (mostly Magpie). Flies around a lo frequently hops on ground with raised tail. Food insects.

IDENTIFICATION Size of Cuckoo, but *wings broader ar blunter*, and *tail somewhat longer and narrower*. Dark abo *bestrewn with small white spots*, pale below. – Adult: Pr maries dark grey. *Crown and ear-coverts silvery-grey*, hind crown with hint of crest. Tertials and secondaries broad tipped white. Throat and breast-sides yellowish-whit – 1st-summer: Variable, usually with *rusty-brown shade o primaries*, also slightly darker head and back than adul – Juvenile: *Primaries bright rusty-brown, tipped black. C dull blackish*. Upperparts blackish with slightly smalle white spots than on adult. Tertials and secondaries with on narrow white tips. Throat and breast ochre-buff.

VOICE Noisy and loud. Usually gives a rattling, cacklin call, 'cherr-cherr-che-che-che-che' or 'ki-ki-ki krie-kri krie...', sometimes recalling Turnstone or alarm of ♀ ha rier. When agitated, a nasal, gruff 'cheh'.

Yellow-billed Cuckoo *Coccyzus americanus* **V**∗

L 29–32 cm. Vagrant from N America. Differs from simila Black-billed Cuckoo in *yellowish bill-base*, paler uppe parts, brighter rusty colour on primaries and *more obviou white terminal spots on tail-feathers*. Orbital ring yellowish

Black-billed Cuckoo *Coccyzus erythropthalmus* **V**∗∗

L 28–31 cm. Very rare vagrant from N America. Told fro similar Yellow-billed Cuckoo by grey-based *blackish bil* darker upperparts, less obvious rusty tone on primari and *narrower, more obscure pale tips to tail-feathers*. Adu has red orbital ring (juv. yellowish).

Senegal Coucal *Centropus senegalensis* –

L 35–41 cm. Breeds in Egypt in open terrain with bush and tall grass; occasionally seen in gardens. Resident. Sh mostly stays hidden in vegetation or runs on ground.

IDENTIFICATION *Big*. Long tail. *Strong, black bill. Wing rusty-brown, rounded*, fligh flappy and slow with brie glides. *Head and tail black back dark brown, unde parts buff-white*. – Juvenile Dark-barred above.

VOICE A series of but bling, hollow notes, lik pouring a bottle of wate 'boo-boo bu-bu-bu-bu...'.

Cuckoo   Oriental Cuckoo

Great Spotted Cuckoo

UCKOO

recently fledged young being fed by foster parent (Meadow Pipit)

flies with regular wingbeats, wings beaten mostly below horizontal plane

ad. ♀ (rufous)

ORIENTAL CUCKOO

long

barred

ad.

white nape patch

variable amount of rusty-buff and dark barring

ad. ♀ (grey)

juv.

ad. ♂ closely attended by two Arctic Warblers

slightly darker than in Cuckoo

ad. ♂

ad. ♂

barring averages sparser, broader and more widely spaced than in Cuckoo

much overlap with Cuckoo but a strong buff colour on the vent and undertail coverts, especially if also barred, is characteristic of Oriental

REAT SPOTTED CUCKOO

at Magpie nest

black

rusty-brown

long

dark

juv.

no rusty panel

ad.

silvery-grey

juv.

1st-summer

rusty on wing reduced or lacking

ad.

rusty-brown

ELLOW-BILLED CUCKOO

strong, mostly yellow

juv.

rufous patch

whit-ish

bright panel on primaries

whitish

black and white tail-side

large white tips and outer edge

BLACK-BILLED CUCKOO

grey

juv.

buffish

subdued panel on primaries

buffish-yellow wash

light tips very small or virtually lacking in juv.

SENEGAL COUCAL

juv.

heavy

ad.

yellow-buff

very long, all dark

M

OWLS *Strigiformes*

Birds of prey, many with predominantly nocturnal habits; 18 species. Catch prey primarily with surprise attack and with the aid of very good hearing and vision. Night vision however not surreal, requires some light from moon or stars. Hunt both from perch and in flight. Plumage dense and soft-surfaced, allowing noiseless flight. Head large, with often characteristic facial pattern. Some have ear-tufts. Sexes similar, ♀ usually slightly larger than ♂. Nest in cavity, in old twig nest, or on ground; eggs white and rounded.

(Eurasian) Eagle Owl *Bubo bubo*   [rB5] / V***

L 59–73 cm, WS 138–170 cm. Resident in mountains and forests, preferring areas with rocks, steep cliffs and mature trees (preferably conifers); often in rocky archipelagos. Scattered distribution, generally rare and local. Sedentary. Only a handful of genuine records in Britain, all in 19th century. Nocturnal and partly crepuscular. Day roost in dense mature spruces or firs, or in cave or crevice. Food mammals (voles, rats, hedgehogs, hares) and birds (corvids, gulls, wildfowl, etc.). Nests on inaccessible cliff-ledge or, less often, on the ground by rock, uprooted tree or among boulders, exceptionally in abandoned raptor's nest or in barn, abandondoned factory, etc. In spite of its size and its impressive claws is by and large a non-violent and retiring owl near its nest, and aggressive behaviour towards humans is rare. Will at times perform wader-like distraction display near nest with small young, feigning injury on the ground, squeaking woefully, completely different from the unhesitating attack of e.g. Ural Owl.

IDENTIFICATION *Largest* owl, strong build enhanced by dense, 'fluffy' plumage; large-headed. When relaxed can appear *barrel-shaped*, when alert may display surprisingly long (but thick) neck. *Ear-tufts long*, visible except in flight, held flattish when relaxed or anxious, more erect when calling or disturbed. Flight powerful and steady, wingbeats rather shallow, glides straight, recalling large buzzard (but head of course huge, and wings more arched when gliding). *Eyes orange-red, large*. Main colour below *yellowish-brown with dark streaks*, broad on breast. Upperparts darker brown, boldly streaked and vermiculated blackish. Throat white, exposed when calling. In flight, primary bases only slightly paler (yellowish-brown) than rest (cf. Great Grey Owl), boldly and evenly cross-barred. – Juvenile: Downy young recognized by proportionately huge bill and talons, and by nest site. Fledged young have fully feathered body at a few weeks of age, but told first few months by partly downy, rounded head with only small ear-tufts.

VOICE Call of ♀ a harsh barking '**rhaev**'. Alarm a fierce, startlingly loud, nasal barking '**kwa**!', often 3–5 notes quickly repeated, 'kwa-**kwa**-kwa!'. Anxiety-call nasal, muffled, gull-like 'gaw'. Song deep sonorous booming '**oo**-hu', second syllable falling in pitch, voice surprisingly faint at close range, still audible at 1½–4 km, usually repeated at intervals of 8–12 sec. but quicker by some younger ♂♂, every 5 sec. or even quicker; at long range only first note heard. Certain ♂♂ more prone to call, others more quiet. ♀ has higher-pitched, hoarse version. Begging-call of young a loud, husky, scraping 'chu**eesh**' (as when planing wood) heard through calm summer and early-autumn nights.

Pharaoh Eagle Owl *Bubo ascalaphus*

L 38–50 cm, WS 100–120 cm. Very closely related to Eagle Owl, a southern, smaller desert form. Ranges seem to overlap marginally, and the Pharaoh Eagle Owl has some morphological separating traits, so the two are nowadays usually treated as separate species. Breeds in deserts or arid habitats, nests in crevice in rocky outcrop or wadi.

IDENTIFICATION *Considerably smaller* than the Eagle Owl (only *c.* 75% its size) and generally *clearly paler* with fine and sparser streaking making the *pinkish-sandy ground colour* stand out more. However, there is some variation in plumage darkness, and some are almost as dark and brownish as Eagle Owl. Bold *dark streaking on underparts is limited to a narrow zone on upper breast, rest is finely barred rufous*. Facial discs are better outlined by black lines bordered whitish. *Ear-tufts somewhat shorter* than in Eagle Owl. Iris colour variable but often *yellow-orange*, slightly less reddish than Eagle Owl. 'Trousers' thin, often *faintly barred* rufous.

VOICE Song much *higher-pitched* than Eagle Owl, and hardly disyllabic, second note almost swallowed, 'huo'. A call variation has been rendered as a trisyllabic '**doo**-u-ho' with first note stressed and the other two higher-pitched.

Brown Fish Owl *Bubo zeylonensis*

L 50–58 cm, WS 125–140 cm. Extremely rare and local in S Turkey (formerly also in N Israel, and adjacent areas?) at fish-holding streams with rocky shores and protecting trees. Stalks fish from perch at bankside or by wading in shallow water, or will fly low over water with legs dangling. Day roost in crevice or dense tree.

IDENTIFICATION *Large*, with *horizontal, broad ear-tufts*. Looks flat- and broad-headed. *Pale mid brown* above, finely streaked darker brown; paler buffish- or pinkish-cream below, *rather sparsely and finely streaked brown*. On upperparts *pale lower scapulars form a light band*. Eyes yellow. Feet unfeathered, yellowish-grey. *Strong, protruding bill* provides characteristic profile. Flight-feathers brown, barred darker. Tail boldly barred brown. Facial discs poorly developed. Faint wing sound heard from flying bird.

VOICE Imperfectly known. Song is a very deep, muffled hooting (thus not audible far!), a trisyllabic 'hu **whoo** huu' (second 'inhaling' note higher), slowly repeated. Also said to give a strange screaming sound reminiscent of Stone Curlew. Alarm is a hissing sound.

Eagle Owl   Pharaoh Eagle Owl

Brown Fish Owl

EAGLE OWL

breeding habitat

mobbed by Hooded Crow

rather uniform above

tufts folded

'boxing gloves'

impressive, fierce-looking, unmistakable

young, just out of nest, c. 5 weeks old—entering 'roaming' stage

blends perfectly with bark of old pine tree

PHARAOH EAGLE OWL

much lighter build than Eagle Owl—adaptation to smaller prey

short tufts

face framed dark

iris colour variable, often yellow-orange

bold streaking

variation

rather sparse wedge-shaped breast-spots

white throat puffed out when calling in all three species

barred rufous, unstreaked

spotted at day-roost

BROWN FISH OWL

bushy tufts

plain face

obvious white 'braces'

yellow eyes

lightly streaked

shape of wings rather closer to *Strix* owl than Eagle Owl

boldly barred flight-feathers

ambush at waterside

bold bars

thin, bare legs

broad and full 'hand', short tail and heavy, protruding bill give characteristic jizz

Great Grey Owl *Strix nebulosa* —

L 59–68 cm, WS 128–148 cm. Mainly resident in lowland boreal forests, often near bogs or infields; rare and local. Partly nomadic. Mainly crepuscular (incl. on light northern summer nights). Food almost exclusively voles (in spite of size—bulk made up of feathers, not muscles and bones!), caught on ground after watch from perch. Nests in abandoned raptor's nest or on top of broken tree trunk, rarely on ground. Some are aggressive when their young are about to leave nest, others fairly tame and will permit rather close observation.

IDENTIFICATION *Large*, with thick neck and proportionately *very large and rounded head*, looking like a *sawn-off log in profile*. Wings very long, broad and rounded, tail fairly long, well rounded. Impressive-looking in flight, with *measured, slow wingbeats* (almost heron-like), capable of long glides. Basically *dusky grey* with darker grey pattern, underside paler, breast and belly coarsely streaked and finely vermiculated. Facial pattern surrounding *small yellow eyes* and bill striking with white 'eye-bows' and *black chin*. Upperwing shows conspicuous *large pale buff patch at base of primaries* and dark grey carpal area and forewing; shoulders pale grey ('braces'). *Uppertail has broad dark terminal bar* (lacking on Ural and Eagle Owls). When pair seen, ♀ larger and often with proportionally slightly smaller head.

VOICE The call of ♀ is a hoarse, high-pitched 'chiep-chiep-chiep'. Alarm or aggression-call is a deep, subdued growling 'grrook-grrook-grröok', and ♀ has a related remarkable drawn-out growling 'grrrrrrrrrrrrk'; snaps bill loudly when angry. Song is a very deep, pumping series (9–12) of booming hoots, well spaced (*c.* 1½ per sec.), falling slightly in pitch and loudness at end, 'bvoo,bvoo,bvoo,...'; normally audible no farther than 400 m. Begging of young reminiscent of call of ♀, 'psiep-psiep', deeper when older.

Ural Owl *Strix uralensis* —

L 50–59 cm, WS 103–124 cm. Resident in old boreal forests interspersed with bogs, often also open water, clearfellings and small fields. In S Europe also mountain forests, beech woods. Food voles, frogs and insects (taken after watch from low perch), but is strongly built and takes also a variety of birds (incl. other owls!). Nests in tree trunk ('chimney'), nestbox or abandoned raptor's nest. *Caution*: Very aggressive when young about to leave nest and can attack intruder fiercely; keep your eyes fixed on the parents if you stumble on an inhabited nest, and leave area quickly!

IDENTIFICATION Medium-large, head rounded, *tail long* and rather wedge-shaped (noticeable in flight); wings rounded. Flight direct, purposeful, recalling Common Buzzard. Plumage *pale buffish grey-brown* (paler than Tawny),

streaked darker brown; in brief encounters, *paleness* is usually best clue. *Eyes black*, stand out well on *plain buffish-grey facial discs*. Bill yellowish. Upperwing evenly barred dark, *lacking conspicuous pale patch* on inner primaries of Great Grey Owl and Short-eared Owl. *Uppertail evenly and boldly barred dark* (cf. Great Grey).

VOICE Call of ♂ raw, very harsh (recalling both Grey Heron and Eagle Owl) disyllabic 'kre-ef!'. Alarm a loud nasal barking 'wak'. Song of ♂ a deep cooing-like hooting, 7 notes in constant pattern, first two notes, then pause of *c.* 4 sec., then two notes immediately followed by three more: 'wo-ho......... wo ho uhwo-ho', audible at 2 km in calm weather; ♀ has higher-pitched, hoarse version. Alternative song of ♂ (courtship, nest-showing, anxiety) a rapid series of 6–8 short, deep hoots, slightly increasing in pitch, pace and stress but decreasing at end, 'po po po po po po'; ♀ has harsher version. Begging-call of young much as for Tawny Owl (though deeper), a high-pitched, throaty 'peechep'.

Snowy Owl *Bubo scandiacus* V**

L 53–65 cm, WS 125–150 cm. Breeds on tundra or high plateaux above tree-limit, preferring areas with scattered rocks and good view. Apparently nomadic, fluctuating with food supply. One pair bred in 1960s–70s in Shetland; otherwise irregular and extremely rare winter vagrant in Britain. Food mainly lemmings and other voles, also birds and rabbits. Nest scrape on ground, in lemming-years with large clutches and lined with store of dead prey.

IDENTIFICATION *Large, strikingly white* in most plumages. *Eyes yellow*. Flight powerful, and wings have proportionately shorter 'arm' and more pointed 'hand' than other large owls. – Adult ♂: Pure white (except for a few scattered tiny dark spots). – Adult ♀: White, with dark spots on crown, dark cross-barring below (except white centre of breast and dark 'scalloping' on back and shoulders; flight-feathers and wing-coverts coarsely barred dark on white ground. – 1st-winter ♂: Very similar to adult ♀ but slightly smaller plumage rather off-white when fresh; upperparts more distinctly cross-barred, on average slightly finer barring below, *tips of tertials, inner greater coverts, and flight-feathers diffusely vermiculated grey*. Bleached to predominantly white in 1st summer. – 1st-year ♀: Like adult ♀, but dark spots and bars heavier and denser, at distance creating strong contrast of white face and neck, and almost uniform dark body and crown; also, more dark barring on crown and breast.

VOICE Alarm a loud, grating bark, 'krek-krek-krek-krek-krek-...' (Mallard-like!), ♂ lower-pitched than ♀. ♀ also has a mewing call, 'pyeey, pyeey, pyeey,...', like a falsetto bark and a thin whistling 'seeuuee'. Song a far-carrying, deep 'gawh', repeated at *c.* 5-sec. intervals, audible at 1–3 km, apparently somewhat variable in tone (soft, clean voice like ♂ Eagle Owl, or more hoarse recalling Great Black-backed Gull) and pattern (single note or a combination of notes). Begging-call of young a high-pitched whistling squeal, like ♀ but more feeble.

Great Grey Owl   Ural Owl   Snowy Owl

GREAT GREY OWL

in profile like a steamship flue funnel

typical face with white 'eye-bows' and black chin

stately look

swarthy face

grey

young, c. 5 weeks old

hunting voles in 'taiga barn country'

pale buff primary base

broad dark tail-band

huge, square

broad band

URAL OWL

typical nest site in hollow tree trunk ('chimney')

deceptively gentle look

pale, plain face, straw-yellow bill

buff-grey

iris black

rounded profile

young, c. 4–5 weeks old

lacks paler patch

evenly barred

long (cf. Tawny Owl)

SNOWY OWL

densely barred

dusky tertial tips marked grey

1st-w. ♀

ad. ♂

young, c. 7 weeks old

white tertials distinctly barred

narrowly barred

ad. ♀

ad. ♀

DZ

(Northern) Hawk Owl *Surnia ulula*   **V***

L 35–43 cm, WS 69–82 cm. Mainly resident in boreal forests, often in upper tree zone on mountain slopes (mixed conifers and birch), preferring vicinity of bog, meadow or clearfell. Fluctuating in numbers, some years locally fairly common. In some autumns considerable numbers move south. About five records in Britain in 20th century. Partly diurnal. Food voles (main prey, taken on ground after watch from treetop) and birds (e.g. thrushes; capable of catching prey as large as Willow Grouse). Nests in tree-hole ('chimney' or vertical) or abandoned raptor's nest. *Caution:* Can fiercely attack intruders when young leave nest; do not go near, and keep your eyes fixed on the parents while in sight of young just out of a nest!

IDENTIFICATION Medium-sized with distinctive proportions: *very long tail* and *rather narrow, bluntly pointed wings*, thus recalling a hawk more than other owls, this enhanced by *direct, agile flight* with short series of wingbeats relieved by brief glides. Perches in treetops, also fearlessly exposed in bare dead spruces. Head rather large and often *flat-crowned* (alert, alarmed). Upperparts dark brown with *pale hindneck and scapulars*, latter forming prominent V on perched bird in rear view ('braces'). Underparts whitish and finely cross-barred except on *upper breast*, which is *pure white*. Face *whitish, strongly outlined black*, expression 'stern'. Eyes yellow. Nape has a pale and dark pattern ('false face', defence purpose), and side of head has a black patch. – Juvenile: Similar to adult but a little greyer and duller, with dusky eye-surround and diffusely patterned throat and upper breast.

VOICE Alarm shrill, like ♀ Merlin, 'ki-ki-kikikikiki'. Food-begging call of ♀ a hoarse, squeezed, drawn-out 'ksheee-lip', ending abruptly. Song (early spring, dark night) a very long and rapid bubbling, almost ventriloqual trill, 'lülülülülülülülü...', 8–9 sec., pauses 8–25 sec., audible to at least 1 km. When agitated (anxiety, courtship) utters a gargling, short cooing, almost like Black Grouse. Begging of young resembles that of ♀ but is more hissing, 'psssss-lip'.

Tengmalm's Owl *Aegolius funereus*   **V***

L 22–27 cm, WS 50–62 cm. Breeds in dense forests with small bogs and glades. Fairly common in N Europe; about 50 records in Britain. Mainly sedentary, but in some autumns many move south, possibly predominantly ♀♀ and young. Strictly nocturnal, difficult to see other than in vicinity of nest. Food mainly voles. Nests in tree-hole (old nest of Black Woodpecker) or box.

IDENTIFICATION Medium-small (about as Little Owl), large-headed (flattish crown). Medium-sized eyes yellow, facial expression 'astonished'. Upperparts brown, with whitish spots and diffuse blotches on shoulders; underparts whitish, diffusely blotched brownish. Wings rounded, *flight rather direct* with series of quick wingbeats and brief straight glides. Greatest risk of confusion Pygmy Owl (small, same range and habitat), but this is only the size of a Starling, has proportionally smaller head, has different, 'stern' facial expression, shows tendency to bounding flight, and has proportionally shorter wings; also, Pygmy often perches in treetops, whereas *Tengmalm's prefers canopy*. – Juvenile: Mainly *chocolate-brown*, with variable amount of white on face.

VOICE Rather vocal. Commonest call squirrel-like 'chiak'. Song a rapid series of deep whistling notes (5–8; when excited, drawn out to much longer series), initially slightly rising in pitch and pace, surprisingly loud (audible to well over 3 km on calm nights), 'pu-po-**po**-**po**-po-po'; pitch and speed vary somewhat; possible to mimic (lowest possible whistling). Begging-call of young a thin, abrupt 'ksi!'.

(Eurasian) Pygmy Owl *Glaucidium passerinum* —

L 15–19 cm, WS 32–39 cm. Breeds in coniferous or mixed forests in boreal zone or in C European mountains, with a preference for mature spruce or fir forests. Mainly sedentary. Crepuscular habits (night vision poor in complete darkness). Rather fearless, and can sometimes be attracted by imitation of song. Food birds, voles; very bold for its size, and capable of killing thrushes (which are larger). Stores food supply in holes. Nests in hole in a tree (often old nest of Great Spotted Woodpecker).

IDENTIFICATION *Very small* (Starling size), round-headed (vestigial ear-tufts rarely seen). *Head is proportionally smaller* than on Little and Tengmalm's Owls. Flight dashing, over longer distances *obviously bounding* (in woodpecker fashion). Perches in spruce tops. At times waves tail or raises it slowly in flycatcher fashion. Facial discs poorly developed, but *short, narrow white supercilia prominent*, and yellow *eyes small* and closely set, giving *'stern' expression*. Lower hindneck has diffuse light pattern creating suggestion of 'false face' (for defence purpose). Grey-brown above, speckled with *tiny white dots*; whitish below, with brown breast (more extensive on sides) and thinly streaked belly. – Juvenile: Very similar to adult, but lacks white spotting on crown, back and wing-coverts.

VOICE Call of ♀ a very thin, drawn-out 'tseeeh' (recalling Blackbird or Robin, but more 'determined'). In autumn (but also at other times) commonest call is so-called scale-song, a series of 5–10 sharp, squeezed whistles, rising in pitch, 'chuuk-chüük-cheek-chiihk-...' (like bicycle pump with finger over hole). Song is repeated mellow, piping whistle, 6–7 per 10 sec., recalling Bullfinch and Scops Owl, but is a fraction more drawn out, slightly higher-pitched and has a hint of accentuated end, 'pyük', audible at 500 m–1 km. In excitement, a fine, subdued stammering is interfoliated between regularly spaced whistles, 'pyük...(popopo) pyük...(popo-po)...pyük...', etc. Begging-call of young is a fine 'tseeh', recalling ♀'s call but shorter.

Hawk Owl   Tengmalm's Owl   Pygmy Owl

HAWK OWL

pale nape pattern creates 'false face'

grim look

on the lookout from top of a dead tree

large pale shoulder patches

finely barred, like a hawk

young just out of nest, c. 4 weeks old

long

hawk-like silhouette in flight, but head is proportionately bigger

TENGMALM'S OWL

♀ looking out of nest hole in old aspen tree

three young just out of nest, c. 4 weeks old

plain dark brown

'astonished' look

PYGMY OWL

well-defined 'false face' on hindneck, esp. in ad.

no white spots

plain grey-brown flank

austere look

short white eyebrows

barred

more distinct pattern than juv.

with a captured Willow Tit

juv.

ad.

flight woodpecker-like, smoothly bounding

DZ

Long-eared Owl *Asio otus* rB4 / W4

L 31–37 cm, WS 86–98 cm. Breeds in forests in vicinity of open country, in copses among arable fields, in plantations on moors, in larger parks with conifers, tall hedgerows, etc. Migratory in northern part of range, sedentary in S and W. Not scarce in Britain in optimum habitats, but declining. Migrants and wintering birds often gather in small flocks at favoured sites for communal roosts in dense trees or bushes. Nocturnal and crepuscular. Food mainly voles, but also small birds, hunting both from perch and in flight, latter probably by far commonest method (but also the one easiest noted). Will at times hunt passerines dazzled by lighthouse light. Nests in old nest of other bird in tree (usually crow), often high up in fir or spruce.

IDENTIFICATION Clearly *smaller than Tawny Owl*, slightly smaller than Short-eared Owl and has proportionately somewhat shorter wings. *Wings are still long and rather narrow*. Long ear-tufts often visible (in courtship; when alarmed; in erect camouflage posture), but can be practically invisible (in flight; when relaxed). *Flight rather slow and wavering*, a few rowing wingbeats (markedly slower than Tawny, very similar to Short-eared but perhaps slightly quicker) interspersed with glides. Similarity to Short-eared Owl enhanced by similar wing pattern: upperwing has *large yellowish-buff patch on base of outer primaries* accentuated by dark carpal patch and dark wing-tip, and underwing has dark 'comma' at wingbend on white ground; separated by several subtle but important differences: *lack of whitish trailing edge to 'arm'* present in Short-eared; *wing-tip being evenly barred dark*; *underbody rather evenly streaked*; tail densely and indistinctly barred; lack of contrastingly dark tertials below; eyes orange; and wings proportionately slightly shorter and broader. – Sexes similar, ♂ being on average paler, less heavily streaked below, face paler and less buff.

VOICE Adults are rather silent. Call of ♀ is a weak, nasal, somewhat cracked '**peh**-ev', repeated. Alarm a rather impetuous, hoarse, nasal 'wrack, wrack-wrack'. Song a series of deep hooting 'oh', repeated every 2½ sec., at first more feeble and deep, audible 500 m–1 km but already faint at 200 m. At times, ♂ and ♀ perform duets (call of ♀ muffled, more nasal). Can also wing-clap below body singly and irregularly during display flight. Begging-call of young *loud* and 'heartbreaking', a disyllabic, drawn-out, *plaintive*, high-pitched '**pee**-eh', audible over 1 km.

Short-eared Owl *Asio flammeus* rB4 / W3–4

L 33–40 cm, WS 95–105 cm. Breeds on heathland, in scrub among meadows, on bogs, esp. in upper tree zone in open boreal forests. In Britain local in N, sporadic in S. Mainly migratory in N Europe, sedentary in rest. Often aggregates in winter in small flocks for communal roosts at favoured sites. Partly diurnal, and in light boreal summers of course mainly active in full 'daylight' even if it is in the night. Food mainly voles located through hearing in flight. Nest a simple scrape on ground.

IDENTIFICATION Medium-sized, *wings very long, narrow* and rather pointed, *head comparatively small, rounded*. Ear-tufts minute and rarely seen. Plumage *pale yellow-brown* and buff-white, heavily streaked. In certain lights, e.g. dusk or overcast weather, can look surprisingly whitish, but otherwise rather similar to Long-eared Owl. Palish face with *distinct black patches around yellow eyes*, expression 'mean'. Flight buoyant and wavering, a trifle slower than Long-eared Owl, *wingbeats rowing in slow-motion fashion*, upstrokes a bit jerky and quicker, downstrokes softer, yet *wings look rigid*. Glides freely, wings then slightly lifted. Often perches on ground. In flight, distinguished from similar Long-eared Owl, apart from slightly slower wingbeats, by *white trailing edge to 'arm'*; wing-tip almost solidly black with only one bold bar inside tip; *contrast between streaked breast and largely unstreaked belly*; tail coarsely barred; tertials contrastingly darker below; wings proportionately somewhat longer and narrower. – Sexes similar, ♂ on average paler below and on face, and less heavily streaked.

VOICE Call of ♀ a hoarse, baleful '**cheh**-ef' (drawn out when begging, '**cheeeh**-op'). Alarm a harsh 'chef-chef-chef'. Song given in flight (often high up), a quick series of subdued, deep hoots, 'uh-uh-uh-uh-uh-uh-...', 6–20 notes at a time, audible to c. 1 km. Short, quick wing-claps also given during song-flight. Begging-call of young like call of ♀ but more hissing and initially somewhat feebler.

Marsh Owl *Asio capensis* —

L 30½–37 cm, WS 80–95 cm. Breeds in marshland or wet meadows with long grass, within region only locally and rarely in NW Morocco; declining. Sedentary. Partly diurnal. Nest on ground, a hollow in tuft of grass.

IDENTIFICATION Like a small Short-eared Owl (and habitat much the same, too) with the following distinctions: size a trifle *smaller* even than Long-eared Owl; *wings shorter, broader and blunter-tipped*; flight more like that of Long-eared than Short-eared Owl, *glides on slightly arched wings*; upperbody, chest and upperwing-coverts almost uniform dark brown, making *cream-white face contrast strongly*; *eyes black*; secondaries dark above, coarsely barred below; belly pale, finely barred. – Juvenile: Has a diffuse, pale panel on upperwing-coverts. At closer range, breast and upperparts can be seen to be coarsely barred.

VOICE Call a harsh, croaking bark, 'krark', frog-like. Song apparently a series of croaking syllables in jerky rhythm, 'kra-kra **krrek**-kra-kra', repeated a few times during circling display flight. In anxiety at nest utters a fine squeaky whistle. Can wing-clap like its relatives. Begging-call of young a disyllabic rather musical 'too-ee' with rising inflection.

Long-eared Owl

Short-eared Owl

Marsh Owl

ONG-EARED OWL

white 'eye-bows' and long wings extending beyond tail differ from Eagle Owl when size not evident

adopts erect, sleek posture with closed eyes when disturbed

wingtip blunt

finely barred wing-tip

tufts can be folded!

finely speckled

orange eyes

belly streaked

several narrow bands

young just out of the nest, 4 weeks old

lacks white edge

finely barred tail

tufts invisible

SHORT-EARED OWL

hunting

can raise short tufts

slightly more slender than Long-eared

black wingtip

black mask, yellow eyes

coarse pattern above

young c. 4 weeks old

pale belly

broad white edge

boldy barred tail

MARSH OWL

juv.

pale-blotched wing-coverts

blunt

underwing darker and more heavily patterned than in congeners

plain

dark eyes, pale face

ad.

stubby rear

plain dark brown

broad white edge

barred

confined to damp meadows

ad.

ad.

ad.

DZ

Tawny Owl *Strix aluco* r**B**3

L 37–43 cm, WS 81–96 cm. Breeds in forests, parkland, wooded farmland, preferring old broadleaf trees (in particular with ancient oaks, frequently providing large holes); habitually found near humans in gardens and towns, and even roost on buildings and hunt rodents around farmhouses. Fairly numerous in Britain, with an estimated stock of 20,000 pairs. Strictly sedentary, rarely moving more than a few kilometres. Nocturnal. Food, mainly mice, voles and insects, taken on ground after patient watch from perch. Said to also hunt in flight but this probably only exceptional. Nests in hole, readily accepts nestboxes if tree-hole are missing. Aggressive when the still greyish and whoolly young, hardly able to fly, leave nest, and can fiercely attack intruder.

IDENTIFICATION A medium-sized, compact owl with *broad, rounded wings* and *large, rounded head*. Flight direct with fairly quick wingbeats; often makes long, straight glides. Ground colour variable from rufous-brown (predominant in Britain) to greyish-brown, whole plumage mottled, finely streaked and vermiculated dark. Facial discs rather plain, but typically has a darker narrow wedge on centre of forehead down to bill, between paler 'eye-brows'. *Thin whitish 'extra eyebrows' on forecrown* add to 'kind' expression. Eyes black. (Note that yellow-eyed species, e.g. Long-eared Owl, can look dark-eyed at night in headlights owing to large pupils.) Scapulars lined with white spots, suggesting 'braces'. No prominent pale patch on inner primaries above (cf. Long-eared and Short-eared Owls, Great Grey Owl). Tail finely and indistinctly barred. – Juvenile: Recognized through first few weeks on remnants of pale down on head, and incomplete facial pattern. – Variation: In NW Africa (*mauritanica*) a rather large, dark race occurs with coarse barring below; only a grey morph.

VOICE Vocal and loud-voiced. Most common call the well-known 'kew**ick**', shrill and repeated; fierce variations used as alarm. Song a hooting with 'mourning' ring, tone ocarina-like, can be mimicked by blowing in hands, first a drawn-out note falling in pitch, followed by *c.* 4-sec. pause, then an abrupt note quickly followed by a rapid series of shivering notes ending with a drawn-out note falling in pitch like the initial one,'**hoooo**uh.......ho, ho'ho'ho'**hoooo**uh'. A hoarse and more wailing variation of the song is used by the ♀ (and allegedly rarely by ♂, too). During courtship either sex utters a low-pitched, shivering tremolo (the so-called 'xylophone trill'), 'o'o'o'o'o'o'o'o'o'o'o'o'o...', audible only to *c.* 50 m. Begging-call of young a squeaking '**psee**-ep'.

Barn Owl *Tyto alba* r**B**4

L 33–39 cm, WS 80–95 cm. Breeds in farmland with scattered copses, gardens, rarely in villages close to fields. In Britain has declined, is not scarce, but local, with an estimated 4000 pairs. Sedentary. Nocturnal and crepuscular, often seen hunting in first dusk. Food voles, frogs, insects. Nests in hole in tree, building or ruin, often using roomy nestboxes inside farmhouses. Pellets characteristic, dark and glossy as if they have been varnished.

IDENTIFICATION Medium size, *slim body*, long wings, *long legs*. *Face* pale and characteristically *heart-shaped*. *Eye dark*. Flight recalls Long-eared Owl, wavering and elegant but *wingbeats* in normal hunting flight *markedly quicker* Appears *long-necked and short-tailed in flight, feet often dangling*, not least when hovering briefly before pouncing on prey. *Plumage typically very light* (♂ normally palest) notably so in flight. – Variation: Underparts practically pure white in W and S Europe, and on W Canaries (mainly ssp. *alba*) and in N Africa and Middle East (*erlangeri*) whereas they are yellow-orange in N, C and E Europe (*guttata*) and on E Canaries (*gracilirostris*) and Madeira (*schmitzi*). Upperparts grey in entire range.

VOICE Call of ♀ (rarely ♂) a drawn-out, purring shriek (Nightjar voice), repeated. Alarm, often in flight, a shrill hoarse and rather spooky squeal. 'Song', or territorial call is a drawn-out (*c.* 2 sec. long) rattling or gargling shriek 'shrrreeee', often repeated. Begging-call of young a drawn-out, wheezy snoring.

Hume's Owl *Strix butleri* –

L 29–35 cm, WS 70–80 cm. Resident in arid lower mountains, in deserts with deep, rocky wadis or cliffs. Breeds within treated area very rarely in Sinai, S Israel, and Jordan. Nocturnal habits. Food voles, mice, insects. Nests on cliff in crevice or hole.

IDENTIFICATION Medium-sized, large-headed, shaped like a small Tawny Owl, but *eyes are yellow* (in juv. more orange-yellow), and *plumage pale and rather plain*, underparts not streaked, only finely and diffusely barred pale ochrous-buff Flight-feathers boldly banded. Face pale and more or less unpatterned. Like Tawny Owl, has a slightly darker narrow wedge on forehead pointing to bill, between paler 'eye-bows'. Flight as Tawny, but due to smaller size has slightly quicker wingbeats and not as steady a flight path.

VOICE A pumping 'do-do-do-do-do-du', rising slightly in pitch at end, appears to serve as defensive call. ♀ begs with a fine, fluty, monosyllabic but slightly upward-inflected hooting 'dooa'. Song a rhythmical, sonorous hooting of five syllables with a voice more reminiscent of Collared Dove than Tawny Owl (voice is 'kind', soft and recalling toy ocarina, lacking vibrato), first a single note falling slightly in pitch, then *c.* 1½-sec. pause followed by four short notes almost in Ural Owl rhythm, 'hoou... **ho**-hu ho-ho', audible to *c.* 300 m and repeated 1–4 times/min. ♀ has deeper-voiced slightly muffled version. An exalted, shivering or stammering series of notes, 'hohohohohoho...', is heard in connection with courtship. Young said to have fine 'sneezing' begging-calls.

Tawny Owl   Barn Owl   Hume's Owl

TAWNY OWL

long glides on arched wings

(grey)

tail short with diffuse barring (cf. Ural Owl)

compact shape

day-roost

adopts erect, sleek posture when disturbed

ad. (rufous)

pale crown-pattern

mauritanica (NW Africa)

dark

ad. (grey-brown)

(common)

barred

young, just out of nest, 4 weeks old

BARN OWL

alba et al. 'pale-bellied' (e.g. W and S Europe, N Africa)

pale heart-shaped face

lacks dark 'comma'

guttata et al. 'dark-bellied' (e.g. N, E & C Europe)

young, 4 weeks old

very pale and 'ghost-like' appearance; legs often dangling when hunting

HUME'S OWL

spotted in torch-light

yellowish eyes

boldly barred

golden-ochre

PHARAOH EAGLE OWL

boldly barred wing- and tail-feathers

beware risk of confusion with Pharaoh Eagle Owl—♂ only fractionally larger than Tawny Owl!

DZ

Little Owl *Athene noctua* r**B**3–4

L 23–27½ cm, WS 50–57 cm. Breeds in Europe in open lowland country with mixture of fields, vineyards, orchards, meadows, copses, cliffs, gardens, parks, hedgerow trees. Further south it is commonly found also in semi-deserts. Sedentary. Introduced in Britain 19th century, where today 5000–10,000 pairs breed. Partly diurnal; this and habit of *perching fully exposed* mean that it is often seen. Food insects, birds, small amphibians and snakes. Nests in hole usually in tree or building, but sometimes also in cliff, quarry, stone wall or directly in ground.

IDENTIFICATION A *rather small and compact* owl, with large, broadly rounded crown and comparatively flat crown, *long legs* and short tail. Often squat posture when alarmed, bobbing (moving body up and down) in excitement. *Flight* fast and over longer stretches *bounding* like woodpecker, but more direct with continuous wingbeats on shorter flights. Brown above, speckled white, finely on crown, more boldly on back; whitish below, densely streaked brown. *Whitish oblique eyebrows* give 'stern' expression. Eyes yellow. Bill greyish-yellow. – Juvenile: Plumage pattern duller, lacking white spots on crown. – Variation: In S Middle East a pale grey-brown race (*lilith*), which in spite of a certain variation in darkness generally comes through as *strikingly pale*. Intermediate populations between this and darker birds of W Europe occur in SE Europe and W Turkey (*indigena*) and in Central Asia (*bactriana*).

VOICE Commonest call a sharp, complaining '*kee-ew*' or '(k)weew', in falsetto, usually falling in pitch, often eagerly repeated. Alarm short, explosive, high-pitched 'chi, chi, chi-chi,...'. Various subdued screeching notes, with or without rolling sound, uttered by ♀ when begging. Song a full and mellow, *slightly drawn-out rather low-pitched hoot with ending upward-inflected*, repeated about every 5–10 sec., 'goooek' ('querying' tone), each song note longer than song of Scops Owl; ♀ has higher-pitched, more nasal, less mellow version. Begging-call of young a hissing 'shree'.

(Eurasian) Scops Owl *Otus scops* V***

L 19–21 cm, WS 47–54 cm. Breeds in broadleaved and mixed open woodland, copses in farmland, churchyards, town parks, larger lush or neglected gardens; also found in wooded mountains well up to 1500 m, sometimes higher. Migratory, normally wintering in Africa S of Sahara, but a minority, perhaps local birds, stay in southernmost Europe and N Africa. In Britain only a few records per decade. Nocturnal. Food mainly insects. Usually hunts when it is dark, avoiding crepuscular hours. Nests in hole. Adopts nestbox if natural tree-holes are lacking. Like Long-eared Owl adopts stiff, elongated posture to avoid detection.

IDENTIFICATION Small, clearly *smaller than Little Owl, only big as a Starling, body thus about as large as a clenched fist*. Sits upright, showing little of ear-tufts when relaxed (then merely as *sharp corners of crown*), more when alert (adopting erect camouflage posture like Long-eared Owl). In flight, shows proportionately long and narrow wings, with flight fast and path hardly bounding (cf. Little Owl). Except at close range, appears *rather uniform brown* (rufous or more greyish; variable) with *pale* (grey-white) '*braces*', slightly paler face and underbody; seen well, shows intricate pattern of black streaks and vermiculations, diffuse whitish spots and rufous patches. Eyes yellow. Pallid Scops Owl (which see) is extremely similar.

VOICE A subdued 'drrr-drrr' by adults when presenting food, in connection with mating, etc. Song most commonly heard call, a constantly repeated short, deep, whistling 'tyuh' every 2–3 (4) sec., easy to mimic, audible to *c.* 1 km; ♀ has slightly higher-pitched version, often duetting with ♂. Told from midwife toad by slightly longer note and fuller, more musical clanging quality (not completely straight and mechanically whistling like the toad).

Pallid Scops Owl *Otus brucei* —

L 20–22 cm. Habitat as Scops Owl. Partly migratory, some wintering in S Israel. Often roosts in dense acacia. Nocturnal. Food mainly insects, also small mammals, birds. Nests in hole or in old nest of Magpie.

IDENTIFICATION *Small*, a trifle larger than Scops Owl (although difference normally impossible to perceive in the field), but in shape and habits identical to that species: *broad thick ear-tufts*, hardly discernible on relaxed bird; erect posture; long wings; short tail. Main distinctions apart from song (see below) are *paler, more dull sand-grey plumage*, especially notable *in face*; distinct, thin streaking below (though a few streaks on breast bolder), enhanced by pale ground colour, and is only *evenly and finely* vermiculated below (lacks bold white blotches and prominent dark barring of Scops Owl); pale '*braces*' *invariably buff* (not whitish as on some Scops); *no rufous* in plumage, and pale spotting above reduced (though some Scops Owls in S Europe and Middle East are rather similar). Rudimentary feathering on bases of toes (Scops unfeathered) difficult to make out in the field. – Juvenile: Finely cross-barred below but *without distinct streaks*, which separates it also from juvenile Scops.

VOICE Song provides best distinction from similar-looking Scops Owl. Voice is *softer and much less loud*, usually audible only to *c.* 300 m. Song *a series of c. 10 low cooing notes*, 'whuo whuo whuo whuo...'; tone subdued and ventriloquial (not loudly whistling with mellow voice as that of Scops), and pace rather like in Stock Dove. The first and last notes are slightly weaker, in Great Grey Owl fashion. At times the cooing is prolonged and delivered a little faster. A high-pitched, squeaky 'tzir tir tir ir' is uttered in excitement.

Little Owl

Scops Owl

Pallid Scops Owl

·TTLE OWL

unmistakable in its habitat

'false face' on nape

often spotted in daylight on prominent perch

bounding flight when moving far

ground colour varies, with tendency to be paler in S and E (but much variation within races)

pale

crown and breast plain

juv.

noctua et al. (Europe except in SE)

lilith (Cyprus, Middle East)

·OPS OWL

white spots

whitish 'braces'

(greybrown)

streaked

juv.

(rufous)

streaked dark, spotted white

streaked and with pale crossbars

(grey)

tufts sometimes folded and invisible, e.g. when in hunting mode

PALLID SCOPS OWL

buffish 'braces'

barred

merely streaked, no white spots

pale face

juv.

frequently shorter wing than Scops Owl

streaked, barring faint

NIGHTJARS *Caprimulgidae*

Medium-sized birds specialized for catching largish flying insects in flight at dusk and at night: have large gape, strong bristles at bill-corners (to enlarge diameter of 'trapping funnel'), long, narrow wings and tail, large head, short neck and legs, small bill. Plumage soft and brown-mottled, often with white spots ('signal flashes') on wings and tail. By day rest motionless on branch or ground. Noticed mostly at night, by calls. Nest on ground.

(European) **Nightjar** *Caprimulgus europaeus* m**B**4
L 24–28 cm, WS 52–59 cm. Breeds in open pine forest on sandy soils, often with some drier, sparsely vegetated bogs and clearings with pine saplings; also in more open sandy mixed and deciduous woods with glades and felled areas. Summer visitor (mostly May–Sep), wintering in Africa.

IDENTIFICATION *Mottled brown, buff-white, grey and black*, and with screwed-up eyes, the Nightjar resting lengthwise on a branch is hard to detect. Adult ♂ has *snow-white spots on wings and tail-corners*, which ♀ and 1st-autumn ♂ lack. ♂ also has a small narrow white patch across lower throat-side (lacking or indistinct on ♀). When hunting insects, flight noiseless, light and pitching with slight climbs, brief hovers, sudden fast glides disappearing from view etc.

VOICE Normally heard only during breeding season and at night. Call a frog-like but sonorous 'krru**it**'. Song a far-carrying (often audible at 1 km) hard reel, at close range amazingly rattling and intense, which, with only brief pauses, carries on 'in two gears' for hours on end, 'errrrrrrurrrrrrr-urrrrrerrrrrrrr...', from late dusk to dawn. When ♀ is in the vicinity, reeling sometimes changes into a hacking 'fi**orr**, fi**orr**' and ends with a halting rattle (sounds like engine breakdown!); wing-claps also form part of courtship.

Red-necked **Nightjar** *Caprimulgus ruficollis* —
L 30–34 cm, WS 60–65 cm. Breeds on sandy heath with tree clumps and bushy vegetation or in closed stone-pine forest. Summer visitor (end Apr–mid Oct), winters W Africa.

IDENTIFICATION The region's largest nightjar, *bigger* and *longer-tailed* than Nightjar. Plumage is similar in pattern, but has *rusty-ochre neck-band, throat and upper breast.* All wing-coverts broadly pale-tipped. Both sexes have pale wing and tail spots (most obvious and whitest on ♂) and rather *large white throat patch* (ditto). – Variation: In N Africa (ssp. *desertorum*) much paler and more rusty.

VOICE A hoarse 'tsche-tsche-tsche...' (like steam engine) is given by ♀. Song a far-carrying disyllabic knocking sound with hollow, slightly nasal quality, repeated in long series, 'kyo**tok**-kyo**tok**-kyo**tok**-...'; volume waxes and wanes a little during course of song. Also wing-claps in courtship.

Egyptian **Nightjar** *Caprimulgus aegyptius* **V**∗
L 24–27 cm, WS 53–58 cm. Resident in deserts and sem deserts with limited vegetation, e.g. sand dunes with lc shrubs and scattered trees, often near spring or stream.

IDENTIFICATION Size of Nightjar, with *long, narrow wir* as latter and rather long tail, but is *paler and more sanc coloured*. Finely patterned above, *without Nightjar's hea black streaking. Lacks obvious white wing patches* (has mu white on inner web of outer primaries, so *under primar conspicuously white*, but on upper side white glimpsed or when wing fully spread). In flight, upperwing pattern c recall a ♀ Kestrel, with '*hand' darker than yellow-brown 'arr Tail pale*, on ♂ with *distinct white corners*. Flies with *ratl slow*, almost Long-eared Owl-like jerky wingbeats. (Duri winter months in Middle East beware small, pale race *ı wini* of Nightjar on passage from Iran; this, however, h fully barred primaries below and heavy black streaking back, like European birds.)

VOICE Song long series of rapidly repeated hollow a mechanical-sounding short churrs, 'krroo-krroo-krroo at distance like an engine (e.g. old-fashioned 2-stroke eng on fishing sloop); tempo drops somewhat towards end each series.

Nubian **Nightjar** *Caprimulgus nubicus*
L 20–22 cm, WS 46–50 cm. Resident in barren, arid, op terrain, in semi-desert with trees (acacia, tamarisk, bes low-growth palm groves). Keeps late hours, but can st singing before dusk. Often sits on roads and bare fields.

IDENTIFICATION The region's *smallest* nightjar. Propc tionately *shorter tail* than Nightjar, and *somewhat broac and blunter wings*. Colours much as Red-necked Nightj predominantly *grey with rusty-ochre neck-band* and reddis brown bars on inner remiges. ♂'s *white wing and tail sp* very conspicuous (the former contrast with otherwise *alm black wing-tips*), ♀'s almost as big.

VOICE Song a hollow, clanging sound repeated in coupl barely once per second, 'kyau kyau' (or 'trül trül'); audi to only *c.* 200 m. Also wing-claps.

(Common) **Nighthawk** *Chordeiles minor* **V**
L 23–25½ cm, WS 54–60 cm. N American species; very ra vagrant Sep–Oct, mainly in Britain. Size of Nightjar, b has *more pointed wings* and *medium-length, shallowly forl tail*. Seen head-on, glides with *wings in V and bent at carp* The unbarred *blackish primaries* have a *distinct broad, wh cross-band* (halfway between carpal and tip) on ♂; less d tinct on ♀. ♂ has *broad white subterminal band on tail*, a a clear *white throat patch* (smaller and buff-tinged on Immatures have white trailing wing-edge.

Nightjar

Red-necked Nightjar

Egyptian Nightjar

Nubian Nightjar

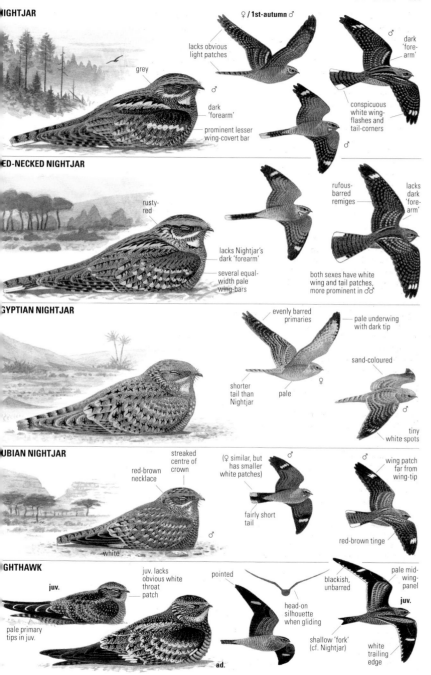

NIGHTJAR

grey

♀ / 1st-autumn ♂

lacks obvious light patches

dark 'forearm'

♂

dark 'forearm'

prominent lesser wing-covert bar

♂

conspicuous white wing-flashes and tail-corners

♂

RED-NECKED NIGHTJAR

rusty-red

rufous-barred remiges

lacks dark 'forearm'

lacks Nightjar's dark 'forearm'

several equal-width pale wing-bars

both sexes have white wing and tail patches, more prominent in ♂♂

EGYPTIAN NIGHTJAR

evenly barred primaries

pale underwing with dark tip

sand-coloured

shorter tail than Nightjar

pale

♀

tiny white spots

♂

NUBIAN NIGHTJAR

streaked centre of crown

red-brown necklace

(♀ similar, but has smaller white patches)

♂

♂

wing patch far from wing-tip

fairly short tail

♂

red-brown tinge

white

NIGHTHAWK

juv.

juv. lacks obvious white throat patch

pointed

blackish, unbarred

pale mid-wing-panel

juv.

pale primary tips in juv.

head-on silhouette when gliding

shallow 'fork' (cf. Nightjar)

white trailing edge

ad.

SWIFTS *Apodidae*

Streamlined small to medium-sized birds with long, pointed, scythe-shaped wings, accomplished and indefatigable flyers. Have 'clinging feet' which (unlike those of swallows) are unsuited for perching. Sexes alike. Can copulate and even sleep (!) in flight. Food mainly insects. Nest in cavities.

(Common) Swift *Apus apus*   m**B**3 / **P**2

L 17–18½ cm, WS 40–44 cm. Breeds in towns and villages and, rarely, in deserted wooded areas or cliffs. Summer visitor (mostly May–Aug), winters in S Africa. Nests in ventilation shafts, cracks etc. in walls, under convex roof tiles or in church towers, sites used regularly year after year. Pairs stay together throughout their lives, and the same pair can reuse the same nest site for over 15 years.

IDENTIFICATION Seen incessantly hunting insects in the air, often with swallows (and confused with these by the layman); told by *dark underparts* (only throat pale), somewhat larger size, different wing shape and flight. *Wings scythe-shaped* with long 'hand' and very short 'arm' (carpal joints indiscernible, right next to body). Flight one minute with *frenziedly fast wingbeats* (action then so rapid that wings appear to 'beat alternately'), the next with *long glides* or sailing, motionless against the wind (swallows flutter, or have more backward-clipping action and brief glides between series of wingbeats). – Juvenile: Blacker ground colour; pale forehead and lores (white feather fringes); whiter throat patch; finely white-edged wing-coverts.

VOICE Various shrill, monotone, ringing screams, 'srriiirr', sometimes faintly downslurred (or upslurred). Especially striking are the choruses of screams from tight flocks flying around low over the rooftops on summer evenings.

Pallid Swift *Apus pallidus*   **V*****

L 16–18 cm, WS 39–44 cm. Breeds in towns and on rock faces, preferably near the sea, nesting in cavities as its relatives. Migrant, winters S of Sahara.

IDENTIFICATION Very like Swift; marginally *broader wings* (seen best on outer wing), *broader head/neck* and *plumper body*. Flies with (subtly) *slower wingbeats* and longer glides. *Forehead and throat* are on average *paler*, and a *dark eye patch* normally shows up against paler brown head-side. Upperwings show *slightly paler brown secondaries, inner primaries and primary-coverts* contrasting with almost blackish outer primaries and darker brown mantle. *Back/ rump subtly paler brown than mantle* (faint hint of dark 'saddle effect'; European Common Swifts are more evenly dark above, but note that Asian populations (*pekinensis*) can have even more obvious dark 'saddle' than Pallid). Underparts are somewhat *paler and more scaly* than on Swift. Note that light and background can give misleading impressions.

VOICE Like Swift's (and at times hard to separate), but commonest call almost always has *lower pitch*, is *more clearly downslurred* (is thus *disyllabic*), and voice is somewhat *drier, more mechanical and hard*, 'vrrüü-e'.

Alpine Swift *Apus melba*   **V****

L 20–23 cm, WS 51–58 cm. Breeds colonially, usually in tall buildings, also in rock faces. Summer visitor (mostly Apr–Sep), winters in S Africa. Pairs for life.

IDENTIFICATION Immediately identified by its *large siz[e]* and for a swift its *rather slow wing action with deep beats* (ca[n] momentarily be taken for a Hobby), also by the *white bell[y]*. Also has *throat white*, but this (surprisingly) not always [s]o easy to see (can be in shadow or have restricted white[...]). Colours above and on breast-band *drab grey-brown*.

VOICE Easily recognized, loud, drawn-out twittering s[er]ries which either accelerate or (normally) slow down an[d] drop a little in pitch at end, e.g. 'ti ti titititititititititi-ti-t[i]-ti-ti ti tü tü'.

Plain Swift *Apus unicolor*

L 14–15½ cm, WS 36–39 cm. Breeds in Madeira and t[he] Canary Islands. Summer visitor (Mar–Aug), but many als[o] stay through the winter; winter area of migrants large[ly] unknown but is thought to be mainly coastal NW Africa.

IDENTIFICATION Very like Swift, but is *slightly smaller* an[d] a little *slimmer* and *narrower-winged. Tail proportionate[ly] somewhat longer with narrower base. Pale throat patch sma[ll]er, not so bright*, and lower edge to upper breast less clea[rly] demarcated. A shade paler grey-brown than Swift (n[ot] sooty-black), but this difference perceptible only under o[p]timal conditions. Flight often appears supremely fast, wi[th] even more ferocious turns, and unlike Swift frequently fl[ies] right through the canopy of trees.

VOICE To all intents inseparable from Swift's.

White-rumped Swift *Apus caffer*

L 14–15½ cm, WS 33–37 cm. Main distribution in sub-S[a]-haran Africa. Breeds in old Red-rumped Swallow nes[ts.] Summer visitor (May–Oct), winters S of Sahara.

IDENTIFICATION Very *dark* with *narrow white rump patc[h]* white and *well-defined small throat patch* and *diffusely pa[le] tipped secondaries*. Wings pointed with long 'hand', *fas[t] agile flight with quick wingbeats. Slim rear end* and *lon[g] deeply forked tail*. Tail often kept folded in flight, formin[g] long pointed end of body.

VOICE A jolting, hard, staccato series which may chan[ge] into a trill, 'chüt-chüt-chüt-chüt-ürrrrrrrr'.

Little Swift *Apus affinis*   **V***

L 12–13½ cm, WS 32–34 cm. Breeds in towns and villag[es] but also on cliff faces. Partially resident.

IDENTIFICATION *Dark*, with quite *broad white rump patc[h]* which *extends onto flanks* (and so can be seen from below[).] In flight, characteristic silhouette with *broad tail-base* a[nd] *short but quite broad, square-cut (rounded when spread) t[ail]* and rather broad inner primaries but short secondaries, *wing looks ample in centre* but *trailing edge looks pinched near body* (effect amplified in spring when inner primar[ies] have been shed, as species starts moult early). Flight often fluttering almost like House Martin, not that fast.

VOICE A high, bouncing twitter pulsating a little in volume (and vaguely recalling Great Snipe display at distance).

Swift

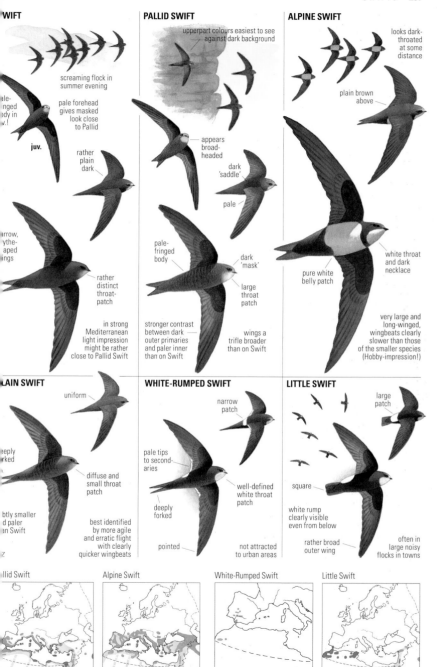

WIFT

screaming flock in summer evening

pale forehead gives masked look close to Pallid

pale-winged body in v.!

juv.

rather plain dark

arrow, ythe-aped ings

rather distinct throat-patch

in strong Mediterranean light impression might be rather close to Pallid Swift

PALLID SWIFT

upperpart colours easiest to see against dark background

appears broad-headed

dark 'saddle'

pale

pale-fringed body

dark 'mask'

large throat patch

stronger contrast between dark outer primaries and paler inner than on Swift

wings a trifle broader than on Swift

ALPINE SWIFT

looks dark-throated at some distance

plain brown above

white throat and dark necklace

pure white belly patch

very large and long-winged, wingbeats clearly slower than those of the smaller species (Hobby-impression!)

LAIN SWIFT

uniform

eeply rked

diffuse and small throat patch

btly smaller d paler an Swift

best identified by more agile and erratic flight with clearly quicker wingbeats

z

WHITE-RUMPED SWIFT

narrow patch

pale tips to secondaries

well-defined white throat patch

deeply forked

pointed

not attracted to urban areas

LITTLE SWIFT

large patch

square

white rump clearly visible even from below

rather broad outer wing

often in large noisy flocks in towns

llid Swift   Alpine Swift   White-Rumped Swift   Little Swift

(Eurasian) Hoopoe *Upupa epops* P5

L 25–29 cm (incl. bill 4–5), WS 44–48 cm. Breeds in farming districts and open, grazed country with copses, hedges and bushes; often seen in vineyards and orchards. Summer visitor (mostly end Apr–Sep), winters in Africa. Spends much time on ground, and needs some short-grass or bare areas for feeding (food mostly worms, insects). Wary (but not exactly shy), keeps a certain distance from man. Nests in hole in tree, stone wall, nestbox, house foundations.

IDENTIFICATION One of the most striking and distinctive birds of the whole region: *buffy-pink with black- and white-striped, broadly rounded wings*, and crown with an erectile *crest like Indian chief's* (though normally raised only momentarily on landing, otherwise rarely). Bill long, narrow, slightly decurved. Tail black with broad white band. Flight flappy and rather unsteady with short undulations, hint of gliding, and uneven rhythm, often low over ground. Moves energetically and jerkily on ground, like a starling.

VOICE When agitated and excited a high, noisy 'scheer' with traits of Collared Dove and distant Black-headed Gull. Also a dry rolling 'cherrr' in mate/brood-feeding and other situations. Song a trisyllabic hollow, muffled 'oop-oop-oop', repeated several times; weak at close range yet carries.

KINGFISHERS *Alcedinidae*

Small and medium-sized, compact birds with proportionately large head and bill, short neck and small legs. Most live by water and are accomplished in plunge-diving for fish.

(Common) Kingfisher *Alcedo atthis* rB4

L 17–19½ cm (incl. bill *c*.4). Breeds at small and moderate-sized fish-rich slow-flowing rivers with some trees and suitable nesting banks; occasionally lakes; also estuaries and coasts in winter. Perches motionless on lookout for small fish, caught following vertical dive. Also hovers occasionally when scanning. Rather shy and restless. Excavates nest in sandy bank, a good metre-long tunnel leading to nest chamber, where young are reared on bed of piled fish bones.

IDENTIFICATION A small, *plump, short-tailed* and short-legged bird with *big head* and disproportionately *long bill*. Perches upright. Beautifully bright colours: *crown and wings greenish-blue* (look more greenish from some angles depending on how light falls), *back and tail bright blue* (shifting from azure to cobalt!), *underparts and cheek patch warm orangey brownish-red*, throat and a patch on neck-side snow-white. Despite this display of colour, can be difficult to pick out if perched motionless in shadow on a waterside branch. More often noted as it flies straight and fast, low over water, calling. Sexes similar, but in breeding pairs ♂ has all-black bill and ♀ reddish base to lower mandible.

Legs light red. – Juvenile: Like adult, but plumage somewhat duller and greener. Legs greyish.

VOICE Normal call, often given in flight, is a short, shar whistled 'zii', sometimes with afternote, 'zii-ti', and i excitement repeated in brief series. Song, seldom heard, simple series of call-like notes in jerky, irregular rhythm.

White-throated Kingfisher *Halcyon smyrnensis* –

L 26–30 cm (incl. bill *c*. 6). Unlike its relatives not strictl tied to water but can be seen among trees in drier farmland in palm groves and in town parks, sometimes even in pur forest with smallish glades. Usually, however, lives near wa ter. Takes fish, amphibians, lizards, insects.

IDENTIFICATION The size of a large thrush. Vivid colour make it unmistakable, with head, belly, flanks and lesse coverts chestnut-brown, *throat and upper breast pure white back, uppertail and most of upperwings blue-green. Bi coral-red*. In flight, gleaming sky-blue primary patche above and *large white wing-panels below*. – Juvenile: Some what duller, and white breast often with fine dark vermicu lation. Bill yellow-brown or orange with darker tip.

VOICE Very *noisy and loud*, its voice can dominate a local ity. The alarm is a croaky, metallic series, 'krix krix krix-ix… Song a very loud and aggravatingly repeated rapid trillin whistle, dropping in pitch and sounding indignant an bleating, 'ti-ti-tü-tü-tu-tu-…', often delivered from top of tall eucalyptus tree.

Belted Kingfisher *Megaceryle alcyon* V**

L 31–34 cm. Very rare vagrant from N America. Recorde e.g. in Iceland, Ireland, Netherlands and Britain.

IDENTIFICATION *Large* and powerful. *Upperparts lead-gre with a white collar*. Breast white with either a grey band (♂ or *both a grey and a rufous band* (♀). Elongated crown- an nape-feathers can be erected, then forming *straggly crest*.

Pied Kingfisher *Ceryle rudis* –

L 25–27 cm (incl. bill *c*. 5). Breeds at rivers, lakes, rive mouths, canals and fishponds; also on coasts, where it ma be seen hovering a fair way out above the surf. Resident.

IDENTIFICATION Large and lively. Hovers and perches in oper easy to see. Plumage is entirely *black and white*. Underpar white, ♂ has *two black bands across breast* and ♀ one. Whi supercilium between black crown and black eye-stripe. Ta fairly long, white with black centre and terminal banc Outer wing black with *white primary-base patches*. Bill black

VOICE Loud, sharp and chirpy whistled notes, often de livered at furious tempo and without any clear pattern (ca recall Great Snipe display) but now and then crystallizin into a rhythmic 'titi-**tütt**-titte**ritt**'.

Hoopoe

Kingfisher

White-throated Kingfisher

Pied Kingfisher

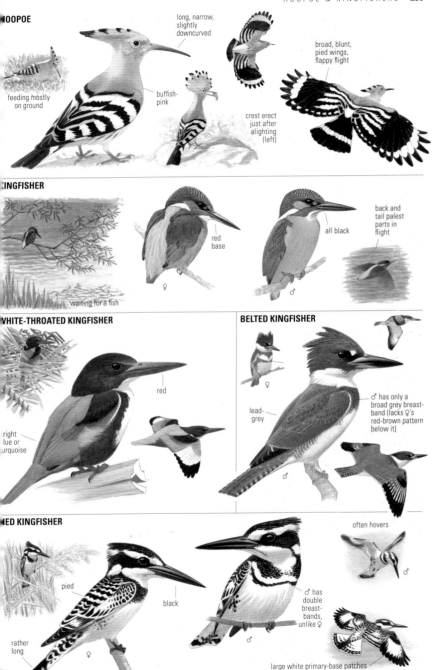

HOOPOE

long, narrow, slightly downcurved

buffish-pink

broad, blunt, pied wings, flappy flight

feeding mostly on ground

crest erect just after alighting (left)

KINGFISHER

red base

all black

back and tail palest parts in flight

♀

♂

waiting for a fish

WHITE-THROATED KINGFISHER

red

bright blue or turquoise

BELTED KINGFISHER

♀

lead-grey

♂ has only a broad grey breast-band (lacks ♀'s red-brown pattern below it)

♂

PIED KINGFISHER

often hovers

♂

pied

black

♂ has double breast-bands, unlike ♀

rather long

♀

♂

large white primary-base patches

M

BEE-EATERS *Meropidae*

Rather small or thrush-sized, slim birds with slender, pointed, downcurved bill, long, pointed wings and long tail, usually with tail projection. Three species breed in the region, many more in rest of Africa and in S Asia. Warmth-loving, favouring southern lowlands and sunbaked open mountain slopes. Specialists in catching flying insects, not least *Hymenoptera*. Sociable habits, often seen in flocks, breed in colonies. Nest in ground holes, often in river bank or sand-pit.

(European) **Bee-eater** *Merops apiaster*   V*

L 25–29 cm (incl. tail projection 0–3), WS 36–40 cm. Breeds in open country in warm regions, in cultivated areas or in broken, open terrain with pastureland, bushes, odd trees or copses, often by rivers with steep banks, but also commonly in sand-pits. Summer visitor (mostly May–Aug), winters in S Africa. Wary.

IDENTIFICATION Medium-sized bee-eater with exotically rich and gaudy plumage colours. Adult should be unmistakable with *bright yellow throat, bluish underbody, yellowish-white shoulder patches* and *red-brown crown / back and inner wing-panel* above. Juvenile on the other hand can at distance be confused with Blue-cheeked Bee-eater (which see), owing to duller and more greenish colours above. Flies in long, shallow undulations. Hunts often high up with straight-winged glides and brief periods of fast wingbeats (much as House Martin). Seen in flight from below most bee-eaters have similar appearance with *pointed, rufous-tinged wings* narrowly bordered black at rear. Note on European Bee-eater the somewhat *broader dark rear edge along secondaries* compared with primaries.

VOICE Call a frequently repeated, soft but abruptly given, rolling 'prüt' with lilting tone, given in chorus from flocks in flight, carries far. When agitated at nest a purer, short, whistled 'wüt'.

Blue-cheeked Bee-eater *Merops persicus*   V***

L 28–32 cm (incl. tail projection 4–8), WS 35–39 cm. Breeds in dry, open terrain, but often near watercourses, at times also in open woods and at glades. Summer visitor (mostly Apr–Sep), winters in Africa.

IDENTIFICATION An *almost entirely green* bee-eater, same size as Bee-eater but adult has *longer and narrower tail projection*. Like Bee-eater has *rusty-red underwings*, so resembles latter from below in flight (unless shape of tail projection discernible). Note *uniform green back* (lacks any hint of pale shoulder patches), *red-brown throat patch* with only *little yellow on chin* and *narrower dark trailing edge to arm below* (evenly narrow on both primaries and secondaries). – Juve-

nile: Duller green. Only rudimentary tail projection. Ha almost no yellow on chin and lacks blue-green on head.

VOICE Call like Bee-eater's but is drier and a bit harde also somewhat higher-pitched, 'prri prri prri...', lacking th more tuneful voice of Bee-eater and therefore does not carr as far.

Little Green Bee-eater *Merops orientalis*   –

L 20–25 cm (incl. tail projection 2–8). Breeds in arid, ope terrain with scattered trees, also in palm groves and garden etc. Often seen perched on wires and bushtops. Not shy.

IDENTIFICATION Smallest bee-eater in region, no bigge than a wheatear, with almost *entirely green* plumage an *black eye-stripe* and a *black cross-band between throat an breast*. Sexes alike. Juvenile lacks elongated tail projection and colours are slightly duller. – Variation: In Israel an Jordan (ssp. *cyanophrys*) entire *throat and a narrow superci ium bright blue*, black band below throat usually broad an diffuse, also *tail projection short*. In Nile Valley (*cleopatra* has *throat pale green*, throat band always narrow, and *ta projection long*.

VOICE Call or alarm short, hard 'kitt' notes or rollin metallic 'krrit'. Song seems to consist of various hard, rol ing, shrill whistles in rapid series lacking a clear structur e.g. 'krrih krrih kru-kru krieh kr-krü'.

(European) **Roller** *Coracias garrulus*   V*

L 29–32 cm, WS 52–57½ cm. Breeds in dry, warm, ope country with scattered trees, copses and open woods (main oak, locally also pine) and plenty of largish ground-dwellin insects (beetles, grasshoppers etc.). Summer visitor (most May–Aug), winters in S Africa. Nests usually in tree-hole.

IDENTIFICATION A Jackdaw-sized, heavily built, green tinged *pale blue* bird with *brown back*. Large head an *strong black bill*. In flight, shows *contrast between blue u| perwing-coverts and blackish remiges*, and even strong contrast on underwing, where coverts are pale blue an may appear white. Leading wing-edge and rump ultrama rine-blue. Over longer stretches flies with rather slow pc powerful, slightly clipped beats and on straight course, bt at times makes minor turns and swerves. Compared wit e.g. Jackdaw, wings appear large on slim body, and are ke| more angled. – Juvenile: Duller and more green-gre' breast and median upperwing-coverts tinged brown.

VOICE Call and alarm a hard '**rack**-ack' (sometimes mon or trisyllabic). When nervous etc. raucous series, 'reehr-eeh eehr-eehr-...' (voice like scolding Jackdaw or Rook). Ca during almost Lapwing-like display-flight starts with a fe hard notes followed by 1–4 fast, hoarse drawn-out 'kraa notes. Structure of call varies somewhat.

Bee-eater

Blue-cheeked Bee-eater

Little Green Bee-eater

Roller

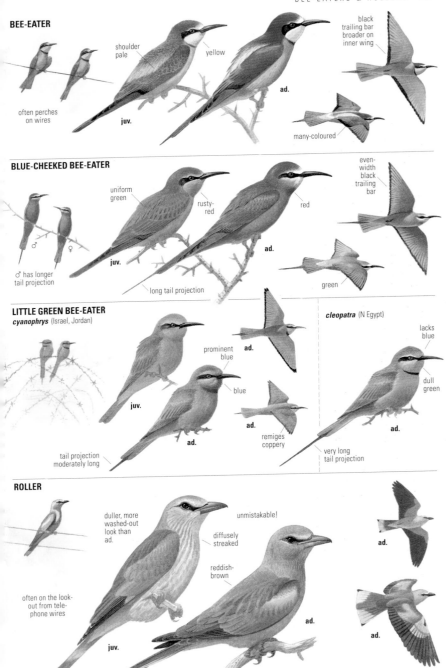

BEE-EATER

shoulder pale

yellow

black trailing bar broader on inner wing

ad.

often perches on wires

juv.

many-coloured

BLUE-CHEEKED BEE-EATER

even-width black trailing bar

uniform green

rusty-red

red

juv.

ad.

♂ has longer tail projection

long tail projection

green

LITTLE GREEN BEE-EATER
cyanophrys (Israel, Jordan)

prominent blue

ad.

blue

ad.

cleopatra (N Egypt)

lacks blue

dull green

juv.

ad.

remiges coppery

ad.

tail projection moderately long

very long tail projection

ROLLER

duller, more washed-out look than ad.

unmistakable!

diffusely streaked

reddish-brown

ad.

often on the look-out from telephone wires

ad.

juv.

ad.

ad.

DZ

WOODPECKERS *Picidae*

Eleven species, all but the aberrant Wryneck specialists in climbing and excavating nest holes in vertical tree trunks. Anatomical adaptations include strong feet with mobile toes (species with four toes have two directed backwards) and sharp claws, stiff tail-feathers which serve as support on vertical surfaces, also powerful awl- or chisel-shaped bill and 'shock-absorbent' braincase. Food includes wood-boring insects; have greatly elongated tongue base for scouring and emptying deep insect burrows. Most species use drumming as a 'song' (both sexes drum). Only the Wryneck is a long-distance migrant, others largely residents.

Black Woodpecker *Dryocopus martius* —

L 40–46 cm, WS 67–73 cm. Nests in mature forest, often pine and mixed forest, but also beech. Expanding in NW Europe. Wary but also inquisitive, can be called up by imitating its voice. Nest entrance oval, *c.* 9×12 cm.

IDENTIFICATION Crow-sized and *entirely black* with *whitish eye* and all-*red crown* (♂) or just a red patch on hindcrown (♀). Compared with e.g. Green Woodpecker, *narrower neck* and longer head with *angular nape*. Flight *flappy* and rather clumsy with *head held up*, with mostly downward jerks of *broad, rounded wings, on straight course* (not dipping as other woodpeckers), recalling Nutcracker if anything.

VOICE Rich repertoire of loud calls. Often heard year-round is the flight-/alarm-call, 'krrück krrück krrück...'. On landing it may utter a shrill **'kliii**-eh'. Song a shrill whistled series, quite distinct from Green Woodpecker's laugh in its more tentative intro, even pitch throughout and 'wilder' voice,'kuih kvi kvi-kvi-kvi-kvi-kvi-kvi-kvi'. In spring drums very loudly in open bursts (like machine-gun salvos; audible at 2–4 km), 1.75–3 sec. long (♂ longer than♀).

(European) Green Woodpecker *Picus viridis* rB3

L 30–36 cm, WS 45–51 cm. Breeds in open deciduous and mixed woods, mature farmland with pasture and trees, parkland and commons, also in large gardens. Common. As all woodpeckers, absent Ireland. Food mainly ants; spends much time on ground. Shy, wary. Nest entrance *c.* 6×7 cm.

IDENTIFICATION Often seen passing in *strongly undulating flight*, and so long as you see *green upperparts with yellow rump* confusion is possible only with Grey-headed Woodpecker (which see for differences); ♀ Golden Oriole, also greenish above and with dipping flight, is much smaller and slimmer, with narrower wings, longer tail. Once twwhe bird perches, on the alert, and can be watched through binoculars, note the *red crown* and *black-enclosed white eye*. Sexes differ in ♂ having *red centre to black moustache*, ♀ not. – Juvenile: Much as adult but whole plumage is spotty and

vermiculated. – Variation: Iberian birds (ssp. *sharpei*) have almost no black on head, are dusky grey around eye, and ♂ has red moustache with black border below only.

VOICE Usually noticed by calls. Often utters a shrill, explosive 'kyü-kyü-**kyück**' in flight, sometimes drawn out in long series with stress on e.g. every second or fourth syllable. Song a loud, laughing series of 10–18 'klü'notes falling somewhat in pitch and accelerating slightly at end, 'klüh-klüh-klü-klü-klü-klü-klü-klü-...'. Drums only rarely, a fast but soft roll *c.* 1½ sec. long. Young beg with hoarse, rasping series of notes.

Grey-headed Woodpecker *Picus canus* —

L 27–30 cm. Breeds in several quite different habitats, e.g. in swamp-forests along lakes and lakeshores with plenty of insect-rich decayed deciduous trees; in open or park-like mature deciduous forest; in open woodland in uplands (often to 600 m, at times higher) and with good coniferous element. Food insects, often ants. Nest-entrance diameter *c.* 5½ cm.

IDENTIFICATION Between Green and White-backed Woodpeckers in size, but confusable of course only with former. Compared with Green, the *head is slightly smaller and more rounded*, this reinforced by the slightly *shorter and more slender bill* ('kinder'appearance), the different, more *uniform grey head pattern* with less black around eye (only lores black) which is *amber-coloured*, the *narrower dark stripe on chin-side*, also *red crown patch restricted to forecrown* (♂) or *no red at all on crown* (♀). Back is moss-green (lacks Green Woodpecker's yellow-green tone), underparts unvermiculated *light grey* (with hint of green), not yellow-tinged. – Juvenile: Very like adult, only somewhat more subdued colours.

VOICE When agitated, series of choking 'chk'. Song a series of 6–9 mechanical-sounding straight whistles which gradually drop in pitch, 'kii kii küü küü, küü, kuu', and often slow down markedly after fast opening; easy to imitate by whistling; lacks Green Woodpecker's laughing tone, sounds more desolate. Drums more often than Green, rolls loud and rather fast, *c.* 1½ sec. long.

Levaillant's Green Woodpecker *Picus vaillantii* —

L 30–33 cm. Breeds in mountain forests of oak, poplar, cedars and pine, extending to treeline at around 2000 m.

IDENTIFICATION Very like Iberian ♀ of Green Woodpecker (which see), but has *pale border above the all-black moustache* (*both sexes*) right up to the bill, also is *less dusky grey behind and above eye*. ♂ has fully red crown, ♀ is red only on hindcrown (black-spotted on rest of crown).

VOICE Like Green Woodpecker's, but song often slightly faster in tempo, with more even pitch and more bubbling in tone. Drums more often than Green, rolls 1–1½ sec.

Black Woodpecker  Green Woodpecker  Grey-headed Woodpecker  Levaillant's Green Woodpecker

BLACK WOODPECKER

flight flappy and slightly clumsy, course straight

all-red crown

angular hindcrown

thin neck

♂

red only on hindcrown

♀

sexes differ already as nestlings

♀

GREEN WOODPECKER

black

juv. ♀

red

juv. ♂

spotted

yellow-green

♂

undulating flight, folds wings completely between wingbeat bursts

black eye-surround

red centre

black

♀

no black around eye

♀   ♂

sharpei
(Iberia)

GREY-HEADED WOODPECKER

red patch on forehead in ♂

narrow

grey

♂

♀

looks neckless, snub-nosed and rather short-billed in flight

plain

lacks black and white bars on tail-sides

often visits old aspen trees

LEVAILLANT'S GREEN WOODPECKER

♂

obvious pale border above all-black moustache

♂

lacks red

♀

prefers ancient oak and cedar woods at high altitudes in Atlas mountains

DZ

Great Spotted Woodpecker *Dendrocopos major* ⌐**B**3

L 23–26 cm, WS 38–44 cm. Breeds in all kinds of woodland, especially with stands of spruce and pine (conifer seeds important winter food), also larger parks and gardens; in Britain mostly in deciduous / mixed woods, often with good element of aspen and waterlogged regrowth. Alert and cautious, but in winter may visit birdtables, suet bags etc., can cling upside-down like a tit. Food insects, conifer seeds, at times also bird eggs and nestlings. To extract seeds wedges cones firmly in special bark crevices ('anvils'); piles of empty cones lie beneath. In some autumns, when cone crop fails in northern taiga, makes invasion-like migrations to S and SW (some reach Britain). Nest entrance *c.* 5×6 cm.

IDENTIFICATION The commonest of the 'pied' woodpeckers, i.e. those with basically *black and white plumage*. Usually easily identified by *saturated red vent* sharply demarcated from whitish belly, by *unstreaked flanks*, and by its two *large, white, oval shoulder patches*. The black wings are barred white. *Flies in deep undulations*, in straight direction. – Adult ♂: Sugar-lump-sized *red patch on hindcrown*. – Adult ♀: Black crown *with no red*. – Juvenile: Combination of mostly red crown (red patch larger and brighter on ♂), more poorly developed black stripe between moustache and nape, paler red vent and sometimes faint streaks on flanks makes it confusable with Middle Spotted and White-backed Woodpeckers; note *white ovals on shoulders* as on adult (eliminating White-backed), *black moustachial stripe reaching bill*, *black sides to red crown patch* and *black hindcrown* (eliminating Middle Spotted). – Variation: Birds in W and S Europe usually more dusky (grey-tinged) brownish-white below and on forehead, also have slightly more slender bill. In Algeria and Tunisia (ssp. *numidus*) breast has a mottled black and red band, and red of vent extends well (but to variable degree) up onto belly.

VOICE Call a short, sharp 'kick!', sometimes slowly repeated (*c.* 1 per sec.) in longer series. When agitated gives a very fast series of thick-voiced chattering notes, 'chrett-chrett-chrett-chrett-...'. Drums in spring, rolls being characteristically short (0.4–0.8 sec.), very fast (difficult to hear individual strikes) and ending abruptly. Young beg with incessant series of thin, high twitters, 'vivivivivi...'.

Syrian Woodpecker *Dendrocopos syriacus* —

L 23–25 cm. Breeds in open, cultivated country, in orchards, gardens, parks, lines of trees, vineyards etc. Has expanded NW during 20th century. Food insects, also fruit and berries. Nest entrance *c.* 5 cm in diameter.

IDENTIFICATION Very like Great Spotted Woodpecker: a black and white 'pied' woodpecker of same size and with *only small red patch on hindcrown* on ♂ and all-black crown

on ♀, also with *large white shoulder patches* and *red vent*. Close observation needed for safe distinction; note: *no black line joining black central nape band with black angled stripe on neck-side* (safest feature on adults); *somewhat longer bill*; *pale nostril feathering* (Great Spotted black); *less white on outer tail-feathers* (only a few white spots at tip of black outer rectrices; Great Spotted has narrowly barred white tail-corners); on average cleaner white head-side and paler forehead (Great Spotted usually has dirty white cheek and brownish-white forehead, but occasionally identical); sometimes a few faint grey streaks on lower flanks and belly (never shown by post-juv. Great Spotted); on average less intensely red vent (but odd birds are similar); often slightly bigger red hindcrown patch on ♂ than corresponding one on Great Spotted (but a few are the same). – Juvenile: Streaked flanks. Often *reddish on breast*.

VOICE Common kick-call is softer than Great Spotted's and often surprisingly like Redshank's alarm-call, 'gipp'; when highly agitated this can be repeated in rapid series, 'gip-gip-gip-...', and also intermixed with a 'chirrr'. Also has thick chattering 'chre-chre-chre-...' (like Great Spotted's). Drumroll resembles Great Spotted's, but can usually be separated by being longer, 0.8–1.2 sec., and by decreasing somewhat in volume towards end (Great Spotted's short roll is more abruptly cut off); faster strike frequency and shorter duration than White-backed Woodpecker's. Young quieter than Great Spotted Woodpecker young.

Middle Spotted Woodpecker *Dendrocopos medius* —

L 19½–22 cm. Breeds in lofty deciduous woods with some old oak, hornbeam and elm and mixture of clearings, pasture and denser parts. Warmth-loving. Food insects and sap. Spends much time high up in tree crowns, and often hops along horizontal, thick branches in search of insects. Nest often excavated in decayed, rotten trunk or thick branch, sometimes strongly sloping or almost horizontal; entrance-hole diameter *c.* 4 cm.

IDENTIFICATION Only negligibly smaller than Great Spotted Woodpecker, but still looks clearly smaller owing to the *short, slender bill* and the *rounded, pale head*. Sexes similar, with: *red crown* (reaches further back on ♂, and colour is brighter red); *lack of black moustachial stripe*; white forehead and head-side, on which eye stands out as a dark spot; *white oval shoulder patches* (cf. White-backed Woodpecker); *pinkish vent* which fades into yellowish-brown belly; and fine *dark streaks on flanks*. Often perches across branches in slightly hunched posture and with drooped tail.

VOICE Kick-call rather weak; little used; very like Lesser Spotted Woodpecker's. More often gives a series of such calls *at fast trotting pace* with slightly different first syllable higher-pitched, 'kick kück-kück-kück-kück-kück-...'. Territorial assertion by song, 4–8 (or more) whining, nasal notes at slow pace (*c.* 2 per sec.), 'gvayk gvayk gvayk gvayk gvayk'. *Does not drum* for territorial purposes (only *very rarely* as auxiliary action).

Great Spotted Woodpecker

Syrian Woodpecker

Middle Spotted Woodpecker

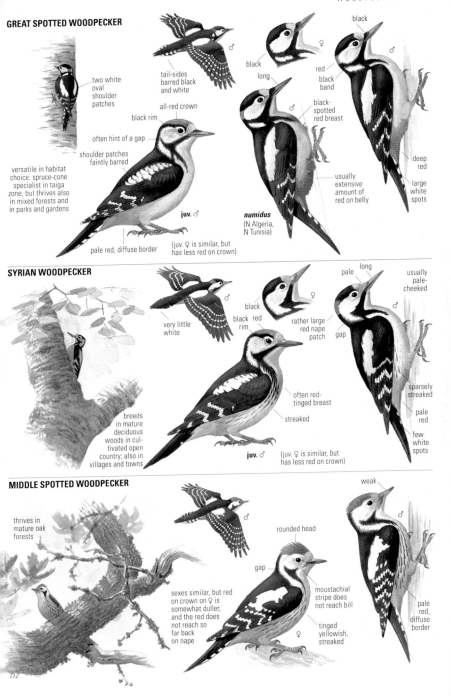

GREAT SPOTTED WOODPECKER

two white
oval
shoulder
patches

tail-sides
barred black
and white

black

red

black
band

black

black-
spotted
red breast

all-red crown

black rim

often hint of a gap

shoulder patches
faintly barred

versatile in habitat
choice: spruce-cone
specialist in taiga
zone, but thrives also
in mixed forests and
in parks and gardens

deep
red

large
white
spots

usually
extensive
amount of
red on belly

juv. ♂

numidus
(N Algeria,
N Tunisia)

pale red, diffuse border

(juv. ♀ is similar, but
has less red on crown)

SYRIAN WOODPECKER

very little
white

black

black
rim

red

rather large
red nape
patch

pale

long

usually
pale-
cheeked

gap

sparsely
streaked

pale
red

few
white
spots

often red-
tinged breast

streaked

breeds
in mature
deciduous
woods in cul-
tivated open
country; also in
villages and towns

juv. ♂

(juv. ♀ is similar, but
has less red on crown)

MIDDLE SPOTTED WOODPECKER

thrives in
mature oak
forests

weak

rounded head

gap

moustachial
stripe does
not reach bill

sexes similar, but red
on crown on ♀ is
somewhat duller,
and the red does
not reach so
far back
on nape

tinged
yellowish,
streaked

pale
red,
diffuse
border

DZ

White-backed Woodpecker *Dendrocopos leucotos* —

L 25–28 cm. Breeds in wet mixed forest, often by lakes and rivers, with plenty of dead and dying deciduous trees (aspen, sallow, alder, birch); thus requires areas undisturbed by forestry operations, so now greatly threatened. Food insects, including larvae of longhorn beetles. Spends much time near ground pecking at willow and alder bases etc., often leaving traces in form of large, deep craters (almost as Black Woodpecker). Mobile and unpredictable, but not particularly shy. Large territory (esp. winter). Nest hole *c.* 5½×6½ cm.

IDENTIFICATION Slightly *bigger* than Great Spotted Woodpecker, longer-necked, with more angular head profile and *longer bill. Vent light red*, poorly demarcated, *belly faintly tinged buffy-pink. Flanks streaked.* Black band on head-side does not reach crown, leaves white 'throughway'. Despite its name, white back can be hard to see, but at least in flight *lower back is conspicuously white.* Perched bird in side view told by *white median coverts forming broad horizontal patch,* broader than white remex bars. ♂ has *all-red crown,* ♀ black. – Variation: Birds in SE Europe and Turkey (ssp. *lilfordi*) have *vermiculated* white back and can also have barred flanks.

VOICE Kick-call is lower-pitched, 'thicker' and not so sharp as Great Spotted's, 'byück', when excited repeated in irregular series, 'byück, byü-byück...'. Drumming usually identifiable by powerful, 'loose' opening and weaker and accelerating ending ('ping-pong ball bouncing to halt'), also by being normally clearly longer (often 1.6–2.1 sec.) than Three-toed Woodpecker's, the most likely confusion risk. (Beware: ♀ White-backed often gives slightly shorter drum, sometimes approaching Three-toed.)

Three-toed Woodpecker *Picoides tridactylus* —

L 21½–24 cm. Breeds in coniferous and mixed forest with some older spruce stands and (often dying) deciduous trees. Food mostly insects; specializes on larvae of spruce bark-beetle, often strips dead spruces; also drills rings of holes in spruces to get at the sap. Usually shy. Nest entrance 4½×5 cm.

IDENTIFICATION Almost as big as Great Spotted Woodpecker. Rather dark-looking because *wings are very dark* and *flanks are diffusely vermiculated grey.* Head-side, too, looks dark with its *broad black bands.* Note *white panel from nape and right down entire back.* ♂ has pale *lemon-yellow crown,* crown of ♀ is black-streaked. – Variation: In C Europe (ssp. *alpinus*) white back-panel is vermiculated.

VOICE Kick-call softer than Great Spotted's, 'bick'. Drumming powerful, well articulated, of medium length (often 1.1–1.4 sec.), with often hint of rise in volume at start (and faint acceleration towards end), most like half a Black Woodpecker roll (and can be confused with that!). Begging-call of young sounds like an insect or distant engine.

Lesser Spotted Woodpecker *Dendrocopos minor* r**B4**

L 14–16½ cm, WS 24–29 cm. Breeds in deciduous woods, old orchards, parkland and, particularly in Britain, river valleys with alders. Nest entrance 3×3½ cm.

IDENTIFICATION *Smallest* woodpecker, a real dwarf with *short, plump body, round head* and *short, pointed bill.* Black above with *white bars across wings and back.* Black bar on head-side does not reach crown, leaving narrow white 'throughway' (as on White-backed Woodpecker). Flanks usually weakly streaked. ♂ has *red crown patch* with black sides; ♀ lacks all red in plumage, has small black-edged dirty brownish-white crown patch. Superficially like White-backed Woodpecker if size not apparent, but differences include *lack of red on vent.* Flies in deep undulations.

VOICE Kick-call short and sharp, generally more feeble than Great Spotted's, though still at times confusingly similar. Territory proclaimed with both drumming and song. Song a series (8–15) of piping, straight notes sometimes slowing at end, 'piit piit piit piit piit piit piit, piit'; lacks Wryneck's whining tone. Drumming rather weak but typically open, more rattling than whirring; tempo constant throughout, and length often 1.2–1.8 sec.; often two drumrolls are given in succession with just a microsecond pause in between.

(Eurasian) Wryneck *Jynx torquilla* P4–5

L 16–18 cm. Breeds in open country with orchards, scrubby pasture, open woodland with fields, copses etc. Summer visitor (mostly May–Sep), winters in Africa. In Britain now ± extinct, rare on passage. Ants are favourite food. Nests in existing tree-hole or nestbox (does not excavate own hole).

IDENTIFICATION Size about that of a small shrike or a Barred Warbler. *Mottled brown and grey above* (merges well with bark of trees), *vermiculated dark below on pale ground* like a small hawk (dirty white on belly, buffish-ochre on breast). *A dark line runs through eye and down neck-side,* with another *dark line along centre of crown and back.* The long *tail is sparsely barred.* Bill rather short and pointed, with strong base. Unlike other woodpeckers, does not climb using tail as support, but behaves more like a passerine; often perches crosswise on horizontal branches, chisels rather than digging with its bill. Flight, too, is more passerine-like, fast with only moderately undulating course. Frequently hops on ground (with slightly raised tail) in search of ants; otherwise rather discreet, quiet habits, and easily escapes detection but for its voice. Sexes alike.

VOICE Alarm a series of hard 'teck' notes. Hisses (and twists head snake-like) if discovered in nest hole or otherwise startled. Young beg with a fast, shuttling 'zizizizi...'. Song a series of 12–18 loud, whining notes, 4 per sec., '**tie-tie-tie-tie**...'; rather long intervals between phrases.

White-backed Woodpecker

Three-toed Woodpecker

Lesser Spotted Woodpecker

Wryneck

WHITE-BACKED WOODPECKER

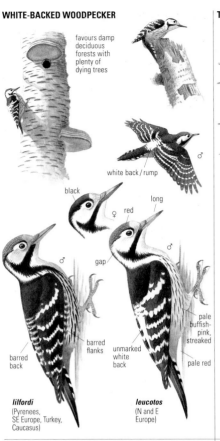

favours damp deciduous forests with plenty of dying trees

white back / rump

black

red

long

gap

♀

♂

pale buffish-pink, streaked

barred flanks

unmarked white back

pale red

barred back

barred back

♂

lilfordi
(Pyrenees, SE Europe, Turkey, Caucasus)

leucotos
(N and E Europe)

THREE-TOED WOODPECKER

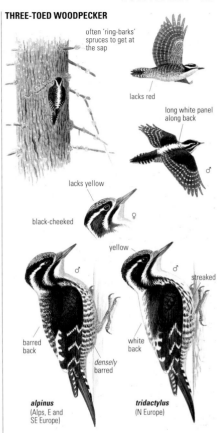

often 'ring-barks' spruces to get at the sap

lacks red

long white panel along back

♂

lacks yellow

black-cheeked

♀

yellow

♂

streaked

barred back

densely barred

white back

♂

alpinus
(Alps, E and SE Europe)

tridactylus
(N Europe)

LESSER SPOTTED WOODPECKER

climbs quietly on underside of thick branches like a large treecreeper

♂

red

weak

♂

gap

barred

black

♀

lacks red

lacks red

WRYNECK

throat ochrousbuff

can be confused with a ♀ Redbacked Shrike in flight

frequently searches for ants on ground

short, pointed

dark band

sexes alike

barred

long, sparsely barred

LARKS *Alaudidae*

Ground-living passerines, mostly brown and off-white. 20 species. Hindclaw long and straight. Many have sustained, hovering or fluttering song-flight and are excellent singers, capable of mimicry, too. Nest on the ground in tuft of grass. Young often leave nest before fledged. Juveniles recognized by pale-fringed feathers on back and wings.

(Common) Skylark *Alauda arvensis*   rB1

L 16–18 cm. Breeds commonly in open, cultivated land, also on meadows and heaths. Mainly resident, but influx in winter from N. Large flocks may congregate on stubble fields and grass steppe in winter.

IDENTIFICATION *Greyish-brown, streaked* above and on breast; breast buff, belly white. Streaking of breast ends rather abruptly above unmarked belly. ♂ can erect *blunt crest* on crown (Crested Lark has longer and more pointed crest). Often seen on fields and along roadsides; when approached, it first squats, then 'catapults' up, retreating a short distance low over ground with *fluttering wings* and landing with half-spread, lowered tail, the brown bird showing *thin, diffuse* (brownish-white) *trailing edge to wing* and *white sides of tail*. Sings at times (often at dawn) from perch but usually in *typical song-flight*, climbing higher and higher on fluttering wings, and eventually stays at one spot at 50–100 (150) m, difficult to detect. At end of song-flight descends while singing, but in the final stage falls to ground with folded wings.

VOICE A variety of calls, all rather dry rolling sounds, e.g. 'prreet', 'prrlyh', 'prrüt-üt' and 'prreeh-e'. Sometimes, often when anxious, a more piping 'p(r)eeh'. Sings from late winter until midsummer, from first light to evening. Song is an incessant outpour of rolling, chirruping and whistling notes at fast pace and for periods lasting 3–15 min., some notes being repeated and varied, and with a few imitations (of e.g. Green Sandpiper, Barn Swallow) woven in.

Oriental Skylark *Alauda gulgula*   —

L 14–16 cm. Breeds on cultivated fields and natural steppe in S Asia and Central Asia (W to Iran); northern populations short-range migrants. Rare winter visitor in Israel.

IDENTIFICATION Very similar to Skylark. In flight small and short-tailed. On ground note following: *shorter primary projection*; more *uniformly pale buff underparts*; on many, *primaries edged red-brown*; outer tail-feathers usually light buffish-grey (but at times almost white, and difficult to judge against the sky); trailing edge of wing not obviously pale; and on average slightly longer bill and shorter tail than Skylark. Can erect small crest like Skylark.

VOICE One flight-call very different from any of Skylark (mainly when taking off), a short, strident, buzzing or 'me-

chanical' 'bzrü'. Other calls more like Skylark, but also a somewhat Barn Swallow-like 'plip'. Song like Skylark's but drier and more repetitive, delivered in drifting song-flight.

Crested Lark *Galerida cristata*   V***

L 17–19 cm. Breeds in fields, on open industrial sites, among railway tracks in harbours, etc. (with easily accessible weed seeds); in S parts of range also in more arid areas. Resident. In Britain & Ireland extremely rare vagrant.

IDENTIFICATION A fraction larger than Skylark, greyish-brown with *long, pointed crest*, visible at rear crown even when folded (short crest of Skylark invisible when folded). Compared with Skylark, usually a slightly more greyish and dark bird (though some races of N Africa paler and rustier) with more *diffuse streaking on breast* (though some are more distinctly streaked). Lacks pale trailing edge of wing (cf. Skylark), and has *buff-brown outer tail-feathers*, not white. Underwing, and often uppertail-coverts, tinged red-brown. *Bill is long and pointed with straight lower edge* (a certain variation in bill length: longest in N Africa, shorter in N Europe and Middle East; cf. very similar Thekla Lark).

VOICE Vocal. Commonest call consists of variable combination of 2–4 whistling, straight notes, one or several drawn out, with desolate ring, e.g. 'treelee**püü**' or 'vü tee **vüü**'. This call used as 'eager' contact-call, but also as alternative song, uttered repeatedly in the territory. Other calls are e.g. cheerful mewing, soft but usually slightly cracked, upward-inflected 'dvuuee', often in flight and repeated, and a more strident, whimpering 'brshü'. The real song is long and often given in flight, richly varied, containing quite a lot of the melancholy, straight whistling notes (cf. above); often difficult to separate from song of Thekla Lark!

Thekla Lark *Galerida theklae*   —

L 15–17 cm. Breeds in more natural and arid habitats than Crested Lark, often at higher altitudes on barren mountain slopes or dry steppes; sometimes also in orchards, in cork oak savanna or shrubbery on sandy soil.

IDENTIFICATION Very similar to Crested Lark, and reliable identification generally requires close view. Marginally smaller. *Bill is shorter* and usually *not so pointed—lower edge slightly convex* on most (exception: Balearics), which creates 'more cute' appearance; bill is often slightly darker, too. *Streaking of breast* generally *more distinct* and finer on paler ground (though some are very similar). Plumage is *greyer*, *mantle and back* on average somewhat *more distinctly streaked*, *tertials more contrasting*, and *uppertail-coverts more reddish-brown* than on Crested. Now and then perches in bushes or trees (as Crested Lark only exceptionally does).

VOICE Similar to Crested Lark's, but softer and more melodious (not such melancholy piping with straight notes), and often more syllables, e.g. 'tu-telli-**tew**-tilli-**tee**'; certain calls, however, are more like Crested's. Flight-song is similar to Crested's, but is slightly softer, *more varied* and pleasing, not so sharply piping. Both use mimicry.

Skylark   Crested Lark   Thekla Lark

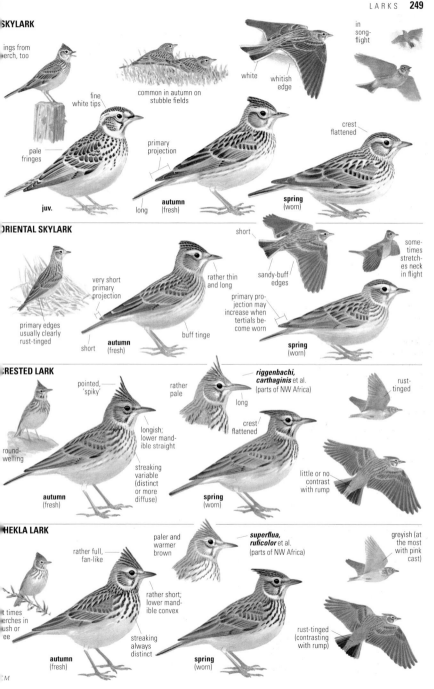

SKYLARK

ings from
erch, too

fine
white tips

pale
fringes

juv.

common in autumn on
stubble fields

primary
projection

long

autumn
(fresh)

white

whitish
edge

in
song-
flight

crest
flattened

spring
(worn)

ORIENTAL SKYLARK

very short
primary
projection

primary edges
usually clearly
rust-tinged

short

autumn
(fresh)

short

rather thin
and long

buff tinge

sandy-buff
edges

primary pro-
jection may
increase when
tertials be-
come worn

some-
times
stretch-
es neck
in flight

spring
(worn)

CRESTED LARK

pointed,
'spiky'

round-
welling

autumn
(fresh)

rather
pale

longish;
lower mand-
ible straight

streaking
variable
(distinct
or more
diffuse)

***riggenbachi,
carthaginis*** et al.
(parts of NW Africa)

long

crest
flattened

spring
(worn)

rust-
tinged

little or no
contrast
with rump

HEKLA LARK

rather full,
fan-like

t times
erches in
ush or
ee

autumn
(fresh)

paler and
warmer
brown

rather short;
lower mand-
ible convex

streaking
always
distinct

***superflua,
ruficolor*** et al.
(parts of NW Africa)

spring
(worn)

greyish (at
the most
with pink
cast)

rust-tinged
(contrasting
with rump)

M

Woodlark *Lullula arborea* r(m)**B**4

L 13½–15 cm. Breeds in open forests, preferring pine on sandy soil, but also in mixed or broadleaved forests with clearings and on heathland with scattered copses. Resident in Britain; increasing. N European birds migrate to S Europe. Rather shy and difficult to approach.

IDENTIFICATION Rather small, brown and *short-tailed*. Often perches exposed in trees, bushes or on wires, unlike other larks (though spends most of the time on ground). *Can recall small woodpecker in undulating flight with short, broad wings* and stubby tail! *Tail tipped white* (adding to the short impression when seen against a bright sky). Sides of tail not white (as on Skylark) but pale brown, and trailing edge of wing on adult not light. When perched, note characteristic *light-dark-light pattern near bend of wing* (primary-coverts dark with buff-white tips, and white patch at their base); also *broad buffish-white supercilia reaching far back* (almost joining on hindneck). Song mostly delivered in drifting song-flight high up (100–150 m), often at very first light, or even beneath stars on pitch-dark night.

VOICE Call a soft whistling yodel, '**tlewee-tlew**ee' or more feebly '**dü**d**lu**ee', often revealing overflying small parties in autumn or spring. Song is one of the most attractive, consisting of pleasant and 'sweet but melancholy' notes in series, opening hesitantly, accelerating, often falling in pitch and gaining in loudness, e.g. 'lee, lee-lee-leeleelee**leelülu**... **ee**-lü **ee**-lü **ee**-lü **ee**-lü-**eelu**-**eelu**eelu... tluee, tluee tluee vi vi vi **tellellellell**...', etc.

(Greater) **Short-toed Lark**
Calandrella brachydactyla **V**∗∗

L 14–16 cm. Breeds in open, dry areas, on cultivated fields or more arid plains. Annual vagrant to Britain late spring or autumn, favouring sandy wasteland, arable fields or open seashores.

IDENTIFICATION Rather small and pale, greyish-brown and off-white. Underparts of adult practically *unmarked*, with only *a dark patch on side of breast* ('dwarf Calandra Lark'; some have faint patch, others have several streaks rather than one patch, but never streaked on *whole* breast; cf. Lesser Short-toed Lark). Important distinction from Lesser Short-toed is *long tertials*, reaching to or very near tip of wing. Median coverts dark with broad pale tips (in pipit fashion). Broad off-white supercilium and largely unstreaked grey-brown cheek, outlined in pale, give distinct facial expression. Bill pointed and rather pale. Some birds (both sexes; more commonly in Spain and Africa) have faint red-brown tinge on crown (often with reduced streaking), others are more evenly grey-brown and obviously streaked (normal plumage in E Europe, Turkey). Song-flight usually slowly drifting

around at some height, undulating, tail folded, *song phrase generally delivered in phase with bursts of quick, fluttering wingbeats* (end of descent, whole ascent). More prolonged song phrases (from agitated bird) or strong winds (forcing singer to beat wings in longer and quicker bursts) at times alter typical pattern. – Juvenile: Apart from pale fringes on upperparts, has sparse and diffuse blotching across whole breast; separated from juvenile Lesser Short-toed by primary projection, bill shape and calls.

VOICE Call a dry chirruping 'drit', drier, shorter and more 'bouncing' than Skylark's, like House Martin call; or more full 'trilp', a bit reminiscent of Tawny Pipit. At times tantalizingly similar to Lesser Short-toed's, 'drrr-t-t'. Song of two types, the simpler being characteristic: brief phrase (1–2 sec.), dry chirruping voice, faltering opening, acceleration and clattering or shuttling end, pauses between phrases 1–3 sec. A more ecstatic song prolongs phrases to 5–30 sec, includes imitations and displays a greater variation, making it easily confusable with Lesser Short-toed (and other larks) unless the call and faltering sections are woven in.

Lesser Short-toed Lark *Calandrella rufescens* **V**∗∗

L 13–14½ cm. Breeds on dry plains (clay), by salines and on desert-like steppe. Mainly resident, but eastern populations migrate to Middle East.

IDENTIFICATION Small, usually fractionally smaller than Short-toed Lark (but some eastern races equal in size), *greyer* and *more evenly streaked*. Best separated by: *streaked breast* (incl. centre of breast, recalling dwarf Skylark); *shorter tertials* (ending 10–15 mm short of wing-tip); *shorter, more bulbous bill*; and voice. Often 'cute-looking' from *rounded head shape*, small bill and wide pale eye-ring. Supercilium usually less obvious. Cheek finely streaked. Song-flight often drifting around with fluttering wingbeats and partly spread tail, in suitable winds now and then shifting to clearly *slower wingbeats* (never seen from Short-toed). – Marked geographical variation: Birds of Spain (ssp. *apetzii*) small and rather grey-brown, breast heavily streaked, flanks streaked, too. Birds in S Middle East and N Africa (*minor*) small but brown, less greyish, streaking of breast fine, flanks unstreaked. Birds of E Europe (*heinei*) and Turkey (*aharonii*) large and greyish, breast finely or diffusely streaked, flanks faintly streaked.

VOICE Most common call a dry, trilling or buzzing 'drrrrd', often slightly falling in pitch or slowing at the end, usually repeated a few times; voice reminiscent of Sand Martin. NB: Short-toed Larks can at times have very similar calls! Song consistently richer, more varied, and paced quicker than Short-toed's; excellent imitations (incl. of other larks, making identification so much more problematic—can mimic Crested or Short-toed for several seconds!); will mix imitation with own drier trills and softer, piping notes in Theklas fashion. Some phrases shorter, more like Short-toed Lark; mixed-in dry trilling calls and 'frizzling' fast pace are then best clues.

Woodlark

Short-toed Lark

Lesser Short-toed Lark

WOODLARK

sings from perch, too

white tips

lacks white edge

striking head pattern

slender

short

pale-dark-pale pattern

broad-winged and short-tailed in song-flight

pale bar

supercilia join at rear, form angle

red-brown

SHORT-TOED LARK

fairly extensive plumage variation noticeable even in small flocks in winter

long tertials cloak primary tips

comparatively pale

rufous-tinged crown on some (variable, and more common in Spain and N Africa)

in undulating song-flight

dark patch on breast-side on most

dark

generally contrast-ingly dark

strong but pointed

practically unmarked

a few lack dark patch

autumn (fresh)

spring (worn)

LESSER SHORT-TOED LARK

LESSER SHORT-TOED

SKYLARK

compare with much larger and longer-billed Skylark, which otherwise has rather similar plumage

note primary projection

short and stubby

streaked

autumn (fresh)

often raises crown-feathers to form tiny crest

sometimes diagnostic display-flight with slow-motion wingbeats

generally only moderate contrast

usually whole breast distinctly streaked

large

rather fine streak-ing

spring (worn)

apetzii (Spain)

spring (worn)

aharonii (C Turkey)

M

Desert Lark *Ammomanes deserti* —

L 15–16½ cm. Breeds in arid (desert-like), open, often sloping or rocky terrain, in wadis or on mountainsides with boulders and smaller stones. Generally avoids flat and sandy desert. Resident. Usually seen singly, in pairs or a few together. Discreet habits; often confiding, and allows quite close approach; unpredictable, often flies off silently.

IDENTIFICATION Chaffinch-sized, sand-coloured, robust lark, front-heavy with tapered rear. *Bill fairly heavy* and long, culmen curved (bill size varies geographically to some extent, e.g. larger in S Morocco, smaller in Sinai and Israel), *base pale brownish-yellow* with slightly darker culmen and tip. Plumage evenly grey-brown above, pinkish grey-buff below, usually with *obvious diffuse streaking on breast.* Short supercilium and ill-defined eye-ring pale; loral streak faintly darker. Tertials brown-grey, tinged red-brown, when fresh obviously pale-edged. *Tail rather dark,* tail-feathers edged rufous-buff. Outer webs of primaries edged red-brown. Sexes alike. (For geographical variation, see plate.) Resembles Bar-tailed Lark, but most reliably distinguished by size and colour of bill, and lack of *distinct* black band on outer tail. Further helpful differences: Desert Lark's usually more streaked breast and neck-sides, slightly longer primary projection, and different voice and choice of habitat.

VOICE Rather silent, still has several calls, some perhaps only variations of same: a common call is a rolling 'churrr'; also, a more subdued whistling, slowly repeated 'cheealp', 'chu-ül', 'chü(u)' or 'chup'. Song rather deep-voiced, loud and resonant with desolate ring, a quickly repeated phrase of 3–6 syllables, often uttered in undulating song-flight, e.g. 'chu-we-chacha, **wooee**' or 'chu-**weeü** chuwe-trutru'.

Bar-tailed Lark *Ammomanes cinctura* —

L 13–14 cm. Breeds in flat, sandy deserts, but also in semi-desert with pebbles and scattered low vegetation. *Not found in mountains* or in rocky, broken terrain. Resident, locally nomadic. Often seen in larger flocks outside breeding; restless, an industrious runner, making brief stops only.

IDENTIFICATION Recalls Desert Lark, but distinguished by the following (apart from habitat and voice): *smaller size* and more chubby, with *rounded head* and *shorter bill*—has a 'cuter' look, though shape rather variable, being fluffed up and rounded on chilly mornings but remarkably slender and noticeably 'leggy' in midday heat; in autumn, upperparts (mantle/scapulars) *often* contrastingly *greyer* than head/nape and wing; tail pale red-brown with *distinct black terminal band,* obvious when alighting (when the tail is briefly spread); *bill off-white* in strong sunlight, faintly *pinkish* in overcast weather; tertials *pale rufous* (best judged from behind); and *outer primaries tipped blackish,* sometimes notice-

able in flight, and *contrasting with tertials* on folded wing.

VOICE Some calls resemble Desert Lark's. Often a dry somewhat strained and hard trilling 'cherr'; also a short nasal 'chüp'. A possibly more diagnostic call is a harsh 'bshee' (recalling autumn call of Reed Bunting). Song characteristic, a thin squeaky note repeated rhythmically a few times, preceded by one or two low notes (audible only at close range), rather like the sound of a distant, old, rusty pump '(tleo) weeeh, (tleo) weeeh, (tleo) weeeh, ...'; often delivered in song-flight, undulating in phase with the song.

Dunn's Lark *Eremalauda dunni* —

L 14–15 cm. Local and scarce breeder in flat and sand desert and semi-desert in Middle East (e.g. Jordan); a different population (possibly a separate species) has recently been found in S Morocco. Resident or nomadic; some scattered winter and breeding records in S Israel. Runs readily but will also make longer stops than Bar-tailed Lark to 'rough up' some plant with its strong bill.

IDENTIFICATION Between Desert and Bar-tailed Larks in size. Note: *heavy bill pinkish* (sometimes with orange cast at tip) but *lacking markedly dark culmen and tip; bill thick even near tip,* culmen strongly curved; red-brown tinge in sand brown plumage; red-brown *streaking on crown, mantle and sides of breast; long tertials* practically reach tip of wing (virtually *no primary projection*); grey, *narrow moustachial and malar stripes; tail-sides black,* central tail-feathers buff; rather obvious light buff-white eye-ring and short supercilium; wings (incl. coverts) fairly uniform and pale.

VOICE Call a throaty upward-inflected 'dshrooee'; also more sparrow-like 'chilp'. More subdued 'prt' and trilling 'drrree-ü' also heard from foraging flocks. Song a rapid stanza of sparrow-like chirpings and drier lark-type notes.

Black-crowned Sparrow-lark
Eremopterix nigriceps

L 11½–12½ cm. Not a regular breeder within treated region, occurs further S (Cape Verde, S Morocco, Sahel region, SE Egypt, S Iraq, etc.); occasional breeder in Israel. Recorded in Algeria. Typically seen in flocks, usually in flat semi-desert and on arid plains with scattered bushes or trees.

IDENTIFICATION *Very small,* Serin-sized, but with proportionately rather large head. Bill large and *conical,* finch-like comparatively rather pale grey-white. *Underwing dark* (blackish on ♂). Outer tail-feathers are black, central pale, like back. –Adult ♂: *Black underparts* (from chin to undertail), sand-coloured above, and *boldly pied head pattern.* –Adult ♀: Entire upperparts pale rufous-buff (rufous tinge decreases with abrasion), crown, scapulars and sides of breast finely streaked. Rest of underparts pale buffish-white. Median coverts a little darker brown, often forming paler band. Colour and tail pattern recall Dunn's Lark, but note smaller size, dark underwing, heavy conical bill (lacking pinkish tinge), and shorter, blunter wings in flight. Alights again surprisingly quickly after having been flushed.

VOICE Call a sparrow-like 'chep'. Song, heard mostly at dawn and dusk, a rhythmically repeated (every 4–5 sec.) phrase of 2–4 notes with melancholy ring, e.g. 'wit ti-wee' last note slightly lower-pitched and drawn out. Usually delivered in drifting, low song-flight with fluttering wings.

Desert Lark

Bar-tailed Lark

▌ESERT LARK

refers sloping,
ony terrain

examples of
geographical
variation:

deserti
(NE Egypt, Israel,
S Jordan)

payni
(S Morocco)

annae
(N Jordan)

plain sand-
brown

longish and strong,
yellowish with dark
culmen and tip

dark grey-
brown

diffusely
streaked

diffusely
set-off black
terminal band

autumn
(fresh)

spring
(worn)

▌AR-TAILED LARK

prefers
flat
desert

plain sand-
brown

small,
pinkish

can adopt rounder
shape in cold
weather

clear-cut black
terminal band

pale rufous

black tip

autumn
(fresh)

spring
(worn)

▌UNN'S LARK

streaked
red-brown

strong head pattern,
e.g. with 'mascara tear'

stout, 'swollen',
pinkish

some worn spring
birds have less strik-
ing head pattern

black

very short
primary
projection

black
sides

autumn
(fresh)

spring
(worn)

▌LACK-CROWNED SPARROW-LARK

unmistakable!

♀

♂

more diffuse head pattern
than Dunn's Lark, and is
less rufous-tinged above

dark

dark centres

heavy,
triangular,
greyish

o primary
projection

black
sides

**1st-winter /
♀ autumn**

ad. ♂ spring
(worn)

Calandra Lark *Melanocorypha calandra* V***

L 17½–20 cm. Breeds in open cultivated country and on natural steppe. Migrant in E, mainly resident in W. In winter often in larger flocks.

IDENTIFICATION Strongly built and large lark, with *heavy yellowish-brown bill* and characteristic *large black patch on side of breast* (varying in shape and size: sometimes very large, sometimes running across breast, sometimes narrow). Pale supercilium and pale eye-surround. Typically has *blackish underwing* with a *wide, distinct white trailing edge* (obvious in flight, and glimpse of white trailing edge at times detectable even on folded wing). Upperparts roughly as Skylark, greyish-brown, streaked dark. White sides to tail. Legs brownish-pink. Song-flight often characteristic, the lark 'hanging' (at times drifting around) at some height (25–100 m) with, provided there is some wind, *slow-motion wingbeats* with stiff, straight wings and folded tail; slow wingbeats may give impression of larger than actual size.

VOICE Call either very characteristic dry rolling, almost frizzling 'schrrrreep', or more Skylark-like 'treeh', 'triptrip', etc. Song like Skylark's, protracted, full of chirping notes at fast pace, mixed up with a few imitations; recognized by slightly slower pace, hinted pauses, and abundance of dry rolling call-like notes woven in. (Cf. Bimaculated Lark.)

Bimaculated Lark *Melanocorypha bimaculata* V***

L 16–18 cm. Breeds on arid, stony plains (semi-desert) and mountainsides; also on arable fields like Calandra Lark, but then usually at higher elevation, near upper limit of cultivation. More migratory habits than Calandra Lark.

IDENTIFICATION Resembles Calandra Lark (heavy bill and dark breast patch the same), but differs in following: slightly smaller (and with a little shorter tail); *underwing brownish-grey; trailing edge of wing just a little paler brown* (not white); *tip of tail white* (sides not); on average more *contrasting and well-marked head pattern* with more obvious supercilium, and dark lines through and below eye; upperparts on average with lighter feather edges in fresh plumage. In song-flight, wingbeats are usually *continuously fast and fluttering*, but in brisk wind slower (bat-like), and in very strong wind can keep wings motionless for long spells. Stiff wingbeats with straight wings in slow motion apparently not executed. *Tail* also often *kept well fanned*, unlike Calandra's.

VOICE Calls resemble Short-toed Lark's, e.g. short, dry twittering 'tripp, tripp', but voice is coarser and calls more varied, including rasping 'tcher' and the like. Song confusingly similar to Calandra's, consisting largely of fast, dry, rolling twitter, but pitch slightly deeper and voice harder, almost 'bouncing' (not so 'high-pitched frizzling'), and is more monotonous with fewer imitations and soft notes.

Thick-billed Lark *Ramphocoris clotbey* –

L 17–18 cm. Breeds in arid, open terrain in NW Africa (e.g. S Morocco) and locally in Middle East (Jordan?). Residen

IDENTIFICATION Large, with *huge bill. Wide white trailin band on wing* and *dark underwing.* – Adult ♂: *Sides of hea black with distinct white spots*; bill pale bluish-grey; unde parts heavily blotched black. – Adult ♀: Resembles ♂ but ha on average less dark grey on head, pale spots less well mark ed; bill pale horn with at most only faint shade of blue; un derparts less heavily blotched. – Winter: Pale fringes cove most of dark areas in plumage. – Juvenile: Head almost un form, without dark pattern, and bill somewhat less heavy.

VOICE Call whistling 'tsu-ee?' ('querying') or open, rollin 'zrrroüh' (rising). Song a rapid stanza of twittering notes

White-winged Lark *Melanocorypha leucoptera* V**

L 17–19 cm. Breeds on natural steppe, dry heaths and t some degree on cultivations. Partly migratory.

IDENTIFICATION Characteristic *wing pattern of white, blac and brown* (sexes largely alike). Wide white trailing edge o wing makes wings look peculiarly narrow against a light sk Most similar to Mongolian Lark (*M. mongolica*; not treated but that is larger, has heavier bill, and has a Calandra-lik black patch on the breast side. – Adult ♂: Crown unstreake red-brown. Breast usually sparsely streaked only. – Adult ♀ Crown streaked, grey-brown with only tinge of rufou Breast generally prominently streaked.

VOICE Call a dry, hard twitter, 'drrit-drrit', recalling B maculated Lark. Song most of all recalls Skylark, but recog nized by sections of faltering notes (as from Short-toed Horned Larks) and by somewhat harder, drier voice.

Black Lark *Melanocorypha yeltoniensis* –

L 18–20½ cm. Breeds on natural steppe, often near marsh and salines. Locally common. Partly migratory.

IDENTIFICATION Adult ♂: Large, all black, with bill heav and straw-coloured; in autumn, pale fringes cover much o black. Song-flight often with *wingbeats in slow motio* Display-flight at low level with very high, stiff wingbeats an dove-like glides; tumbles down to court ♀ with raised ta and lowered wings and neck. – Adult ♀: Brown-gre blotched dark; plumage rather variable, some resembl Calandra Lark, having a larger dark patch on side of breas Note, however, dark legs, no wide white trailing edge o wing, and usually obvious blotching below. Underwin coverts almost black, clearly darker than wing-feathers.

VOICE Song like Skylark's but slightly higher-pitched ar even more frizzling and frantically twittering (like 'quarre ling young Starlings'); loudness frequently varied. So miaowing, 'pleading' notes are frequently mixed in.

Calandra Lark

Bimaculated Lark

White-winged Lark

Black Lark

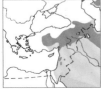

◀CALANDRA LARK

blackish

white sides

black patch (size variable)

obvious white edge

compare Calandra (left) with smaller Short-toed Lark (right)

pink

autumn (fresh)

spring (worn)

◀BIMACULATED LARK

dark

brown-grey

white tail-tip

strong head pattern

in song-flight

CALANDRA   BIMACULATED

short tail

lacks pale edge

autumn (fresh)

spring (worn)

◀THICK-BILLED LARK

dark

very heavy, bluish grey-white

broad white edge ♂

pale

grey

autumn ♀ (fresh)

spring ♂ (worn)

dense dark patches

◀WHITE-WINGED LARK

grey-brown, streaked

red-brown

red-brown

three-coloured wing ♂

white wing patch

autumn ♀ (fresh)

spring ♂ (worn)

looks quite narrow-winged against a light sky, when white rear edge 'disappears'

◀BLACK LARK

lacks pale rear edge

coverts blacker than flight-feathers

rather Starling-like in flight

autumn ♂

♀ (worn)

summer ♀ (worn)

rather 'anonymous' head pattern

summer ♂ (worn; abraded to all black)

dark

dark grey

autumn ♀ (fresh)

spring ♂ (still fairly fresh)

spring ♂

M

Horned Lark (Shore Lark)
Eremophila alpestris **P+W**4

L 16–19 cm. Breeds in mountains above tree-limit, on alpine moors and dry stony ground, often on crown of lowest fells, in extreme north on tundra at sea-level. Migrant in north. Scarce and local in Europe, has decreased.

IDENTIFICATION Slightly smaller than Skylark. In all plumages, recognized by characteristic *head pattern of black and yellow* (Europe, N Turkey, Caucasus, N Africa) *or white* (rest of Asia). Along sides of crown elongated black feathers form tiny, narrow 'horns' (not always visible in the field). Sexes and ages similar, but when pair-members seen together on breeding ground adult ♂ is: larger; more well marked on head; has longer 'horns'; practically unstreaked pinkish-brown crown and nape; and less streaking on mantle. In winter flocks, immature ♀ can sometimes be picked out: small size; head pattern more ill-defined (lores grey, not black, and dark cheek patch small); 'horns' missing; crown, nape and mantle rather distinctly streaked dark. Still, many autumn birds are difficult to sex or age. In flight, note *blackish tail*. – Juvenile: Upperparts and breast with pale and dark dots, and hint of dark cheek patch. – Variation (apart from yellow or white ground colour on head; see above): Birds in N Europe (ssp. *flava*) have black on head separated from black breast patch; birds in SE Europe (*balcanica*), N Turkey and Caucasus (*penicillata*) have black on head merging with black breast patch. In Middle East (*bicornis*), upperparts are warmer rufous-tinged.

VOICE Call fine, short squeaky notes, single 'eeh', or with 'echo', '**eeh**-dü' or '**eeh**-deedü', rather metallic, 'rippling' ring, and calls fairly anonymous. Sometimes more harsh 'prsh'or'tsrr'. Song, delivered from rock or in flight, usually promptly repeated brief phrases (1–2½ sec.) with characteristic acceleration after faltering, 'fumbling' opening (same song pattern as Short-toed Lark) and desolate chirruping, jangling voice (rather like Lapland Longspur).

Temminck's Lark *Eremophila bilopha* —

L 14–15 cm. Breeds in semi-desert, on drier steppe, lower mountainsides (< 1000 m), etc., at lower altitude than Horned Lark. Mainly resident. Discreet habits; not shy.

IDENTIFICATION Like a small Horned Lark of Middle East (ssp. *bicornis*), but separated by following: upperparts, incl. wing-coverts and whole tertials, vividly rufous (lacking dark centres); primary projection short; primary tips blackish (fringed pale in fresh plumage), contrasting strongly with rufous tertials. Note that black breast patch never merge with black of cheek (as in some races of Horned Lark), and that ground colour of head is white (never yellow).

VOICE Call a fine, squeaky jingle as of keys. Song slightly

finer than Horned and contains some cracked, thin note phrases frequently short, without much variation. Pace al more even (accelerations not so obvious as with Horned).

Dupont's Lark *Chersophilus duponti*

L 17–18 cm. Breeds on dry, sandy soil with tufts of grass, natural steppe or in semi-desert and flatter mountainside high plateaux as well as low plains near sea. Will also feed o arable fields. Mainly resident. Extremely shy and elusiv quickly running for cover. Nervous and jerky movements.

IDENTIFICATION Size as Skylark but slimmer; slimness e hanced by *narrower, longer neck, longer legs* and *long, sligh downcurved bill*. Sometimes adopts upright stance. Th plumage is rather like Skylark's (brown above, brea streaked, white sides of tail), but is slightly darker with (fresh plumage) fine pale feather tips, and has *thin pale centr crown-stripe* (best visible head-on), lacks pale trailing ed of wing, and has thin dark malar stripes (much as o Crested Lark). – Variation: The description refers to birds i Spain and NW Africa north of Atlas (ssp. *duponti*). In S A geria and in Libya (*margeritae*), the plumage is distinctl more red-brown, and the bill is longer.

VOICE Sings mostly at dawn and dusk, a slowly repeate brief, melancholy phrase, opening with a few short whistl followed by a nasal, miaowing, drawn-out note, 'wu-tle tre-weeüüih'. Song-flight at great height, also at night.

(Greater) Hoopoe Lark *Alaemon alaudipes*

L 19–22½ cm. Breeds in flat desert, semi-desert, in ope wadis, etc. Resident. Usually seen singly or in pairs; larg flocks rare. Fairly tame. Prefers to run away rather than fl

IDENTIFICATION Large, slim, pale sand-coloured (colou variation: some are more grey) lark with rather *broad win with black and white pattern* (vaguely recalling Hoopoe flight). *Legs and bill long* (bill-tip slightly downcurved Head pattern well marked, with black moustachian strip and eye-streak. Breast usually finely but distinctly blotche Starts singing from top of bush or ground (or in Hoopo like, flappy, low flight), continuing in *spectacular song-flig* with steep climb for 1–5 m with spread tail, followed b vertical collapse, wings folded, landing on ground or endin with a short, low flight.

VOICE Commonest call a strong, rolling, upward-inflecte 'zrrruee'. Song consists of characteristic thin piping an melancholy-sounding notes, starting tentatively, accelera ting and then slowing down again, e.g. 'voy voy... **vüüü**(cha **vüüü**(cha) swe-swe-swe-swe-swe sisisi... svee, sveeh', whe '**vüüü**' is piercingly thin and most far-carrying. In Midd East, an almost Black Woodpecker-like 'kri-kri-kri-kri-k is inserted in the phrase (apparently not heard in Africa).

Horned Lark

Temminck's Lark

Dupont's Lark

Hoopoe Lark

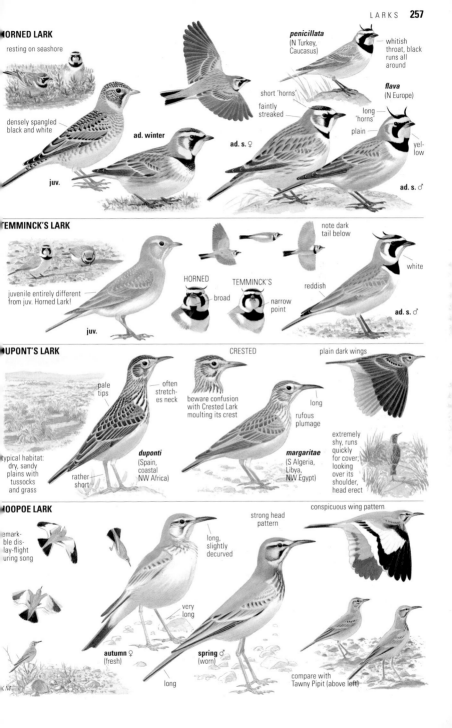

HORNED LARK

resting on seashore

densely spangled black and white

ad. winter

juv.

ad. s. ♀

penicillata (N Turkey, Caucasus)

whitish throat, black runs all around

short 'horns'

faintly streaked

flava (N Europe)

long 'horns'

plain

yellow

ad. s. ♂

TEMMINCK'S LARK

juvenile entirely different from juv. Horned Lark!

juv.

HORNED

TEMMINCK'S

broad

narrow point

note dark tail below

white

reddish

ad. s. ♂

DUPONT'S LARK

pale tips

often stretches neck

CRESTED

beware confusion with Crested Lark moulting its crest

long

rufous plumage

plain dark wings

typical habitat: dry, sandy plains with tussocks and grass

rather short

duponti (Spain, coastal NW Africa)

margaritae (S Algeria, Libya, NW Egypt)

extremely shy, runs quickly for cover; looking over its shoulder, head erect

HOOPOE LARK

remarkable display-flight during song

strong head pattern

conspicuous wing pattern

long, slightly decurved

very long

autumn ♀ (fresh)

spring ♂ (worn)

long

compare with Tawny Pipit (above left)

K M

SWALLOWS and MARTINS *Hirundinidae*

Relatively small passerines with long, pointed wings, adapted for a life largely in the air, where they catch insects by means of fast, agile flight and their large mouth. Bill small, legs short. Most are long-distance migrants which winter in the tropics. Sexes alike. Nest either in excavated holes in sandy banks or in cup skilfully built of mud, plants and saliva.

(Common) **Sand Martin** *Riparia riparia*   m**B**2

L 12–13 cm. Breeds colonially in vertical sandy or earth banks, e.g. in gravel-pits and river banks, where nest is excavated (often a good metre horizontally into the earth). Rather tied to water, often seen in numbers hunting insects in low flight over lakes and rivers. Summer visitor (in Britain & Ireland Mar–Oct), winters in tropical W Africa. Prior to and during migration roosts communally in large reedbeds.
IDENTIFICATION One of the smallest swallows. Dull *grey-brown above* (primaries slightly darker) *and white below*, with *grey-brown breast-band* separating white throat from white belly. *Underwings dark* (grey-brown, wing-coverts somewhat darker). Tail shallowly but clearly forked and without white 'windows' (cf. Crag Martin). Differs from closely related Brown-throated Martin in somewhat larger size, dark breast-band, white throat and *contrastingly darker under-wing-coverts*. Flight rapid and light. When accelerating, wings beaten backwards and in towards body, so then look narrow; does not glide that often. – Juvenile: Pale, rusty-buff or whitish fringes, especially on tertials and upper-wing-coverts, are visible at close range; at distance these areas look a little paler than on adult.
VOICE Call a dry, voiceless rasp as from coarse sandpaper, 'trrrsh'; when excited, e.g. in throng around colony entrance holes, this harsh call is uttered in long series at rapid and often quickening rate, 'trrrsh, trre-trre-trre-rrerrerre...'. These series, or very similar combinations of calls, seem to constitute the song. Alarm-call a higher, excited 'chiir'.

Brown-throated Martin *Riparia paludicola*   —

L 10½–11½ cm. Breeds in winter in small colonies and groups in sandbanks, much as Sand Martin, but, within the region treated here, only in Morocco (more numerous in S Africa). Habits as for Sand Martin, but is a resident.
IDENTIFICATION The *smallest* hirundine in the region. Has proportionally slightly *shorter wings* than Sand Martin, and these therefore look somewhat broader and blunter. The different shape combined with *lack of dark breast-band*, and diffusely dark-spotted or *dusky throat*, can recall Crag Martin unless the small size is apparent. The underwing is rather plain (wing-coverts only moderately darker than flight-feathers). Flight lighter and more fluttering than

Sand Martin's, with more frequent changes in direction and speed. – Juvenile: Same differences as for Sand Martin.
VOICE As Sand Martin's, but higher, softer and less rasping and hard, 'chirr', 'prri' and the like.

(Eurasian) **Crag Martin** *Ptyonoprogne rupestris*   **V**＊＊

L 14–15 cm. Breeds in caves and cavities on cliff faces, rarely also in holes in buildings and walls. Found at all levels, even high up (>2000 m) in remote mountain areas. Migrant in N, resident or short-distance migrant in S.
IDENTIFICATION Rather large and compact swallow with *heavy body, broad neck and broad but pointed wings*. Drab *grey-brown above* with flight-feathers a shade darker, and dirty brown-tinged greyish-white below on breast and *gradually darkening rearwards* towards undertail-coverts. A moderate range and in good light, *throat* is seen to be *faintly streaked or dusky*. Underwing pattern characteristic, with *rather pale brown-grey remiges and clearly contrasting sooty black coverts*. An important character is visible only when the shallowly forked tail is spread: *small white 'windows' on most tail-feathers*. Flight powerful and agile, often glides on flat wings held straight out, twists and turns, sweeps rapidly along, gives a few quick beats, and so on. Often seen high up in air like a swift, or patrolling back and forth along vertical cliff high in mountains. – Juvenile: Same plumage differences as for Sand Martin.
VOICE Rather discreet repertoire. Calls often heard include a short, high, clicking 'pli', a cheerful Linnet-like 'piieh', also short, chirpy, somewhat House Martin-like 'tshir' and 'trit'. Song consists of a quiet, fast, twittering series.

Rock Martin *Ptyonoprogne fuligula*   —

L 12–13 cm. Breeds in desert or desert-like terrain on cliff faces, in ravines, locally also on ruins and other buildings. Resident. Widely distributed in Africa and Middle East.
IDENTIFICATION Like Crag Martin, with same shape and basic pattern, including *white 'windows' on most tail-feathers* (visible only if tail spread), but is slightly but *clearly smaller* and a shade *paler grey-brown above*, especially on lower back/rump (discernible in direct comparison or with sufficient experience) and *paler dirty white below*. *Throat is pale*, not washed grey-brown as on Crag Martin, and *undertail-coverts are fairly pale*, not dark as on Crag. An important distinction in flight is that *only the carpal area on underwing is darker than rest of wing* (on Crag Martin all coverts are dark, creating a larger area dark), and that the *contrast is less strong* than on Crag Martin.
VOICE Unobtrusive. Soft, rather dry 'trrt' like House or Sand Martin, also more Barn Swallow-like, slightly nasal 'vick'. Song a muffled, somewhat raucous twitter.

Sand Martin   Brown-throated Martin   Crag Martin   Rock Martin

SAND MARTIN

pale fringes in juv.

grey-buff

well-marked breast-band

brown

breeding colony in gravel-pit

juv.

ad.

BROWN-THROATED MARTIN

rusty-buff fringes in juv.

shaded brown

squarer tail than Sand Martin

brown-grey

broad 'arm'

white

small and compact shape may attract attention

juv.

ad.

CRAG MARTIN

dusky, diffusely streaked

medium grey

prominent white 'windows'

rather dark

dark vent

strongly contrasting dark underwing-coverts

breeds in steep cliffs

ROCK MARTIN

like a small, bleached version of Crag

pale

pale

paler back / rump than Crag Martin

rather pale

note how the appearance of the underwing changes with different angles and light

small white 'windows'

paler than Crag Martin

rather pale

dark carpal area (but paler coverts on inner wing)

Barn Swallow *Hirundo rustica*   m**B**2 / **P**1

L ad. 17–21 cm (incl. tail projection 3–6½), juv. 14–15 cm. Breeds commonly in cultivated areas with farmyards, small villages etc. Summer visitor (in Britain & Ireland mainly Apr–Oct), winters in Africa. Often roosts communally in reedbeds outside breeding season. Nest an open mud cup reinforced with plant material, placed on roof beam or projection inside barn, boathouse, under bridge, in culvert etc. Commonly hunts insects low over ground, often around legs of grazing cattle, but also at treetop height.

IDENTIFICATION The symbolic swallow, well known to one and all through its breeding habits and through its characteristic appearance with *long, pointed wings* and *deeply forked tail with wire-thin elongated streamers*. Blue-glossed black above, white or buffish-white below (in most of Europe; see Variation below) with *blue-black breast-band and blood-red throat and forehead*. The red colour is surprisingly difficult to see on flying bird and at a little distance, when the swallow then looks mostly all dark on head and breast. When tail is spread while braking and turning, *small white 'windows' are visible on outer tail-feathers*. Flight fast and powerful with *clipped beats*, passing back and forth and often low above ground or water surface (may then drink in flight); less inclined than House Martin to make long, slow, curving glides. Sexes alike (but ♂ has on average narrower and longer tail-streamers). – Juvenile: Short, blunt tail-streamers. Forehead and throat rusty buffish-white or brownish-pink (not blood-red). Upperparts with minimal blue gloss. – Variation: Birds along E Mediterranean coast (ssp. *transitiva*) are reddish-buff below, and those in Egypt (*savignii*) deep rusty-red.

VOICE Noisy, its loud calls enlivening farmsteads and small villages. In 'itinerant flight' gives cheerful sharp 'vit!', often repeated two or more times. Mates preen each other and entertain the barn livestock with cosy chatter almost like Budgerigars. Cats are announced with sharp 'si**flitt**' notes and birds of prey with similar 'flitt-flitt!'. The rather loud song consists of a rapid twitter now and then interrupted by a croaking sound which turns into a dry rattle.

Red-rumped Swallow *Cecropis daurica*   V**∗∗**

L 14–19 cm (ad., incl. tail projection 3–5). Breeds on cliffs in mountain areas and along steep coasts, sometimes also in cavities on ruins, under bridges etc. Migrant, winters in tropical Africa. Builds a closed nest of mud with entrance tunnel, fixed to roof of cave, recess or similar site. Rare but regular vagrant to Britain & Ireland, mostly Mar–May and Aug–Nov. Often hunts insects high up in remote mountain districts.

IDENTIFICATION Resembles Barn Swallow, having similar size and shape with *long, pointed tail-streamers*, but immediately told by *pale rump*, which at distance can look white (cf.

House Martin) but at close range is *light rusty-red* at top and somewhat paler (sometimes whitish) to rear. Close views also reveal that *nape has a narrow rusty-brown band* and that *head-sides are pale*, not blue-black. Wings and tail dark brownish-black (tail without Barn Swallow's white 'windows'), mantle, scapulars and crown shiny blue-black. Red-rumped Swallows in flight are also identifiable from below at distance by their *pale throat and breast* and *squared-off black undertail-coverts* (white on Barn Swallow). At close range, underparts are seen to be pale rusty-buff or buff-white with very discreet, thin streaks. Flight intermediate between Barn Swallow and House Martin, clipped wing-beats but also a lot of slow gliding on straight wings. – Juvenile: Shorter tail-streamers, less blue gloss above, buff-white tips to tertials and many wing-coverts.

VOICE Noisy, but not so loud as Barn Swallow. Call, often given by flocks hunting insects, a rather soft but spirited, nasal 'tveyk' (recalling Tree Sparrow in tone). Alarm a sharp 'kiir!' Song allied to Barn Swallow's in structure, contains croaking sounds and rattles, but the introductory twitter is much lower in pitch and harsher, slower and shorter (with rhythm and ring almost like song of Black-headed Bunting).

(Common) House Martin *Delichon urbicum*   m**B**2 / **P**

L 13½–15 cm. Breeds commonly and colonially in villages, farms, towns, also all kinds of open country; attracted to houses, but also cliff faces in undeveloped areas. Summer visitor (Apr–Oct), winters in Africa. Outside breeding period roosts communally in trees (not reeds). Confiding toward humans, building its closed, convex mud nest beneath eaves of house walls, on bridge girders etc., sometimes even on ferry boats in regular service, oblivious to the boat's movements or to rather intrusive human activities; those breeding in the species' original sites attach their nests to rock walls on faces or mountain precipices. When young have fledged, birds often perch in numbers on telephone wires ('music score'). Hunts insects at all levels, but often high in air.

IDENTIFICATION Easily recognized by *pure white rump* contrasting sharply with otherwise *black upperparts*, with crown, mantle and scapulars *glossed blue*. Underparts white. Black tail short and moderately forked (lacks streamers). On ground, e.g. at pool of water to collect mud for nest, the white-feathered feet (as on a grouse) are noticeable. Flight not so swift as Barn Swallow's, more *fluttery* with frequent and at times *long glides on straight wings*, often in gentle curves at slow speed. – Juvenile: Like adult but with hardly any blue gloss above, and has throat and upper breast sullied brownish-grey. Slight touch of yellow at base of lower mandible (adult has all-black bill).

VOICE Noisy, especially at colonies. Gives incessant dry but pleasing twitter, 'prrit' with variations according to mood and requirements. When agitated, utters a higher, emphatic and drawn-out '**chi**err'. Song little more than a chatty burst of chirps with no clear structure, but the whole sounds rather sweet and 'eager'.

Barn Swallow   Red-rumped Swallow   House Martin

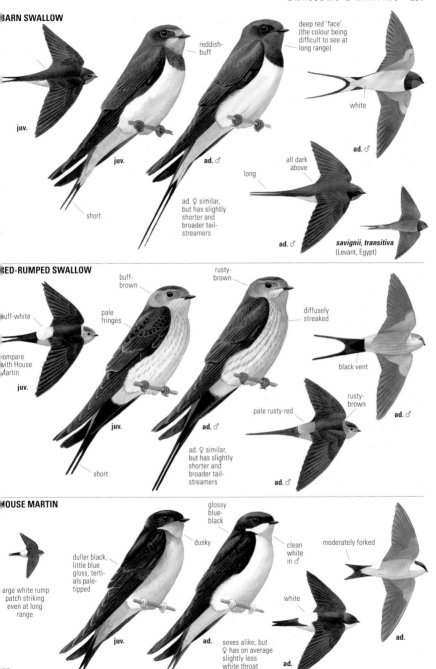

BARN SWALLOW

juv.

juv.

reddish-buff

ad. ♂

short

ad. ♀ similar, but has slightly shorter and broader tail-streamers

deep red 'face' (the colour being difficult to see at long range)

white

ad. ♂

all dark above

long

ad. ♂

savignii, transitiva (Levant, Egypt)

RED-RUMPED SWALLOW

buff-white

compare with House Martin

juv.

buff-brown

pale fringes

juv.

rusty-brown

diffusely streaked

ad. ♂

ad. ♀ similar, but has slightly shorter and broader tail-streamers

short

pale rusty-red

ad. ♂

black vent

rusty-brown

ad. ♂

HOUSE MARTIN

large white rump patch striking even at long range

duller black, little blue gloss, tertials pale-tipped

juv.

glossy blue-black

dusky

ad.

sexes alike, but ♀ has on average slightly less white throat

clean white in ♂

white

ad.

moderately forked

ad.

PIPITS and WAGTAILS *Motacillidae*

Rather small, slim, long-tailed passerines with pointed bill which spend much time on the ground. Migratory in N and E. Insectivores. Pipits (ten breeding species, three vagrants) are brown and white with varying degree of streaking on breast and upperparts; have characteristic songs, usually delivered in typical song-flight. Nest on ground among grass tufts. Wagtails (four species) have more contrasts in plumage and simpler songs, also very long tail which they wag up and down. Nest in holes, recesses or cover of vegetation.

Tawny Pipit *Anthus campestris* **V***

L 15½–18 cm. Sparse breeder on sand dunes, sandy open ground, at gravel-pits and in clearings, in S Europe also on barren mountain slopes. In Britain & Ireland rare but regular on passage (esp. Aug–Oct). Winters in Africa.

IDENTIFICATION A large, slim pipit with relatively uniform *sandy-coloured* plumage. *Dark loral stripe* in all plumages (occasionally less obvious owing to angle of light or wear). *Hindclaw relatively short* (cf. Richard's Pipit). Legs pinkish or light brown. – Adult: *Crown to back almost unstreaked* (only hint of diffuse spotting), *underparts unstreaked buff-white*, sometimes with a few faint narrow streaks on upper breast-side. Distinct *pale supercilium*; usually dark moustachial stripe and very thin lateral throat-stripe. *Median coverts contrastingly dark, in fresh plumage broadly tipped buff-white.* – Juvenile: Crown to back heavily dark-patterned, feathers finely fringed white. *Upper breast dark-spotted* (varying in extent; at times some streaks on flanks, too). Resembles Richard's Pipit (which see).

VOICE Call a full 'tshilp' (soft ending; slightly House Sparrow-like, but can also recall Short-toed Lark) or shorter 'chüp'. Song simple, usually delivered in undulating song-flight, two or three syllables with ringing tone, often stressed and drawn out at end, slowly repeated in time with peaks in flight, e.g. 'tsirliih...tsirliih...tsirliih...'. (In the Balkans it may sound like 'sr'r'riih', trembling and dropping in pitch.)

Long-billed Pipit *Anthus similis* **—**

L 16–17½ cm. Breeds in N Israel, W Syria and Lebanon (also W Jordan?) on bare, open slopes with plenty of (often flat) rock outcrops and grass, herbs and other low vegetation (± garrigue). Resident or short-distance migrant.

IDENTIFICATION Like Tawny Pipit and roughly the same size (marginally larger), but slightly *longer-tailed* and *longer-billed*. Plumage somewhat *greyer* and *less contrasty* than Tawny Pipit's, uppertail darker. *Supercilium* often rather narrow and *reaches far back towards nape*, longer than on Tawny. Breast lacks clear dark streaking, is merely obscurely spotted. Belly and flanks often tinged warm buff.

Median coverts proximally dark. Bill-base light pink.

VOICE Call resembles Desert Lark's, a short 'chupp'. Alarm 'siih'. Song delivered from rock or in fluttering, slow song flight, simple, vaguely like Tawny Pipit's but a bit more varied, three or four chords randomly strung together in slow drawl, e.g. 'siu...chürr...siu...chivü...srrüi...siu...' and so on

Richard's Pipit *Anthus richardi* **V**

L 17–20 cm. Rare but regular Siberian vagrant in (mostly W Europe, primarily in autumn (mid Sep–Nov), extremely rare at other seasons; thus does not breed within region.

IDENTIFICATION A touch bigger than Tawny Pipit, this reinforced by *longer legs and tail* and by habit of *standing more upright and stretching neck*. When perched on wire against pale background (sky), a useful pointer is *very long hindclaw* (longer than hind toe). Bill averages somewhat longer and stouter than Tawny Pipit's (like small thrush, but difference not always striking). *Flight powerful with long dips.* Often *hovers* briefly before landing in grass (only exceptionally a sign of this in Tawny). Plumage similar to juvenile Tawny Pipit (dark-streaked on back and breast), but has *pale lore* (at certain angles, however, e.g. in oblique front view, can appear dusky). – Adult: *Breast and flanks usually warm buff* (even rusty-tinged), contrasting with whiter belly-centre; *upper breast distinctly streaked*; often streaks coalesce to form dark wedge on lower throat-side. Greater and median coverts and tertials rather *broadly tipped rusty-brown* when fresh. – Juvenile (often even into Oct): Like adult, but upperpart streaking darker, and the blackish median coverts and tertials *narrowly and sharply edged white*.

VOICE Call typical, esp. when flushed or on migration, frothy, grating or hoarse, drawn-out and uninflected (or faintly downslurred) 'pshee!' (or 'shreep'; some House Sparrows can sound similar, but on other hand is very different from Skylark's calls). Straight, less typical, weaker 'chee calls heard at breeding site, and sometimes given by vagrants Song simple, delivered in deeply undulating song-flight, grinding 'tschivü-tschivü-tschivü-tschivü-tschivü'.

Blyth's Pipit *Anthus godlewskii* **V****

L 15½–17 cm. Rare autumn vagrant in NW Europe from Mongolia and adjacent regions.

IDENTIFICATION Very like Richard's Pipit in plumage, but is *slightly smaller* and has *shorter legs, shorter and somewhat more pointed bill, shorter tail* and tendency to more distinct wing-bar. In the hand, differs in certain measurements (e.g. shorter hindclaw) and in amount of white on second outermost tail-feather (short, broad white wedge on Blyth's, long and narrow on almost all Richard's), but safe identification in the field requires detailed study at close range and preferably close check of call. *Underparts somewhat more uniform buff* (belly not so white). Many identified on having adult-type median coverts (often odd ones present even in 1st autumn): these have *broad and sharply defined light ochre-buff tip*, against which the dark centre has blunt, broad edge (no darker rusty-brown, somewhat more diffuse and narrower tip, with dark centre projecting in broad wedge).

VOICE Call like Richard's Pipit's but slightly higher in pitch, a little purer in tone, somewhat reminiscent of Yellow Wagtail, 'pshiu', often interspersed with low 'chip-chüp'.

Tawny Pipit

Long-billed Pipit

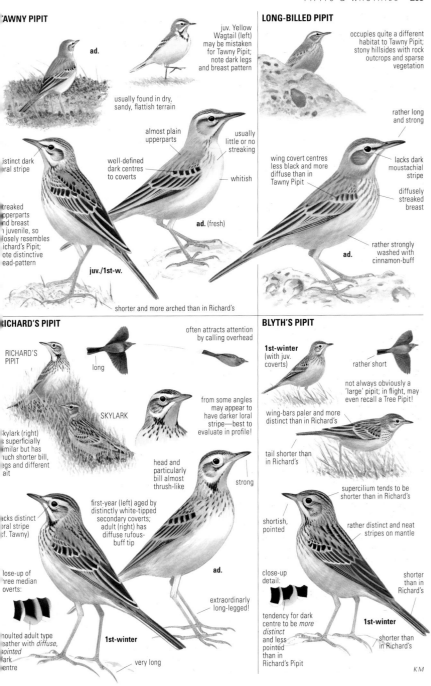

TAWNY PIPIT

ad.

juv. Yellow Wagtail (left) may be mistaken for Tawny Pipit; note dark legs and breast pattern

usually found in dry, sandy, flattish terrain

almost plain upperparts

usually little or no streaking

well-defined dark centres to coverts

whitish

distinct dark loral stripe

streaked upperparts and breast in juvenile, so closely resembles Richard's Pipit; note distinctive head-pattern

ad. (fresh)

juv./1st-w.

shorter and more arched than in Richard's

LONG-BILLED PIPIT

occupies quite a different habitat to Tawny Pipit; stony hillsides with rock outcrops and sparse vegetation

rather long and strong

wing covert centres less black and more diffuse than in Tawny Pipit

lacks dark moustachial stripe

diffusely streaked breast

rather strongly washed with cinnamon-buff

ad.

RICHARD'S PIPIT

RICHARD'S PIPIT

long

often attracts attention by calling overhead

from some angles may appear to have darker loral stripe—best to evaluate in profile!

SKYLARK

Skylark (right) is superficially similar but has much shorter bill, legs and different gait

head and particularly bill almost thrush-like

strong

lacks distinct loral stripe (cf. Tawny)

first-year (left) aged by distinctly white-tipped secondary coverts; adult (right) has diffuse rufous-buff tip

ad.

close-up of three median coverts:

extraordinarily long-legged!

moulted adult type feather with diffuse, pointed dark centre

1st-winter

very long

BLYTH'S PIPIT

1st-winter (with juv. coverts)

rather short

not always obviously a 'large' pipit; in flight, may even recall a Tree Pipit!

wing-bars paler and more distinct than in Richard's

tail shorter than in Richard's

supercilium tends to be shorter than in Richard's

shortish, pointed

rather distinct and neat stripes on mantle

shorter than in Richard's

close-up detail:

tendency for dark centre to be *more distinct* and less pointed than in Richard's Pipit

1st-winter

shorter than in Richard's

KM

Water Pipit *Anthus spinoletta*   **W**4

L 15½–17 cm. Breeds in mountain regions on bare alpine slopes and upland plains. Short-distance migrant, winters in lowland areas at freshwater marshes, ponds, watercress-beds, flooded fields etc., also at coasts on waterlogged meadows with fresh water. Very closely related to Rock and Buff-bellied Pipits (the three formerly regarded as conspecific).

IDENTIFICATION In winter plumage very like Rock Pipit (incl. usually *dark legs*) but *a shade paler and browner above* (not olive-grey), and *ground colour of underparts is purer white with more distinct streaks. Supercilium and wing-bars generally paler* and better marked. *Pure white on outer tail-feathers* (but note that Rock Pipit's light dusky tail-sides can appear white in strong light). – Summer: Moults into characteristic plumage in Mar, with *almost unmarked pale pinkish underparts* (streaked only on flanks) and dark *ash-grey head and nape* which usually contrast with *brown mantle, distinct white supercilium* and *whitish wing-bars*. – Variation: Birds in the Caucasus (ssp. *coutellii*), whose winter range includes Middle East, are *more distinctly streaked above* and more buff below (on whole of underparts; *spinoletta* has pink hue strongest on breast).

VOICE Call like Rock Pipit's or often faintly upslurred, 'vüisst'. Song extremely similar to Rock Pipit's, but to the practised ear sometimes recognizable by certain more rhythmic motifs and thin, fine, drawn-out notes (vaguely recalling Red-throated Pipit), e.g. 'zrü zrü zrü-zrü-zrü-zrü-zü-züzü-züzüzüzüzü svirir**irr**-svirir**irr**-svirir**irr** suu**ü**-suu**ü** psiii**eh**-psiii**eh**-psiii**eh**' (though note that certain geographical variations occur!).

Rock Pipit *Anthus petrosus*   r**B**3 / **W**3 (2?)

L 15½–17 cm. Breeds on rocky coasts and islands. Most N European breeders migrate to W European coasts in winter.

IDENTIFICATION A relatively sturdy pipit with predominantly *dark plumage* and *smudgy markings*. Best characters are usually *dark legs* (sometimes not-so-dark reddish-brown but never pink as e.g. Meadow Pipit) and *relatively long, dark bill*. Outer tail-feathers pale greyish, not pure white (except sometimes at tip). *Supercilium usually indistinct and short.* Narrow white eye-ring. Underparts dirty white, often tinged buff; breast and flanks *heavily but diffusely streaked dark*. Upperparts brown-grey with *faint olive tinge*, mantle indistinctly dark-streaked. – Variation: Nordic and Russian birds (ssp. *littoralis*) virtually identical to W European (*petrosus*) in winter, but some acquire distinctive summer plumage in which underparts tinged pink and dark streaking somewhat reduced. Extreme examples may closely resemble Water Pipit but have greyer mantle and rump area and bold dark lateral throat stripe.

VOICE Call a sharp, explosive, drawn-out 'viisst', usually given singly or repeated less quickly than in Meadow Pipit. Song confusingly similar to Water Pipit's (and not that unlike Meadow Pipit's!), series of repeated, sharp, fine notes with three or four theme changes during the song, the full version always delivered in typical climbing and then descending song-flight on rigid wings, 'zrü-zrü-zrü-zrü-zrü-zrü-zre-zre-zre-zre-zre-zre-sui-sui-sui-sui-zri-zri-zri...' or similar. Told from Meadow Pipit's by slightly sharper, almost 'electric' voice and harder and less pure introductory 'zri' series at lower pitch.

Buff-bellied Pipit *Anthus rubescens*   **V**∗∗

L 15–16 cm. Very rare vagrant from America or Asia. Ssp. *rubescens* breeds in N America and W Greenland and has been recorded in autumn in W Europe; ssp. *japonicus* (very different in appearance; see below) breeds in E Asia (so far as known not W of Lake Baikal) and is regular in winter in e.g. Israel. Very closely related to Rock and Water Pipits (all three until recently regarded as conspecific).

IDENTIFICATION A *touch smaller* than Water Pipit, with somewhat *more slender bill* and *paler lores* which may suggest Meadow Pipit, but Buff-bellied always *much less distinctly streaked on upperparts* than Meadow. – Winter: Ssp. *rubescens* is clearly *buff below* with rather narrow but distinct dark streaks and, usually, *dark legs*. Ssp. *japonicus* *whiter below* with *distinct blackish streaks* (incl. large dark patch on side of throat, roughly as on young Red-throated Pipit), *darker and greyer above*, often has neat eye-ring and distinct supercilium, sometimes with dark border above (can recall Olive-backed Pipit!); legs reddish-brown, often paler than on *rubescens*.

VOICE Call like Meadow Pipit's 'psipp' but slightly more squeaky and emphatic; usually uttered singly.

Berthelot's Pipit *Anthus berthelotii*   –

L 13–14½ cm. Breeds as only pipit commonly in Canary Islands and Madeira on dry plains, stony and sandy terrain with low vegetation and also mountain slopes. Resident.

IDENTIFICATION *Small* as a Meadow Pipit, but with somewhat different proportions: *short rear end* (legs thus appear placed far back), *rather big head* but slender bill. Adult is predominantly *grey and dirty white*, whereas juvenile often tinged rufous above. Dirty white *distinct supercilium* characteristic. Pale eye-ring and diffuse pale area beneath eye and in centre of ear-coverts, together with *dark eye-stripe* (incl. lores), *dark moustachial stripe* and *narrow dark lateral throat-stripe*, give head a bold and almost striped appearance. *Breast clearly streaked on dirty white ground colour* (no buff tones). Upperparts unstreaked (nape) or with very meagre fine streaks. Pale tips to median coverts form obvious wing-bar. Actions more like a Desert Lark than a pipit, makes rapid mouse-like dashes interrupted by fleeting stops, hard to flush, prefers to run away.

VOICE Call not unlike Yellow Wagtail's, 'tsri(e)', also shorter and 'chup'. Song, given in undulating song-flight, rather energetically delivered, somewhat Tawny Pipit-like call slowly repeated 4–7 times, 'tschilp...tschilp...tschilp...' at times a bit faster and with almost prinia-like rattling and rolling quality, 'tsivirr, tsivirr, tsivirr, tsivirr'.

Water Pipit   Rock Pipit

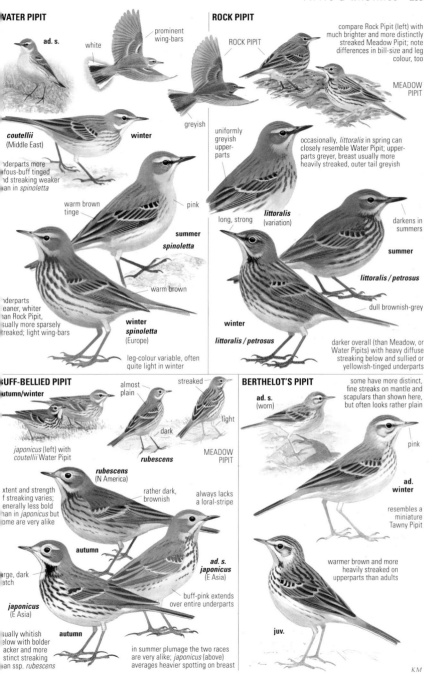

WATER PIPIT

ad. s.

white

prominent wing-bars

ROCK PIPIT

ROCK PIPIT

compare Rock Pipit (left) with much brighter and more distinctly streaked Meadow Pipit; note differences in bill-size and leg colour, too

MEADOW PIPIT

greyish

coutellii
(Middle East)

winter

underparts more
rufous-buff tinged
and streaking weaker
than in spinoletta

uniformly
greyish
upper-
parts

occasionally, *littoralis* in spring can closely resemble Water Pipit; upper-parts greyer, breast usually more heavily streaked, outer tail greyish

warm brown
tinge

pink

littoralis
(variation)

long, strong

summer
spinoletta

darkens in summers

summer

warm brown

littoralis / petrosus

underparts
cleaner, whiter
than Rock Pipit,
usually more sparsely
streaked; light wing-bars

winter
spinoletta
(Europe)

dull brownish-grey

winter

littoralis / petrosus

leg-colour variable, often
quite light in winter

darker overall (than Meadow, or Water Pipits) with heavy diffuse streaking below and sullied or yellowish-tinged underparts

BUFF-BELLIED PIPIT

autumn/winter

almost
plain

streaked

light

BERTHELOT'S PIPIT

ad. s.
(worn)

some have more distinct, fine streaks on mantle and scapulars than shown here, but often looks rather plain

dark

japonicus (left) with
coutellii Water Pipit

rubescens

MEADOW
PIPIT

pink

rubescens
(N America)

extent and strength
of streaking varies;
generally less bold
than in *japonicus* but
some are very alike

rather dark,
brownish

always lacks
a loral-stripe

**ad.
winter**

resembles a
miniature
Tawny Pipit

autumn

large, dark
patch

ad. s.
japonicus
(E Asia)

buff-pink extends
over entire underparts

warmer brown and more
heavily streaked on
upperparts than adults

japonicus
(E Asia)

usually whitish
below with bolder
blacker and more
distinct streaking
than ssp. *rubescens*

autumn

juv.

in summer plumage the two races
are very alike; *japonicus* (above)
averages heavier spotting on breast

KM

Meadow Pipit *Anthus pratensis*   r**B**1 / **P**1 / **W**1

L 14–15½ cm. Breeds in open country, on heaths and moors (very common), coastal meadows, pastures and bogs. Resident in Britain & Ireland, where also large passage (Mar–May, and esp. Sep–Oct) and many winter visitors from NE.

IDENTIFICATION Typical pipit, with olive-tinged grey-brown and buff-white or dirty white, ± streaked plumage. Like Tree Pipit, but usually distinguished from latter without difficulty by calls and habitat. Both species can, however, occur in same places (wooded pasture, edges of raised bogs, on passage on meadowland), and close study of appearance then needed unless calls heard. Note *somewhat slimmer bill* (which often looks all dark at distance, lacking Tree Pipit's pink base to lower mandible), slightly *more diffuse*, less contrasty head pattern with *less marked supercilium* and slightly narrower and not so pale submoustachial stripe. A *narrow dirty white eye-ring* is often palest part of face. *Upperparts more heavily streaked* than on Tree Pipit (but some in fresh plumage are rather similar). Hindclaw is long and not so curved (cf. Tree Pipit), but close views and good observation conditions needed to confirm this. Legs pinkish. Ground-dwelling, prefers to land *on ground* after flight excursions but often also on fence wires and posts, at times also on bushtops, low isolated trees and telephone wires. Song-flight characteristic, first climbing with rapid wingbeats, then dropping on stiff wings.

VOICE Call, from bird rising with jerky springing flight, a few heated, thin 'ist ist ist' notes; birds flying over give similar or shorter 'ist', 'ist üst' or the like. Alarm a tremulous double note, 'sitt-itt' (or 'tirri'). Song like Rock Pipit's: series of rapidly repeated thin, piercing notes with motif changing three or four times during verse, which is fairly constant in structure, 'zi zi zi zi zi zi zi zi zi zi zü-zü-zü-zü-zü-zü-zü-svisvisvisvisvi **tüü tüü tüü tüü** tii-svia' (note that initial notes are higher and less 'jangling' than Rock Pipit's, and that verse often ends with a slight flourish). Complete verse only in song-flight; perched bird reels off intro after intro, 'zi zi zi zi zi…'.

Tree Pipit *Anthus trivialis*   m**B**2 / **P**2

L 14–16 cm. Breeds locally in open woodland, in Britain often young conifer plantations, sparsely timbered heaths and commons, downland with scrub, woodland glades, but in N and W Britain mainly older oak and birch woods in uplands. Summer visitor (mainly Apr–Sep), winters in Africa.

IDENTIFICATION Like Meadow Pipit, but has *somewhat heavier bill*, rather more bold head markings with slightly stronger supercilium and broader and paler submoustachial stripe, as well as suggestion of dark eye-stripe (though differences often minor and hard to see). Underparts show *more contrast between warm buff breast and white belly* and between

heavy streaking on breast and thin streaks on flanks (Meadow Pipit more evenly yellowish- or off-white below, with equal coarse streaks both on breast and on flanks). Mantle/back normally less distinctly dark-marked than Meadow Pipit (but difference not always evident). At close range, short and curved hindclaw can sometimes be seen. Legs and bill-base pink. Slightly heavier in build than Meadow Pipit, and flight is a little more direct, not quite so light and skipping. Rather shy and difficult to approach. Often forages on ground, but when startled is more likely to fly up and *perch in tree*; in open country it tends to fly off farther than Meadow Pipit. Song-flight characteristic, starts from tree, climbs up, *then parachutes down on rigid wings, legs dangling near end of descent*, to perch in different tree. Often pumps tail downwards when perched.

VOICE Call is a drawn-out, uninflected, hoarse 'spihz'. Alarm a discreet but penetrating, clear 'sütt'. Song repeated frequently, even in middle of day when many species take siesta, a jaunty, loud verse made up of trills and repeated notes with varied tempo, e.g. 'zit-zit-zit-zit cha cha-cha-cha sürrrrrrrrrrrrr **siiiii**-a tvet-tvet-tvet-tvet **siiva siiiva siiihva** cha-cha-cha'; the sequence of the various components varies.

Olive-backed Pipit *Anthus hodgsoni*   **V**

L 14–15½ cm. Breeds in open woodland in Siberia and in far NE of Europe, winters in S Asia. Rare but annual autumn vagrant in W Europe (a few spring and even winter records).

IDENTIFICATION Very like Tree Pipit (with similarly short curved hindclaw; all other pipits have slightly longer and less curved hindclaw), but has somewhat different head pattern, more *green-tinged* brown-grey and *less clearly marked mantle and crown*, also on average *heavier black spotting on breast* (extremes, however, the same with regard to breast markings). *Supercilium bold, whitish and unstreaked* (Tree Pipit has finely streaked pale yellow-brown supercilium of variable prominence), *distinctly tinged rusty-buff in front of eye* (at best faintly buff on Tree Pipit). Above supercilium usually shows a *dark border to crown-side* (never obvious on Tree Pipit). Behind ear-coverts, diagonally below rear end of supercilium, almost invariably has an *isolated pale spot and below this a dark spot* (sometimes a faint hint of such spots on Tree Pipit). In autumn always shows *fresh tertials with broad olive-brown edges* (imm. Tree Pipits have narrow grey-white edges to often slightly worn tertials).

VOICE Call very like Tree Pipit's, a straight, slightly hoarse, high-pitched 'spiz' (doubtfully distinguishable from Tree Pipit, but Olive-backed Pipit's call on average little less hoarse, somewhat finer and less drawn out). Alarm a fine, sharp 'sitt'. Song as Tree Pipit's, but the verses are often a bit shorter, the voice softer and higher, the tempo fast throughout (does not slow down intermittently like Tree Pipit's) and the trills are drier (which together with some very thin, high notes can bring to mind song of Red-throated Pipit); the voice at times recalls Dunnock (which Tree Pipit's song never does).

Meadow Pipit   Tree Pipit

Olive-backed Pipit

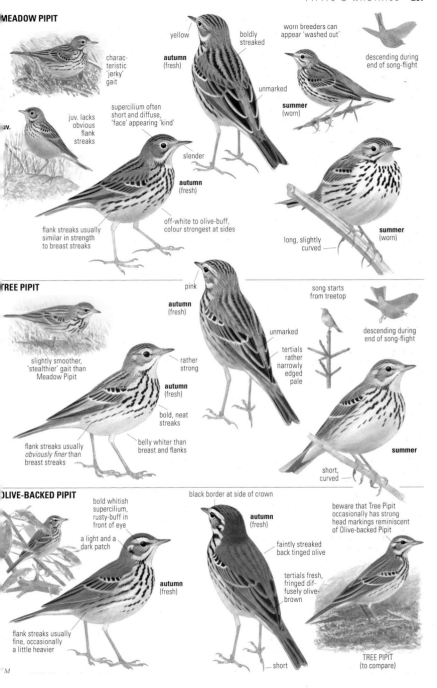

MEADOW PIPIT

charac-
teristic
'jerky'
gait

JUV.

juv. lacks
obvious
flank
streaks

yellow

boldly
streaked

autumn
(fresh)

worn breeders can
appear 'washed out'

descending during
end of song-flight

unmarked

summer
(worn)

supercilium often
short and diffuse,
'face' appearing 'kind'

slender

autumn
(fresh)

flank streaks usually
similar in strength
to breast streaks

off-white to olive-buff,
colour strongest at sides

long, slightly
curved

summer
(worn)

TREE PIPIT

pink

autumn
(fresh)

song starts
from treetop

unmarked

tertials
rather
narrowly
edged
pale

descending during
end of song-flight

slightly smoother,
'stealthier' gait than
Meadow Pipit

rather
strong

autumn
(fresh)

bold, neat
streaks

flank streaks usually
obviously finer than
breast streaks

belly whiter than
breast and flanks

summer

short,
curved

OLIVE-BACKED PIPIT

bold whitish
supercilium,
rusty-buff in
front of eye

a light and a
dark patch

black border at side of crown

autumn
(fresh)

beware that Tree Pipit
occasionally has strong
head markings reminiscent
of Olive-backed Pipit

faintly streaked
back tinged olive

tertials fresh,
fringed dif-
fusely olive-
brown

autumn
(fresh)

flank streaks usually
fine, occasionally
a little heavier

short

TREE PIPIT
(to compare)

M

Red-throated Pipit *Anthus cervinus* V**

L 14–15 cm. Breeds sparsely on bare mountains and tundra in northernmost Europe, on swampland in willow zone and open, low upland birch forest with grassy patches. Summer visitor, winters in Africa and locally in Middle East. Vagrant Britain & Ireland (mostly Apr–Jun, Aug–Nov).

IDENTIFICATION Resembles Meadow Pipit in size and shape, and occurs on passage (at times also breeds) in same habitats as that species. Normally, however, only autumn immatures present identification challenge. In all plumages *heavily black-streaked above, including back and rump* (Meadow Pipit only poorly marked on rump), with *brown ground colour* and some rufous and whitish fringes (lacks olive tinge shown by most autumn Meadow Pipits). – Adult: Easily told by *rusty-red throat*. Some (prob. mostly ♂♂) also have rusty-red on supercilium, forehead and upper breast, as well as having these areas completely unpatterned (thus lack dark lateral throat stripe); others (prob. mostly ♀♀) have more restricted rusty-red and are heavily streaked. The rusty colour is retained in autumn (winter plumage), but is initially more yellow-brown in tone. – 1st-autumn: Lacks rusty-red on throat, which is buff-white. Note *heavily dark-streaked upperparts* with brown rather than olive-grey ground colour, *distinct pale tertial edges, pale lores, complete narrow whitish eye-ring*, thin bill with usually yellowish base to lower mandible, *whitish stripes along mantle*, and often strong black wedge-shaped streak on lower throat-side.

VOICE Call a fine, drawn-out, squeaky 'pssiih', with hint of hoarseness at close range (but much higher, purer and longer than Tree Pipit's call). Alarm a 'chüpp' like Ortolan Bunting's. Song very characteristic in its rhythmic theme with ringing tone, its needle-fine, drawn-out notes (with call-note quality) and its redpoll-like dry buzzing sound, e.g. 'svü-svü-svü-**svü**, svü-svü-svü-**svü**, svü-svü-svü-**svü**, psiü psiiiü psiiiüh sürrrrrrrrr wi-wi-wi-wi tsvü-tsvü-tsvü'.

Pechora Pipit *Anthus gustavi* V***

L 14–15 cm. Breeds in boggy areas with open conifer forest and willows, on damp tundra with scattered trees and bushes, by swampy river banks with rushes, willows and isolated trees. Summer visitor (Jun–Sep), winters in SE Asia. Rare vagrant in W Europe, mostly Sep–Oct; odd spring records. Terrestrial. Hard to flush, scampers mouse-like in vegetation.

IDENTIFICATION Owing to boldly streaked upperparts, with *whitish stripes along mantle-sides* (often a hint of a secondary stripe) and frequently dark patch on throat-side, most resembles young Red-throated Pipit, but differs as follows: somewhat *heavier bill* with usually pinkish or light brown base; *shorter tertials*, which do not cover primaries right to the tips; more prominent pale, almost *whitish wing-bars*; contrast between *white belly and buff breast* (Red-

throated Pipit more uniform yellowish-white below); *crown and ear-coverts rather light rufous-brown, distinctly streaked dark*; tendency to have dark lores but indistinct moustachial and lateral throat stripes; supercilium short and diffuse finely streaked; tail-sides not pure white.

VOICE Can call frequently on passage, *but much less inclined to call when* (reluctantly) *flushed* than other pipits, a short, sharp, clicking 'dzepp' with almost electrical overtone, repeated a few times. Song soft with strained, 'mechanical' voice, hard, dry trills broken by double note, 'turrrrrr tirrrrrrr-**chu**chi-turrrrrrrr-**chu**chi-tirrr-turrrrrr...'.

White / Pied Wagtail *Motacilla alba* r(m)**B**2 / **P**

L 16½–19 cm. Common breeder in various habitats, often in open cultivated country close to habitation and water, e.g. farmyards and lakesides, as well as villages, towns and cities, in N also far from habitation in open forest or on bogs. Partial resident in W and S Europe, summer visitor in N (Apr–Oct), wintering in Middle East and NE Africa. Nests in stone walls, beneath roof tiles, in ventilation shafts, banks, among ivy, under stones etc. Attracted to bare areas such as grass lawns, golf courses, flat rocks, roofs and asphalt roads, where the fast-running bird can easily see and catch insects.

IDENTIFICATION Slender, with *long, narrow black and white tail* which *is constantly wagged up and down*. Walks with jerking head movements, rushes after prey only to pull up suddenly with tail pumping excitedly. – Adult ♂ summer (WHITE WAGTAIL, ssp. *alba*): Black crown and nape with *sharp border against ash-grey mantle*. White forehead/forecrown. Jet-black bib (bill to breast). White wing-bars and white tertial fringes. – Adult ♀ summer (*alba*): Like ♂, but less black on nape and *diffuse border against grey back*, or only a little black on crown, or even *all-grey crown and nape*. – Winter: White chin and throat, on some tinged yellow; black breast-band. Forehead white, white with greenish-yellow tinge, or light grey with greenish tinge (adult ♂ whitest; immatures may have no white at all). – Juvenile: Dark crown-sides. Wing-bars greyish. Breast patch grey, not black. – Variation: In Britain & Ireland, and sporadically on adjacent Atlantic coasts, PIED WAGTAIL, ssp. *yarrellii* breeds, summer-plumaged adult ♂ of which has *jet-black upperparts* and ♀ (and imm. ♂) *dark sooty-grey with black rump*; flanks dark olive-grey. Winter and immature plumages somewhat less dark above. Breeding in rest of Europe (passage migrant in Britain & Ireland) is ssp. *alba* with light grey upperparts. In Morocco *subpersonata*, with distinctive head pattern.

VOICE Call is easily recognizable, a two- or three-syllable cheerful 'tsli-**vitt**' or 'zi-ze-**litt**'. Normal song very plain and simple, a few twittering notes, a pause, then a few more notes followed by further pause etc.; the whole gives impression of being casual and reflective and does not always sound like 'proper' song. The opposite is so with the ecstatic song variant, used in excitement, in territorial conflicts and when mobbing Cuckoo or Sparrowhawk, a long, very fast series of indignant chirping notes.

Red-throated Pipit

Pechora Pipit

White / Pied Wagtail

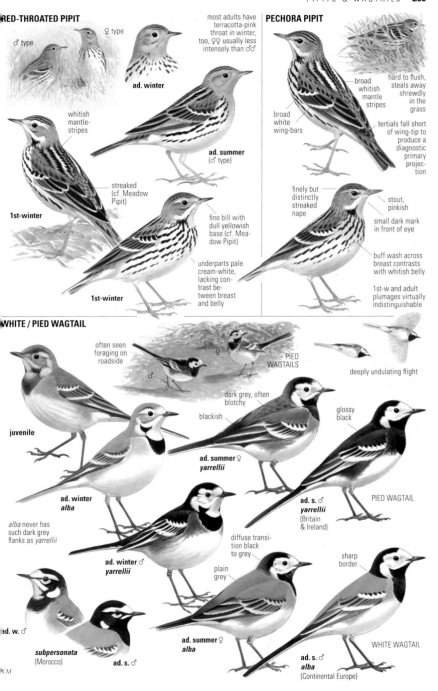

RED-THROATED PIPIT

♂ type

♀ type

most adults have terracotta-pink throat in winter, too, ♀♀ usually less intensely than ♂♂

ad. winter

whitish mantle-stripes

streaked (cf. Meadow Pipit)

1st-winter

ad. summer (♂ type)

fine bill with dull yellowish base (cf. Meadow Pipit)

underparts pale cream-white, lacking contrast between breast and belly

1st-winter

PECHORA PIPIT

broad whitish mantle stripes

broad white wing-bars

hard to flush, steals away shrewdly in the grass

tertials fall short of wing-tip to produce a diagnostic primary projection

finely but distinctly streaked nape

stout, pinkish

small dark mark in front of eye

buff wash across breast contrasts with whitish belly

1st-w and adult plumages virtually indistinguishable

WHITE / PIED WAGTAIL

often seen foraging on roadside

PIED WAGTAILS

deeply undulating flight

juvenile

dark grey, often blotchy

blackish

glossy black

ad. winter *alba*

alba never has such dark grey flanks as *yarrellii*

ad. summer ♀ *yarrellii*

ad. s. ♂ *yarrellii* (Britain & Ireland)

PIED WAGTAIL

ad. winter ♂ *yarrellii*

diffuse transition black to grey

plain grey

sharp border

ad. w. ♂

subpersonata (Morocco)

ad. s. ♂

ad. summer ♀ *alba*

ad. s. ♂ *alba* (Continental Europe)

WHITE WAGTAIL

KM

Yellow Wagtail *Motacilla flava*   mB3 / P3

L 15–16 cm. Breeds in lowland areas on marshy pastures, waterlogged meadows, beside lakes and at sewage-farms, in N Europe also on forest bogs and mires and in wet clearings. Summer visitor (mainly Apr–Sep/early Oct), winters in Africa. Migrants on pastures and golf courses on autumn passage (roost in reedbeds). Nests in tussocks on ground. Complex geographical variation in appearance, especially of ♂♂, and this cannot be treated in detail here; at least eight different races just within the region covered.

IDENTIFICATION The least extreme of the wagtails in proportions, with tail slightly shorter than White Wagtail's and clearly shorter than that of Grey. ♂ is *saturated yellow below*, ♀ a little duller, especially on breast. Autumn immatures are pale below, *buff-white with only faint yellow tinge on vent* (the yellow at times difficult to see in the field, or can even be lacking altogether; cf. Citrine Wagtail). *Mantle and back greyish-green or grey-brown*, with *olive tinge barely discernible* in the field. *Black legs*. Wing brownish-grey, with rather distinct but narrow wing-bars pale yellow or dirty white. – Variation in ♂♂: In greater part of Continental Europe (ssp. *flava*), the head is blue-grey with prominent white supercilium and white submoustachial stripe between dark grey cheek and yellow throat; small white patch below eye. In Britain (*flavissima*), crown, nape and ear-coverts are yellow-green, and rest of head bright yellow like underparts. Typical ♂♂ of other races illustrated opposite. Note that intermediate forms from border areas between two races occur.

VOICE In much of Europe call a fine 'psit' or slightly fuller 'tslie' (also 'tsrlie'), but in Balkans and in parts of E Europe (*feldegg*) usually a more grating, frothy 'zrri(e)' (with more obvious 'r'; cf. Citrine Wagtail). Song is one of the most modest of all bird songs, usually consisting of two (occasionally one or three) scraping notes, with the last generally slightly stressed, 'srrii-**srriiht**'; often delivered from top of a post, low bush or fence-wire perch.

Citrine Wagtail *Motacilla citreola*   V**

L 15½–17 cm. Breeds on waterlogged meadows, open bogs in forest with low vegetation and sparse willow bushes, and soggy river banks. Migrant, winters in SE Asia (a few in Middle East). Rare but increasingly numerous vagrant to W Europe. Expanding westwards (now breeds e.g. Poland).

IDENTIFICATION Like Yellow Wagtail in size and proportions, but tail on average somewhat longer. *Double broad white* (or whitish) *wing-bars* important character in all plumages (exception: heavily worn breeding birds). – Adult ♂ summer: Unmistakable, with entirely *lemon-yellow head, black nuchal band, ash-grey mantle/back*, yellow belly (but usually *whitish undertail-coverts*; always yellow on Yellow

and Grey Wagtails). Flanks often dusky grey (as on White Wagtail, unlike the other wagtails). – Adult ♀ summer, adult ♂♀ winter: Crown and cheeks have grey tone, nape lacks black, yellow of underparts is usually paler. Note *yellow supercilium* and *yellow-framed dirty grey ear-coverts*, as well as *white wing-bars*. In winter plumage especially, some dark spots on breast. – 1st-winter: Lacks pure yellow in plumage (but breast and forehead often have tinge of yellow-buff) and can be confused with pale immature Yellow Wagtails. Note following: *broad white wing-bars; grey or brownish-grey upperparts* (never olive-toned); *pale lores* and often forehead (Yellow Wagtail dark); *light-framed ear-coverts*; *all-dark bill* (Yellow Wagtail has slightly paler base to lower mandible); *never any yellow on undertail-coverts; flight-call.*

VOICE Contact-call often very like Yellow Wagtail's, 'tslie'. Alarm and flight-call (often heard from vagrants in W Europe) a characteristic loud, ripping, straight (voice as in Yellow Wagtail song) 'tsriip', repeated several times. (Black-headed Yellow Wagtail, *feldegg*, also have r-sound in call, but call not so loudly ripping and straight, more soft, frothy and with faint terminal upslur.) Song most like White Wagtail's, a few uncomplicated syllables at a time, delivered from top of a willow bush (slightly longer phrases in song-flight); individual notes also not unlike Yellow Wagtail call (*flava*).

Grey Wagtail *Motacilla cinerea*   rB3 / P3

L 17–20 cm. Breeds in scattered pairs on small, fast-flowing watercourses with plenty of exposed rocks in and alongside the water and with woodland or at least rows of trees along the banks; at times also on lakeshores and slower rivers. Mostly resident in Britain & Ireland; also migrants and winterers from N Europe (passage Aug–Oct). Nests in rock crevice, cavity in stone bridge, mill foundations etc. beside water.

IDENTIFICATION Compared with the other wagtails, Grey Wagtail has *longest tail* and *shortest legs*. Constantly pumps its long tail, and so strongly that *whole rear end rocks* with it. In all plumages *bright yellow vent* and also *yellow-green rump*. Legs brownish-pink (other wagtails black). *Grey upperparts* contrast with *black wings*. Tertials edged (yellow-)white. Distinct pale supercilium. In *deeply undulating flight*, when long tail is very striking, shows a *broad white wing-bar*, most evident from below. – Adult ♂ summer: *Black bib*. Pure *white supercilium and submoustachial stripe*. Rest of underparts yellow (but flanks sometimes a little paler). – Adult ♀ summer: Throat white with some greyish-black; sometimes more ♂-like with much black on throat, but always shows some white, and has buff-white supercilium and indistinct submoustachial stripe. – Winter: Pale throat with no dark feathering. Older birds are yellower on breast and whiter on throat, younger ones paler yellow on breast and buff-tinged on throat.

VOICE Call basically like White Wagtail's but sharper and higher-pitched, 'zi-zi'. Alarm a repeated 'süiht', often interspersed with call. Song a short, mechanical series of sharp notes, 'ziss-ziss-ziss-ziss', often mixed with alternative phrase of higher notes, 'si si si siü'.

Yellow Wagtail

Citrine Wagtail

Grey Wagtail

YELLOW WAGTAIL

flavissima (Britain) ♂

flavissima ♀

lutea (SW Siberia) ♂

beema (N Kirghiz steppe) ♂

thunbergi (N Fenno-Scandia) ♂

cinereocapilla (Italy) ♂

iberiae (Iberia, NW Africa) ♂

feldegg (Balkans, Turkey, Caucasus) ♂

feldegg ♀

'superciliaris' (SE Russia) ♂

'dombrowskii' (Romania) ♂

juv.

dark lores

olive-brown

pale base

1st-w.

pale yellow

♀ / ad. w.

ad. s. ♂

flava (C Europe)

CITRINE WAGTAIL

pale 'cheek-surround'

grey-ish

pale lores

all dark

broad wing-bars

white

1st-winter (Oct)

1st-w. (Dec–Jan)

little black

ad. s. ♂

werae (SE Russia, SW Siberia)

deep yellow

♀ / ad. w.

ad. summer ♂

citreola (N Russia, NW Siberia)

GREY WAGTAIL

prefers fast-flowing water-courses with exposed stones

wide white wing-bar

yellowish-green

very long tail; frequently wags tail and whole rear body

bright yellow

1st-winter

pink

ad. s. ♀ / 1st-s. ♂

some ♀♀ have much whiter throat

brownish-pink, short

black bib

ad. summer ♂

bright yellow

(White-throated) **Dipper** *Cinclus cinclus* [r]**B**3 / (**W**)

L 17–20 cm. Breeds along shallow watercourses in upland regions, often forested, preferably beside fast-flowing torrents with exposed boulders as convenient perches, less on lowland rivers with weirs. Dives for aquatic invertebrates, *swims underwater* using wings and can walk on bottom (helped by skeletal bones that are uniquely solid among flying birds); is also a 'winter bather', hopping in from edge of ice. Also swims on surface, body very low. Nest of straws and moss, domed and with side entrance, placed out of reach on wall of rock or mill over water, or even behind waterfall. Mostly resident, but northern populations move south (Norwegian birds usually southeast).

IDENTIFICATION *Compact build* with strong legs and *short tail, which is often held somewhat cocked.* Plumage *dark with big white bib*, either entirely sooty-black (birds in north, east and large parts of France and NW Iberia) or brownish on head and rusty-brown beneath bib (mainly Britain & Ireland; less obviously in C Europe). Nictitating membrane white. Looks *compact* in flight, about Starling-sized. *Flight straight*, with *fast whirring beats of the rounded wings.* Often curtseys repeatedly when perched. – Juvenile: Dull grey with pale wavy bars above and dark ones below. Shows hint of pale bib, but this is sullied grey and is poorly outlined on belly.

VOICE Call a short, sharp and 'electric' 'zrik!' penetrating through the roar of the rushing water, often given by bird flying past along watercourse. Song, often heard even in depths of the coldest winter, a slowly delivered, rather subdued series of alternating hard, harsh, squeaky and throaty notes, some repeated a few times, which can recall song of both Common Crossbill and Bullfinch. Both sexes sing.

(Bohemian) **Waxwing** *Bombycilla garrulus* **W**5(4)

L 18–21 cm. Breeds in northern parts of the coniferous belt, often in remote, lichen-rich, mature forest in damp, mossy terrain. Nest high up on branch, often in pine. In some years large flocks move south in winter, reaching W Europe, when often seen eating rowan berries (in Britain mainly hawthorn) in gardens and suburban roads. Nevertheless, remains alert and rather nervous, retreating to higher treetops at first sign of disturbance. In winter can eat frostbitten and semi-fermented berries, which may intoxicate the bird and render it temporarily incapable of flight. Has apparently developed a highly efficient liver (better than humans) to cope with this, since it usually recovers quickly.

IDENTIFICATION Starling-sized, compact in build with thick neck. *Large crest on crown* combined with *reddish-buff* colour eliminates all other species. Narrow black eye-mask and *black bib, yellow tail-tip* plus *yellow and white pattern on wing* combine to make it unmistakable. Flight straight with

long undulations; flocks keep tight formation, thus recall a flock of Starlings (enhanced by same size and similar shape) – Adult ♂: Tips of primaries edged yellow on outer web and *broadly white-edged on inner web*, forming *V-shaped marks* or 'angles' on each tip; *broad yellow terminal tail-band*; many long red waxy appendages on secondaries, and sometimes also some on tail-feathers; all-black bib with well-demarcated lower edge. – Adult ♀: Like adult ♂ but has thinner white edges to inner webs of tips of primaries, or even incomplete white 'angles' on primary tips; thinner yellow tail-band; often fewer and shorter waxy appendages; and blurred lower edge to bib. – 1st-year: Lacks any white tips to *inner* webs of primaries, has just yellowish-white tips to outer webs (thus no bright yellow, no 'angles').

VOICE Contact-call a pleasant ringing 'sirrrrr' as from a small bell, often in pealing chorus from flock before rising. Song slow and halting, a series of call notes mixed with hard, raucous sounds, 'sirrr sirrrrr chark-chark chi-chark sirrrr sirrrr...' etc.

(Grey) **Hypocolius** *Hypocolius ampelinus* —

L 21–24 cm. Breeds in Middle East, mainly in Iraq and S Iran, in or near semi-deserts with scattered trees or bushes, in date palm plantations and groves and gardens. Periodically seeks diet of berries, often in small flocks, like close relative Waxwing. Winters S to the Gulf, in W and N Saudi Arabia and in UAE. Vagrant or scarce winterer in Oman, vagrant in Israel and Yemen.

IDENTIFICATION At distance appears like a shrike but is rather slimmer and shorter-winged with a proportionately *longer tail.* When seen close *unmistakable* with *very pale bluish-grey plumage* and a prominent *black end-band to tail.* Close relationship to Waxwing revealed by similar-shaped bill, *shoulderless shape* and *habit of stretching neck* when alarmed. Flight usually straight with continuous wing-beats, thus not undulating. – Adult ♂: Has *large black patches on sides of head*, forming a 'mask', which widen at rear and join at nape. *Primaries* largely *jet black* but *broadly tipped pure white*, striking in flight both from above and from below; the black parts are largely concealed when perched while the white tips are the more prominent. – Adult ♀: Like adult ♂ but less bluish-tinged, more dull and buff-tinged grey, and *lacks the black marks on head*, has only *narrow subterminal black marks on primaries* and a *narrower black end-band on tail.* – 1st-year: Very similar to adult ♀. Differs in having fine pale tips to tertials and tail-feathers, less broad and dark end-band to tail, and slightly looser, 'woollier' plumage, but safe separation in the field requires close views.

VOICE Contact-calls include mewing 'meee' and variants, and a downwards-inflected 'wheeoo' (almost Wigeon-like). Has also a mellow, trilling, rapid 'tre-tur-tur', where the first note is slightly higher-pitched than the following two. Apparently has no real song.

Dipper   Waxwing   Hypocolius

DIPPER

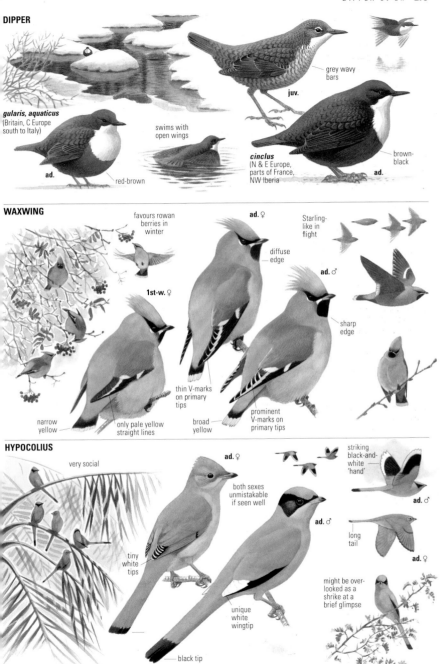

gularis, aquaticus
(Britain, C Europe
south to Italy)

ad.

red-brown

swims with
open wings

grey wavy
bars

juv.

cinclus
(N & E Europe,
parts of France,
NW Iberia

ad.

brown-
black

WAXWING

favours rowan
berries in
winter

ad. ♀

diffuse
edge

ad. ♂

Starling-
like in
flight

1st-w. ♀

sharp
edge

narrow
yellow

only pale yellow
straight lines

thin V-marks
on primary
tips

broad
yellow

prominent
V-marks on
primary tips

HYPOCOLIUS

very social

ad. ♀

both sexes
unmistakable
if seen well

ad. ♂

striking
black-and-
white
'hand'

ad. ♂

long
tail

ad. ♀

tiny
white
tips

unique
white
wingtip

might be over-
looked as a
shrike at a
brief glimpse

black tip

DZ

ACCENTORS *Prunellidae*

Five small passerines with features of sparrows as well as warblers and small thrushes, spending much time on ground or in shrub layer. Bill rather thin and pointed (but has fairly deep base), plumage mostly brown, buff, grey and black. Wings rather short and rounded. Sexes and ages very similar. Forage mostly on ground, hopping in crouched posture.

Dunnock *Prunella modularis* ⌐B1 / W
L 13–14½ cm. Breeds in parks, gardens, open woodland, heaths and commons with gorse or scrub, areas with hedges and thick shrubbery, also young conifer stands; but in N Europe mainly in dense spruce plantations, also in juniper country and upland forest. Resident; migratory in N. Rather shy and retiring in summer. Food mostly insects. Nests in thick bush or low down in conifer.
IDENTIFICATION Robin-sized, with warbler-like *thin bill*. Rather House Sparrow-like *brown and streaked upperparts.* Characteristic *blue-grey head* (ear-coverts and crown tinged grey-brown) *and breast.* Legs red-brown. Lacks white in tail. Greater coverts have fine pale spots at tips, at times appearing as thin wing-bar. Overall impression, however, is of a *uniformly coloured and rather dark* bird. Flight jerky, light and springing. Often perches in view when singing, but otherwise mostly keeps hidden, and is rather difficult to catch sight of; dives down into cover at slightest disturbance.
– Juv.: Browner, *more boldly streaked*, and has pale chin.
– Variation: Birds in Ireland and NW Scotland (ssp. *hebridium*) are darker, especially above and on breast. Birds in rest of Britain (*occidentalis*) something between those and birds of Continental Europe (*modularis*).
VOICE Alarm-call (also at times as contact) a rather strong, uninflected pipe with discordant tone, 'tiih'. Call on migration, when often heard at first light or in late evening, uttered especially in flight, a ringing or shivering, very thin 'tihihihi'. Song, usually given from top of bush or medium-sized conifer, clear and quite loud, an irresolute shuttling or patchwork of sounds *c.* 2 sec. long at quite even pitch, '**tü**telli**ti**telleti**tü**tell**ü**to**te**litell**e**ti' or similar.

Alpine Accentor *Prunella collaris* V***
L 15–17½ cm. Breeds in high alpine country, usually at 1800–3000 m. Found mostly in barren areas with boulders and low-growing plants. Rather sociable, and small groups can be seen even during breeding season. Largely resident. Food mostly insects and seeds. Nests in rock crevice.
IDENTIFICATION When walking in the right terrain and one or more Skylark-sized, compact and rather dark birds fly up, the alternatives are few: Alpine Accentor, Horned Lark, Water Pipit or some bunting. Of these, Alpine Accentor is the strongest flyer, fast and in shallow dips almost like a small thrush. If seen perched, it shows *light yellow patch at bill-base*, a *dark panel across wing* formed by the greater coverts and, in good light, *heavy red-brown flank streaking*. At really close range, some pale spotting is visible on throat-centre and fine white tips on wing-coverts. – Variation: Birds in Turkey (ssp. *montana*) are somewhat more pale brown and less heavily streaked above.
VOICE On rising, gives muffled lark-like rolling calls, e.g. 'drrü drrü, drrip', 'tschirr'. May also utter stifled click like

a chat. Song a drawn-out, creaking and trilling phrase with no clear structure: some bubbling notes recall Crested Tit, others are squeaky. Generally delivered from perch, occasionally in fluttering song-flight.

Siberian Accentor *Prunella montanella* —
L 13–14½ cm. Breeds in N Siberia and on both sides of N Urals in willows and birch forest bordering bogs, tundra and rivers, also in open, impoverished conifer forest. Migratory, winters in SE Asia. Rare vagrant in W Europe. Shy, keeps mostly in cover.
IDENTIFICATION Roughly the size of Dunnock. Characterized by *broad and long buff supercilium* (widest behind eye) accentuated by *brownish-black crown* (darkest at sides) and *dark cheeks. Entire underside yellowish rusty-buff* (some with intimation of dark feather centres on breast). Above, varies somewhat in colour: *usually distinctly reddish-brown* with darker red-brown streaks, but sometimes more cold dark brown with mere hint of red-brown tinge and heavily black-spotted (then more like Black-throated Accentor).
VOICE Call a fine 'ti-ti-ti'. Song a rather monotonous Dunnock-like verse, high-pitched and with quite *hard r-sound*; minimal range on scale.

Radde's Accentor *Prunella ocularis* —
L 13–14 cm. Breeds in high alpine habitats at 2000–3000 m, in boulder and shrub terrain only just below treeline. Resident or short-distance migrant. Lively and restless, scampers on ground, flicks wings, flies between low bushes, rarely stays in one spot for long.
IDENTIFICATION As Dunnock in size. Seen briefly, plumage most like Siberian Accentor, with *broad, pale supercilium, dark crown* and dark cheeks, and warm buff underparts. Differs from Siberian Accentor in that supercilium is dirty white or creamy (not saturated yellow-buff), and buff colour below is *tinged orange and concentrated on breast*, with some contrast against whiter belly (as on Black-throated Accentor) and throat. At close range, throat-sides show fine dark spots.
VOICE Call Dunnock-like, a fine 'ti-ti-ti'. Song *crystal-clear* in tone, a fast verse in which some notes are repeated several times and short *high trills* are interwoven.

Black-throated Accentor *Prunella atrogularis* —
L 13–14½ cm. Breeds in spruce thickets and bushy country, also in deciduous forest, often on upper slopes (Urals). (In C Asia, also found in tall conifer forest in mountain valleys.) Very rare vagrant in W Europe, mainly in autumn.
IDENTIFICATION Like Siberian and Radde's Accentors, with *brown-black crown* and *broad buff supercilium*. Distinguished by *black throat*, easy to see in breeding plumage but harder with autumn immatures, when broad pale fringes conceal some (rarely, on 1st-winter ♀, almost all) of the dark colour. Note that *back is dull dark brown* (not strongly tinged reddish-brown) with dark spotting, also that *warm orange-buff breast contrasts with whiter belly*. Breast has some dark feather centres.
VOICE Call like that of other smaller accentors, a fine 'ti-ti-ti'. Song Dunnock-like and shuttling with little change in pitch, though verses somewhat shorter and volume a bit weaker; contains fewer clear notes than Dunnock's.

DUNNOCK

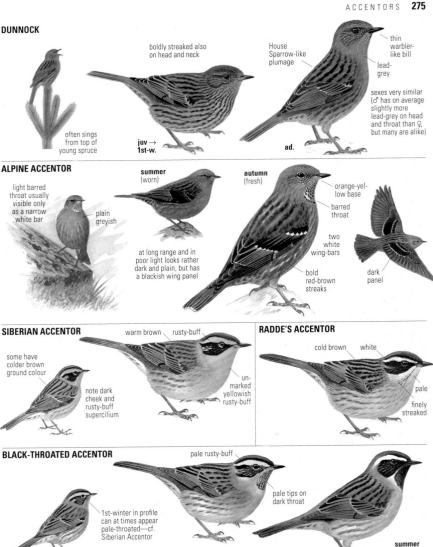

often sings from top of young spruce

boldly streaked also on head and neck

**juv →
1st-w.**

House Sparrow-like plumage

thin warbler-like bill

lead-grey

sexes very similar (♂ has on average slightly more lead-grey on head and throat than ♀, but many are alike)

ad.

ALPINE ACCENTOR

light barred throat usually visible only as a narrow white bar

plain greyish

summer (worn)

autumn (fresh)

orange-yellow base

barred throat

two white wing-bars

at long range and in poor light looks rather dark and plain, but has a blackish wing panel

bold red-brown streaks

dark panel

SIBERIAN ACCENTOR

warm brown   rusty-buff

some have colder brown ground colour

note dark cheek and rusty-buff supercilium

unmarked yellowish rusty-buff

RADDE'S ACCENTOR

cold brown   white

pale

finely streaked

BLACK-THROATED ACCENTOR

pale rusty-buff

1st-winter in profile can at times appear pale-throated—cf. Siberian Accentor

pale tips on dark throat

autumn

summer

DZ

Dunnock

Alpine Accentor

Radde's Accentor

Black-throated Accentor

THRUSHES *Turdidae*

Large family of small to medium-sized passerines with well-developed songs. Juvenile plumage in most cases spotted. The true thrushes (pp. 294–301) are bigger, with strong legs and bill; 18 species. The smaller thrushes (chats) comprise rather dissimilar, often brightly coloured species, such as nightingales, redstarts, wheatears, rock thrushes; 34 species.

(European) **Robin** *Erithacus rubecula*   rB1 / W

L 12½–14 cm. Breeds in woodland, gardens, parks, forest edge, generally with some dense vegetation and open areas; in N Europe prefers spruce forest with some deciduous element (e.g. birch), also mixed forest. Migratory in N Europe. Wary but by no means shy. Food insects, snails, worms. Nests in hollow stump, bank, crevice etc.

IDENTIFICATION Small and brownish with *rusty-red 'bib'* (covers entire breast and face). *Narrow yellow wing-bar* (most obvious on juveniles). Often looks rather big-headed. Has *thin and rather long legs*. At times ruffles up feathers so that it looks rounded and compact, but more often looks rather slim. Droops wings, hops rapidly on ground with feet together, curtseys and cocks short tail. Perches motionless, makes a sudden movement, stands still again and so on. – Juvenile: Whole plumage finely spotted. Lacks red bib.

VOICE Call a short, hard 'tick'. When nervous, not least at roosting time and 'reveille', the ticking calls are protracted into long rapid series, 'tick-ick-ick-ick-...', sounding like a watch being wound up. Alarm an extremely thin and sharp, ventriloquial 'tsiiih'. On nocturnal migration a thin, hoarse 'tsi'. Song often begins with a few high, drawn-out, thin notes, then the verse drops in pitch and speeds up with fast runs of trembling and excited clear squeaky notes; tempo and volume vary, no two verses are the same.

(Common) **Nightingale** *Luscinia megarhynchos* mB4

L 15–16½ cm. This is the 'real' nightingale, with the more beautiful song, which has found a place in literature and in the minds of everyday people. Breeds in woods and groves with rich undergrowth, often by water but also in drier habitats with dense shrubbery; at times in gardens and orchards. Summer visitor (mainly Apr–Sep), winters in S Africa.

IDENTIFICATION More often heard than seen, and when seen its appearance is, for many, surprisingly plain: *brown above* with *rusty-red tail and rump*, grey-buff below with diffusely paler throat. The *rather big black eye* is set off by an *indistinct whitish eye-ring*. Very like Thrush Nightingale, and sometimes inseparable in field, but at close range a shade redder on uppertail and *lacks obvious grey vermiculation on lower throat/breast* (but can have light grey shading). – Variation: Birds in Central Asia (*golzii*) are

paler, longer-tailed and have a hint of a pale supercilium.

VOICE Has no conspicuous calls. Alarm a whistled, slightly upslurred 'üihp', as well as a creaking 'errrr'. Song powerful and melodic, varied throughout, phrases rather short (2–4 sec.) with often equal-length pauses in between, consisting of trilling sounds, fluted whistles and rippling or gurgling notes; the song is best recognized by recurring whistles in crescendo, 'lu lu lü lü li li'.

Thrush Nightingale *Luscinia luscinia*   V**

L 15–17 cm. The nightingale of N and E Europe, a less vocally pleasing substitute for the 'real' nightingale of W and S Europe. Breeds in thick, damp, shady deciduous forest, often in hazel stands, shrubbery and waterside scrub; also in thickly wooded parks, gardens. Summer visitor, winters in Africa. Vagrant in Britain (Apr–Jun, Aug–Oct).

IDENTIFICATION Confusingly like Nightingale. On average *somewhat less reddish-brown on tail and rump*, a shade greyer brown on back, and as a rule has rather distinct *grey vermiculation on lower throat/breast* (but none of these features is absolute or easy to judge in field).

VOICE Lacks conspicuous calls. Alarm a piercing, ventriloquial, straight whistle, 'ihp' (like alarm of Collared Flycatcher), also a creaking hard 'errrr'. Song more loud than beautiful (still remarkable!), series of tongue-clicking and gurgling notes, the verses often introduced by a few pensive sharp whistles. Audible for kilometres on still nights.

Rufous Bush Robin *Cercotrichas galactotes*   V***

L 15–17 cm. Breeds in dry, rather open country with dense shrubbery, hedges, fruit orchards; fond of tamarisk, prickly-pear cactus, pistachio trees; often near human habitation. Summer visitor (mid May–Sep), winters S of Sahara.

IDENTIFICATION Brownish above and dirty white below, with *long red-brown tail with black and white tips* to outer feathers. Bold head markings, with *white supercilium* and *narrow dark eye-stripe* plus often *dark lateral crown-stripe*. Spends much time on ground. Often spreads tail and jerks it upwards, then closes and slowly lowers it. Also flicks and opens wings. Sexes and ages alike. – Variation: In Spain, NW Africa, Egypt and Israel (ssp. *galactotes*) the back is warm brown and contrasts little with tail, and breast is buff, whereas birds in Greece and Turkey (*syriaca*) have slightly more grey-brown back, grey tone to breast and flanks and more contrasted head markings.

VOICE Call a hard tongue-clicking and a soft straight whistle, 'üh'; also a peculiar, insect-like buzzing sound, 'bzzzzz'. Song thrush-like and clear with rather melancholy tone, may be likened to lethargic song of Song Thrush (short phrases, short pauses), but higher and more chirping.

Robin

Nightingale

Thrush Nightingale

Rufous Bush Robin

ROBIN

brown

spotted

juv.

ad.

NIGHTINGALE

often sings hidden in cover

compare ♀ Redstart (right) in flight

pale eye-ring

warm brown

plain red-brown

rather bright red-brown above

pale rust-red with dark centre

long

hint of pale super-cilium

usually plain sandy-buff

unspotted rusty-buff

1st-winter

golzii (Central Asia)

ad.

THRUSH NIGHTINGALE

dull grey-brown, but some are warmer brown, more like Nightingale

usually 7 primary tips visible

NIGHTINGALE

hint of a stripe

darker brown with reddish tinge

long 1st primary

breast variably diffusely mottled grey-brown

usually 8 primary tips visible

THRUSH NIGHTINGALE

pale buff-white, often spotted dark

1st-winter

short 1st primary

ad.

RUFOUS BUSH ROBIN

often jerks tail upwards

black and white spots visible when tail is spread

strong 'lace' pattern

grey-brown

rufous-buff

tail spots not obvious (from above) when tail is closed

often perches in prickly-pear cactus

syriaca (SE Europe, Turkey, N Syria)

galactotes (SW Europe, N Africa)

KM

Bluethroat *Luscinia svecica* **P**4

L 13–14 cm. Breeds in N Europe among willows and soggy upland birch forest, in rest of Europe in swampland by fens and rivers overgrown with bushes, reeds, alder etc. Summer visitor, winters mainly from NE Africa to W India. Passage migrant in Britain (Mar–May, Aug–Oct). Food insects. Nests in tussock or low in dense willow bush.

IDENTIFICATION A rather small and slim bird with long thin legs, in all plumages told by *obvious whitish supercilium* and *rusty-red patch on sides of tail-base* (often seen in flight, or when perched bird flicks tail up). – ♂: *Bright blue bib*, bordered below by a narrow black and white band and a broader rusty-red one. Inlaid in centre of the blue, birds in chiefly N Europe (ssp. *svecica*) have a *rusty throat patch*, those in rest of Europe, E Turkey and Caucasus (incl. *cyanecula*, *magna*) a *small white patch* or none at all. In autumn, some of the bright throat colours are replaced by yellowish-white areas. – ♀: Variable throat markings; usually just an arc of black spots on creamy-white ground, older birds occasionally with some blue, plus black and rusty border across breast (those with most blue can look quite ♂-like). – Juvenile: Finely spotted like juvenile Robin, but note tail markings.

VOICE Commonest call a dry, throaty clicking 'track' (like flag halyard flapping in the wind). When agitated, also a wheatear-like whistled 'hiit' (and at times a Redstart imitation, **'hu**it'). In autumn a hoarse, cracked 'bzrü' may also be heard. Song powerful and clear, often starts with a much-repeated, loud, metallic 'zrü', or a polysyllabic 'zri-zri-**zrütt**' or the like, which slowly speeds up and suddenly turns into a cascade of melodious or hard and squeaky notes, often mixed with good imitations of other species (even reindeer bells may be mimicked!).

Red-flanked Bluetail *Tarsiger cyanurus* **V*****

L 13–14 cm. Breeds in taiga, mainly in rolling, upland terrain in undisturbed, damp, mossy spruce forest with some birch. Rare breeder in E Finland (arrives from late May). Long-distance migrant, winters in SE Asia. Rare vagrant in W Europe. Rather shy. Food insects. Nests in hollow trunk or stump, in bank among roots etc.

IDENTIFICATION Seen well in good light, adult ♂ is a beautiful bird with *dull blue upperparts* (only rump, tail-base, wing-bend and crown-side are brighter blue), *orange flanks* and white underparts. Wing blue or (usually) olive-tinged grey-brown. ♀, 1st-summer ♂ and autumn immatures more modest in olive-grey and off-white and can easily be overlooked, are blue only on uppertail, but this normally looks just dark in field. Instead, note *orange patch on flanks*, dusky breast and *grey cheeks* framing *narrow whitish bib*, also *whitish eye-ring*. Often flicks wings and tail.

VOICE Alarm-calls somewhat recall Black Redstart's: an often repeated, whistling, straight 'viht', and a muffled, hard, slightly throaty 'track'. Song, often delivered from treetop at very first light, a rather constant, fast, short, clear, melancholy verse, not unlike Redstart's but deeper and clearer, 'itrü-**chürr**-tre-tre-tru-trurr'.

White-throated Robin *Irania gutturalis* **V*****

L 16½–18 cm. Breeds on dry, rocky slopes, often above 1000 m, with dense bushes and scattered trees. Summer visitor (mostly end Apr–Aug), winters in E Africa. Very rare vagrant in N Europe. Rather shy, but sings from open perch, and occasionally in gliding song-flight. Food mostly insects. Nests in bush or low down in thick-foliaged tree.

IDENTIFICATION Rather big with slimness of a thrush, with long neck, long legs and *long, heavy bill*, and *long black* (♂) or *dark grey* (♀) *tail* which is often raised. – Adult ♂: Unmistakable, with *orange breast*, *black head-sides* framing *narrow white bib*, with *white supercilium* and *lead-grey upperparts*. – ♀ / 1st-winter ♂: Grey-sullied buff-white below (breast weakly vermiculated grey) with *pale orange flanks*. Pale eye-ring and forehead. Brownish-grey above.

VOICE Call disyllabic, 'chi-**litt**', not that unlike Pied Wagtail. When agitated, a clicking 'check'. Alarm also a creaking 'churrr'. Song *very fast and twittering*, full of sharp whistles and hard, creaky calls, often given at such high speed that the notes seem to stick in the throat; can recall an ecstatically singing *Sylvia* warbler or shrike.

Siberian Rubythroat *Luscinia calliope* **V*****

L 14½–16 cm. Breeds in mixed coniferous forest with rich element of birch, willow and undergrowth, often in damp spots; at times also in well-wooded parks. Summer visitor (mainly late Apr–early Sep), winters in SE Asia. Shy, tends to keep well concealed. Nests low in bush or in tussock.

IDENTIFICATION Like a cross between Bluethroat and Nightingale, long-legged but rather short-tailed. In all plumages *dark lores* and *pale supercilium*. – Adult ♂: *Red bib* narrowly edged black; *submoustachial stripe white*; *lores black*. – ♀ / 1st-winter ♂: More diffuse head markings (*grey lores*; *smudgy submoustachial stripe*) and either partly light red bib (older ♀) or off-white chin without black frame.

VOICE When nervous, a whistling, disyllabic '**ii**-lü' and a tongue-clicking 'chack' (as Fieldfare but weaker); sometimes also a muffled creaking 'arrr'. Song rippling, *chatty and calm*, like a Garden Warbler if anything, but recognized by sprinkling of harder, lower and more strained notes (voice like Finsch's Wheatear) as well as higher whistles (thus rather wide amplitude); some individuals interweave skilful imitations of a wide range of species.

Bluethroat

Red-flanked Bluetail

White-throated Robin

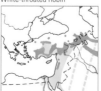

Siberian Rubythroat

BLUETHROAT

1st-w. ♂

ad. ♀ (bird with blue)

(variation)

ad. ♂ *cyanecula* (S and C Europe)

prom- inent

red

singing from willow

many ad. ♀♀ are quite similar to 1st-winter ♀

ad. ♂ *svecica* (N Europe)

rusty-red tail-base

1st-winter ♀

RED-FLANKED BLUETAIL

orange

neat whitish throat patch

1st-winter / ♀

grey-blue

grey-brown

blue shoulder

ad. ♂ autumn

blue

often jerks tail downwards

ad. ♂

prefers ancient spruce forests in hilly taiga

WHITE-THROATED ROBIN

grey-black

ad. ♂ (variation)

orange- buff

long

narrow white throat patch

deep orange- brown

orange

long-legged

black

1st-winter / ♀

ad. ♂

SIBERIAN RUBYTHROAT

rather compact, long-legged and short-tailed

ad. ♀ (variant with red throat)

buff-white diffuse edge

striking head pattern

warm brown tinge

ad. ♀ (typical) / 1st-winter

ad. ♂

plain brown

DZ

(Common) Redstart *Phoenicurus phoenicurus* mB3 / P2

L 13–14½ cm. Breeds in woodland, often old deciduous and mixed, in gardens and parks, but in N (incl. Scotland) also in older, derelict pine forest. Passage migrants in coastal scrub. Summer visitor (Apr– Sep), winters in Africa. Rather unobtrusive. Keeps mainly to trees and bushes, with mostly only fleeting visits to ground. Nests in tree hole or nestbox.

IDENTIFICATION Rather slim. Usually perches fairly erect and *vibrates tail* so that the *rusty-red colour* is exposed. – Adult ♂: *Black throat, orange-red breast, white forehead, ash-grey crown and back*. In autumn the bright colours are partially concealed by buff fringes. – 1st-winter ♂: Rather ♀-like, but has greyish-black elements on throat, orange-mottled breast and sometimes a little white showing on forehead. – ♀: Grey-brown above and buff-white below. Like ♀ Black Redstart, but is *paler* and *less grey in tone*, and has distinctly *paler throat, lower breast and central belly*. Rarely more ♂-like with orangey breast. – Variation: ♂ in E Turkey and Caucasus (ssp. *samamisicus*) is subtly greyer and darker above, with white wing patch (variable in size), and ♀, too, often has hint of paleness on secondaries.

VOICE Call a soft whistled, upslurred 'huit' (in *samamisicus* a straight 'ih' like Collared Flycatcher), often followed by clicking 'tick-tick-tick'. Song from very first light, a short, soft, melancholy verse at high pitch, often with standard beginning but with different ending in terms of details, e.g. 'sih **trüi-trüi-trüi** si-si pli si**veuy** si'.

Black Redstart *Phoenicurus ochruros* mB5 / P4 / W5

L 13–14½ cm. Breeds both in uplands, on slopes with boulders, cliffs and scattered bushes, and at lower levels in towns, docklands and industrial areas, where the natural mountains are substituted by church towers, chimneys and factory buildings. Very local in Britain, in lowlands, also passage / winter visitors on coasts. Nests in recess or hole in building.

IDENTIFICATION Size and shape as Redstart, and has similar behaviour: perches upright, constantly vibrates its *rusty-red tail*, is rather shy and nervous, often flits about. – Adult ♂ (Europe): *Greyish-black* with *bold white wing patch*. – 1st-summer ♂: Lacks white wing patch (some have at most hint of narrow stripe). Colour of body varies, many as grey-brown as ♀, some grey-black like adult ♂. – ♀/1st-winter: *Sooty brown-grey* on head and body, clearly darker than ♀ Redstart (dark also on throat, lower breast and belly, where Redstart obviously paler). – Variation: ♂ in greater part of Europe (ssp. *gibraltariensis*) has grey upperparts, but in Iberia ('*aterrimus*') back is often blacker; ♂ in Turkey and Caucasus (*ochruros*) usually has some rusty-red on belly; ♂ in Middle East (*semirufus*) is quite distinct, has entire belly and lower breast rust-red with sharply demarcated black bib, has black back and lacks white wing patch.

VOICE Call a straight, slightly sharp whistle, 'vist', often repeated impatiently. When highly agitated, a discreet clicking is added, 'vist, tk-tk-tk'. Song loud, frequently given at first light from high perch, usually consists of four parts: starts with a few whistles and a rattling repetition of same note, followed by a pause *c.* 2 sec. long, then a peculiar crackling sound (not very far-carrying), after which the verse terminates with some brief whistled notes, e.g. 'si-srü **till-ill-ill-ill-ill**....... (krschkrschkrsch) **srü**visvi'; the sequence of the four components may sometimes be switched around.

Moussier's Redstart *Phoenicurus moussieri* [V***]

L 12–13 cm. Breeds in open country with bushes, scattered trees and small woods, often on mountain slopes; also in open, bushy coastal forest, in cultivations and gardens. Predominantly resident. Rather bold. Nests in hole or recess.

IDENTIFICATION Small and compact, *tail and wings fairly short*. Constantly vibrates *rusty-red tail*, flicks wings, often perches in open. – Adult ♂: Unmistakable, with *entirely rusty-red underparts* and rump, rest of *upperparts black* with *large white wing patch* and *white forehead-band that runs back over eye and well down neck-side*. In autumn plumage, black areas have brown fringes. – ♀: Like a short-tailed ♀ Redstart with *more saturated orange below* and with *a faint pale panel on wing* (where ♂ has white wing patch). – 1st-winter ♂: Like adult ♂, but black areas are brown-grey (broadly fringed pale) and remiges are browner.

VOICE Call a thin whistle, 'hit', and a soft 'chirrr'. Song a thin twitter, shuttling and slightly irregular in composition, roughly like Dunnock, squeaky notes with occasional short 'chirr' notes inserted; verses 2–5 sec. long.

Güldenstädt's Redstart *Phoenicurus erythrogastrus* —

L 15–16½ cm. Breeds in high alpine terrain on bare, rocky slopes with scant vegetation right up to the summer snowline, from 2200 m (more often above 3000 m) to *c.* 5000 m. Winters in valleys down to 1500 m (at times 900 m).

IDENTIFICATION *Quite big* (but size not always easy to determine at distance in alpine habitat!), sturdy and broad-chested with large head. Often perches openly on jutting rock. Restless, often changes perch. Flight strong, almost thrush-like. – Adult ♂: Unmistakable: *white crown / nape* (in fresh plumage tinged greyish-yellow), *very big white wing patch, dark rust-red underparts*, and *black bib*. (Note that local race of Redstart has white wing patch, but that Black Redstart lacks this.) – ♀: Like an oversized, pale ♀ Black Redstart but with *hint of small pale wing patch* at bases of flight-feathers.

VOICE Call 'tsi' or 'tsi tek-tek'. Song a brief, clear Rock Thrush-like verse with hard sounds interwoven.

Redstart

Black Redstart

Moussier's Redstart

Güldenstädt's Redstart

REDSTART
phoenicurus
(Europe)

smudgy cheek

♂♂ characters subdued by pale fringes

samamisicus
(Caucasus, Turkey)

many ♀♀ are duller with just a hint of buff on the breast

hint of a pale panel

1st-w. ♂

♀

ad. ♂

white panel

red

♀

ad. ♂

BLACK REDSTART
gibraltariensis
(Europe)

southeastern subspecies:

brown, worn wing

1st-summer ♂
(advanced)

ad. ♂
ochruros
(Turkey)

most 1st-summer ♂♂ wear a plumage identical to ♀

♀ / imm. / 1st-s. ♂

ad. ♂

mouse-grey

white panel

ad. ♂
semirufus
(Middle East)

MOUSSIER'S REDSTART

in habits and habitat a mixture between a chat and a redstart

white tiara

short tail

♀

rich rufous

1st-w.
♂

ad. ♂
spring

large white wing patch

GÜLDENSTÄDT'S REDSTART

drab and featureless

dark rufous with no dark centre

♀

ad. ♂
spring

high-alpine species

large white wing patch

DZ

WHEATEARS Oenanthe

A rather homogeneous group of smaller thrushes, all members of the genus *Oenanthe*. Ground-dwellers, preferring open terrain with grass fields and rock outcrops, rarely perching in trees or taller bushes. Insect eaters, hole nesters. All have characteristic black-and-white tail pattern, sometimes also rufous, often with inverted black 'T' on white ground.

(Northern) **Wheatear** Oenanthe oenanthe   m**B**2 / **P**2

L 14–16½ cm. Breeds in open, stony country with meadowland, often on moorland, coastal grassland, pasture and farmland with stone walls, downland, locally coastal shingle. In S Europe mostly high up in alpine country. Summer visitor to Britain & Ireland (mainly Mar–Oct), winters in Africa; Greenland and Canadian breeders (ssp. *leucorhoa*) also winter in Africa, are thus among the world's real long-distance migrants (moreover cross wide oceans, estimated to involve 2400 km non-stop flights in 30 hours). Food insects. Nests in hole in rock crevice, cairn, stone wall, rabbit burrow, roof, etc.

IDENTIFICATION *Tail black and white* in all plumages (black *T-pattern* at tip, white base, slightly variable, width of black end-band tending to be narrower in S Europe, invariably black protruding on central tail-feathers as long as width of black end-band, or longer). *Upperparts grey or grey-brown.* White or *pale buff supercilium.* – Adult ♂ spring–summer: *Ash-grey crown and upperparts,* white supercilium, black eye-mask, black wings, and pinkish-buff colour on throat and breast (can fade to almost white in late summer, then can vaguely recall pale-throated morph Black-eared Wheatear, which see). – ♀/autumn: Grey-brown above and buff below with darker wings and ear-coverts, lighter buff supercilium. Adult ♂ in autumn identified by black wings (with narrow pale fringes), black lores and some black on ear-coverts. ♀ and 1st-winter similar, have dark brown wings, lack black on head; paler birds confusable with Isabelline Wheatear (which see). – Variation: In Greenland and Iceland (*leucorhoa*), breeders are slightly larger and somewhat darker rufous in fresh plumage. However, nominate birds from Fenno-Scandian and Russian tundras are similarly saturated in early spring or autumn, and safe identification of genuine *leucorhoa* generally involves trapping and measuring.

VOICE Call a straight whistle, like indrawn 'hiit', as well as a tongue-clicking 'chack'. Song, often delivered from elevated perch (top of a rock, wire or the like) but sometimes in short song-flight, an explosive, fast, hard, chirpy and crackling verse with interwoven whistling 'hiit', varying in details and hard to transcribe; sometimes contains one or two imitations of other birds. Often sings at first light and just before, but also during day.

Seebohm's Wheatear Oenanthe seebohmi   —

L 14–16 cm. Often treated as a race of Wheatear, but due to distinct ♂ plumage and allopatric range here treated as a separate species. Breeds on grassy, stony slopes of Atlas range and other mountains in NW Africa, generally at 1500–2350 m. Summer visitor (usually Mar–Oct), short-distance migrant wintering in SW Mauritania and rest of W Africa. Nests in hole in ground or among rocks.

IDENTIFICATION Like a Wheatear but adult ♂ summer has *black throat, black underwing-coverts* and on average *more white on forehead.* Black end-band on tail rather narrow, like Wheatear in S Europe. ♀♀ and immatures are often inseparable from Wheatear unless showing hint of mottled grey throat (rare though).

VOICE Similar to Wheatear, but insufficiently studied. Call 'heet', a bit like Horned Lark. On some recordings the song sounds weaker and less scratchy than Wheatear, more softly warbling.

Isabelline Wheatear Oenanthe isabellina   **V*****

L 15–16½ cm. Breeds on short-grass plains or slopes in warm, dry climates, e.g. at border between natural steppe and semi-desert, often with scattered boulders. Frequently on slopes at foot of mountain ranges. Migratory, wintering in wide sub-Saharan zone. Insect eater. Nests usually in ground hole (e.g. rodent burrow).

IDENTIFICATION Slightly larger on average than Wheatear, with proportionately a little shorter wings and tail, often adopting a *more upright stance with vertical legs and more stretched neck*, although some variation and overlap in these respects between the two species. Like autumn-plumaged Wheatear, and the two at times very tricky to separate, but apart from general characters given above differs in: *broader black terminal tail-band* (intermediate between Wheatear and Desert Wheatear, tail-band broader than length of black protruding basally on central tail-feathers); *paler wings* with broader buff fringes, so that *wing-coverts appear closer in tone to upperparts* than to dark brown primaries, and in flight against dark background wing looks paler; often, *alula stands out as much blacker than paler brown primary-coverts and greater coverts* (less contrast in Wheatear due to rather dark centres to all wing-coverts); *supercilium whitest before eye, buff and narrower behind* (whitest and broadest over and behind eye on Wheatear, thus the opposite pattern); on average *paler brown ear-coverts*; slightly longer tarsi. Sexes similar, but ♂ has on average blacker lores.

VOICE Call a fairly sharp 'chip', sometimes with hint of downslur, 'chiü'; also a stifled clicking 'chack'. Song distinctive, drawn out (verses often 10–15 sec., pauses brief), 'chatty', and including diagnostic 'Harpo Marx whistles' (fast series of short wolf-whistles), 'vi-vi-vi-vi-vü-vü-vü-vü-vuy-vuy-vuy-...', hard, crunchy sounds and also more or less good imitations.

Wheatear

Seebohm's Wheatear

Isabelline Wheatear

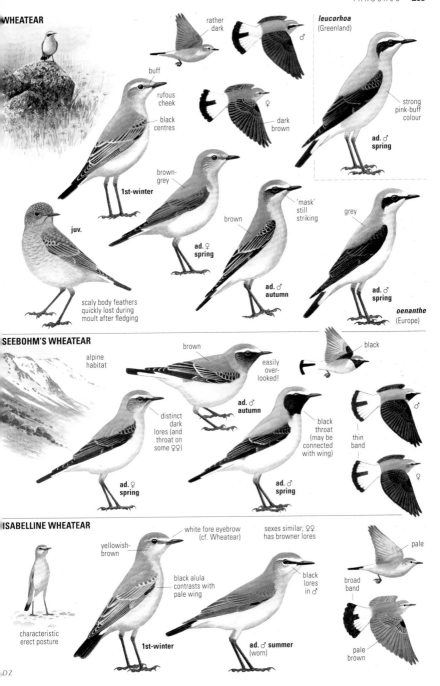

WHEATEAR

rather dark

buff

rufous cheek

black centres

leucorhoa (Greenland)

strong pink-buff colour

ad. ♂ spring

♂

♀

dark brown

1st-winter

brown-grey

'mask' still striking

brown

grey

juv.

scaly body feathers quickly lost during moult after fledging

ad. ♀ spring

ad. ♂ autumn

ad. ♂ spring

oenanthe (Europe)

SEEBOHM'S WHEATEAR

alpine habitat

brown

easily over-looked!

black

ad. ♂ autumn

♂

distinct dark lores (and throat on some ♀♀)

black throat (may be connected with wing)

thin band

♀

ad. ♀ spring

ad. ♂ spring

ISABELLINE WHEATEAR

white fore eyebrow (cf. Wheatear)

sexes similar; ♀♀ has browner lores

pale

yellowish-brown

black alula contrasts with pale wing

black lores in ♂

broad band

characteristic erect posture

1st-winter

ad. ♂ summer (worn)

pale brown

DZ

Black-eared Wheatear *Oenanthe hispanica* **V****

L 13½–15½ cm. Breeds in open country with scattered bushes, trees and rocks, usually below 600 m but locally higher; often in barren pastureland, along streamsides, on stony slopes. Migrant, winters S of Sahara. Food insects. Nests under boulder or in ground hole or thick tussock.

IDENTIFICATION Rather small, slim and long-tailed. Often adopts crouched posture. As Pied Wheatear (but unlike Northern, Isabelline, Finch's, Mourning and others), does not have uniform-width terminal tail-band but has *more black along outer edges of tail* and *less black in centre of each half of tail* (on some, black terminal band is fully broken by white). Two morphs in both sexes, one pale-throated (commonest in W) and one dark-throated (dominates in E). – ♂: *Mantle/back white*, tinged yellow-ochre in fresh plumage. Black of head/throat does not meet black of wing (cf. Finch's Wheatear). – ♀: Like ♀ Pied Wheatear, but is *warmer yellowish grey-brown above* and *more uniformly yellow-orange on breast*. – Variation: Western race (*hispanica*) ♂ has less extensive eye-mask or bib, also ♂♀ in fresh plumage are more yellow-ochre on mantle and breast; rather short primary projection. Eastern, from S Italy and Balkans eastward (*melanoleuca*), has more extensive black on head of ♂, which is more black and white in summer, and ♀ is more a dull greyish yellow-brown above (and thus more like ♀ Pied); long primary projection (± as Pied).

VOICE Call usually a broken, almost hissing 'brsche', sometimes also a more Wheatear-like clicking 'tshack'. Alarm a descending whistle, 'viü'. Song a short, chirping and dry, rather explosive verse (can recall song of Short-toed Lark, or speeded-up Whitethroat), e.g. 'chuchürrche-chuchirr-tri'. Often a few syllables or entire verse replaced by good mimicry of e.g. Barn Swallow or Red-rumped Swallow, Linnet or Goldfinch. Song is rather variable, and doubtful whether it can be safely distinguished from that of Pied. Also has a chattering subsong with harsh and hoarse sounds mixed with chirpy and clearer whistled notes and mimicry.

Pied Wheatear *Oenanthe pleschanka* **V****

L 14–16½ cm. Breeds on barren mountain slopes with low vegetation, high plateaux with boulder-strewn grassland, at times adjoining grazed or cultivated but bare areas; generally ascends higher and selects more precipitous habitats than Black-eared Wheatear. Migrant, winters mostly in E Africa; main passage thought to pass E and S of Jordan. Nervous and restless, making constant movements.

IDENTIFICATION Like Black-eared Wheatear in size, shape and tail pattern (i.e. *black terminal band* not uniform in width, but *broader along tail-sides and narrower in centre of each half of tail*), and some ♀ and autumn plumages can be

extremely difficult to separate. – ♂ summer: *Black mantle patch, meets black throat at sides. Wings dark in flight* (ligh on Mourning Wheatear). *Breast buff in fresh plumage, un dertail-coverts light yellowish-white* (Mourning always pur white on breast, rusty on undertail-coverts). Adult ha blacker wings, 1st-summer browner; in addition, 1st-summer is not so perfectly black and white, retains traces of winter plumage. (A pale-throated morph, '*vittata*', occur very rarely.) – ♂ winter: Since black mantle has broad grey brown fringes, broader on 1st-winter than on adult, simila rity to eastern Black-eared Wheatear arises. Is, however *colder brown-grey above* with *more distinct pale feather tips* (scaly). – ♀: Very like eastern Black-eared Wheatear bu *darker and colder earth-brown above*, in fresh plumage *pale scaled*, and is *pinkish rusty-brown or pinkish-grey on breast* often with *irregular lower border* against pale belly and with tendency towards *diffuse mottling* (♀ Black-eared more evenly yellow-orange on breast with more regular lower border and no mottling). *Throat* in spring variably *dark,* most showing hint of this also in autumn (but some then look rather pale-throated); dark of throat extends a bit further towards breast than on Black-eared Wheatear (but when head sunk between shoulders this becomes hard to judge).

VOICE Strongly reminiscent of Black-eared Wheatear's, and the two species seem not easy to distinguish by voice. Main call a cracked, frizzling 'brsche', at times also throaty clicking 'tshak' notes. Alarm a whistling 'vih(e)'. Song verses are often somewhat shorter, and intervening pauses longer, than Black-eared's, but song is similarly explosive, twittering and dry, e.g. 'surtu-**shirr**-echu'; some variation in intonation, can at times resemble Whinchat's song. Verses occasionally protracted and some mimicry included.

Cyprus Wheatear *Oenanthe cypriaca* —

L 14–15 cm. Formerly considered an isolated race of Pied Wheatear, now usually regarded as a full species. Breeds only in Cyprus, where common mid Mar–early Oct in barren grassy and rocky areas at all levels, often in and around villages; often perches in treetops, especially when singing. Winters in S Sudan and Ethiopia, stops off on passage in e.g. Israel and Egypt.

IDENTIFICATION Like Pied Wheatear, but differs in: *slightly shorter wings* (shorter primary projection) *and tail*, looks compact and large-headed; *broader black terminal tail-band*; *less white on nape*, not continuing onto mantle (does on Pied); *smaller white patch on tail-base* above; *underparts more saturated yellow-ochre* in autumn, and breast never fades to fully whitish in summer. Sexes similar, i.e. ♀ always has *dark* (grey-black or dark brown-grey) *throat and back*, and has *some white on nape*; *crown* in autumn *sooty-grey* on ♀ and 1st-winter ♂, but some white usually visible on crown-sides (can create white supercilium), wearing to more white in summer. 1st-summer ♂ has contrast between brown-grey remiges and black back/rest of wing.

VOICE Calls a frizzling 'brzü', a hoarse, hard 'tschick', an upslurred whistle, 'jüi' (occasionally downslurred 'jiü'). Song totally diagnostic among the wheatears, a series of similar cracked, hoarse notes, 'bizz-bizz-bizz-bizz-bizz...', completely lacking dry twittering and varied character of its relatives; verses usually 3–10 sec. long.

Black-eared Wheatear

Pied Wheatear

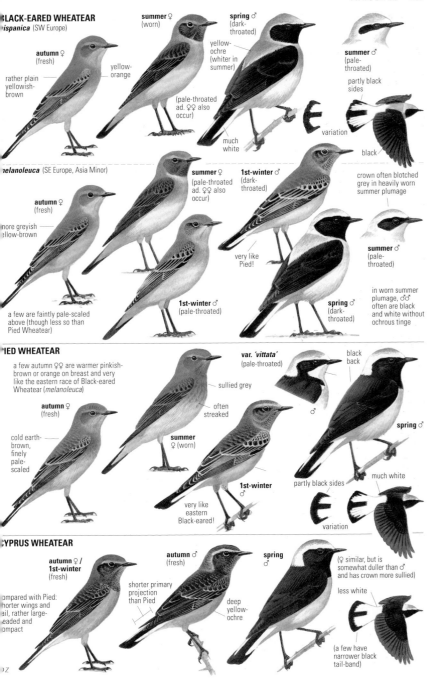

BLACK-EARED WHEATEAR
hispanica (SW Europe)

autumn ♀
(fresh)

rather plain
yellowish-
brown

yellow-
orange

summer ♀
(worn)

spring ♂
(dark-
throated)

yellow-
ochre (whiter in
summer)

(pale-throated
ad. ♀♀ also
occur)

much
white

summer ♂
(pale-
throated)

partly black
sides

variation

black

melanoleuca (SE Europe, Asia Minor)

autumn ♀
(fresh)

more greyish
yellow-brown

summer ♀
(pale-throated
ad. ♀♀ also
occur)

1st-winter ♂
(dark-
throated)

crown often blotched
grey in heavily worn
summer plumage

1st-winter ♂
(pale-throated)

very like
Pied!

summer ♂
(pale-
throated)

spring ♂
(dark-
throated)

in worn summer
plumage, ♂♂
often are black
and white without
ochrous tinge

a few are faintly pale-scaled
above (though less so than
Pied Wheatear)

PIED WHEATEAR

a few autumn ♀♀ are warmer pinkish-
brown or orange on breast and very
like the eastern race of Black-eared
Wheatear (*melanoleuca*)

autumn ♀
(fresh)

cold earth-
brown, finely
pale-
scaled

**summer
♀ (worn)**

sullied grey

often
streaked

var. 'vittata'
(pale-throated)

black
back

**1st-winter
♂**

spring ♂

very like
eastern
Black-eared!

partly black sides

much white

variation

CYPRUS WHEATEAR

**autumn ♀/
1st-winter**
(fresh)

compared with Pied:
shorter wings and
tail, rather large-
headed and
compact

autumn ♂
(fresh)

shorter primary
projection
than Pied

**spring
♂**

deep
yellow-
ochre

(♀ similar, but is
somewhat duller than ♂
and has crown more sullied)

less white

(a few have
narrower black
tail-band)

Finsch's Wheatear *Oenanthe finschii* —

L 15–16 cm. Breeds on barren mountainsides and rocky plateaux with low vegetation. Short-distance migrant. Winters and makes stopovers at higher and cooler altitudes than most of its relatives. Defends territory aggressively and successfully against other wheatears. Restless and fidgety. Shy, flies far away, watches intently. Food mostly insects, but also some plant material. Nests usually in hollow in rock or in crevice.

IDENTIFICATION A trifle bigger than Black-eared Wheatear, and has proportionately *somewhat bigger head* and *thicker legs and bill*. Tail has *uniformly wide dark terminal band* (like Mourning Wheatear, unlike Black-eared and Pied). In flight, has rather pale remiges. – Adult ♂: Black shoulders and wings but *white back* (rather narrow area, can be hard to see in side view, but is obvious on bird flying away). Much *black on throat*, extends well down and *joins at sides with black of wings*. Remiges finely tipped white. White of crown and mantle tinged *buffy-pink* in fresh plumage, crown whiter with *dusky grey centre* when worn. Breast white (negligible buff tinge), undertail-coverts with hint of pinkish-buff. – 1st-year ♂: As adult ♂, but remiges and, especially, *primary-coverts browner*, the latter *with pale tips*. – ♀: Predominantly *brown-grey and dirty white* (lacks warmer buff-brown tone on breast shown by Black-eared and Pied) with rather *contrastless greater coverts*; distinct pale tips to primary-coverts. Ear-coverts usually reddish-brown. Off-white supercilium. Throat colour variable, majority is *dusky grey* (in worn plumage some almost as dark as ♂), often darkest at sides and lower edge, others apparently all pale. Note tail pattern.

VOICE Call a short 'tsit', often repeated; also has a pebble-clinking 'tsheck-tsheck' and a rasping 'bsheh'. Song varied throughout, short, clear and loud verses at slow tempo with 'bent whistles' and occasional interwoven scraping, creaky, almost 'electrical' notes (voice thereby very different from that of Black-eared Wheatear's), e.g.'tk-tk su**iiih**-va chi-chi **trrüü**-u........ chu-chu kretrü-chiiih-a...', etc. Mimicry rare in song. Performs song-flight.

Maghreb Wheatear *Oenanthe halophila* —

L 14½–16½ cm. Closely related to Mourning Wheatear, and these two often included as races of the same species, but here the NW African population is tentatively separated as a separate species on account of clear sexual dimorphism and allopatric range. Breeds like Mourning Wheatear in barren mountain tracts, on rocky slopes and in ravines or wadis, sometimes in flatter, rocky terrain with meagre vegetation. Mainly resident. Food insects. Nests in rock crevice or in ground hole.

IDENTIFICATION In many respects like Mourning Wheatear but sexes differ clearly. Note tail pattern with *evenly broad dark terminal tail-band*, and somewhat *paler flight-feather bases* visible in flight. – ♂: Very difficult to separate from Mourning Wheatear, having *back, throat and wings black*, but has on average *narrower black across back*, *less rusty or buff tinge on undertail-coverts* (often so little as to appear near white in the field), and *somewhat less pale bases to remiges* (though still some contrast). Rarely, black on back broken up by white, appearing streaked rather than solidly black. – ♀: *Back fairly pale buff-brown*. Throat pale or, more often dusky grey, becoming quite dark in worn plumage. Some dark-throated ♀♀ similar to 1st-year ♂, having brown-grey cast on nape (pale) and back (darker). Exceptionally has black-streaked back, then usually inseparable from ♂.

VOICE Both calls and song very similar to Mourning Wheatear.

Mourning Wheatear *Oenanthe lugens* —

L 14½–16½ cm. Like Maghreb Wheatear, breeds in barren mountain tracts, on steep slopes and ravines or poorly vegetated wadis, or at times in flatter, boulder-strewn or rocky terrain with scant vegetation. Short-distance migrant or resident. Food mostly insects. Nests in rock crevice, under rock or in ground hole.

IDENTIFICATION Like Pied Wheatear, but differences include: *uniformly broad dark terminal tail-band*; in flight *pale bases to flight-feathers* clearly contrasting with narrow darker tips and with darker wing-coverts (Pied has more uniformly dark wings in flight); slightly heavier bill and legs. Compared to ♂ Pied Wheatear, also has *smaller black bib* which covers only throat and does not reach onto upper breast; always *pure white breast* but distinct *rusty-buff colour on undertail-coverts*; *white of nape* less extensive, *does not reach onto upper mantle*; 1st-year has *white tips to primary coverts* and often some outer greater coverts, *forming small white patch* (not seen on Pied). Sexes alike. For difference from very similar ♂ Maghreb Wheatear, see that species. – Variation: In S Syria/N Jordan a rare morph occurs, 'BASALT WHEATEAR', with all-black body and head (but still with pale remex bases). Interestingly, this morph has white not rusty-buff undertail-coverts, and maybe it deserves more study.

VOICE Calls are a straight whistled 'hiit' and a clicking 'tschack'; sometimes also an almost crossbill-like clipped 'chüp-chüp'. When disturbed, often gives a cracked 'bzrr' on rising. Song short verses, well spaced, with clear and melancholy ring, jingling and rolling (closest to *Blue Rock Thrush* voice), e.g. 'tritra-**chürr**-tru'; often a 'tschack' as prelude or interwoven, and mimicry may also be inserted at times. Relatively little variation; the song is sometimes protracted into longer verses. Often performs song-flight.

Finsch's Wheatear

Maghreb Wheatear

Mourning Wheatear

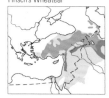

FINSCH'S WHEATEAR

greater coverts
rather plain
(cf. ♀ Pied)

autumn ♀
(pale-
throated)

pale tips

1st-winter ♂

evenly
broad
band

white
back

rather
pale

white
wedge

brown-
grey

short
primary
projection

white
back

autumn ♀
(dark-throated)

spring ♂

white back some-
times concealed

MAGHREB WHEATEAR

tiny

darker wings
than ♀
Finsch's

autumn ♀
(pale-
throated)

1st-winter ♂

rather
pale

buffish-
grey

longer primary
projection
than
Finsch's

autumn ♀
(dark-throated)

faint buff
(difficult to
see)

spring ♂

1st-winter showing pale
tips to primary-coverts

MOURNING WHEATEAR

often found on
boulder-strewn
mountain slopes

black back
(cf. Finsch's)

evenly
broad
band

very pale—
has 'Magpie
pattern'!

rounded
veil

1st-w. ♂

sexes
similiar,
♀ only
slightly duller

pale tips
to primary-
coverts

strikingly
rusty-buff

'BASALT WHEATEAR'
(black morph)

this very odd
colour morph
is restricted to
an area of dark
basalt desert in N
Jordan and S Syria

white!

Hooded Wheatear *Oenanthe monacha* —

L 15½–17 cm. Breeds in unvegetated mountain deserts, in steep rocky ravines and wadis, often with boulders. Usually resident. Often takes larger flying insects in characteristic manner after long chases high in air. Often perches rather unobtrusively on a rock, only to disappear the next moment down into a ravine. Nests in rock crevice.

IDENTIFICATION *Rather big* with *large head* and *long bill*, long wings and long tail but *rather short legs*. Often adopts crouched posture. – ♂: *Tail has much white* and lacks proper black terminal band, has *black only on central feathers and minimally at corners*, which reinforces long-tailed impression. (Same tail pattern is seen only on White-crowned Wheatear, but this has black breast and belly, black nape, smaller bill and longer legs.) *Black* mantle, throat and *upper breast*, black even on upper flanks (the black merges with black of wing-edge). The white of crown narrows on nape. – ♀: Same tail pattern as ♂, but the white is replaced by *light rusty-brown*. Entire plumage *pale grey-brown* with somewhat paler underparts and *rusty rump*. Wings with darker centres and sandy fringes. Note *long bill*.

VOICE Less well known. Calls a straight whistled 'vit', a clicking 'tschack' and a rolling, thick 'prrü'. Song *chatty and composed*, the verses often rather long, consisting mostly of crackling calls and hard, knocking sounds but also occasional clearer and softer notes, also with plenty of mimicry of other birds.

White-crowned Wheatear *Oenanthe leucopyga* [V***]

L 17–18½ cm. Breeds in desolate mountain deserts, on barren stony plains, in ravines, wadis and at fault-line precipices; also on or next to houses at oases. Is a true desert bird, thriving in terrain totally devoid of vegetation. Boulder country seems also to attract it. Resident. Often shy (locally more bold). Food insects. Nests in rock crevice or hole.

IDENTIFICATION About as *large* as Black Wheatear, but is *slimmer* and has longer wings and tail and narrower bill. Flight fast and strong. *All-black* plumage with *white crown*, but the white 'coronet' is not acquired until one year of age; occasional younger birds have scattered white feathers in otherwise black crown. *Tail has much white* with black central line and only small black markings on outer corners (same tail pattern as ♂ Hooded Wheatear). In flight, shows *all-dark wings*. Sexes alike.

VOICE Rich and varied repertoire. Powerful voice, good mimic ('the Hill Myna of the desert'). Calls a shrill whistled 'hiit', a high cracked 'bizz' and a hoarse, throaty 'tschreh'. Song often loud with many clear whistled sounds; for long periods repeats short verses with e.g. a few *descending whistle notes* as major component, 'si-sü-so-trui-trü'; sur-

prisingly *expert imitations* are often inserted as parts of th song. A chatty, softer subsong also occurs. Sometimes pe forms song-flight with fluttering wings and fanned tail.

Black Wheatear *Oenanthe leucura* —

L 16–18 cm. Breeds on steep dry slopes with rock outcrops landslip precipices, bare areas, bushes and scattered tree from sea-level to 3000 m (mostly below 2000 m). Avoid flatter terrain, denser woodland and usually proximity o humans. Resident. Shy, flushes early, usually flies uphill Food mostly insects. Nests in rock crevice or other hollow.

IDENTIFICATION Big and *hefty*, with rather *big head* an *heavy bill*, somewhat more plump than often similar-looking White-crowned Wheatear. Pumps tail upwards, then slowl lowers it. Flight strong, wings broadly rounded. *All dark* (♂ black, ♀ sooty-brown) with white tail-base, and the typica wheatear tail markings of black T-pattern on white back ground. *Terminal tail-band strongly developed and uniform broad*, roughly as on Wheatear. Flight-feathers dark (bu against the light they contrast somewhat paler against dar wing-coverts). Sole confusion risk is younger White crowned Wheatear, but that is slimmer and has incomplet tail-band.

VOICE Calls a descending, whistled 'piüp', often repeate in series, and a slightly thick clicking 'chett-chett'; someti mes a rolling 'cherr'. Song of 'coarsely twittering' verses, bit lower-pitched than other wheatears in same region somewhat thrush-like (but without e.g. Blue Rock Thrush melancholy ring and clear, rolling voice). Has undulatin song-flight.

Blackstart *Cercomela melanura* —

L 14–16 cm. Breeds in rocky desert, in wadis and on boul der-strewn mountain slopes with scattered shrubbery an acacias. Resident. Bold, will approach very close withou showing any fear. Spends much time on ground. Ofte perches low in acacias. Food insects. Nests in rock crevice.

IDENTIFICATION *All-grey* plumage (lighter below) wit *all-black tail*. Black alula contrasts with lighter grey wing coverts. Often fans tail and wings and closes them agai especially on landing or when changing perch. Also pump tail downwards. Sexes and ages alike.

VOICE Single phrases of song (see below) seem to be use as call. Alarm a whistled 'vih'. When agitated, a crackly o frizzling 'bshreh', somewhat recalling e.g. Pied Wheatea Song consists of 2–3 quite deep, clear fluting syllable with accent on the first, repeated at 4–5 sec. intervals 'chürre-lu...... trrü-troo...... chürrlü...... chürrchur......' tone is desolate and melancholy, and volume is ofte amplified by the bare rock faces.

Hooded Wheatear   White-crowned Wheatear   Black Wheatear   Blackstart

HOODED WHEATEAR

confined to mountain desert

pale-tipped primary-coverts in 1st-w.

black breast obscured by pale fringing

1st-winter ♂

only a hint of a tail-band

♀

may bleach to white

slender, long

note typical shape: attenuated, long-billed and short-legged

black reaches far down onto breast (cf. Mourning Wheatear, p. 286)

jizz like Spotted Flycatcher!

white reaches far up

♂

buff

♀

ad. ♂ spring

beware confusion with ad. White-crowned Wheatear shown below; tail-pattern very similar. Look for the white belly!

WHITE-CROWNED WHEATEAR

ad.

glossy black

1st-year

unmistakable!

variation; not related to sex or age

the predominantly white tail always separates from Black Wheatear

1st-year

black reaches to the legs

adult

1st-y.

broken-up band, never solid as in Black Wheatear

BLACK WHEATEAR

sooty brown

strong

♀

dull black

♂

evenly broad tail-band

pale

evenly broad black tail-band evident from both above and below

black reaches behind legs

often pumps tail up, then down

BLACKSTART

watches for insects from the lower branches

ash-grey

often opens wings and fans tail

velvet-black tail striking

all black

DZ

Desert Wheatear *Oenanthe deserti*   **V****

L 14½–15½ cm. Breeds on dry, sandy steppe-like heath or in semi-desert with scattered bushes, low vegetation and rocks (but shuns pure sand desert). Short-distance migrant. Food insects. Nests in hole in ground or in rock crevice.

IDENTIFICATION Rather compact with comparatively big head, short neck and tail. *Tail almost all black*, white only at very base of sides, sometimes showing in flight as tiny white corner. Rump buffy-white, conspicuous in flight. Remiges in flight rather light grey-brown, but note that *underwing-coverts in both sexes are contrastingly dark* (blackish in ♂). – ♂: Invariably *black bib*. Resembles black-throated morph of Black-eared Wheatear, but differs in black tail and in having *whitish scapulars*, not black, so that folded wing shows narrower area of black. Also, *black of throat merges narrowly with black of wing*, normally not seen in Black-eared Wheatear. In autumn, black areas have pale fringes. 1st-summers have browner wings than adults. – ♀: Note, besides tail pattern, quite *pale and grey-brown* colours including wing (usually appearing darker in other ♀ wheatears); breast buff-brown (not orange-tinged); *upperparts light* yellow-tinged grey-buff; *scapulars very pale buff-brown*, at least as pale as mantle; reddish-brown ear-coverts and dirty yellow-white supercilium. Occasionally has darkish throat.

VOICE Call a squeaky, drawn-out 'viieh', a hard clicking 'tsack' and a thin, muffled rattle, 'tk-tk-tk'. Song characteristic, a rather constantly repeated, always *short*, clear fluting and tremulous verse *dropping* in pitch and sounding desperately *mournful* in tone, 'trüü-trururu'. Song-flight before dawn, sometimes in chorus.

Kurdish Wheatear *Oenanthe xanthoprymna*   —

L 14–15 cm. Closely related to Persian Wheatear, and previously often treated as a race of that. However, latter is morphologically distinct, has slight sexual dimorphism and separate range, hence here given species status. Breeds on barren mountain slopes with low, meagre vegetation, often quite high up (1200–4000 m). Largely migratory; also seen on passage and in winter at lower levels. Rather bold. Usually perches on low bushes, lowest branch of trees or on rocks, does not pump tail much. Food mostly insects.

IDENTIFICATION A rather *compact* wheatear with proportionately large head, fairly short neck, tail and legs. Often perches rather horizontally. Shares with Persian Wheatear characteristic of having *rusty rump and vent*, but unlike that has in adult plumage *white tail-base*. 1st-year birds usually have rusty tail-base, though. Sexes similar, although ♂ is on average slightly smarter than ♀ (but easy to mix them up; best marks are that ♂ has darker remiges, greyer back and deeper rusty colour on vent than ♀). *Black bib* (dark grey in ♀), dirty white breast, *white supercilium* (dusky in ♀), grey crown, mantle and back (tinged brownish in ♀), some pale fringes to wing-feathers so that these look rather variegated. Differs from Red-rumped Wheatear e.g. in *narrower pale fringes to wing-coverts* (wings hence rather dark in spring), *greyer mantle/back* and slightly smaller size and slimmer shape.

VOICE Calls a whistle on descending scale, 'eehp', more unobtrusive than corresponding call of Wheatear, and a thick-voiced clicking 'tchack'; the latter may run into fast rattling series, 'chr'r'r'r'. Song short and fairly deep-voiced, a rather constantly repeated, twittering verse of 4–6 dry, rolling notes, 'chu-ey chu **churr**' or variants.

Persian Wheatear *Oenanthe chrysopygia*   —

L 14½–16 cm. Formerly often lumped with Kurdish Wheatear but differs consistently in sufficiently many ways and has allopatric range, hence here given species status. Breeds in E Transcaucasia, Iran and adjacent parts of Central Asia, wintering in E Africa and migrating through Arabia and E parts of Middle East. Habitat and habits as for Kurdish Wheatear.

IDENTIFICATION Sexes alike. Appears like a ♀ Kurdish Wheatear, but *rump and base of tail invariably rusty-red* (no white), and *throat off-white* (never dark). Wings rather *featureless brown-grey*, tertials edged rufous when fresh; *underwing-coverts pale* (not contrastingly dark). *Upperparts rufous grey-brown* (never predominantly grey). *Insignificant short pale supercilium*, quickly fading off behind eye.

VOICE Call a harsh 'kersch', rather like Pied Wheatear. Song a prolonged, rather lark-like, meditative warble interfoliated with the harsh call in twos, 'kersch-kersch'.

Red-rumped Wheatear *Oenanthe moesta*   —

L 14½–16 cm. Breeds mostly at lower levels on flat, dry, sandy or clayey heaths or in more barren, stony semi-desert, at times among tamarisk bushes and scanty low vegetation; avoids pure sand desert. Spends much time on ground. Resident. Food insects. Nests in ground hole.

IDENTIFICATION Resembles Kurdish Wheatear (which see), but note: slightly *larger size, longer bill and legs*; nearly all-*dark tail* with rusty-red rump and narrow rufous base on ♂ but extensive rufous base on ♀; *white edges to greater coverts* form pale panel, mostly on ♂; *pale wing-bars* along both greater and median coverts; *mantle, back and scapulars of* ♂ *black in worn plumage*, occasionally causing confusion with other wheatear species (Mourning, Pied, Finsch's). – ♀: Typically has obvious *rusty-buff colour on crown, nape and head-sides*, and *light buffish-white panel on wing*. Frequently pumps tail upwards and slowly lowers it.

VOICE When agitated, a repeated short 'prrit', almost like Little Bustard. Song remarkable, soft and protracted, scratchy, *low-pitched*, structure indefinite and *voice trembling*, e.g. 'churr-urr chorr-orr chürr-ürr chirr-irr...'. Shorter, simpler verses also uttered.

Desert Wheatear   Kurdish Wheatear   Red-rumped Wheatear

DESERT WHEATEAR

contrasting pale panel on inner wing

dark throat obscured by pale fringes

nearly all black

a few ad. ♀♀ wear to grey on throat

scapulars white

black 'bridge' (cf. Black-eared Wheatear)

dark under-wing-coverts

pale

1st-winter ♂

ad. summer ♂

♀

KURDISH WHEATEAR

rufous

1st-w. ♀

ad. spring ♂

dark wing-feathers

white

dark coverts

subdued darkish throat

brown-grey

tail rufous in many 1st-years

ad. ♀ / 1st-winter ♂

1st-winter ♀

rufous vent

ad. spring ♂

PERSIAN WHEATEAR

pale coverts

pale wing-feathers

white supercilium in front of eye

greyish

sexes alike

general impression like a small, bright Isabelline Wheatear

1st-winter

pale orange-buff

rufous lower flank

RED-RUMPED WHEATEAR

1st-year ♂ in spring with worn wings may cause confusion

♀ unmistakable; only wheatear with rufous head

grey

broad band

♀

rufous head

pale wing in both sexes

1st-w. ♀

winter ♂

♂

always rufous tail

♀ spring

ad. spring ♂

white-fringed wing-coverts characteristic

tail mainly dark with rufous base and reddish-buff rump

DZ

Whinchat *Saxicola rubetra* mB3 / P3

L 12–14 cm. A bird of open lowlands with a wide range. Breeds in uncultivated, usually damp areas, e.g. rough pastureland, lake-sides, commons, tussocky grassland, water meadows, also bracken-covered slopes. Summer visitor (Apr–Sep, rarely Oct), winters in tropical Africa. Food insects. Nests in tussock.

IDENTIFICATION *Small, short-tailed* and plump with rather long dark legs. Perches upright, twitches or wags tail slowly, flies off low in energetic flight and swoops up to new perch in top of weed or on fence. Although thus often keeps to low perches will sometimes take higher song post in treetop. Prominent light supercilium characteristic of all plumages. *Rump dark-spotted yellow-brown*, and rest of upperparts streaked dark. A little white on tail-base, glimpsed in flight. – Adult ♂: Black lores and *brownish-black head-side framed by pure white stripes. Throat and breast orange-buff.* Wing-feathers rather dark, with one or two *white wing patches*, easiest to see in flight. – ♀/1st-winter: Buffish-white supercilium, light brown head-side, normally no white wing patch. Ageing in autumn difficult except for very bright adult ♂♂. Both adults and young can have finely spotted breast.

VOICE When agitated, utters a soft, short whistle and a clicking call note, 'yu tek, yu tek-tek...'. Song (often at night) a varied, loud, fast, short verse with mix of hoarse rasping sounds, clear deep notes and mimicry; verse begins a little haltingly, accelerates and is abruptly cut short; a recurrent variant resembles a compressed version of Corn Bunting song. Readily separated from song of Wheatear, another nocturnal songster, by its more mellow, less scratchy sound, its explosive delivery and the frequent inclusion in the song of the short anxiety whistle 'yu'.

(Common) Stonechat *Saxicola torquatus* rB3 / P

L 11½–13 cm. A widespread polytypic species, sometimes treated as two or more separate species but here kept together due to partly intergrading forms and still uncertain situation in area of contact. Regardless of taxonomy adopted, obviously two rather distinct groups of subspecies, here afforded separate accounts, the first in Europe and W Turkey (mainly ssp. *rubicola*; resident except in N). As to the other group, see below. Breeds in open areas with low vegetation, often heather and gorse. Found both at low level near sea and on higher moorland (incl. above treeline in alpine zone).

IDENTIFICATION (*rubicola* et al.) Adult ♂: *Entire head* (incl. chin) *black* (in autumn plumage partly concealed by pale fringes); *white patch on neck-side. Breast reddish-orange* extending to upper belly and flanks. *Rump brown and dark-mottled* (bleached in summer, can then show small white

patch, rarely even a large white patch). *Underwing-coverts grey with fine white edges*, rarely if ever giving blackish and strongly contrasting impression. – ♀: Throat dusky brown (may be largely concealed by pale fringes in autumn).

VOICE Alarm rather like Black Redstart's, a shrill, sharp whistle and a throaty clicking, 'vist trak-trak, vist...'. Song a short, high-voiced, twittering or squeaky verse that recalls both Horned Lark and Dunnock; song rather monotonous, lacks Whinchat's variation and wider tonal range.

'Eastern Stonechat' *Saxicola torquatus maurus* et al. V

L 11½–13 cm. A subspecies group of Stonechat, here given separate account for practical reasons. Breeds in Siberia and N Central Asia (*maurus, stejnegeri*). Two more races generally referred to this group, but apparently unconnected with the rest: *armenicus* (E Turkey, Armenia, possibly lower parts of Caucasus) and *variegatus* (plains W of Caspian Sea). Habitat and habits similar to those of Stonechat in Europe (see above), with a tendency to select lower altitude and avoid the most barren alpine habitats.

IDENTIFICATION (*maurus*) Adult ♂: Similar to Stonechat but has *larger white neck-side patch reaching further back on nape*, and *orange-red breast patch on average smaller*, leaving *rest of breast, belly and flanks white* (though rarely, tricky intermediates occur). *Rump unstreaked buff-white* (bleached in summer to large white patch). *Underwing-coverts black*, strongly contrasting with rest of underwing (but can still be difficult to judge). – ♀: Throat pale, buffish-white, never dark-looking. – 1st-winter: Stragglers to W Europe show hint of *pale buff eyebrow* and *large buff unstreaked rump patch*. – (*variegatus*) Adult ♂: Similar to *maurus* but has much white basally on tail with inverted 'T' like in Wheatear. – (*armenicus*) Adult ♂: Similar to *maurus* but somewhat larger and has a little white sometimes visible at base of tail.

VOICE Probably identical to Stonechat.

Fuerteventura Stonechat *Saxicola dacotiae* —

L 11–12½ cm. Breeds only on Fuerteventura (Canary Islands; prob. max. total 1300 pairs) on barren slopes and in ravines with bushes and odd trees. Resident. Food insects.

IDENTIFICATION Very slightly smaller than Stonechat, but has similar shape and general behaviour including wagging of tail and flicking of wings. Has very subtly thinner and longer bill (a common tendency for island-living taxa) and longer and narrower tail. – Adult ♂: *Head brown-black with narrow white throat* (continues at lower edge onto neck-side) and *short white supercilium*. On upper breast an orange-buff patch, but this is paler orange and smaller than in any Stonechat. Small white wing patch. *Rump dark.* – 1st-winter ♂: Duller than adult. Usually no white wing patch (a hint at most). In fresh plumage a pale panel along secondaries and a faint pale wing-bar along greater coverts. – ♀ Paler and greyer than ♂, head brown-grey, faintly spotted; supercilium diffuse. Throat invariably pale.

VOICE Quiet. Voice very similar to Stonechat.

Whinchat

Stonechat (both race groups)

Fuerteventura Stonechat

Madeira

Canaries

WHINCHAT

perched in lush meadow

broad, distinct supercilium in all plumages

1st-w. (fresh)

♀ **spring**

striking white in ♂

orange throat and breast

ad. ♂ spring

ad. ♂

white base

white tail-base glimpsed when takes off low over ground

ad. ♀

heavily spotted

STONECHAT

ad. ♂

ad. ♂

— variation —

1st-w. ♀ (fresh)

spotted

short wing

spotted

grey

ad. ♂

♀ **spring**

half-collar

ad. ♂ spring

juv.

plain white

variation

rufous rump

spotted

spotted and streaked all over

almost entire under-parts orange

ad. ♂ spring

rubicola (Europe)

'EASTERN STONECHAT'

large white rump

ad. ♂

hint of eyebrow

plain, pale

♀

very large white patch

slightly longer

1st-w. ♀ (fresh)

plain, pale

jet-black

ad. ♂

small orange patch

ad. ♂ spring

maurus (Siberia)

variegatus (W of Caspian Sea)

very similiar to *maurus* except for distinctive tail-pattern

ad. ♂ spring

ad. ♂

white tail-sides

tail pattern like Wheatear

FUERTEVENTURA STONECHAT

(ad. ♀ often with hint of white wing patch)

long and thin

short white eyebrow

♂

1st-w. ♂ (fresh)

1st-w. ♀ (fresh)

throat white

dark tail and rump

♀

favours dry river beds—'Barrancos'

ad. ♂ spring

z

Song Thrush *Turdus philomelos* r(m)**B**2 / **P**+**W**

L 20–22 cm. Breeds in lush woodland, parks and well-vegetated gardens. N populations migratory, winter in W and S Europe. Often rather tame. Food snails, insects, worms. Nest, often well concealed in e.g. ivy, hedge, on bank, a moss-clad bowl with smooth grey interior (of clay and decayed wood).

IDENTIFICATION Small and rather compact, *fairly short-tailed*. Fast, slightly jerky flight. Brown above, and yellowish-white or *white below, densely sprinkled with black spots* (cf. Mistle Thrush). *Underwing rusty-buff*, sometimes perceptible in flight. Often has *slightly warmer ochre tone on flanks* next to wing-edge (and on breast), but never rusty-red like Redwing, and lacks latter's white head-stripes. Sexes alike.

VOICE Call a rather discreet, fine, sharp 'zit' (not so hard and clicking as Robin's, and a little softer than Rustic Bunting's). Alarm an excited series of 'electrical' scolding sounds, 'tix-ix-ix-ix-ix-...' (sharper, higher-pitched than Blackbird's). Song *loud and proclaiming*, sounds 'cocksure' and 'dogmatic'; strongly varied, often squeaky and shrill cascades of notes, pauses few and very brief; characteristic is recurring repetitions 2–4 times of same group of notes, e.g. 'kücklivi kücklivi, tixi tixi tixi, **pii**-eh, trrü-trrü-trrü tixi**fix**, chü-chü-chü, ko-ku-ki**klix** ko-ku-ki**klix**,...'.

Redwing *Turdus iliacus* m**B**5 / **W**1

L 19–23 cm. A characteristic bird of N Europe's conifer forests, extending into upland birch forest, even into willow zone; at southern limits also mixed woodland, regenerating scrub, often near water. In Britain & Ireland very rare breeder (mostly Scotland), but abundant winter visitor to fields, woodland edge, parks etc. Rather shy.

IDENTIFICATION *Small* (Song Thrush size), rather short-tailed and compact, with fairly big head. Note *whitish stripes above eye and beneath cheek*, also *rusty-red flanks* when perched, and *rusty-red underwing* in flight (redder colour than on Song Thrush). Sexes alike.

VOICE Call on migration a protracted, indrawn 'stüüüf' (hint of hoarseness); often heard on October nights, even over towns and cities. Also has a nasal 'gack'. Alarm a hoarse, rattling scold, 'tret-tret-tret-tret-...'. Song highly variable among individuals but constant within each one, recognized by general tone and structure: rather short verses, with 3–6 sec. intervals, consisting of a loud, easily transcribed section immediately followed by a soft squeaky twitter; among common introductory themes e.g. descending 'chirre chürre chorre', rising 'tru-tra-tro-trü-tri', soft Common Rosefinch-like 'vi**dje**-vi**djü**', short and mournful 'trüi trai', clear ringing 'tüllüllüllüllüllüll...' etc. Flocks at migration stopovers sometimes give noisy chorus of squeaks.

Mistle Thrush *Turdus viscivorus* r**B**2 / **W**2

L 26–29 cm. Breeds in open woodland, parks, orchards, large gardens, scrub, groups of conifers. In S Europe often found on mountain slopes in treeline zone, and forages on alpine meadows and bare mountain. In Britain & Ireland mostly resident, with immigrants from N Europe Sep–Apr in winter also on fields. Wary and rather shy in much of range, less so in W Europe. Nest grass-lined, in tree.

IDENTIFICATION Big, and patterned basically as Song Thrush, but normally easily told by calls, *larger size*, *more elongated* and *long-tailed* shape, *more powerful flight with longer undulations* (not jerky and hopping), also *white underwings*. At times, *white tips to outer tail-feathers* can be made out. Perched bird seen in gap in foliage can be more difficult to distinguish from Song Thrush. Note: *pale brown-grey head-side and neck-side*; tendency towards *bigger black patch on breast-side* and irregular *dark markings on neck-side*; *more rounded black spots* on breast and belly; *greyer wing-feathers with paler edges*.

VOICE Call a characteristic drawn-out, wooden dry rattle 'zer'r'r'r'r'. Alarm a hard rattling call resembling Fieldfare's but slightly drier. Song most like Blackbird's: short, varied verses with clear and loud voice; differs in more desolate and slightly harder tone, faster tempo and shorter pauses, smaller tonal range (i.e. sounds a little more monotonous), no (or fewer) squeaky notes at end of each verse 'truitrüvu... churichuru... chüvutru... churuvütru...'.

Fieldfare *Turdus pilaris* m**B**5 / **W**1

L 22–27 cm. Breeds in various types of woodland and bush, scrub, also in parks, avenues of trees and gardens. Food in summer worms and insects. In Britain & Ireland, very rare local breeder; abundant passage and winter visitor, large flocks on open fields and other grassy areas, also taking rowan berries, fallen fruit etc. Often nests in small colonies (for extra protection against corvids, which are actively pursued in flight and sometimes bombarded with excrement); twig nest rather high up in treetops, often in a main fork.

IDENTIFICATION Big, *long-tailed*, stocky thrush. In flight apart from tail length, identified by *light grey rump* (shows well from behind), *white underwings* (shown otherwise only by Mistle Thrush) and rather *flapping, less undulating flight*. Red-brown back and grey crown/nape. Heavily spotted below, and breast has rusty-yellow tinge. Sexes similar – Juvenile: Identified by pale spots on wing-coverts.

VOICE Migrants utter a squeaky 'gih' and also chattering 'schack-schack-schack'. When pursuing crows, gives furious chatter. Song simple, a few chattering notes with no clear structure, short pause, more chatter and so on. In song-flight, gives ecstatically chattering, drawn-out, faster song

Song Thrush

Redwing

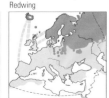

Mistle Thrush

Fieldfare

SONG THRUSH

habitually smashes snail shells on stone 'anvil'

compare juv. Blackbird (above), having darker reddish-brown breast

plain brown

ageing in the field difficult; sexes alike

singing

short

rusty-buff

yel-low-buff

'arrow-head' spots

REDWING

compare silhouettes of three flying Redwings (left) with three Starlings (right)

at a distance, head markings usually most striking feature

singing

rusty-red

short

strong head pattern

bold white supercilium

rusty-red

1st-winter

juvenile tertials and retained outer greater coverts have white tips, revealing age

white submous-tachial stripe

ad.

MISTLE THRUSH

compared with Song Thrush, Mistle is bigger, stands more upright and is paler

MISTLE   SONG

white corners

slightly paler than back

singing

fairly long

white

rather plain dark

pale vertical cheek patch

often dark patch

paler outer coverts

1st-winter

in autumn, young birds told by retained juvenile outer greater coverts, these shorter and paler

rounded black spots

ad.

FIELDFARE

hardy, sometimes retreats only to snow-limit

fairly long

white

white   dark-breasted

all dark

pale grey

1st-winter

grey

ochre

ad.

M

(Common) **Blackbird** *Turdus merula*   r(m)**B**1 / **P**+**W**1

L 23½–29 cm. Breeds commonly in woodland, parks and gardens, also in juniper country. N populations migratory, winter in W Europe Sep/Oct–Mar/Apr. Confiding. Sings from rooftop, TV aerial, or treetop in wood. Food earthworms, insects, berries. Nests in bush, on trellis, behind logs, etc.

IDENTIFICATION ♂: *All black*, in spring and summer with *yellow bill* and *narrow yellow eye-ring*, making it unmistakable to any birdwatcher. But the beginner seeing a black bird with yellow bill on the lawn must eliminate the Starling: the Blackbird is *not a flocking bird*, has *long tail* (which is often jerked upwards and slowly lowered), often makes quick *two-footed bounds* or walks a few steps before *standing dead-still* for a few seconds (looking out for worms), takes a few more steps, etc. Also, plumage lacks pale spots. – ♀: Sooty-brown with slightly *lighter brownish-white throat* and *diffuse mottling on breast*. Bill dark. – Juvenile: Like ♀, but with small pale spots on upperparts.

VOICE Rich repertoire. Noisy. Common calls: a deep 'pok'; on migration, a quiet, fine rolling 'srrri' (more jingling and 'looser' than Treecreeper's). When agitated: a very fine, Robin-like, straight 'tsiih'; hard clacking 'chack-ack-ack-ack-...'. Alarm: at the sight of a cat or owl, or prior to going to roost, series of metallic, high 'pli-pli-pli-pli-pli-...', which (often on take-off) turn into a crescendo. Song well known for its melodic, mellow tone, a clear, loud fluting (almost in the major key) at slow tempo and on wide, often sliding scale, with soft twitter appended; verses rather short, repeated at 3–5-sec. intervals. Some have desolate tone, sound quite like Mistle Thrush.

Ring Ouzel   *Turdus torquatus*   mB3 / P

L 24–27 cm. Breeds in uplands in gulleys or on steep boulder-strewn slopes, on bare mountainside or, in Continental Europe, in upland forest. Migratory in much of range; in Britain & Ireland, moorland breeders arrive Apr and leave Sep, with migrants from Fenno-Scandia Mar–May and Sep–Nov. Shy. Food insects, worms, berries. Nests in rock crevice, hollow stump, hole in wall, etc.

IDENTIFICATION Size and shape roughly as Blackbird. – Adult: Predominantly *black* or sooty brownish-black with *white half-moon across breast* (on ♀♀, but also some ♂♂, the white has narrow brown feather fringes, most obvious in fresh plumage) and *yellow bill* with dark tip. Wing-feathers have pale edges, and in flight *wings appear much paler* than Blackbird's. Sexes similar, but ♂ on average blacker with whiter half-moon. – 1st-winter: ♂ resembles adult ♀, has hint of pale half-moon. ♀ usually completely lacks pale half-moon, can be confused with Blackbird but has somewhat paler wings than latter (though by no means as pale as adult

Ring Ouzel!). – Variation: Birds in S and C Europe (ssp *alpestris*) have pale-scaled underparts and paler wings.

VOICE Call a stony clicking 'tück', harder and somewha higher-pitched than Blackbird's call. Also a Fieldfare-lik squeak. Song loud and desolate-sounding, a few repeate notes or motifs at slow tempo; several types included, e.g 'trrü trrü trrü trrü.... si-**vütt** si-**vütt** si-**vütt**.... chuvü chuvüü chuvüü chuvüü....' etc.

Blue Rock Thrush   *Monticola solitarius*   — [V***

L 21–23 cm. Breeds in rocky, precipitous habitats on coasts in mountain ravines and on mountainsides with boulder and rock outcrops, locally also in stone quarries, on ruins churches and even inhabited buildings in flat terrain; thus unlike Rock Thrush, also inhabits lower levels. Mainly resi dent. Wary, often takes cover. Food insects, plant matter small lizards. Nests in cavity or recess in cliff or wall.

IDENTIFICATION *Large* and *slim* with *long bill*. (Against th light and at distance beware similarly long-billed Mistl Thrush, which also breeds in the S European mountains!) ♂ dull *blue-grey* with dark wings (1st-winter even duller blu and vermiculated dark, primary-coverts tipped pale). ♀ dar brown, vermiculated below, with *dark brown tail*, rarely wit faint blue tinge on back, uppertail, sides of breast and flanks.

VOICE Call a disyllabic squeaky pipe, 'üh-vih'; also a dee clicking 'chook'. Song varies somewhat, often like Roc Thrush's (genuine confusion risk!), is loud but not that sof has clear and rolling or tremulous tone (r-sound), strongl melancholy ring, often shuttling back and forth.

(Rufous-tailed) **Rock Thrush**   *Monticola saxatilis*   **V**∗∗

L 17–20 cm. Breeds on steep, dry, rocky mountain slopes o higher-lying alpine meadows. Found above 1500 m (od exceptions), and thus usually chooses higher sites than Blu Rock Thrush. Summer visitor (mostly Mar / Apr–Sep), win ters S of Sahara. Shy. Food insects, plant matter, smal lizards. Nests in rock cavity, under boulder or in ground.

IDENTIFICATION ♂ in spring unmistakable (but garish co ours remarkably camouflaged in strong mountain light). ♂ winter plus ♀ and immatures all similar, more anonymou brownish and *densely vermiculated dark*, told by *short rust red tail*, stocky build and rather long bill; ♂♂ often reveale by intimation of *pure white spotting on back* (♀ at the mos buffish-white) and touch of blue-grey on crown and throat.

VOICE Call a short, squeaky 'viht'; also a clicking 'chak Song, sometimes in undulating song-flight with termin glide, a *soft*, clear and melodic verse (with melancholy bas tone) which often drops in pitch; at distance can recall bot Redwing and Blackbird; like Blue Rock Thrush's, but softer and not quite so melancholy in tone.

Blackbird

Ring Ouzel

Blue Rock Thrush

Rock Thrush

BLACKBIRD

imm. ♂♂ have darker bill and orbital ring than adults

pale spots

♂

ad. ♀

brown, streaked

ad. ♂

all yellow

all dark

bill partly darkens in winter

juv.

typical juv.; some, presumably ♂♂, are darker than this

faintly brown-tinged wing

all-black wing

ad. summer ♂

1st-winter ♂

RING OUZEL

♂

(odd ad. ♀♀ are more ♂-like)

pale

ad. ♀

brownish-white

pale edges

rather long

barely visible paler brown crescent (1st-w. ♀ has even darker crescent, often not perceptible in the field)

alpestris (S & C Europe)

pale-scaled

ad. ♂

white crescent

1st-winter ♂

mostly black

torquatus (N & W Europe)

ad. ♂

KM

BLUE ROCK THRUSH

note the drawn-out silhouette; so different from Rock Thrush

usually looks all dark and featureless in flight

ad. ♂

found in rocky habitat down to coastline

long bill

♀

long-tailed

blue colour not always easy to see at long range

ad. ♂ summer

ROCK THRUSH

♂ in song-flight

note typical short rear

white back

rufous tail

ad. ♂

confined to higher altitude then Blue Rock Thrush

short bill

ad. ♀

♀

vermiculated above

short-tailed

ad. ♂ summer

DZ

NORTH AMERICAN THRUSHES
Some North American thrushes have repeatedly been found in Europe as stragglers, and those most commonly encountered are shown here. Thrushes are strong and capable flyers, but for a North American passerine to safely reach Europe it needs to be helped in autumn by often quite strong tailwinds connected with low pressure over the N Atlantic. One species is a member of the genus *Turdus*, richly represented in Europe, while the remaining four belong to genus *Catharus*, a typically American group of smaller thrushes.

American Robin *Turdus migratorius* V***
L 22–25 cm. (British emigrants' red-breasted substitute for the European Robin, much loved and much missed favourite garden bird. Rare but recurring vagrant in Europe, mainly in autumn and winter. Confiding, often a garden or park bird in America, and stragglers in Europe usually occur in same habitats, behaving much like a Blackbird or Fieldfare.

IDENTIFICATION Rather heavily built, with shape and behaviour recalling Fieldfare, thus long-winged and a good flyer. Has also rather long tail and strong feet, hops or walks a few steps on lawns, then freezes and watches attentively in search of earthworms. *Entire breast, flanks and upper belly tomato or rusty red* (buff-white tips in fresh plumage, narrowest on ad. ♂). *Dark grey above* (faint brown tinge on ♀ and 1st-winter), *head almost black* (duller on ♀ and 1st-winter) with white 'eyelid spots' above and below eye. Throat white, with coarse, dense black spotting, on some so densely as to look black at distance. Adult has almost *all-yellow bill* in spring, but darker in winter months; 1st-winter has all-dark bill. Odd ♂♂ have sooty or densely black-smudged mantle and deeper red underparts (var. '*nigrideus*').

VOICE Blackbird-like, including deep chacking 'chok-chok-chok' and more excited 'kli-kli-kli-...'.

Swainson's Thrush *Catharus ustulatus* V***
L 16–18 cm. Rare vagrant in Europe, mainly in autumn. Quite vigilant.

IDENTIFICATION Fairly small, *considerably smaller than Song Thrush* (but without comparison this can all the same be hard to ascertain). Basically like Song Thrush in pattern, but *dark spots on underparts smaller* and *restricted to throat and breast*, while spotting on flanks is weaker and stops far ahead of legs (Song Thrush heavily spotted down to belly and flanks, continuing past legs), and in flight has typical *Catharus* (and *Zoothera*) *underwing markings of broad white and dark bands*. Narrow dark lateral throat stripe. Flanks greyish. Resembles Grey-cheeked Thrush, but has *buff-tinged and more distinct eye-ring* (not greyish-white and narrow), as well as *weak buff tinge to cheeks* and *clear buff-white streak above lores*. Olive-tinged grey-brown above (but note that more rusty-toned birds breed in NW America).

VOICE Flight-call, often at night, a high, soft, upslurred 'kvüüi'. When nervous, a sharp, resonant 'chipp'.

Grey-cheeked Thrush *Catharus minimus* V***
L 15–17 cm. Closely related to Swainson's Thrush. Rare vagrant in Europe, mainly in autumn. Shy, preferring to remain in cover.

IDENTIFICATION Small. As on other *Catharus* thrushes, *underwing* has *white–dark–white pattern*. Resembles Swainson's Thrush (finely spotted dark on lower throat and breast, narrow dark lateral throat stripe, greyish flanks, uniformly grey-brown above) but differs in *more greyish-white ground colour on head-side and throat*, usually *narrower and less prominent grey-white eye-ring* (can be slightly stronger on 1st-winter), and *lack of distinct pale supraloral stripe*. – Variation: Race *bicknelli* ('BICKNELL'S THRUSH', considered separate species by some), in NE North America, is small, with *faint* rusty tone to tail, and half of lower mandible is yellow.

VOICE On migration, often at night, gives a shrill piping, drawn-out 'tsii(ü)' (hint of drop in pitch).

Hermit Thrush *Catharus guttatus* V***
L 15–17 cm. Very rare vagrant in Europe. Shy, almost always remaining concealed.

IDENTIFICATION Small. In flight, as on other *Catharus* thrushes, *underwing* has *characteristic pattern of broad white and dark bands*. Breast only *sparsely and sparingly spotted*, but *spots are large, black and distinct*. Flanks usually entirely unspotted (diffusely grey-mottled at most). Grey-brown above, with *rusty-red tone to rump and tail* (like Nightingale, but is slightly bigger, more compact and shorter-tailed than that species, besides other obvious plumage differences). *White eye-ring*. Often jerks tail upwards and slowly lowers it.

VOICE Call a Blackbird-like though higher and more stifled 'chück', repeated in series. Also has a strained, nasal, tit-like 'veeüh' (faintly upslurred).

Veery *Catharus fuscescens* V***
L 16–18 cm. Very rare vagrant in Europe (a few autumn records). Shy, prefers to stay in cover.

IDENTIFICATION Small. As on other *Catharus* thrushes, *underwing has typical markings of broad white and dark bands*. Very similar to both Grey-cheeked and Swainson's Thrushes, but *throat / breast spotting is less distinct* and less extensive. In addition, *upperparts* are *entirely rusty-toned* (only rump and tail red-brown on Hermit Thrush; but note that western race of Swainson's Thrush is similarly reddish-brown above). Eye-ring always narrow and indistinct. With more fleeting observations quite possible to mix up with Thrush Nightingale, but note that entire upperparts (not just tail) are rufous-tinged, and that legs are pinkish-brown, not grey.

VOICE Call a whistled, hoarse 'pyiu' (downslurred).

Unlike Song Thrush, all of the species treated opposite (apart from American Robin) have a striking underwing pattern.

AMERICAN ROBIN

vagrants in Europe mostly recorded in suburban parks and gardens, where they behave much like Blackbirds

1st-w. ♀

♂♂ average blacker on head than ♀♀

white corners

pale tips characteristic of fresh plumage (autumn–winter)

ad. ♂ winter

underparts more intensely coloured in ♂♂

contrast between replaced inner greater coverts and retained (juv.) outers best means of ageing 1st-year birds; sexing more difficult, ♀♀ generally browner on head and have lighter colour on underparts

SWAINSON'S THRUSH

much smaller than Song Thrush (left)

SONG THRUSH

more heavily marked ear-coverts

Swainson's (right) lacks Song Thrush's muted supercilium behind the eye

buff-tinged 'face', more prominent eye-ring than Grey-cheeked

flanks soft greyish with much more subtle spotting than in Song Thrush

Swainson's lacks song Thrush's heavily patterned ear-coverts, buff-tipped median coverts and buff flank

HERMIT THRUSH

habitually cocks and slowly lowers tail, more than the other *Catharus* thrushes

prominent white eye-ring

upperparts rather warm olive-brown

breast spots comparatively large and distinct

bright reddish tail

Hermit (unlike other *Catharus* thrushes) often has light-tipped median coverts

GREY-CHEEKED THRUSH

best identified by differences in face pattern

SWAINSON'S

eye-ring usually reduced to thin half-crescent behind eye

GREY-CHEEKED

most vagrant American thrushes recorded in Europe are 1st-years, but adults and 1st-year are very similar

greyish cheeks and face, lacks distinct buff eye-ring and supraloral stripe of Swainson's

NE populations (ssp. *bicknelli*) are more reddish-brown on upperparts

VEERY

vagrant Veeries in Europe are liable to be mistaken for Thrush Nightingale, especially if views are fleeting

VEERY

THRUSH NIGHTINGALE

inconspicuous eye-ring

comparatively bright reddish overall

more subdued spotting than other *Catharus* thrushes

silky-white

KM

ASIAN THRUSHES

Of the seven Asian thrush species presented here, two breed marginally within the treated range, whereas the other five are stragglers from Siberia, usually seen in W Europe in autumn and winter.

White's Thrush *Zoothera dauma* V***

L 27–31 cm. Breeds locally just W of Ural mountains, but main distribution in Siberian taiga. Summer visitor, winters in SE Asia, returning to breeding sites at end of May. Rare vagrant to W Europe. Usually forages on ground. Very shy; when spotted, flies up into dense tree crown and perches motionless, or deftly takes cover in undergrowth.

IDENTIFICATION *Large* and *elongated*, with *heavy bill*. In powerful flight, undulating over longer distances, *broadly white-framed black band on underwing* may be glimpsed. Plumage otherwise olive-tinged grey-brown at distance, but at close range handsomely *black-scaled, including on entire upperparts*. Black scaly markings at times coalesce on breast, which can then appear rather dark. *Wings fairly variegated* (incl. bicoloured remiges). Sexes alike.

VOICE Silent except for its song, which is striking: monotonous, drawn-out, ventriloquial, second-long whistles with pauses of 5–10 sec. between, 'tüüü....... tüüü....... tüüü...'; at times alternates between two pitches; audible at 1 km.

Siberian Thrush *Geokichla sibirica* V***

L 20–21½ cm. Siberian taiga species, very rare vagrant in Europe, mainly in autumn and winter. Shy.

IDENTIFICATION Size of Song Thrush. In all plumages, *broad white-bordered dark band on underwing* and *white-tipped outer tail-feathers*. – Adult ♂: Mostly *dark lead-grey* (looks black in field) with *white supercilium*. Centre of belly and flanks show much white. – 1st-year ♂: Like adult ♂, but has *some white on throat and head-sides*, supercilium not pure white, and *browner wing-feathers and primary-coverts*. Many have some distinct white streaks and dark vermiculations on breast and flanks. – ♀: *Olive-brown above* and on flanks. Narrow yellowish-white wing-bar on greater coverts. Supercilium buff. Breast yellowish-white with *dark crescent-shaped spots*. Rarely has bluish tinge on rump, upperwing and flanks.

VOICE Call a fine, discordant 'tsii' or a short 'zit'.

Black-throated Thrush *Turdus atrogularis* V***

L 23–25½ cm. Closely related to Red-throated Thrush, now frequently treated as separate species. Breeds in Siberian taiga, westward to the W slopes of Urals; occasionally strays to W Europe, especially in late autumn and winter.

IDENTIFICATION Fieldfare-shaped. *Plain brownish-grey upperparts*. Marked contrast between *black on throat/breast* and *relatively unmarked belly and undertail-coverts* (the latter mainly whitish but can have rufous element). *Flanks only diffusely spotted grey*. *Rufous-red underwing-coverts* visible in flight. Adult ♂ has uniformly black bib, finely pale-fringed in autumn. *Outer tail blackish*. Adult ♀ and 1st-winter ♂ heavily dark-spotted and -streaked on breast (markings partly merging) and throat-sides (♂ finely streaked on entire throat, ♀ usually mainly on breast). 1st-winter has pale-tipped outer greater coverts. 1st-winter ♀ has sparser dark streaking on breast and throat-sides. *Tail grey-brown*.

VOICE Call a Fieldfare-like whining 'gvih gvih'. When uneasy, a Redwing-like 'gyack' and hard Ring Ouzel-like 'tack tack'. Song well-spaced series of simple chattering sounds, 'chip-chip-chip' and 'chat-chat-chat', etc., interfoliated now and then by brief, husky warbling or fluting phrases, e.g. 'trro-**uu** trre-**vee**', like 'hoarse Blackbird'.

Red-throated Thrush *Turdus ruficollis* V***

L 23–25½ cm. Closely related to Black-throated Thrush. Breeds in SC Siberian taiga, Altai and Transbaikalia; very rarely straggles to W Europe.

IDENTIFICATION A close copy of Black-throated Thrush, with the black on *throat, breast and tail replaced by rufous-red*. *Rufous-red underwing-coverts* visible in flight. Ageing and sexing much as in Black-throated. 1st-winter ♀ has sparse dark streaking on breast and throat-sides.

VOICE Calls apparently quite similar to Black-throated. Song poorly known, possibly mainly consists of well-spaced simple fluty phrases, not of chattering notes.

Eyebrowed Thrush *Turdus obscurus* V***

L 20½–23 cm. Very rare but near-annual autumn vagrant in Europe from Siberian taiga.

IDENTIFICATION Size of Song Thrush, slim, rather long-tailed and short-billed. *Sides of breast and flanks orange-buff*. Prominent *white supercilium, dark lore* and beneath this a *white stripe under eye* and across chin. Underwing greyish-white. Adult ♂ is almost uniformly grey on throat, cheeks and neck, whereas other plumages have off-white throat with black spots or streaking, and white-spotted dark cheeks. Sexing not always possible except perhaps for very neatest birds being adult ♂♂ (blackish lores, dark grey upper cheeks, no streaking on throat), and dullest 1st-years being ♀♀ (almost no orange tinge below, much streaking on throat).

VOICE Migration call a Redwing-like, fine 'ts(r)iih'.

Dusky Thrush *Turdus eunomus* V***

L 21–24 cm. Closely related to Naumann's Thrush, now frequently treated as separate species. Breeds in C Siberian taiga N to tundra border. Very rare vagrant in Europe.

IDENTIFICATION Shape of Song Thrush. Adult ♂ attractive with *black-and-white head pattern* and *extensive bright copper-red on upperwing*. Very dark, black-spotted back; whitish underparts with *heavy brownish-black spots on flanks* and an often complete *black band across breast*. *Supercilium white, cheeks black*. Sexes similar, but ♂ much more neat.

VOICE Calls are rather Fieldfare-like, e.g. a squeaky 'giieh' and a clicking 'chak-chak'.

Naumann's Thrush *Turdus naumanni* V***

L 21–24 cm. Closely related to Dusky Thrush. Breeds in SC Siberian taiga. Very rare vagrant in Europe.

IDENTIFICATION Shape and size as Dusky Thrush but plumage differs markedly. *Underparts extensively and heavily blotched rufous-red*, feathers in fresh plumage tipped buff. Adult ♂ neatest and reddest with no or only very few dark streaks on sides of throat, and some *red also on scapulars and rump* of otherwise brown upperparts, to young ♀ with paler and less red below and prominent dark lateral throat stripes.

VOICE Calls are apparently very similar to Dusky Thrush.

WHITE'S THRUSH

a shy species of stealthy habits, including a peculiarly sneaking gait, and cryptic plumage pattern

pale corners

long

bold underwing pattern

MISTLE THRUSH, juv. (to compare)

large, elongated

bold scaly markings

sexes and ages alike in the field

BLACK-THROATED THRUSH

1st-winter ♂

ad. ♂

supercilium may be more prominent on some birds

rather drab, only rufous in plumage is on underwing coverts

no rufous in tail

1st-winter ♀

mainly whitish, never strongly marked

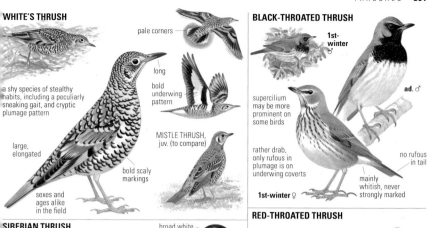

SIBERIAN THRUSH

broad white supercilium

boldly banded underwing

dark slate-grey

♂

♂

boldly patterned

ad. ♂

♀

(♀♀ are tricky to age)

lighter throat and patch on ear-coverts than in ad. ♂

less pure white than in ad. ♂

1st-year ♂

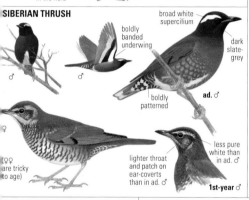

RED-THROATED THRUSH

1st-w. ♂

ad. ♂

supercilium more prominent on some

rusty-buff tinge

greyish

diffusely streaked grey

white

rufous outer tail

1st-winter ♀

EYEBROWED THRUSH

compare with Redwing (right)

REDWING

1st-w.

(variation)

a few have less white above and behind eye

ad. ♂

ad. ♂

1st-winter

ad. ♀ is very similar, but lacks white tips to greater coverts

DUSKY THRUSH

1st-w. ♀

dark brown

rusty-red

black breast-band

sexes similar—
♀♀ average duller and less contrastingly marked than ♂♂

ad. ♂

red-brown

1st-winter ♂

aged on retained, light-tipped juvenile greater coverts

NAUMANN'S THRUSH

rusty-red

sexes similar—♀♀ often have more black streaking on throat and paler, more reduced red spotting on underparts

ad. winter

1st-winter ♂

always more patterned than in Red-throated Thrush

KM

WARBLERS *Sylviidae*

In traditional bird taxonomy, the warblers are a large family of small insect-eaters. 63 species are treated here, of which 14 breed in Britain & Ireland. They have attractive songs, or at least marked vocal powers. All species inhabit woodland, shrubbery or tall vegetation, and most are, because of their diet, pronounced migrants.

The warblers are lively and nimble in their movements. They constantly hop or fly from one perch to another, and often nervously twitch or flick their wings or tail. They spend long periods concealed in the vegetation, and contacts with them are usually no more than brief glimpses. Fortunately, most species have characteristic calls.

A general overview of this great variety of small and elusive birds may more easily be gained by learning the five main genera. These are presented below.

NB: Recent research has shown that the traditional arrangement of warblers in one family does not reflect evolution and true relationships. Warblers are part of a superfamily, *Sylvioidea*, containing also babblers, tits, larks and swallows. However, the traditional treatment has pedagogic merits and is kept for this edition. Some recent splits at the species level have been covered, however.

Sylvia

(19 species.) Often live in dense, thorny thickets, or else in woodland. Sturdily built, with rather heavy bill and strong feet. The tail is narrow and square-ended. Their plumage colours are

Whitethroat

brown or grey above and white or buff below, males sometimes with rusty-red or pinkish on throat and breast and black on head (the only warbler genus in which the sexes are often somewhat dissimilar). The song of most is fast, uneven and twittering, and usually a great deal of experience is required to distinguish the species by calls. Males of several species have a short song-flight.

Acrocephalus

(11 species.) Live mostly in open swampland (reeds, bushes etc.). Slim, with long, pointed bill and flat forehead. Quite broad and bluntly rounded tail. Agile and active, they hop up reed stems, climb nimbly, sometimes with head pointed

Reed Warbler

downwards, then dive back down into the vegetation. Plumage brown above, buff-white below, some having pale supercilium. No white on tail. Sexes alike. Song varied, often diagnostic, verses long and often containing excited sounds which are repeated several times, in some species also masterly mimicry. Often heard at night.

Locustella

(5 species.) Resemble *Acrocephalus*, but have more rounded and broader tail and more streaked plumage which is often darker, keep more to the ground and low vegetation, and often scamper away unseen on the ground rather than flying off. Typically nocturnal singers. Four species have a monotonous, mechanical, insect-like song.

Grasshopper Warbler

Hippolais

(8 species.) Live in woods or in bushes with trees nearby. Closely related to *Acrocephalus*, differing in more square-ended tail and broader bill-base. Plumage brownish-grey or grey-green above, dirty white or yellowish-white below (those with green and yellow elements can recall *Phylloscopus*, see below). Sometimes the secondaries have pale edges. Sexes alike. Song loud, drawn out, excitable or chattering, in a couple of species including mimicry.

Icterine Warbler

Phylloscopus

(16 species.) Breed mostly in woodland, spending much time in treetops (but nest placed on ground). Most species are small, with slender bill and legs. Tail rather narrow and short, and square-ended (or even slightly forked). Tirelessly on the go and nimble, flutter and move about in the foliage, the smallest even hovering. Plumage usually grey-green above and white or yellowish-white below. Some species have one or two narrow pale wing-bars, and most have a pale (yellow or white) supercilium. Song loud and usually clear and pleasing, varying greatly among species.

Willow Warbler

JUDGING COLOUR IN VARYING LIGHT

Warbler identification involves assessing colours correctly. Backlighting or strong direct light impart entirely different effects to the same colour, and birds living in the aquarium-green light of the canopy foliage easily become 'mis-coloured'. Here, try to watch the bird for as long as possible and from several different angles against the light, to see it both in shadow and in sunlight.

Below:
Blyth's Reed Warbler against the light: upperpart colour becomes dark and hard to judge, contrasts in head markings increase. Note also the flat forehead!

Above:
Singing Blyth's Reed Warbler in strong direct light: upperparts look pale and brown. Note also the peaked crown!

Above:
Blyth's Reed Warbler in sunlit foliage: green tones in plumage intensified. Head markings become indistinct.

SCRUB WARBLER
saharae (N Africa, east to Libya)

inquieta (Middle East, Egypt)

GRACEFUL PRINIA

prefers shrubbery and dense undergrowth in more lush areas than Scrub Warbler

long tail and short rounded wings

dark with pale tips

occurs in more arid environment than Graceful Prinia

spread tail mainly blackish

strongly graduated tail

weak eye-stripe and buffish, indistinct supercilium

prominent eye-stripe and whitish supercilium

pale eye

dark eye

winter pale

lacks supercilium

black

buffish

pinkish

ad. summer

DZ

Scrub Warbler *Scotocerca inquieta* —

L 10–11 cm. Breeds in semi-desert and on dry, sandy or stony steppe with low bushes or grass (wadis, garrigue), sometimes in same habitat as Spectacled Warbler with rich but low undergrowth but more often alongside Desert Lark and Mourning and Maghreb Wheatears in rather mountainous country or slopes with stony deserts. Resident. Bold, lively, restless. Nests low down in vegetation.

IDENTIFICATION A small *pale sandy-coloured* 'ball of feathers' with *long, narrow, dark tail* which is often held raised in Wren fashion. *Head relatively big*, with characteristic *broad, pale supercilium* set off by *narrow dark eye-stripe* and finely dark-streaked crown. Legs long and appear to be set far back. – Variation: Birds breeding in Algeria east to Libya (*saharae*) are *pale* with rather plain pinkish-buff plumage, *poorly marked buffish-white or buff supercilium* and *pale eye*. Birds in Egypt and Middle East (*inquieta*) are *slightly darker* and more pinkish-tinged, have *well-marked whitish supercilium* and *dark eye*. Breeders in SW Morocco (*theresae*) are *darkest*, quite *saturated brown, heavily streaked dark on crown and nape*, and have *rather pale eye*; forepart of supercilium rufous-tinged.

VOICE Calls variable. When agitated, a metallic, dry descending trill, 'prrrrrrr'. Highly characteristic is a disyllabic whistle with second note lower, 'wii-wew'; a four-syllable variant with notes successively falling seems to have mainly a song function, 'sii-sü-su-so' (at distance almost like Common Sandpiper call). True song often introduced by differing first syllable, e.g. 'tsrisrü si-sü-su' or 'trsirr-vuy-vuy-vuy'.

Graceful Prinia *Prinia gracilis* —

L 10–11 cm. Breeds in shrubbery and tall grass in both wet and drier habitats, often along ditches, riversides and pools, on sandy ground, in agricultural areas and near buildings, wherever there is dense undergrowth, tamarisk bushes, reeds, rushes, etc. Resident. Bold, active. Nests low in vegetation.

IDENTIFICATION Recalls Scrub Warbler, but note following points: *head smaller, more rounded* and *more uniform in colour* (lacks both pale supercilium and dark eye-stripe); plumage *more greyish-tinged*, not so warm brownish; more *dark-streaked on mantle and back*; long *tail is strongly graduated* and each feather has *black-and-white tip*. Breeding ♂ has *black bill*. Eye yellow-brown.

VOICE Call an explosive 'tlipp!' or a metallic rattling trill, 'srrrrrrt', often repeated and with somewhat 'swaying' rhythm. Alarm a sharp 'tsiit', not unlike that of Tree and Red-throated Pipits, but harder. Song a frenzied grinding 'srr**lip** srr**lip** srr**lip** srr**lip**...'.

Scrub Warbler

Graceful Prinia

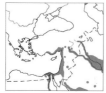

Barred Warbler *Sylvia nisoria* **P**4

L 15½–17 cm. Breeds in tall bushes with scattered trees in open country; shares habitat with Red-backed Shrike. Summer visitor; passage migrant in Britain, mainly on E coast (Aug–mid Oct, rare in spring). May be seen flying between bushes, and in spring in near-level song-flight.

IDENTIFICATION A large, *long-tailed Sylvia.* Wing-coverts and tertials ± pale-edged. White at tips of tail-feathers. – Adult ♂: *Iris bright yellow. Upperparts lead-grey,* darkest on head; *distinct white fringes to wing-coverts and tertials.* Entire *underparts closely and heavily vermiculated* dark grey. – Adult ♀/1st-year ♂: Like adult ♂, but iris slightly duller yellow, upperparts more brown-grey, barring below not so obvious and complete. (1st-year ♀ nearly unbarred, has browner tone above, and yellow-brown iris.) – 1st-autumn: Grey-brown above, buffish-grey and unbarred below. Iris dark. Told by *light buffish-grey tertial and covert fringes.*

VOICE Call characteristic, a loud rattle, dying towards end, 'trrrrr't't't-t', somewhat recalling scolding House Sparrow; sometimes also added to song; tempo varies, is occasionally slower. Alarm a hoarse, muffled 'chaihr'. Song like Garden Warbler's (at times confusingly so!), but almost always audibly *harder and more rasping* and with slightly *higher pitch.* Even more similar is the ecstatic song of some Whitethroats, and at times the rattling call must be heard or the bird seen before certain identification is made.

Garden Warbler *Sylvia borin* **m**B2 / **P**2

L 13–14½ cm. Breeds in woods with clearings, in groves, overgrown parks and larger gardens with tall trees (not content with just shrubbery and solitary trees) with rich undergrowth. Summer visitor (in Britain & Ireland mainly mid Apr–Sep, passage migrants to Oct), winters in tropical Africa. Discreet in habits, difficult to see other than briefly. Nests low down in bush or dense herbage.

IDENTIFICATION 'Anonymous' appearance, an olive brown-grey and rather plump warbler with no obvious features. *Bill rather thick and stubby,* bluish grey-buff with slightly darker tip. Legs light greyish, rather strong. Diffuse paler eye-ring (around dark eye) and even less prominent pale, short supercilium. *Often an inkling of a purer grey tinge on neck-side.* Sexes and ages similar, but 1st-autumn birds have fresher wings with paler tips to tertials and remiges.

VOICE Call (heard from foliage, usually without bird being seen) a cackling series of slightly nasal, 'thick' clicks, 'chek, chek, chek, chek,...'; tempo quickens with increasing unease. (Blackcap's anxiety-call a harder and more distinct clicking.) Song beautiful, 3–8-sec.-long verses of rapid, rather deep notes, one moment thrush-like and the next harsher, not forming any clear melody but *shuttling*

irresolutely up and down: it sounds like a *rippling brook.* Lacks Blackcap's flute-like final notes. Some are harder-voiced, can easily be confused with Barred Warbler.

Blackcap *Sylvia atricapilla* **m**B2 / **P**2 / **W**4

L 13½–15 cm. Breeds in shady woodland with dense under-storey, in parks and gardens with rank undergrowth. In Britain & Ireland common in deciduous and mixed wood-land. A common bird over large parts of C and S Europe. Migratory in N and E, short-distance migrant or resident in W and S. Nests low down in bush.

IDENTIFICATION Rather stocky build, roughly as Garden Warbler. Dirty grey above and light olive-grey below, ♂ has *small black cap* (reaches only to upper part of eye; cf. the small-bodied and big-headed tits, p. 344, which have larger black cap as well as a black bib), ♀ and juvenile with *red-brown cap.* 1st-winter ♂ has cap brown, black or a mixture of both colours. Tail invariably lacks white. Bill and legs grey.

VOICE Call a tongue-clicking 'teck' (harder and louder than Lesser Whitethroat's 'tett'), when uneasy repeated in long series, 'teck-teck-teck-...', occasionally with drawn-out, hoarse 'schreh' notes interposed. Song one of the finest; begins like Garden Warbler's with an irresolute chattering (can be shortened or omitted, often so during midday heat and in S Europe), *turning into clear, slightly melancholy flute-like notes at end.* As other *Sylvia* warblers, has a drawn-out subsong of mimicry and squeaky, rasping sounds.

(Common) Whitethroat *Sylvia communis* **m**B2 / **P**1

L 13–15 cm. Breeds in scrubby areas, in farmland with hedge-rows, at woodland edge. Summer visitor (in Britain May–Aug, passage migrants to Oct), winters S of Sahara. Often sings from bushtop, wire or in short ascending song-flight.

IDENTIFICATION Rather big, *sturdy* and *long-tailed,* some-what heavy and slow in movements. *Bill strong* with dif-fusely paler base. *Legs* rather strong, *yellow-brown.* Distin-guished in all plumages by *broad rusty-brown edges to tertials and greater coverts* (a feature shared only with Spectacled and Tristram's Warblers), which contrast with grey-brown back. White tail-sides. – Adult ♂: Grey head, *white eye-ring* and throat, pink-tinged breast. Iris reddish-ochre. – Other plum-ages: Grey-brown head, diffuse eye-ring, off-white throat, buffy-white breast, iris grey-brown or yellow-brown.

VOICE Call a hoarse 'vaihd vaihd vaihd' with nasal tone. Alarm a drawn-out, hoarse 'chaihr', recalling Dartford War-bler. Song from perched bird normally very characteristic, a short, fast verse with *scratchy, hoarse, gruff voice,* delivered in *jerky* and *jolting rhythm.* In song-flight, verses are drawn out and voice becomes more like other *Sylvia* species, at times confusingly like Barred Warbler's! Sings frequently.

Barred Warbler   Garden Warbler   Blackcap   Whitethroat

BARRED WARBLER

eye dull yellow

adult ♀/ 1st.-s. ♂

white tips to tail (striking in flight)

eye bright yellow

scaly

pale-fringed tertials

imm. autumn

some show barring on rear flank

pale wing-bars

dark-spotted

adult ♂

barred below like a hawk

GARDEN WARBLER

usually sings well concealed in foliage

short bill and 'mild' expression

grey neck-side

nondescript and featureless—a good character!

plain

BLACKCAP

rufous

moulting

imm. ♂ autumn

black

♀/ juv. ♂

MARSH TIT (for comparison): note pale cheek and black bib

ad. ♂

rather dark dusky grey

WHITETHROAT

greyish

sexing frequently difficult except for fully developed ♂♂, like the one shown

brownish

bright rufous wings contrast with rest of plumage

contrasting dark-centred tertials

buff

pink

♀/ imm. autumn

♀/ 1st-s. ♂

ad. ♂ spring

DZ

Lesser Whitethroat *Sylvia curruca* mB3 / P2

L 11½–13½ cm. Breeds in farmland and parkland with mature hedgerows, in young conifers, gardens with hedges and berry bushes, even in pure thorn scrub (but not so often as Whitethroat, and prefers some trees). Summer visitor, winters in NE Africa. Very numerous in E Mediterranean on passage (esp. early Apr). Unobtrusive at breeding site, its presence revealed mainly by its voice. Nests in dense bush.

IDENTIFICATION *Small* but *compact* and rather short-tailed *Sylvia*. Within the region, always *grey-brown above* with greyer crown and tail. Best feature is *darker grey ear-coverts*, darker than crown, combined with small size and *dark grey legs*. Some have suggestion of greyish-white supercilium. *Iris mainly dark* (grey or brown). *Bill shortish, dark* with paler blue-grey base. – Variation: East of Caspian Sea, paler and browner birds occur in semi-deserts (ssp. *halimodendri*), rarely straggling to W Europe. In mountains of Central Asia, large and dark birds (*althaea*) breed, but no definite records of these exist from Europe, and field identification tricky. Likewise, smallest and palest brown desert form (*minula*) has not been confirmed yet from Europe.

VOICE Call a quiet, dry, clicking 'tett'. On migration, also a chattering, scolding 'che-che-che-che' (superficially like Blue Tit). Song in Europe a *rattling, loud series*, 'tell-tell-tell-tell-tell', preceded by a short scratchy warble. In E Turkey, Caucasus, Central Asia and Siberia, emphasis usually on the scratchy warble, while the rattle sequence is shortened or omitted. Intermediate versions also occur.

Eastern Orphean Warbler *Sylvia crassirostris* V***

L 15–16½ cm. Very closely related to Western Orphean Warbler, and only recently separated, mainly on account of different morphology and song. Breeds in the same kind of habitats, requiring at least some trees, not only bushes. Shy like its relative, too. Nests in tree or dense bush.

IDENTIFICATION Very similar to Western Orphean Warbler, and without song or close and prolonged views often impossible to tell from this. Eastern Orphean tends to have somewhat *blacker and better defined hood in* ♂ (though many similar!), *upperparts more greyish* (especially in ♂), *underparts whiter* (only moderately buff-grey wash on flanks and vent), and always *dark-centred longer undertail-coverts* (but these often difficult to see). Bill somewhat stronger and longer (but much overlap); perhaps more useful is tendency in adult ♂ to have *more and better defined bluish-white at base of lower mandible*, often reaching gonys angle (much less and more diffuse pale base in most Western).

VOICE Calls are not thought to differ from those of Western Orphean Warbler (which see). Song, however, is very different, being longer and more varied, often elabo-

rate phrases containing several motifs, some even recalling Nightingale, e.g. 'trü trü trü shivü shivü, yu-yu-yu-brü-triüh'.

Western Orphean Warbler *Sylvia hortensis* V***?

L 14½–16 cm. Breeds in deciduous woods (e.g. in oak forest, olive groves or luxuriant riverine vegetation), in tall scrub with some trees (maquis), often on rocky, sunny mountain slopes, also locally in pine forest. Migrant, winters S of Sahara. Rather shy and elusive at breeding site, keeps mostly in treetops. Nest often at breast height, in bush or low tree.

IDENTIFICATION Large, somewhat bigger than Blackcap, and has *longer bill* and slightly *bigger head*. Dull *brown above*, dirty white below with *pinkish-buff tinge on flanks and vent*, and sometimes faintly on breast, creating *contrast to white throat*. *Tail square-ended, with sides white-edged*. – Adult ♂: *Greyish-black on ear-coverts, lores and forehead*, sometimes dark also on crown (but always indistinct border with grey hindcrown/nape, unlike Sardinian Warbler). *Iris pale*. – Adult ♀: Head less black than ♂'s (often only ear-coverts and lores darker grey, giving impression of outsized Lesser Whitethroat), and upperparts are brown. Iris variable, often with narrow pale ring, sometimes a little darker grey. – 1st-winter: Like adult ♀, but *head lacks any black*, being grey-brown, and *iris always dark*. Frequently a *pale supercilium from forehead to eye*.

VOICE Call a tongue-clicking 'teck', barely distinguishable from Blackcap. Also has a buzzing rattle when agitated, 'trrrrr', somewhat recalling Rüppell's Warbler, and a Barred Warbler-like hoarse 'chaih'. Song a simple composition of clear motifs repeated a few times in Ring Ouzel fashion, e.g. 'türu türu türu türu..... liru liru liru trü...'.

Arabian Warbler *Sylvia leucomelaena* —

L 14½–16 cm. Breeds in acacia stands in semi-desert. Resident. Relatively shy; restless and mobile, often disappears without a trace, despite openness of breeding terrain.

IDENTIFICATION Most like Eastern Orphean, but has *longer and round-tipped tail* with *white tips to outer feathers* (not white sides as on Eastern Orphean), also always *dark eye* with *trace of white eye-ring, shorter bill* and *blacker cap*. Brilliant white throat. Often perches quite upright, but also hops about in crouched posture in dense acacia crowns. *Constantly flicks tail downwards.* Overall impression in fact not at all unlike that of a bulbul (which is often found in same territory). Sexes similar. Immature has brown-grey crown.

VOICE Call a rather 'thick' clicking 't(r)ack', often repeated. Song short verses at slow pace with clear, attractive voice and wide tonal range, can recall Blackcap, Eastern Orphean Warbler, Rustic Bunting or, at distance, even Blackbird.

Lesser Whitethroat   Eastern Orphean Warbler

Western Orphean Warbler   Arabian Warbler

LESSER WHITETHROAT

some have pale 'spectacles'

earth-brown

halimodendri
(Central Asian semideserts)

pale sandy-brown

imm. autumn

extensive white in tail (not visible in this position)

unmarked

dark ear-coverts

lead-grey

autumn

back and wing same colour (cf. Whitethroat)

ad. summer

althaea
(Central Asian mountains)

dark slaty-brown

very dark

strong

this large and dark form has not yet been recorded whitin our region

autumn

EASTERN ORPHEAN WARBLER

pale eye

white tail-sides (unlike Arabian Warbler)

grey

outer tail-feathers have much white

faint pinkish-buff breast-sides and flanks

spotted

ad. summer ♂

WESTERN ORPHEAN WARBLER

pale eye

brownish-grey

strong

whole underparts suffused pinkish-buff in typical birds

diffusely spotted (cf. Lesser Whitethroat)

imm. autumn

dark-eyed ♀♀ and imm. surprisingly similar to Lesser Whitethroat when size not evident!

plain buffish

identification on plumage characters outside range very difficult

ad. summer ♂

ARABIAN WARBLER

found in stands of acacia

♀ winter

eye-ring subdued in some

constantly dips tail downwards

dark eye

looking all-black in flight

often adopts typical crouched, stealthy posture

pale eye-ring

tail slightly graduated with pale tips to outer tail-feathers

plain

ad. summer ♂

only faint pale tips, unlike Eastern Orphean Warbler

DZ

Sardinian Warbler *Sylvia melanocephala* V**

L 13–14 cm. Breeds in tall bushes and open woodland with dense thickets, locally also in mere waist-high vegetation; sometimes in gardens or tree clumps close to habitation. Primarily resident. Active and restless, is not shy, often reveals itself. Nests in bush, usually rather low down.

IDENTIFICATION About size of Lesser Whitethroat, *compact* and rather large-headed. Flicks wings and tail, often adopts crouched posture with head lowered. In all plumages: *white throat* but *rather dark flanks*; red-brown eye-ring (reddest on ♂, duller on ♀) outside red orbital ring; rounded, *rather long tail*; on adult, dark tail-feathers with white tail-corners. Iris ochre (adult) or grey-brown (imm.). – ♂ (from Aug): *Black cap* with usually diffuse border against *dark grey mantle*; mid-grey flanks. – ♀: Grey head (may have some greyish-black on forecrown and ear-coverts), *brown back*, greyish-brown flanks. – Variation: E Mediterranean birds (ssp. *momus*) slightly smaller and paler.

VOICE Call a loud, very hard 'tseck'. More characteristic is the 'rattle-call', a series of mono- or disyllabic rattling notes at rapid speed, 'trr-trr-trr-trr-trr' or 'tü-**trru** tü-**trru** tü-**trru** tü-**trru**', or with first note deviating, 'tu**rett** trett-trett-trett-trett'. Song a *Sylvia* chatter very familiar in S Europe, usually short (2–5 sec.), harsh, rattling verses at high speed with very brief whistled notes inserted; includes a great deal of the 'trr-trr' rattling sound. Observers should learn this species' song to enable easier comparisons.

Ménétries's Warbler *Sylvia mystacea* —

L 12–13 cm. Breeds in bushes, often tamarisk, or in open wood with undergrowth (maquis), in drier, open and high-lying terrain, but also along watercourses, on fringes of palm groves etc. Migrant, winters around Red Sea.

IDENTIFICATION Very like Sardinian Warbler, especially of ssp. *momus*, but neater and has slightly shorter tail. – ♂ told by *dark cap* not being velvety black but more *dull greyish-black* and *less extensive* at rear; also, flanks are not so grey but *whitish*; majority in Turkey have white throat and breast-centre, but some (and most of those in Caucasus and Central Asia) have *pinky or light brick-red tone to throat/breast*. – ♀ told from Sardinian (*momus*) by: *paler* and *more uniform sandy-brown upperparts* (without contrastingly greyer head); more uniform buff-washed grey-white underparts; usually *white eye-ring*; more uniform grey-brown tertials (darker with more distinct pale edges on Sardinian); paler bill with light grey or pinkish base (blue-grey on Sardinian).

VOICE A tongue-clicking 'tseck' like Sardinian Warbler's, also a sparrow-like chattering 'cher'r'r'r' quite unlike Sardinian's rattle-call, more like Rüppell's Warbler (but shorter and more feeble). Song like Sardinian's, with 'simmering',

chattering notes interspersed with clear whistles, but tempo somewhat slower and rhythm a little more jerky.

Cyprus Warbler *Sylvia melanothorax* —

L 12½–13½ cm. Breeds in dense shrubbery in open terrain (maquis) in Cyprus, where fairly common. Summer visitor (Mar–Oct), winters in Israel, Jordan and Sinai; holds winter territories in barren wadis. Shy. Nests low down in bush.

IDENTIFICATION Looks like a Sardinian Warbler with *dark-spotted underparts* and *narrow white eye-ring* (outside indistinct brick-brown orbital ring). In all plumages, greyish upperparts and *contrasting pale edges to tertials, secondaries and greater coverts* (as Rüppell's Warbler); *undertail-coverts* have *dark centres*; flanks greyish(-buff); bill-base pinkish. – Adult ♂ has black cap and is heavily dark-spotted below; 1st-winter ♀ has few or almost no spots visible below; adult ♀ and 1st-year ♂ are intermediate between those two, have some greyish-black on forecrown and cheeks.

VOICE Call a throaty clicking 'zreck'. When agitated, gives hoarse drawn-out 'tschreh tschreh...'. Alarm a rattling, high-pitched 'ze-ze-ze-ze-ze-...'; sometimes Sardinian-like 'tri-tr'tr'tr'tr'tr...' (though faster). Song a dry, rugged warble, rather monotonous, virtually lacking clear whistled notes. Tempo slow and slightly jerky and uneven.

Rüppell's Warbler *Sylvia rueppelli* V***

L 12½–13½ cm. Breeds in dense, often thorny or prickly shrubbery with some low trees on arid, rocky mountain slopes, sometimes in open oak forest with shrub layer (tall maquis). Summer visitor, winters S of Sahara. Rather shy at breeding site, but ♂ often seen in song-flight in spring (at times 'butterfly-flight'). Nests low down in dense bush.

IDENTIFICATION Size of Sardinian Warbler, but slimmer and with slightly smaller head and more obvious neck. Characteristic of all plumages is *grey upperparts* with *contrasting pale edges to dark tertials, secondaries and often greater coverts*; alula often strikingly black. Bill narrow and longish, on some (not all) with clearly decurved tip. – ♂ (from Jan): *Black* crown, *head-sides and throat* (unique), distinct white moustachial stripe. Orbital ring and legs dull brick-brown. – ♀/juvenile ♂: Crown and mantle greyish (faintly tinged brown). Orbital ring rather weak, brownish-white. Some spring ♀♀ have dark grey on throat. Told from immature ♀ Cyprus Warbler by *unmarked greyish-white undertail-coverts, light greyish-white flanks* and *pale blue-grey bill-base*.

VOICE Call a distinct clicking 'zack', most like Blackcap's, also a dry rattling 'zrrr' and intermediates. Alarm a sparrow-like rattling series, often more drawn out, faster and feebler-voiced than Barred Warbler's, 'zerrrrrrrrr'r'r...'. Occasionally a 'bubbling' nasal scold, 'zett-ett-ett-ett-...'. Song resembles Sardinian's most, but has slightly *lower pitch* and more 'bouncing', *shuttling* character between lower, 'chatting' warbled notes and high, clear whistles; the warble is also more pleasing and sonorous than Sardinian's.

Sardinian Warbler

Ménétries's Warbler

Rüppell's Warbler

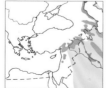

SARDINIAN WARBLER

momus

contrasty pale tertial edges

grey

warm brown

grey

deep buff-brown

♀

melanoce-phala

♀

darker

momus (Middle East)

ad. ♂

white corners

black cap

melanocephala (Europe, Turkey)

grey

grey-white

paler grey above and below than European breeders

ad. ♂

MÉNÉTRIES'S WARBLER

head and back uniform

diffuse tertial edges

often sweeping tail movements

buff

♀

yel-low-pink

rubescens (SE Turkey, Middle East)

ad. ♂

whitish, or only faint pink tinge

black forecrown grades smoothly into grey hindcrown

mystacea (NE Turkey, Caucasus)

yel-low-pink

pink

ad. ♂

CYPRUS WARBLER

distinctive pale edges

spotted

cold grey-brown

yel-low-pink

imm. ♀ autumn

grey

grey-brown

♀ spring

spotted, mostly on throat and breast

whitish narrow eye-ring outside brick-red orbital ring

black

grey

yel-low-pink

boldly spotted

♂ spring

RÜPPELL'S WARBLER

touch of dark 'Mexican moustache'

distinctive pale edges

culmen slightly curved

(variation)

pale throat

♀ spring

dark-spotted throat

♀ spring

immatures in autumn usually similar to pale-throated ♀♀, but have slightly browner crown; a few (probably ♂♂) have a little dark on throat

unmistakable!

♂ spring

DZ

Spectacled Warbler *Sylvia conspicillata* V***

L 12–13 cm. Breeds in low bushes and scrub (garrigue) on mountain slopes, in brush and herbaceous vegetation on dry heaths or saltflats. Short-distance migrant or resident. Rather shy, yet often seen in the open. Nests low in bush.

IDENTIFICATION Like a *small* and *slim* Whitethroat, but has *shorter primary projection*, proportionately smaller body/ *bigger head*, and *narrower bill* with paler inner part of lower mandible; also consistently *quicker* and 'more nervous' in movements. – Adult ♂: Differs from Whitethroat in: *greyish-black lores* and forecheeks; *darker brownish-pink breast* and touch of grey on lower throat. – ♀/1st-winter: Resembles both Whitethroat and Subalpine Warbler. Told from former by smaller size, short wing, slimmer bill, broader rusty-brown wing-covert fringes (hardly any grey-black centres visible); from latter by shorter primary projection, brighter rusty colour on wing and *paler yellow-pink legs*.

VOICE Call a high, dry buzzing 'drr', often strung together in long series with short syllables inserted, 'drrrrrr-dr-dr-dr-drrrrrrr...' ('rattlesnake rattle'). Song a usually high, fast warble which is often introduced by several clear whistled notes (which can sound a bit like Crested Lark).

Subalpine Warbler *Sylvia cantillans* V**

L 12–13 cm. Breeds in various habitats, often in bushes on dry slopes or sandy heaths, also in open woodland (often holm oak) or in lush shrubbery beside watercourses etc. Summer visitor (mostly Apr–Sep), winters S of Sahara. Discreet habits, but has short song-flight. Nests in thick bush.

IDENTIFICATION Size as Lesser Whitethroat, but *more slender* in build; fairly short-tailed. – Adult ♂: Easily identified by *lead-grey head* and grey mantle, and by having *brick-red on throat*, bordered by *white moustaches*, and on breast. Eye-ring and orbital ring brick-red. Legs brownish-pink. – Adult ♀: As ♂, but throat/breast usually only lightly washed brownish-pink (cf. ♂ Ménétries's Warbler); more *brownish above* and not lead-grey on head; faintly brick-coloured or *whitish eye-ring* (outside indistinct red-brown orbital ring). Legs greyish brown-pink. – 1st-winter: As ♀, but browner above, buff-white below, and initially with greyish yellow-brown iris (adult more reddish). – Variation: Complex, not yet fully resolved. Birds in Balkans and Turkey (*albistriata*) morphologically rather distinct, adult ♂ having deep brick-red breast, largely whitish belly and prominent white submoustachial-stripe. Breeders in SW Europe (*cantillans*) and NW Africa (*inornata*) are more evenly orange-buff or pink-buff below in adult ♂, but birds of W Mediterranean islands and parts of N Italy (*moltonii*) separated, sometimes as species, mainly on account of different call and genetic distance, whereas plumage is similar.

VOICE Call a Lesser Whitethroat-like dry clicking 'tett', often rapidly repeated in long series (unlike e.g. Sardinian Warbler); some variation: on W Mediterranean islands and NW Italian mainland a dry trill 'trrrrt', in SE Europe and Turkey a rough 'trek' (often doubled). At times gives muffled, creaky 'ehd' like Trumpeter Finch, and when agitated a hoarse 'tscheh tscheh tscheh...'. Song *higher-pitched* than Sardinian's, and verses on average a bit longer; almost Linnet-like in 'rippling' and *bouncing* nature, a rambling mix of harsh twitters and short, high squeaky notes.

Asian Desert Warbler *Sylvia nana* V***

L 11½–12½ cm. Closely related to African Desert Warbler, but now commonly treated as a separate species. Breeds in semi-desert or on dry steppe in Central Asia. Migrant, wintering west to Levant and Arabia. Rare autumn vagrant to W Europe. Active and restless, often runs on ground. Not hard to see owing to openness of habitat. Nests low in bush.

IDENTIFICATION *Small*. Habitat, *light grey-brown upperparts* and size give immediate pointers to species, but in brief glimpse other pale brown species are possible contenders. *Ochre fringes to tertials and greater coverts* and broad off-white *eye-ring* can recall Spectacled Warbler. Bear in mind that: *eye is yellow; legs light yellow-brown; uppertail bright rusty-brown. Bill slender, dull yellow with just dark tip* and culmen. Like African Desert Warbler but is *somewhat greyer above* on head and back, has *darker primaries* and more *vivid rufous tinge on uppertail;* best character offered by *black central streaks to tertials and central tail-feathers.* Can recall a ♀ Whitethroat but is much smaller, has yellow eye and paler yellowish legs.

VOICE Call a chattering high 'cherr'r'r' (can recall both sparrow and Blue Tit). Song a spirited, clear signal, introduced by a few subdued, chatty notes, followed by a silvery-clear trill falling in pitch, 'che-tre-zerr **srrrihr'r'r'r'**. The chattering alarm is sometimes annexed to the song.

African Desert Warbler *Sylvia deserti* —

L 11–12 cm. Formerly treated as a race of Asian Desert Warbler, nowadays separated on account of slightly different song and distinct morphology. Breeds in semi-deserts in NW Africa. Resident. Habits as for Asian Desert Warbler.

IDENTIFICATION Same size and shape as Asian Desert Warbler, which see for differences. *Light ochrous-buff upperparts* of a small warbler in NW Africa eliminates all other options. In poorer views, ♀ or immature Spectacled and Tristram's Warblers need to be eliminated; then note: *pale yellow eye; legs light yellow-brown* (not pinkish-brown); *uppertail with a rusty tinge; sides of head very pale*, running into throat without contrast. Fine *dull yellow bill with dark tip* and culmen.

VOICE Chattering call, 'kerrrr', resembles Asian Desert Warbler. Song a clear, jerky verse, best recognized by *high, clear voice* and by containing so few harsh notes. Each song opened with a dry trill (call-like) followed by a series of loud notes in irregular rhythm, 'krrrr-**ti-tu-ti-ti-tew-ti**'.

Spectacled Warbler

Subalpine Warbler

Afr. Desert W.   Asian Desert W.

SPECTACLED WARBLER

♀ like a small cute Whitethroat, but rufous wing plainer and even more striking

very short primary projection and rufous tertials with narrow, *pointed* dark centres (cf. Whitethroat)

♀/ imm. autumn

♀/ imm. autumn

comparatively large, pure grey head with black lore

rufous wing

dull pinkish

ad. ♂

SUBALPINE WARBLER

ad. ♀ spring

pale eye-ring outside red orbital ring

pinkish-buff

lead-grey

pale legs (cf. Lesser Whitethroat)

dark tertials with distinct tawny-brown edges

♀ autumn/ imm.

broader white 'moustache' in *albistriata*

pale rufous below

ad. ♂ spring

albistriata (SE Europe and Turkey)

white

ad. ♂ spring

ASIAN DESERT WARBLER

has the peculiar habit of sometimes 'tailing' Desert Wheatear!

dark shafts on tertials and central tail-feathers

pale grey-brown

pale eye

AFRICAN DESERT WARBLER

plain tertials and central tail-feathers

pale eye

golden-buff

DZ

Dartford Warbler *Sylvia undata* r**B**4

L 13–14 cm. Breeds on bushy coastal heaths or in higher sites (low maquis, garrigue), sometimes also in open pine or oak wood with whin and heather. In Britain mostly on heaths with gorse and heather. Mainly resident. Rather inquisitive and fearless, often reveals itself in the top of a bush. Nests low in bush.

IDENTIFICATION *Small*, smaller even than Spectacled Warbler, but has *longer tail* which is often *cocked*. Flight fluttering and jerky, bobbing tail. *Dark grey* or grey-brown above, *dull wine-red below* (but red element of underparts frequently surprisingly difficult to see; often appears all dark at distance and in poor light, which can invite confusion with Marmora's and 'Balearic Warbler') with fine off-white spots on throat (close-range views) and dirty-white belly-centre. Iris red-brown, *orbital ring red*. Sexes similar, but ♀ browner above and not so saturated wine-red below. – Juvenile: Lacks red orbital ring, has dark iris, and is even browner than ♀; very like juvenile Marmora's Warbler (which see).

VOICE Call distinctive, a *drawn-out, harsh*, 'chaihhrr' (sometimes with extra note, 'chaihhrr-chr'). Alarm a chattering 'trü-tr'r'r'r', rather like Sardinian Warbler but higher in pitch. Song short *high-speed* bursts of a rattling warble with occasional whistled notes inserted; has *faster tempo, weaker voice* and *lower pitch* than Sardinian Warbler.

Marmora's Warbler *Sylvia sarda* V***

L 13–14 cm. Breeds on hillsides in low maquis and garrigue (up to *c.* 1800 m altitude) and in rocky coastal scrub (heather, broom, etc.); avoids woodland. Resident or short-range migrant, wintering in N Africa and rarely S Italy. Nests low down in bush or scrub.

IDENTIFICATION Size and shape are as Dartford Warbler (though on average not quite so long-tailed), as is behaviour. Distinguished from that species by *grey underparts* (but beware that Dartford Warbler, too, can look greyish below in poor light) and *lack of distinct, pale spots on throat* (ad. ♂ can show hint, ♀ and juv. have entire throat paler greyish-white). Sexes similar, but adult ♂ often has darker (grey-black) forehead and lores. – Juvenile: Very like juvenile Dartford Warbler, but has *whiter throat, breast-centre and belly* (not so warm grey-buff), and *flanks are more greyish-buff* (less warm brown). See also very similar and closely related 'Balearic Warbler'.

VOICE Alarm and contact call a rough, throaty 'chreck' or 'tscheck' (recalling as varied and unrelated species as Stonechat, White-winged Tern and Dipper!). Also a chattering series. Song resembles both Sardinian and Spectacled Warblers, a short, fast, liquid warble or outburst of notes, which usually ends with a rather *clear trill* or *flourish*.

'Balearic Warbler' *Sylvia sarda balearica* —

L 13–14 cm. Differs from Marmora's Warbler very subtly morphologically and on vocalization, and by some regarded as a separate species, although here kept as a distinct local race until better studied. Resident on Balearic Islands (except Menorca). Apparently only few confirmed records (non-breeders) on Spanish mainland. Habitat and habits as Marmora's Warbler.

IDENTIFICATION In most respects very similar to Marmora's Warbler, and without vocalization, above all song, and close views of preferably adult ♂ frequently impossible to separate these two. A fraction *smaller* but proportionately *longer-tailed* than Marmora's Warbler, and adult ♂ has *whitish throat*, not grey. On average *slightly paler grey above* and *tinged pinkish-buff on flanks* (Marmora's Warbler usually rather dark grey above and all-grey on underparts). There is a tendency for the legs to be brighter orange-brown, especially in juveniles (often slightly duller grey-brown in Marmora's Warbler), but much overlap. Very similar juvenile Dartford Warbler is best separated on calls.

VOICE Similar to Marmora's Warbler, but apparently usually separable with practice. Call both a nasal short, subdued 'tset' or 'tret', and a nasal, muffled 'cherr' (or 'catch'; the latter may recall a distant Snipe being flushed), and is rather different from the call of Marmora's Warbler. Song is at least as fast and short as Marmora's Warbler, possibly faster, but less pleasing, is more mechanical and repetitive, each stanza shuttling up and down in pitch like a rattle, at a distance somewhat recalling rattling contact call of Sardinian Warbler.

Tristram's Warbler *Sylvia deserticola* —

L 11½–12½ cm. Breeds on bare mountain slopes or heaths with bushy vegetation (low maquis, garrigue); scattered holm oaks and cedars tolerated. Mostly resident, but some move short distance south in winter, and often found at slightly lower level and on flatter ground, e.g. in semi-desert or dry steppe with low, dry bushes. Nests in bush.

IDENTIFICATION Small and rather long-tailed (although tail somewhat shorter than Dartford Warbler's), plumage intermediate between Spectacled Warbler (*rusty-brown on wing, white eye-ring*) and Subalpine Warbler (*brick-red on throat, breast and flanks*, at least hint of white moustachia stripe). Note: entire throat red-brown, finely scaled white when fresh; grey on head and back. Autumn birds, however, are browner above, and throat appears pale in side view, so can be very like Spectacled Warbler.

VOICE Call a very hard metallic 'chick', often repeated in series. Alarm a sparrow-like chatter, 'chett-ett-ett-ett-ett-ett-...', with slight nasal tone. Song fast and twittering with a few short whistles inserted, thus like several other *Sylvia* warblers; as fast as song of Dartford Warbler, but on average more varied, at times a little recalling Subalpine Warbler.

Dartford Warbler

'Balearic W.'   Marmora's W.

Tristram's Warbler

DARTFORD WARBLER

attracted to areas with gorse and broom

true colour of underparts sometimes difficult to detect due to poor light-comditions

♀ shows faint brown tinge

warm earth-brown

grey

white spots

duller vinous-red than ♂

tail long and narrow, often raised

dusky, brownish-white throat

dark vinous-red

imm. autumn

warm brown hue

yellow-brown

♀

♂ spring

MARMORA'S WARBLER

grey

sexes very similar but ♀ shows paler throat and no dark lore

grey

dark lore

dark

only faintly tinged brown

paler throat than juv. Dartford

♀

dusky grey-buff

lead-grey

imm. autumn

orange-brown

♂ spring

'BALEARIC WARBLER'

spends much time foraging on the ground, like its congeners

dark lore

pale

buffish-pink hue

brownish-orange

♂ spring

TRISTRAM'S WARBLER

long tail is often raised (unlike Spectacled)

white eye-ring outside red orbital ring

pale central throat

ad. ♀/ winter ♂

lateral throat stripe pinkish-buff

ad. ♀/ winter ♂

long

buff tinge

rufous-brown when fresh; worn birds in spring may have only traces of rufous

beware confusion with similiar-looking Spectacled Warbler

usually somewhat warmer buff tinge than on juv. Spectacled Warbler

imm. autumn

ad. ♂ spring

rusty-red

DZ

Sedge Warbler *Acrocephalus schoenobaenus* m**B**2 / **P**

L 11½–13 cm. Breeds in dense vegetation in marshy areas: in reeds (preferably with some bushes), riverside willows, swampy bushland or along ditches with lush herbage, grass, reeds, etc.; locally also drier habitats. Summer visitor (in Britain & Ireland mid Apr–Sep/Oct), winters S of Sahara. Mobile, bold, often in open. Nests low in dense vegetation.

IDENTIFICATION Has a *distinct and quite long off-white or buff-white supercilium* which contrasts with *rather dark, diffusely streaked crown* and cheeks. Mantle/back diffusely dark-streaked, but in the field often appear almost uniform brown. In flight, from behind, shows *warmer yellow-brown rump*. Crown darker at sides, creating lighter brown broad central band, on some juveniles so distinct that confusion can arise with Aquatic Warbler (which see). – Juvenile: Often distinct dark spots on breast, and fresher wing than adult.

VOICE Anxiety-call a muffled, dry rolling 'errrrr'. Alarm a sharp 'tsek!'. Song, often in brief song-flight, loud, long sequences of not very varied *excited notes* now and then relieved by *rapid cascades of trills and whistles* and occasional interwoven mimicry (e.g. of Coot, Wood Sandpiper, Yellow Wagtail), e.g. 'zrüzrü-trett zrüzrüzrü-trett zrüzrüzrü psit trutrutru-pürrrrrrrrrrrurrrrrr **vi-vi-vi** lülülü ze**tre** ze**tre**...'.

Aquatic Warbler *Acrocephalus paludicola*　　**V (P)**∗∗

L 11½–13 cm. Breeds only in open, waterlogged sedge meadows and prefers just foot-high vegetation, so now rare (after so much draining); main known stronghold in E Poland. Summer visitor, winters in W Africa, passes through Holland and Morocco; regular autumn vagrant in Britain. Rather shy and retiring. Nests in sedge tussock.

IDENTIFICATION Like Sedge Warbler in shape and general appearance; differs as follows: *narrow, distinct yellow median crown-stripe* (much more prominent and narrower than on any juvenile Sedge); two clear *yellow-buff bands along mantle-sides*, and *heavier black streaking on upperparts*; normally *paler and more yellowish-white overall impression* (but at distance and in evening light nevertheless surprisingly dark and Sedge Warbler-like); *pale lores*, so supercilium looks broad at front (juveniles and some adults; other adults have slightly darker loral stripe). Silhouette when singing distinctive: tail pointing down, *neck extended* to the limit with each verse. – Sexes alike. Adult almost always has breast and flanks finely streaked, whereas juvenile does not.

VOICE Call a clicking 'chack'. When anxious, a Sedge Warbler-like 'errrr'. Alarm short, hoarse 'tscht' notes (like a grasshopper!). Song (mostly in evening, up to dusk; mainly perched, only seldom in short song-flight) like a sleepy Sedge Warbler; almost all verses start with a churring 'trrrr' (like anxiety-call), immediately followed by a series

of whistles, 'jüjüjüjü' or 'didididi' (a bit like Wood Sandpiper, but shriller and 'looser'); for long periods, utters nothing but churr and whistle... pause... churr and whistle... etc.; at times more complex verses with 4–5 motifs.

Moustached Warbler *Acrocephalus melanopogon*　**V**∗∗∗

L 12–13½ cm. Breeds, mostly locally and sparingly, in reedbeds, often adjoining small open patches and mixed with bulrush stands. Fond of dense areas of fallen reed, where it often hops along low down. Mainly resident in W, migratory in E (present at Austrian breeding sites mostly Mar–Sep). Does not perform song-flight. Nests low down in reeds.

IDENTIFICATION Quite similar to Sedge Warbler, yet distinctive when seen well. Is marginally larger and more compact. Often hops on ground or near water, frequently flicks tail and raises it slightly. Differs from Sedge Warbler in: *short primary projection* (at most a third of tertials, two-thirds in Sedge); *whiter supercilium, which is squarer-ended and broader at rear*; *more uniformly dark crown* (can appear unstreaked brownish-black in field) and cheeks, lower edge of cheeks with *hint of narrow black moustachial stripe*; *reddish-brown tone in fresh plumage* (spring), especially on rump, nape, neck- and breast-sides; whiter underparts (not buffy yellow-white). – Variation: In E Turkey, Caucasus and Middle East (ssp. *mimicus*) more like Sedge Warbler, with greyer-brown, less rusty plumage and paler, more streaked crown.

VOICE Call a muffled, 'throaty' clicking 'treck' (thus *thicker voice* than Sedge Warbler, rather like Stonechat). Variations include shorter 'trk', which in excitement turns into fast clicking series, 'tk-tk-tk-tk-...'; also rolling 'trrrt'. Song like Reed Warbler's but more animated, a little faster, softer and more varied, and best recognized by recurring *series of Nightingale-like rising whistled notes*.

Zitting Cisticola *Cisticola juncidis*　　　　**V**∗∗∗

(Alt. name: Fan-tailed Warbler.) L 10–11 cm. Breeds in open country in warm climates in tall grass or in fields; prefers drier terrain, avoids trees. Resident. Nests low in vegetation.

IDENTIFICATION *Small*, podgy and *short-tailed*, sandy-brown and *dark-streaked*. *Undulating song-flight* can at times reveal that *tail is strongly rounded* and below dark and *broadly tipped white*, and *wings short and rounded*. On perched bird, the short tail eliminates confusion with Graceful Prinia (which it otherwise most resembles). In breeding season, sexes separable: ♂ has black bill and uniform brown crown, ♀ light brown bill and more streaked and pale crown.

VOICE Call a loud 'chipp!'. Song a monotonous repetition of a short, sharp note, uttered for long periods (usually in wide circling song-flight) with regular barely second-long pauses, 'dzip... dzip... dzip... dzip... dzip...'.

Sedge Warbler　　　　　　Aquatic Warbler

Moustached Warbler　　　　Zitting Cisticola

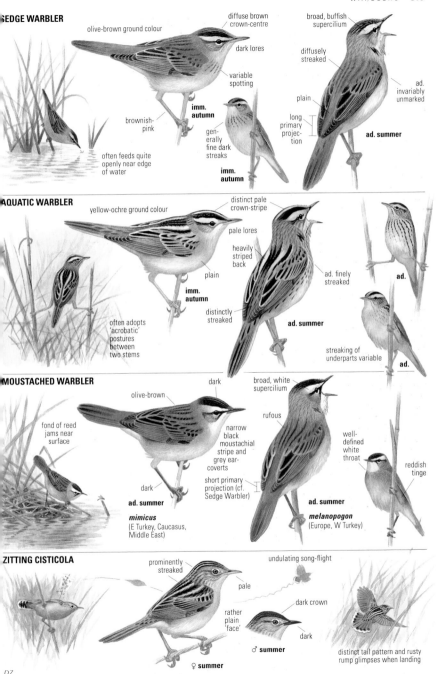

SEDGE WARBLER

olive-brown ground colour

diffuse brown crown-centre

dark lores

variable spotting

broad, buffish supercilium

diffusely streaked

ad. invariably unmarked

plain

long primary projection

ad. summer

brownish-pink

imm. autumn

generally fine dark streaks

imm. autumn

often feeds quite openly near edge of water

AQUATIC WARBLER

yellow-ochre ground colour

distinct pale crown-stripe

pale lores

heavily striped back

plain

imm. autumn

ad. finely streaked

ad.

distinctly streaked

ad. summer

often adopts 'acrobatic' postures between two stems

streaking of underparts variable

ad.

MOUSTACHED WARBLER

dark

olive-brown

broad, white supercilium

rufous

narrow black moustachial stripe and grey ear-coverts

well-defined white throat

reddish tinge

fond of reed jams near surface

dark

short primary projection (cf. Sedge Warbler)

ad. summer

mimicus
(E Turkey, Caucasus, Middle East)

ad. summer

melanopogon
(Europe, W Turkey)

ZITTING CISTICOLA

prominently streaked

pale

undulating song-flight

dark crown

rather plain 'face'

dark

♂ **summer**

distinct tail pattern and rusty rump glimpses when landing

♀ **summer**

DZ

(Common) Grasshopper Warbler *Locustella naevia* m**B**3

L 12½–13½ cm. Breeds in habitats with low, thick vegetation, often in tussocky marshland beside lakes (where tussocks tallest), in young conifer plantations or clear-felled areas, among tall grass and herbage with scattered bushes, often along riversides. Summer visitor (in Britain & Ireland May–Jul), winters so far as known in tropical Africa. Except when singing, very hard to see; keeps well concealed and creeps in grass like a mouse. Nests low in dense vegetation.

IDENTIFICATION A Sedge Warbler-sized, grey-brown bird with no striking plumage features. *Dark-spotted above* on *olive-tinged grey-brown* ground; off-white or sometimes warmer buffy yellow-white below, *unspotted* or with variable amount of *small dark spots on throat* and upper breast (many have a few spots only; rarely, showing dense heavy spotting, and in addition some diffuse streaking on flanks, producing similarity to Lanceolated Warbler, which see). Distinguished from Sedge Warbler by: *lack of prominent, whitish supercilium* (has just faint suggestion); *off-white, diffusely streaked undertail-coverts*; olive-tinged grey-brown and weakly dark-spotted *rump not contrasting appreciably with back*; evenly fine-streaked crown without darker sides. Legs pinkish, bill rather dark. Tertials dark, with bases broadly and diffusely edged brown. Sexes alike. In autumn, juveniles are often yellowish below while adults are off-white, although there seems to be a certain overlap as to this.

VOICE Call a sharp, piercing 'psvitt'. Song peculiar, an insect-like, *mechanical, dry ringing or whirring* reel, 'sirʳrʳrʳrʳrʳrʳr...', heard mostly from late dusk and at night, and which often continues for several minutes. The sound may be likened to an alarm clock with muffled clapper; volume varies somewhat depending on how bird turns its head. (Three congeners have similar song: Lanceolated, Savi's and River Warblers. See those for differences.)

Lanceolated Warbler *Locustella lanceolata* V**

L 11½–12½ cm. Breeds in swampy open forest with rank undergrowth, among willows and herbage in waterlogged clearings in the taiga, also in more open habitats such as extensive marshes with rushes and scattered bushes. Habits as Grasshopper Warbler, but winters in SE Asia and does not arrive at breeding sites until end of May/Jun.

IDENTIFICATION *Small and compact*, looks *shorter-necked* and *shorter-tailed* than Grasshopper and Pallas's Grasshopper Warblers, at times also a mite bigger-headed. Coloration much as Grasshopper Warbler, i.e. usually has grey tone to the *dull brown* plumage, but is always *more heavily black-streaked*, both above and below. In typical plumage, *throat, breast and flanks distinctly and narrowly streaked*

black (least obvious on throat); occasional heavily streaked Grasshopper Warblers have *dense spotting on* central/lower throat and upper breast (almost forming bib) and have some *smudgy long streaks* on flanks. Other Lanceolated (often juvs.) have only sparse streaking on throat/breast and can be unstreaked on flanks, much resembling Grasshopper Warblers; identified by markings being in form of *sparse, distinct streaks* (not more diffuse, rounded spots) and these being *more evenly distributed both on throat and on breast*. Note for all Lanceolated: *tertials are almost black with narrow sharply defined and complete brownish-white fringes* (on a few, outer web has slightly broader fringe, but this always sharply defined, not diffuse as on Grasshopper Warbler) *undertail-coverts rusty yellow-buff* or buffish-white, appearing unstreaked or with distinct black shaft spots.

VOICE Call a clicking 'chick'. Song like Grasshopper Warbler's but differs in: slightly higher pitch; slightly faster tempo; on average shorter verses (rarely over 1 min. Grasshopper often 5–10 min.); *sharper, more piercing* sound; *hint of River Warbler-like shuttling*, 'zizizizizizi...' unlike Grasshopper's more uniform and dry reeling. (Although, rarely, aberrant Grasshopper Warblers with vocal defect occur which can sound very similar!)

Pallas's Grasshopper Warbler *Locustella certhiola* V***

L 13–14 cm. Breeds in Siberia, parts of Central Asia, Mongolia and farther eastwards, in soggy grassland with dense shrubbery or in thick riverside vegetation (reeds, rushes, willows), in damp forest clearings or bog margins etc. Very rare autumn vagrant in Europe. Mostly remains concealed low down in vegetation, difficult to see.

IDENTIFICATION Can be described as something between Grasshopper Warbler and Sedge Warbler. Has *pale supercilium* which is a touch more prominent than Grasshopper's, but nothing like Sedge's long and well-marked one. Ground colour of *upperparts* usually a shade more *reddish-brown*, not so olive-grey as Grasshopper's, and dark spotting on crown, back and uppertail-coverts averages heavier (though some similar), while rump is often contrastingly reddish-brown and unstreaked (quite like Sedge Warbler). Most important feature: *tail-feathers have white tips* (can be worn off, or be slightly more diffuse on some juveniles) with dark subterminal marks; tertials dark brown with sharply defined light brown edge to outer web and (almost without exception) a *white spot near tip of inner web*; undertail-coverts entirely (or almost so) *unstreaked, rusty-buff* (with whiter tips). Underparts plain off-white with only faint yellow-buff wash to breast and flanks (adult), or distinctly yellow-buff with warmer tone to throat, breast and flanks and with breast spotted (mainly, but not exclusively, immatures).

VOICE Several different calls, including clicking 'chick', shorter ticking 'pt', dry rolling 'trrrrrt'. Song not monotonously mechanical and insect-like as with most congeners, but a verse made up, as with an *Acrocephalus*, of fast repetitions of different motifs, e.g. 'tri-tri prt-prt chiv-chiv-chiv srrrrrt **sivih-sivih-sivih**'; song rather soft, but the final 'sivih' notes are quite loud and carry far.

Grasshopper Warbler

Lanceolated Warbler

GRASSHOPPER WARBLER

olive-brown, streaked dark

usually unmarked breast on both imm. and ad.

some have long and diffuse streaks on flanks

often tinged yellow

imm. autumn

off-white

summer

diffuse olive-brown partial fringes, widest on outer web

some have throat and upper breast streaked (cf. Lanceolated Warbler)

climbs nimbly like a mouse in the grass

long, ill-defined dark streaks

LANCEOLATED WARBLER

distinctly streaked

streaking of under-parts variable

rusty yellow-buff

summer

distinct, evenly narrow, pale tertial fringes

rusty yellow-buff, finely tipped white, longest coverts often without black spots

immatures have more diffuse and discreet streaking below

short

summer

short-tailed   **imm. autumn**

distinct narrow streaks or small spots, variably extensive

PALLAS'S GRASSHOPPER WARBLER

fairly distinct pale supercilium

PALLAS'S GRASSHOPPER

rounded tail tipped dark and white

warm red-brown

greyish or dark crown

rusty-brown

white spots to tips of inner webs of tertials

blackish

usually unmarked rusty-buff

yellowish

summer

GRASSHOPPER

imm. autumn

long

rather plain

DZ

River Warbler *Locustella fluviatilis*   V***

L 14½–16 cm. Breeds in soggy young deciduous growth along rivers and at swamp edges; can thrive in small stands of dense and shady swamp forest (birch, willow, alder), but not in low shrubland. Summer visitor (mostly mid May–Aug), winters in E Africa. Shy and elusive (except when singing). Nests near ground in dense vegetation.

IDENTIFICATION A rather large, dark, elongate warbler with broad, rounded tail. *Upperparts are uniform dark grey-brown*, sometimes with slight olive-green tone. Underparts dirty white with olive-tinged brown-grey flanks. The only salient features are a *short and indistinct, off-white supercilium* and a contiguous pale eye-ring, *diffuse grey mottling* (of varying strength; some are poorly marked) *on upper breast* (sometimes lower throat, too), and *olive-brown undertail-coverts with broad whitish tips* (produce spotted impression). At close range, outer web of outermost long primary off-white, edge of wing curved, and undertail-coverts cover almost entire length of tail (features common to all *Locustella* species). Bill rather dark, legs pink. Sexes and ages usually inseparable in the field.

VOICE Calls not very distinctive, a 'zrr' and a throaty 'tschick'. Song, fortunately, much more characteristic, a remarkable machine-like shuttling which carries on with only short breaks from dusk to morning hours, 'zre-zre-zre-zre-zre-…' or 'dze-dze-dze-dze-dze-…'; some variation in tempo (usually rather sluggish) and pitch. The sound can be likened to that of a 'giant wartbiter' (i.e. a bush cricket) or a powerful sewing-machine; at close range, a metallic overtone is audible.

Savi's Warbler *Locustella luscinioides*   mB5 / V**

L 13½–15 cm. Breeds in extensive, tall reedbeds, less often in bulrushes or rushes or other dense waterside vegetation. Summer visitor (mid Apr–Aug/Sep), winters S of Sahara; very rare breeder in Britain & Ireland, only handful of localities. Moves about discreetly in cover, but not really shy. Nests low down in dense vegetation.

IDENTIFICATION Bearing in mind habitat (dense, tall reeds) and its uniform brown and off-white plumage, can be confused with Reed Warbler; like that species, it is red-brown above and has short, indistinct, pale supercilium. Differs in: *long pale reddish-brown undertail-coverts* with diffusely paler tips (usually look uniformly coloured in field, but some have pale tips distinct enough to create similarity to River Warbler; see latter); *curved wing-edge* (wing straighter on *Acrocephalus*) with *whitish outer web to outermost long primary*; *more uniform and slightly darker reddish-brown upperparts* (lacks Reed's usually contrasting warmer red-brown rump); brown-grey wash on breast and

tinge of reddish brown-grey on flanks (lacks Reed's buffy yellow tones); broader tail with fine dark bars on upper side (requires good light and close view). Legs brownish grey-pink. Sexes and ages alike. – Variation: E populations (especially from Caspian Sea eastwards, ssp. *fusca*) have olive-grey cast above and are generally less red brown, often have slightly stronger greyish tinge to breast and usually have more distinct pale tips to undertail-coverts, thus in several respects are closer to River Warbler.

VOICE Call a sharp, metallic 'pvitt!'. Song most like Grasshopper Warbler's, i.e. consists of an endless reeling insect-like sound, given mainly at night. Differs, however (apart from fact that singer perches out in tall, dense reeds!), in *higher frequency, lower pitch* and in the noise 'surrrrrrrrr…', sounding more of a *hard and noteless buzzing* than a high whirr. (A possible confusion risk at distance and at night in S and C Europe is the mole-cricket Song verses begin with an accelerating series of click which turn into the reel, 'pt… pt pt-ptptptsurrrrrrr…'.

Cetti's Warbler *Cettia cetti*   rB

L 13–14 cm. Breeds in dense, rather tall and often well delimited vegetation, preferably near but not in water, e.g. in drier tall reeds with scattered bushes, stands of papyrus, willow, bamboo, various thick bushes etc.; sometime breeds near human habitation, in thickly wooded parkland, beside reservoirs and canals, etc. Mainly resident but migratory in E. Generally keeps concealed in vegetation, often hops on ground or low down in shrubbery Nests low down in dense vegetation.

IDENTIFICATION A medium-sized, rather compact warbler, *short-necked* and *broad-tailed*, with *short, strongly rounded wings*. Bill pointed. Plumage *uniform red-brown above*, dusky greyish-white below with rusty tinge on flanks and belly. Head pattern roughly as on Reed and Savi's Warblers, the two species which it superficially most resembles i.e. has a *narrow and not particularly distinct pale supercilium* which is set off by dark lores and dark eye-stripe. Best identified by: body shape and *short primary projection longer supercilium* than Reed and Savi's; faint *grey tone on ear-coverts* and neck- and breast-sides but slightly warmer (greyish) *red-brown, rather dark tone on lower flanks* and belly; *dark brown undertail-coverts* with narrow pale fringes relatively *short undertail-coverts* reaching only halfway along tail. Lively and active, flicks wings and tail, often cocks tail. Sexes and ages alike.

VOICE Call an explosive, metallic 'plitt!' which may be repeated in series and turn into rattling 'plir'r'r'r'. Song a *sudden and loud outburst* of metallic, clanging notes rhythm characteristic: first 1–4 slightly tentative notes usually with the last one stressed, then a half-stop followed by a rapid series of groups of similar notes, at times dying at end, e.g 'plit, plit-**plüt**!… tichut tichut-tichut chütt, chutt' (or, why not: 'Listen!… What's my name? … Cetti Cetti-Cetti—that's it!').

River Warbler   Savi's Warbler

Cetti's Warbler

IVER WARBLER

not confined to water-courses—requires lush brushwood or tall bushes

dark olive grey-brown

indistinct, short supercilium and slightly paler eye-ring

diffusely grey-mottled upper breast

a few have very faint breast streaking only

olive-brown, prominently tipped pale

the tail is rather long, broad and well rounded

undertail-coverts reach far out towards tail-tip

AVI'S WARBLER

linked to vast reedbeds

olive grey-brown

note the proportionally small head (cf. Reed Warbler)

unmarked

red-brown

fusca (Transcaspia, Central Asia)

rusty-buff with a hint of paler tips

wing-edge is greyish-white and is smoothly curved (unlike in *Acrocephalus*)

rufous-grey flanks

a hint of pale tips at most

CETTI'S WARBLER

crown often slightly peaked

pale supercilium and eye-ring create a slightly Chiffchaff-like expression

warm red-brown

often raised tail

CETTI'S

usually keeps in cover, giving its explosive, loud song from a dense thicket

side of head and breast tinged grey

REED

undertail-coverts brown, often with pale tips

compare with posture of a Reed Warbler

(European) Reed Warbler *Acrocephalus scirpaceus* mB2

L 12½–14 cm. Breeds in reedbeds; locally, e.g. Britain, extends into willow herb, rape fields, migrants also in scrub, etc. Summer visitor (in Britain mid Apr–Sep/Oct), winters in tropical Africa. Rather bold and inquisitive, easy to see. Nest basket-shaped, woven around a few reed stems.

IDENTIFICATION Pointed head with *flat forehead* and *long, thin bill*. Climbs nimbly on reed stems. *Uniform brown above* and *buff-white below* with warmest tone on flanks and under-tail-coverts. Has a short and rather indistinct pale *supercilium*, which usually *does not extend past eye*; lores dark. Brown colour of crown and mantle/back usually has slight olive-grey tinge, while *rump is somewhat lighter and warmer rusty-brown*. Sexes alike. In late summer, adult is somewhat worn and more grey-brown above and whitish below, juvenile warm rusty-brown above and saturated buff on flanks. (See also similar Marsh and Blyth's Reed Warblers.) – Variation: Eastern birds (esp. E of Black Sea, ssp. *fuscus*) are slightly darker and greyer above and paler below, i.e. are more like Marsh, Blyth's Reed, Olivaceous and Savi's Warblers.

VOICE Call a short, unobtrusive 'che', sometimes slightly harder, almost clicking 'chk'. When agitated, drawn-out hoarse 'chreeh', thick rolling 'chrrrre'and disyllabic 'trr-rr'. Song, heard most at dusk and dawn, is 'chatty' and slow-paced, consists mostly of jittery notes which are repeated 2–3 times, interrupted by occasional mimicry or whistles, e.g. 'trett trett trett **tirri tirri** trü trü **tie** tre tre vi-vü-vu tre tre trü trü **tirri tirri**...'. Now and then tempo is raised, but Reed Warbler never achieves the real crescendo of Sedge Warbler.

Marsh Warbler *Acrocephalus palustris* mB4–5 / P4

L 13–15 cm. Breeds in rank herbaceous vegetation, often in damp stands of meadowsweet, nettle, cow parsley etc., often beside ditches or soggy wasteland, sometimes on fringe of reedbed if growing on slightly drier ground and mixed with herbage. Summer visitor (in Britain rare and local, mid Apr–Sep), winters in tropical Africa. Unobtrusive habits.

IDENTIFICATION Very like Reed Warbler. Told by: *song*; breeding habitat; slightly *shorter bill* and *more rounded head shape*, a 'kinder' appearance; often less distinct pale supercilium, but slightly more prominent pale eye-ring; in spring, crown and mantle rather light grey-brown with faint *green cast*; somewhat more *yellowish below*; on average somewhat longer primary projection and more distinct white tips to each primary (but quite a few Marsh and Reed Warblers similar in this respect). – Juvenile: Warmer brown above with rusty-yellow rump, thus *extremely like Reed Warbler*; legs light yellowish-pink (light brown-grey on juv. Reed).

VOICE Calls similar to Reed Warbler's. Song, heard most from nightfall to dawn, a stream, broken by only brief pauses, of whirring, excitable and whistling notes with f[...] the most part *high voice* and *furious tempo*, which on clos[...] study proves to consist almost exclusively of *expert mimic[...]* (often e.g. Blue Tit, Blackbird, Chaffinch, Magpie, Bar[...] Swallow, Linnet, Common Gull, Quail, Bee-eater, Jack[...]daw). Now and then longer passages of fast, dry trills occu[...] 'prri-prri-prrü-prri-...', and hoarse 'ti-**zaih**, ti-**zaih**'. Ten[...]po varies, slower passages of listless repetitions can at time[...] recall Blyth's Reed Warbler, but speed soon picks up again[...]

Blyth's Reed Warbler *Acrocephalus dumetorum* V**[...]

L 12½–14 cm. Breeds in overgrown clearings in deciduo[...] forest, in bushes along riversides etc. Not attracted to ree[...] or waterlogged ground. Summer visitor (in Finland most[...] late May–Aug), winters in India. Habits as Marsh Warble[...]

IDENTIFICATION Resembles Reed and Marsh Warblers. Tol[...] by: *song*; habitat; songpost in bush or lower tree (not i[...] reeds or herbage; few exceptions); *more obvious, long pal[...] supercilium, extending just beyond eye*; rather *short prima[...] projection*; *legs rather dark*, brown-grey (never light pink[...]ish); *long bill* and *flat forehead*; *greyish-brown upperparts* flanks more olive-grey than buff. Some have markedly un[...] form brown wing, without contrasting dark centres t[...] wing-feathers. – Juvenile: Rusty-toned above (like Reed).

VOICE Call a clicking 'zeck', often repeated, and a rollin[...] 'zrrrrt'. Song characteristic (safest indication in field ider[...] tification), mostly at night, varied groups of notes, ofte[...] with masterful mimicry inserted, which are *repeated at slo[...] pace* 3–5 times, sometimes even up to 10 times; *the clickin[...] call is stuck in between each repetition*. Neat 'scale exercise[...] such as 'loh-lü-**lii**-a' (steps up the scale) recur regularly.

Paddyfield Warbler *Acrocephalus agricola* V**

L 12–13½ cm. Breeds in reeds, seemingly having broadly th[...] same habitat requirements as Reed Warbler. Summer visito[...] (mostly May–Aug), winters in India. Nests low in reeds.

IDENTIFICATION *Smaller* than Reed Warbler, with somewha[...] *shorter bill* and *longer tail*. Colour varies but is usually *pale[...] brown* than Reed, rusty-coloured in fresh plumage, mor[...] grey-brown when worn. Facial expression different owin[...] to short bill which has *black distal area on straw-yellow lowe[...] mandible*, also to *uniformly broad, long pale supercilium[...] sometimes accentuated by hint of dark crown-side edge. Mos[...] have warm *rusty yellow-brown rump* (worn birds can be grey[...] brown; cf. Booted Warbler) and *short primary projection*.

VOICE Call slightly feebler than e.g. Reed Warbler. Alarr[...] a weak rolling 'cherrr'. Song most like Marsh's, a fast strearr[...] consisting mostly of mimicry; voice, however, weaker, tem[...]po constantly high, and the song more 'bouncing', lackin[...] hoarse drawn-out sounds and dry trilling runs.

Reed Warbler

Marsh Warbler

Blyth's Reed Warbler

Paddyfield Warbler

REED WARBLER

scirpaceus
(Europe,
N Africa)

rich buff

**imm.
autumn**

summer

warm
brown

red-
brown

warm
buff

grey-brown

hops among reed stems
in typically crouched
posture, body rather
'spool-shaped'

fuscus
(Middle East,
Central Asia,
SE Russia)

note: very similar to Marsh!

**imm.
autumn**

pale

MARSH WARBLER

long primary projection
and dark primaries
with pale tips

note: imm. Marsh and
Reed Warblers are
extremely similar

grey-brown
tinged olive
when fresh

off-white tinged
yellowish

compa-
ratively
short and
blunt

summer

pale
yellow-
ish-pink

**imm.
autumn**

just a touch
paler tawny-
brown

prefers lush stands of meadowsweet, nettles etc.

BLYTH'S REED WARBLER

plain
tertials

greyish-brown

lower
mandible
usually
diffusely
tipped
darker

olive-
buff

short primary
projection

**imm.
autumn**

summer

brown-grey,
rather dark

resembles Marsh Warbler,
but is shorter-winged, longer-
billed and has darker legs

PADDYFIELD WARBLER

rather long and
rounded tail

contrasty
tertials

long, pale supercilium, often
edged dark above

rather
pale
brown

distinct
dark tip

short primary
projection

rusty-
brown

same favourite
habitat as
Reed Warbler:
extensive
reedbeds

**imm.
autumn**

summer

beware: a few have more
diffuse head pattern and
less rusty-brown on rump
(imm. or abraded ad.) and
can be very similar to richly
coloured Booted Warbler

ad. has
pale eye

Great Reed Warbler *Acrocephalus arundinaceus* **V**∗∗

L 16–20 cm. Breeds in tall, dense and preferably extensive reedbeds; if necessary makes do with smaller reedbeds, e.g. along canals and dykes. Summer visitor (May–Aug in N of range), winters in tropical Africa. Basket-shaped nest woven around strong reed stems at medium height above water.

IDENTIFICATION Like a *larger version of Reed Warbler*, similar in colour (brown above, buff-white below) and with same habitat and same type of song. Apart from size and gruffer voice, differs in: proportionately *larger head and bill* (thick and long like thrush's); usually *dark spot at tip of lower mandible*; slightly longer tail; somewhat longer primary projection (with clear white tips to primaries); often slightly more distinct, *broader pale supercilium* and on average *darker lores and eye-stripe*; often a trifle paler and more grey-brown nape/hindneck. At closest range, a few fine grey streaks on lower throat/upper breast. Legs pinky-brown or brown-grey. Sexes alike. – Adult late summer: Usually somewhat worn, with lighter brown-grey cast above and whiter below. Primaries worn, brownish-grey. – Juvenile: Plumage uniformly fresh, warm rusty-brown above and buffish below. Primaries fresh, dark with pale tips.

VOICE Call a coarse clicking with slightly 'thick' voice, 'kshack', or with more audible r-sound, 'krrack'. In anxiety a hard rolling 'krrrrr'. Song roughly as Reed's in composition, with various repeated hoarse notes, but is very loud, far-carrying, recognized by gruff, croaking voice and by recurring series of repeated shrill falsetto sounds, e.g. 'trr trr karra-karra-karra **krie-krie-krie** trr-trr-**kie-kie**'.

Clamorous Reed Warbler *Acrocephalus stentoreus* —

L 16–18 cm. Breeds in dense stands of papyrus, secondarily also in reedbeds, bulrushes, maize and other dense vegetation tall enough to hide a person, in association with water or at least marshland. Within the region mostly resident. Nest similar to Great Reed Warbler's.

IDENTIFICATION Resembles Great Reed Warbler in size, song and habitat (although, unlike Great Reed, prefers papyrus if offered a choice), but differs in several ways in terms of shape and plumage details. Clamorous Reed gives a different impression owing to its *shorter, rounded wings* (i.e. has *short primary projection*), *longer and more rounded tail* and *somewhat longer and*, especially, *narrower bill*. Differences in proportions are visible not least in flight, when it looks *shorter-bodied*. Brown above, roughly similar to Great Reed, but generally darker below. In addition, a less common dark (melanistic) morph occurs in Middle East, as well as intermediates. Usually *upperparts are rather dark red-brown* and *underparts shabby brownish-buff with just slightly paler throat* (colours can recall Savi's Warbler rather than

Great Reed). The buff-white *supercilium* is at times shorter and in particular narrower and thereby less prominent tha on Great Reed, and *lores and eye-stripe are not so dark*. Dar individuals often lack supercilium altogether, also hav darker bill (almost like young Starlings). Legs greyish.

VOICE Similar to Great Reed's, especially calls, a slight 'thick' 'track' and, when agitated, a hard rolling 'trrrrr' Song gruff-voiced like Great Reed's, but verses *more ten tative* in composition and have *more uncertain patter* ('chacking' like Fieldfares before dawn!), and *shrill falsett* notes are mostly given singly, not in fast series as in Great Reed; a typical verse might go 'track, track, track karra **kru**-kih karra-**kru**-kih chivi **trü** chivi **chih**'.

Thick-billed Warbler *Acrocephalus aedon* **V**∗∗

L 16–17½ cm. Breeds in S Siberia in dense vegetation o willows, young birch etc. beside taiga bogs. Winters in India Extremely rare autumn vagrant in Europe.

IDENTIFICATION Slightly smaller than Great Reed Warble with *short, stubby bill* and few plumage features. Greyish brown above with slightly *warmer red-brown rump and tai base* (in flight can recall nightingale, since tail is not onl rusty but also long and rounded), and dirty white belo with buff-tinged breast, flanks and undertail-coverts. Lack prominent supercilium, has just *light grey lores* and diffus buffy grey-white eye-ring. Crown slightly warmer browr on many creating *suggestion of brown cap*. Primary projec tion medium-long. Legs greyish. Rather sluggish in action often hops in forward-crouched posture with tail lowered.

VOICE Call a clicking 'chack', often given in series. Occa sionally a hoarse Red-backed Shrike-like 'veht' in alar (mimicry?). Song a *fast, loud twittering warble with lots o mimicry* added; most like Marsh and Icterine Warblers; lack Marsh's harsh, repeated 'ti-zaih' and very fastest runs of dr whirring trills, and is often *harder-voiced*; lacks Icterine whining nasal notes.

Basra Reed Warbler *Acrocephalus griseldis* —

L 15½–17½ cm. Breeds in vast beds of papyrus or reeds, for merly restricted to S Iraq, Kuwait and SW Iran but recently ha spread to N Israel and possibly Syria. Winters in E Africa.

IDENTIFICATION *Smaller and slimmer than Great Reed Wa bler*, but larger than Reed Warbler. *Dark-tipped bill long, thi and pointed*, lower mandible with *yellowish-pink base* (the bill is not as all dark as in Clamorous Reed Warbler). *Uppe parts rather cold grey-brown*, not as warm rusty-brown o rump as usually Great Reed. *Underparts decidedly whit* than in congeners, any colour being restricted to a fai cream tinge on flanks and vent. *Legs greyish. Pale superc ium rather well marked*, reaching to just behind eye; *dar lores and eye-stripe* rath prominent. Note long wing and *long primary projection*

VOICE Call a harsh 'chaar and a hard 'chak'. Song sul dued, slow and disrupt, wit hard, low and guttural voic consists of a prolonged seri of simple notes, a little lik bulbul chatter.

Great Reed Warbler

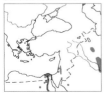
Clamorous Reed Warbler

Basra Reed Warbler

REAT REED WARBLER

often sings from exposed perch

young birds tinged warm brown in fresh plumage

diffuse, broad pale supercilium continues behind eye

thrush-sized bill

warm grey-brown

faint grey streaking

yellow-brown tinged flanks

moderately rounded

long primary projection (as long as tertial length)

imm. autumn

summer

legs strong, usually dark (though sometimes pinkish-brown)

LAMOROUS REED WARBLER

a melanistic variety, dark as juv. Starling, occurs in Middle East

distinct narrow eyebrow

long, narrow

lacks pale above eye

dark olive-brown

long, narrow

dusky olive-brown breast and flanks

often feeds near water surface in open reeds

strongly rounded

short primary projection (about half the tertial length)

imm. autumn

summer

HICK-BILLED WARBLER

mild expression with eye standing out

pale lore

stout

short primary projection

often crouched posture, reminiscent of a babbler; not a wetland species

BASRA REED WARBLER

supercilium reaches just behind eye

narrow, long and pointed

cold grey-brown

paler on belly and flanks than Great Reed

long primary projection

z

Icterine Warbler *Hippolais icterina* **P**4

L 12–13½ cm. Breeds in woodland, incl. in dense oak forest, birch or pine clumps in pastureland, or in dense parks. Summer visitor, winters in tropical Africa; passage migrant in Britain & Ireland, mostly E and S coasts (Aug–Oct, few in spring). Keeps high in treetops, hard to see. Nests in tree.

IDENTIFICATION Moderately large warbler with *broad-based bill*, quite large head (often with crown-feathers raised), long wings (*primary projection about equal to tertial length*), and fairly short, *square-ended tail*. *Greyish-green above, uniform light lemon yellow below* (rarely paler). Note *pale lores* and contiguous with these rather indistinct yellow-white supercilium, pale eye-ring and fairly *pale, yellowish-pink bill-side*; pale edges to tertials and secondaries create *pale panel on folded wing*, most obvious in spring, can be worn off on autumn adult. *Legs* greyish, often *blue-grey*. Sexes alike. – Juvenile: Brown-tinged above, not quite so green as adult, and pale wing-panel more buffy-yellow (and therefore not so prominent). Underparts often slightly paler yellow.

VOICE Call a cheerful trisyllabic 'teh-teh-**lüü**it'. Also a clicking 'teck', in anxiety 'te te te...' in short series. Song *loud, fast*, drawn out, varied, spiced with imitations of other species. Best recognized by recurrent *nasal, shrill notes*, e.g. '**gie gie**...'. Also recognized by the typical call being interwoven into the verse. Repetitions (often 2–4) are more common than in Marsh Warbler, and tempo slower.

Melodious Warbler *Hippolais polyglotta* **P**4

L 12–13 cm. Replaces Icterine Warbler in SW Europe and breeds in similar habitats, but also in lower vegetation such as shrubbery with scattered trees. Summer visitor, winters in W Africa; passage migrant in Britain & Ireland, mostly S and W coasts (Aug–Oct, few in spring). Nests in tree.

IDENTIFICATION Like Icterine Warbler, and solitary birds on passage require close study. In shape a trifle more plump and 'kindly-looking', with *subtly shorter bill* and clearly shorter and less pointed wing (*primary projection between half and two-thirds of tertial length*). Upperparts a shade more brownish-green, not so greyish, and underparts often pale yellow with a faint buff-brown suffusion. 'Face' and bill pattern as Icterine. *Wing more uniform*, and pale edges to tertials and secondaries are usually almost entirely abraded. *Primaries often slightly worn and brown-grey*, generally lack distinct pale tips. *Legs* greyish as Icterine *or slightly tinged brownish*. Sexes alike. – Juvenile: Pale wing-panel brownish-yellow and moderately contrasting. Primaries fresh and slightly darker brown-grey.

VOICE Various short clicking calls, e.g. 'tett', 'tre-te-te-tü', drawn-out series of 'te-te-te-te-te-...' and faster sparrow-like chattering 'tr'r'r'r'r'r'rt'. Song *drawn-out verses at high speed*,

at times lightning-fast and with fragments of mimicry (then Marsh Warbler-like), sometimes not quite so fast and with somewhat shrill and nasal tone (when Icterine-like), sometimes simpler in structure, more shuttling, excitable and varied (like a small *Sylvia* species). Listen long, it will help.

Olive-tree Warbler *Hippolais olivetorum* —

L 16–18 cm. Breeds in open forest with clearings, in tall maquis (holm oak, cork oak) or olive or almond groves, etc. Summer visitor (early May–Aug), winters in tropical Africa. Keeps well hidden. Nests in dense bush or tree.

IDENTIFICATION *Very big* and elongated, with *long, powerful bill* and *sloping forehead* (unless crown-feathers raised). Wings *long* (primary projection = tertial length, much longer than on Upcher's Warbler), with *dark greyish-black primaries*. Grey above and *off-white below* (often dusky grey on breast and flanks), but told by: more distinct *pale wing-panel* formed by whitish edges to longest tertial and secondaries (the white concentrated at middle of feathers) and *almost complete lack of pale supercilium* (has merely whitish spot above lores and a diffuse pale eye-ring). Legs strong, (blue-)grey. Sexes alike.

VOICE Call deep tongue-clicking 'chack'. Alarm a throaty, slightly nasal chattering 'kerrekekekekek', like a giant Blue Tit. Song gruff-voiced and raucous, the verses rather primitive in structure with cyclical repetition, can sound like an amplified Olivaceous Warbler at slow speed, e.g. '**chak** chi chi-**chak** chira **chuk** chi-chi **chak**-era **chak** chü chi-**chak**...'.

Upcher's Warbler *Hippolais languida* —

L 14–15 cm. Breeds in bushy areas and cultivations in dry, barren terrain, on slopes in river valleys or at higher levels. Summer visitor (May–Aug/Sep), winters in E Africa.

IDENTIFICATION A trifle bigger than Olivaceous Warbler with *somewhat longer bill, legs and tail*. Head shape a little more rounded, with steeper forehead. Tail looks a little broader, too. Plumage is similarly grey above and off-white below, but *tail-feathers and wing-tips somewhat darker greyish-black*; longest tertial and inner secondaries have pale greyish-white edges which create *suggestion of pale wing-panel*, and lores a shade paler and more smudgy grey. Has habit of *swinging tail*, including *sideways*, and *fanning it little*. Like Olive-tree Warbler, often glides the final stage before landing. Olive-tree is bigger and lacks supercilium.

VOICE Call a hard clicking 'zack' and in anxiety a fast, dry 'trrrt'. Alarm a long series of 'soggy', slightly impure clicking calls, 'scheck scheck scheck...'. Song energetic, nasal, repeating phrases several times (Blyth's Reed Warbler fashion), often sticks on high, 'bent' whistle, interspersed with rattle '...vieh vieh vieh trrrt vieh vieh vieh chechecheche vieh...'.

Icterine Warbler 
Melodious Warbler 
Olive-tree Warbler 
Upcher's Warbler

CTERINE WARBLER

usually sings well concealed in lush foliage

pale lores and short supercilium

grey-green

often well-marked pale panel

light yellow

summer

lead-grey

long primary projection; as long as the visible tertials

imm. autumn

usually pale fringes to the greater coverts

MELODIOUS WARBLER

often sings from exposed perch

weak panel

yellow with buff-brown cast

summer

variable, brownish or grey

some individuals show distinct wing-panel!

imm. autumn

short primary projection; half the length of the tertials

never shows distinct pale fringes to the greater coverts

LIVE-TREE WARBLER

usually sings from perch in cover, in orchards and groves

both species often glides before alighting (unlike Olivaceous)

short and rather thin supercilium

contrasting pale wing-panel

long and stout (but pointed)

blackish wing- and tail-feathers

distinct pale fringes to the greater coverts

summer

long primary projection; as long as tertial length

UPCHER'S WARBLER

often sways tail, partly sideways, and fans it slightly

pale edges to secondaries form hint of paler panel

dark tail and primaries (cf. Olivaceous Warbler)

rather stout

never distinct fringes to greater coverts

summer

medium-length primary projection, just less than ¾ of tertial length

Olivaceous Warbler *Hippolais pallida* V***

(Alt. name: Eastern Olivaceous Warbler.) L 12–13½ cm. Breeds in open woods, parks, orchards, bushes with scattered trees, in vegetation along rivers, etc. Summer visitor (mostly May–Aug), winters S of Sahara.

IDENTIFICATION Rather like a washed-out, greyish Reed Warbler; same size and has same pointed head with *flat forehead* and *long, narrow bill*, and virtually the same 'facial expression' with *dark lores, short pale supercilium* ending at rear edge of eye, and *pale eye-ring.* Told from Reed Warbler (incl. from worn, greyish individuals of eastern race *fuscus*) by: habit of regularly *dipping tail downwards* when hopping among vegetation; basically *greyish rump* (*never rusty yellow-brown*); practically *square-ended tail* with narrow *grey-white outer edges and tips* to outer feathers; always *greyish upperparts* and *whiter underparts; broader bill-base* seen head-on; usually somewhat shorter primary projection. Very like Sykes's and Upcher's Warblers (which see).

VOICE Call a clicking 'chack', slightly 'thick' and a bit nasal, like cross between Blackcap and Garden Warbler. Rapid series of tongue-clicks (nasal undertone) also heard, 'zet-zet-zet-zet-...', and a sparrow-like, muffled rattle, 'chr'r'r'r'r' and quiet 'chrrre'. Song a rather monotonous *babbling* at moderate pace, shifting from lower *hoarse* notes to higher-pitched squeaky ones in a recurring, *cyclic pattern*, rising and falling in pitch; notes tend to be 'blurred' together and scratchy. Does not mimic other species.

Isabelline Warbler *Hippolais opaca* —

(Alt. name: Western Olivaceous Warbler.) L 13–14 cm. Breeds in maquis, open woods, large gardens, orchards, along rivers, etc. Summer visitor (mostly late Mar–Aug; arrives Spain only in May), winters W Africa S of Sahara.

IDENTIFICATION A rather *large, slim and long-tailed*, drab brown-and-white warbler recalling Reed Warbler, and could be confused with Melodious Warbler, too, if light is poor. In NW Africa needs to be separated from very similar race of Olivaceous Warbler (*reiseri*). Note *uniformly pale brown upperparts* ('milky tea') *without contrastingly lighter edges to secondaries, no downward* (dipping) *movement of tail*, and *long and broad bill*, often with slightly convex ('swollen') sides when seen from below or in front.

VOICE Calls very similar to Olivaceous Warbler and probably not separable. Song is best means of separation, being less hoarse and scratchy, slightly slower and better-articulated. 'Talking' quality somewhat recalls Reed Warbler, but does not repeat notes in twos or threes, is more varied, and frequent nasal notes show affinity with related Melodious and Icterine Warblers. Each song often opens with a few call notes, 'chek... chek... chek...'.

Booted Warbler *Hippolais caligata* V*

L 11–12½ cm. Breeds in low scrub (spiraea, pea species) in natural steppe, in low brushwood on meadows, overgrow in pastureland, etc. Summer visitor, winters in India.

IDENTIFICATION *Smallest* of its group, with *short,* usual rather dark bill, fairly dark lores* and rather distinct but *shor pale supercilium;* can recall a *Phylloscopus* rather than *Hip polais.* In fresh plumage, buffish-brown above, more greyis brown when worn. *Flanks* often tinged *rusty-buff.* Has pal edged outer tail-feathers (but this not always obvious). A appear to have darkish sides of crown, above superciliu Bill heavier than Chiffchaff's, and with paler brownish-pin base. *Legs brownish-pink with darker, greyer toes.* Sometim twitches wings and tail, but does not repeatedly dip ta downwards like Olivaceous Warbler. Very similar to Sykes Warbler (which see for differences). Can also be confuse with Paddyfield Warbler (which see).

VOICE Call a dry tongue-clicking 'chrek', a little 'con pound', with 'r' in, slightly recalling Moustached Warble In anxiety, a dry rolling, muffled 'zerrrr'. Song a fast, twi tering verse of *c.* 2–6 sec. which 'boils over' with energy, chirping, nasal and chattering very rapid stream of shor almost bouncing and 'trembling' notes. Typically *opens lo and tentative* but *quickly picks up speed and stregth.* Do not contain mimicry.

Sykes's Warbler *Hippolais rama* V*

L 11½–13 cm. Breeds in dry plains and semi-deserts, nestin in low trees (saxaul) and tall bushes (often tamarisk). Sur mer visitor, winters Pakistan and India. Straggler to Europ

IDENTIFICATION Very similar to both Booted and Oliv ceous Warblers; separation without help of song requir close and prolonged observation. Averages *paler and plain* than Booted, being more *milky-tea-coloured above* and *wh ish below* (not rusty-tinged), tertials being rather plain wi concolorous shafts. Usually *slightly longer tail and bill* (b some overlap). Tip of lower mandible with insignificant da smudge (Booted has more dark, Olivaceous nothing Twitches tail nervously but does not repeatedly dip it dow wards as Olivaceous. Does not seem to twitch wings at t same time, as Booted then does. Rather *short primary pr jection*, and *primary tips brownish* rather than dark grey as most Olivaceous. Secondaries never finely white-tipped.

VOICE Call a dry tongue-clicking 'chek', similar to Boote but cleaner, without 'r' sound. Also has fuller 'tslek' recallin Bluethroat. Song hurried, similar to Booted but loude both from start and all over; contains harder and mo scratchy notes, and some quickly *repeated a few tim* creating vague *resemblance with Sedge Warbler*, further e hanced by interwoven brief trills and whistling notes.

Olivaceous Warbler | Isabelline Warbler | Booted Warbler | Sykes's Warbler

OLIVACEOUS WARBLER

rather tolerant; inhabits several types of bushy and scrubby habitats, even gardens

darkish lore

olive-grey

narrow, pointed

sometimes white-tipped second-aries

primary projection medium-long, fully half the tertial-length (cf. Sykes's Warbler)

imm. autumn

regularly dips tail down-wards

compare with Sykes's, below right; sometimes extremely similar and difficult to separate

ISABELLINE WARBLER

often found in tamarisk stands

pale lore

sandy-brown

heavy, broad

broad and 'swollen', sides convex

no downward dipping of the tail

BOOTED WARBLER

head-pattern may recall *Phylloscopus* warblers

broad-based bill when seen head-on

prefers low, dense shrubbery in open terrain, such as low bushes on steppe

grey-brown

primary projection short

often short with distinct dark tip

rusty-buff tinge

usually brownish-pink with darker toes—booted!

SYKES'S WARBLER

saxaul forest, the main habitat for this vagrant species in its Central Asian stronghold

both species flick tail, and Booted also wings, but do not dip tail downwards repeatedly like Olivaceous Warbler

pale grey-brown

supercilium usually faint or absent behind eye

fairly long with only faint dark tip to lower mandible

primary projection very short ($^1/_3$ or less of tertial length)

imm. autumn

Willow Warbler *Phylloscopus trochilus*   mB1 / P

L 11–12½ cm. Breeds commonly wherever a few trees or taller bushes exist, is one of N Europe's commonest birds, with over 2 million pairs in Britain & Ireland; found in upland birch and willow zone, in all types of woodland and in copses and more wooded parks and gardens. Lively and restless, flits about in canopy in search of insects. Summer visitor (in Britain & Ireland Apr–Sep), winters in tropical Africa.

IDENTIFICATION Most distinctive feature a *pale supercilium*. *Greyish brown-green above* (green tone obvious in W Europe, birds in N Fenno-Scandia often more grey-brown), sometimes a shade paler and brighter green on rump. *Yellowish-white on throat and breast*, whiter on belly. *Legs* usually *brownish-pink* (occasionally dark brown-grey!). Commonest confusion risk is Chiffchaff, but note, in addition to *voice*: normally brownish-pink or *light brown legs* (Chiffchaff's darker); on average *longer and bolder pale supercilium* and *darker lores and eye-stripe*, giving more patterned 'face'; *ear-coverts* olive-grey but usually rather *pale just below eye* (Chiffchaff's more uniformly dark, so white lower eye-crescent stands out); on average somewhat *stronger bill* with more obvious *yellow-pink at side/base*; somewhat more elongated shape, not so compact and chubby; *longer primary projection* (often *c*. three-quarters of tertial length; Chiffchaff half to two-thirds, but often difficult to judge). – Juvenile/1st-winter: More *saturated pale yellow below* and on supercilium. Sometimes faint buff tinge to breast-side.

VOICE Call a soft whistle, disyllabic and upslurred, '**hu**-itt'; varies a little (and some are more like Chiffchaff's faster and differently stressed '**hweet**'); very like Redstart's call, but a bit weaker. Song a frequently repeated soft whistling, somewhat descending verse *c*. 3 sec. long, e.g. 'sisisi-**vüy-vüy-vüy** svi-svi-vi tuuy tuuy tuuy si-si-**sviiy-sü**'; recognized by delightfully sweet voice with softly inflected notes.

Wood Warbler *Phylloscopus sibilatrix*   mB3

L 11–12½ cm. Breeds mainly in closed woods. Favoured habitats beech wood with at least sprinkling of younger trees, oak forest, generally with minimal undergrowth (in N Europe, also mixed spruce and deciduous forest). Summer visitor (in Britain mid Apr–Aug; rare Ireland), winters in tropical Africa. Rather unobtrusive in habits, but not shy.

IDENTIFICATION Size of Willow Warbler and similarly *elongated* in shape, but has *even longer and more pointed wings* (primary projection ≥ tertial length), which makes it look *shorter-tailed*. Told by saturated *dark moss-green upperparts* (at times tinged grey) with *contrasting pale tertial edges*, *yellow-green edges to remiges* and greater coverts, very *long and prominent yellow supercilium*, and *well-marked grey-green eye-stripe* from lores backwards. *Throat, ear-coverts and upper

breast pale lemon-yellow usually *abruptly demarcated fro[m] silky-white rest of underparts*. (Rarely lacks much yellow an[d] green pigments, then looks brown and white; needs to b[e] identified by voice, size, shape and distinct head markings.

VOICE Call a sharp 'zip'. When agitated, a muffled flu[te] note with slightly melancholy quality, 'tüh'. Song high[ly] distinctive, an accelerating series of sharp, metallic, call-lik[e] notes ending in an almost pulsating trill,'zip... zip... zip, zip zip zip zip zip-zip-zip-zipzipzipzvürrrrürrrr'(often likene[d] to a spinning coin on a marble slab). An alternative song interposed now and again between the reeling song verses, series of melancholy, initially intensified, soft notes, 'tüh tü[h] tüh-tüh-tüh-tüh' (vaguely like Willow Tit, but sequence faster and the notes straighter).

Western Bonelli's Warbler *Phylloscopus bonelli* V[*]

L 10½–11½ cm. Breeds in woods, in N usually at low[er] levels in more open forest with some understorey, in S mo[re] often on mountain slopes in pine or oak forest. Summ[er] visitor (mostly end Apr–Aug), winters in tropical W Afric[a].

IDENTIFICATION Not quite size of Willow Warbler. Typic[al] plumage features: *yellowish rump* which contrasts wi[th] duller greenish-brown back; *whitish underparts*, with obvious yellow or buff on throat, breast or belly; *primari[es] and tail-feathers edged light yellowish-green*; *tertials da[rk] with contrasting yellow and white edges*; pale-edged secon[d]-aries produce *bright green panel on folded wing*; *unbrok[en] pale eye-ring*; *pale lores* and *light grey ear-coverts*, givin[g] pale 'face' (but note that in field some show rather da[rk] lores and distinct supercilium); darkish, brown-grey legs.

VOICE Call a loud, distinctly disyllabic, up-turned, whi[s]tled 'hü-**eef**'. Song a simple repetition of one high note, e.[g.] 'svi-svi-svi-svi-svi-svi-svi-svi'; voice 'silvery' and 'laughing', a bit like Wood Warbler's in tone.

Eastern Bonelli's Warbler
Phylloscopus orientalis   V[**]

L 11–12 cm. Breeds in much the same habitats as its relativ[e] Western Bonelli's Warbler; frequently found in closed oa[k] woods. Summer visitor, winters in tropical E Africa.

IDENTIFICATION Very similar to Western Bonelli's, and in[ ]matures, and some worn autumn adults, not always separa[ ]ble except on calls. Eastern is a fraction larger and longe[r]-winged. In spring, note: tertials and most greater cover[ts] rather bleached and worn, often creating *hint of pale, greyis[h] wing panel* (Western typically fresher and has greener win[g] in spring). Upperparts subtly more grey-brown (but usual[ly] not an obvious difference). Bill a little stronger and showin[g] a little more pink-brown at base (some overlap). Autum[n] birds depressingly similar in plumage.

VOICE *Call a monosyllab[ic] flat* 'chip', almost like Hou[se] Sparrow fledgling. On m[i]gration this call, or a rela[t]ed, slightly muffled 'iss(t[)]'. Song quite similar to Wes[t]ern Bonelli's, but separabl[e] structure the same but voi[ce] flatter, *drier and mo[re] mechanical* or insect-like.

Willow Warbler

Wood Warbler

W. Bonelli's W.   E. Bonelli's W.

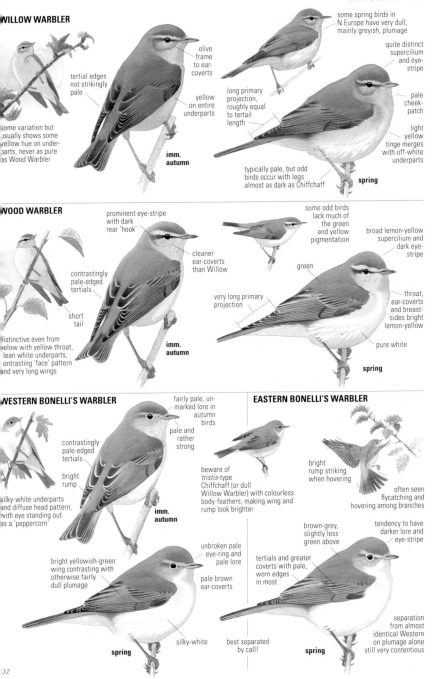

WILLOW WARBLER

some variation but usually shows some yellow hue on under-parts, never as pure as Wood Warbler

olive frame to ear-coverts

tertial edges not strikingly pale

yellow on entire underparts

imm. autumn

long primary projection, roughly equal to tertial length

typically pale, but odd birds occur with legs almost as dark as Chiffchaff

some spring birds in N Europe have very dull, mainly greyish, plumage

quite distinct supercilium and eye-stripe

pale cheek-patch

light yellow tinge merges with off-white underparts

spring

WOOD WARBLER

distinctive even from below with yellow throat, clean white underparts, contrasting 'face' pattern and very long wings

prominent eye-stripe with dark rear 'hook'

cleaner ear-coverts than Willow

contrastingly pale-edged tertials

short tail

imm. autumn

very long primary projection

some odd birds lack much of the green and yellow pigmentation

green

broad lemon-yellow supercilium and dark eye-stripe

throat, ear-coverts and breast-sides bright lemon-yellow

pure white

spring

WESTERN BONELLI'S WARBLER

silky-white underparts and diffuse head pattern, with eye standing out as a 'peppercorn'

contrastingly pale-edged tertials

bright rump

bright yellowish-green wing contrasting with otherwise fairly dull plumage

fairly pale, un-marked lore in autumn birds

pale and rather strong

beware of *tristis*-type Chiffchaff (or dull Willow Warbler) with colourless body-feathers, making wing and rump look brighter

imm. autumn

unbroken pale eye-ring and pale lore

pale brown ear-coverts

silky-white

spring

best separated by call!

EASTERN BONELLI'S WARBLER

bright rump striking when hovering

often seen flycatching and hovering among branches

brown-grey, slightly less green above

tendency to have darker lore and eye-stripe

tertials and greater coverts with pale, worn edges in most

separation from almost identical Western on plumage alone still very contentious

spring

(Common) Chiffchaff *Phylloscopus collybita* m(r)**B**2/**W**4

L 10–12 cm. Breeds in woodland, normally open and with tall deciduous trees and moderate scrub layer (locally also in conifer forest with some broadleaved mixed in). Mainly summer visitor to Britain & Ireland (mid Mar–Oct), winters around Mediterranean and partly S of Sahara; scarce in Britain in winter, when also in scrub, reeds, gardens, etc. Active, confident. Nests on ground in domed cup.

IDENTIFICATION Quite similar to Willow Warbler, but slightly smaller, more compact and with duller colours. Grey-tinged brownish-green above, off-white below with variable yellow and buff tinge on throat and breast. Often more distinct element of *buffy-brown on breast-side*, especially in autumn, than shown by Willow Warbler. Distinguished by: *dark legs; fine, often rather dark bill*; usually *rather short and indistinct pale supercilium*; moderately *dark eye-stripe* which *divides whitish eye-ring at front and rear*; rather uniformly dark below eye and on ear-coverts, so that *white lower eye-crescent* shows well; short primary projection. Habitually *dips tail downwards* when moving in canopy (not seen from Willow Warbler). – Variation: Birds in the greater part of Europe (ssp. *collybita*) and in N Fenno-Scandia (*abietinus*) are very similar, but *abietinus* averages a shade paler, and greyer above; in the extreme NE, green and yellow elements in the plumage gradually decrease.

VOICE Call a soft whistled 'hweet', faintly upslurred and with emphasis at end. In late summer–autumn, juveniles occasionally emit straight calls, 'hiip' (which can then be confused with Siberian Chiffchaff). Also, birds of Turkey, Caucasus and Middle East have straight calls (cf. Siberian). Song a series of well-spaced, clear, forceful, *monosyllabic* (rarely disyllabic) notes on *two or three pitches*, 'sílt sült sült sült silt silt sult sült sült sült sült silt...'. Birds newly arrived at breeding site often add a muffled 'perre perre' between verses.

Siberian Chiffchaff *Phylloscopus collybita tristis* **V***

L 11–12½ cm. Traditionally regarded, as here, as an eastern race of Chiffchaff, but could also be an incipient separate species based on different vocalization and plumage. Breeds in coniferous taiga forest from the Ural region eastwards. Migratory, winters mainly in India; odd birds stray annually to W Europe in autumn (end Sep–Oct).

IDENTIFICATION Differs from Chiffchaff in Europe in being *grey-brown on crown, nape and mantle* (no trace of green), and *light rusty-buff on supercilium, ear-coverts and neck-/breast-sides* (no trace of yellow). Back / rump, however, show weak olive-green tone, and *remiges and rectrices are narrowly edged green* when fresh. *Bill and legs always black*. In fresh autumn plumage, quite a few immatures have light grey-brown tips to greater coverts enough to create suggestion of pale wing-bar. (It should be stressed that certain N European birds of race *abietinus* can be similarly grey-brown, also that it is usually very difficult *in the field* to confirm for certain that all trace of yellow is lacking on head and breast.)

VOICE Call a mournful, piping, straight (or negligibly downslurred) 'hii(e)p', but note that a straight call is no automatic ticket to this race, since some other eastern populations also have it. Song clearly *faster* and more *varied* than Chiffchaff in Europe, the notes run together, *pitch higher* and more *multisyllabic* than monosyllabic notes.

Iberian Chiffchaff *Phylloscopus ibericus* **V*****

L 11–12 cm. Formerly treated as a race of Chiffchaff but recently separated as full species on account of different vocalization, genetics, morphology and migration habits. Breeds in mixed and deciduous woods, often on foothills or low mountain slopes. Summer visitor (Mar–Sep), apparently mainly wintering in tropical W Africa. Nests on ground.

IDENTIFICATION Would often be overlooked were it not for the song. Small as a Chiffchaff, with wings nearly as rounded, but differs in having *cleaner green upperparts, yellow breast* and *whiter belly*, more like Willow Warbler. In spring virtually *no buff or brown tinge on head, neck or breast*, in autumn a trace of brown on sides of head at the most. *Supercilium* often *vividly lemon yellow*, especially *in front of eye*. An often rather *distinct dark eye-stripe* adds to similarity with Willow. On average, *legs slightly paler* than Chiffchaff, and base of bill has more pale brown (but overlap in both respects). *Dips tail downwards* repeatedly when feeding, just as Chiffchaff (but unlike Willow Warbler).

VOICE Call a characteristic *downslurred* soft whistling 'wee-uu'. Song basically resembles Chiffchaff on tone, is somewhat variable, but often consists of three motifs, a Chiffchaff-like part (only a little quicker and more monotonous), a series of fast, dry, stammering notes, and a few drawn-out whistles; a common phrase is 'chief chief chief chief tr-tr-tr-tr-tr sweet sweet sweet' (order of motifs may be swapped).

Canary Islands Chiffchaff *Phylloscopus canariensis* —

L 11–12½ cm. Recently separated as a local species rather than subspecies of Chiffchaff. Breeds on W Canary Islands in forests and copses with rich undergrowth. Resident.

IDENTIFICATION Quite rounded wings give *short primary projection* (and long-tailed look). Upperparts a rather *dark olive-brown*. Underparts *sullied and tinged buff when fresh*, wearing to purer yellow in spring. *Pale supercilium long, narrow and distinct*. Bill *rather long*, dark with pale cutting edges, tip often appearing downcurved. *Legs long*.

VOICE Call either Chiffchaff-like 'hweet', more straight 'heep' or disyllabic 'vüsst-eest'. Song like Chiffchaff but penetrating and 'explosive', and rhythm more uneven, together giving some songs a remarkable resemblance to Cetti's Warbler song; others are fairly similar to Iberian Chiffchaff.

Caucasian Chiffchaff *Phylloscopus lorenzii* —

L 10–11½ cm. Often regarded as a race of Sind Mountain Chiffchaff (*Ph. sindianus*) but here treated as full species. Breeds on mountain slopes, usually at 1500–2500 m, in willow, tall scrub or in more open mixed forest near treeline.

IDENTIFICATION Differs from similar Siberian Chiffchaff in *bolder whitish supercilium, darker brown crown* (often giving capped look), *pink-buff sides of head and beast* and more *contrasting white throat*, relatively *longer and broader rather square-ended tail*, *strong and quite black feet* with strong claws. The mainly brown upperparts can have trace of green on back, scapulars and wing-coverts.

VOICE Call a slightly melancholy pipe, faintly downslurred, 'tüü(u)', rather like Siberian Chiffchaff. Song like Chiffchaff's, 'pyit pyüt pyet pyüt pyet pyit pyet pyüt...'; tempo a bit faster, verses shorter and tone slightly more squeaky. Sometimes stifled 'te-ti' thrown in between verses.

CHIFFCHAFF

all species in the Chiffchaff group habitually dip their tail downward

a certain variation in e.g. amount of yellow on underparts, even in W and C Europe, makes subspecific identification of individual birds difficult or impossible

collybita
(W and C Europe)

extensive overlap between these forms

WILLOW WARBLER

(to compare; a short-winged ♀ shown)

supercilium distinct

pale cheek, diffuse frame

cleaner below than Chiffchaff

longer primary projection than Chiffchaff

pale

primary tips evenly spaced (unlike Willow Warbler)

supercilium subdued

dark cheek, making eye-ring stand out

duller flank than Willow

short primary projection, half to two thirds the length of the tertials

dark

abietinus
(N and NE Europe)

IBERIAN CHIFFCHAFF

many are impossible to identify in the field on plumage characters; song and calls are the best clue!

greenish, no brown hue

supercilium bright yellow in front

often paler bill

slightly longer primary projection, than Chiffchaff

usually clean yellowish

underparts cleaner yellow than Chiffchaff

medium-dark

SIBERIAN CHIFFCHAFF

call and plumage must be used in combination for safe identification

brown-grey

supercilium buff, no yellow

buff 'face', breast-sides and flanks; no trace of yellow

typical *tristis* shows no green hue on head or mantle, but greenish on wing, rump and tail is normal

tristis
(Siberia)

CANARY ISLANDS CHIFFCHAFF

vocalizations differ in many respects from those of other Chiffchaffs!

dark brownish-green

supercilium long

long and strong, slightly decurved

very short primary-projection

rather pale

strong buff tinge in fresh plumage

CAUCASIAN CHIFFCHAFF

brown-capped

brown

brown continues down to upper-tail-coverts

lacks obvious greenish fringes to wing and tail (unlike Siberian Chiffchaff)

supercilium white

white throat

pinkish-brown

DZ

Chiffchaff

Iberian Chiffchaff

Canary Islands Chiffchaff

Madeira

Canaries

Caucasian Chiffchaff

Arctic Warbler *Phylloscopus borealis* V**

L 11½–13 cm. Breeds in northernmost forest belt, in Fenno-Scandia usually in pure upland birch forest and often on sloping ground near a brook, farther E also in coniferous forest with plenty of birch and willow. Summer visitor (in Sweden end Jun–Aug), winters in SE Asia (incl. S Thailand, Burma, Indonesia) and thus has an almost record-long migration route. Active, spends much time in tree crowns, but is not really shy. Nests on ground, no feather lining.

IDENTIFICATION Just larger than Willow Warbler. Most like Greenish Warbler owing to *greenish-grey upperparts, off-white underparts*, very *distinct and long supercilium, long and broad dark eye-stripe* and a *short, narrow whitish wing-bar*, but distinguished by: bigger size; rather plump body but proportionately *not so big head*; *thicker and more wedge-shaped bill* (heavy base, pointed tip), with *dark tip to lower mandible* (lacking on Greenish); tendency towards *olive-grey tinge* (at times in form of diffuse spotting) *on breast and flanks*; *dark lores extend to bill-base*, but *pale supercilium ends short of forehead*. Primary projection fairly long. Tertials uniformly greyish-green. Primaries often have brownish edges (just as on Radde's Warbler). Ear-coverts are usually mottled dark. Exceptionally, wing-bar is worn off on one wing or even on both; occasionally shows hint of a second, shorter wing-bar on median coverts. Legs brownish-pink, sometimes darker brown-grey. Often looks 'snub-nosed', with flat forehead and slightly peaked hindcrown.

VOICE Call a short, sharp, scratchy 'dzri' (a bit like Dipper, easily penetrates through noise of rushing brook), quite unlike other *Phylloscopus* species. Song a fast, rather hard whirring trill at low pitch, 'sresresresresresresre...'; recalls Cirl Bunting song; now and then pitch drops ('vowel change') or tempo changes halfway through song.

Greenish Warbler *Phylloscopus trochiloides* V*

L 9½–10½ cm. Breeds in Europe (ssp. *viridanus*) in lowland deciduous or mixed forest (oak, elm, lime), often in tall groves and parks, also in small dense copses (alder, birch, rowan, oak) along coasts or in spruce forest mixed with deciduous. Summer visitor (broadly May–Sep, varies with latitude), winters in India; a few reach Britain annually (mostly Jun–Jul, often sing). Nests usually on ground.

IDENTIFICATION Small (as Chiffchaff), with proportionately *rather large head* and *rounded crown*. Greenish-grey above, off-white below with faint hint of yellow on throat and breast. Distinguished by *long, distinct, whitish supercilium* (which often extends onto forehead, past nostril; cf. Arctic Warbler) and a *short, narrow, slightly diffuse whitish wing-bar*. Lacks second (median-covert) white wing-bar (cf. Two-barred Greenish and Green Warblers), but in fresh plumage

sometimes has a few slightly paler grey-green covert tips which can look like a hint of a wing-bar in the field. Dark eye-stripe normally not quite so well marked as on Arctic lores usually with only a *dark spot in front of eye and not reaching bill-base* (few exceptions). *Lower mandible pale.*

VOICE Call a frothy, faintly disyllabic 't'sli', or more clearly disyllabic (almost White Wagtail-like) 'tisli'. Song which has sharp, high call-like quality, is a short, vaguel shuttling verse, often with slightly jerky rhythm and a half stop inserted, e.g. 'tisli-zizi-tisli-züt-sitzlie-zi, t'sli-sli-z sli-sli'; sometimes the verse ends with a trill (can be striking ingly Wren-like), sometimes it sounds more like Coal Tit.

Two-barred Greenish Warbler
Phylloscopus trochiloides plumbeitarsus V**

L 10–11 cm. An eastern race of Greenish Warbler. Breeds i C and E Siberian taiga; very rare autumn vagrant in Europe

IDENTIFICATION Very like European Greenish Warble Safest features are that *wing-bar* on tips of greater covert is *longer* (extends to scapulars), *broader* (rule of thumb broader than tarsus) and *better demarcated*, also that *short, narrow second wing-bar* is usually present on tips o median coverts. Other differences are only averages an insignificant: a shade greener, less grey above; rather pure white below; slightly stronger yellow tinge on eyebrow an head-/neck-side; slightly heavier (but not longer) bill somewhat darker eye-stripe behind eye. Told from Yellow browed Warbler by more *uniformly coloured tertials, al pale lower mandible* and *size*.

VOICE Very like that of European Greenish, and ofte impossible to separate with certainty. Call usually slightl fuller and more clearly disyllabic, 'tsi-z'li' (may even be tr syllabic). Song on average longer, faster and more liqui with smaller tone-steps and no half-stops as in *viridanus*.

Green Warbler *Phylloscopus nitidus* –

L 10–11 cm. A close relative of Greenish Warbler, recentl elevated to being a full species. Breeds on mountain slopes i N Turkey, Caucasus and N Iran in lush deciduous and mixe forest, at times close to treeline among birch, alder, beech juniper, willow, etc. Summer visitor (early May–Aug/Sep) winters in S India. Habits otherwise as Greenish.

IDENTIFICATION Resembles Greenish Warbler (*viridanus*) but differs in *Wood Warbler-green upperparts* and in ver *prominent and clear yellow supercilium, yellowish cheek throat, neck-sides* and *upper breast.* (Beware that yellow an green tints easily disappear among sunlit foliage, so similar ity to *viridanus* often amplified in field.) *Yellow-white win bar* obvious, and at times short trace of a second one vis ble. The yellow supercilium ends short of bill-base, does no reach forehead as on mos Greenish. Proportionatel *large head* and *strong bill.*

VOICE Call like Two-bar red Greenish. Song very lik Greenish but often separabl on inclusion of dry trillin or 'buzzing' phrase, 'tsee tserrr'; 1–3 such phrases in serted in each song.

Arctic Warbler

Greenish Warbler

Green Warbler

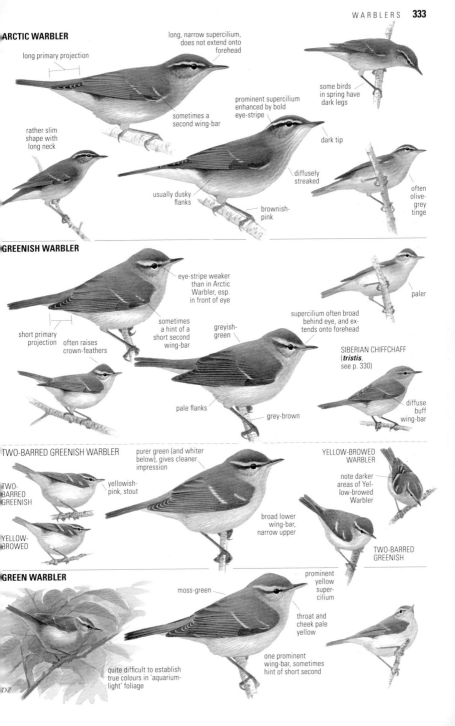

ARCTIC WARBLER

long primary projection

long, narrow supercilium, does not extend onto forehead

some birds in spring have dark legs

rather slim shape with long neck

sometimes a second wing-bar

prominent supercilium enhanced by bold eye-stripe

dark tip

diffusely streaked

usually dusky flanks

brownish-pink

often olive-grey tinge

GREENISH WARBLER

eye-stripe weaker than in Arctic Warbler, esp. in front of eye

paler

short primary projection

often raises crown-feathers

sometimes a hint of a short second wing-bar

greyish-green

supercilium often broad behind eye, and extends onto forehead

SIBERIAN CHIFFCHAFF (*tristis*, see p. 330)

pale flanks

grey-brown

diffuse buff wing-bar

TWO-BARRED GREENISH WARBLER

TWO-BARRED GREENISH

YELLOW-BROWED

purer green (and whiter below), gives cleaner impression

yellowish-pink, stout

broad lower wing-bar, narrow upper

YELLOW-BROWED WARBLER

note darker areas of Yellow-browed Warbler

TWO-BARRED GREENISH

GREEN WARBLER

moss-green

prominent yellow supercilium

throat and cheek pale yellow

one prominent wing-bar, sometimes hint of short second

quite difficult to establish true colours in 'aquarium-light' foliage

DZ

Yellow-browed Warbler *Phylloscopus inornatus* V*

L 9–10½ cm. Breeds in Siberian taiga and reaches just to edge of treated area in N Ural range, to upper reaches of Pechora. Normal winter range SE Asia. Visitor in autumn to W Europe, and although a rarity still the most numerous eastern vagrant, with several hundred records in some years in Britain alone; most records in late Sep–Oct in trees and scrub along coasts. Agile and restless, frequently flicks its wings. Often associates with other *Phylloscopus* warblers and Goldcrests.

IDENTIFICATION Often first attracts attention by its call. Smaller than Chiffchaff, with quicker action. Greyish-green above, off-white below, legs medium brown. *Long and very prominent supercilium* pale yellow above dark eye-stripe. *Broad yellowish-white wing-bar* on tips of greater coverts, and a shorter, less conspicuous second bar on tips of median coverts; prominence of wing-bar on greater coverts enhanced by dark surround on the wing. *Tertials* (esp. shorter two) dark-centred and *edged yellowish-white*, which eliminates Greenish Warbler. Sides of crown can be a shade darker grey-green, often giving a hint of a pale crown-stripe, esp. on hindcrown. (See also Hume's Leaf Warbler.)

VOICE Vocal. Commonest call a loud, penetrating, high-pitched 'sweeet' or 'tso**eest**', the quality of the call often recalling Coal Tit, though higher-pitched and more clearly rising at end, apart from being uttered in a consistent way (Coal Tit calls often vary in pitch and details). Song a few hesitant, very fine notes, 'tsewee, sese-wee...sewees' (somewhat recalling Hazel Grouse).

Hume's Leaf Warbler *Phylloscopus humei* V***

L 9–10 cm. Closely related to Yellow-browed Warbler but has profoundly different vocalization and slightly different plumage, hence nowadays commonly treated as a separate species. Very rare autumn or winter visitor from Central Asian mountain forests.

IDENTIFICATION Very similar to Yellow-browed Warbler, and at times inseparable on plumage. In fresh autumn plumage, slightly more dull and greyish-green above; *supercilium* (fore-part), *cheeks, neck- and breast-sides faintly buff-tinged*; flight-feathers and wing-coverts not quite so dark grey; *light tips to median coverts rather dusky and ill-defined*, not so distinct and pale; wing-bar on greater coverts often slightly tinged buff, not pure light yellow; bill and legs on average somewhat darker. In worn spring plumage usually inseparable from Yellow-browed Warbler.

VOICE *Best told from Yellow-browed by voice.* Call outside breeding a forceful disyllabic whistling 'dsu-**weet**', shorter, softer and lower-pitched than Yellow-browed, or a slightly descending 'dsee wo' (rather similar to alternative song; see

Yellow-browed Warbler

below). Song a drawn-out, at end slightly falling, buzzing note, 'bzzzzzzzeeo' (like a drawn-out migration call of Redwing). Alternative song is a downcurled and rather low-pitched 'veeslo' (or transcribed '**tiss**-yip'), frequently doubled or repeated several times.

Pallas's Leaf Warbler *Phylloscopus proregulus* V*

L 9–9½ cm. Rare autumn visitor (mostly Oct–Nov) from Asia, but among Siberian stragglers to W Europe one of the most numerous (up to *c.* 300 in one year in Britain alone). Very active and quickly moving about, constantly changing perch, fluttering and hovering among bushes or in tree canopy, and may even hang upside-down in tit fashion.

IDENTIFICATION *Very small*, about the size of a Goldcrest, with which it is often seen, but clearly differs from crests in its *much quicker movements* and restless feeding actions. Differs from Yellow-browed in the narrow but striking *pale yellow* (or whitish), *well-marked rump patch*, obvious when hovering or when feeding on the ground, but otherwise may be difficult to see. *Crown very dark* olive-green, divided by narrow but *clear-cut pale yellow central crown-stripe*. Often looks *large-headed* and *short-tailed*. Seen from below, note *long dark eye-stripe* and *vividly yellow supercilium with orangey forepart*. Shares with Yellow-browed and Hume's Leaf dark margins to pale wing-bar on greater coverts.

VOICE Less vocal than Yellow-browed. Commonest call a quiet, slightly rising 'chuee', rather modest, soft and nasal. Also, a finer, straighter 'pih'. Song surprisingly loud and trilling in Wren-like fashion ('Canary of the taiga').

Dusky Warbler *Phylloscopus fuscatus* V**

L 10½–12 cm. Breeds in Siberia in willows and other shrubbery on taiga bogs or on wet meadows, invariably preferring wet ground. Migratory, winters in SE Asia; very rare autumn vagrant in W Europe and then often found in weeds and scrub in open habitats.

IDENTIFICATION Rather Chiffchaff-like in shape but has rounded wings with *shorter primary projection*, and a trifle longer legs. In colour closest to a Caucasian Chiffchaff, *rusty-tinged dark grey-brown upperparts* and dusky grey-brown underparts, darkest on sides of breast and flanks, sometimes lightly tinged rusty. Identified by: *thin red-brown legs*; *narrow bill* with a little red-brown at base; *long, distinct, pale supercilium, well defined, white and usually narrow before eye* and usually broader and sometimes *tinged rusty-buff behind eye*; *broad dark eye-stripe*. Lacks any yellow on flanks and vent as in Radde's Warbler.

VOICE Call a hard clicking 'teck', like Lesser Whitethroat.

Radde's Warbler *Phylloscopus schwarzi* V**

L 11½–12½ cm. Breeds in Siberia in the taiga; requires glades beside watercourses with rank undergrowth. Migratory, winters in SE Asia; rare autumn vagrant in Europe.

IDENTIFICATION Willow Warbler size but sturdier, with slightly *larger head* (often with steeper forehead in profile), *stronger bill and legs*. Rather *dark brown with olive-grey cast* above, off-white below with *warm yellow-buff* (or more olive-brown) *tone* on breast, flanks and, especially, *vent and undertail-coverts*. *Strong pale supercilium* and *dark eye-stripe*. Told from Dusky Warbler by: *thicker bill*; *thicker and* on average *paler pinky-brown legs*; in front of eye *supercilium dorsally more diffusely defined, broad and in fresh plumage yellow-buff*, behind eye usually whitish. A bit *more sluggish* in movements than Dusky, not so active and restless.

VOICE Call a soft, slightly nasal, 'thick' clicking, 'chrep' or 'chett', sometimes in 'simmering' series, 'chett-et-et-et-et'.

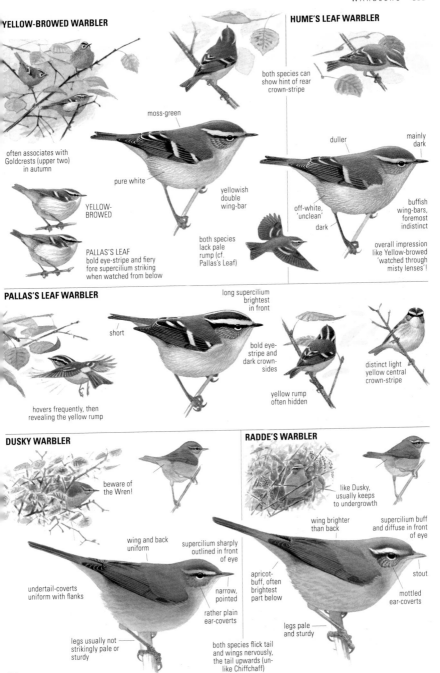

YELLOW-BROWED WARBLER

HUME'S LEAF WARBLER

both species can show hint of rear crown-stripe

often associates with Goldcrests (upper two) in autumn

moss-green

pure white

YELLOW-BROWED

PALLAS'S LEAF
bold eye-stripe and fiery fore supercilium striking when watched from below

yellowish double wing-bar

both species lack pale rump (cf. Pallas's Leaf)

duller

mainly dark

off-white, 'unclean'

dark

buffish wing-bars, foremost indistinct

overall impression like Yellow-browed 'watched through misty lenses'!

PALLAS'S LEAF WARBLER

long supercilium brightest in front

short

bold eye-stripe and dark crown-sides

distinct light yellow central crown-stripe

yellow rump often hidden

hovers frequently, then revealing the yellow rump

DUSKY WARBLER

RADDE'S WARBLER

beware of the Wren!

like Dusky, usually keeps to undergrowth

wing brighter than back

supercilium buff and diffuse in front of eye

wing and back uniform

supercilium sharply outlined in front of eye

apricot-buff, often brightest part below

stout

undertail-coverts uniform with flanks

narrow, pointed

rather plain ear-coverts

mottled ear-coverts

legs pale and sturdy

legs usually not strikingly pale or sturdy

both species flick tail and wings nervously, the tail upwards (unlike Chiffchaff)

DZ

CRESTS *Regulus* and WREN *Troglodytes*

Crests are traditionally kept together with the warblers, in particular with the superficially similar *Phylloscopus*, but newer genetic research has shown that they are not close to the warblers. Their true relationship is still to be established; for convenience they are kept here and with the Wren (related to flycatchers and thrushes), all being small arboreal species.

Goldcrest *Regulus regulus* rB1 / P+W1

L 8½–9½ cm. Breeds in coniferous or mixed woods, preferring dense stands of spruces or firs; on Continent often at higher elevations. Partly resident, partly migratory. On migration often shows up in open country. Associates with tits in winter. In woods, keeps high in canopy, restlessly moving and hovering among branches. Confiding.

IDENTIFICATION Attention to presence of a Goldcrest usually attracted by thin, high-pitched calls. *Minute size* (smallest bird in Europe) and combination of being pale green above and off-white below eliminates all but a few. Short neck and proportionately large head give compact, rounded impression. Dark 'peppercorn' *eye looking large* in otherwise pale, bland 'face'. *Crown-stripe yellow, bordered black* at sides. Dark bill very fine. – Adult ♂: *Much orange admixed in yellow crown-stripe*, generally hidden on relaxed bird but shown in display or when feathers are fluffed up. – Adult ♀: Usually *no orange on crown*, but a few have a very little (normally not seen in the field) in centre. – Juvenile: Crown plainer than adult, but dull conspicuously pale. Acquires adult-like 1st-winter plumage by end of summer. – Variation: Breeders on outer Canary Islands, i.e. Tenerife, La Palma, La Gomera, Hierro (ssp. *teneriffae*), have obvious *black band across forecrown*, are deeper pink-buff on underparts, and tertials are less pale-tipped. Bill subtly longer than on European Goldcrests.

VOICE Vocal. Commonest call a thin, high-pitched, reedy 'zree-zree-zree' (3–4 syllables). Alarm is a sharp, high-pitched, straight 'tsiih'. Song a cyclic repetition (4–6 times) of a high-pitched, rhythmic 'piteetilü', ending with a little trill or Treecreeper-like flourish, 'zezesuzreeo'. Difficult to hear for many elderly people.

Firecrest *Regulus ignicapilla* rB4 / P3

L 9–10 cm. Breeds in deciduous and mixed woods, but will also be found in conifers (incl. pines, cedars). On average selects lower-growing woods than Goldcrest, and can often be found in tall bushes and hedgerows, or in more mixed woods with richer sprinkling of deciduous trees, including in varied older parks and larger gardens. In Iberia and NW Africa found in cork oak and beech woods. Short-range migrant. Behaviour as Goldcrest. Nests in conifers.

IDENTIFICATION Like Goldcrest, best located and identified by its calls. *Minute size* and general coloration narrow possibilities down to a few. Unlike Goldcrest, has *prominent white supercilium* and *black eye-stripe making the head pattern look more striped and 'sharp'*. Also, *side of neck bright greenish-yellow*, tinged ochrous, not plain grey-green as on Goldcrest. Crown as Goldcrest, but black lateral crown-stripes join across forehead. – Adult ♂: *Much orange on crown*, usually visible also when relaxed. – Adult ♀: *No or very little orange* visible on centre of crown, those with most

being difficult to sex in the field. – Juvenile: Lacks distinct crown pattern until autumn moult, but has striped 'face' pattern reminiscent of adult but duller.

VOICE Contact-call similar to that of Goldcrest but slightly lower-pitched, usually opens with a longer and more stressed note followed by 2–3 shorter and slightly accelerating and rising notes, 'züü zü-zi-zi'. Song a *repetition of the same fine, high-pitched note*, slightly rising in pitch initially and in strength, and ending in a brief trill; it lacks the cyclic rhythm of Goldcrest's song.

Madeiran Firecrest *Regulus madeirensis* —

L 9–10 cm. Breeds fairly commonly on Madeira in forests. Resident. Separated as a local species rather than a subspecies of Firecrest on account of partly different vocalization and genetics, and on subtle differences in morphology.

IDENTIFICATION Very similar to Firecrest but differs in having a *deeper golden-bronze patch on sides of neck*, more black on wing at base of primaries forming a *rather large black patch outside broad wing-bar*, and on having *shorter white supercilium*, fading off just behind the eye. Also, has on average *longer bill* and *slightly longer legs*, which combine to give a subtly more lanky look. ♂ has less extensive orange, more yellow, on crown than in Firecrest.

VOICE Among several fine calls reminiscent of Firecrest there is also a distinctive straight, shrill 'wheez', and a similarly straight whistle, 'peep', almost like call of Siberian Chiffchaff (ssp. *tristis*). Song rather similar to Firecrest's, a hurried, slightly accelerating series of fine notes.

(Winter) Wren *Troglodytes troglodytes* rB1 / W

L 9–10½ cm. Breeds in woodland with dense undergrowth, in overgrown clearings and scrub, often in rank streamside growth and gardens, also on barren islands with hedges, shrubbery, walls etc. Fond of dense patches created by up-rooted trees, piles of branches from forestry, dense bushes of bramble, roses, etc. Resident in much of Europe, but migratory in Fenno-Scandia, wintering in W Europe.

IDENTIFICATION *Very small*, and this reinforced by ludicrously *small tail that is usually raised vertically*, also by short neck. *Reddish-brown above and sullied brownish-white below with fine dark vermiculations*. A rather subdued pale brownish-white supercilium is about the only plumage feature that sticks out. Bill quite long, pointed and slightly decurved. Sexes and ages alike. – Variation: Rather slight in spite of wide distribution, but birds on northernmost British Isles (*zetlandicus*), Faroes (*borealis*) and Iceland (*islandicus*) are progressively darker and have longer bill and legs.

VOICE When nervous, a rattling, hard 'zerrrr' as well as single hard clicking 'zeck!' notes, the two sometimes combined in long series. The 'zerrr' call can be mixed up with the call of Red-breasted Flycatcher but is harder and more metallic. Song, delivered in most seasons of the year, amazingly loud for so small a bird, a rather consistently repeated series of metallic ringing notes and trills roughly like a Canary or Tree Pipit, e.g. 'zitrivi-si **svi-svi-svi-svi-svi** zivüsu **zü-zü-zü-zü** si-**zirrrrr svi-svi-svi** siyu-**zerrrrr** sivi!'. Some of the high-pitched and fine notes in the song do not carry far, making the song sound a bit disrupted at distance. Sings mostly from cover but at times from exposed perch on a branch.

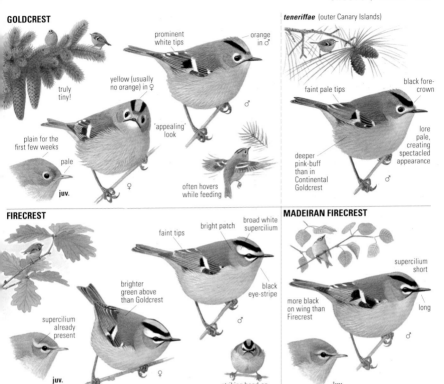

GOLDCREST

truly tiny!

prominent white tips

orange in ♂

yellow (usually no orange) in ♀

'appealing' look

plain for the first few weeks

pale

juv.

♂

♀

often hovers while feeding

teneriffae (outer Canary Islands)

faint pale tips

black fore-crown

lore pale, creating spectacled appearance

deeper pink-buff than in Continental Goldcrest

♂

FIRECREST

brighter green above than Goldcrest

faint tips

bright patch

broad white supercilium

black eye-stripe

♂

supercilium already present

juv.

♀

striking head-on appearance!

MADEIRAN FIRECREST

supercilium short

more black on wing than Firecrest

long

♂

juv.

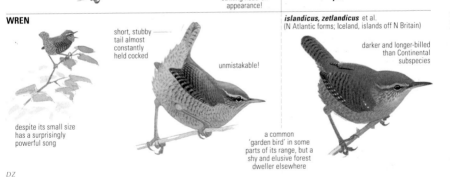

WREN

despite its small size has a surprisingly powerful song

short, stubby tail almost constantly held cocked

unmistakable!

islandicus, zetlandicus et al.
(N Atlantic forms; Iceland, islands off N Britain)

darker and longer-billed than Continental subspecies

a common 'garden bird' in some parts of its range, but a shy and elusive forest dweller elsewhere

DZ

Goldcrest

Firecrest

Madeiran Firecrest

Madeira

Canaries

Wren

FLYCATCHERS *Muscicapidae*

Small passerines, related to thrushes, seven native species within region. Characteristic habit of choosing prominent perches, where they remain quite still between dashing forays to catch insects in the air, on the ground or in foliage. Short legs, delicate feet and broad but pointed bill (gape wide). Often flick tail or wings. Juvenile plumages spotted.

Spotted Flycatcher *Muscicapa striata* mB2

L 13½–15 cm. Breeds in gardens, parks and in forests, often with small glades and openings. Summer visitor, wintering in Africa. Nests in recess on tree trunk or building, often against a wall on branch of climbing shrub.

IDENTIFICATION Medium-small, greyish-brown, rather slim passerine, lacking obvious features. Bill dark, fairly long and strong. *Tail and wings long*, lower body rather slim. Black *legs short*. Dull *grey-brown upperparts* and *off-white underparts* unmarked except for *dark streaking of breast*, *sides of throat, forehead* and *forecrown* (best seen head-on; ground colour of forecrown slightly paler grey-brown than rest of upperparts). 'Peppercorn' eye dark, with indistinct, narrow palish eye-ring. In fresh plumage, secondaries diffusely edged pale and may form indistinct lighter panel on folded wing, but this still very different from wing patch of ♀ Pied Flycatcher. Sexes alike, ages similar and generally not separable in the field. Sits rather upright, often quite exposed, flicking its tail and watching for flying insects, then makes quick sally to snap prey, alighting on same or new perch. Flight fast and agile. When hunting insects sometimes rises quickly several metres and hovers, at other times may glide a couple of metres during long, sweeping dives.

VOICE Vocal, but repertoire unobtrusive and primitive, and often overlooked. Call a short, shrill 'zee', not unlike what can be heard from other species such as Chaffinch, Hawfinch and Robin. Alarm more characteristic, a call-like note immediately followed by one or two short, dry, clicking ones, 'eez-tk(-tk)'. Song a series of simple, *quiet*, high-pitched, squeaky or scratchy notes, often mixed with a few soft trills, rhythm a little uneven with *notes well spaced*.

(Asian) Brown Flycatcher *Muscicapa dauurica* [V***]

L 12–13 cm. Very rare vagrant from Siberia, where it breeds in taiga (recorded only in Denmark and Sweden). Behaviour and habitat much as for Spotted Flycatcher.

IDENTIFICATION Much like a small Spotted Flycatcher, differing in having *unstreaked forecrown* and usually a *uniform pale brown-grey wash across breast* only; prominent *whitish eye-ring* (around rather large dark eye) and *whitish lores*. Bill strong, long and flat in profile, broad-based with slightly convex sides seen from below, and with pinkish or *straw-coloured base to lower mandible*, visible from the side.

VOICE Call recalls Spotted's, a high-pitched, piercing 'tzi'.

Red-breasted Flycatcher *Ficedula parva* V* / (P4–5)

L 11–12 cm. Breeds in forests in a variety of habitats, mostly in lush, dense patches with rich undergrowth and a brook or other water, preferring deciduous woods but not shunning mixed. Summer visitor (May–Sep), in W Asia.

IDENTIFICATION Small, restless, dashing around in canopy like small warbler (unlike other European flycatchers). When

perched it often reveals its flycatcher identity by flicking its wings and cocking its *black and white tail*. Mainly brown above and whitish below, with narrow pale eye-ring around dark eye. Rather small bill dark with pinkish-brown base to lower mandible. – Adult ♂: *Orange-red throat patch* with *diffuse lower border* but sharply defined laterally against *lead-grey sides to neck*; lead-grey colour also variably invades head and nape. Keeps red throat in winter. Uppertail-coverts black. – Subadult ♂: Like adult, but red colour on throat and grey on head less developed. – 1st-summer ♂/ ♀♀: No orange-red on throat (exceptions very rare), and no lead-grey on head. Uppertail-coverts brown or blackish, *not blacker than central tail-feathers. Underparts tinged creamy-buff*. – 1st-winter: Like 1st-summer ♂/♀♀, but underparts rich buff, and wing-coverts tipped buff and forming wing-bar on greater coverts; tertials edged buff.

VOICE Vocal. Common call on migration a slurred rattle, 'serrrt', like a soft Wren. In anxiety/alarm a short, soft disyllabic whistle, 'deelü'. Other calls a short, shrill 'zree' and dry clicking 'tek' notes. Song opens with a few high-pitched, sharp notes followed by a rhythmic middle section in Pied Flycatcher fashion, and ends with highly characteristic *slow-paced series of descending clear notes*, 'teh tüh tuh taa taa'.

Taiga Flycatcher *Ficedula albicilla* V***

L 11–12½ cm. Breeds in dense Siberian taiga and rarely W of Urals in easternmost European Russia, preferring patches with plenty of birch and other deciduous trees mixed in. Rare vagrant to NW Europe, perhaps often overlooked.

IDENTIFICATION Very similar to Red-breasted Flycatcher, but adult ♂ summer has orange-red throat patch *neatly outlined below* by *lead-grey breast-band*; tendency to have *brownish cap* and ear-coverts, but lores and supercilium grey. ♀♀ and immatures have invariably *jet-black uppertail-coverts, blacker than central tail-feathers*. Note also that *bill is* darker, *nearly all black* with only limited brown base below (although this character often difficult to assess in normal side view). *Underparts tinged greyish-brown*, usually duller and colder than Red-breasted. Often contrasting white bib. – 1st-winter: Like Red-breasted, but wing-coverts and tertials usually *edged and tipped whitish* rather than buff (some overlap). – 1st-summer ♂: Very similar to adult with developed red throat-bib. – Adult ♂ winter: Moults to ♀-like plumage, more or less losing the red throat-bib.

VOICE Common call a very fast, hard rattle, 'trrrt' like a branch in the forest creaking in the wind. Song completely different from Red-breasted, more recalling a bunting or pipit, consisting of a rapid series of dry trills and sharp notes at varying pitch, shuttling up and down the scale a bit, e.g. 'zri-zri-**da** zri-zri-**da** tü-tü-tü zrri-**daa**-zi'.

Spotted Flycatcher

Red-breasted Flycatcher

SPOTTED FLYCATCHER

streaked forehead

diffusely streaked breast

watches from exposed perch

ad.

juv. →
1st-wint.

ad.

all flycatchers wear a pale-spotted juv. plumage for a short period after fledging

note characteristic shape with short legs and long rear

BROWN FLYCATCHER

plain forehead

broad, pale base to bill

plain

ad.

palish spectacles

pale wing-bar in 1st-w.

dark lateral throat patch

1st-wint.

note compact shape with large head and short tail

RED-BREASTED FLYCATCHER

buff

1st-w.

white tail-sides (more obvious in flight)

small orange throat-patch

2nd-s. ♂

very rarely, ad. ♀♀ may attain small red throat-patch, too

pale buff

ad. ♀

orange-red continues onto breast

ad. ♂

TAIGA FLYCATCHER

pale-edged tertials

uppertail-coverts blacker than uppertail

tail-pattern similar in both species

dark

grey-brown

1st-w.

pale throat contrasts with dark breast (reminiscent of Red-flanked Bluetail!)

lead-grey

ad. ♀

brown

grey

ad. ♂

DZ

Pied Flycatcher *Ficedula hypoleuca*   m**B**3

L 12–13½ cm. Breeds in gardens, parks, mixed and deciduous open forests, even in remote taiga habitats. Summer visitor, wintering in W Africa. Particularly attracted to mature oak trees providing airy surroundings and rich insect life. Active and restless. Nests in tree-hole or box.

IDENTIFICATION Most spring ♂♂ black and white, other plumages brown and off-white. All have *white or buff-white wing patch* on otherwise dark wing, narrow white sides to tail (few exceptions). Note amount of *white visible on primary bases* on folded wing: usually *none* or *only narrow wedge*, barely extending to tips of primary-coverts. Often cocks tail nervously and *flicks one wing upwards.* – Adult summer ♂: *Upperparts blackish* or dark brown-grey (more frequent in E Europe) with usually *small white patch on forehead* (divided in two, one spot on each side). Primaries rather dark and fresh. Brownest very similar to ♀♀ but usually told by white on forehead, *extensive white on outer webs of tertials*, and *blacker uppertail.* A few have paler grey rump. – 1st-summer ♂: Similar to adult ♂, but in close view primaries are more abraded and browner. – Adult ♀: Upperparts brown. No white patch on forehead (very little indistinct buff-white at most). White on tertials covering only outer margin distally. – Adult autumn ♂: Like adult ♀, but sometimes recognized by *blacker uppertail* and rather blackish wing-coverts. – 1st-winter: Like ♀, but has a 'step' between broad white edge on outer web and narrow one on inner web on tips of tertials. – Variation: Adult summer ♂ ssp. *iberiae* (C Spain) looks like a ♂ Collared Flycatcher without white collar, is very black with large white patch on forehead, much white in wing, and has all-black tail; some are even semi-collared. Many adult ♀♀ develop a white patch on forehead.

VOICE Vocal. Alarm a persistently repeated short, metallic 'pik'. Call a quiet clicking 'tec'. Song loud, rhythmic, a 2-sec.-long phrase with repetitive elements and sudden changes of pitch, 'zi **vree**zi **vree**zi tsu tsu chu-**vee** chu-**vee** zi zi zi'; at times inserts a pleasing, melodious figure.

Atlas Flycatcher *Ficedula speculigera*   —

L 12–13½ cm. Breeds in woods and ochards in the Atlas mountains in NW Africa. Summer visitor (arrives late Apr–May), winters in W Africa. Closely related to Pied Flycatcher.

IDENTIFICATION Adult summer ♂ like Collared Flycatcher, *jet-black above* with *much white on forehead and wing*, and with *all-black tail*. Differs on *lack of complete white collar* (but may rarely have a narrow and near-complete one), and on voice. – Adult ♀: Like Collared but *uppertail on average blacker*, contrasting against *pale grey rump* and *greyish back*. Large white primary patch. – 1st-summer ♂: Very similar to Pied Flycatcher, and some inseparable. Typical

birds have rather *large white forehead patch* (never divided in two), and are deeper black on crown, nape and mantle.

VOICE Call a repeated 'veet', subtly different from call of Pied Flycatcher. Song, too, resembles Pied but is slightly slower and deeper-voiced, and includes large tone steps. It sounds less cheerful, more 'pensive' than song of Pied.

Semicollared Flycatcher *Ficedula semitorquata*   —

L 12–13½ cm. Breeds in deciduous or mixed woods in mountains at lower and middle levels, often in copses along rivers. Summer visitor, wintering in Africa.

IDENTIFICATION Intermediate between Pied and Collared. On summer ♂, white on side of neck usually reaches further back than on Pied (but a few are confusingly similar); look for prominent *white tips to median wing-coverts* (may merge with large white wing patch, though), more *white on primary bases* than on Pied (about as much as on 1st-summer ♂ Collared), and *extensive white on outer tail-feathers* (sometimes visible from below on perched bird). Adult ♀ very similar to ♀ Collared; typically has *white-tipped median coverts*, but these may rarely be absent (and rarely there is a hint of such pale tips on Collared, too); white on primary bases somewhat less extensive on average.

VOICE Call a clear, straight piping note, 'tüüp', lower-pitched than in Collared, almost like call of Siberian Chiffchaff (*tristis*) at distance. Sometimes a thinner, slightly shrill and rising 'tüühp'. Contact-call also a quiet clicking 'tec'. Alarm a hoarse 'shah'. Song *weak*, easily missed in forest chorus, strained notes like Collared but rhythmically more akin to Pied. The song has more even pitch than Collared, e.g. 'sree sü-sü, sru-tee sru-tee srreeh see-vee sree'.

Collared Flycatcher *Ficedula albicollis*   **V*****

L 12–13½ cm. Breeds in deciduous woods, in gardens and parks. Summer visitor, wintering in Africa. Behaviour much as for Pied Flycatcher, with which it can hybridize.

IDENTIFICATION Adult summer ♂ instantly recognized by *broad white collar* around neck. Otherwise similar to Atlas Flycatcher: *large white patch on forehead, very black upperparts* incl. primaries, *large white patch on base of primaries*, and *all-black tail.* Rump often pale grey or even white. – 1st-summer ♂ has browner primaries, smaller white areas, but has the white collar. – ♀ and 1st-winter very similar to Pied, but spring ♀ slightly paler and tinged more greyish above, and *white primary patch larger and club-shaped or rectangular*, reaching closer to outer edge of wing than on any Pied. – Adult autumn ♂: Brown above like autumn ♀♀ and 1st-winter, but recognized by *blackish wing* with *large white primary patch*. Oddly, much white on outer tail-feathers.

VOICE Contact and alarm a drawn-out, straight, thin whistle, almost as if 'inhaling', 'eehp' (like a note from the song). Also a quiet clicking 'tec'. Song completely different from that of Pied, a series of rather slowly delivered drawn-out, harsh or strained whistling notes, often with marked changes in pitch.

Pied Flycatcher   Semicollared Flycatcher   Collared Flycatcher

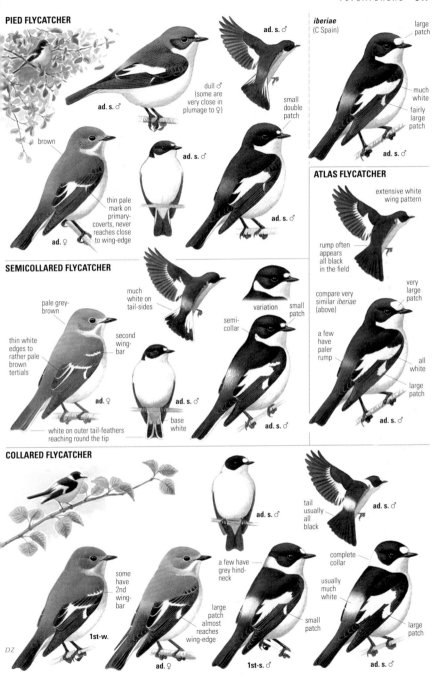

PIED FLYCATCHER

ad. s. ♂

dull ♂
(some are
very close in
plumage to ♀)

ad. s. ♂

small
double
patch

brown

ad. s. ♂

thin pale
mark on
primary-
coverts, never
reaches close
to wing-edge

ad. ♀

ad. s. ♂

iberiae
(C Spain)

large
patch

much
white

fairly
large
patch

ad. s. ♂

ATLAS FLYCATCHER

extensive white
wing pattern

rump often
appears
all black
in the field

compare very
similar *iberiae*
(above)

a few
have
paler
rump

very
large
patch

all
white

large
patch

ad. s. ♂

SEMICOLLARED FLYCATCHER

much
white on
tail-sides

pale grey-
brown

variation

small
patch

semi-
collar

thin white
edges to
rather pale
brown
tertials

second
wing-
bar

ad. ♀

ad. s. ♂

base
white

ad. s. ♂

white on outer tail-feathers
reaching round the tip

COLLARED FLYCATCHER

ad. s. ♂

tail
usually
all
black

ad. s. ♂

some
have
2nd
wing-
bar

1st-w.

a few have
grey hind-
neck

large
patch
almost
reaches
wing-edge

ad. ♀

complete
collar

usually
much
white

small
patch

1st-s. ♂

large
patch

ad. s. ♂

DZ

TITS *Paridae* et al.

Small, lively, rather squat passerines with short bill. Most are woodland-dwellers, nest in cavities and in winter gather in mixed flocks (roving tit bands). Mainly residents. Of the species grouped here, Penduline Tit, Long-tailed Tit and Bearded Reedling are not so closely related to the 'real' tits.

Great Tit *Parus major* rB1 / W4

L 13½–15 cm. Breeds in woodland (both in desolate taiga and close to humans in parks and gardens). Frequent visitor to birdtables and seed-dispensers. Bold, at times plain audacious, may take seed from outstretched hand. Food insects and seeds, suet, beech nuts etc. Nests in nestbox or tree-hole.

IDENTIFICATION Largest tit, easily identified by *yellow underparts* with black central band, glossy *black head* with *big white cheek patches*, moss-green back and narrow white wing-bar on blue-grey wing. Sexes similar, but ♂ is stronger yellow below with *broader black median band* (forms all-black patch in centre of belly), while ♀ is often somewhat paler yellow and has narrower and often broken black line. – Juvenile: Resembles adult, but has yellowish cheek patch without complete black border along lower edge.

VOICE Rich repertoire. Main calls cheerful, include a somewhat Chaffinch-like 'ping ping', a cheery 'si **yu**tti **yu**tti' and a chatty 'ti **tuu**i'. A kind of inquiring, more timid 'ti ti tüh' is often heard in autumn. Scolds with harsh 'che-che-che-che-che-...' (Magpie fashion). Fledged young fill the summer wood with insistent, shrill 'tete**te** tete**te**...'. Song, heard as early as late winter, a simple, seesawing ditty with slightly mechanical intonation, e.g. '**ti**-ta **ti**-ta **ti**-ta...' or trisyllabic and with different stress, 'ti-ti-**ta** ti-ti-**ta** ti-ti-**ta**...'.

Coal Tit *Periparus ater* rB2 / W4

L 10–11½ cm. Breeds mainly in conifer woods, often with some taller spruces. Resident, but N populations in some years move south in large numbers. Forages much in tops of trees and in outer branches. Nests in hole.

IDENTIFICATION Like a small and colourless cousin of Great Tit, with similarly *black head, large white cheek patches* and *narrow white wing-bar*. However, has bigger head and *fuller nape* which merges more into mantle. *Underparts* are *dusky greyish-buff* rather than yellow and lack black central band. Views from behind reveal best feature, an oval *white patch on nape*. When a trifle agitated may raise a small crest like a tiny 'spike' on hindcrown. Back blue-grey. Besides white wing-bar on tips of greater coverts, has a second bar in the form of a shorter 'string of beads' along median coverts. – Variation: Birds in Britain and in Iberia are more olive-toned, not so blue on back, and flanks are tinged reddish-brown. In Ireland, underparts and cheeks are faintly yellowish.

VOICE Call is various fine, clear notes with melancholy ring, e.g. 'tüüh', 'tih tüh-e'. Song, often delivered from top of tallest tree in territory, like Great Tit's though 'played at too fast a speed', a high-voiced, very *fast* scraping series with call-like tone, 'sit**chu**-sit**chu**-sit**chu**-sit**chu**-...'.

(European) Blue Tit *Cyanistes caeruleus* rB1 / W3

L 10½–12 cm. Breeds in woodland, preferring deciduous or mixed woods, and in parks and gardens. Resident, but N populations move SW in some autumns, sometimes in huge numbers. Often seen at birdtables in winter. Readily hangs upside-down on nut-holders. Nests in hole or nestbox.

IDENTIFICATION A smallish tit with a *small but rounded head* compressed into shoulders (often looks neckless). Shape combined with its lively disposition and strong head pattern, with *black eye-stripe* and small *blue skull-cap* on otherwise white head, give an attractive and captivating appearance. *Underparts yellow* with narrow greyish-black central stripe on belly. *Wings bluish*, with bright blue on greater coverts and wing-bend. Sexes similar (♀ usually a bit duller).

VOICE Vocal. Main call a fast, high 'sisisüdu' with final note lower, sometimes just 'sisisi' (can sound almost as sharp as Grey Wagtail). Alarm a scolding series with ending stressed and somewhat halting, 'ker'r'r'r'rek-ek-**ek**'. Song a couple of drawn-out, sharp notes followed by a silvery trill on lower pitch, 'siiih siiih, si-sürrrr'; sometimes two rapidly repeated short verses, 'si-si-sürrr, si-si-sürrr' (Treecreeper-like voice).

African Blue Tit *Cyanistes teneriffae* —

L 11–12 cm. Breeds in N Africa and on Canary Is in forests and gardens. Resident. Recently separated as a local and allopatric species rather than part of Blue Tit.

IDENTIFICATION Similar to Blue Tit but is *smaller* and *darker* with partly different vocalization. *Crown is dark, blackish-blue*, and *white supercilium* and band across hindcrown *narrow*. *Back is greyish-blue*, lacking green element of Blue Tit. Wing *deeper ultramarine*, and populations in Canary Is *lack white wing-bar*, have uniform wing.

VOICE Some calls recall Great Tit, others Crested Tit.

Azure Tit *Cyanistes cyanus* —

L 12–13 cm. Breeds in woods, often in riverside willows and in scrub bordering wetlands. Rare vagrant in W Europe.

IDENTIFICATION Unmistakable, with *white and blue plumage*. Sole confusion risks are albinistic Blue Tit or hybrid with Blue Tit ('*pleskii*'). Note *white crown* (without blue cap), *long tail* with *white corners, very broad white wing-bars*, and *much white on tertials and tips of flight-feathers*.

VOICE Calls most similar to Blue Tit's, e.g. 'tsi-tsi-tserrr de-de-de'. Song like a mixture of Blue and Crested Tit.

Great Tit   Coal Tit

Blue Tit   Azure Tit

GREAT TIT

dull cap and yellowish cheek

juv.

ad. ♂

ad. ♀

broken-up belly-band

ad. ♂

wide black belly-band

white sides

COAL TIT

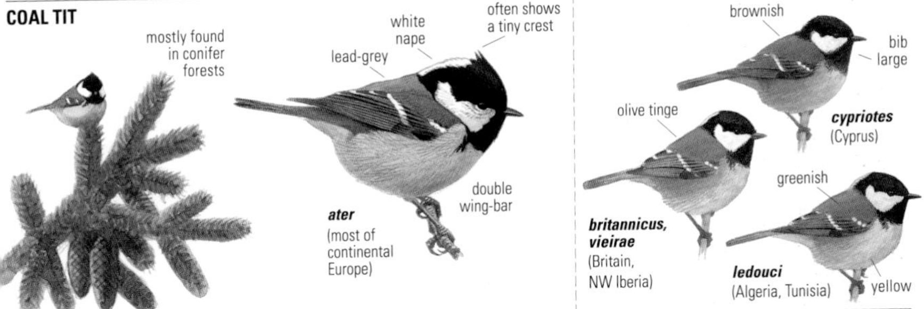

mostly found in conifer forests

white nape

lead-grey

often shows a tiny crest

double wing-bar

ater (most of continental Europe)

brownish

bib large

cypriotes (Cyprus)

olive tinge

greenish

britannicus, vieirae (Britain, NW Iberia)

ledouci (Algeria, Tunisia)

yellow

BLUE TIT

dull blue 'beret'

no white

♀

bright blue 'beret'

♂

often hangs upside-down

dull cap, yellow cheek

juv.

AFRICAN BLUE TIT

ultramarinus (NW Africa)

indigo-blue

♂

teneriffae (C Canaries)

lacks wing-bar (except in juv)

♂

AZURE TIT

hybrid Blue Tit × Azure Tit ('*pleskii*')

extensive white in wing

white cap

striking white in spread tail

DZ

Crested Tit *Lophophanes cristatus* rB4

L 10½–12 cm. Breeds in coniferous forest, often in older moss- and lichen-rich spruce (in Britain confined to Caledonian pine forest of Scotland), locally in S Europe in deciduous woodland. Resident. In N Europe joins winter tit bands. Often forages on ground or low branches. Not exactly shy, though usually difficult to approach. Visits birdtables only rarely. Nest excavated in rotten trunk or stump.

IDENTIFICATION Rather small, with brownish upperparts and dirty white underparts. *Head* neatly *patterned in black and white* and furnished with a *pointed, triangular crest* which is always visible (but which may be raised to varying extents). Sexes and ages alike in field. – Variation: Minimal. In W and S Europe upperparts warmer brown, head-sides have faint buff tone and flanks reddish-brown tinge.

VOICE Advertising-call a characteristic 'bubbling', cheery trill, 'burrur**ret**', often repeated, quite different from calls of other tits (but a bit like Snow Bunting's). Also short sharp conversational 'zit' notes. Song a series of calls and sharp notes, delivered alternately at rapid pace, 'zi-zi-**züt** burrurre zi-zi-**züt** burrurre zi-zi-**züt** burrurre...' and so on.

Marsh Tit *Poecile palustris* rB3

L 11½–13 cm. Breeds in unthinned, often damp deciduous woods with plenty of dead and dying trees; also in larger gardens and parks with older fruit trees and understorey. Resident. Frequently at birdtables in winter. Fearless. Dominant over Willow Tit where ranges overlap. Nests in cavity (tree-hole made by Willow Tit, nestbox, natural hole etc.).

IDENTIFICATION Very like Willow Tit (which see), and *best told by voice*. Has proportionally somewhat smaller head (but difference not striking). Most important plumage difference is *lack of obvious pale panel on secondaries* on folded wing (fresh autumn birds can show suspicion of a diffuse pale area). More subtle differences are that *black cap is glossy* (though duller on immatures), *cheeks are on average less clean white* and *black bib somewhat smaller* (though some overlap). Marsh Tit in N Europe is a shade browner above and not so whitish below as Willow, but in W and C Europe the two species are similar in general colour.

VOICE Call an explosive, spirited 'pi**chay**', sometimes slightly longer 'zee-zee-**chay**', or with (vaguely) Blue Tit-like scold appended, 'pi**chay**-de-de-de-de-de-de-de-de-**det**'. Several types of song occur, and same bird can vary; recognized by vocal tone. Common song type is a repetition of a single loud note, 'chüpp chüpp chüpp chüpp...' (fast variants become Greenfinch-like, 'chipchipchipchip...'); another type more like Coal Tit, '**veeta-veeta-veeta-veeta**-...', or is a mixture of the two, 'tee-chüpp tee-**chüpp** tee-**chüpp**...'.

Willow Tit *Poecile montanus* rB3

L 12–13 cm. Breeds in coniferous forest, often at higher elevation, and in upland birch forest, as well as (mainly in W and C Europe) at lower levels in mixed woods with much birch, alder and willow, along rivers, in scrub in overgrown marshland, etc. Resident. Not so frequently seen at birdtables in winter. Nest excavated in rotten, usually narrow trunk.

IDENTIFICATION Rather big head, full nape. Plumage *grey-brown and off-white* with *black cap* and *black bib*. Confusingly like Marsh Tit; *best told by voice*, also by *pale panel on*

secondaries (on summer adults occasionally lost through abrasion, and a few newly moulted autumn Marsh Tits can have suggestion of pale secondary edges) contrasting with the otherwise brown-grey folded wing. Further differences exist (but are difficult to see in field, or are not absolute): black *crown is* always *dull* (glossy on adult Marsh); *bib is* on average *larger*, and cap extends further down nape; *head-sides average whiter*. Sexes and ages alike. – Variation: Fenno-Scandian, Russian and E European birds (*loennbergi, borealis*) are whitish on head-sides and below and brown-grey above, while British birds (*kleinschmidti*) are darker, distinctly brownish above, more soiled on cheeks and buff on flanks. Continental Europe harbours a few more races of intermediate appearance.

VOICE Call 1–2 short notes followed by 2–4 lower, hoarse, harsh, drawn-out notes, 'zi-zi **taah taah taah**', quite unlike Marsh Tit's (but very like that of Siberian, which see). Also fine conversational 'zi' notes. Song of two main types: a series of pensive, melancholy, Wood Warbler-like notes, 'tiu tiu tiu tiu tiu tiu' (in Alps and parts of E Europe dialectal difference with straighter note, 'düü düü düü...'), and a rather variable high-pitched short warble ending in a trill.

Siberian Tit *Poecile cinctus* —

L 12½–14 cm. Breeds in old undisturbed conifer forest in N, mostly lichen-rich ancient pine forest, but also in the lower upland birch forest. Resident; rare S of breeding area.

IDENTIFICATION Resembles Willow Tit (which is found in same habitat), but has fluffier feathering, appears to have somewhat bigger head, and has much *bigger dark bib* (often with broken lower edge), *rusty-buff colour on flanks* and *dull dark brown cap* (not black). Back is darker and much browner than northern Willow Tit's.

VOICE Call confusingly similar to Willow Tit's, 'zi-zi **tah tah tah**', but the harsh notes are not so drawn out and as a rule not so firmly stressed. Contact-call a fine 'tsih'. Song a grinding, buzzing series, 'chi-**ürr** chi-**ürr** chi-**ürr**...', as well as a more Marsh Tit-like simple verse, 'che-che-che-che-...'. Function of a more melodic brief warble, 'zi-zi-yut**vuy**', which may be repeated, is more uncertain.

Sombre Tit *Poecile lugubris* —

L 13–14 cm. Breeds in open deciduous forest on mountain slopes or lower down in rocky terrain with trees and bushes; often in fruit orchards. Resident. Unobtrusive in behaviour.

IDENTIFICATION Proportions and size roughly as Great Tit. Much as Willow Tit in coloration, but with (brown-tinged) *greyish-black cap* drawn further down onto head-sides and with *much bigger dark bib*, with result that the *off-white colour of cheeks* becomes a *typically narrow wedge*. Bill heavy.

VOICE Calls Blue Tit-like 'si-si-si' and scolding, sparrow-like 'kerr'r'r'r', often combined; or more grating, 'zri-zri-zri', and scold more like Great Tit's, 'checheche-che...'. Song a fast, unsophisticated series in Marsh Tit style, with impure, grating voice, 'chriv-chriv-chriv-...'.

Crested Tit

CRESTED TIT

only European tit having a fully developed crest

crest can be folded back

MARSH TIT

palustris
(Nordic countries, Continent)

glossy black

sullied brown

brown

usually small bib

throat bib small and usually does not expand at lower edge

dresseri
(Britain, W Brittany)

WILLOW TIT

grey

bull-necked

whole cheeks white

dull black

brown

pale panel along secondaries

borealis, loennbergi
(Fenno-Scandia, Russia)

pale

grey-brown

buff tinge

intermediates are found on the Continent

kleinschmidti
(Britain)

usually rather large bib

buff-brown

bib often expands at lower edge

SIBERIAN TIT

throat bib is very large and broad at lower edge

dull grey-brown   black mask

tawny-brown

SOMBRE TIT

dark cap extends far down, white cheek forms narrow wedge

dark brown-grey

very large, bib

lacks brown tinge

rusty-brown

DZ

Marsh Tit

Willow Tit

Siberian Tit

Sombre Tit

Long-tailed Tit *Aegithalos caudatus*   rB2

L 13–15 cm (incl. tail 7–9). Not closely related to the 'true' tits (pp. 342–345), but a member of the same larger grouping which also includes warblers, swallows and larks. Breeds in deciduous and mixed woods with rich undergrowth (often hazel and goat willow) and some dead trees, also bushy places. Mainly resident. Usually seen in small family parties which move quickly through woods and gardens. Confiding but restless, normally allows only brief views. Builds a closed, oval nest of moss in branch fork, held together with spiders' webs and cleverly camouflaged with birch bark and lichen.

IDENTIFICATION A very *small and rotund tit with long tail*; in flight, looks like a tiny pale ball with a tail. Flight is skipping and with short undulations. Clings acrobatically to thin twigs, can hang upside-down. *Head and underparts whitish*, with (Britain & Ireland, ssp. *rosaceus*) broad black band *from side of forehead back along crown-side*, finely dark-streaked ear-coverts, light *reddish-brown flanks and belly*, and often gorget of dark spots across breast. *Back black, with wine-red or reddish-brown colour* on scapulars. Narrow tail is black with white edges. Bill short, black 'peppercorn eye', upper orbital skin yellow-orange. Sexes alike. – Juvenile: *Forehead and entire side of head dark*. Orbital ring dull red. Moults into adult plumage Jul–early Oct. – Variation: In N Europe (*caudatus*), adult has all-white head (no dark bands) and is whiter below, with pink flank. Continental birds (mostly *europaeus*) as British & Irish, but head-side either cleaner or more soiled (in S Spain, Italy and Turkey even grey-striped). A zone of intermediate forms exists between birds with all-white head and those with dark band on crown-side. Turkish birds (*alpinus*) have dark throat patch, grey back and shorter tail than European.

VOICE Restlessly moving parties utter loud trisyllabic, sharp 'srih-srih-srih' and slightly explosive rippling calls, 'zerrrr'. Also has chatty clicking calls, 'pt' (or 'zepp'). Drawn-out, high trills sometimes heard, e.g. when nervous and excited. Song rather soft, twittering, seldom heard.

Bearded Reedling *Panurus biarmicus*   rB4

L 14–15½ cm. The Bearded Reedling is (rather surprisingly) most closely related to the larks, but is shown here due to its superficial similarity to the tits. Breeds colonially in large reedbeds at lowland lakes and swamp margins. Very local in Britain, almost entire population confined to E and S coasts of England. Mainly resident, but dispersive, undertakes sporadic eruptions in autumn (signalled by high 'towering flights' over breeding lake in dense, highly excited flocks), some at least reaching more southerly winter grounds. Numbers may be severely reduced by cold winters. Feeds on insects and reed seeds. Nest of reed stems low in reeds.

IDENTIFICATION A small, *light yellowish-brown* bird with *long pale yellow-brown tail* glimpsed among the dense jungle of reeds should always be a Bearded Reedling. Flight a little unsteady, with irregular, shallow undulations and whirring wingbeats, almost as if billowing forwards. Climbs reed stems nimbly. Often twitches tail or raises and fans it. Occasionally spreads tail in flight, too. – Adult ♂: *Head light blue-grey* with *long, black drooping moustache* (in other words, despite name, does not have beard!). Throat white. *Undertail-coverts black*. – Adult ♀: Head buffish-brown without black moustache. Throat off-white. Undertail-coverts buff. – Juvenile: Like adult ♀, but has *black back* and *black areas on tail*, and plumage is a bit more yellowish-buff. ♂ separated from ♀ by all-black lores and orange-yellow bill (♀: greyish or brown lores and grey-brown to greyish-black bill). Moults to adult appearance Jul–Oct.

VOICE Usually reveals its presence by flight- and flock-call, a lively, sort of twanging and slightly impure 'psching', which is repeated and echoes in chorus from the reedbeds when a flock decamps and flies a short distance. Also has a rolling 'chirrr' and softer, clicking 'pett'. Song 3–4 discordant, squeaky notes, e.g. 'pshin-dshick-tschreeh'.

(Eurasian) Penduline Tit *Remiz pendulinus*   V**

L 10–11½ cm. Placed in a separate family, one reason being its aberrant breeding biology. Breeds in deciduous trees bordering rivers and lakeshores, in young regrowth beside overgrown swamps etc. Requires access to suitable nesting trees with thin, hanging branches (birch, willow, alder). Migratory in N, resident in S. Has spread towards NW in recent decades. Food mostly insects and spiders. Polygamous breeding habits. Several nests may be sited close to each other, but is not a true colonial nester; the same ♂ can start more than one nest. Nest unusual, fixed skilfully at end of a thin hanging branch, is pouch-shaped and furnished with entrance tunnel; exterior is pale and downy owing to interwoven seed-hairs of willow, aspen and bulrush.

IDENTIFICATION *Very small*, with *conical and sharply pointed* bill. Adult almost like a ♂ Red-backed Shrike in coloration, with *black 'mask' through eye*, *light grey crown/nape* and *reddish-brown mantle/back*. ♂ has broader eye-mask than ♀ and darker red-brown back, which is noticeable when pair seen together. ♂ also has stronger sprinkling of diffuse red-brown spots on breast than ♀. Flight is light and a little springy. Climbs nimbly on thinnest branches, freely hangs upside-down. – Juvenile: Duller. Eye-mask grey-brown and poorly indicated (never black), back drab yellowish grey-brown, crown greyish-buff with slightly paler forehead.

VOICE Call a fine, whistled 'tsiiü', faintly downslurred; sounds tender and almost 'dreamy' in tone (more drawn out than Reed Bunting's 'siu', softer and more downslurred than Robin's 'tsiiih'). Song a simple ditty with the call (or fragments of it) interwoven and with high trills, well spaced, e.g. 'tsiiü... sirrrr... tvitvitvi... tsiiü... zver'r'r'r'... tsiiü...' (like a soft, high-voiced Greenfinch).

Long-tailed Tit   Bearded Reedling   Penduline Tit

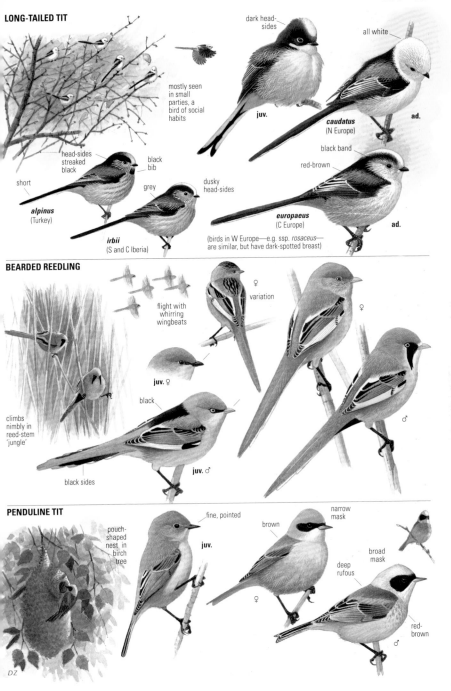

LONG-TAILED TIT

mostly seen in small parties, a bird of social habits

dark head-sides

all white

juv.

caudatus
(N Europe)

ad.

head-sides streaked black

black bib

grey

dusky head-sides

black band

red-brown

short

alpinus
(Turkey)

irbii
(S and C Iberia)

europaeus
(C Europe)

ad.

(birds in W Europe—e.g. ssp. *rosaceus*—are similar, but have dark-spotted breast)

BEARDED REEDLING

climbs nimbly in reed-stem 'jungle'

flight with whirring wingbeats

♀ variation

♀

juv. ♀

black

♂

black sides

juv. ♂

PENDULINE TIT

pouch-shaped nest in birch tree

fine, pointed

juv.

brown

narrow mask

broad mask

deep rufous

♀

♂

red-brown

DZ

NUTHATCHES *Sittidae*

Six small to medium-sized, compact passerines with big head and strong feet, specialized in climbing on sloping surfaces, including head-first downwards or upside-down on underside of branches. Long, awl-shaped bill and short tail. Food insects, seeds, nuts; store food reserves. Nest in hole or crevice, sometimes wholly or partly walled with mud.

(Eurasian) Nuthatch *Sitta europaea* rB3

L 12–14½ cm. Breeds in mixed and deciduous woods with some older trees and in larger parks and gardens. Resident. Commonest nuthatch, and widespread. Food insects, seeds and nuts. In winter often visits birdtables in worst Rambo style, chases off others, scatters seeds in all directions. Nests in tree-hole; if entrance is too big, it may be plastered with mud until size suitable. Nest chamber of pine-bark flakes.

IDENTIFICATION Silhouette distinctive: *neckless*, with *big head* and *long pointed bill*. Quick and active, makes sudden movements, climbs jerkily but nimbly on branches and trunks. Ability to *climb downwards head-first* eliminates treecreepers. Flight on short, rounded wings is straight over shorter distances, in short undulations over longer stretches. *Blue-grey above* and paler below, with *black eye-stripe*. Sexes similar, but ♂ has more intense *red-brown colour on flanks* (near leg). – Variation: Continental and British birds (ssp. *caesia* and others) have warm rusty-buff tone below. Those in Fenno-Scandia (*europaea*) are whitish on breast and belly-centre and lack white on forehead. In E Russia and Siberia (*asiatica*) small, paler blue-grey birds with white on forehead occur; occasionally seen in winter in Finland.

VOICE Rich repertoire of loud calls, used frequently; is therefore usually first located by voice. When foraging, commonly utters sharp conversational 'zit' and associated more drawn-out 'ziit'; sharper than similar tit calls. When excited and as warning, uses a forceful and very loud 'twett' (or slightly upslurred 'chuitt'), which is often repeated in short groups with brief pause in between. Several song types, all loud; delivered from high perch. Often heard is a slightly slower series of single whistling notes which are bent either up or down, 'vuih, vuih, vuih, vuih...' or 'viiu, viiu, viiu, viiu...'; other song variants are a fast, almost trilling 'vivivivivi...' with clear tone, also a slower and more rhythmically grouped 'jujuju jujuju...'.

Corsican Nuthatch *Sitta whiteheadi* —

L 11–12 cm. Breeds in tall pine woods in Corsica (endemic, total only c. 2000 pairs), especially at 1000–1500 m altitude, mostly in steep-sided valleys; in winter sometimes at lower levels. Food pine seeds, insects. Nests in hole in pine.

IDENTIFICATION *Small*, with *small head* and *short bill*. *White supercilium* contrasts with ♂'s *black crown* (blue-grey on ♀) and *black eye-stripe* (grey on ♀). *Underparts greyish-buff*, throat whiter. Lacks reddish-brown in plumage. Forages high up in pines, mostly at ends of branches like a tit.

VOICE When nervous or agitated, gives hoarse drawn-out 'pschehr', slowly repeated, somewhat reminiscent of anxious Starling. Song a clear and loud, fast trilling 'dididididididi...', almost like Alpine Swift (but not dropping in pitch and slowing down like latter); tempo varies a bit. Similar trills can be used as contact-call.

Krüper's Nuthatch *Sitta krueperi* —

L 11½–12½ cm. Breeds in forest at variable level, mostly in coniferous at higher altitude. Prefers pine forest on Lesbos and in Turkey, but evidently spruce in other places. Mainly resident. Food seeds and insects. Nests in tree-hole.

IDENTIFICATION *Small*, with *small head* and *relatively short bill*. Blue-grey upperparts and dirty greyish-white underparts as other nuthatches, but identified by combination of *reddish-brown breast patch*, *white supercilium* and *black forecrown*. Sexes similar, but on ♂ black crown patch is bigger and more sharply defined at rear, lores and eye-stripe blacker, flanks more blue-grey (tinged brownish on ♀).

VOICE When nervous or agitated, hoarse, harsh 'eehch', sometimes in almost Great Spotted Cuckoo-like fast scolding series, 'zreh-zreh-zreh-zreh...'. Call Greenfinch-like, gently upslurred, 'dvui'. Song a fast shuttling series with shrill, nasal tone and alternate high and low notes, sounds like an old-fashioned ball-horn on a bicycle or like a toy trumpet, 'tutitutitutituti...'; speed varies: in highest gear, it sounds like a trill, in lowest, the series is spelt out. Monosyllabic 'ti-ti-ti-ti-...' series are also heard occasionally.

Algerian Nuthatch *Sitta ledanti* —

L 11½–12½ cm. Breeds locally in N Algeria (not discovered until 1975; c. 100–200 pairs known) in oak forest (350–1120 m) or mixed forest of e.g. oak, maple, poplar and conifers (2000 m). Resident. Nests in tree-hole.

IDENTIFICATION Very like Corsican Nuthatch, but ♂ has (as Krüper's) just *black forecrown* (♀ greyish-black or dark grey) sharply demarcated from *blue-grey hindcrown/nape* and back (border diffuse on ♀), and *underparts are paler and more warm pinky-buff*. Supercilium broad as in Corsican.

VOICE When nervous or agitated, a harsh 'scheeh', repeated a few times. Song a slowly repeated, nasal whistling or fluting call on rising pitch and with a short coda, 'vuü-di vuü-di vuü-di...'; occasionally rapid disyllabic trills, 'di-du-di-du-di-du...'.

Nuthatch

Corsican Nuthatch

Krüper's Nuthatch

Algerian Nuthatch

NUTHATCH

reduced amount of red-brown

all nuthatches can climb head-first downwards

asiatica
(E Russia, Siberia)

white supercilium, often white forehead

small, pale subspecies

♂

dark red-brown

europaea
(Fenno-Scandia, W Russia)

♂

caesia
(Britain, Continental Europe)

♂

breast and belly rusty-buff

♂♂ invariably told by dark red-brown vent

♀

europaea

rusty-buff vent

CORSICAN NUTHATCH

confined to pine forest

♀

diffuse grey centres

grey

♂

whole crown black

pale greyish-buff

KRÜPER'S NUTHATCH

less extensive black on crown, rear edge ill-defined

♀

singing ♂

buffish grey-white

red-brown

blue-grey

♂

black cap on forecrown, rear edge well marked

large rusty-red patch

ALGERIAN NUTHATCH

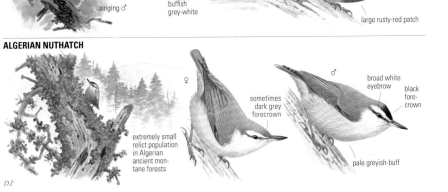

extremely small relict population in Algerian ancient montane forests

♀

sometimes dark grey forecrown

♂

broad white eyebrow

black fore-crown

pale greyish-buff

DZ

(Western) Rock Nuthatch *Sitta neumayer* —

L 14–15½ cm. Breeds in rocky terrain (often limestone) with scattered bushes and trees at variable level (sometimes coastal cliffs, more often 1000–2500 m). Nest, in rock crevice, made of mud, closed and with entrance tunnel.

IDENTIFICATION A trifle bigger than Nuthatch (marginal, barely noticeable in field), but has *slightly longer bill, paler blue-grey colour above* and *lacks red-brown on flanks* and undertail-coverts. (Has faint tinge of reddish-buff on flanks/belly, roughly as local S European Nuthatches, but is *whiter on breast*.) Tail is uniformly grey without white and black markings, but this more difficult to see in the field. Pale upperparts, long bill and *habitat* often best features for species. Resembles Eastern Rock Nuthatch. Active, restless.

VOICE Often heard, loud. Song (by both sexes, at times in duet) drawn-out series of clear whistles; the series speed up often and drop in pitch, at distance can momentarily be taken for Woodlark song (but is much more cutting, not so soft), e.g. 'vi-**yu** vi-**yu** vi-**yu**... tui-tui-tui-tui-... vivivivivi... trrrrtititi-ti-ti-ti-tü tü tü tu... **ti**-vü **ti**-vü **ti**-va, **ti**-vü...'.

Eastern Rock Nuthatch *Sitta tephronota* —

L 16–18 cm. Breeds in similar habitats to Rock Nuthatch, but normally not below 1000 m, and does not shun forest. Often perches in trees, and found in oak forest with some rocks and boulders. Mud-plastered nest on rock or in tree.

IDENTIFICATION *Distinctly bigger* than Rock Nuthatch, and appears to have proportionately even *bigger head* and *heavier bill*. Plumage differs in that the *black eye-stripe expands behind eye into a broad band* and *runs further down neck-side*; both *black bands clearly visible from behind* on nape of perched bird (can only just be made out on Rock Nuthatch). Often perches upright.

VOICE Resembles Rock Nuthatch's, but voice somewhat more powerful and has a *clearly deeper pitch*, and *tempo* is on average *slower*. Song e.g. '**tu**-ti **tu**-ti **tu**-ti... vitt-vitt-vitt-vitt-vitt-... choo-choo-choo-choo-... tvui tvui tvui tvui...'.

WALLCREEPER *Tichodromidae*

Only one species in this family, which is very closely related to the nuthatches and often treated as a subfamily of those.

Wallcreeper *Tichodroma muraria* **V***

L 15½–17 cm. Breeds in rocky terrain, on sheer cliff faces with some vegetation and nearby water, at 1000–3000 m altitude. Lives in inaccessible sites, is hard to see. Found at lower levels in winter, sometimes on buildings and in quarries; then not too shy. Food insects. Nests in cliff crevice.

IDENTIFICATION The Hoopoe of the rock face! When creeping over a cliff wall it can, with its grey and black colours

(blood-red of wing only glimpsed), elude observation, but when shifting position the *broad, rounded wings* are spread, gaudily *marked with red, black and white* above. Flight fluttering and jerky, with brief glides. – ♂ summer: Throat and breast all black. Much red on wing. – ♀ summer: Throat greyish-white with smaller grey-black central patch. Less red on wing. – Winter: Throat / breast pale greyish-white.

VOICE Rather silent. Song drawn-out, kind of strained whistles with glissando and with typical pattern in which one note rises, then one falls, 'tu... rruuüh... ziiüu...'.

TREECREEPERS *Certhiidae*

Small brown-spotted and white birds with pointed, slightly decurved bill and stiff tail-feathers. Two similar species. Quiet and unobtrusive, creep jerkily up trunks and thicker branches (like mouse), then fly down to base of next tree and so on. Food insects. Nest under bark flake on trunk.

(Eurasian) Treecreeper *Certhia familiaris* r**B**2

L 12½–14 cm. Breeds in woods, often with coniferous element. Prefers denser patches. Mostly resident, but populations in N and E periodically wander southward.

IDENTIFICATION Has on average *somewhat shorter bill* than Short-toed Treecreeper, and at least birds in N Europe are *whiter below*, have *whiter supercilium* and *whiter, more contrasting pale spots on crown/mantle*. (Continental birds of both species are, however, very similarly coloured.) Fine differences exist in wing pattern (see plate) but require photograph or extremely close view. Voice is best clue!

VOICE Call repeated, fine, 'buzzing' whistles, 'srri' (softer and more unobtrusive than Blackbird call). Sometimes also thin, 'pure' notes (without r-sound), 'tiih', loosely repeated. Song a high 2–3-sec.-long verse (Blue Tit voice) on *falling pitch* and with a slight flourish or *trill at end*. (Beware: some can imitate Short-toed Treecreeper song!)

Short-toed Treecreeper *Certhia brachydactyla* **V***

L 12–13½ cm. Breeds in woods. Habits and habitat choice as Treecreeper, but is more a bird of lowland and deciduous woodland. Also found in parks and gardens in towns.

IDENTIFICATION Confusingly similar to Treecreeper. *Lower flanks are on average more brownish* and *bill a touch longer*, but individual variation and minor differences render this of little help in the field. Voice, however, is important!

VOICE Call a strong, Coal Tit-like 'tüüt' with clear and penetrating quality, often repeated several times at slightly increasing trotting pace, '**tüüt**, tüüt, tüüt tüüt...'. A more Treecreeper-like 'srri' heard rarely. Song a *short verse* with *jolting rhythm*, *even pitch* and call-like voice, '**tüüt** e-to e-ti**titt**'. (Beware: some can imitate Treecreeper song!)

Rock Nuthatch

Wallcreeper

Treecreeper

Short-toed Treecreeper

Eastern Rock Nuthatch

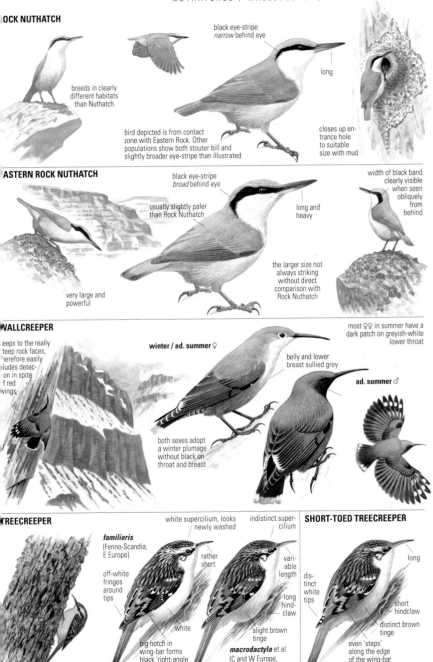

OCK NUTHATCH

breeds in clearly
different habitats
than Nuthatch

black eye-stripe
narrow behind eye

long

bird depicted is from contact
zone with Eastern Rock. Other
populations show both stouter bill and
slightly broader eye-stripe than illustrated

closes up en-
trance hole
to suitable
size with mud

ASTERN ROCK NUTHATCH

black eye-stripe
broad behind eye

usually slightly paler
than Rock Nuthatch

long and
heavy

width of black band
clearly visible
when seen
obliquely
from
behind

very large and
powerful

the larger size not
always striking
without direct
comparison with
Rock Nuthatch

WALLCREEPER

eeps to the really
teep rock faces,
herefore easily
ludes detec-
on in spite
f red
vings

winter / ad. summer ♀

most ♀♀ in summer have a
dark patch on greyish-white
lower throat

belly and lower
breast sullied grey

ad. summer ♂

both sexes adopt
a winter plumage
without black on
throat and breast

TREECREEPER

familiaris
(Fenno-Scandia,
E Europe)

white supercilium, looks
newly washed

indistinct super-
cilium

SHORT-TOED TREECREEPER

rather
short

vari-
able
length

long

off-white
fringes
around
tips

dis-
tinct
white
tips

distinct white
tips

long
hind-
claw

short
hindclaw

white

big notch in
wing-bar forms
black 'right-angle
indentation'

slight brown
tinge

macrodactyla et al.
(C and W Europe,
Corsica, Italy)

distinct brown
tinge

even 'steps'
along the edge
of the wing-bar

SHRIKES Laniidae

Eight medium-sized passerines with long tail, strong bill with hooked tip, strong feet, sharp claws and generally a broad dark band through or behind eye. Spend long periods perched motionless at top of bush or in tree on lookout for prey (insects, lizards, small rodents, small birds). Majority are long-distance migrants. Nest in branch fork in bushes.

Great Grey Shrike Lanius excubitor   P+W4–5

L 21–26 cm. Wide distribution, with habitat and habits different in S and N. N European birds (ssp. *excubitor*) breed in upland birch forest, thinly pine-clad bogs, in clearings; short-range migrant. Southern populations (several races) live in more open, often dry country provided there are lookout posts (e.g. wires, trees) and nest sites (thick bushes); resident. Widespread, fragmented range and partly quite noticeable variation might be sufficient grounds in future for subdivision into two or three species, but morphological and genetic evidence is partly contradictory and more research is required before a robust taxonomy can be proposed.

IDENTIFICATION A *big, light grey and white* shrike with large head, *black eye-mask*, black, rather short and *blunt wings* with *relatively large white primary-base patch*, and *long, rounded tail*, black with white sides. Scans from exposed perch, jerks tail sideways, *flies in deep undulations* when moving longer distances; can hover. Sexes similar but generally differ subtly when seen close, at least in breeding season: ♂ has *all-black bill*, *black* or very dark *lores* (complete 'mask'), and on average blacker wing-feathers, whereas ♀ usually has slightly paler base to lower mandible, paler lores and not quite so blackish wings. N European populations further differ in that adult ♀ is finely vermiculated grey on sides of breast and on flanks, adult ♂ not (frequently difficult to assess in the field, though). – Juvenile: Dirty grey above, virtually wholly unvermiculated, but off-white below with close brown-grey vermiculations. – Variation: In SE Russia and S Siberia (ssp. *homeyeri*), sometimes wintering in Black Sea region, vagrant to W Europe: paler with more white in wings and tail, including large primary patch and broad white wing-bar also on secondaries; grey-white forecrown and pale grey mantle. – In Middle East and Arabia (*aucheri*): like *excubitor* of N Europe but ♂ has broad black 'mask' reaching onto forehead (though much less so than in Lesser Grey Shrike), and flanks are ash-grey. Juv. of this race and all following are completely unbarred below, unlike birds of N Europe. – In N Africa both a coastal dark form with slate-grey upperparts and grey breast and no white supercilium (*algeriensis*), and a paler desert form with much white on wing and tail (*elegans*). In both, ♂ often has a little black on forehead. – On Canary Is (*koenigi*): similarly dark

as *algeriensis*, only smaller and with hint of white supercilium in many. – In deserts of Central Asia (*pallidirostris*): very large white wing patch on 'hand' only, buff tinge below in many, narrow rather short tail. Juv. and 1st-winter have *pale-based bill, pale lores* and *broad buff-white tips to greater coverts*. Note that adult ♂ has all-black bill and black lores in spring and summer.

VOICE Call usually a drawn-out trill, 'prrrih'. In anxiety, hoarse 'vaihk vaihk vaihk'. Song quiet, a simple phrase, e.g. a metallic or squeaky double note, or various short notes followed by hoarse trill, repeated rhythmically once per 1–sec.; also has a soft subsong with mimicry.

Iberian Grey Shrike Lanius meridionalis   –

L 23–25 cm. Here treated as a separate species based on genetic evidence and distinct morphology. Breeds in open arid plains, often on limestone heaths or stony wasteland with scattered trees and bushes. Mainly resident. A shy bird which will often retreat to cover long before close views have been obtained. Food mainly insects and small reptiles.

IDENTIFICATION Size similar to N European Great Grey Shrike, but head perhaps very slightly larger, bill stronger and *tail a little narrower*. Plumage differs clearly from Great Grey Shrike in being *much darker grey above*, in particular on crown, nape and mantle, and having a *greyish-pink flush on breast and belly*. Broad black 'mask' and *narrow white well-marked supercilium*, which often runs up over base of bill. White on scapulars narrow and restricted. Moderately large primary patch but *invariably no white on secondaries*. – Juvenile: Head pattern less distinct and no pink on underparts, which are often finely vermiculated grey. Attains nearly adult-like plumage in late autumn.

VOICE Both calls and song appear very similar to those of Great Grey Shrike, but no detailed study attempted.

Lesser Grey Shrike Lanius minor   V***

L 19–21 cm. Breeds in open terrain with cultivations, fruit orchards, avenues of poplars and scattered trees and groves (but not in woodland). Warmth-loving, is most at home in lowland. Often seen scanning from roadside telephone wires.

IDENTIFICATION Size between Great Grey and Red-backed Shrikes. Large rounded head, stout rounded bill. Like Great Grey (similar colours and *white primary-patch*), but adult differs in *black forehead and forecrown* (solid black and more extensive on ♂, smaller and often mottled grey on ♀), *longer primary projection* and proportionately slightly *shorter and less strongly rounded tail*. *Breast and belly pale salmon-pink*. Many adults in autumn (perhaps mainly ♀♀) have grey forehead. – Juvenile: Lacks black on forehead. Barred on crown and mantle, but not below (reverse of juv. Great Grey).

VOICE Rather silent. Call a Magpie-like double chacking, 'tsche-tsche'. Song a hard and screaming parakeet-like 'tschilip!', given at intervals from treetop or doubled in display-flight. A rasping drawn-out song full of mimicry is also heard, apparently mostly from unpaired ♂♂.

Great Grey Shrike

Iberian Grey Shrike

Lesser Grey Shrike

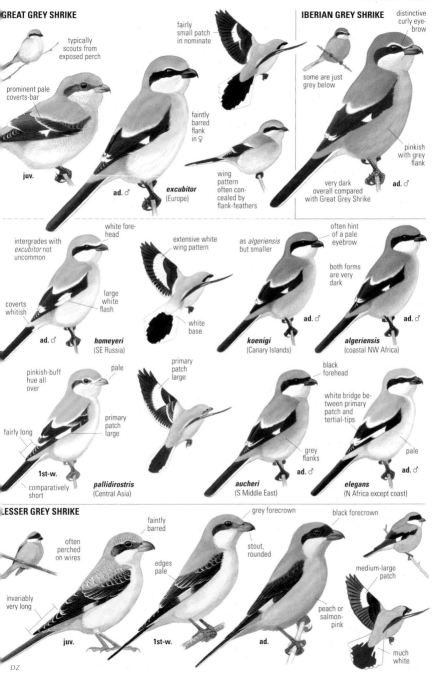

GREAT GREY SHRIKE

typically scouts from exposed perch

prominent pale coverts-bar

fairly small patch in nominate

faintly barred flank in ♀

juv.

ad. ♂

excubitor
(Europe)

wing pattern often concealed by flank-feathers

IBERIAN GREY SHRIKE

distinctive curly eye-brow

some are just grey below

pinkish with grey flank

very dark overall compared with Great Grey Shrike

ad. ♂

---

intergrades with *excubitor* not uncommon

white fore-head

large white flash

coverts whitish

ad. ♂

homeyeri
(SE Russia)

extensive white wing pattern

white base

as *algeriensis* but smaller

often hint of a pale eyebrow

both forms are very dark

ad. ♂

koenigi
(Canary Islands)

ad. ♂

algeriensis
(coastal NW Africa)

---

pinkish-buff hue all over

pale

primary patch large

fairly long

1st-w.

comparatively short

pallidirostris
(Central Asia)

primary patch large

aucheri
(S Middle East)

ad. ♂

black forehead

white bridge between primary patch and tertial-tips

grey flanks

black forehead

pale

ad. ♂

elegans
(N Africa except coast)

LESSER GREY SHRIKE

often perched on wires

faintly barred

edges pale

grey forecrown

stout, rounded

invariably very long

juv.

1st-w.

peach or salmon-pink

ad.

black forecrown

medium-large patch

much white

DZ

Red-backed Shrike *Lanius collurio* P5

L 16–18 cm. Breeds in open cultivated country, often on heaths and pastures with hawthorn, sloe and dog-rose, also in juniper stands. Summer visitor which winters in tropical Africa and returns in May; autumn migration mainly in Aug. Has declined in recent decades. Insect specialist. Some set up food stores by spearing surplus on bush thorns.

IDENTIFICATION Often perches upright, well visible; when nervous, jerks tail sideways. Short flights rapid and straight, longer undulating. – Adult ♂: Breast and belly light brownish-pink; *no vermiculation* on flanks. Throat white. *Crown light ash-grey, mantle reddish-brown.* Broad *black eyestripe* ('bandit's mask'), sometimes reaching a little above base of bill and often with narrow white border above and on forehead. *Tail black, with white sides at base* ('wheatear pattern'). Rarely a small white patch at primary-bases. – Adult ♀: Creamy off-white below with *coarse vermiculations. Crown brown or brownish-grey,* nape greyer, mantle duller brown than ♂; variably vermiculated above (some being unbarred, others with much vermiculations). *Eyestripe brown*; lores often pale. *Tail dark brown with narrow white edges.* Rarely attains advanced plumage, more like ♂, still invariably with barred breast-sides. – Juvenile/1st-winter: Like ♀, but usually *more rufous-tinged and heavily vermiculated above.* Tertials have buff tips and black subterminal bars. A few with warmer brown uppertail confusingly like young of both Brown and Turkestan Shrikes.

VOICE Alarm a nasal, hoarse 'vehv', loosely repeated. When highly agitated, series of tongue-clicking 'tschek'. Song of two types, either the alarm note 'vehv' used as a territorial signal, or a prolonged quiet warble (can be construed as subsong), latter rather seldom heard; harsh and squeaky sounds mixed with expert mimicry of other birds.

Isabelline Shrike *Lanius isabellinus* V***

L 16½–18 cm. Closely related to Turkestan Shrike, breeding in Mongolia and W China tablelands. Rare vagrant in Europe, mostly in autumn, in same habitat as the Red-backed.

IDENTIFICATION Often noted on *rust-red uppertail,* which is marginally longer than in Red-backed Shrike. – Adult ♂ sandy grey-brown above (often with pink-buff sheen when fresh but wearing to purer grey in summer) and *strongly tinged buff below.* Complete black eye-mask (which does not reach above bill) and *short, ill-defined and buff-tinged supercilium.* Has a small *white primary-base patch.* ♀ and immatures have incomplete eye-mask (lores often pale), no or very insignificant primary patch, and ♀♀ as a rule faintly vermiculated on breast and sides. – Although classical Isabelline and Turkestan Shrike ♂♂ are quite distinct, ♀♀ and immatures with intermediate appearance are sometimes met with,

difficult to identify. For separation of immatures from Red-backed Shrike, see under Turkestan Shrike.

VOICE Calls and songs are presumed to be inseparable from Turkestan Shrike, but a closer study is lacking.

Turkestan Shrike *Lanius phoenicuroides* V***

L 16½–18 cm. Often treated as race of Isabelline Shrike, but adult ♂♂ usually differ quite clearly, hybridization limited and genetic difference sizable, hence here treated as separate species. Closely related also to Red-backed Shrike, and hybridizes with it in a few limited areas of its Central Asian range. Rare autumn vagrant in N and W Europe.

IDENTIFICATION Resembles both Isabelline and Brown Shrikes with *rust-red uppertail,* and a black 'mask' in ♂. – Adult ♂ slightly darker and duller grey-brown above than Isabelline and *much whiter below,* pink-buff tinges restricted to flanks and sides of breast. Also, has a *pure white and prominent supercilium,* often a *rusty-tinged crown/nape,* and the *black eye-mask has tendency to reach above base of bill.* A variety has grey crown/nape/mantle ('*karelini*'). *White primary-base patch* as in Isabelline. – ♀ and young are like Isabelline (and often doubtfully separable) but on average whiter below with more obvious dark vermiculations. – Young vagrants in Europe told from superficially similar young Red-backed on slightly *paler and greyer ground colour above* with less or *no vermiculation on mantle and nape, paler* and brighter rusty-red colour on uppertail (but a few are similar in this respect!) and on *shorter primary projection.*

VOICE Calls and songs are very similar to those of Red-backed Shrike. The alarm, also used as territorial signal, a repeated harsh 'vehv'. Has a prolonged subsong when close to ♀ containing mimicry and a variety of scratchy notes.

Brown Shrike *Lanius cristatus* V**

L 17½–20 cm. Rare vagrant from Siberia and Altai (ssp. *cristatus*), where it breeds in taiga bogs or in glades in forest or in forest steppe. Overlaps in range with Red-backed Shrike in W Siberia, and hybrids between the two have been reported but are apparently very rare.

IDENTIFICATION Similar to Red-backed and Turkestan Shrikes but a trifle larger with bigger head, *stronger bill* and *narrower tail-feathers.* Adult ♂ is rather *uniformly rufous-brown above* incl. uppertail, lacking contrast between more rusty-red uppertail and duller brown rest of upperparts of Turkestan. Normally *no white primary-base patch* (rarely hint at the most). *Complete black 'mask'* and *prominent white supercilium.* Underparts unbarred and tinged *yellowish-buff* (lacking pinkish hue of many ♂♂ Turkestan Shrike). Sexes similar but often differ subtly: ♂ has blacker 'mask', ♀ nearly invariably faintly vermiculated on breast and sides. – Vagrant immatures seen in autumn in Europe strongly resemble immature Red-backed Shrikes. Note on Brown: *strong bill; somewhat shorter primary projection* (still, slight overlap with three, not two primaries with emarginated outer webs; *narrower tail-feathers* and *strong tail graduation; solidly dark centres to tertials* (sharply fringed buff-white) without obvious black subterminal marks; and more *obvious pale supercilium* and dark grey (not red-brown) ear-coverts.

VOICE Not closely studied but apparently similar to that of Red-backed, Turkestan and Isabelline Shrikes.

Red-backed Shrike   Isabelline/Turkestan Shrike

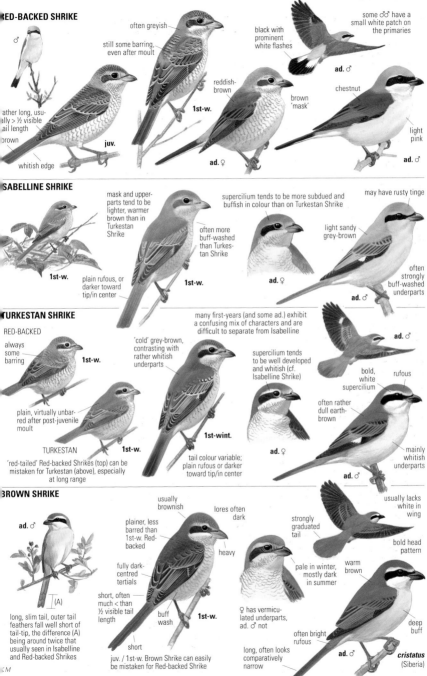

RED-BACKED SHRIKE

♂

often greyish

still some barring, even after moult

black with prominent white flashes

some ♂♂ have a small white patch on the primaries

ad. ♂

reddish-brown

brown 'mask'

chestnut

rather long, usually > ½ visible tail length

1st-w.

brown

light pink

whitish edge

juv.

ad. ♀

ad. ♂

ISABELLINE SHRIKE

mask and upperparts tend to be lighter, warmer brown than in Turkestan Shrike

supercilium tends to be more subdued and buffish in colour than on Turkestan Shrike

may have rusty tinge

often more buff-washed than Turkestan Shrike

light sandy grey-brown

1st-w.

plain rufous, or darker toward tip/in center

1st-w.

ad. ♀

often strongly buff-washed underparts

ad. ♂

TURKESTAN SHRIKE

RED-BACKED

many first-years (and some ad.) exhibit a confusing mix of characters and are difficult to separate from Isabelline

ad. ♂

always some barring

1st-w.

'cold' grey-brown, contrasting with rather whitish underparts

supercilium tends to be well developed and whitish (cf. Isabelline Shrike)

bold, white supercilium

rufous

plain, virtually unbarred after post-juvenile moult

often rather dull earth-brown

TURKESTAN

1st-w.

1st-wint.

ad. ♀

mainly whitish underparts

'red-tailed' Red-backed Shrikes (top) can be mistaken for Turkestan (above), especially at long range

tail colour variable; plain rufous or darker toward tip/in center

ad. ♂

BROWN SHRIKE

usually brownish

lores often dark

usually lacks white in wing

ad. ♂

plainer, less barred than 1st-w. Red-backed

strongly graduated tail

bold head pattern

fully dark-centred tertials

heavy

warm brown

pale in winter, mostly dark in summer

(A)

short, often much < than ½ visible tail length

buff wash

1st-w.

deep buff

long, slim tail, outer tail feathers fall well short of tail-tip, the difference (A) being around twice that usually seen in Isabelline and Red-backed Shrikes

short

juv. / 1st-w. Brown Shrike can easily be mistaken for Red-backed Shrike

♀ has vermiculated underparts, ad. ♂ not

often bright rufous

long, often looks comparatively narrow

cristatus (Siberia)

ad. ♂

KM

Woodchat Shrike *Lanius senator*   **V***

L 17–19 cm. Breeds in open wooded areas with glades, interspersed cultivations and scattered trees, preferably in association with fruit orchards. Usually requires some sandy or bare ground in territory. Tropical migrant (Africa).

IDENTIFICATION Slightly larger than Red-backed Shrike. –Adult ♂ (ssp. *senator*): Easily identified by *red-brown crown and nape, black eye-mask, forehead and mantle*. Wings black above with white *primary-base patch*. Large *white scapular patch* and uppertail-coverts add further to 'showy' appearance, especially in flight. Lower back ash-grey. –Adult ♀: As ♂ but duller, with *dark brown-grey mantle*, and often *some buffy-white mixed with black of forehead and eye-mask*. Some are borderline cases, impossible to sex in the field. – Juvenile/1st-winter: *Grey-vermiculated* above, on breast and on flanks. Basic colour grey, with variable elements of brown; almost always distinctly *reddish-brown on neckside, greater coverts and tertials*. Note *suggestion of pale scapular patch* (white with grey scalloping) and *pale rump*; pale primary patch diffusely defined and tinged rusty. – Variation: On W Mediterranean islands (*badius*) white primary patch lacking (though a few have a hint), and forehead has less black. Also, lower back is darker grey, and sexes differ less, ♀ being almost as neat as ♂. – In SE Turkey and eastward (*niloticus*) breeders have extra big white primary patch (still, some overlap with *senator*), white tailbase and much black on forehead. While adult may be difficult to separate from *senator* in the field, the immature is the more characteristic in having *large, pure white and distinct primary patch*; early hatching and moult means often incipient adult features as early as 1st autumn.

VOICE Alarm a series of short, hoarse calls, 've-ve-ve-...', or faster, like a hoarse trill, 'dsherrrrr'. Song rather loud, sometimes fast stream of squeaky and clicking notes, sometimes slower with expert mimicry; phrases sometimes repeated several times, but as a rule a bit varied in detail.

Masked Shrike *Lanius nubicus*   —

L 17–18½ cm. Breeds in open woodland with bushes and glades, usually associated with smallish cultivations, often on hillsides with stone pines, oak and thorn bushes. Often uses less exposed perches than its relatives, takes cover in treetops or larger shrubberies. Winters in Africa.

IDENTIFICATION *Smallest* shrike. *Slimmer* than Red-backed and has *longer, narrower tail*. Black (♂) or grey (♀) above, with *white forehead* and *white scapular patches*. *Large white primary-base patches* striking in flight. *Flanks orange* (more saturated on ♂). A certain amount of individual variation in ♀ plumage, some having very dark grey-black crown/nape, almost like ♂. – Juvenile: Grey, white and black, heavily

WOODCHAT SHRIKE

1st-w.

always has light rump

niloticus has larger and more well-defined white patch, on darker primaries than *senator*, thus resembling Masked Shrike

niloticus (SE Turkey, Middle East, W Central Asia)

never rufous

dark

slim bill

1st-w.

slender build, proportionately longer-tailed than Woodchat

MASKED SHRIKE

vermiculated and with hint of pale scapular patch and large white primary patch, like eastern juvenile Woodchat (*niloticus*); told from latter by smaller size, *lack of reddish-brown on wing and nape*, by *dark uppertail-coverts, narrow tail, small bill* and more uniformly dark tertials and greater coverts. Some that hatch early mature to plumage resembling adult as early as Sep, but majority retain barred and greyish juvenile plumage through autumn.

VOICE When nervous, a snipe-like 'chaihr'; alarm a dry rattle. Song (surprisingly enough) superficially like both Olive-tree and Olivaceous Warblers, a rather flat verse with cyclic repetition and harsh voice; more rugged and jolting in structure, with slower tempo than Olivaceous, not quite so grating and low as Olive-tree.

BUSH-SHRIKES *Malaconotidae*

A rather large and varied family of mainly African species. Most have long, strong bill, rather long legs, long tail and strongly rounded wings. Only one species within treated range. Nests in fork in dense bush.

Black-crowned Tchagra *Tchagra senegalus*   —

L 21–24 cm. Breeds in NW Africa in low-lying, dry, open bushy country. Resident. Frequently hops on ground, often perches low down in bushes. Rather shy and elusive, but loud calls reveal its presence. Food insects, small lizards etc.

IDENTIFICATION Strongly patterned head, with *black crown, broad, long creamy-white supercilium* and *black eye-stripe*. Scapulars and tertials black, edged ochre. Heavy black bill. Wing bright *rusty-brown. Tail long*, black and broadly tipped white, central tail-feathers greyish barred dark. Sexes alike, ages very similar and often impossible to separate in the field. Juvenile has black portions on crown, eye-stripe, scapulars and tertials tinged brown, scapulars being diffusely tipped buff.

VOICE Song remarkably loud, powerful glissando whistles, rising or falling sound like human wolf whistles. ♀ has rattling trill 'trrrrrrr...', sometimes in duet with ♂.

Woodchat Shrike

Masked Shrike

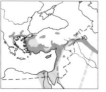

Black-crowned Tchagra

WOODCHAT SHRIKE

senator (Europe to W Turkey)

WOODCHAT

juvs.

RED-BACKED

light-centred scapulars, more variegated wing than juv. Red-backed

lighter, greyer head than juv. Red-backed; less obvious 'mask'

juv.

prominent light primary flash often extensive, and buff-washed

♀♀ (and imm.) usually more extensively pale on lore than ad. ♂♂

dark grey-brown

usually prominent flash

white

red

much black

black or blackish

often difficult to sex, generally easiest when viewed as a pair

ad. ♀

ad. ♂

---

badius (W Mediterranean islands)

senator

beware that the primary flash in *senator* may at times be largely concealed by overlaying secondaries

black band averages narrower than in *senator*

slightly stronger bill than *senator*

old, faded juv. primaries and their coverts may become very pale at their bases

lacks obvious white flash; at most very small white or buff spots on innermost primaries

ad. ♂

sexes more alike than in *senator*

lacks white at base of primaries

juv.

2nd cal.-yr (spring)

lacks obvious white primary flash

ad. ♂

MASKED SHRIKE

juv. → **1st-wint.**

often develops a whitish forehead and supercilium

bold white primary flash

dark rump

rump and uppertail darker than Woodchat

slim

dark greyish-brown

black

long, narrow black tail!

more delicate build than Woodchat, proportionately longer, narrower tail

juv.

ad. ♀

deep orange

ad. ♂

K M

BLACK-CROWNED TCHAGRA

other members of the genus are confined to Africa S of the Sahara

unmistakable!

rufous wings striking in flight

unobtrusive habits, hops quietly about in dense bushes

sexes alike

DZ

BULBULS *Pycnonotidae*

A family of small or moderately large passerines, according to recent new insights most closely related to warblers, larks and swallows & martins, although here placed with the sunbirds for practical reasons – both groups occur in Middle East, Arabia and N Africa. Bulbuls thus live in southerly, warm habitats, are social in their habits and lively and loud-voiced. Spend most of the time in the canopy of trees (often palms) and bushes, or are seen in gardens. Nest in tree.

Common Bulbul *Pycnonotus barbatus* —

L 19–21 cm. Breeds in gardens, around cultivations, by watercourses in mountains and at oases, so long as vegetation present. Does not shun proximity of humans, and before long is noticed: noisy, flies about a lot, often seen in small groups, often perches openly. Resident. In behaviour very similar to White-spectacled Bulbul.

IDENTIFICATION A barely thrush-sized, rather *long-tailed, dark grey* and dirty greyish-white bird. *Head is sooty-grey* (lores almost black, no pale eye-ring), upperparts almost equally dark grey, and breast slightly paler brown-grey with *whitish belly and vent*. Tail dull black, wings dark grey. Bill black, thrush-sized but slightly downcurved. Sexes and ages alike. *Flight straight and a little unsteady, wings rounded.* Often opens and closes tail in flight and frequently also when perched. Can erect crown-feathers, although they do not form a full crest but more make head profile peaked.

VOICE Very loud and noisy. Flocks often utter a repeated, hoarse, slightly casual 'chahr chahr chahr…'. When uneasy, bulbuls give sharper but similar 'tshirr' in persistent series. Song powerful with characteristic jerky, uneven rhythm, varies in length and detail, e.g. '**chick** chille**wee, chuwü**', delivered most frequently at dawn and dusk.

White-spectacled Bulbul *Pycnonotus xanthopygos* —

(Alt. name: YELLOW-VENTED BULBUL.) L 19–21 cm. Breeding habitat and habits very like those of Common Bulbul, to which it is closely related (ranges of the two species meet at Suez Canal). Lives in palm groves, wadis with some vegetation, gardens etc. Very social, usually seen in small flocks, appears always to have something to argue about or sort out among the palm leaves. Resident.

IDENTIFICATION Like Common Bulbul, but differs in having *lemon-yellow vent*, darker, almost *black head* which contrasts against a *body that is paler grey* than in Common. Further, has a *narrow grey-white eye-ring*, pale grey (not so whitish) belly and slightly more square-cut tail.

VOICE Appears to be very like Common Bulbul's. However, song appears to be consistently delivered at a slower pace with a few notes at a time and lacking the characteristic jolting rhythm of Common Bulbul.

White-eared Bulbul *Pycnonotus leucotis* —

L 17–19 cm. Closely related to Himalayan Bulbul (*P. leucogenys*; not treated), but these two now usually kept as separate species. Has recently spread towards west from Iraq and the Gulf region, and into Syria. Found in similar habitats as its relatives. Very bold and confiding, will approach people closely in gardens. Resident.

IDENTIFICATION Like White-spectacled Bulbul in shape but is marginally smaller. Differs in having *somewhat more developed blunt crest*, which it often erects. Most obvious character is *large white black-surrounded cheek patch* Great Tit fashion. Like White-spectacled it has *yellow vent* (only more orange-tinged). The *dark tail is white-tipped* clearly visible on birds flying away or perched. Bill a little longer and stouter than on its relatives.

VOICE Appears to be the same as in White-spectacled Bulbul, although a closer study of this is lacking.

SUNBIRDS *Nectariniidae*

A large family (there are over a hundred species in Africa and Asia) of tiny, dapper birds which, with their thin and pointed downcurved bill, often lustrous plumage colours and habit of feeding on flower nectar, can be said to be the humming birds of the Old World. Hover only rarely (and not with rapid insect-like wingbeats), but perch to inspect flowers. Also feed on insects, spiders, etc. Nest in tree crowns.

Palestine Sunbird *Cinnyris osea* —

L 11–12 cm. Breeds especially in dry, savanna-like acacia terrain but also alongside cultivations, in parks, gardens and such places where nectar-producing flowers exist. Resident, or makes just local movements. Food nectar and insects.

IDENTIFICATION Due to its very short tail looks almost chubby and cute. Full of life, *fast-moving and agile*. Often flicks tail nervously. *Flight fast and jerky. Bill black, slender, pointed and downcurved.* – ♂ breeding (*c.* Dec–Jul): All black with, at close range and in correct light, metallic blue and lilac sheen. Otherwise appears just dark. – ♀: Grey-brown above, grey-white below with faint yellowish cast. Wings contrastingly dark. – ♂ eclipse: Like ♀, at times with a few scattered black feathers.

VOICE Rich repertoire of loud calls. Whistled 'viyu', like something between Nuthatch and Hume's Leaf Warbler (!), also a rising 'tvüit' with cheerful merry tone; also has 'electric' tongue-clicking 'züt-züt-züt züt züt…'. Song a few whistled notes followed by a rattling trill, 'tvui tvui tvu **tirrrrrr**', at distance a bit like song of Black Redstart.

Nile Valley Sunbird *Anthodiaeta metallica* —

L 9–10 cm (plus tail projection of ♂ 4½–6½). Within the region breeds only near Cairo. Habitat as Palestine Sunbird. May perform minor migratory movements. Breeding starts in Apr.

IDENTIFICATION Tiny. Nervous actions, flicks wings and fans tail. *Bill* as with all sunbirds slightly downcurved but differs in being *markedly short*. ♂ unmistakable in breeding plumage (*c.* Dec–Jul), with *green-glossed upperparts, black breast, yellow belly* and *long, thin tail projection* which has club-shaped end. Black breast will show purple metallic gloss in the right light and angle. Other plumages grey-brown above, with off-white throat and *yellowish-white belly*; eclipse somewhat stronger yellow and can have a few glossy coverts and some dark on throat. The long tail-feathers are not worn in autumn.

VOICE Rather similar to Palestine Sunbird's. Has a high-pitched song with some repeted, hard and trilling notes.

COMMON BULBUL

bulbuls are noisy and conspicuous species, often found in urban areas, so easily noticed

dark 'face'

white

WHITE-SPECTACLED BULBUL

loud chorus from 'untidy' gatherings

dark hood

white eye-ring

yellow

WHITE-EARED BULBUL

white in tail obvious in front view and in flight

striking white cheek

yellow-orange

PALESTINE SUNBIRD

on most occasions ♂ appears all black in the field

grey with only slight yellowish-green tinge

♀

ad. ♂ **breeding** (Dec–Jul)

DZ

NILE WALLEY SUNBIRD

yellow eyebrow

pale yellow

♀

very long 'streamers'

ad. ♂ **eclipse**

ad. ♂ **breeding** (Dec–Jul)

Common Bulbul

White-spectacled Bulbul

White-eared Bulbul

Palestine Sunbird

Nile Valley Sunbird

BABBLERS *Timaliidae*

A very large family (over 300 species in the world) of small to moderately large passerines with mostly plain colours, rounded wings and strong feet. Recent genetic research has (surprisingly) shown them to be closely related to the *Sylvia* warblers.

Fulvous Babbler *Turdoides fulva* —

L 23–25 cm. Resident breeder in arid bush country, in semi-desert with scattered trees and bushes, in palm groves with undergrowth, by oases, etc. Lives in flocks, often of 5–10 birds. Nests in thickets.

IDENTIFICATION Barely thrush-sized, *long-tailed*. Rather *uniformly pale reddish sandy-brown* (warmer in colour than Arabian Babbler) with poorly defined dark feather centres on crown, nape and mantle. Throat whitish. *Bill strong*, slightly decurved, usually *greyish-black* (touch of yellow at gape on ♂), but straw-yellow with dark tip in NE Africa (ssp. *acaciae*). *Iris* usually *dark*. Legs thick, light brown.

VOICE Noisy. Call short 'chitt' notes and various drawn-out trills, e.g. a rather hard rolling sound with metallic tone, often on higher note, 'priür'r'r'r'r'r'r'r....'. Song a series of piping diphthongs, 'piü piü-piü-piü piiay'.

Arabian Babbler *Turdoides squamiceps* —

L 25–29 cm. Resident breeder in arid bush country, in semi-desert with some acacia and tamarisk, near cultivations and gardens. Lives in flocks. Nests usually in thick bush.

IDENTIFICATION Thrush-sized, *light grey-brown, long-tailed*. Head and neck somewhat greyer than body, and *crown and nape finely spotted dark*. Bill strong, slightly decurved, greyish-black (ad. ♂♂, young) or yellowish with diffusely darker tip (ad. ♀♀). Iris pale yellow (♂), ochre (♀) or grey-brown (juv.). Flight fluttering and direct; glides on extended wings. Hops two-footed, and often jerks its usually cocked tail.

VOICE Utters chatty whistles with piping tone; more characteristic is a *high, piercing trill*, 'zvir'r'r'r'r'r'r'r...'. Alarm a trill and hoarse 'ksherh' notes, like a small Jay. Song a series of piping whistles, 'piay piay piay piay piay...'.

Iraq Babbler *Turdoides altirostris* —

L 20–23 cm. Resident Middle East breeder which has recently expanded from Iraq into Syria and to SE Turkey. Mainly found in riverine reedbeds. Lives in flocks.

IDENTIFICATION Smallest babbler in area, resembling extralimital Common Babbler *T. caudata* (Iraq, Iran), but has *pale lores, warm brown crown, less streaking on crown and breast*, whiter throat and *stronger rusty tinge on sides of breast and flanks*. Legs dull dark brown, *dark bill stout*.

VOICE Drawn-out descending high-pitched trills.

CROWS *Corvidae*

Medium-sized to large, heavily built passerines. Legs sturdy, bill strong. Highly evolved, alert, quick to learn. Social, often seen in flocks. Sexes alike. Omnivores; food comprises insects, seeds, nuts, berries, refuse, offal, eggs and young of other birds, etc. Usually build stick nest in tree crown.

Azure-winged Magpie *Cyanopica cyanus* —

L 31–35 cm (incl. tail 16–20). Resident breeder in Iberia in woods mainly of stone pine but also deciduous trees. Odd gap in distribution all the way to Far East. Outside breeding season, roams about in family parties or larger groups. Alert and rather shy. Dashes around nimbly in dense treetops when foraging. Food mostly insects and berries. Nest often at edge of crown of holm oak.

IDENTIFICATION Medium-large, rather slim and *long-tailed*. In shadow of treetops, often seen only as a grey-brown bird with strong contrast between *black cap* and *whitish throat*. But better light reveals the beautiful *light blue colour on wings and tail* and the brownish-pink tone to back and buff tone to breast and belly. Primaries edged white. Flight fluttering like Magpie's, but lighter and more elastic.

VOICE Several calls; most commonly heard (e.g. frequently repeated from moving flocks) is a high, grinding, slightly nasal, gently sluurred 'vrrüiih'; sometimes high 'kui' and whining 'vih-e' heard. Alarm a drawn-out, high-frequency, hard rolling or rattling 'krrrrrr...'. Song seldom heard and poorly known, a soft, squeaky twitter.

(Common) Magpie *Pica pica* rB2 (1?

L 40–51 cm (incl. tail 20–30). Breeds commonly around farms and in urban areas. Resident. At times gathers in noisy flocks of 5–25 birds, known as magpie parliaments. Sounds the alarm against cats. Vigilant but not timid, swoops down and patrols lawns and flower beds in centre of noisy cities in search of food. Has false reputation of being a silver thief. Builds roofed stick nest in tree crown.

IDENTIFICATION Unmistakable—probably the bird which most people recognize. *Black and white*, with *very long*, green-glossed *tail*. In flight, short, rounded *wings show large white panels on 'hand'*. Flight fluttering, on straight course, with occasional sweeping glides, sometimes from high rooftop right down to ground. Walks confidently and slightly jerkily with tail often raised, also makes strong bounds.

VOICE Most calls hoarse and unmusical. Perhaps best known is the alarm against cat or owl, long-drawn-out, fast, very hoarse staccato series, 'tsche-tsche-tsche-tsche-...'. Other calls include hard, hoarse and whining sounds; disyllabic, clicking 'cha-ka!', 'chiah-cha' etc. in conversation. Song more rarely heard, a quiet, harsh twittering 'subsong'.

Fulvous Babbler

Arabian Babbler

Azure-winged Magpie

Magpie

Iraq Babbler

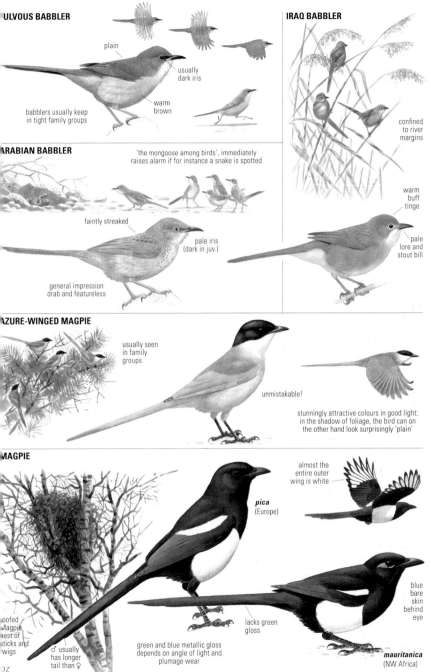

FULVOUS BABBLER

plain

usually
dark iris

warm
brown

babblers usually keep
in tight family groups

IRAQ BABBLER

confined
to river
margins

warm
buff
tinge

pale
lore and
stout bill

ARABIAN BABBLER

'the mongoose among birds', immediately
raises alarm if for instance a snake is spotted

faintly streaked

pale iris
(dark in juv.)

general impression
drab and featureless

AZURE-WINGED MAGPIE

usually seen
in family
groups

unmistakable!

stunningly attractive colours in good light;
in the shadow of foliage, the bird can on
the other hand look surprisingly 'plain'

MAGPIE

almost the
entire outer
wing is white

pica
(Europe)

blue
bare
skin
behind
eye

roofed
Magpie
nest of
sticks and
twigs

♂ usually
has longer
tail than ♀

lacks green
gloss

green and blue metallic gloss
depends on angle of light and
plumage wear

mauritanica
(NW Africa)

(Eurasian) **Jay** *Garrulus glandarius*   r**B**2 / (**W**)

L 32–35 cm, WS 54–58 cm. Breeds in various types of woodland, both coniferous and deciduous, and in larger wooded parks. Prefers areas with acorns (secondarily beech nuts, hornbeam seeds), which are cached in autumn as winter food; shipments high up with crop full of acorns can extend over several kilometres. Mostly resident, but N populations migrate S and SW in some autumns. Vigilant and shy, difficult to approach. Omnivore; summer diet includes a good many eggs and young of small birds. Nests usually in tree.

IDENTIFICATION Plumage striking, yet a bird which few city-dwellers are acquainted with. Mainly *pinkish grey-brown* with whitish throat and vent. *Above, on wing-bend, a light blue panel*, finely vermiculated black. Head has a *broad black moustachial stripe* at throat-side, also black-spotted whitish crown (in Europe; race *glandarius*). Crown-feathers can be raised, creating pointed head shape. In flight, conspicuous *large white rump patch* and broad white wing-band on 'arm'. Flight fluttering and a little uneven, on straight course. Told at long range from similarly flapping Nutcracker by *rather long tail* and *short bill.* – Variation: British & Irish birds (*rufitergum, hibernicus*) are darker red-brown. In Caucasus (*krynicki*), Middle East (*atricapillus*) and central N Africa (*cervicalis*), crown is all black and bill thicker. In NE Russia (*brandtii*) head is reddish-brown.

VOICE Most often heard is the characteristic, loud and intense, hoarse scream, 'kschaach!', which normally functions as warning but is sometimes used also as advertising-call; often repeated a couple of times in quick succession; when a flock of Jays catches sight of an owl, Goshawk or marten, a real uproar can break out in the wood! Sometimes gives a descending mew, 'piyeh', very like Buzzard's, but confusion seldom arises since the Buzzard calls almost invariably from the skies whereas the Jay is always heard from dense woods. Mimicry of its arch-enemy the Goshawk's cackle is often practised, too, 'kya-kya-kya-...'. Song is heard at times during late winter, a rather odd mixture of clucking, knocking, mewing and raucous sounds; quiet, does not carry far.

Siberian Jay *Perisoreus infaustus*   —

L 26–29 cm. Breeds in the northern coniferous forests, with some preference for denser, older forest with element of beard lichen. Resident. Caches insects, seeds and berries for winter requirements, hides the food in trees (in bark crevices, among needles etc.). Omnivore, forages partly at camp fires and at elk-slaughtering sites, and in summer robs some small birds' nests. Often goes about its business quietly and confidently even when forest walkers intrude quite closely upon it.

IDENTIFICATION Roughly thrush-sized, fairly *long-tailed, dark* and *uniform.* Plumage is grey-brown with element of

brownish-pink, and with *darker brown crown* and somewha paler brownish-white throat. Bushy nostril featherin forms a *small buff patch on forehead.* In flight, which i noiseless, smooth and fast with glides and sharp turn among the spruce branches, shows *rusty-brown panel nea wing-bend above* and *rusty-brown tail-base and tail-sides* Sexes and ages alike.

VOICE For the most part rather silent. Occasionally give short mewing 'geay' notes, upslurred 'kui' and thinne 'kiiy'. In momentary excitement, Siberian Jays may erup into louder outburst with Jay-like screams (though feebler) 'kreh', and longer mewing 'geeaih'. Song seldom heard, low 'subsong' (audible at only *c.* 20 m) of twittering, whist ling and mewing sounds.

(Spotted) **Nutcracker** *Nucifraga caryocatactes*   **V**∗∗

L 32–35 cm, WS 49–53 cm. Breeds almost exclusively i areas with easy access to spruce forests for nesting and ric supply of either hazel or arolla pine for winter foo stores. Nuts and seeds are cached in the ground layer durin late summer–autumn, and nut depots are memorized i detail and can be re-found in winter with great accuracy even beneath thick snow cover. European breeders (ssp *caryocatactes*), which are vigilant and shy, are mainl residents. In some years, when in Siberia the arolla pine seed crop is poor, large flocks of the slender-billed rac (*macrorhynchos*), however, migrate in late summer and earl autumn from NE Russia and Siberia all the way to C Eu rope, where they are relatively tame in behaviour; no concen trated return passage of Siberian birds noted. Omnivorou in summer. Nests in thick spruce (occasionally pine), usuall against trunk. Early breeder.

IDENTIFICATION Size of a Jay but with quite differen proportions: *tail is short* (best feature in flight) and *bi long and powerful.* Head is more elongate than Jay's an lacks hint of crest. *Plumage is dark brown* and *sprinkle with small white spots*; crown and greater part of wings are by contrast, unspotted brownish-black. Wings broad an rounded. Flight direct, often high up, with fluttering slightly hesitant wingbeats recalling Jay. From below shows contrastingly *white vent* and *white tail-tip.* Sexes alik and ages similar. – Variation: In Europe, bill is heavie (thicker base) and tail has narrower white tip. In Siberia race *macrorhynchos*, bill is more uniformly narrow an averages a trifle longer, and tail has broader white tip.

VOICE A rather silent bird on the whole. Does, howeve have a characteristic call heard recurrently, at least seasona ly (early spring, summer), a drawn-out, hard rolling soun with an almost machine-like quality, 'krrrrreh', which i often repeated a few times in rapid succession; it is longer feebler and at higher fre quency than Hooded Crow usual call, and the pitch i more even and mechanica Soft Jackdaw-like calls, 'yail and 'kya', sometimes given Song seldom heard, a quie 'subsong' of squeaky, hars and twittering sounds, o same type as Magpie's.

Jay   Siberian Jay   Nutcracker

JAY

mobbing a day-roosting Tawny Owl

streaked

pinkish grey-brown

black moustache

typically fluttering, slow flight and open flock formation

pale blue patch

white

black

rusty-red

dark

black

white

cervicalis
(N Algeria, N Tunisia)

atricapillus
(Middle East)

SIBERIAN JAY

favours ancient northern, coniferous forest rich in lichens

grey-brown

rusty-red

often scans from exposed perch

slightly darker cap

rusty-red tail-sides and rump

NUTCRACKER

hazel nuts favourite food where available, stored in autumn for winter supply

macrorhynchos
(E Russia, Siberia)

slender-billed, and has on average longer bill

caryocatactes
(Europe, W Russia)

thick-billed, and has on average shorter bill

fluttering flight with rounded wings somewhat recalls Jay, but note short tail and bigger head

white vent

dark rump

densely spotted white

in the Alps, Nutcrackers are fond of the seeds of the arolla pine

white tip (widest on the slender-billed race)

DZ

(Western) **Jackdaw** *Corvus monedula*   r**B**2 / **W**

L 30–34 cm, WS 64–73 cm. Breeds both in immediate vicinity of humans, i.e. in cavities in houses and other buildings, and in mature parks and deciduous woods with hollow trees; also locally in mountains and on sea-cliffs. Short-range migrant in N, otherwise resident. Pairs for life, and pair-members almost always seen together, often perch close together, look 'amorous'. Also very social, forages in flocks and, especially in autumn, gathers at dusk in large throngs to roost in favoured copses or town parks. Can become quite tame where not persecuted. Omnivore. Nests in chimney, air shaft, tree-hole, duck nestbox, cave, rock crevice, etc.

IDENTIFICATION At distance *all dark*, and often included in the term 'crow' by the layman. At closer range, however, *dark grey*, not black, and has *lighter grey neck-side and nape*. Eye, moreover, is *greyish-white*, and bill is much more slender than on Hooded/Carrion Crow and Rook. On ground, struts around quickly with upright posture. In flight, told from Hooded/Carrion Crow by *faster and slightly deeper beats* of the proportionately somewhat longer and narrower wings, also by broad but short neck, and by short bill that gives bird a somewhat 'docked' appearance at front. As a rule also flies in *denser flocks* than crows (almost as pigeons), but flock formation can be similar. Sexes and ages alike. – Variation: Birds from NE Europe (ssp. *soemmerringii*) have paler neck-side, especially in fresh autumn plumage, producing greyish-white patch on lower neck; intergrades into nominate race in Fenno-Scandia, so racial determination risky. Jackdaws in S and W Europe (*spermologus*) are darker.

VOICE Voluble. Conversational- and advertising-calls rather short and cutting and quite pleasing, some also hoarse and harsh. Often utters jolting 'kya', readily repeated in energetic series, harder 'kyack!', drawn-out 'kyaar' and slightly harsher 'tschreh', but details and volume vary with mood; often the birds chatter quietly together, when perched on a chimney and billing and cooing. Alarm-call is a furious, hoarse, drawn-out 'chaiihr'. A cackling noise is heard from large roosting flocks, before they settle for the night.

(Red-billed) **Chough** *Pyrrhocorax pyrrhocorax*   r**B**4

L 37–41 cm, WS 68–80 cm. Breeds in mountains (at 1200–3000 m, locally even higher) with steep precipices and ravines, as well as along steep rocky coasts (thus around sea-level) with caves and deep clefts. Resident. Often fearless and approachable. Social in habits except when breeding, but colonies may occur, sometimes mixed with Alpine Choughs, when suitable nest sites are few. Forages on ground, for insects and other small invertebrates and berries. Nests on cliff-ledge on precipice or in cave, in crevice, at times also on or in cavity in building.

IDENTIFICATION A *glossy metallic black*, fully Jackdaw-sized bird with distinctive proportions: *wings are rather long* and *uniformly broad right up to body*, has *blunt wing-tips* which are *deeply 'fingered'*, has *rather short tail* (equals wing width) which is *straight-ended*, and the *red bill is long, pointed and decurved*. The red legs are rather long. On perched bird, wing-tips reach tail-tip (tail protrudes clearly beyond wing-tips on Alpine Chough). Acrobatic flyer—the Chough may be likened to an air-show pilot in an old biplane: despite the broad and 'frayed' wings, the bird rolls and turns over with the greatest of ease, and it plunges in tail-chases on *folded wings* (only carpals stick out) with a whizzing sound, only immediately to shoot skywards again. Sexes alike. – Juvenile: Bill is dull orange-yellow and shorter than adult's, plumage duller sooty-black. Legs are often duller red.

VOICE The most typical call, often uttered as advertising-call in flight, is a cutting, almost whizzing, slightly descending 'chiach', akin to some calls of Jackdaw but more piercing, higher-pitched and with thicker-voiced ending. The call varies in details, so that a whole spectrum of related calls occurs, 'chiaa', 'chrai', 'chi-ah', 'tschraah' etc.

Alpine **Chough** *Pyrrhocorax graculus*   —

L 36–39 cm, WS 65–74 cm. Breeds only in mountains (at 1500–3900 m) with steep, inaccessible precipices. Resident. Gregarious, and may be seen all year in large flocks (several hundred). Often visits top-station restaurants at ski resorts in search of food remains, and can then be quite fearless. Restless, and before long moves on again over the mountain peaks. Feeds largely on insects in summer, on berries, seeds and food scraps in winter. Nests on cliff-ledge or in cavity.

IDENTIFICATION Resembles its relative the Chough in being *all black* and in having red legs, 'fingered' primaries and *contrastingly jet-black underwing-coverts*, also in acrobatic flight with headlong plunges and playful tail-chases. Differs, however, even at long range, in: *longer tail* (longer than wing width) *with narrower base and rounded tip*; somewhat shorter wings with *bulging rear edge of 'arm'* and slightly *narrower wing-base*, and *more rounded wing-tip with fewer and not such long 'fingers'*. At closer range, the diagnostic short and *bright yellow bill* is visible. At distance can also be confused with roughly same-sized Jackdaw, but that has shorter tail (barely equal to wing width), broader tail-base, broader head/neck and uniformly grey underwing. – Juvenile: Sooty-black somewhat duller than adult, also dark legs and often dark tip to yellow bill.

VOICE At least 90% of repertoire is characteristic and cannot be confused with Chough's. Commonest call from wandering flocks, e.g. around top stations at ski resorts, is a high, rolling 'zirrrrr' with an almost electric quality, also a piercing, whizzing 'ziieh' and something between the two, e.g. 'zrr rieh'. Anxiety-call a slightly deeper and fuller 'krrrrü'. Occasional calls are more like Chough's.

Jackdaw

Chough

Alpine Chough

JACKDAW

'chimney-sweep bird', fond of chimneys

distinct pale half-collar (shown by some birds in Scandinavia as well)

soemmerringii
(Finland, Russia, E Europe)

grey

dark face

often seen in pairs

uniformly grey under-wing

spermologus
(S & W Europe)

birds in Scandinavia (*monedula*) somewhat intermediate

CHOUGH

ferocious dives with wings folded back

broad outer wing, deeply 'fingered'

darker coverts

short

tail-base broad

brownish-yellow

juv.

short

ad.

long, red

ALPINE CHOUGH

rounded wing-tip with fairly short 'fingers'

darker coverts

tail length longer than wing width

tail-base narrow

yellow

feeding on mountain slope

gregarious, roves in dense flocks, restless and constantly on the move, one moment high in the sky, the next sweeping low over the ground

long

short-legged

DZ

Rook *Corvus frugilegus* rB1 / W

L 41–49 cm, WS 81–94 cm. Breeds in colonies (rookeries) in agricultural areas. Builds loose stick nests close together in crowns of tree clumps near farms and in villages; in spring, before trees in leaf, nests look like large witches' brooms. Forages in flocks on ploughed fields, on pasture and along ditchsides, often accompanied by Jackdaws. Relatively bold where not persecuted. Omnivore, but takes mostly insects and earthworms. Migratory in far NE.

IDENTIFICATION Size of Carrion Crow (and often miscon-strued as one), with *all-black plumage* which from certain angles shows reddish-lilac gloss. Adult told by *bare greyish-white skin around bill-base* and lack of nostril feathering, but usually also has a different profile from Carrion Crow, with *flat forehead, peaked crown* and *'short nape', flattened breast* and *ample belly* (ruffled, drooping belly-feathers). Bare bill-base develops Feb–May of 2nd calendar-year; up to then immatures are very like Carrion Crow, but at close range told by somewhat different bill shape with *straighter culmen and more pointed tip*, also usually by calls. Experienced observers may be able to tell that Rook has *slightly more flexible*, faster and deeper *wingbeats*, plus *more rounded*, almost wedge-shaped *tail-tip* and on average somewhat longer 'hand' and narrower wing-tip (subtle!).

VOICE Hoarse, nasal, noisy croaks without open rolling r-sound of Carrion Crows, more grinding and irascible 'geaah', 'geeeh', 'gra gra grah...' and the like. Noise from a rookery at nest-building and mating time can be deafening.

Hooded Crow *Corvus cornix* rB2 / (W)

L 44–51 cm, WS 84–100 cm. Breeds in open woodland, on moors and wooded shores, in tree clumps in farming areas and in larger town parks. Resident in large parts of Europe, but many Finnish and Russian birds migrate in winter to Sweden and the North Sea countries. Hybridizes with Car-rion Crow (see below) in a narrow zone of contact. Vigi-lant and shy with good reasons, for is an outlaw in most countries. Omnivore, robs other birds' nests, takes refuse and carrion, insects and other invertebrates on fields and sea-shores (incl. small fish, mussels etc.), berries, seeds etc. Builds open stick nest in tree crown, usually well concealed.

IDENTIFICATION Size of Rook and similar in shape, but recognized immediately by *bicoloured plumage* with dirty *light grey body* and black wings and tail along with *black head and straggly black bib down to breast* (cf. House Crow, p. 368). Flight is rather apathetic, almost sloppy, the wing-beats a little hesitant and without any bite (unlike Jackdaw's somewhat deeper and more resolute beats). Often flies high up. Flock formation generally loose.

VOICE Noisy. Calls are mostly hoarse, hard and croaking.

Commonest call, also functioning as song, is a hard rolling croak repeated 3–4 times, 'krrah krrah kraah'. Birds bicker among themselves, or mob Sparrowhawk and owls, with sti-fled but stubborn grating 'krrrr krrrr...'; but, when it comes to the dreaded Goshawk, it gives full vent to loud, hard and indignant 'krrah' notes.

Carrion Crow *Corvus corone* rB2 / (W)

L 44–51 cm, WS 84–100 cm. Closely related to Hooded Crow (above), which it replaces in W and SW Europe, the two hybridize in a narrow zone of contact, producing var-iable intermediates. Habits and habitat as for Hooded.

IDENTIFICATION Shape and size just as for Hooded Crow, but differs in having *entire plumage black* (with faint metal-lic green and bluish-lilac gloss when fresh). Told from Raven by considerably smaller size, by broader, shorter wings, by *more listless flight* with shallower wingbeats, and by fact that *rear edge of tail is rounded*, not wedge-shaped. Harder to separate from immature Rook, but at close range shows *thicker bill* which is *blunter, with more curved culmen* at tip; flight is somewhat lazier, and wings a little shorter and broader (though differences minute!).

VOICE Very like Hooded Crow's, but often sounds a bit harder and more 'malevolent' in tone, not so rolling and open. Difference of doubtful use in field, however, bearing in mind individual variation in, especially, Hooded Crow.

(Common) Raven *Corvus corax* rB3–4

L 54–67 cm, WS 115–130 cm. Breeds in deserted woods, in uplands and on coastal cliffs. Has large territory, roams widely. Pairs for life. Resident. Very shy and wary. Omni-vore. In winter, frequent visitor at carrion. Builds stick nest on inaccessible cliff-ledge, in tree or occasionally on power pylon. Breeds early in spring.

IDENTIFICATION Largest passerine, *bigger than a Buzzard*. Plumage *all black* with metallic sheen (green, bluish-lilac). *Bill very thick*, this often visible in flight, as is heavy, some-times *shaggy throat feathering*. Typical flight silhouette, with *long and proportionally narrow wings* with 'fingered' but *rather long and narrow 'hand'*; also *well projecting neck area* and *wedge-shaped tail-tip*. Flight distinctive, too, often rather high up with *slow but deep and driving, elastic wing-beats* with slightly *backswept 'hand'*, speed fast. Often occurs in pairs. Often soars, when wings held flat or even slightly lowered, not raised as with a buzzard or Golden Eagle.

VOICE Loud, clanging, jarring with rolled r-sound. Call a deep 'korrp', sometimes repeated 3–4 times. Alarm a faster, hard 'krack-krack-krack'. Shows spring feelings with vari-ous rather odd calls, knocking and clucking sounds and a resounding 'klong'.

Rook

Hooded Crow

Carrion Crow

Raven

ROOK

often peaked crown

pointed, rather straight culmen

pale base

narrow wing-bases

feeding in flock on fields

juv.

ad.

HOODED CROW

innovative—empties unattended fishing-gear on spring ice

pale grey and black plumage renders the species unmistakable

CARRION CROW

careful observations required to separate from juv. Rook

stout; culmen curved

rather evenly broad wings

RAVEN

teasingly close attendance on a Golden Eagle with prey, hoping to get some scraps

very heavy

large— the size of a Buzzard!

rather long, narrow and pointed wings

wedge- shaped

DZ

Brown-necked Raven *Corvus ruficollis* —

L 48–56 cm, WS 103–120 cm. Breeds in desert and dry steppe, and at times seen patrolling mountain deserts and rocky wadis but is not a mountain bird. Resident. Wary and shy, often rather difficult to approach. Nests in tree crown, on power pole or cliff-ledge, exceptionally on ground.

IDENTIFICATION Very like Raven, and not easily separable in field. Calls often the best clue (see below). Is *a trifle smaller* than Raven, but difficult to assess without direct comparison. Shape and flight resemble Raven's, but has proportionately *somewhat smaller* (thinner) *bill*, somewhat less bushy throat feathering and *slightly narrower 'hand'* (one 'finger' fewer at wing-tip). Quite often flies with bill pointing downward, and not uncommonly has less obvious wedge-shape to tail-tip (can show slightly projecting central rectrices as on Red-footed Falcon), but these features do not always hold, or are hard to make out. At closest range may show *bronze-brown colour on nape* (Raven can show indication). Toes often yellowish (at least in Middle East).

VOICE Calls include a surprisingly *crow-like* croaking 'krii-eh', a bit hoarse and high-pitched, not deep and re-sounding or with rolling 'r' as Raven's. Another call, again more like Hooded Crow than Raven, is a strained, grumbling 'grreu grreu grreu', perhaps with territorial function.

Fan-tailed Raven *Corvus rhipidurus* —

L 45–52 cm, WS 103–110 cm. Breeds in mountain desert and locally lower down in rocky, barren tracts. Resident. Social, often seen in flocks. Frequently soars, and performs aerobatics. Omnivore; visits refuse tips, cultivations, oases etc. Nests on cliff, on ledge or in cave.

IDENTIFICATION Size of Brown-necked Raven, smaller than Raven. Has very short tail; on ground, wing-tips extend far beyond tail-tip (wings and tail equal in length on Brown-necked). Bill rather short and thick but head fairly small (can give vulture impression). Often seen walking around on ground with half-open bill. In flight, immediately recognized by most peculiar silhouette with very broad wings, broadest next to body, and docked tail.

VOICE Call a Raven-like, 'loosely' rolling and gentle-sounding 'korr korr'. In irritation, e.g. when mobbing a raptor, utters a high, rolling 'trrrü trrrü...' with dry wooden ring, somewhat recalling voice of Rook.

House Crow *Corvus splendens* —

L 37–42 cm, WS 68–80 cm. Indian species which has relatively recently found its way to the region via ships (or introduced?). Breeds colonially in vicinity of humans, close to ports and towns with ready access to trees and food in form of refuse (fish offal, edible items at rubbish tips etc.).

Omnivore. Social, almost always in flocks. Builds stick nest high in tree.

IDENTIFICATION A real character; between Jackdaw and Hooded Crow in size, but slimmer and more elegant in shape than either. On ground, is *rather slim and lanky*, with *strikingly long legs* and long wings, *small head*, but proportionately *large bill* which is often emphasized by steep forehead. The plumage recalls Hooded × Carrion Crow hybrids, i.e. *head with bib attached and back, wings and tail are black* while *rest is dark grey* (nape, neck and upper breast a shade paler, medium grey). Close views reveal that the *black on head ends just behind eye* and is thus less extensive than on Hooded Crow. In flight, contrast between black and grey often difficult to see at range, when the bird just looks all dark. *Wings are slim like Rook's and slightly angled at* carpal, and so, with its rather long, heavy bill and *long tail* it looks at distance like a miniature Raven.

VOICE Quiet. Has a muffled, somewhat Hooded Crow-like croak, 'kraar'. Mobs raptors with harder 'krao-krao'.

STARLINGS *Sturnidae*

Four mid-sized passerines, compact, with short, strong feet and pointed bill. Often seen in large, dense flocks. Spend much time walking on ground, gait rapid. Varied diet. Nest in hole. One Asiatic species is treated below, while the three European species are described on the next spread.

Tristram's Starling *Onychognathus tristramii* —

L 24–27 cm. Breeds in barren mountain tracts, especially in or near ravines, wadis and valleys with some vegetation. Resident. Gregarious, often seen in smallish parties (flocks of 10–50 not unusual) except when breeding. Food insects, fruits and berries. At times visits lower-lying cultivations to feed. Nests in cavity in rock face.

IDENTIFICATION Size approximately as Starling, but somewhat *longer-tailed* and *slimmer in build*. Plumage broadly all *black*, at certain angles with bluish-lilac lustre. Main feature a *big, bright orange* (copper-coloured) *wing-panel* on 'hand', conspicuous on flying birds but otherwise visible only as narrow band along edge of folded wing. Eye (brownish-)red. Sexes separable by ♀ having greyish head and neck with diffuse darker streaking or spotting.

VOICE Noisy and *loud*. Voice, moreover, very characteristic: flocks moving along mountainsides emit loud glissando calls, sounding like *wolf-whistles* or like tuning in an old radio and getting distorted tones at full amplification, 'viyvuviy' or 'vooviiyuu'. Alarm a hoarse scream, 'veeech', recalling Golden Oriole but with more malevolent tone. Supposed song a soft series of squeaky and rasping notes.

Brown-necked Raven  Fan-tailed Raven

House Crow  Tristram's Starling

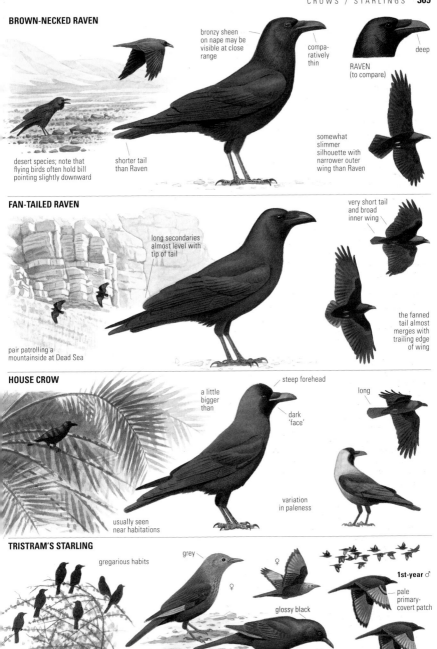

BROWN-NECKED RAVEN

bronzy sheen on nape may be visible at close range

comparatively thin

deep

RAVEN (to compare)

desert species; note that flying birds often hold bill pointing slightly downward

shorter tail than Raven

somewhat slimmer silhouette with narrower outer wing than Raven

FAN-TAILED RAVEN

long secondaries almost level with tip of tail

very short tail and broad inner wing

the fanned tail almost merges with trailing edge of wing

pair patrolling a mountainside at Dead Sea

HOUSE CROW

a little bigger than

steep forehead

dark 'face'

long

variation in paleness

usually seen near habitations

TRISTRAM'S STARLING

gregarious habits

grey

♀

♀

glossy black

1st-year ♂

pale primary-covert patch

ad. ♂

♂

DZ

(Common) **Starling** *Sturnus vulgaris* ⌐B1 / **W**+**P**1

L 19–22 cm. Familiar breeder in farmland and suburban areas, and in woodland (esp. with oak). Post-breeding flocks, sometimes of enormous size, forage on fields, in gardens, on cherry trees, on shoreline seaweed, etc. In Britain & Ireland largely resident, but massive influx from N and NE Europe *c.* Aug–Mar, some roosting in urban sites.

IDENTIFICATION When a *black* bird with *yellow bill* lands on the lawn, beginners have to decide between Starling and Blackbird. On closer inspection the differences are considerable: Starling has some *yellowish-white spots* in the plumage, the feathers have a *metallic green and violet sheen*, the *tail is short* and the *bill rather long and pointed*, it does not make two-footed hops but *walks with jerky, energetic movements* and so on. Flight fast and slightly undulating, silhouette distinctive with *rather pointed wings* and *short tail*. Legs brownish-pink. Sexes similar (♂ in spring has wholly unspotted breast and blue-grey at base of lower mandible, ♀ retains some pale spots below and has yellowish-white lower-mandible base). – Winter: Bill dark. Entire plumage densely sprinkled with yellowish-white spots. Chin whitish. – Juvenile: All *dull dirty brown*. Adult-like 1st-winter plumage acquired end Jul–Sep; immatures with pale-spotted body but retaining dull brown head are easily recognized.

VOICE Rich repertoire. On take-off and in flight, often a short buzzing 'chürrr'. Alarm at nest a hoarse, drawn-out 'steeh' and on sight of hawk a sharp 'kyett!'. Young beg with rippling, hoarse buzz, which from fledged young can at times become almost a metallic rattle. Song in vicinity of proposed nest site in spring, but also occasionally on warm sunny days in autumn and winter, consists of loud glissando whistles and rather soft, knocking sounds, squeaks and fine imitations of other birds, e.g. often Lapwing, Curlew, Coot, House Sparrow, Magpie and Hooded/Carrion Crow.

Spotless Starling *Sturnus unicolor* —

L 19–22 cm. Resident breeder around habitations and farmyards, sometimes in olive groves or on cliffs with cavities available. Habits otherwise as Starling's.

IDENTIFICATION Very like Starling, but in spring and summer is *entirely unspotted black* (♀ and 1st-year ♂ can still have some fine spots on undertail-coverts) with *more uniformly purple gloss* (green from certain angles) than Starling. In winter plumage, ♀ and immatures have pale spots on back, belly and undertail-coverts. *Crown* usually *unspotted*, unlike normal winter-plumaged Starlings. Bill often short-looking. *Legs light pink* (on average paler than in Starling).

VOICE Very like Starling's, but whistles of song sharper in tone, as well as clearer and louder and often also upward-bent, and trills are often stronger and more rolling.

Rose-coloured Starling *Pastor roseus* **V****

L 19–22 cm. Breeds colonially in agricultural country or on steppe. Occurrence erratic, adapts opportunistically to mass emergences of grasshoppers and other insects. A few stray to NW Europe, mostly in summer.

IDENTIFICATION *Body pink*, head, wings and tail black, on adult ♂ with metallic lilac sheen on head and green sheen on wings; *long drooping crest on nape*. ♀ somewhat duller, with shorter crest and brown-toned back. Immatures have pink of back and belly soiled grey-brown and generally lack metallic sheen and nuchal crest. Juvenile like a washed-out young Starling, but has *shorter and blunter, yellowish bill*, some contrast between darker wings and paler body and, in particular, *pale rump* in flight (beware aberrantly pale young Starlings, though these always have dark, pointed bill).

VOICE Calls short, harsh. Song, often given with trembling 'penguin wings', a jingling, noisy chorus of knocking, squeaky and silvery sounds, like older immature Starlings.

ORIOLES *Oriolidae*

Medium-sized passerines with quite elongate, thrush-like shape. Mostly hide in tree crowns. Only one species in region.

(Eurasian) Golden Oriole *Oriolus oriolus* m**B**5 / **P**5

L 22–25 cm. Breeds in lofty deciduous forest (few conifers admixed not a hindrance), in agricultural areas with deciduous copses, often near rivers or lakes, in larger parks etc. In Britain rare, in commercial poplars. Summer visitor (May–Aug), winters in tropical Africa. Shy, restless, mobile. Difficult to catch sight of on breeding grounds, keeps concealed high up in foliage (even yellow ♂ blends into the sun-dappled leaves and escapes detection!). Food insects, berries. Nest suspended in branch fork (Walt Disney style) high in canopy.

IDENTIFICATION Flight fast and sweeping through foliage; over longer stretches thrush-like, in long, shallow undulations. – Adult ♂: Unmistakable, *bright yellow with black wings and black tail* and reddish bill. Lores and tertials black. – ♀/1st-summer ♂/juvenile: *Green above, off-white with fine streaking below*. Variable amount of yellowish-green or yellow on flanks and belly. (Note that occasional ad. ♀♀ can be very yellow and unstreaked below, and hard to separate from ad. ♂ in field, but have dark grey lores and partially olive tertials and central tail-feathers.)

VOICE Often emits a screaming, hoarse 'veeaahk', somewhat reminiscent of Jay but with more nasal and more strained voice. Also has a falcon-like, fast 'gigigigigi'. Song a beautiful, loud, fluting whistle, confusable only (at distance) with Blackbird song, e.g. 'foh-flüo-fiih-fiioo' or shorter 'fo **flüh**-füo' or just 'fiiooh', the details varied.

Starling

Spotless Starling

Rose-coloured Starling

Golden Oriole

STARLING

singing ♂

short tail and pointed wings in flight

grey-brown

paler

still juv. feathers

begging young and feeding adults

pale

ad. s. ♀

blue-grey

ad. summer ♂

juv.

juv. → 1st-winter

winter

dark

SPOTLESS STARLING

at nest entrance

even a young ♀ (right) appears blacker than ad. ♂ Starling (left)

entirely spotless from c. Mar; plumage has some purple gloss (green at certain angles)

wing-feathers have no buff edges

no pale spots

only tiny pale spots

ad. winter

ad. summer ♂

often quite light pink

ROSE-COLOURED STARLING

juv. Rose-coloured (centre) flanked by young Starlings moulting to winter plumage

STARLING, juv.

ROSE-COLOURED, juv.

ad. s. ♂

head glossy black

long crest

pale

pale bill-base

dark

sullied grey

short

bright pink

contrastingly dark wing

juv.

winter

ad. summer ♂

GOLDEN ORIOLE

ad. ♀ / 1st-s. ♂

ad. ♂

ad. ♀ / 1st-s. ♂

bright yellow with black wings

prominent streaks

a few ad. ♀♀ are more yellow, approaching ad. ♂

faint streaks

juv.

ad. ♀ / subad. ♂

ad. ♂

KM

SPARROWS *Passeridae*

Sturdily built passerines with conical bill well suited for consuming seeds. Closely related to the weavers *Ploceidae*, which occur mainly in Africa. Have primitive vocal repertoire, often build bulky, 'twiggy' nests and have complete moult from juvenile to 1st-winter plumage in summer.

House Sparrow *Passer domesticus* r**B**1
L 14–16 cm. Breeds in proximity of humans both in rural and in urban areas. Is therefore well known, and often thought of as the most numerous bird, which is far from the case (commonest in Britain are Wren and Chaffinch). Resident. Social, even when breeding, occurring in dense flocks. Nests under roof tiles, in air duct, recess, sometimes in tree.

IDENTIFICATION Robust, with broad body and fairly big head with *stout bill*. Plumage rather bushy and 'loose', and often looks matted and unkempt. Main features are *heavily black-streaked brownish back*, *squat posture* with legs drawn in when perched, and laborious and clumsy, 'drilling' flight with *continuously whirring wingbeats*. – ♂: *Black bib*, lores and eye region. *Grey crown* with chestnut sides. Throat-sides whitish, but *cheeks usually dusky grey*. *Broad white wing-bar*. Bill black when breeding. – ♀/juvenile: Dusky brownish grey-white below, and dirty brown above with black-streaked back. Often a distinct light dirty buff supercilium. Juvenile ♂ grows the first black throat-feathers in Aug–Sep.

VOICE During courtship, long series of well-spaced monosyllabic chirps slightly varied throughout, e.g. 'chilp chev chilp chelp chürp...'. In irritation, a rattling 'cher'r'r'r'r'.

Italian Sparrow *Passer italiae* —
L 14–16 cm. Intermediate between House and Spanish Sparrows. Formerly regarded as a 'stable hybrid swarm' as a result of mixing between these two, but recently more often regarded either as a separate species (followed here), or as a race of Spanish Sparrow. Breeds in mainland Italy, on Corsica and partly on Sicily. Mixes with House Sparrow in S Alps. Similar-looking birds in N Africa are more likely hybrids between House and Spanish Sparrows than real 'Italians'!

IDENTIFICATION Combines *chestnut crown* and *whitish cheeks* of Spanish with *lack of bold black spotting on underparts* of House Sparrow. Often a *hint of short white supercilium* as in Spanish. ♀ usually inseparable from the other two species.

VOICE Usually not separable from House Sparrow's.

Spanish Sparrow *Passer hispaniolensis* V***
L 14–16 cm. Unlike House Sparrow, usually breeds in tall shrubbery or in clumps of trees, often in large colonies. Closed nests of straw and twigs. Also builds in stick nest of larger birds (e.g. White Stork). Mainly resident in W, migratory in

E; in E Mediterranean in spring and autumn, seen *migrating by day in typically dense flocks*, which move in purposeful, billowing low flight and at surprisingly high speed. Often hybridizes with House Sparrow, and intermediates can be seen where the two species meet, mainly in N Africa.

IDENTIFICATION Same shape as House Sparrow. – ♂ easily recognized by combination of: *wholly chestnut crown*; *big, broad black bib* which *meets blackish shoulders*; *coarsely black-spotted lower breast and flanks*; *whitish cheeks*; *narrow white supercilium* (sometimes broken just above eye); in spring and summer due to wear largely *black on mantle and shoulders*. – ♀: Normally impossible to separate in the field from House Sparrow (has on average slightly bigger bill, paler belly and sometimes hint of grey streaking on flanks, but differences as a rule too vague for safe identification).

VOICE Like House Sparrow's, but a bit *higher* and more *metallic* in tone. *Fast, monotonous* series, e.g. 'chili-chili-chili-...', often heard. The hum from larger colonies has a typically musical ring, is audible at several hundred metres.

(Eurasian) Tree Sparrow *Passer montanus* r**B**2–3
L 12½–14 cm. Breeds in open woodland, usually near cultivated land or human habitation, locally around houses and in gardens like House Sparrow. Resident. Nests in hole.

IDENTIFICATION Smaller and neater than House Sparrow, with ♂ of which it is still often confused by the layman. Like that species, has black-streaked brown back, distinct white wing-bar, black bib, lores and bill, and dusky underparts, but told by: *wholly vinaceous red-brown crown*; *pure white headside* with *big black cheek patch*; *smaller black bib* which does not expand into large patch on breast; *narrow white neck-collar* (broken only on nape). Sexes and ages alike.

VOICE Rather like House Sparrow's. Distinctive, however, is a 'tsu**witt**', slightly nasal and disyllabic, with cheery tone. In flight, often a dry 'tett-ett-ett-ett'. Song a fast series of 'tsvit' and variants, like House's chatter but a bit higher-pitched.

Dead Sea Sparrow *Passer moabiticus* —
L 12–13 cm. Breeds in dry lowland areas with tamarisk or other bushes and access to open water; not tied to humans. Resident and nomadic. Rather shy.

IDENTIFICATION Small. ♂ has *dark grey crown, nape and cheeks* (tinged brown in fresh autumn plumage), small black bib, *pale supercilium becoming buff at rear*, and characteristic *light yellow neck-side*. ♀ like a *small but pale* House Sparrow; a few have trace of yellow on neck-side.

VOICE Feebler voice than House Sparrow. Flight-call repeated high, disyllabic 'chi-vit'. Song consists of fast, rolling 'trirp-trirp-trirp-trirp' or slower series of sonorous but more varied chirps, 'chilp chrilp chrirp chilp...'.

House Sparrow

Spanish Sparrow

Tree Sparrow

Dead Sea Sparrow

HOUSE SPARROW

common both in urban areas and near humans in the countryside

dusky, sullied look

♀

winter ♂

grey

dusky

summer ♂

ITALIAN SPARROW

♀♀ inseparable from the closely related House sparrow

red-brown

whitish

summer ♂

SPANISH SPARROW

not depen-dent on vicinity of humans

♀

winter ♂

pale

♀ usually impossible to separate from ♀ House Sparrow

sometimes hint of grey streaking on whiter underparts

red-brown

bold black pattern

summer ♂

TREE SPARROW

often nests in nestboxes or tree-holes

juv.

black cheek patch

nearly full white collar

vinaceous red-brown

feeding on a stubble field

DEAD SEA SPARROW

small

♀

autumn ♂

strikingly small, and gives a cleaner impression than House Sparrow

grey

red-brown

spring ♂

DZ

(Common) **Rock Sparrow** *Petronia petronia* —

L 15–17 cm. Breeds in various habitats with bare surfaces: rock faces in alps, ravines, wadis, quarries, rocky desert tracts, open terrain with walls, ruins and even occupied buildings, locally also alpine meadows with boulders, cultivations. Resident. Social, can be seen in large flocks in non-breeding season. Nests in cavity in rocks or in crevice.

IDENTIFICATION Rather big and *very stocky*, with broad head and *hefty bill*. Markings typically heavy and stripy, with patterned wing and *broad pale supercilium* between *dark crown-side* and dark eye-stripe. Greater part of *whitish underparts heavily and openly streaked brownish-grey* (look rather dark at distance). *A yellow spot where throat meets breast* (though often hard to see when bird perched in hunched-up posture). Tail dark with *large white terminal patches on outer feathers*, seen best in flight. *Undertail-coverts dark with broad pale fringes*. Sexes alike.

VOICE Voluble. Call a short nasal 'vüi'. In anxiety, a hard trilling 'tii-tür'r'r'r'. Most typical is a loud, drawn-out call in glissando, repeated and varied as song, 'sle-veeit' or 'tveyu**itt**'; sounds slightly nasal, creaky and 'seductive' in tone; occasionally a shorter 'tvü**it**' is also heard.

Pale Rock Sparrow *Carpospiza brachydactyla* —

L 14½–15½ cm. Breeds in arid and hilly areas on barren slopes with scattered bushes and some ground vegetation. Migratory. Nests in fork on tree or dense bush.

IDENTIFICATION Rather slim, with small head, and *large, short and podgy bill* swollen at very base. Plumage rather *plain grey-brown* with no obvious features (back lacks heavy streaking). On closer inspection, shows a poorly indicated, short, pale supercilium and a diffusely paler eye-ring, as well as two rather indistinct pale wing-bars. *Bill* when breeding is rather *dark grey* with slightly paler base to lower mandible, during winter months somewhat lighter grey-buff. In flight, shows *small white spots on tail-tip*.

VOICE Call a Greenfinch-like 'dvui', also a short, low-pitched trill, 'trrü' (can recall distant Bee-eater). Song a plain, full-second-long throaty buzz, 'bzrüüüüiz', like an insect or electric motor, sometimes prefaced by a few short grace-notes, 'bz-bz-bz-bzrüüüiz' (Corn Bunting-like).

Yellow-throated Sparrow *Gymnoris xanthocollis* —

L 13–14½ cm. Breeds within the region only at a few places in SE Turkey in groves and cultivations, e.g. of almond, olive and date palm, or in dry scrub with scattered trees. Migratory, winters in India; returns in May.

IDENTIFICATION A rather small and *slim* sparrow with small head but *long and strong, conical, pointed bill*, the size of which is accentuated by flat forehead/crown (though may

erect crown-feathers). Plumage brown-grey with somewhat paler belly and *double wing-bar*. ♂ told from ♀ by *reddish-brown lesser coverts, yellow throat patch, dark lores* and, in breeding season, *black bill*. Tail square-ended. Legs grey.

VOICE Resembles House Sparrow's. Simple chirping song, but tempo characteristically fast and tone full and ringing. Call a piercing 'chi**ah**' (like miniature Chough).

Desert Sparrow *Passer simplex* —

L 13–14½ cm. Breeds in sand deserts at oases, near settlements or cultivations. Resident and nomadic. Nests in cavity.

IDENTIFICATION Like a washed-out, small House Sparrow. Plumage *very pale buff*, like sunlit desert sand. Sexes very different once juvenile plumage exchanged for adult type in summer. – ♂: *Black bib, lores and eye region*; when breeding, *black bill*; brownish-black primary-coverts, alula, outer greater coverts, primary tips and tertial centres; *grey tinge to crown and mantle*. – ♀: Mainly pale buffy-white, with warmer *ochre tone on crown/mantle* and ochre tinge to breast; lacks black bib, and dark wing markings are brownish-grey, not brownish-black as on ♂. Normally pale bill darkens when breeding. – Juvenile: As ♀, but wing markings even more lacking in contrast (seen best on tertials in rear view).

VOICE Like House Sparrow's but more metallic, and mixed with White Wagtail-like high 'chi-vi chi-vi'. Often gives Greenfinch-like 'chüpp-chüpp-chüpp' in flight.

(White-winged) **Snowfinch** *Montifringilla nivalis* —

L 16½–19 cm. Breeds on bare mountains, often at 1900–3100 m. Resident, and remains at rather high altitude even in winter (exceptionally down to 1000 m). Often seen foraging at ski-resort restaurants; restless but often quite fearless.

IDENTIFICATION A fairly *big and elongate*, long-winged bird with conspicuous white panels on tail and wings. Hopping on ground, a rather dull grey-brown bird showing one narrow but long white wing-panel. When it takes off, the white is dominant: greater part of *inner wing is white*, and *tail is white with narrow black central band* (and thin black terminal band). At close range, shows dull earth-brown mantle/back, greyish head, and dirty white underparts with a small and often diffuse black bib. Bill black when breeding, ivory-yellow during winter. Flight fast and powerful, undulating over longer stretches. Sexes similar.

VOICE Rich repertoire. Calls include a hoarse 'zyiiih', a creaky mewing 'myaih', short 'ti ti zü' and, often when nervous, a Crested Tit-like rolling 'tir'r'r'r'. Also has a clattering 'chett-chett-chett-chett'. Song, delivered from perch on boulder or in circling gliding flight, has sparrow quality, is variable, halting and jolting with recurrent dry, strained trills. Also has a soft, twittering subsong.

Rock Sparrow

Pale Rock Sparrow

Desert Sparrow

Snowfinch

ROCK SPARROW

variegated pattern

broad stripes on head

strong bill

boldly streaked

pinkish lower mandible

juv.

pale median crown-stripe

small yellow spot (concealed when in hunched-up posture)

ad.

PALE ROCK SPARROW

singing ♂

rather plain pale grey-brown

pale

hint of wing-bar when fresh

sexes and ages alike; 'bland' look with diffuse head pattern

strong bill with decurved culmen

autumn (fresh)

summer (worn)

YELLOW-THROATED SPARROW

yellow throat patch sometimes difficult to see

♂

two white wing-bars

off-white below and grey-brown above

♀

pale

winter ♂

chestnut 'shoulder'

black

yellow patch

summer ♂

DESERT SPARROW

found at oases in sand deserts

very pale, buff and white

pale

winter ♀

sandy-brown

winter ♂

pale base

summer ♀

greyish

summer ♂

pale wing patch

SNOWFINCH

song-flight

greyish head

long white band

ivory

dark bib often concealed in fresh winter plumage

ad. summer ♂ (colours of ♀ somewhat duller)

earth-brown

black

winter

winter

long-tailed

SNOW BUNTING, winter ♂ (to compare)

DZ

FINCHES *Fringillidae*

A large and varied group of relatively small passerines with large, conical bill. Tail of most narrow, with indented tip. Flight powerful, undulating. Most species are accomplished and frequent singers. Often social in habits outside breeding season; migrate and forage in flocks. Most important food is seeds, which are split or husked with the heavy bill; in summer also take some insects. Build open, basket-shaped nests which are usually placed in thick bushes or trees.

(Common) Chaffinch *Fringilla coelebs*   r**B**1 / **W**+**P**1

L 14–16 cm. Breeds commonly in all types of woodland and in parks and gardens, and is Britain's second commonest bird after Wren (estimated *c.* 5.4 million pairs). Prefers rather more open woods to dense ones, and often forages on ground. Resident British & Irish population augmented by large autumn / winter influx from N and NE Europe. Builds neat nest in tree fork, camouflaged on outside with lichens and moss.

IDENTIFICATION Size of House Sparrow, but slimmer and has longer tail. Usually easily recognized by *double white wing-bar, white tail-sides* and *grey-green rump*. – ♂ (once juv. plumage moulted Jul–Sep): *Head-side and breast rusty-red; crown and nape blue-grey* (tinged brown in fresh autumn plumage); mantle reddish-brown. – ♀/juvenile: Grey-green above with faint brown tinge, greyish-white below; narrower wing-bars than ♂. – Flight strong and undulating. On migration forms fairly loose flocks (a bit looser than Brambling, but mixed flocks common). Often alights with a few fluttering sweeping turns, showing white patterns.

VOICE Powerful voice and frequent singing make this one of the birds most heard in woodland and parks. Several calls characteristic once learned. Perched bird gives a spirited, sharp 'fink!' (only Great Tit's 'ping' is confusable), while in flight it has a more unobtrusive 'yupp' (softer than Brambling's 'yeck', weaker than Greenfinch's 'jüpp!'), which is repeated frequently by migrating flocks overhead. Also heard often in N Europe is a forceful, upcurled whistle, 'hüitt', often persistently repeated, which in S Scandinavia, Britain and Continental Europe is commonly replaced by a rolling, straight, discordant whistle, 'rrhü' (and many local variants; in S Europe also a Thrush Nightingale-like uninflected 'hiit'); function varies between advertising, warning and territory maintenance, and the call is often termed the 'rain-song'. When highly agitated, a sharp, fine piping note, 'ziih'. Fledged young beg with a loud 'chripp'. Song highly characteristic, rather constant in delivery and tirelessly repeated, a bright, loud, almost rattling verse introduced by 3–4 rapidly repeated sharp notes which turn into a similar series of lower notes, the whole terminating

in a lively flourish, 'zitt-zitt-zitt-zitt-sett-sett-sett-chatt-chite**rii**dia'; some Chaffinches add an almost out-of-place, Great Spotted Woodpecker-like 'kick' at the very end.

Brambling *Fringilla montifringilla*   **W**+**P**1–3 / (m**B**5)

L 14–16 cm. Common breeder in upland birch forest of Fenno-Scandia; absent from dense, tall forest, prefers more open coniferous with some deciduous growth. Summer visitor to breeding sites (mostly mid Apr–Oct), winters in C and S Europe, but in mild winters many remain in S Scandinavia. Rather shy. Food mainly seeds, in summer also insects and in winter often beech nuts. In some winters when beech mast plentiful, enormous flocks gather around beech trees. Builds nest in tree fork, generally in birch, also in spruce.

IDENTIFICATION Same size and shape as Chaffinch, but in all plumages recognized by *white rump* (tinged yellow on juvenile) and little or *no white on sides of tail*. Like Chaffinch, has pale wing markings, but these are partially rusty-yellow as well as narrower and less conspicuous. Flight as Chaffinch's, but single-species passage flocks are often somewhat more compact. – ♂ summer: *Head, nape, shoulders and mantle glossy blue-black*. Throat, *breast and lesser wing-coverts unspotted rusty-yellow*. Some dark spots on lower flanks. Bill black. – ♀: Crown, mantle/back and nape-sides rather dark grey-brown, head-sides and central nape paler buffy grey-white, *breast buffish*. – ♂ winter: Black areas of summer plumage partly concealed by fairly broad rusty-buff fringes, giving somewhat piebald appearance. Bill straw-yellow with dark tip. Most reliably told from ♀ by *black showing through on entire head-side*. Younger ♂♂ have black-spotted lesser coverts and broader pale fringes to upperparts.

VOICE Has characteristic call, a loud, croaking or hard, nasal 'te-**ehp**', uttered both in flight and when perched. Migrating flocks give hard, short, slightly nasal 'yeck' calls (cf. Chaffinch), oft repeated. In anxiety, utters repeated, fine, silver-clear notes, 'slitt, slitt, slitt,...'. Song very distinctive and with desolate ring, a slowly repeated, straight, barely second-long, crude buzzing note, 'rrrrrhüh', which can recall a distant wood-cutting crosscut saw.

Blue Chaffinch *Fringilla teydea*   —

L 16–18 cm. Breeds in pine forest at 1200–1800 m altitude in Tenerife (relatively numerous) and Gran Canaria (scarce). Resident. Normally forages in pine woods but sometimes in bushy areas on mountainsides, and exceptionally (in severe weather) can be seen in cultivations down to *c.* 500 m. Generally seen singly or in small flocks (<10). Food mainly seeds of Canary pine. Late breeder, end May–Jul (Aug).

IDENTIFICATION Noticeably *bigger* than Chaffinch and more front-heavy, with *heavier breast, head and bill*. Both sexes have only *poorly indicated, narrow greyish-white wing-bars* and *lack white on tail*. The stout, conical bill is rather pale grey. – ♂: *Dark grey-blue on head and mantle, paler lead-grey on throat, breast and flanks*; undertail-coverts white. – ♀: Brown-grey above, greyish-white below.

VOICE Call a disyllabic 'chrooit', with cracked and slightly squeaky voice. Song shorter than Chaffinch's and not so forceful, has decelerating ending of a few repeated, rather harsh, descending notes, the last of which is drawn out, e.g. 'sitt-sitt-sitt rüha-rüha, rrüü**aah**'.

Chaffinch

Brambling

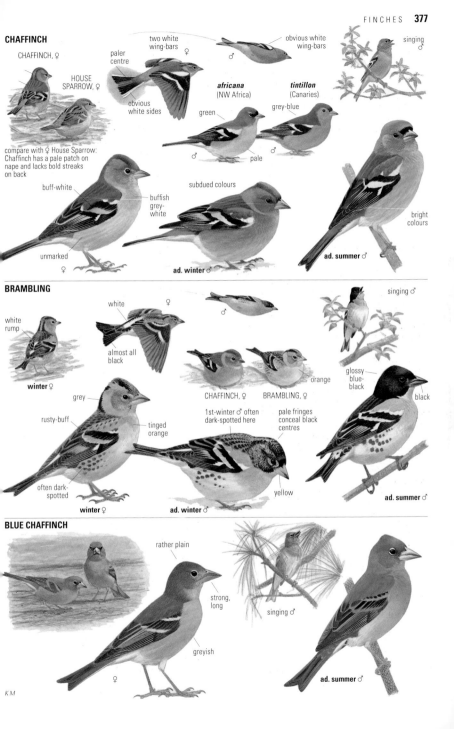

CHAFFINCH

CHAFFINCH, ♀

HOUSE SPARROW, ♀

compare with ♀ House Sparrow: Chaffinch has a pale patch on nape and lacks bold streaks on back

two white wing-bars ♀

paler centre

obvious white sides

obvious white wing-bars ♂

singing ♂

africana (NW Africa)

green

♂

pale

tintillon (Canaries)

grey-blue

♂

buff-white

buffish grey-white

unmarked

♀

subdued colours

ad. winter ♂

bright colours

ad. summer ♂

BRAMBLING

white rump

white ♀

winter ♀

white rump

almost all black

♂

singing ♂

grey

rusty-buff

tinged orange

often dark-spotted

winter ♀

CHAFFINCH, ♀

BRAMBLING, ♀

orange

1st-winter ♂ often dark-spotted here

yellow

ad. winter ♂

pale fringes conceal black centres

glossy blue-black

black

ad. summer ♂

BLUE CHAFFINCH

rather plain

strong, long

greyish

♀

singing ♂

ad. summer ♂

K M

(Common) **Linnet** *Carduelis cannabina*   r+m**B**2 / **P**+**W**

L 12½–14 cm. Breeds in areas with thick bushes, in gardens and especially on coastal heaths with gorse, locally also in orchards. Restless and active, often rises in hopping flight only to drop down again soon. Pair-members stay close together throughout summer.

IDENTIFICATION *Slim* and rather long-tailed, with *short grey bill*. In all plumages, *brown mantle and back* and buffy-white throat with indistinct dark spotting in centre. *Side of head typically marked* with pale above and beneath eye and dusky cheeks with pale spot in centre. *Primaries edged white.* – Adult ♂: *Red on forehead and breast* (in autumn brownish-red and partly concealed by pale fringes). Uniform grey nape. – ♀/juvenile: No red in plumage; brown-grey nape and lightly streaked breast, and streaked crown. Juvenile somewhat more streaked and duller brown above than adult ♀.

VOICE Flight-call a dry but still slightly nasal, bouncing 'tigg-**itt**' or 'tett-ett-**ett**'; uttered with particular intensity on rising. Song a mixture of short, rattling syllables akin to call plus musical whistling notes. Sometimes gives drawn-out verses delivered at rather fast tempo, but more often the song is 'on low heat', and the verses are short and well spaced. Common elements in the song, besides the call, are e.g. 'piuu', 'trrrüh' and 'tuh-kii-**yüü**'.

Twite *Carduelis flavirostris*   r(m)**B**3 / **W**3–4

L 12½–14 cm. Breeds on treeless moors and bare coastal heaths (incl. crofting lands in Scotland and Ireland), many moving in Oct–Nov mainly to coasts of North Sea / S Baltic, returning in Mar–Apr. Often forages at refuse tips and on tilled fields, may then be seen in large, dense flocks of hundreds of birds. Restless and mobile, but not shy.

IDENTIFICATION Resembles a Linnet in size and shape, but on average has a trifle *longer and more deeply cleft tail*, smaller head and slightly bushier, 'looser' feathering. Told from 1st-winter Linnet by: *more heavily dark-streaked*, usually distinctly *buff-tinged plumage*; lack of dark brown (cinnamon) *above*; *no dark spotting on throat-centre*; *bolder pale wing-bar* (buff-white tips to greater coverts); ♂ in addition has *pink on rump*, most obvious in spring but usually visible also in fresh autumn plumage (absent on ♀). Bill *yellow* (with dark tip) in winter, brown-grey when breeding.

VOICE Call a rather short and hard 'yett', somewhere between flight-calls of Brambling, Redpoll and Linnet, and which can be rather hard to pick out in mixed finch flocks, as well as a characteristic *drawn-out and slightly rising, bleating* 'tveeiht' (hence 'Twite'!). Song more like Citril Finch's than Linnet's, fast trills, buzzing sounds and sequences of twitters, recognized by interwoven low rattling 'trrrrrrr' notes; traces of the bleating call often heard also in the song.

(Common) **Redpoll** *Carduelis flammea*   m(r)**B**2 / **W**2

L 11½–14 cm. Breeds in birch forest, young conifers, less often in willows on bare mountains, as well as in small deciduous copses in open terrain; in the Alps often in larch-dominated conifer forest above 1400 m. In Britain & Ireland partial migrant and winter visitor, very numerous in some years (invasions), often in alders in winter. Mobile and restless, dense winter flocks seldom stay in one spot for long, and even during breeding the birds roam widely, often high up. Clings nimbly, even upside-down, to tips of birch twigs. Nests in branch fork.

IDENTIFICATION Greyish and dark-streaked, with small *yellowish finch-type bill with dark culmen, pale wing-bars, black bib, black loral region* and dark forehead plus *red forecrown*. In good view, only confusion risk is Arctic Redpoll. *Rump streaked*, but note that some (esp. ad. ♂) have white (or light red) ground colour with very faint streaks and in field can be confused with Arctic. Only adult ♂ has much bright red on breast, otherwise sexes and ages are difficult to separate in field. – Variation: Birds in British Isles, C Europe and SW Scandinavia (LESSER REDPOLL, ssp. *cabaret*) are smaller and darker brownish than those in Fenno-Scandia (*flammea*). Winter visitors from Greenland (*rostrata*) are big, rather dark and heavily streaked, in fresh plumage brownish like *cabaret*.

VOICE Call, given often in flight, an almost echoing, hard and metallic, repeated 'chett-chett-chett'. When nervous, a somewhat Greenfinch-like but feebler and usually hoarser 'jüih' (used also as advertising-call). Song consists of the metallic call interspersed with dry, reeling 'serrrrrrr'; delivered mostly in undulating, wide-ranging song-flight.

Arctic Redpoll *Carduelis hornemanni*   **V**∗∗

L 12–14 cm. Breeds as a rule in willows and in low, open birch forest on or near tundra or upland heath, rarely also in tall, closed upland birch forest. Migratory habits unclear; some move southward with Redpoll flocks in winter months.

IDENTIFICATION Very like Redpoll, and in certain plumages impossible to separate reliably. *Bill small* and pointed, with *straight culmen*. Rather fluffy, 'loose' plumage. – Adult ♂: Large *unmarked white* (or faintly pink) *rump*. Only *few and narrow dark streaks on flanks*, contributing to pale overall impression. *Ground colour of mantle/back pale.* Forehead usually pale. *Breast light pink* (never bright red). – ♀: Lacks pink on breast and rump. Many have a small unspotted white rump patch, but some (probably imms.) have faintly streaked grey-white rump and are deceptively like Redpolls. Flank streaking varies, some being heavily streaked.

VOICE Very like Redpoll's and difficult to distinguish from latter's. The reeling trill is on average a little softer and has more cracked and almost 'lisping' quality, and the 'chett-chett-chett' call is also softer and higher in tone.

Linnet

Twite

Redpoll

Arctic Redpoll

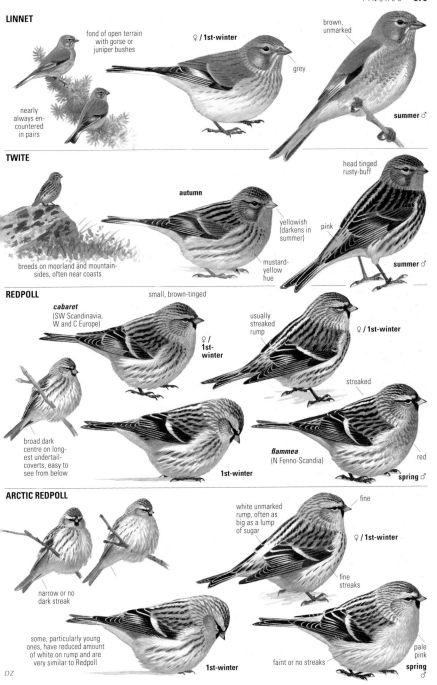

LINNET

fond of open terrain
with gorse or
juniper bushes

nearly always en-
countered in pairs

♀ / **1st-winter**

grey

brown,
unmarked

summer ♂

TWITE

breeds on moorland and mountainsides, often near coasts

autumn

yellowish
(darkens in
summer)

mustard-
yellow
hue

head tinged
rusty-buff

pink

summer ♂

REDPOLL

small, brown-tinged

cabaret
(SW Scandinavia,
W and C Europe)

♀ /
**1st-
winter**

usually
streaked
rump

♀ / **1st-winter**

streaked

broad dark
centre on long-
est undertail-
coverts, easy to
see from below

1st-winter

flammea
(N Fenno-Scandia)

red

spring ♂

ARCTIC REDPOLL

narrow or no
dark streak

some, particularly young
ones, have reduced amount
of white on rump and are
very similar to Redpoll

1st-winter

white unmarked
rump, often as
big as a lump
of sugar

fine

♀ / **1st-winter**

fine
streaks

faint or no streaks

pale
pink

spring
♂

DZ

(European) Goldfinch *Carduelis carduelis* m(r)**B**2 / **P**

L 12–13½ cm. Breeds in low-lying deciduous woodland, pine plantations and orchards. Mainly summer visitor, majority wintering in S and W Europe. Often seen in flocks after breeding. Frequently forages on seeding thistles, burdocks, etc.

IDENTIFICATION Unmistakable with its *red 'face'*, its otherwise *black and white head*, the *jet-black wing with broad yellow wing-bar* (visible also on perched bird) and prominent *white rump* and *black tail with white spots*. Sexes in practice alike in the field. Bill relatively long and very pointed, ivory-white. – Juvenile (up to Aug– Oct): Head grey-brown and diffusely streaked, without adult's red, white and black pattern.

VOICE Call a trisyllabic, skipping and cheery 'tick**elitt**'; conversational monosyllabic 'litt' or disyllabic 'te**litt**' sometimes heard from foraging flocks. Flocks also give rasping 'tschrre' notes. Song rather quiet, akin to Siskin's, consisting of rapid trills, mewing notes and twittering sequences but *always recognized by interwoven call note*.

(European) Greenfinch *Chloris chloris* r(m)**B**2 / **W**

L 14–16 cm. Breeds at woodland edge, in wooded pasture and copses, in bushy areas and in parks and gardens. Common also in villages and towns. Hardy, and many overwinter in N, but some migrate to W and SW Europe. Always wary, at times shy. Nests in tree, bush or trellis.

IDENTIFICATION Has a rather *stout body, head and bill*. Bill pale pink or ivory-coloured. Recognised by *yellow edges to primaries*, forming yellow panel on folded wing, by *greenish tone to underparts* in all plumages and, in flight, by *yellow at base of tail-sides* as well as on primaries. – ♂: *Yellowish-green breast*, grey-green upperparts, greyish head-side and *light grey wing-panel*. Much yellow on primaries and tail. – ♀: Duller grey colours, mantle / back tinged brown, less yellow on wings and tail. – Juvenile: Diffusely streaked.

VOICE Call a short, forceful 'jüpp', usually repeated in fast series and becoming short trills, 'jüp-üp-üp'. Also a loud, upcurled 'ju**it**'. Song of two different types, one an unmusical, wheezy 'dschrüüüüuh' which is repeated with long pauses, the other a pleasing and drawn-out Canary-type song consisting of trills and fast runs of whistles and twitters, a bit like Tree Pipit's song, e.g. 'jüpp-jüpp-jüpp jürrrrrrrrr tuy-tuy-tuy-tuy-tuy ju**it** chipp-chipp-chipp-chipp-chipp dürdürdürdür jürrrrrrrrr...'; sometimes the wheezing song type is woven into the musical song.

(Eurasian) Siskin *Carduelis spinus* r(m)**B**2 / **P**+**W**2

L 11–12½ cm. Breeds in coniferous and mixed forest, mainly in spruce; prefers alder and birch nearby for food. In Britain & Ireland, winter influxes from N and E Europe, sometimes on huge scale. Often large flocks in winter, when

also visits garden nut-baskets. Mobile, roams far and wide, even when breeding. Not that shy. Nests high up in spruce.

IDENTIFICATION Small and neat, with rather small head and short tail. Told by *dark wing with contrasting yellow or yellowish-white markings* and by *yellow bases to tail-sides* (Greenfinch style); in addition has *heavily streaked lower flanks*. – ♂ (from Sep): *Crown and bib black* (partly concealed by light grey fringes when fresh); eyebrow, *breast and rump unstreaked greenish-yellow*. – ♀: Crown grey-green, breast and rump white with yellow-green tinge and diffusely streaked. – Juvenile: Brownish mantle / back; head palish, streaked dark.

VOICE Two similar disyllabic whistled calls with characteristic ring, one descending and the other rising, 'tilu' and 'tluih'; the two are freely mixed together by overhead parties. A stifled, rattling call, 'tetete', is also heard. Song a flowing series of twittering and trilling notes with traces of mimicry (some are masterly mimics), now and then broken by a *drawn-out choking or wheezing note*.

Citril Finch *Carduelis citrinella* —

L 11½–12½ cm. Breeds in mountain forests at *c.* 700 m up to treeline, usually in spruce stands adjoining alpine meadows and clearings. Mostly arboreal, but often forages on ground. Primarily resident.

IDENTIFICATION The size of Siskin, with rather *short, pointed bill*. *Two yellowish-green wing-bars*, but otherwise no pale markings on wings and tail in flight. – ♂: 'Face', *underparts and rump yellowish-green*; *napel side of throat ashgrey*; *back nearly unstreaked olive-grey*. – ♀: Duller and greyer; less yellow-green in plumage, especially on 'face'; grey of neck-side runs in a band across breast.

VOICE Call a monosyllabic short 'teh', e.g. loosely repeated in flight, or fast, bouncing 'te-te-te'. Begging- and anxiety-calls sharp 'ziit' and piping 'tüht'. Song somewhat recalls Goldfinch, often short with pauses, contains a lot of Redpoll-like *buzzing trills* and almost Chaffinch-like passages.

Corsican Finch *Carduelis corsicana* —

L 11½–12½ cm. Closely related to Citril Finch, from which it was recently separated. Breeds on Corsica and Sardinia at sea level to *c.* 1650 m on slopes with low-growing tree-heath and scrub, sometimes also in tree copses. Resident.

IDENTIFICATION Very similar to Citril Finch, differing on *dark-streaked brown back* and slightly *clearer yellow underparts and 'face'* being less green-tinged than in Citril. Uppertail-coverts greyish, not olive-green.

VOICE Calls very similar to Citril Finch, but flight call perhaps slightly feebler and almost 'trembling' in tone. Song possibly on average longer and more structured, and has therefore been likened to that of Wren.

Goldfinch

Greenfinch

Siskin

Citril Finch

GOLDFINCH

grey-white, finely streaked

ad.

black wings with broad yellow bar common to all plumages

red 'face'

very long and pointed

white

ad.

juv.

ad.

ad.

GREENFINCH

♀ ♂

ad. ♂

much yellow

much yellow

display-flight with slow-motion wingbeats and high wing action

stout, conical

brown cast, faintly streaked

unmarked moss-green

grey-white, streaked

♀

yel-low-green

juv.

less yellow than on ♂

greyish, faintly tinged green

ad. summer ♂

SISKIN

climbs nimbly on the thinnest twigs of alder and birch

short, notched

broad yellowish wing-bar in all plumages

ad. ♂

black, in autumn fringed grey

long, pointed

streaked

♀

bright yellow rump and tail-sides

no other small finch has a distinct pale patch at base of primaries

juv.

ad. ♂

CITRIL FINCH

ad. ♂

finely streaked

greenish-yellow

ad. ♂

CORSICAN FINCH

ad. ♂

ad. ♀

virtually unstreaked

dark

♀ duller than ♂ and has grey across breast

comparatively long tail giving longer, more slender shape than Siskin

grey across breast

greyish-green, faintly streaked

warm brown, distinctly streaked

♀

greenish

yellowish-green

greyish

bright yellow

diffusely streaked

juv.

ad. ♂

ad. ♂

KM

(European) **Serin** *Serinus serinus*   **P**5 / (m**B**)

L 11–12 cm. Breeds in tree clumps and woodland edge, also in gardens, parks, churchyards, orchards etc., preferably with some conifers (often silver fir, thuja, cypress etc.). Rare on passage in Britain (mostly May, Sep–Oct), a few breeding (irregularly) in S. Bold. Makes frequent aerial excursions. Nests rather high in thick conifer or in citrus grove.

IDENTIFICATION Small, with proportionately *large head* and *tiny little bill*. Important features are its small size and its *energetic, skipping flight* and often *restless* behaviour. Mantle, back and *flanks* always heavily *streaked*. On side of head *long pale supercilium* extends down onto *pale neck-side*, framing darker *cheek with pale central spot* (Linnet style, but more strongly marked). *Rump pale*, bright yellow and contrasting on ♂, duller greenish-yellow (sometimes not very obvious) on ♀. ♂ has *forehead, facial markings, neck-side and breast bright lemon-yellow*, these areas being paler yellowish-white on ♀. – Juvenile: All areas which are yellow on adult are buffish-white.

VOICE Call is a buzzing or bouncing trill, 'zir'r'r'rl', with typically high and silvery, clear voice. Alarm an upcurled 'tü-ih', much as Redpoll's but more clearly disyllabic. Song a *frantically fast* and almost strained stream of squeaky, sharp and jingling notes (often likened to crushing glass, and call-note voice recognizable) at *even pitch*, delivered from treetop or in song-flight with stiff 'slow-motion' wingbeats.

(Atlantic) **Canary** *Serinus canaria*   —

L 12½–13½ cm. Original form of the domestic Canary. Breeds in Madeira and Canary Islands (though not on Fuerteventura and Lanzarote) in orchards, copses, shrubbery, hedges etc. from sea-level to *c.* 1500 m. Resident. Social habits. Mobile, flies a lot. Nests in bush or thick tree.

IDENTIFICATION Compared with Serin, is a *bigger, more elongate* bird with *longer bill and tail* and shorter wing (only moderate primary projection). Has some grey and pale brown in the plumage and has *more diffuse, more washed-out* pattern than Serin, and lacks Serin's distinctly contrasting dark crown against pale forehead. – ♂: Greenish-yellow facial markings, nape and neck markings, throat, breast-centre and rump; head-side and crown diffusely grey; lower flanks light yellow, streaked. – ♀: Duller, less yellow; pale rump indistinct. – Juvenile: Lacks yellow.

VOICE A rather characteristic call is a slightly descending, tremulous 'ti-ti-türr' with cracked, 'lilting' tone. Also an upslurred, Greenfinch-like 'juit', a somewhat Siskin-like falling 'tiüh', and a silvery 'tvi-vi-vi-vi' rather like Serin's. Song is most like that of Goldfinch and has same tempo; contains nasal, discordant notes, rapid twitters, muffled trills and many high notes.

Syrian Serin *Serinus syriacus*   —

L 12–13 cm. Breeds on lower mountain slopes and in hilly high-lying terrain with scattered trees and bushes, sometimes also in orchards. Some remain in breeding area but often move to lower levels, while others migrate to S Israel and Sinai. Rather shy. Nests in tree.

IDENTIFICATION Only a trifle bigger than Serin, but is a bit *more elongate and long-tailed*, with proportionately slightly smaller head. *Bill small and short*, as Serin's. Plumage somewhat reminiscent of Citril Finch, rather *uniform and pale* with lightly streaked olive-grey mantle, *unstreaked yellowish-green rump* and pale yellow-tinged grey-white underparts with *no or just minimal streaking*. *Wings* distinctly *greenish* in both sexes. – ♂: Bright yellow forehead, eye-surround and throat; grey-toned crown, nape and cheeks; *lacks Serin's dark crown and yellow supercilium and neck-side framing darker cheek*. – ♀: Duller, less yellow, somewhat more olive-grey and streaked. – Juvenile: All areas which are grey on adult are buffish-white.

VOICE Call a dry, slightly nasal, and rolling or tremulous 'pe-re-ret' or 'pü-tü', closest to Citril Finch's but lower in pitch. An almost Greenfinch-like, forceful 'chüpp', at times repeated, and a higher, faster variant, 'chip-ip-ip-ip', are often given, too; both these are heard in the song as well. Song most like Red-fronted Serin's and Citril Finch's, squeaky and twittering notes at fast pace, sometimes with strained buzzing sounds admixed. Verses usually fairly short but delivered in rapid succession with only brief pauses.

Red-fronted Serin *Serinus pusillus*   —

L 11½–12½ cm. Breeds in mountain areas near or just below treeline in coniferous and mixed forest, often along watercourses in high-lying valleys with sparse, low vegetation or by alpine meadows, with some juniper and rhododendron. Moves to somewhat lower levels in winter. Social. Generally not shy. Nests in low bush or dense tree.

IDENTIFICATION *Small* as Serin (but marginally longer tail), and makes similar *restless* aerial excursions in *bounding flight*. Unmistakable in adult plumage, having *black head* with dazzling *red patch on forecrown*. Plumage otherwise is *densely and heavily black-streaked*, and overall appearance is dark. *Rump orange-yellow*, diffusely streaked. Sexes often alike, but some ♀♀ have black on head less solid and extensive (are dark-streaked sooty-grey on hindcrown, behind eye and on breast), and red crown patch is often slightly smaller. – Juvenile (up to Oct/Nov, occasionally still in Mar): Head largely uniform cinnamon-brown.

VOICE Call a simmering, twittering trill, 'tvir'r'r', cheery and 'sweet' in tone, resembling both Citril Finch's and Syrian Serin's; the trill (which is invariably long like Serin's, not short like Citril Finch's, which often also utter monosyllabic calls) drops very slightly at end. Song a long, twittering and frenzied series, like a fast Goldfinch song, and also like Citril Finch's: trills and rapid whistles mixed with short wheezing sounds.

Serin   Syrian Serin   Red-fronted Serin

SERIN

ad. ♂

light yellow

pale patch

variation

ad. ♀

SISKIN, ♀

bright yellow

short and chubby, grey

boldly streaked

ad. ♀
(dull bird)

lacks any yellow

juv.

boldly streaked on white ground

ad. ♂

yellow

dark sides

narrow, short wing-bar

ad. ♀

CANARY

streaked

moderately streaked

grey

stout, rounded shape; pinkish

dull greenish-yellow

ad. ♀

juv.

long

yellow-green

ad. ♂

dark sides

ad. ♀

SYRIAN SERIN

much yellowish-green on wing

faintly streaked

deep yellow

pale green-ish-yellow

ad. ♀

yellowish sides

rather long

unmarked grey

ad. ♂

ad. ♀

RED-FRONTED SERIN

cinnamon

deep red

short and chubby, grey-black

sexes usually alike, but some ♀♀ have less black on head, neck and breast, and have streaked rump

buffish-yellow

juv.

ad.

flock feeding on mountainside

ad.

DZ

(Eurasian) **Bullfinch** *Pyrrhula pyrrhula* ⎡**B**2

L 15½–17½ cm. Breeds in mixed woods, in parks and larger gardens, in churchyards etc. with some conifers, sometimes (esp. in Fenno-Scandia) also in coniferous forest. Mostly resident, but some N European breeders migrate (mainly end Oct) to S Scandinavia and C Europe. Not really shy, but unobtrusive and easily overlooked in summer. Quiet, almost sluggish in behaviour. Often seen in pairs or in small, rather loose flocks. Food various kinds of seeds and shoots of fruit trees; also some insects in summer. Nests in bush or tree, often on sheltered branch.

IDENTIFICATION A rather big and very *compact* finch which looks bull-necked or *neckless*, has *plump body*, quite big head, and short but deep, *'podgy' bill*. Bearing in mind its fairly sedentary lifestyle, has surprisingly long wings and long tail, and flight is fast and in long undulations; at range, still recognized by plump body and broad neck. In rear view, identified in flight by *white rump patch* contrasting with *black tail* and *grey back*. Also has a *broad, off-white wing-bar*. Adult has black crown and 'face', and sexes best separated by underpart colour, ♂ *bright pink-ish-red*, ♀ greyish-buff. ♂ also has pure ash-grey mantle and back, whereas ♀ has the grey tinged brownish. – Juvenile (up to Sep–Oct): Head entirely grey-brown, and sexes are alike (being ♀-coloured below).

VOICE Call a short whistle or fluted note, low-pitched, discreet, usually with melancholy ring, 'phü', locally in Europe with distinctly falling pitch, 'phü-ew'. A variation with toy trumpet tone could have its origin in N Russia. Conversational call, sometimes when flushed, a stifled, repeated 'bütt'. Song soft, slowly recited, halting and tentative, a mixture of call-like, low-pitched fluted notes and choking squeaky and scraping notes; often double-noted 'phü-phü' inserted, or the fluted notes are drawn out into melancholy descending notes, 'pyüüüh'. Some notes are so muffled that they are audible only at very closest range, making the song sound even more discontinuous and hesitant at distance.

Pine Grosbeak *Pinicola enucleator* **V**∗∗∗

L 19–22 cm. Breeds in taiga, usually in mature, undisturbed coniferous forest with some birch and berry-bearing shrubs. Probably mainly resident, but in some autumns mass southward eruptions occur (usually end Oct and in Nov). Unobtrusive and retiring during breeding, but in winter quite fearless (old Swedish colloquial name 'silly fool'), and flocks may then readily feed on rowan berries in gardens and roadside trees in town centres. Food seeds, buds and shoots of e.g. spruce and birch, also berries of rowan, bilberry, cowberry etc. In dense forest sometimes forages quietly down in shrub layer. Nests in tree against trunk, usually of spruce.

IDENTIFICATION *Thrush-sized*, hefty body and thickset neckless silhouette; *long-tailed*. Bill short and deep with curved culmen. In all plumages, *double white wing-bar* (greater and median coverts have white tips and narrow white outer edges) and *white outer edges to tertials* contrasting with otherwise greyish-black wings. Flight powerful and fast, in deep undulations. – Adult ♂: Greater part of plumage *raspberry-red* (with touch of light grey); mantle, back and uppertail-coverts spotted dark. – ♀/1st-year ♂: Red colour of adult ♂ replaced by *greyish yellow-orange*.

VOICE Call in flight clear, loud, fluting 'plüit' and 'ju-jü', or especially when alert before flushing, multisyllabic 'plüijüh-püllidijüh' and the like. Flocks converse with low, fast 'bütt-bütt-bütt...'. Song very pleasing, with desolate ring well tuned to the deserted breeding forests, a very fast, rapidly repeated, high and silver-clear short verse (2–3 sec.) which yodels or shuttles irresolutely on a couple of notes; the verse is often a shade more stressed in middle than at start and end, may perhaps recall Wood Sandpiper display; can sound like 'pilidü-pilüdi-pilidü-pilidipü-pilüpipi' or similar.

Hawfinch *Coccothraustes coccothraustes* ⎡**B**

L 16½–18 cm. Breeds in deciduous and mixed woods, preferring mature lofty deciduous with plenty of oak, hornbeam, beech, ash and elm. Also attracted to fruit trees, especially cherries, kernels of which it cracks with its powerful bill (can generate over 50 kg force!). Diet also includes insects. Very wary and shy and difficult to observe, spends most time up in canopy or seen flying fast high up between woodland-edge trees. Usually nests well up in deciduous tree against trunk or in fork, in fairly exposed site.

IDENTIFICATION A rather big finch with totally distinctive proportions: has *very powerful, triangular bill*, *big head* and thick neck, but *short tail*. (Size of head due to powerful jaw muscles; the Hawfinch is a flying pair of nut-crackers!) Dominant colours are *rusty-brown and buff* with black, white and grey embellishments. Bill in summer greyish-black with blue-grey base, in winter ivory-white or pale yellow-brown. Wings glossy blue-black, inner primaries with club-shaped extensions. Sexes similar, but separable by ♂ having all-black remiges, while ♀ has an ash-grey panel on secondaries. In flight, shows conspicuous *broad white wing-bar on 'hand'*, and the short *tail is white-tipped*. – Juvenile: Breast greyish-yellow, belly coarsely spotted dark.

VOICE Has a very hard and sharp clicking 'pix!' with an almost electric quality, or like the sound made by jabbing a spike into solid granite; with a bit of practice easy to recognize and distinguish from e.g. Robin's ticking 'tic', often repeated at slow pace in undulating flight (one 'pix!' on each rise). Besides this call, has more anonymous 'zrri' and 'zih' which are easily drowned in the varied sounds from the woodland's Chaffinches, Spotted Flycatchers, thrushes and others. Song a rather quiet, stumbling series of 'zih' and 'zrri' notes, rather hard to make out.

Bullfinch   Pine Grosbeak   Hawfinch

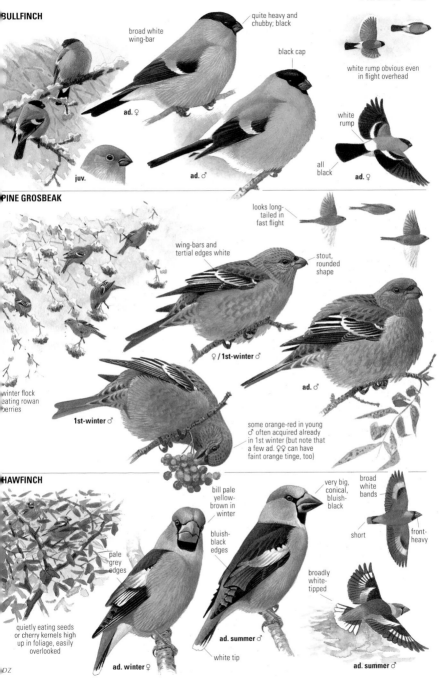

BULLFINCH

broad white wing-bar

quite heavy and chubby; black

black cap

white rump obvious even in flight overhead

white rump

ad. ♀

all black

ad. ♀

juv.

ad. ♂

PINE GROSBEAK

looks long-tailed in fast flight

wing-bars and tertial edges white

stout, rounded shape

♀ / 1st-winter ♂

ad. ♂

winter flock eating rowan berries

1st-winter ♂

some orange-red in young ♂ often acquired already in 1st winter (but note that a few ad. ♀♀ can have faint orange tinge, too)

HAWFINCH

bill pale yellow-brown in winter

very big, conical, bluish-black

broad white bands

bluish-black edges

short

front-heavy

pale grey edges

broadly white-tipped

quietly eating seeds or cherry kernels high up in foliage, easily overlooked

ad. winter ♀

ad. summer ♂

white tip

ad. summer ♂

DZ

CROSSBILLS *Loxia*

Chunky finches with large head and bull-neck, long wings and deeply cleft tail, plus heavy bill with crossed tips, an adaptation for prising open cones and extracting seeds. Often breed early in year (Feb–Mar), but breeding season protracted. Generally resident in northern coniferous forests, but periodic food shortages trigger mass eruptions towards S and W. Often silent when feeding (easily overlooked; watch for falling cones and seed wings discarded by the birds), when they clamber about nimbly like small parrots. High-flying family parties attract attention by their repeated loud, metallic calls. Flight undulating.

Common Crossbill *Loxia curvirostra*   rB(3–)4 / (**W**)
L 15–17 cm. Breeds in conifers, including smallish clumps, prefers spruce. Commonest and most widely distributed crossbill, but numbers fluctuate regionally. Nests high up.
IDENTIFICATION Very similar to Parrot Crossbill (which see), but is a trifle smaller, with proportionately somewhat *smaller head, slimmer neck* and *smaller bill*. Note that *bill is longer than it is deep*, with culmen only moderately curved. *Lower mandible less deep than upper and lower edge lacks obvious S-shape with bulging centre.* – Post-juvenile ♂: Head, underparts and rump brick-red, on some with variable amount of yellow-orange and greyish-green (not strictly age-related). – Post-juvenile ♀: Grey-green or dull yellowish-green, unstreaked below, rump often pale yellow. Back and head-sides brownish, wings and tail dark brown-grey. – Juvenile: Grey, with boldly streaked underparts. Some streaked feathers can remain through 1st winter. (For rare variant with narrow pale wing-bars, see under Two-barred Crossbill.)
VOICE Calls frequently. Call a loud, metallic, fairly high 'glipp', usually repeated in series, 'glipp-glipp-glipp-...'; generally sounds 'clipping', as if an 'l' is inserted after rather soft initial consonant, but some have a harder 'kipp'. A certain variation in pitch and volume occurs, dependent on age, individual, population or mood, but pitch is always higher than in Parrot Crossbill. Song resembles Parrot Crossbill's, but is recognized by interwoven call-notes.

Parrot Crossbill *Loxia pytyopsittacus*   **V**∗∗ / (m**B**5)
L 16–18 cm. Breeds in N and E Europe in pine forests or in conifer woods with plenty of pines; seen also in spruce during irruptions. Not so common as Common Crossbill, but quite numerous in some invasion years; exact status poorly known. Has bred in Britain. Nests high up in pine.
IDENTIFICATION A shade bigger than Common Crossbill and with proportionately somewhat bigger head and bill and thicker neck. Plumages the same as for Common (which see). Best identified by *larger, thicker bill*, roughly *as deep as*

it is long and with *lower mandible almost as deep as upper* culmen more strongly curved, so bill-tip looks blunter; lower edge of lower mandible S-shaped, with *parrot-like bulge in middle*. Tip of lower mandible usally not visible beyond culmen in profile.
VOICE Calls frequently. Call very like Common Crossbill's but on average *deeper and harder*, more echoing, often less 'clipping' (cf. Common Crossbill) and with harder initial consonant, 'tüpp-tüpp-tüpp-...'. Difference can be detected by the trained ear, and is easiest way (outside Britain) to identify the species despite some individual variation in voice. Song a mix of strained trills and soft twitters with call-like notes (latter distinguishing it); some notes are repeated rapidly a few times.

Scottish Crossbill *Loxia scotica*   rB∕
L 15½–17 cm. Breeds in N Scotland in ancient pine forest Resident; any movements poorly known. Nests in pine.
IDENTIFICATION Plumage as Common and Parrot Crossbills; size, structure, bill size and voice intermediate between those two. Best identified by distribution and often discernibly heavier bill than Common Crossbill (though some overlap). Within this species' range in Scotland, Parrot Crossbill probably not safely distinguishable in the field.
VOICE Call intermediate between those of Common Crossbill and Parrot Crossbill.

Two-barred Crossbill *Loxia leucoptera*   **V**∗∗∗
L 14½–16 cm. Probably breeds regularly only in Russia; in some years nests in Finland and Sweden. Prefers larch cones often takes rowan berries in winter.
IDENTIFICATION Plumages similar to those of other crossbills, but adult ♂ often brighter, raspberry-red, and scapular and tail-coverts have darker centres. A bit smaller than Common Crossbill (though some overlap), and bill not quite as heavy. Best identified by voice and by *broad white wing-bars and clear-cut broad white tips to tertials*. – Adult: Very broad white wing-bars and tertial tips make it unmistakable. – Juvenile Underparts greyish, boldly streaked; wing-bars and tertial tips considerably narrower than adult. (Some Common Crossbills have pale/white wing-bars and tertial tips, but usually one or both wing-bars are incomplete, and tertials generally have just narrow pale fringes at tip, not broad, clear-cut and white tips.) – 1st-year: Some are identifiable by retained juvenile wing-coverts and tertials with little white.
VOICE Call weaker and higher than Common Crossbill's and not so metallic and echoing, 'chip-chip-chip-...'. Also has a diagnostic nasal, slightly discordant piping as if from a muted toy trumpet, 'tviiht' (contact-, alarm- and, less often, flight call). Song fast, varied and twittering, Siskin-like.

Common Crossbill   Parrot Crossbill

Scottish Crossbill   Two-barred Crossbill

COMMON CROSSBILL

culmen moderately curved, lower mandible not so deep

grey-green, diffusely streaked

yellowish-green

ad. ♀

ad. ♂

grey-brown, boldly streaked

spruce-cone specialist

juv.

imm. ♂♂ are often incompletely red, but ad. ♂♂, too, can have patches of grey-green

♂

PARROT CROSSBILL

pine-cone specialist

culmen strongly decurved in parrot fashion, lower mandible very deep and bulging down in the centre

plumages just the same as for Common Crossbill

ad. ♀

all measurements intermediate between Parrot and Common

SCOTTISH CROSSBILL

deep

strong head, 'bull-necked'

ad. ♂

TWO-BARRED CROSSBILL

larch-cone specialist

Common Crossbills very rarely have white wing-bars, too, but these bars are narrower, and tertial tips are evenly fringed white, do not have more white at tips

COMMON CROSSBILL, ♀ wing-barred variant to compare

boldly spotted black

broad white wing-bars, the lower widest towards back

ad. ♀

juv.

ad. ♂

more bright red than brick-red

DZ

Common Rosefinch *Carpodacus erythrinus* **V*** / (m**B**5)

L 13½–15 cm. Breeds in deciduous scrub and lush bushy areas, often on soggy overgrown lakeshores, in hazel stands, parks and smaller copses, along watercourses and field margins, and also bushy clearings in deciduous woods. Summer visitor (mostly end May–Aug), winters in India. A comparatively recent colonist from E. Has bred in Britain. Relatively bold, ♂ often sings from exposed perch; species is otherwise mainly quiet and retiring, easily eludes detection. Food mostly seeds, shoots and buds, some insects. Nests low down in bush or thicker small tree.

IDENTIFICATION Barely Bullfinch-sized, somewhat front-heavy finch with rather heavy breast and head/neck, but slim rear end and long tail. *Bill hefty*. Full adult ♂ distinctive with its red colour and its song, but ♀ and majority of younger ♂♂ are more nondescript and harder to identify (and in fleeting view can be confused with Corn Bunting, ♀ House Sparrow, pale young ♀ Yellowhammer etc.). Note: bill shape; *'colourless' brown-grey* plumage with little contrast between upperparts and underparts; rather *uniform brown-grey head*; diffuse dark mottling on crown, back and underparts; *two narrow and rather indistinct pale wing-bars*; pale tertial edges. No white on tail. – Adult ♂: *Bright crimson-red* on breast, crown, cheeks and rump; pale wing-bars often pinkish. – 1st-summer ♂: Majority are ♀-like, brownish-grey with no trace of red (and normally inseparable from ♀♀). Some acquire red feathers, however, and a few are adult-like; usually still identifiable by greyish-white wing-bars, scattered brown-grey feathers among the red, and sometimes by some retained worn wing- and tail-feathers. – ♀: All brownish-grey; wing-bars narrow and greyish-white. – Juvenile: Like ♀, but has fresher plumage with slightly warmer olive-brown tone and heavier dark spotting, and has buffish wing-bars.

VOICE Call an upslurred whistle, 'vüih', at times a bit hoarse and rather Greenfinch-like but otherwise with slightly softer voice (cf. song); sometimes heard also from overhead migrants. Song a typically soft whistling consisting of a few rhythmic notes in glissando, e.g. 'vid**ye**-vüy-viid**ya**' or 'vid**ye**-vüd**ya**-vüy'. Birds can be called up by high wolf-whistles. When excited, e.g. when courting ♀, ♂ also gives quiet, ecstatically chattering, twittering notes between the loud song verses.

Sinai Rosefinch *Carpodacus synoicus* —

L 13–14½ cm. Breeds in mountain deserts in wadis and ravines, locally on ruins. Requires water within reasonable flying distance, otherwise lives in very barren, rocky surroundings. Resident. Mobile, and rather difficult to get close to (but locally less shy). Visits drinking spots in morning (c.

09.00 hrs) and afternoon, but on hot days also several times in the middle of the day. Sometimes seen in small flocks (often family parties). Food mostly seeds. Nests in rock crevice, often high on cliff face, occasionally near ground.

IDENTIFICATION Size and shape much as Common Rosefinch, but *bill is somewhat smaller* and *culmen slightly less curved*. In all plumages, *pale* and without prominent markings, *almost totally lacks dark streaking above* and has *no pale wing-bars*. – Adult ♂: *Red 'face'*, paler *pink underparts, head-side and rump*; brown hindcrown, nape and mantle have tinge of pink, too. *Forecrown and crown-side have silvery-white feather tips*, with slightly fainter silvery-white also on rear cheeks. – ♀/1st-summer ♂: Almost uniform light grey-brown, like dry mud, the head marginally browner; wing-covert fringes a shade lighter than rest of wing. (Can be taken for a young Trumpeter Finch, bearing in mind colour and pale, heavy bill.) Occasional younger ♂♂ (maybe adult ♀♀, too) can have a little pink on 'face' and on underparts.

VOICE Flight-call on rising a surprisingly sparrow-like 'chip... chip...' (or 'chiepp'), with hard ending. Voice is thick (i.e. the sound has overtones), making the call at times, e.g. in an echoing wadi, reminiscent of Bearded Tit. Perched birds give a rather rattling, piping, somewhat House Sparrow-like chatter, 'tviht, cheht tvüt, tviht-tviht...'. Song (imperfectly known?) is sometimes stated to be short, clear and melodic, but according to other sources it consists of a simple verse of rather sparrow-like, chirping notes (cf. above!).

Great Rosefinch *Carpodacus rubicilla* —

L 19–20½ cm. Alpine species breeding on bare mountains at above 2500 m (usually above 3000 m), preferably close to glaciers, on sunny, boulder-strewn slopes with grass and supply of seeds of dandelions. Resident, but in winter descends to c. 1000 m or lower depending on snow situation; then seen in small flocks, and rather mobile. As a rule shy, difficult to approach. Late nester, often not incubating until end Jul. Nests in rock crevice or under boulder.

IDENTIFICATION *Big* as Pine Grosbeak, and similarly stocky. *Bill heavy*. Wings and tail rather long. Flight deeply undulating over longer stretches. Plumage quite *dark*, which together with size and manner of flight can at distance give thrush-like impression. – Adult ♂: *Dark raspberry-red sprinkled with fine white spots* on crown and underparts. Rather *swarthy on lores and around eye*, giving 'stern' facial expression. Rear cheeks have pinkish-white lustre. *Bill rather pale, yellowish-grey*. – ♀/1st-summer ♂: Entirely brownish-grey (no red) and heavily streaked dark; lores, eye region and forecheeks fairly dark. Bill grey-brown.

VOICE Call a high, cheerful 'tvui', often uttered in undulating flight. Alarm a discordant, thick, sparrow-like 'chick'. Song is a plain, clear-voiced verse at slow pace and on somewhat descending pitch, e.g. 'chüih, chü-chü-chü tvutvutvu-tsü'; some verses are longer, and the delivery varies a bit.

Common Rosefinch   Sinai Rosefinch   Great Rosefinch

COMMON ROSEFINCH

ad. ♂

two buff-white wing-bars obvious

♀-type plumage is non-descript—still, chubby bill and dark beady eye typical

juv. is somewhat richer olive-brown than ad. ♀

juv.

short and stout, rounded shape

red head and breast

red

♀ / 1st-summer ♂

ad. ♂

SINAI ROSEFINCH

slightly browner head

pale grey-brown, lacking features

silvery-white crown

one ♂ (left) and two ♀♀ in typical desert habitat

ad.♂

at certain angles the red looks more pink

ad. ♀ / 1st-summer ♂

ad. ♂

confined to mountain deserts

GREAT ROSEFINCH

rather distinctly streaked

big and long-tailed, flight deeply undulating

heavy, pale

rasp-berry-red, finely spotted white

Pine Grosbeak size

ad. ♀

long

ad. ♂

high-alpine species

DZ

Trumpeter Finch *Bucanetes githagineus*   **V*****

L 11½–13 cm. Breeds in rocky and stony desert and semi-desert at lower or moderately high levels; also thrives in barren, unvegetated terrain so long as water exists within flying distance. Rather common in Canary Islands. Resident, but mobile. Rather bold. Social in habits, but flocks usually quite small. Nests on ground in crevice or under vegetation.

IDENTIFICATION *Small*, long-winged finch with rather big head and *short but very thick and stout bill. Narrow pale eye-ring.* Contrastingly *dark* (almost black) *primary tips and tail-feathers* with *narrow pale edges.* – ♂ summer: Distinctive *pale red bill* and *pinkish legs*, vivid *pink tail-sides* at base, pink-tinged rump, belly and lower flanks, *pink edges to flight-feathers and greater coverts* and unstreaked warm brown mantle and back, otherwise plumage is *uniformly light grey* (partly suffused pink). – ♀/♂ winter: Bill pale greyish yellow-brown, often with faint pink tinge; legs light ochraceous-pink. Plumage colours somewhat more subdued, with less pink and pure grey, more buff and brownish-grey tones. – Juvenile: Lacks pink.

VOICE Call short, jolting, nasal 'ahp' notes with lilting or discordant tone (as in the song, though lower-pitched), often repeated; volume and pitch vary somewhat. Song consists mainly of a peculiar nasal and drawn-out buzzing note, a loud hoarse buzz with even and mechanical tone, 'aaaaaahp', like the sound from a small toy trumpet; sometimes two pitches audible. Often the song begins or ends with a few short, clear notes, e.g. 'chu chu zi aaaaaahp'.

Mongolian Finch *Bucanetes mongolicus*   **—**

L 11½–13 cm. Breeds in barren, rocky mountain tracts at over 2000 m. Resident, or makes only short movements down to lower levels after breeding. Seen on ground or in flight, often in small parties. Not shy. Nests on ground.

IDENTIFICATION Size of Trumpeter Finch, and very like that species, but has slightly *smaller bill* (esp. less deep base) which is always *yellowish-brown* or light grey-brown, never pink or light red. Further, *head is light brown like mantle* (never contrastingly grey), *crown to back* in all plumages *lightly but distinctly dark-streaked* (as good as uniform on Trumpeter), and wing pattern is different, this especially noticeable on adult ♂ (but harder to see on some ♀♀). Thus, in flight, shows a *short, broad, white bar on secondaries* and a *hint of second wing-bar at base of greater coverts.* Secondaries are also more broadly tipped white than on Trumpeter. Usually shows *narrow white patch on 'arm'* on folded wing.

VOICE In flight, often utters a rising note immediately preceded by a short grace-note, 'tu-vüit', sometimes also shorter 'tvüi' or 'tui'. Song a pleasing, melodic, short verse, e.g. 'viit-vüah…vreyah', repeated at slow pace.

Crimson-winged Finch *Rhodopechys sanguineus*   **—**

L 13–15 cm. Alpine species which breeds in bare mountain tracts at 1700–3300 m (locally 1100 m), but in winter often found lower down. Often inhabits steep precipices, ravines and boulder-strewn slopes practically devoid of vegetation. Social habits, small parties may be seen throughout year. Mobile, often forages and drinks far from nest site. Rather shy for the most part. Nests in rock crevice or cavity.

IDENTIFICATION *Big head*, *thick neck* and *heavy, triangular bill, shortish tail* but *long wings.* Best plumage features are: *extensive pinkish-white bases to secondaries and primary-coverts* forming striking pale wing-panel, visible also on folded wing; *black crown* (in fresh plumage partly concealed by light brown fringes); throat by and large concolorous with brown on head-sides; *breast, flanks and back dark rufous*, last two *spotted dark.* Sexes similar, but ♀ duller. – Variation: Ssp. *alienus* (Atlas mountains, NW Africa) is subtly larger, has less contrast below between rufous breast/flanks and buff-white belly, and back and flanks lack dark spotting. Rump darker rufous, not so clear pink, and *throat often light*, forming a pinkish-buff bib. Also, nape is more greyish.

VOICE Has a variety of calls: sonorous 'chivli' and 'chuchelütt' (at distance recalls Woodlark), slightly shrill 'yuvü-yuvü', and higher 'picha'. Song, often delivered from a rock pinnacle but at times in deeply undulating song-flight, a rapid, quite short, rhythmic verse repeated fairly constantly throughout and recognizable by its rhythm and its sonorous, musical tone; sometimes a drawn-out and stressed note is heard towards the end of the verse, e.g. 'chodlü-chodlü-udle-friiiih-chodlü'. A shorter verse occasionally given in flight.

Desert Finch *Rhodospiza obsoleta*   **—**

L 13–14 cm. Breeds in open, dry, usually flat country, often by irrigated cultivations, at oases etc. (thus, despite name, not a real desert bird!). Largely resident. Often appears in flocks. Unlike Trumpeter Finch, readily perches in trees; food taken mostly on ground. Nests in fork of bush or tree.

IDENTIFICATION A *sandy-coloured* or grey-buff finch with characteristic wing and tail markings: *black, finely white-edged primary tips*, and on closed wing a *large white and pink panel* broken only by black alula and black primary-covert tips. – ♂: Tertials black, narrowly edged white. *Bill black. Lores black* on adult, pale on 1st-winter. – ♀: Tertial centres, bill and lores grey-brown.

VOICE Three often heard calls: a simmering 'drrr'r' falling at end, a bent 'tvoi' (or 'tvio') somewhat recalling Common Rosefinch, and finally a lilting 'dveüüt' recalling both Twite and Rock Sparrow. Song a soft, halting series of short, uneven notes, sometimes with interwoven drawn-out buzzing notes almost like those of Trumpeter Finch.

Trumpeter Finch   Mongolian Finch

Crimson-winged Finch   Desert Finch

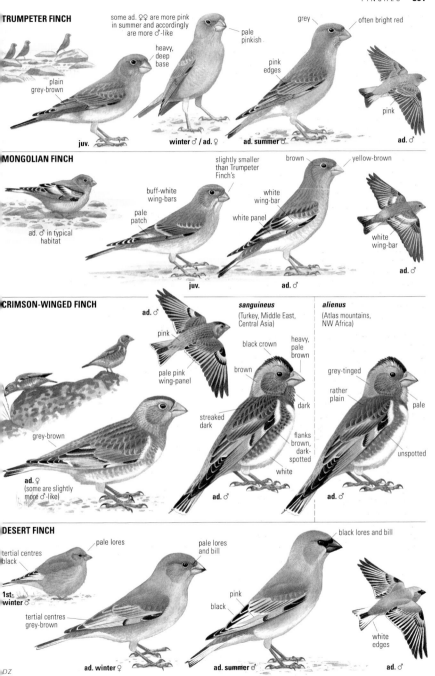

TRUMPETER FINCH

some ad. ♀♀ are more pink in summer and accordingly are more ♂-like

grey

often bright red

pale pinkish

heavy, deep base

pink edges

pink edges

plain grey-brown

pink

juv.

winter ♂ / ad. ♀

ad. summer ♂

ad. ♂

MONGOLIAN FINCH

slightly smaller than Trumpeter Finch's

brown

yellow-brown

buff-white wing-bars

white wing-bar

pale patch

white panel

white wing-bar

ad. ♂ in typical habitat

white wing-bar

juv.

ad. ♂

ad. ♂

CRIMSON-WINGED FINCH

ad. ♂

sanguineus (Turkey, Middle East, Central Asia)

alienus (Atlas mountains, NW Africa)

pink

black crown

heavy, pale brown

pale pink wing-panel

brown

grey-tinged

rather plain

pale

streaked dark

dark

grey-brown

flanks brown, dark-spotted

unspotted

ad. ♀ (some are slightly more ♂-like)

white

ad. ♂

ad. ♂

DESERT FINCH

black lores and bill

pale lores

tertial centres black

pale lores and bill

1st-winter ♂

pink

black

tertial centres grey-brown

white edges

ad. winter ♀

ad. summer ♂

ad. ♂

BUNTINGS *Emberizidae*

More terrestrial than the closely related finches. Majority of species belong to genus *Emberiza*, have triangular bill and characteristic plumage colours and pattern. Song loud, often short and species-specific. Most species are migratory. Favoured food grass seeds. Nest in tussock or low in bush.

(Common) **Reed Bunting** *Emberiza schoeniclus* r**B2**/**W**2

L 13½–15½ cm. Breeds in reedbeds, tall rushes and shrubbery on wet ground or at lake margins, also in drier sites (young conifer plantations, rape fields etc.). In Britain & Ireland resident, with immigrants from N and E Continent late Sep–Apr/May. ♂ easy to see in breeding season, sings from reed stem or bushtop, but species otherwise rather unobtrusive; flies off in springing, slightly uneven and jerky flight, quickly taking cover in vegetation.

IDENTIFICATION Mid-sized bunting. *Bill small* (except in SE Europe and areas around Black and Caspian Seas, where bill is strong and sturdy) and *dark*. *Plumage mostly brown and buffy-white* with *dark streaking and white tail-sides.* – ♂ summer: Easily identified by *black head and throat* with *pure white neck-band* and *narrow white moustachial stripe.* – Other plumages: Buffish supercilium and grey-brown crown and cheeks. ♂♂ usually told by dusky or irregularly black-spotted throat, while ♀♀ have unspotted buffish-white throat-centre with distinct black lateral throat stripes. *Lesser wing-coverts reddish-brown.* Cheeks always slightly swarthy (never uniform red-brown as on Little Bunting). Legs red-brown or black (darker than Little's).

VOICE Calls include a high, softly downslurred 'siü' and a slightly impure, ringing 'bzü', heard mostly from autumn migrants. On territory, sometimes short, sharp 'zi zi...'. Song a brief, simple verse, usually at *slow tempo*, a few sharp rolling notes spelt out and ending with a couple of faster notes or a short trill, e.g. 'sripp, sripp, sria, srisrisirr'; varies in detail, and occasionally faster, but species nevertheless recognized by voice.

Pallas's Reed Bunting *Emberiza pallasi* V***

L 12–13½ cm. Breeds in similar habitat to N populations of Reed Bunting, i.e. on tundra along watercourses in lower vegetation (herbage, brush, dwarf birch, low shrubs), but also e.g. open larch forest bordering tundra. Migrant, winters in SE Asia, returns to European breeding grounds late, in mid Jun. Behaviour rather like Reed Bunting's.

IDENTIFICATION A small version of Reed Bunting, with similar habits. Identified by: *small* size (no bigger than Little Bunting); *narrower and weaker bill* with, in ♀-type plumages, *pinkish base to lower mandible* (Reed Bunting has somewhat heavier and darker bill); *ash-grey* (♂) or *dull brown* (♀) *lesser*

coverts; in ♀-type plumages *almost no streaks on breast and flanks*, so dark lateral throat stripes stand out as only obvious markings on underparts; *two buff-white wing-bars* (more prominent than on Reed); *more uniformly brown crown and cheeks* (Reed: dark border to cheeks and sometimes crownsides); usually discernibly *paler, more buffy-yellow ground colour* (but some eastern Reed are similarly pale). Sexing as for Reed Bunting. Adult ♂ has contrastingly *greyish-white rump* (but shown also by some Reed in SE Russia).

VOICE Call a rather full, sparrow-like, somewhat down-slurred 'tschialp' and a slightly rasping 'tschirp'. Said also to have a more Reed Bunting-like fine 'dsiu'. Song very *plain and flat*, consists of a short series of monotonous, rasping notes, 'srih-srih-srih-srih'; little variation.

Little Bunting *Emberiza pusilla* V*

L 12–13½ cm. Breeds in NE Europe in taiga, often in glades and open coniferous forest with scattering of birch and will low, but also in upland birch forest with some spruce; most territories beside watercourses. Rare but regular vagrant in Britain, mostly in autumn (Sep–Oct) but with increasing frequency of overwintering. Spends much time on ground.

IDENTIFICATION *Small*. In summer plumage, has *red-brown head* with *black lateral crown-stripes* and *white eye-ring*. In autumn, most likely to be confused with ♀ Reed Bunting but told by: *call*; bolder *white eye-ring*; *uniform red-brown or orangey yellow-brown cheeks* with narrow dark border at rear (but without dark edge bordering entire throat-side); *rather light red-brown lores and fore-supercilium*; a *distinct pale* (reddish-brown) *median crown-stripe* between dark crown-sides; often a *small light buff spot on rearmost cheek* proportionately somewhat *longer bill* with *straight culmen*; more prominent buff-white wing-bars; grey-brown lesser coverts; invariably pinkish legs. Sexes alike (individual variation greater than slight tendency for adult ♂ to have more distinct head markings than ♀).

VOICE Call a sharp, short clicking 'zick'; volume varies with mood. Song pleasing and melodic, quite short and high-pitched and containing both clear notes and more rasping, hard, rolling sounds, e.g. 'trütrütrütrü srri tüüy sivi-sivi chu si' or 'si**trü**-si**trü**-si**trü** srisrisri svi-svi-sürrr'.

Yellow-browed Bunting *Emberiza chrysophrys* V***

L 13–14 cm. An extremely rare autumn vagrant from E Siberia; recorded a few times end Sep–Oct.

IDENTIFICATION Barely as large as Rustic Bunting. Rather *big head* and stout bill. *Straight culmen, pink lower mandible*. In summer unmistakable: *bright yellow eyebrow*, (brownish-black crown with narrow white median crown-stripe*. In autumn more like Little and Rustic Buntings, but retains a *tinge of yellow in eyebrow* and *trace of yellow the white crown-stripe. Flanks and breast narrowly and distinctly streaked black*.

VOICE Call a sharp, piercing click, 'zick!' (like Little Bunting's). Song short, has *slow tempo*, often starts with a *drawn-out soft note*, ends more rapidly.

Reed Bunting

Pallas's Reed Bunting

Little Bunting

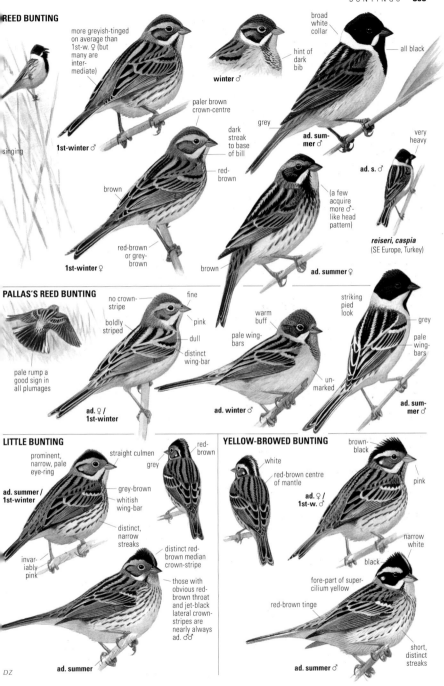

REED BUNTING

singing

more greyish-tinged on average than 1st-w. (but many are intermediate)

1st-winter ♂

winter ♂

hint of dark bib

broad white collar

all black

grey

ad. summer ♂

very heavy

ad. s. ♂

reiseri, caspia
(SE Europe, Turkey)

paler brown crown-centre

dark streak to base of bill

red-brown

brown

red-brown or grey-brown

1st-winter ♀

brown

(a few acquire more ♂-like head pattern)

ad. summer ♀

PALLAS'S REED BUNTING

pale rump a good sign in all plumages

no crown-stripe

fine

boldly striped

pink

dull

distinct wing-bar

**ad. ♀ /
1st-winter**

warm buff

pale wing-bars

ad. winter ♂

un-marked

striking pied look

grey

pale wing-bars

ad. summer ♂

LITTLE BUNTING

prominent, narrow, pale eye-ring

straight culmen

grey

**ad. summer /
1st-winter**

grey-brown

whitish wing-bar

distinct, narrow streaks

invariably pink

red-brown

distinct red-brown median crown-stripe

those with obvious red-brown throat and jet-black lateral crown-stripes are nearly always ad. ♂♂

ad. summer

YELLOW-BROWED BUNTING

white

red-brown centre of mantle

**ad. ♀ /
1st-w. ♂**

brown-black

pink

narrow white

black

fore-part of super-cilium yellow

red-brown tinge

short, distinct streaks

ad. summer ♂

DZ

Snow Bunting *Plectrophenax nivalis* m**B**5 / **P**+**W**3–4

L 15½–18 cm. Breeds in high-alpine habitat in boulder zone of bare mountains, but also along northern rocky coasts and on open, treeless high moors and tundra. Predominantly passage and winter visitor in Britain & Ireland (mostly mid Sep–Apr), mainly from Scandinavia and Greenland, but small numbers breed in Scotland. In winter, often seen in large, dense flocks, mostly along seashores, some on moors and open coastal pasture. Quite confiding, but restless and mobile, continually on the move.

IDENTIFICATION *Large white panels on inner wing* and white at tail-base, flash in flight. Can be confused only with Snowfinch of S Europe (p. 374), but that species stays in its breeding mountains and does not descend to the lowlands where Snow Bunting winters (records of Snow Finch below 500 m exceptional), is bigger, has more white on tail, etc. – ♂ summer: Unmistakable, with *all-white head and breast and all-black back*. – ♀ summer: Diffusely spotted grey on crown, cheeks and breast-sides; back is not solid black. – Juvenile: Very different: *grey head*, and breast diffusely soiled grey. (This plumage not seen away from breeding sites.) – Autumn: All plumages similar, with *rusty yellow-brown tone* on breast-side, cheek, crown/nape and shoulders; back yellow-brown and streaked black. *Bill yellowish with black tip*. ♂ has *more white on wing and on primary-coverts* (latter often white with black tips, forming isolated black spot on white wing-edge) and blacker wing-tip than ♀. – Variation: Compared with Fenno-Scandian birds (ssp. *nivalis*), Icelandic birds (*insulae*), of which some resident in Iceland and some winter in British Isles and Netherlands, have slightly darker plumage with darker primary-coverts (often brown on ♂, too).

VOICE Calls, which often reveal birds overhead, are a rippling, Crested Tit-like 'per'r'r'rit', and a soft but emphatic whistle, 'pyüu', much as Little Ringed Plover; the whistle is given mostly by lone birds. When irritated (e.g. overcrowding in larger flocks, territorial disputes), gives a hoarse 'bersch'. Song a brief twitter with clear voice (almost like Rustic Bunting, but somewhat harder) and desolate ring (as Lapland Bunting), e.g. '**swi**to-**sü**vee-vi**tü**ta-**sü**veh'.

Lapland Bunting *Calcarius lapponicus* **P**+**W**4 / (m**B**5)

(Am.: Lapland Longspur.) L 14–15½ cm. Breeds in upland wet willows and on bare mountains and moors (also tundra) and open bogs with dwarf birch, creeping willow and scrub. Scarce passage and winter visitor in Britain & Ireland (Aug/ Sep–May), often singly or in small parties, on coastal fields and meadowland; has bred in Scotland. Most winter on cultivated steppe in SE Europe. Rather wary, creeps away on ground or freezes, then rises quite high when flushed, flight powerful.

IDENTIFICATION A rather *heavily built* medium-sized bunting with quite *thick but short bill* and *long, straight hindclaw*. Bill *straw-yellow* (slightly darker and greyer on juvenile) *with dark tip*. Legs rather dark brown. – In summer plumage ♂ is unmistakable (black head/throat, yellowish-white band behind eye and down across neck-side to breast-side, red-brown nape, white belly), and ♀, too, identified by *red-brown nape* and by diffuse *greyish-white patch on breast* and pale bill. – In fresh autumn plumage more like Reed Bunting, but told by: *broad red-brown edges* to tertials and greater coverts, the latter forming on closed wing a reddish-brown panel framed by *narrow pale wing-bars*; bill; *paler grey-brown median crown-stripe* (paler and more distinct than on Reed); *grey-brown heavily streaked rump*; sometimes *reddish-brown nape*. Sexes and ages very similar in autumn, but adult ♂ sometimes distinguishable by intimation of red-brown nape and by dense black breast spotting (thus not sparse streaks).

VOICE Rather vocal. A couple of calls correspond to those of Snow Bunting: on migration (often when flushed) a dry rattling 'pr'r'rt', a bit harder and drier than Snow Bunting's, and a short whistled 'chu' (rarely very Snow Bunting-like 'pyu'); also a slightly hoarse or more voiced 'chüp', often from high-flying birds (incl. nocturnal migrants). At nest site, a nervous disyllabic 'tihü' also heard. Song a short, rather constant jingling verse with desolate ring, e.g.'kretle-**krlii**-trr kritle-kretle-trü'; somewhat recalls that of Horned Lark (partly same habitat!) but lacks latter's hesitant introduction. Performs occasional high song-flight with fanned tail and intermittent hovering, when verses protracted.

Rustic Bunting *Emberiza rustica* **V**∗∗

L 13–14½ cm. Breeds in swampy spruce or pine forest with birch, willow and other deciduous trees or in dense, water-logged deciduous forest. Summer visitor (May–Sep), winters in SE Asia; autumn vagrant in Britain. Unobtrusive. On passage, stops off in swampy bushland, easily escapes detection; often singly or in quite small parties on passage, but large flocks occur on coasts of Gulf of Bothnia.

IDENTIFICATION Same size as Reed Bunting. Told by *white belly, plain reddish-brown rump, red-brown spots on flanks*, pinky-brown legs, *longish bill* with *straight culmen* and *pink lower mandible*. – ♂ summer: *Red-brown band on breast and white nape*. Black head with *white supercilium* and *small white spot on nape*. – ♀ summer: Crown and cheeks spotted brown, supercilium off-white, the red-brown on nape narrow. Some are very like ♂. – Autumn: Black and red-brown areas partly concealed by buff-brown fringes; sexing and ageing in field doubtful. Like Little Bunting, often has a dark-edged *buff-white spot on rear cheek* and *distinct pale eyebrow*.

VOICE Call and alarm a short piercing 'zit', sharper than Song Thrush's call, more cutting than Robin's 'tick'. Song, often heard during light night-time hours, unmistakable through its soft, rather gloomy, melodic quality, its lack of hard or sharp sounds, and the jaunty rhythm; a verse may go '**duu**-dele-**düü**do-de**luu**-delü'.

Snow Bunting   Lapland Bunting   Rustic Bunting

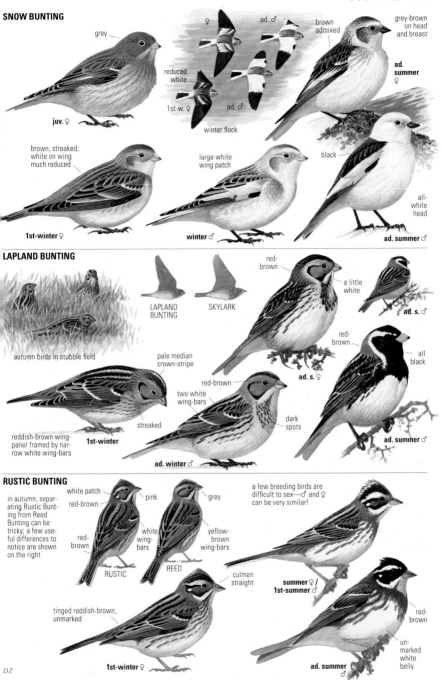

SNOW BUNTING

grey

juv. ♀

♀   ad. ♂

reduced white

1st-w. ♀   ad. ♂

winter flock

brown admixed

grey-brown on head and breast

ad. summer ♀

brown, streaked; white on wing much reduced

large white wing patch

black

all-white head

1st-winter ♀   **winter** ♂   **ad. summer** ♂

LAPLAND BUNTING

autumn birds in stubble field

LAPLAND BUNTING   SKYLARK

pale median crown-stripe

streaked

reddish-brown wing-panel framed by narrow white wing-bars

1st-winter

red-brown

a little white

ad. s. ♂

red-brown   two white wing-bars   dark spots

red-brown

all black

ad. s. ♀   **ad. summer** ♂

ad. winter ♂

RUSTIC BUNTING

in autumn, separating Rustic Bunting from Reed Bunting can be tricky; a few useful differences to notice are shown on the right

white patch   pink

red-brown   grey

red-brown   white wing-bars   yellow-brown wing-bars

RUSTIC   REED

culmen straight

a few breeding birds are difficult to sex—♂ and ♀ can be very similar!

summer ♀ / **1st-summer** ♂

red-brown

tinged reddish-brown, unmarked

1st-winter ♀   **ad. summer** ♂

unmarked white belly

DZ

Ortolan Bunting *Emberiza hortulana* P5

L 15–16½ cm. Breeds at lower levels in agricultural country with some patches of trees and deciduous copses, in open wooded pasture and in clearings, and in S Europe, Turkey and Caucasus in mountain regions, usually in glades or at forest edge, often above 1500 m. Summer visitor (mainly Apr–Sep), winters in tropical Africa. Nocturnal migrant. Forages mostly on ground, often on dry, short-grazed grassy areas. Rather shy and wary, quick to take cover.

IDENTIFICATION Medium-sized, slim bunting. *Bill rather long. Legs and bill buffish-pink*, culmen a shade darker and with straight profile. *Yellowish-white eye-ring* in all plumages, as well as *heavily dark-streaked grey-brown mantle* and *dark lateral throat stripe on light yellow ground* (juv. lacks yellow up to Jul/Aug). *Flanks and belly* have variably distinct *orange-brown tone* (except juv.), and *head and nape* are usually *olive-toned. Rump grey-brown and streaked.* – ♂ summer: Head, broad breast-band and narrow lateral throat stripe *plain olive-grey*; belly and flanks *saturated orange-brown.* – ♀ summer/adult autumn: Crown, breast and lateral throat stripe *finely dark-spotted.* Olive-grey breast-band narrow or indistinct; orange colour below sometimes weaker. Some ♂♂ have unspotted olive-grey breast-band as in summer. – 1st-winter: Head brown-grey (without green tinge) and boldly streaked; lateral throat stripe, breast and upper flanks heavily streaked. Eye-ring buffish-white. Throat light buffy-yellow.

VOICE Call a metallic, almost disyllabic 'sli-e', also a short and slightly jolting, at distance almost clicking, 'chu'; these two are given alternately at *c.* 2-sec. intervals, e.g. on nocturnal passage. Also has a flatter clicking 'plett', e.g. when flushed. Song a plain verse with typically *ringing tone* and with repeated theme changed a good halfway through; can sound *like an echo*, 'srü-srü-srü-srü-dru-dru-dru' or 'sia sia sia drü drü'; the 'echoing' notes are often lower. Mountain-dwelling Ortolans of S Europe have the ringing tone in the first stage but generally cut short the second part to a single, slightly bent, drawn-out final note, 'srü-srü-srü-srü-chuuüy' (so confusion risk with Cretzschmar's can arise: see latter).

Cretzschmar's Bunting *Emberiza caesia* V***

L 14–15½ cm. Breeds on sunny and dry, bare mountain slopes with mixture of rock outcrops, grass and thorn bushes. In regions where Ortolan Bunting occurs in vicinity, Cretzschmar's often (but not always) breeds at lower levels, usually below 1350 m. Mainly coastal. Summer visitor (in Greece mostly mid Apr–Aug), winters S of Red Sea. Night migrant. Perches on ground or low rocks, but also uses low bushes as songpost. Not shy. Migrant flocks, at times quite large, often seen at stopovers in S Israel in Mar.

IDENTIFICATION Somewhat smaller than Ortolan Bunting,

which it otherwise resembles in shape. Legs and bill as on Ortolan, but *bill a trifle shorter.* Bold *eye-ring is white* (not yellowish). *Rump red-brown*, saturated and unstreaked on adult ♂, duller red-brown on adult ♀, only tinged red-brown and also streaked on 1st-winter. Plumages much as Ortolan, but latter's green tone on head, breast and lateral throat stripe are replaced by *lead-grey*, and Ortolan's light yellow throat by *rusty-red* or white. Adult ♂ has *very dark orange-brown colour on underparts*: seen head-on at distance, breast and throat look distinctly paler than the dark belly, an impression never given by Ortolan. 1st-winter is very like Ortolan, but has whiter throat and eye-ring and faint red-brown tinge on rump.

VOICE Calls of Ortolan type, and because of some dialectal variation very similar at times; as a rule, however, recognized both by sharp, rasping note ('tsrip') and by melancholy ring and quite low-pitched, full whistle with faint downslur ('chiiu'). Also has drier 'plett', very like Ortolan's. Song rather like S European Ortolan's, having same structure with (2–)3 repeated short notes and a drawn-out terminal one, 'ji ji ji jüüü' or slightly faster 'si si si-siüüh', but can be told by fact that it *always lacks the pleasant ringing tone* in the first part; some singers alternate this with hoarse (at times really asthmatic!) song variants, 'cha-cha-cha-**cheeeh**'.

Grey-necked Bunting *Emberiza buchanani* —

L 14–15½ cm. Breeds on bare mountain slopes or high plateaux, mostly above 2000 m, with rock outcrops, scant vegetation and scattered low bushes. Has similar habitat requirements to Rock Bunting and southern populations of Ortolan Bunting. Summer visitor (in E Turkey mainly Apr/May–Aug/Sep), winters in India. Quite placid, not shy.

IDENTIFICATION Like Cretzschmar's Buntings, but *somewhat longer-tailed. Bill long and narrow, all pink* (lacks dark culmen). *Eye-ring* white. *Mantle/back* invariably *discreetly spotted*, looks rather *uniform brown-grey* in fresh plumage. *Rump plain brown-grey*, or just faintly streaked. – Adult ♂: Greyish head (dirty grey, sometimes ash-grey); minute off-white throat patch and off-white (longer) submoustachial stripe; *cold red-brown breast* (with fine whitish fringes when fresh, looks vermiculated); *brownish-white belly* and *red-brown shoulders.* – Adult ♀: Like ♂, but *duller* and has *finely streaked breast* and streaked crown. – 1st-winter: Rump dull brown without rusty tone, mantle/back only moderately spotted. Note narrow, pointed bill and white eye-ring.

VOICE Calls similar to those of Cretzschmar's and Cinereous Buntings, a forceful 'chüpp' and a higher and sharper 'zrip'; these are alternated and vary in detail. Song, too, is like those of its congeners, perhaps most like that of Cinereous: plain, short, with slightly scratchy voice (usually not so nice a ringing sound as Ortolan's), 'srü-srüsrü **srih**-srusru' or 'sri-srisri-su**sriih**-sria'; slightly wider scale range also occurs, 'tru-trü-**tri** tre-tra'; often *third syllable from end most stressed.*

Ortolan Bunting Cretzschmar's Bunting Grey-necked Bunting

ORTOLAN BUNTING

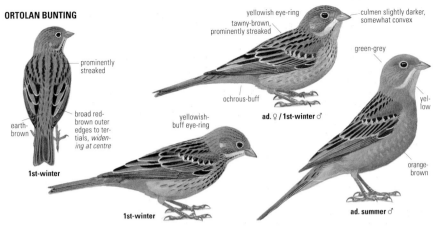

prominently streaked

earth-brown

1st-winter

yellowish eye-ring

tawny-brown, prominently streaked

culmen slightly darker, somewhat convex

broad red-brown outer edges to tertials, *widening at centre*

ochrous-buff

ad. ♀ / 1st-winter ♂

green-grey

yellow

yellowish-buff eye-ring

1st-winter

orange-brown

ad. summer ♂

CRETZSCHMAR'S BUNTING

prominently streaked

tawny-brown

broad *rusty-brown* outer edges to tertials, widening at centre

1st-winter

whitish eye-ring

tawny-brown, prominently streaked

culmen slightly darker, somewhat convex

rusty-tinged

ad. ♀ / 1st-winter ♂

blue-grey

rusty-red

uniformly brown

whitish eye-ring

1st-winter

deep rusty-brown

ad. summer ♂

GREY-NECKED BUNTING

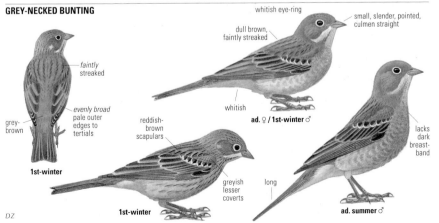

faintly streaked

grey-brown

evenly broad pale outer edges to tertials

1st-winter

whitish eye-ring

dull brown, faintly streaked

small, slender, pointed, culmen straight

whitish

ad. ♀ / 1st-winter ♂

reddish-brown scapulars

greyish lesser coverts

long

1st-winter

lacks dark breast-band

ad. summer ♂

DZ

Yellowhammer *Emberiza citrinella* rB1 (2?)

L 15½–17 cm. Breeds commonly in farmland, in bushy areas, woodland edge and wooded pasture, in glades and clearings, on heaths and coastal meadows. Predominantly resident, but many Scandinavian birds migrate Oct–mid Nov to North Sea countries, returning Mar/Apr. Wary without being really shy, usually flushes early; often gains height, perches high in tree or drops into thick bushes.

IDENTIFICATION *Long-tailed* with *unmarked red-brown rump* and *elements of yellow* in plumage. Mantle / back olive-brown, heavily streaked black. Bill rather small, lower mandible light blue-grey. *Much white on tail-corners*, often seen in flight, when looks longer-tailed than e.g. Reed Bunting. Flight in *long undulations* and slightly jerky. – ♂ summer: *Almost entirely yellow head* with just a few dark markings on crown- and head-sides. *Saturated yellow below*, with much *olive-green and red-brown on breast* and with *flanks streaked red-brown*. – ♀ summer: Greyish-green, streaked crown, at times with a small yellow spot; grey-green head-side with no yellow. Underparts *paler yellow, with greyish-black streaking on breast and flanks*. – Autumn: ♂ often told by intimation of deep yellow on head and underparts, and by olive-green and red-brown colour on breast. ♀ rather like ♂ summer. 1st-winter ♀ often has almost wholly brown and streaked head and buff-white underparts with only faint yellow tinge; note: yellow on underparts can be hard to see in field—cf. Pine Bunting!

VOICE Call a discordant 'stüff'. Also has variety of more stifled, short, clicking calls, 'pt...pt, pt, pitti**litt**...'; also sharp 'tsit' and fine, drawn-out 'tsiih'. Song well known, a run of 5–8 rapidly repeated short notes and a different ending, e.g. 'si-si-si-si-si-si-**süüü**'; the penultimate note is often higher and the last lower, 'sre-sre-sre-sre-sre-sre **siii**-suuu' ('a little bit of bread and no CHEESE!'); voice occasionally much more River Warbler-like, 'dzre-dzre-dzre-...'.

Pine Bunting *Emberiza leucocephalos* V***

L16–17½ cm. Breeds in Siberia (apparently not W of Urals) in similar habitats to closely related Yellowhammer, but more associated with forest than latter. Hybridizes to some extent with Yellowhammer in W Siberia. Migratory, winters in Central Asia and locally W to Middle East and S Europe. Rare vagrant in W Europe, mostly in autumn and winter.

IDENTIFICATION A sibling species of Yellowhammer in which all yellow pigment is replaced by white. Adult ♂ (sometimes ♀, too) in addition has *reddish-brown throat and supercilium*. Autumn birds, especially 1st-winter ♀, often hard to separate from pale young ♀ of Yellowhammer. Note: *ground colour of underparts whitish with no suggestion of yellow*; *mantle / back rather pale, greyish-brown without green tone*; *shoulder* often with *red-brown tone*; light brown-tinged off-

white *supercilium without yellow*; narrow pale *edges to primaries white*, not yellow; *flanks streaked* predominantly *red-brown*, not greyish-black as on tricky Yellowhammers.

VOICE Most calls are just like Yellowhammer's, at least to the human ear. Besides the usual 'stüff' call, occasionally gives a slightly different 'tsick' and a faintly downslurred 'chüeh'. Song sounds very like Yellowhammer's.

Cirl Bunting *Emberiza cirlus* rB4

L 15–16½ cm. Breeds at woodland edge, in glades, in larger parks etc.; prefers slightly hilly terrain with some tallish trees and thick bushes and hedges, often found on dry sunny slopes (south- and east-facing), e.g. with vineyards, but avoids open plains. In far SW often on mountain slopes to *c.*1500 m. Resident. Huge decline in N of range, in Britain now confined to handful of coastal sites in SW England.

IDENTIFICATION Like Yellowhammer but a bit smaller, with proportionately *slightly bigger bill* and shorter tail. ♂ unmistakable, with *black throat and eye-stripe* and *broad olive-green breast-band*. ♀ and 1st-winter are very like Yellowhammer, but distinguished by: *olive-tinged grey-brown* and *streaked rump*; *greater contrast on head-side* between dark markings and pale ground colour; *reddish-brown shoulders*, *plainer brown-grey greater coverts*; *wholly greenish grey-brown, streaked crown* (no hint of paler crown-centre); tendency towards more two-toned bill; voice.

VOICE Call a fine, sharp clicking 'zitt'. Also has somewhat Reed Bunting-like, downslurred 'siiü' and, e.g. in pursuit-chase, fast sharp trills, 'zir'r'r' (can recall Long-tailed Tit). The song is a fully second-long, flat, dry trill, 'sre'sre'sre'sre'sre'sre'sre'sre', rather rattling and metallic; sometimes shifts between two pitches; at distance can to some extent recall song of Arctic Warbler.

Cinereous Bunting *Emberiza cineracea* —

L 16–17 cm. Breeds locally and often sparingly on dry, stony slopes with low vegetation, maquis and scattered trees. Summer visitor, winters around Red Sea.

IDENTIFICATION Rather big and *slim* and *long-tailed*. With its *pale and almost unstreaked* plumage, its *narrow pale eye-ring* and *rather heavy, pale bill*, most resembles a ♀ Black-headed Bunting, but differences include *white tail-corners*. – ♂: *Yellowish throat*; *plain yellowish-grey head*; *grey breast*; *greyish mantle / back* with discreet dark spotting. – ♀/1st-winter: Buffy-white throat; mantle and head brownish; crown finely streaked; breast grey-brown, finely streaked. – Variation: Breeding ♂ in SE Turkey and further east (ssp. *semenowi*) is yellow-tinged also on lower breast and belly.

VOICE Calls quite like those of Cretzschmar's and Grey-necked Buntings, a sharp, rasping 'tschrip' as well as a full 'chülp' (or shorter 'chü'), which are often uttered alternately and varied in detail. Song like Grey-necked's, a plain, rapidly delivered verse with slightly hoarse voice and uneven pace, e.g. 'zre, zrü-zrü-zrü **zrih**-zra'; often ends with a high and a low note in rapid succession.

Yellowhammer

Cirl Bunting

Cinereous Bunting

YELLOWHAMMER

tendency to have pale nape patch

primaries edged yellowish

ad. ♀ / winter ♂

bright yellow

red-brown

pale yellow

1st-winter ♀ generally at least a faint yellow tinge

red-brown

ad. summer ♀

ad. summer ♂

PINE BUNTING

hybrid between Pine Bunting and Yellowhammer

primaries edged white

winter ♂

red-brown

whitish

1st-winter ♀ never tinged yellow

ad. summer ♀

ad. summer ♂

CIRL BUNTING

plain nape

striped head-side

ad. winter ♂

juv.

fine, distinct streaks

olive-grey

yellow tinge

ad. summer ♀

ad. summer ♂

CINEREOUS BUNTING

rather plain brown-grey

grey

semenowi (SE Turkey)

ad. s. ♂

cineracea (W Turkey)

plain breast

yellow

juv.

pale un-marked belly

semenowi (SE Turkey)

ad. s. ♀

white

♂ *semenowi*

cineracea

ad. summer ♂

DZ

Red-headed Bunting *Emberiza bruniceps* [V***]

L 15–16½ cm. Very closely related to Black-headed Bunting, and hybridizes with latter where the two meet (SE Caspian Sea). Breeds on steppe (cultivated or not) with some bushes or scattered trees, in semi-desert, on open mountain slopes or high-lying plains. Summer visitor (by Volga mostly early Jun–early Aug), winters in India. Not that shy. Rare records in W Europe, but all these are generally regarded as involving escaped cagebirds; a few, however, probably refer to genuine vagrants.

IDENTIFICATION *Fairly big* and rather *long-tailed*. *No white on tail.* – ♂: Unmistakable. Besides *red-brown mask*, note *yellowish rump* and *yellowish-green, streaked mantle.* – ♀: Rather nondescript and hard to identify. Note: *long, heavy, light grey bill*; *narrow buffy-white eye-ring*; dull light brown mantle/back and head/nape with only *modest, fine streaking* on mantle and forecrown; usually lightly streaked yellow-green or *yellowish rump*; wholly *unstreaked dirty yellowish-white underparts* and *bright yellow undertail-coverts.* – 1st-winter: Like ♀, but has *buffish-white breast* (sometimes with fine streaks), *more heavily streaked grey-brown mantle/back* and rump without yellow-green tone. Very like Black-headed Bunting, and often inseparable from that species.

VOICE Has many calls, most recalling those of congeners, including rather Ortolan-like 'chüpp' and Cretzschmar's-like 'zrit', also quite Yellowhammer-like 'chüh', sharp 'tsit' and fast series of clicks, 'ptr'r'r'. Song is to human ear identical to Black-headed Bunting's, a slightly accelerating verse with low pitch, jerky rhythm and very rolling r-sound, 'zrit...zrit...srütt, srütt-srütt sütt**e**ri-sütt sütter**reh**'; the pitch drops a little during the verse.

Black-headed Bunting *Emberiza melanocephala* V***

L 15½–17½ cm. Very close relative of Red-headed Bunting, and the two hybridize locally. Breeds in open, dry country with bushes, often in farmland with scattered tree clumps, in vineyards and orchards, in riverine woodland, at forest edge and in glades, as well as on drier mountain slopes with thorn bushes and isolated trees. Summer visitor (in Greece mainly early May–end Jul), winters in India. Not shy.

IDENTIFICATION *Fairly large*, a trifle bigger than Red-headed, and similarly long-tailed. *Lacks white on tail.* – ♂: Unmistakable. Apart from *black hood*, note *unstreaked red-brown upperparts.* – ♀: Very like Red-headed and often difficult to separate in field. Has *slightly longer bill* and sometimes *faint red-brown tone* to the dull grey-brown mantle/back. Rump dull brown or tinged yellow-brown (never has green tinge of some Red-headed). In addition, *crown* is sometimes *more distinctly dark-streaked* and ear-coverts a little darker. – 1st-winter: Very like Red-headed

(which see) and often not safely separable in the field.

VOICE Most calls are very like those of the closely related Red-headed Bunting, and because of some variation in articulation (in both species) safe field separation based on voice alone is impossible. Song as Red-headed's (above).

Yellow-breasted Bunting *Emberiza aureola* V**

L 14–15½ cm. Breeds in NE Europe and large parts of Siberia (where very common) in bushy areas, deciduous scrub and copses, often along watercourses, lakesides and at edge of wetlands. Summer visitor (in Finland mostly mid Jun–end Jul), winters in SE Asia. Rather confiding.

IDENTIFICATION Roughly size of Reed Bunting; has *longer bill* than latter, with *straighter culmen* and *pink* (not greyish-black) *lower mandible*. In all plumages, *element of yellow or buffy-yellow below* and *sparse, narrow streaks on flanks.* – ♂ summer: *Black 'face'*, unmarked *dark red-brown upperparts*, *red-brown breast-band*, *saturated yellow underparts*, *large white wing patch*. 1st-year has *small and dark-spotted white wing patch*, a *sprinkling of brown and white on black 'face'*, narrow or broken breast-band; some are paler yellow below. – ♀/1st-winter: *Buffy yellow-white supercilium*; dark rear eye-stripe and a *dark lower border to light yellow-brown cheek*; *dark crown-sides* frame narrow, paler grey-brown median crown-stripe; *white on tail-sides* (unlike ♀ Chestnut Bunting, which has all-dark tail). *Rump grey-brown* (with faint red-brown tinge), *heavily streaked*. Undertail-coverts whitish.

VOICE Call/alarm a sharp clicking 'tsick'. Song easily recognized by dominant element of clear loud notes repeated in pairs and wide tonal range, often in steps up the scale (but verses commonly seesaw up and down in pitch at end), 'tru-tru trüa-trüa **tri-tri** cha' (last note falling) or 'trü-trü, tra, tro-tro **triih**' (lowest pitch in middle). Wide individual variation.

Chestnut Bunting *Emberiza rutila* [V***]

L 12½–13½ cm. Breeds in E Siberian taiga. Migratory, winters in SE Asia. Very rare autumn vagrant in W Europe, incl. in Norway, Netherlands and Malta. (Some spring records of adult ♂♂ deemed likely to have involved escapes.)

IDENTIFICATION Small. *Rump unmarked red-brown*. Underparts *light yellow*. Bill longish, with pink at base of lower mandible. ♂ unmistakable. ♀/1st-winter is rather like Yellow-breasted Bunting, differs in: *smaller* size; *unspotted red-brown rump*; *more strongly indicated lateral throat stripe*; *white throat* but *yellowish undertail-coverts*; *all-dark tail*.

VOICE Call a short, hard, sharp clicking 'zit', sometimes rapidly repeated. Song a brief, loud, rather fast verse which may be likened to a snippet of Pallas's Warbler or Olive-backed Pipit song; quite often each phrase starts with three slow, pure notes or double notes, followed by a characteristic *scratchy, trilling* section (lacking in rather similar song of Yellow-browed Bunting) and a softer, Red-start-like ending (as Yellow-browed), e.g. 'tvia tvia tvia sre-sre-sre sisicha'. See also Little Bunting.

Red-headed Bunting

Black-headed Bunting

Yellow-breasted Bunting

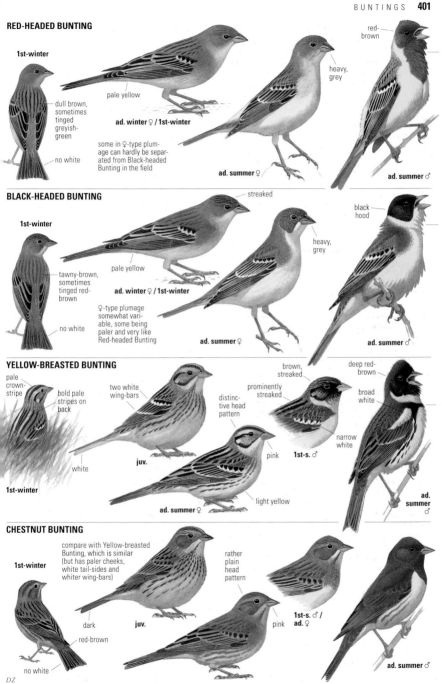

RED-HEADED BUNTING

1st-winter

dull brown, sometimes tinged greyish-green

no white

pale yellow

ad. winter ♀ / 1st-winter

some in ♀-type plumage can hardly be separated from Black-headed Bunting in the field

heavy, grey

ad. summer ♀

red-brown

ad. summer ♂

BLACK-HEADED BUNTING

streaked

1st-winter

tawny-brown, sometimes tinged red-brown

no white

pale yellow

ad. winter ♀ / 1st-winter

♀-type plumage somewhat variable, some being paler and very like Red-headed Bunting

heavy, grey

ad. summer ♀

black hood

ad. summer ♂

YELLOW-BREASTED BUNTING

pale crown-stripe

bold pale stripes on back

two white wing-bars

distinctive head pattern

white

juv.

1st-winter

pink

light yellow

ad. summer ♀

brown, streaked

prominently streaked

narrow white

1st-s. ♂

deep red-brown

broad white

ad. summer ♂

CHESTNUT BUNTING

compare with Yellow-breasted Bunting, which is similar (but has paler cheeks, white tail-sides and whiter wing-bars)

1st-winter

dark

red-brown

no white

juv.

rather plain head pattern

pink

1st-s. ♂ / ad. ♀

ad. summer ♂

DZ

Corn Bunting *Miliaria calandra*   r**B**3

L 16–19 cm. In greater part of its European range breeds in extensive, open farmland, mostly where cereals, fodder plants and leguminous plants are cultivated on a large scale. Requires elevated songpost, but isolated trees and bushes or telephone wires, or even just fence posts, will suffice. In S of range, breeds also on dry mountain slopes with low thorn bushes. Huge decline in NW Europe since 1960s. Mostly resident. Social, often seen in small loose flocks. Complex breeding habits; locally 2–3 ♀♀ per ♂. Not very shy.

IDENTIFICATION Big, with quite *heavy body* and *stout bill*. Often looks rather *large-headed*. Plumage not unlike that of a lark, dark-streaked grey-brown above and buffy-white below with dark streaks on throat-side, breast and (more narrowly) flanks. *Tail lacks white*. Head-side shows no striking pattern, is somewhat irregularly streaked on buff-white ground, with *darker lateral throat stripe* and *lower edge of cheek*, and has a *dark spot on rearmost ear-coverts*. In worn plumage (summer), breast streaks often coalesce to form irregular dark patterns. *Legs and bill-side yellowish-pink*. Sexes alike. Moves short distances often with fairly *ponderous flight and dangling legs* (can give lark-like impression); over longer stretches flies in deep undulations.

VOICE Call a discordant, metallic 'tsritt', also a short clicking 'bitt' or 'bt', often rapidly repeated in 'electrified' series, 'bt'bt'bt'bt...'. Song is characteristic (but can be skilfully imitated by some Whinchats), a brief, little-varied and oft repeated verse with *halting start, accelerating* to a *squeaky, jingling ending*, 'tück tück-zick-zik-zkzkzrrississss'.

Rock Bunting *Emberiza cia*   **V**∗∗

L 15–16½ cm. Breeds on steep, often boulder-strewn or rocky, open mountain slopes immediately above treeline (often below 1500 m in Europe) with grass and herbage along with thorn bushes and scattered trees, sometimes in glades and alpine meadows just below treeline; locally in coastal regions down to sea-level. Chiefly resident, but in winter usually descends to lower levels. Not really shy, but unobtrusive and is easily overlooked when ♂ not singing.

IDENTIFICATION Fairly big, short-winged but *long-tailed*. Identified in all plumages by contrasty head markings, with *dark stripes along crown-side, through eye and framing cheek*, combined with *ash-grey throat and ash-grey breast* (mostly on ♂), *rusty-brown* (or on young ♀ orange-buff) *belly, white outer tail-corners*, and *grey lesser wing-coverts*. Lower mandible light lead-grey. *Rump unstreaked red-brown*. Sexes similar. – Adult ♂: Black head markings, whitish supercilium, unspotted pale lead-grey on throat and breast, and unstreaked flanks. – ♀: Occasionally inseparable from ♂ in field, but many can be told by less

extensive grey on breast, light streaking on lower throat, breast and flanks, and more diffuse head pattern.

VOICE Call a short, sharp 'tsi', also a drawn-out 'tsiii', at times faintly downslurred and then rather Reed Bunting-like, 'tsiii'; when nervous, fine twittering series, 'tir'r'r'r'(like begging young White Wagtails). Song *high-pitched* and *clear*, a melodic verse with distinct scale changes and often typically *tentative introduction* and uneven rhythm, e.g. 'sütt, titt-itt, svi cha-cha-sivi-süasüa, sitt sivisürrr si'.

Striolated Bunting *Emberiza striolata*   —

L 13–14 cm. Lives in uninhabited, barren mountain regions with scant vegetation, where it is shy and rather difficult to observe and breeds only in rock crevices; resident, but in winter occasionally moves to cultivated fields and plains at lower levels.

IDENTIFICATION A *fairly small* bunting with *much rusty-brown on wing*, with *dark spots or shading on throat/ breast*, *straw-yellow lower mandible* and *no white on tail*. Can recall Rock Bunting as head is often greyish-white with dark stripes, but differs in: smaller size; tail lacking pure white (Rock Bunting has white outer corners); yellowish lower mandible (Rock Bunting: light grey); dark-mottled throat and breast (Rock: unspotted light grey); *rusty-brown lesser coverts* (Rock: grey); *no white wing-bars* (Rock: two narrow ones). ♀ is more brownish than grey on head, neck and breast, and head markings are less distinct.

VOICE Calls include a nasal, whining 'chueht', a shorter 'tvett' and hoarse, House Sparrow-like 'chriff'; song a brief, cheerful, short phrase, repeated much the same all the time, with first and last notes lower, '**tru**-ee-ah **tre-tre** trivitri-**trah**'.

House Bunting *Emberiza sahari*   —

L 13–14 cm. Resident breeder in proximity of humans by settlements, on mountain slopes with small cultivations and scattered bushes and trees, at times also in village gardens, where it is anything but shy (even flies unconcernedly into houses, hence name) and nests in cavities, on ledges or in recesses in houses. Sometimes treated as race of Striolated Bunting, which is anyway closely related to.

IDENTIFICATION Very similar to Striolated Bunting but has much *richer rusty-brown colour on belly and wings* and is as good as *unspotted on rusty-brown mantle* (has only fine shaft streaks), whereas Striolated is *paler below* and *heavily dark-spotted on brown-grey upperparts*. In addition, ♂ Striolated has on average more contrasty greyish-white and black head markings, while head of ♂ House Bunting is more swarthy and diffusely patterned.

VOICE Calls include a Red-rumped Swallow-like, merry 'dvueet', a desolate, moaning 'chuoo' with falling pitch, and a somewhat Greenfinch-like 'chüpp'. Song a short, rather consistently repeated, high-pitched stanza with rather jolting rhythm, e.g. 'wiss-to süss-to she**viss**-chu'.

Corn Bunting

Rock Bunting

House Bunting   Striolated Bunting

CORN BUNTING

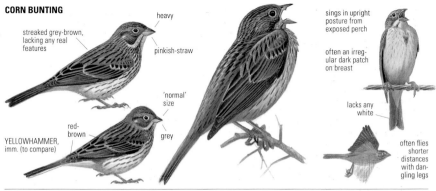

streaked grey-brown, lacking any real features

heavy

pinkish-straw

'normal' size

red-brown

grey

YELLOWHAMMER, imm. (to compare)

sings in upright posture from exposed perch

often an irregular dark patch on breast

lacks any white

often flies shorter distances with dangling legs

ROCK BUNTING

grey

two white wing-bars

prominently streaked

unmarked red-brown

white tail-sides

greyish lesser coverts

♀ / 1st-winter ♂

ad. s. ♂

striking head pattern

plain

unmarked red-brown

long

ad. summer ♂

STRIOLATED BUNTING

yellow-pink

lacks white wing-bars

buff

earth-brown, lightly streaked

red-brown lesser coverts

dark covert-bar

♀ / 1st-wint. ♂

diffuse transition

streaked

pale buff-grey

ad. summer ♂

HOUSE BUNTING

plain-looking back and wings characteristic

red-brown, plain

wing uniform

♀ / 1st-wint. ♂

sharp border

streaked

deep rusty-buff

ad. summer ♂

DZ

NORTH AMERICAN PASSERINES

Every year a number of North American passerines are recorded in Europe, primarily in Britain, Ireland and Iceland. They are assisted by prevailing westerly winds, and a strong tail-wind is probably a prerequisite for a successful crossing. Two spreads show some species which have now become reasonably regular visitors to Europe. See also pp. 298–299, where a few American thrushes are treated.

VIREOS Vireonidae

The vireos constitute a large family resembling both warblers and flycatchers but with no real counterpart in Europe. Most are fairly small, grey and green above and off-white below. They have a strong, flycatcher-like bill.

Red-eyed Vireo Vireo olivaceus V∗

L 13–14 cm. Annual vagrant in autumn, mostly end Sep–mid Oct. Arboreal; vagrants in Europe often seek mature gardens, but in windswept coastal areas may be found in hedgerows, willows, etc. Frequently keeps well hidden or quiet, will 'disappear' for long spells.

IDENTIFICATION The size of an Icterine Warbler, and somewhat recalls one, having rather *large head* and *heavy bill, fairly short and straight-ended tail, olive-green upperparts* and lead-grey legs. Recognized by striking head markings: *long, well-marked, whitish supercilium bordered dark above* (distinctly) *and below* (shorter), and *blue-grey crown*. The *red* (or red-brown) *iris* sometimes discernible. Rather sluggish and slow when not moving far, more like a Garden Warbler than a Willow Warbler, but can certainly make dashing sallies now and then. Call is a slightly squeaky, nasal 'chway'.

NEW WORLD WARBLERS Parulidae

A large family of small arboreal insect-eating birds. Although they have a thin, pointed bill and warbler-like habits, they are most closely related to the buntings (pp. 392–403). Often distinctly patterned in bright colours. Most have *large white patches on the tail*, often visible from below.

Northern Parula Parula americana V∗∗

(Alt. name: Parula Warbler.) L 10½–11½ cm. Rare autumn vagrant, mostly end Sep–mid Oct. Goldcrest-like behaviour, moves restlessly, will hover or hang upside-down.

IDENTIFICATION *Small*. Fairly dark *bluish grey-green above* (head and edges of wing-feathers more bluish, esp. on adult, mantle always green), with *two* narrow, distinct *white wing-bars*. Thin white eye-ring, broken in front and behind. *Throat and breast yellow, belly white*. ♂ has hint of a dark breast-band, missing on ♀. In Europe usually silent.

Yellow-rumped Warbler Dendroica coronata V∗∗∗

(Alt. name: Myrtle Warbler.) L 12½–13½ cm. Rare autumn vagrant, mostly in Oct. Vagrants may spend much time foraging in short grass, often joining flocks of pipits.

IDENTIFICATION In all plumages recognized by *yellow rump patch* and *whitish throat* (lower throat sometimes tinged buff), being the only wood-warbler with this combination (Magnolia Warbler, p. 420, has pale yellow throat). Autumn birds are *brown above, moderately streaked dark*, and have *two*

light wing-bars, white underparts with dark streaks on breast and flanks, and a small *yellow patch on breast-side* near wing-bend (can be missing on 1st-winter ♀). Sometimes a *small yellow patch on centre of crown* (often concealed by brown fringes, or missing, esp. on 1st-winter ♀). Ear-coverts and lores fairly dark grey-brown, indistinct supercilium, usually rather prominent whitish eye-ring. Uppertail-coverts and tail-feathers edged blue-grey. Vagrants often vocal; call a hard, liquid 'chick', often uttered in flight.

Blackpoll Warbler Dendroica striata V∗∗

L 12–13 cm. Rare autumn vagrant, mostly end Sep–Oct.

IDENTIFICATION Autumn birds are *yellow-tinged on cheek, eye-surround, throat and breast*. Belly and *undertail-coverts* are *white*. Crown, back and indistinct eye-stripe dull olive-green, *back rather thinly streaked dark*. Wing dark with *two well-marked but rather thin white wing-bars* and *white edges to tertials*. Breast and flanks finely streaked on most. *Feet yellowish-brown*, soles orange-yellow. Call a hard but liquid 'chip'.

(American) Yellow Warbler Dendroica petechia V∗∗∗

L 11½–13 cm. Rare autumn vagrant, mostly end Sep–Oct.

IDENTIFICATION Autumn birds are *yellow-tinged* with slightly darker *grey-green upperparts* and *hint of yellow wing-bar*. Adult ♂ has strong yellow on head and underparts and reddish-brown streaks below. ♀ and immatures are duller with greener crown/nape and only faint or non-existent red-brown streaks below. Spread tail shows yellow panels on inner webs. Call a distinct 'tsip'.

American Redstart Setophaga ruticilla V∗∗∗

L 13–14 cm. Very rare autumn vagrant, mostly in Oct. Has very active, darting movements.

IDENTIFICATION Long-tailed. Often raises and spreads tail and droops wings. ♀/1st-winter ♂ have head grey, back and wing olive-grey, underparts off-white, and have *large yellow patches on tail, a small yellow wing patch* and a *yellow* (♀) or *orange-yellow* (1st-winter ♂) *diffuse patch on side of breast* at wing-bend. Call a clicking 'chik' and a thin 'tsiit'.

Black-and-white Warbler Mniotilta varia V∗∗

L 11½–13 cm. Very rare vagrant in autumn.

IDENTIFICATION Peculiar habit of *creeping along tree trunks* and thicker branches, at times with head down almost like a nuthatch. *Broadly striped plumage in black and white*, with a *white median crown-stripe* and a *white supercilium*. Bill very *long and narrow*. Sexes similar; adult ♂ has some black on throat and breast. 1st-winter ♂ *boldly streaked blackish on flanks and undertail-coverts*; 1st-winter ♀ much *less distinctly streaked* on underparts and has *buffier-tinged ear-coverts and flanks* than ♂. Call a clicking 'tic' or a thin 'ziit'.

Northern Waterthrush Seiurus noveboracensis V∗∗∗

L 13–14½ cm. Very rare vagrant in autumn. Ground-dweller, often seen near water, feeding on insects along edge of pond or stream.

IDENTIFICATION Not unlike a pipit, but has *longer and more prominent eyebrow, heavier head and bill*, shorter tail, and habit of often *bobbing rear end*. Bill dark, legs pale. Colour of underparts varies, being either whitish or yellowish.

RED-EYED VIREO

adult and 1st-w. very similar; iris bright red in adult, duller in 1st-w.

lead-grey

strong, hooked at tip

1st-winter

size and proportions similar to Icterine Warbler but shorter-tailed

NORTHERN PARULA

1st-w. ♀

1st-w. ♂

small and brightly coloured, prominent eye-ring; ♂ has some chestnut on breast

YELLOW-RUMPED WARBLER

small yellow spot, often hidden

yellow

1st-w. ♂

yellow patch

prominent white spots

1st-w. ♀-type, autumn

often forages on ground

brighter birds with blue-grey in wing coverts and upper-tail-coverts probably ♂♂, but reliable age/sex determination requires critical judgement

BLACKPOLL WARBLER

much larger than Yellow-browed Warbler (p. 334), streaked above and below, more subdued supercilium

dull yellow

light tail-spots

muted streaks

prominent white wing-bars and tertial edges

extremely difficult to age/sex in autumn

pale

YELLOW WARBLER

1st-w. ♂ (bright)

conspicuous eye-ring, completely lacks dark eye-stripe

1st-w. ♀

variable, some 1st-w. ♀♀ are rather dull

spreads tail momentarily revealing yellowish inner webs

AMERICAN REDSTART

some lack wing-bar

1st-w. ♀

delicate build, striking tail pattern

fans and flirts long tail constantly

variable wing-bar, more extensive in ♂♂ than ♀♀

1st-w. ♂

BLACK-AND-WHITE WARBLER

1st-w. ♂

♂ has bolder black stripes below than ♀

Nuthatch-like foraging habits

1st-w. ♀

diffuse, brownish streaks in ♀♀

1st-w. ♂

NORTHERN WATERTHRUSH

bobs tail, like Common Sandpiper; often skulks in the shadows

size of Rock Pipit but shorter-tailed

KM

Dark-eyed Junco *Junco hyemalis* **V∗∗**

L 13½–14½ cm. An American bunting which, like White-throated Sparrow (below), is part of the subfamily *Emberizinae*. Rare vagrant in Europe, chiefly in spring but occasionally at other seasons, too. Forages almost invariably by hopping on ground. Jerks tail. Takes cover in trees and bushes.

IDENTIFICATION *Pink finch-type bill*, *sooty-grey upperparts*, *head and breast* and *white belly* give unmistakable appearance. Tail long with *white outer edges*, obvious in flight. Legs reddish-brown, toes darker. Sexes rather similar, but ♀ lighter grey with brown tinge above, and with belly not pure white. Call a clicking 'tick', which may be repeated in fast series, and a frothy tongue-clicking 'check'.

Indigo Bunting *Passerina cyanea* **[V∗∗∗]?**

L 11½–13 cm. An American bunting (placed in the subfamily *Cardinalinae*, cardinals, New World grosbeaks and others, more like European finches than buntings in appearance). Rare vagrant in Europe. Recorded at all seasons, with some preponderance for summer. That some records involve escaped cagebirds cannot be ruled out; the species is often kept in captivity.

IDENTIFICATION Adult ♂ summer is an elegant bird, *all blue* with *dark lores* and *light blue-grey lower mandible*. Can be confused only with Blue Grosbeak *Passerina caerulea* (not treated here), but that is bigger and has red-brown wing-bars. At distance just looks 'dark'. In winter plumage the blue is largely concealed by brown fringes. – ♀/1st-winter: More nondescript, being *reddish grey-brown above* with faint dark streaking, and yellowish grey-brown below with grey breast spotting. *Tail-feathers, uppertail-coverts* and at times primaries *edged blue-green* (often absent on 1st-winter ♀). 1st-winter has slightly more distinct light brown wing-bar and heavier breast spotting than adult ♀. Call a sharp 'tsick'.

TROUPIALS, COWBIRDS et al. *Icteridae*

A family occurring only in America. Most species show features of both starlings and finches. Plumage often black, sometimes with some bright colours, and have either powerful finch-like bill or long, narrow and pointed bill and sturdy, rather long legs. Tail often long, with rounded tip.

Bobolink *Dolichonyx oryzivorus* **V∗∗∗**

L 16–18 cm. Rare autumn vagrant in Europe; all records Sep–Nov but for one in spring. Habits in part lark-like, inhabits low crops and stubble fields. Also perches in bushes and reeds.

IDENTIFICATION Adult ♂ summer unmistakable. ♀ and winter plumages not unlike ♀ House Sparrow but *yellower in tone*; are brown and dark-streaked above and yellowish-white below, with buff supercilium, *dark lateral crown-stripe* and *buff median crown-stripe*, *pale lores* but *distinct dark eye-stripe* above *pale, buff cheek*. *Two pale bands on mantle* (roughly as on Aquatic Warbler). Surprisingly similar to ♀ Yellow-breasted Bunting (p. 400), but is much *bigger* and *lacks dark lower border to cheek*. Often stands upright on ground and extends neck, when head looks quite small. *Tail-feathers narrow* and *pointed*. *Flight path straight*, not undulating. Call in flight a sharp 'pink'.

Baltimore Oriole *Icterus galbula* **V∗∗∗**

(Alt. name: Northern Oriole.) L 17–19 cm. Rare autumn vagrant in Europe; most in Sep–Oct; occasional winter and spring records.

IDENTIFICATION Large and *long-tailed*, with *long, pointed bill*. Adult ♂ summer unmistakable. ♀ and winter plumages recognized by: *two white wing-bars*, broadest and most obvious on median coverts; *underparts vivid yellow*, strongest on throat, breast and undertail-coverts; *greyish-yellow tail and rump*. Mantle and back somewhat variable, olive-brown with heavy dark spotting (indicative of ♂) or more uniformly grey (indicates ♀). Adult ♀ and 1st-winter ♂ can show some black on head and throat. Call a disyllabic, clear and full whistle, 'pyoo-li'; also has a nasal chatter.

Rose-breasted Grosbeak *Pheucticus ludovicianus* **V∗∗∗**

L 17½–20 cm. An American bunting belonging to the subfamily *Cardinalinae*. Rare autumn vagrant in Europe, almost exclusively in Oct.

IDENTIFICATION *Big* and stocky, with *heavy head and bill* almost of Hawfinch proportions. *Bill* predominantly *pale*, often with obvious *pink* tone. Adult ♂ summer is handsome (white belly, black upperparts with white rump and white wing markings and a large red breast patch), but has not yet been recorded in Europe. All records here have involved birds in 1st-winter plumage. These resemble adult ♀ and, besides stout, conical, pink bill, are identified by: very *broad and distinct whitish supercilium* and *whitish neck-side* almost completely framing brown ear-coverts; dark brown, finely dark-streaked crown with narrow *whitish median crown-stripe*; a *white patch below eye* and a diffuse pale spot on rearmost ear-coverts; *two white wing-bars* and *white spots at tips of tertials*. Above, brown and dark-streaked; below, off-white with buff tinge, or occasionally more obvious buffy-yellow tone, and with dark streaking on throat-side, breast and flanks. 1st-winter ♂ can often be told from ♀ by much *broader white wing-bar on median coverts, bigger white patch on primary bases* and *light red tinge on breast*, or by fact that a bit of red from axillaries is just visible. Call a hard 'chick'.

White-throated Sparrow *Zonotrichia albicollis* **V∗∗∗**

L 15–17 cm. An American bunting (see p. 392), included in the subfamily *Emberizinae*. Rare vagrant in Europe, chiefly in spring but occasionally at other seasons. Forages mostly by hopping on ground. Often takes cover in bushes.

IDENTIFICATION Recognized by *white throat* (can be finely edged black at lower edge), *grey breast, broad pale supercilium*, and *broad dark lateral crown-stripes* sandwiching a *pale median crown-stripe*. Also has black eye-stripe and *uniform grey cheeks*. Head markings usually black and white, typically with *yellow tone to supercilium in front of eye*. Some, including most 1st-winters, have the pale head markings tinged light brown. 1st-winter also has slightly less clear-cut, off-white throat (instead of pure white and sharply defined as on adult). Call a sharp, fine, 'ziit' and a hard 'chink'. – Can be confused with same-sized and closely related White-crowned Sparrow *Z. leucophrys* (se p. 417) recorded a few times in Europe, but that has light grey throat (tinged buff-white on 1st-winter), light grey head- and neck-sides, is grey around nape, and has broad white crown-band.

DARK-EYED JUNCO

conspicuous white, in flight

dark grey all over

white belly

♀/ **1st-winter** ♂

pinkish

INDIGO BUNTING

rather nondescript

ad. ♂ **spring**

pale

rusty-buff wing-bars

♀/ **imm.** ♂

winter ♂

in autumn, young birds of both sexes may look like ♀♀; blue in the ♂ plumage indicates a ♂

BOBOLINK

alert posture with stretched neck

pointed tips

occasionally mistaken for autumn Yellow-breasted Bunting (pp. 400); feral or escaped weaver sp. may also be misidentified as Bobolink; pointed tips to tail feathers diagnostic

prominent crown-stripes

pinkish

1st-winter / ad. autumn

difficult to age and sex in autumn, when virtually all records occur

bright, yellowish-buff head and breast; lacks the moustachial and lateral throat stripes of most European buntings

BALTIMORE ORIOLE

♀♀ usually duller than ♂♂

1st-w.

1st-w.

bright birds are usually ♂♂

beware: sexing young birds on 'brightness' not always reliable

long, pointed

1st-wint.

rather long

brightness and blackish-centred scapulars indicate a ♂

ROSE-BREASTED GROSBEAK

vagrants often found foraging on black-berries

prominent white flash in ♂♂

1st-winter ♂

bold head markings

1st-year ♀♀ lack white

slightly larger than Hawfinch!

huge bill

1st-winter ♀

WHITE-THROATED SPARROW

'tan-striped' variety

adults of either sex may be 'tan-striped' or 'white'striped'

develops yellow in 1st-autumn

weaker wing-bars than adult

streaked in juv./1st-w.

adult

**juv. →
1st-w.**

'white-striped' variety

yellow in front of eye

forages mostly on the ground like a Dunnock, but sometimes joins flocks of House Sparrows

adult

KM

Vagrants

Those species which have been recorded only a few times within the region treated, which largely corresponds to the West Palearctic (see definition p. 8), are briefly described below. As a rule, this means 4–15 records in total since 1900 (but strict consistency over numbers has not been sought, and new records of rarities continue to be reported). Those

seeking fuller descriptions for the majority of these species are referred to Alström, Colston & Lewington (1991), *A Field Guide to the Rare Birds of Britain and Europe*, and to general handbooks or specialist literature on rare birds (see list of references on p. 427–428). Accidentals (generally 1–3 records in all) are simply listed on pp. 418–421.

Falcated Duck *Anas falcata* [V***]

ad. ♂

Falcated Duck

L 46–54 cm. Breeds in East Asia. Has been encountered in a great many European countries, e.g. Belgium, Britain, Denmark, Finland, France, Germany, Netherlands, Spain and Sweden. Records as a rule thought to involve escapes (the species is common in bird collections), but at present at least records from Jordan, Netherlands and Turkey are regarded as genuine. – At distance, appears all grey with dark head. ♂ unmistakable when seen closer, has *dark green head* (save sharply defined white chin) with *gently sloping forehead* and *full nape*. A Teal-like *yellow patch on sides of undertail*, but this may be entirely or partly hidden by *overhanging, long and scythe-shaped tertials*. – ♀ may recall a *small, broad-winged* Pintail, has *narrow, dark bill* and vermiculated head-side. In flight, shows *dark speculum*.

Cape Teal *Anas capensis* [V***]

Cape Teal

L 44–48 cm. Breeds in sub-Saharan Africa. A few spring and summer records each from Israel and Libya. Seen in many European countries but then regarded as most likely an escape. – Roughly the size of Wigeon. All plumages basically the same, all pale grey-brown with few eye-catching features other than a *pinkish, rather stocky, upturned bill* and a *dark green speculum broadly lined with white*. Whereas the *head is plain*, the *body is sparsely blotched* darker brown, most markedly on flanks. The head often appears rather large with rounded, almost crested hindneck.

Spectacled Eider *Somateria fischeri* —

ad. ♂

Spectacled Eider

L 52–58 cm. Breeds in Alaska and NE Siberia, wintering in holes and cracks in the ice far north. A few winter records from N Norway. – Size of Velvet Scoter, thus *somewhat smaller than Eider*, although shape and proportions similar. Plumage, too, recalls Eider, but ♂ has *black breast* (and belly), large, *rounded 'white spectacles'* (formed by large oval white patches outlined black) on predominantly *green head*, and *orange bill* partly concealed by velvety-textured, green and white feathering; ♀ has *light buff 'spectacles'* on somewhat darker rufous-brown head, and body is rufous densely barred black. ♂ in eclipse plumage like ♀ but told by white tertials and partly white back.

Pacific Loon *Gavia pacifica* V***

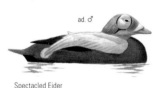

juv.

Pacific Loon

L 60–68 cm, WS 95–115 cm. Breeds in Alaska, C and W northern Canada and NE Siberia. A few recent records from Britain. An Italian record has been questioned. – Closely related to and very similar to Black-throated Loon. Winter- plumage birds differ from that species on *lack of white patch at rear flanks* ('anterior thigh patch') when swimming (narrow white visible only when floating high), on more regularly showing a *thin dark chin-strap* across upper throat (perhaps mainly an adult winter feature, but many 1st-winters have it also; note that a few Black-throateds have a hint of this mark, too). A *dark transverse bar runs across vent* (visible sometimes when preening underparts, or in flight) separating white belly from small white central vent patch. Often appears rather *small-billed* and has on average *more rounded head and full hindneck*, which is also slightly paler grey.

Red-billed Tropicbird   *Phaethon aethereus*   **V***

L 45–50 cm (plus tail projection on ad. *c.* 45), WS 100–115 cm. Breeds in Cape
Verdes and in South Atlantic, Red Sea and Persian Gulf. Recorded in Britain,
Canary Islands, Israel, Madeira and Portugal. – A pale pigeon-sized bird, *white*
with *black outer primaries*, black spots and a long black patch above on innermost 'arm', a *black eye-stripe* and greyish-black vermiculation on mantle/back
and lesser upperwing-coverts. Adult has *bright red bill* and *very long, thin tail-
streamers*. Juvenile lacks streamers, and has yellowish-brown bill with dark tip
and black terminal tail-band. Often flies high up with mechanical wingbeats,
much as a Sandwich Tern. Often hovers before plunging, and makes brief glides
in searching flight.

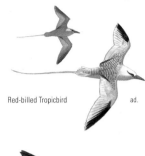

Red-billed Tropicbird   ad.

Masked Booby   *Sula dactylatra*   —

L 80–90 cm, WS 165–185 cm. Breeds in Indian Ocean, S Carribean area and
South Atlantic. Recorded in France, Israel, Morocco and Spain. – Resembles
Gannet but is slightly smaller and has less wedge-shaped or pointed, a little
more bluntly rounded tail. Adult is white with *all blackish flight-feathers*, hence
can be confused with either immature or subadult Gannet retaining dark
flight-feathers longer and more extensively, or with Cape Gannet (*Morus cap-
ensis*; not treated). Masked Booby has *black tertials and longest scapulars*
(white in older Gannets, and in Cape Gannet), and the *'face' is dark bluish-
grey*. Bill shape differs slightly, the *base being deep and heavy* (and tinged
blue-green) and the yellowish tip more attenuated.

Masked Booby

ad.

Pink-backed Pelican   *Pelecanus rufescens*   —

L 125–135 cm, WS 225–260 cm. Breeds in Africa. Recorded occasionally in
several countries within treated region, although frequently difficult to tell wild
birds from free-flying escapes from parks. The following countries or areas usually regard the species as being a genuine vagrant: Canary Islands, Egypt, Israel,
Italy, Morocco and Spain. – Looks like a subadult White Pelican, is not clean
white but *dirty greyish-pink*. Is somewhat *smaller* and has slightly *shorter bill*; in
addition, *lacks dark on culmen and gape*. In spring, a *dark patch by eye*. Feathers
of underparts rather elongated and of loose structure.

Magnificent Frigatebird   *Fregata magnificens*   **V***

L 90–114 cm, WS 215–245 cm. Breeds in Cape Verdes and in Central and South
America. A few definite records in Europe (Britain, Denmark, France, Italy,
Spain), plus several records of indeterminate frigatebirds. – *Very big*, with *long,
narrow, pointed wings, long, deeply forked tail* (normally held closed in flight and
then appears very long and narrow) and retracted head with long, powerful,
cormorant-like bill. Often flies with slow, intermittent wingbeats with slightly
backward-angled 'hand' (pointed carpal), frequently glides and makes deft manoeuvres. ♂ is *all black* with light grey bill (red throat-sac is rarely shown away
from breeding sites), ♀ has *white area on crop* between breast and lower belly,
while immature has in addition a *whitish head*. Separation from other frigatebird
species requires care and can prove impossible at sea; two other species have been
found within treated region, Ascension and Lesser (see p. 418). Magnificent is
largest of them, with green sheen on head but purple on mantle, scapulars and
breast; no white below in adult ♂ (whereas e.g. Lesser has narrow white patch in
'armpit').

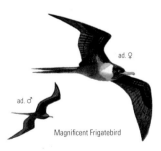

Pink-backed Pelican

ad. ♀

ad. ♂

Magnificent Frigatebird

Dwarf Bittern   *Ixobrychus sturmii*   —

L 27–30 cm. Medium-small bittern, related to Little Bittern of Europe. Breeds
in a wide range in sub-Saharan Africa. A few accepted records from Canary Islands. – Same size and shape as Little Bittern, with short and rounded wings.
Combines *lead-grey upperparts, boldly dark-striped underparts* and a *yellow bill*.
Head and much of upperparts plain bluish-grey. Underparts light yellowish-
buff with broad blackish streaks. The yellow bill has a narrow dark culmen.
Legs pale yellow with often *bright orange long toes*, as a rule clearly visible in
flight behind tail-tip. (Not illustrated.)

Snowy Egret

Intermediate Egret

Goliath Heron

juv.

Yellow-billed Stork

juv.

Pallas's Fish Eagle

DZ

Green-backed Heron *Butorides virescens*   **V***

L 40–47 cm. American close relative of near-cosmopolitan Striated Heron (p. 80), and sometimes lumped with that. The two species are in contact in Central and N South America, with some interbreeding. Has straggled several times to Azores and to Britain, Iceland and Ireland, presumably from E North America. – Size as Striated Heron, and very similar to that. Main differences are overall *darker colours* and *deeper rufous* (almost purplish-tinged) *sides of head, neck and breast* with *more contrasting white streaking on centre of throat and along sides of chin*, and *darker slaty upperparts with hint of blue-green tinge* (Striated paler and greyer above and less deep rufous below). (Not illustrated.)

Snowy Egret *Egretta thula*   **V***

L 55–65 cm, WS 90–105 cm. American species which has been recorded in Britain and Iceland. – Very like closely related Little Egret and, like that, has *dark tarsi with yellow toes*, but told by: almost invariably distinctly *yellowish rear to lower tarsi*, sometimes so extensive that on much of tarsus only front is dark (Little invariably has all-dark tarsus); *more saturated yellow toes* including soles (Little: duller yellow toes with greenish-tinged soles); *bright yellow lores* outside courtship period (blue-grey or greenish on Little); on adult summer *bushier and shorter white nape plume*; slightly smaller, and has shorter neck and legs.

Intermediate Egret *Mesophoyx intermedia*   —

L 60–72 cm, WS 105–115 cm. Breeds in sub-Saharan Africa and in South and East Asia. A few records from Egypt, Israel, Italy and Jordan. – Only slightly larger than Little Egret, but due to the *short and generally all-yellow bill* (only extreme tip dark) recalls more a Great Egret. If *smaller size* not evident, single birds often difficult to tell reliably from Great. *Gape ends below centre, or at least inside rear of eye* (extends behind rear edge in Great), *head is more rounded* (crown peaks at centre) and *neck* usually less kinked, more *smoothly S-curved*. Feet all-dark. Adult lacks head plumes (unlike Little Egret).

Goliath Heron *Ardea goliath*   —

L 135–150 cm, WS 210–230 cm. Breeds in tropical Africa and in E Iraq. Occasional records within treated region, including in Israel (several, spring and autumn), Jordan and Syria (late 19th century). – *Very big* heron, as Grey Heron with *blue-grey upperwing* and white foreneck with black stripes, but with *red-brown crown and hindneck*, dark *red-brown underbody*, *powerful black bill* and black legs. Adult has a *red-brown crest on hindcrown*. Juvenile lacks crest and is somewhat paler.

Yellow-billed Stork *Mycteria ibis*   —

L 95–105 cm, WS 150–165 cm. Breeds in tropical Africa. Seen occasionally in Canary Islands, Egypt, Israel, Jordan and Spain. – Recalls a White Stork, but has *red facial mask* (can be partly orange) and *yellowish bill* (with slightly darker yellowish-orange tip) with slightly downcurved tip. Plumage white, *wing-coverts, shoulders and mantle/back tinged pink*. Legs orange-red. In flight very like White Stork; best told, even at long range, by *black tail*. Immature is sullied grey-brown, has grey-green legs and pinkish-yellow mask.

Pallas's Fish Eagle *Haliaeetus leucoryphus*   —

L 73–84 cm, WS 185–210 cm. Breeds locally and rarely on lakes and rivers in Central Asia and eastward in S Asia. Declining; formerly bred at N Caspian Sea. Vagrants recorded in e.g. Finland, Israel, Norway and Poland. – Adult easily identified by *white, evenly rounded, medium-long tail with broad black terminal band* (as young Golden Eagle), dark brown body and *all-dark wings*, and contrastingly *buffy-white head*. Rather long, bare, light grey tarsi. Immature pale cinnamon-brown on head and body, with dark brown band through and behind eye; in flight, can recall Steppe Eagle at quick glance considering size and *white band along underwing*, but note that this band runs along median coverts, not greater, and that *'armpit' is white* and *inner primaries have very large whitish patches*. Also, all-dark tail is more square.

Bateleur *Terathopius ecaudatus* —

L 50–60 cm, WS 170–190 cm. Breeds in Africa. Vagrants recorded in N Egypt and in Israel. – Unmistakable owing to its long, peculiarly shaped wings with *bulging rear edge to 'arm'* and *constricted inner 'hand'*, so that wing-tip is narrow but still deeply 'fingered'. In addition, *tail is 'docked'*, on adult so short that *feet clearly stick out behind* in flight and head/neck seem disproportionately long and markedly protruding. Flight often fast with elastic wingbeats, holds wings raised in V-shape when gliding, performs flight-rolls, etc. While the young bird has a plumage much as a dark buzzard, the adult develops a more striking appearance with black head, underbody and scapulars, rufous back and tail. *Cere and feet orange-red.* ♂ has all-black remiges, ♀ white with black trailing edge.

American Kestrel *Falco sparverius* **V***

L 23–27 cm. American species, recorded as a vagrant in Britain, Denmark and Malta. – Looks much like a *small* Kestrel, but *wings are subtly shorter and broader.* Along rear edge of wing in flight usually pale subterminal spots, forming *suggestion of paler band.* Often hovers with fanned tail, which is *reddish-brown with broader black terminal band* (broader on ♂). Irrespective of plumage, identified by very contrasty head pattern with *pointed black moustachial stripe* and *two black marks on cheek and nape-side.* Both sexes have *blue-grey crown with reddish-brown centre* and black-barred red-brown mantle/back, but ♂ has *blue-grey wings*, ♀ *red-brown.* Juvenile ♂ has bolder barring on back than adult ♂.

Amur Falcon *Falco amurensis* **V***

L 26–30 cm, WS 63–71 cm. Breeds in SE Siberia, E Mongolia and E China, wintering in SE Africa; main movements involve crossing the Indian Ocean. Recorded in Britain, Hungary, Italy and Sweden. – The Far Eastern counterpart of Red-footed Falcon, closely related to this but a fraction smaller and more compact in proportions. Adult ♂ differs in having *pure white underwing-coverts* and on average slightly darker crown, mantle and leading edge to wing. Adult ♀ differs from Red-footed by having *white* (not rufous-buff) *underparts with bold black streaking on breast and barring on flanks* (Red-footed: finely streaked or spotted only). Also, *crown is dark grey*, not rufous-buff. Juvenile very similar to both juvenile Red-footed and Hobby, differing only in *hint of dark barring on lower flanks* sometimes visible.

Allen's Gallinule *Porphyrio alleni* **V***

L 22–25 cm. Breeds in tropical Africa. Vagrants recorded in several countries, incl. Canary Islands, Finland, France, Germany, Greece, Israel, Morocco, Poland and Spain (and pre-1950s in Britain). Frequents dense, wet swamps, difficult to see. – *Small*, no bigger than a Water Rail. *Short all-red bill* and *dark blue and green* plumage. Like Purple Gallinule (below), but clearly smaller, has *red legs* and *dark central undertail-coverts* (as on Moorhen), so that white at stern is divided in middle. Juvenile looks like cross between young Ruff and ♀ Little Crake (!).

Purple Gallinule *Porphyrio martinica* **V***

L 29–33 cm. Breeds in North America. Vagrant in Britain, Canary Islands, Iceland, Norway and Switzerland. – Somewhat Moorhen-like, but has *no white line along flanks*, legs are a bit longer, and *underparts are deep ultramarine-blue* and *upperparts greenish.* Bill red with yellow tip, but *frontal shield blue-grey.* Like Allen's Gallinule but *bigger*, with *all-white undertail-coverts* (lacking dark centre) and *yellow legs.*

American Coot *Fulica americana* **V***

L 31–37 cm. Breeds in North America. Recorded in, among others, Britain, Iceland, Ireland and Portugal. – Like European Coot, but differs in *dark band across whitish outer bill* (bill is tinged bluish, and dark band is purplish when seen close), *white sides to undertail-coverts* and a small *dark* (reddish) *spot on uppermost frontal shield.* Is also somewhat slimmer, has *straighter back profile* and has *paler grey body plumage* than European Coot.

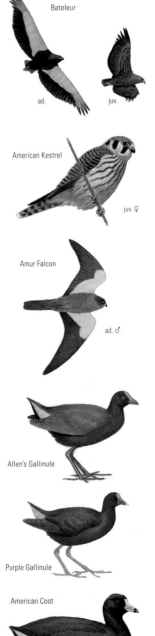

Bateleur
ad. juv.

American Kestrel
juv. ♀

Amur Falcon
ad. ♂

Allen's Gallinule

Purple Gallinule

American Coot

DZ

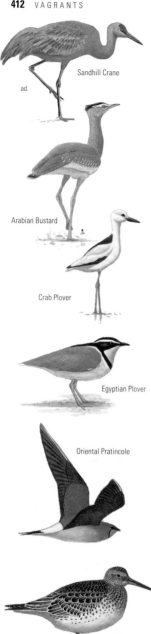

Sandhill Crane

ad.

Arabian Bustard

Crab Plover

Egyptian Plover

Oriental Pratincole

Great Knot

juv.

DZ

Sandhill Crane *Grus canadensis*   V***

L 80–95 cm, WS 155–180 cm. Breeds in North America and NE Siberia and winters in S North America and Mexico. Vagrant in Britain, Faeroes, Ireland and Netherlands. – Nominate race, the one which has been recorded in Europe, is *smaller than Crane*. Plumage is *grey* with diffuse spotting; breeding birds acquire rusty tone above, as Crane. Adult has *white throat and side of head* and *red forehead/forecrown*. Juvenile harder to separate from juvenile Crane, but is *more distinctly rusty-brown above on back and wing-coverts* and on head.

Arabian Bustard *Ardeotis arabs*   —

L 70–90 cm. Breeds in Africa S of Sahara. Formerly bred in Morocco, where still found occasionally. – *Grey-brown* and *big*, with *finely vermiculated neck feathering* (thick and bushy on adult ♂). Nape shows a *short crest*, enhanced by a black stripe along crown-side; crest mostly held flat, not very obvious except during display. Has *less white on wing* (when seen in flight) than other bustards.

Crab Plover *Dromas ardeola*   —

L 38–41 cm. Breeds in Persian Gulf, Oman and Gulf of Aden. Vagrant in Egypt, Israel, Syria and Turkey. – *Large, black and white*. Resembles if anything an Avocet whose thin upcurved *bill* has been replaced by a *straight, thick, dagger-shaped* one. Often stands erect like a Cream-coloured Courser, but can crouch like a plover, too. Actions plover-like: runs, stops and watches, runs, and so on. In flight note all-white upper forewing but rest of upperwing black (whereas distant flying Avocet has white upperwing with black wing-tip and a black covert-bar on forewing). Flight-call somewhat Gull-billed Tern-like, '**wedd**e-vek, **wedd**e-vek,…'.

Egyptian Plover *Pluvianus aegyptius*   —

L 19–21 cm. Breeds in tropical Africa. Recorded in Libya and Egypt, and formerly bred near Cairo. Reported from Israel and Canary Islands. – A rather small and *compact* wader with *broad wings* and shortish legs. Plumage unmistakable, exceptionally smart with *broad black and white stripes on wings*, black and white head pattern and *orange underparts* with *black breast-band*. Frequently bobs head and fore body up and down when walking. Tame, allowing close approach. Almost always seen by rivers.

Oriental Pratincole *Glareola maldivarum*   V***

L 23–27 cm. Breeds in South Asia from India eastward. Vagrant in Britain, Cyprus, Israel, Netherlands and Sweden. – *Underwing reddish-brown* (as on Collared Pratincole), but *upperwing dark* and *without clear white trailing edge to 'arm'* (as on Black-winged Pratincole). Note: *tail-streamers short*, terminate 1–3 cm *short of wing-tips* on standing bird, give stub-tailed appearance in flight; often *warmer orange-buff tone to upper belly* (below brown-grey breast patch); rounded or oval-shaped nostrils (relatives have narrowly slot-shaped, although this requires close range to be established); on average *broader black and white frame to bib, blacker lores* than on Collared and *less red on base of lower mandible* than on Collared. Identification pitfall exists in that odd Collared Pratincoles may wear off much of white trailing edge to 'arm', and immatures and subadults rarely have shorter tail-streamers than usual. Important to secure as complete a description as possible of any potential Oriental.

Great Knot *Calidris tenuirostris*   V***

L 24–27 cm. Breeds in NE Siberia. Vagrant in Britain, Denmark, Israel, Morocco, Norway, Poland and Sweden. – Marginally bigger than a Knot, and with different proportions: longer neck, *smaller head, longer and slightly decurved bill with thick base but thinner tip*, long wings giving *attenuated, protracted rear end*. In summer plumage distinctive, with *coarse dark breast spotting, arrowheads on flanks, contrasty upperparts* with broadly white-fringed dark feathers, and *rusty-brown on shoulders*. In winter plumage resembles Knot, but more distinctly streaked above and spotted on breast.

Wilson's Snipe *Gallinago delicata*   **V**∗∗∗

L 23–28 cm. American counterpart of European Snipe, but recently separated on account of different vocalization and slight but apparently consistent morphological differences. Breeds in North America. Recorded in Britain, France and Ireland. – Very similar to Snipe, with same size and shape. Display-sound differs clearly, is deeper (almost recalling Tengmalm's Owl!), but vagrants to Europe will most likely be silent. Safe identification by plumage characters requires close views and use of as many characters as possible. *White trailing edge to secondaries narrower* (often < 2 mm), but some overlap; underwing-coverts more narrowly tipped white resulting in *darker underwing*; more and denser barring of *outer tail-feathers* with typically 3–4 distinct dark bars visible on rather white ground, dark bars about same width as pale (Snipe: often 1–2 indistinct bars on greyish-buff ground, dark bars narrower than pale). (Not illustrated.)

Swinhoe's Snipe *Gallinago megala*   **V**∗∗∗

L 26–31 cm. Breeds in taiga in C and E Siberia, but possibly overlooked further west. At least a Russian record from European N Ural, and recently a ♂ displayed in SE Finland. So, although still only two records known from covered range, included here anticipating several more in years to come. – Slightly larger and bulkier than Snipe but somewhat smaller than Great Snipe. Very similar to both Pin-tailed Snipe and Snipe, best told by *display-sound* (a whinnying sound given in flight, with increasing loudness and strange hollow tone, ending in pulsating flourish often followed by subdued harsh notes, 'krrek-krrek-krrek-krrek', like from wooden rattle). Identification using structure and plumage requires close range and careful judgement, and *plumage characters apparently largely overlap with Pin-tailed Snipe*. Differs from latter on *lack of 6–7 pin-like outer tail-feathers* (2–5 are narrower but not pin-like), on *slightly greater bulk* and *longer tail with toes not protruding much* in flight, from Snipe on *lack of broad white trailing edge to secondaries*, *darker underwing* and on average *shorter bill*. Flight call might be useful but differences poorly known. (Not illustrated.)

Hudsonian Godwit *Limosa haemastica*   **V**∗∗∗

L 37–42 cm, WS 67–79 cm. North American species, recorded a few times in Britain, Denmark, Norway and Sweden. – Not unlike Bar-tailed Godwit in size and shape (but has even *more obviously upturned bill* and is somewhat more *long-legged*), but closer to Black-tailed in plumage, having (hint of) *white wing-bar*, *white rump* and *all-black tail*. The wing-bar is well-marked only on inner primaries. In non-breeding plumages, *breast is unpatterned grey-brown*. In flight, shows *diagnostic blackish 'armpit' and underwing-coverts* in strong contrast to paler rest of underwing. Call a quick double-sound, 'toe-wit', or single 'wit'.

juv.

Hudsonian Godwit

Willet *Tringa semipalmata*   **—**

L 31–36 cm, WS 54–62 cm (eastern nominate race). Breeds in North America. Vagrants observed in Finland, France, Italy and Norway. – The size and general shape of a Greenshank but differing in *shorter, straighter and bulkier bill* (culmen straight, only lower mandible slightly upturned); *shorter, thicker neck*; more *extensively patterned and cross-barred breast and flanks*; and *very bold white wing-bar* on upperparts contrasting with *blackish primaries and primary-coverts* and, below, with blackish rest of underwing. Western race *inornata* (possibly a separate species) is larger with longer legs and bill. Flight-call a fast, clear trilling 'kli-li-li' or a bisyllabic 'kee-lit'. (Not illustrated.)

Grey-headed Gull *Chroicocephalus cirrocephalus*   **—**

L 39–43 cm, WS 100–115 cm. Breeds in tropical Africa and in South America. Recorded in Israel, Jordan, Morocco, Spain and Tunisia. – A bit *bigger than Black-headed Gull*, with slightly larger head. Has *more black on wing-tip*, and in addition has *two white spots in the black* near tip. *Underwing dark*. Adult summer has *light grey hood*. Wings *broader* and somewhat *blunter* than on Black-headed. Bill rather long, and *feathering of forehead at bill-base drawn out and shallowly sloping* as on Slender-billed Gull. Adult has *whitish eye*, neat red orbital ring and white eye-ring.

Grey-headed Gull

1st-winter

DZ

Crested Tern

(Greater) **Crested Tern** *Sterna bergii* —

L 50–54 cm (incl. tail-streamers 7–10). Breeds from S Africa to Pacific Ocean, incl. Red Sea and Persian Gulf. Frequently moves up to Suez region, and straggles to Eilat in Israel. – *Very big*, long-winged. *White forehead* even in breeding plumage (though narrow right at front). Black crown with *bristly, loose crest*. *Bill very long and fairly narrow* with slightly decurved culmen, usually *pale greenish-yellow* (a bit brighter yellow during courtship). In Middle East (ssp. *velox*), mantle, *rump and upperwing rather dark grey* (esp. seen in overcast weather) without light blue tinge of e.g. Lesser Crested Tern (p. 202). Commonest call a deep croaking 'karrack'.

Elegant Tern

Elegant Tern *Sterna elegans* V***

L 39–43 cm. Breeds in W Central America. Remarkably, one took up residence during 1974–85 in a Sandwich Tern colony in France, where it paired with a Sandwich and produced young; also seen in 1987, and two different individuals in 1984. Vagrants recorded also in Belgium, Denmark, Ireland and Spain. – Very like Lesser Crested Tern, but has *somewhat longer and more decurved bill* which is often *slightly brighter orange* but with tip sometimes a bit yellower. *Shaggier crest* than Lesser Crested. In winter plumage, white forehead but with *black neatly around eye* and backwards. Elimination of aberrant yellow-billed Sandwich Tern (very rare, still a possibility) requires careful observation and good judgement. Also confusable with Royal Tern (p. 202), but differences include longer, narrower bill.

Chestnut-bellied Sandgrouse

Chestnut-bellied Sandgrouse *Pterocles exustus* —

L 28–32 cm (excl. tail projection of ad. 4–8), WS 50–56 cm. Breeds in Africa. In 19th century bred at Suez, and occurred sparingly there as late as 1929; now extinct in N Egypt. Vagrant in Hungary. – *Brownish-black belly*. Told from Black-bellied Sandgrouse by smaller size, *all-dark underwing* and *narrow tail projections*, from Spotted Sandgrouse by e.g. *dark primaries above*. When perched, ♂ looks like dirty or washed-out Black-bellied. ♀ has very boldly spotted throat and breast. In flight, utters Willow Ptarmigan-like grating 'vitt-karr-arr'.

Pallas's Sandgrouse

Pallas's Sandgrouse *Syrrhaptes paradoxus* V***

L 27–32 cm (excl. tail projection of ad. 4–11). Breeds on Asian steppes. Rare and irregular vagrant in W Europe (in Britain most recently 1990), at times in minor invasions which can lead to breeding (e.g. 1888–89). – *Black belly patch*, rather *pale wings above and below*. ♂ told from similar Spotted Sandgrouse by having *mantle to rump heavily barred black* while *upperwing-coverts are uniform* in colour. Diagnostic needle-like pointed outer primaries rarely seen in the field. ♀ has more black spots on crown, nape and wing-coverts than ♂. Flight-call a deep trisyllabic 'ku-ke-**rik**', repeated a few times.

African Collared Dove

African Collared Dove *Streptopelia risoria roseogrisea* —

L 28–30 cm. Breeds in sub-Saharan Africa and Arabian peninsula. Occasional or rare autumn/winter vagrant in S Israel and Jordan. See also Barbary Dove (p. 424). – Very like Collared Dove, but *a bit smaller* and has slightly *shorter tail*. Plumage is a shade *paler* than Collared's, with *whiter belly*, but *flight-feathers* and *uppertail* on the other hand are *darker*, giving slightly greater contrast. The multisyllabic song is recognized by the first syllable being higher and drawn out, followed by hint of a pause and then a falling section with *rolling r-sound*, 'kaaw, kurroo-ooh'.

(American) **Mourning Dove** *Zenaida macroura* V***

L 28–33 cm. Breeds abundantly in North America in most habitats, including cities. Recorded in Britain, Denmark, Germany, Iceland and Ireland, but three of these refer to same individual. A Swedish record was most likely an escaped bird. – Size about as a Collared Dove but perhaps slightly *slimmer* and shape more elegant. *Long pointed tail*, with *white sides* visible when spread in flight. Colours *pale pinkish-brown* with grey cast above and on nape, and *a few scattered large black blotches on upperparts.* Eye and bill dark, legs pink. ♂ has more purplish iridescence on nape and sides of neck than ♀. Both sexes have a small dark spot on lower cheek. Whistling wing-noise in flight. (Not illustrated.)

Chimney Swift

White-throated Needletail

Pacific Swift

Yellow-bellied Sapsucker

Cliff Swallow

Chimney Swift *Chaetura pelagica*   V***

L 12–13 cm. Breeds in North America. Autumn vagrant in Britain (several), Canary Islands, France, Ireland, Sweden. – *Small*, with *thickset body* and broad wings, in particular with *broad outer wing* (secondaries shorter, and wing can appear almost pinched-in where primaries meet secondaries). Plumage rather *uniform grey-brown*, just *a bit paler on throat and upper breast*. Fine tail-spines seen only at close range. Flight is one moment fast on backswept wings, the next slower with fast, shallow wingbeats relieved by glides. Call a rapid twitter.

White-throated Needletail *Hirundapus caudacutus*   V***

L 19–21 cm, WS 50–54 cm. Breeds in S Siberia and Central Asia. Vagrant in Britain, Faeroes, Finland, Ireland, Netherlands, Norway and Sweden. – *Big*, with *heavy, compact body*, neckless, stub-tailed (shape something between fat cigar and 'flying barrel'). *Flight impressively fast*, the bird seems to draw easily away from other swifts (though these are still fast flyers!). Identified otherwise by *white vent* (which extends a bit onto flanks) and *brownish-white back*.

Pacific Swift *Apus pacificus*   V***

L 18–19 cm. Breeds in Asia, from C Siberia eastwards. Vagrant in Britain and Sweden. – Like a Common Swift with *narrow white rump patch*. To eliminate possibility of a partially albinistic Swift, look for: slightly *deeper tail-fork*; longer wings; slightly *more projecting, heavier head*, which can give impression of *slimmer body*; often *somewhat larger white chin patch*; pale-scaled underbody; coarser and *harsher call clearly dropping at end*, 'vriiüh' (can recall Pallid Swift).

Yellow-bellied Sapsucker *Sphyrapicus varius*   V***

L 18–21 cm. North American species. Migratory. Somewhat recalls Three-toed Woodpecker, e.g. in habit of drinking sap. Vagrant in Britain, Iceland and Ireland. – Basically *black and white*, adult with *red crown*, and ♂ also with *red throat* (white on ♀). *Side of head broadly striped black and white.* Throat framed by black band, which *on breast* widens into *broad black patch* (lacking on 1st-winter). There is a *white rectangular patch on wing*. Mantle and back finely mottled black and white. Underparts tinged yellowish (sometimes most strongly on breast-side).

(American) Cliff Swallow *Petrochelidon pyrrhonota*   V***

L 13–14 cm. Breeds in North America. Vagrants recorded in autumn in Britain, Canary Islands, France, Iceland and Ireland. – Most closely related to House Martin and similar in shape (but with *square-ended tail* and *broader wings*). Distinctively coloured head, with *cheeks, throat and neck-sides red-dish-brown, nape pale* so that blue-black crown forms isolated skull-cap, and has *creamy-white forehead*. *Rump orange.* At distance, confusion with Red-rumped Swallow conceivable, but *lacks tail-streamers* and *back has diffuse white streaking*. Greyish-black patch between throat and breast. Call a nasal, sparrow-like 'chiev'.

Cedar Waxwing *Bombycilla cedorum*   V***

L 15–19 cm. Breeds in North America, with similar habits and habitat selection as Waxwing, although also breeds in more open country. Vagrant in Britain and Iceland. – *Smaller* than Waxwing with plainer plumage pattern. Main difference is *absence of yellow or white tips to primaries, absence of white tips to primary-coverts* and *dull grey-white vent* (Waxwing: deep rufous vent). Also, belly is more tinged yellowish, not greyish-pink, and secondaries are all-dark, lacking white tips or edges. Black 'mask' has *narrow white outline* also above, *running across forehead* (white only below 'mask' in Waxwing). Call like Waxwing but higher-pitched and feebler. (Not illustrated.)

Catbird

Black Bush Robin

ad. ♂

Pied Bushchat

Long-tailed Shrike

ad.

Daurian Jackdaw

ad.

Evening Grosbeak

ad. ♂

DZ

(Grey) Catbird *Dumetella carolinensis*   V***

L 19–22 cm. Relative of Northern Mockingbird. Breeds in North America. Vagrant in Belgium, Britain, Channel Islands, Germany and Ireland. – Like a small thrush with long tail. Adult is *uniform ash-grey*, with *black crown* and *dark reddish-brown vent*. Hops on ground or keeps to low vegetation.

Black Bush Robin *Cercotrichas podobe*   —

L 20–23 cm. Breeds in Africa S of Sahara, and in Arabia. First recorded in Israel in 1981, bred at Eilat in 1994 and is regular there since. – *Entirely sooty-black*, with *long tail* which is often *held straight up* and quickly spread or flicked. *Tail-feathers broadly tipped white*, undertail-coverts narrowly tipped white. Long-legged. Perches mostly on ground or in low bushes. Sings from bushtop; song rather like that of Rufous Bush Robin, often accompanied by tail-raising.

Siberian Blue Robin *Luscinia cyane*   V***

L 12–13 cm. Breeds in Siberia. Vagrant in Britain, Channel Islands and Spain – *Small*, compact thrush with *short tail*, *strong pale pinkish feet* and strong, *pointed bill*. Adult ♂ *dark blue above, pure white below* with dusky or blackish border along lores, cheeks and sides of breast. ♀ and immature brown above except for *blue-tinged tail*, buffish-grey-white below. (Not illustrated.)

Pied Bushchat *Saxicola caprata*   —

L 13–14½ cm. Breeds in Central Asia and South Asia. Vagrant in Cyprus and Israel. – Somewhat bigger than Stonechat. Adult ♂ is *all black, with white rump*, white belly and small *white shoulder patch*. ♀ and immature ♂ are brownish-grey like ♀ Black Redstart, but differ from latter in *shorter all-dark tail* with only rump rusty-coloured; told from dark ♀ Stonechat of race *rubicola* by *all-dark wing* (just hint of pale patch on some) and *paler throat*.

Eastern Crowned Warbler *Phylloscopus coronatus*   V***

L 11–12 cm. Breeds in E Asia. Vagrant in Finland, Germany, Netherlands and Norway. – Resembles Arctic Warbler but differs on *palish central crown-stripe* (prominent at rear, more diffuse at front), *darker olive-grey eye-stripe and lateral crown-sides* and *supercilium being yellow in front of eye* but less so behind. *Underparts rather whiter* than in Arctic, and *upperparts purer green*. (Not illustrated.)

Long-tailed Shrike *Lanius schach*   V***

L 21½–24 cm. Breeds in Asia. Vagrant in Britain, Denmark, Israel, Jordan, Sweden and Turkey. – Rather big and *long-tailed*, tail strongly rounded. Black facial mask and forehead (smaller than on ♂ Lesser Grey Shrike), light grey on crown and mantle. *Rump, back and scapulars are rusty yellow-brown*, and vent, too, is rusty-yellow. Small white primary-base patch on adult. Deep bill.

Daurian Jackdaw *Corvus dauuricus*   —

L 30–34 cm. Breeds in SE Siberia, Mongolia and NE China. Vagrant in Denmark, Finland, France, Germany, Netherlands and Sweden. – Same size and shape as Jackdaw, but differs in all plumages in having *dark eye* (ad. Jackdaw has whitish iris, juv. grey) and in adult plumage *pale grey-white nape, neck-ring and belly* (not unlike a well-marked 'dwarf Hooded Crow'). Silvery-white feather tips at rear of head-side. 1st-year plumage either similar (light morph; less common) or dark grey like Jackdaw (dark morph; seems to be the commoner), with only hint of paler, slate-grey belly; dark morph lacks Jackdaw's paler grey nape.

Evening Grosbeak *Hesperiphona vespertina*   V***

L 18–20 cm. North American species. Vagrant in Britain and Norway. All records in spring, several involving ♂♂. – Starling-sized. *Big, triangular, light greenish-yellow bill*. ♂ is *yellow and olive-brown* with black tail and *black wings with white tertials*; forehead and crown-sides yellow, crown-centre black. ♀ more reminiscent of Hawfinch, buffy greyish-white and grey-brown with white patches on wings and tail; *usually has yellow tinge on neck-side*. Yellow axillaries.

Tennessee Warbler *Vermivora peregrina*   **V***

L 10½–12 cm. Breeds in North America. Autumn vagrants recorded in Britain,
Faeroes and Iceland. (New World warblers, family *Parulidae*, are like thin-billed
buntings and fill niche corresponding to that occupied by warblers in Europe.)
– In autumn plumage can be taken for a *Phylloscopus* warbler owing to its *green
upperparts, yellowish underparts* and *yellow supercilium*. Legs dark grey, pointed
bill rather long and light grey with dark culmen. Call a sharp 'tsiit'.

Tennessee Warbler   winter

Ovenbird *Seiurus aurocapilla*   **V***

L 13½–15 cm. North American species. A *ground-dwelling*, forest bird. Autumn
vagrants recorded in Britain, Ireland and Norway. – *Uniform greenish-brown
above*, pale below with *bold black spots* and narrow black lateral throat stripes.
Pale eye-ring on otherwise uniform brown side of head. *Reddish-brown median
crown-stripe bordered by narrow black stripes*. Trips along on ground like a pipit.

Ovenbird

(Common) Yellowthroat *Geothlypis trichas*   **V***

L 12–13 cm. North American species, recorded in Britain and Iceland. – Told
by *unmarked* plumage, *olive-green above* and off-white below with *yellow throat*
and *yellowish undertail-coverts*; flanks brownish-white. Adult ♂ has a *black
mask* extending over forehead, bordered at rear by light grey. 1st-winter ♂ has
dusky ear-coverts and a little dark colour on forehead. ♀ has olive-brown head.

Yellowthroat   1st-winter. ♂

Scarlet Tanager *Piranga olivacea*   **V***

L 15½–17 cm. North American species; belongs to family *Thraupidae*, found
only in New World. Vagrants recorded in Britain, France, Iceland and Ireland,
nearly all in Oct. – *Bill heavy, rather long* and pointed. In autumn plumage, *head
and body greyish yellow-green*, palest below. Adult ♂ has brightest yellow colour
below, yellow tone to crown and greener mantle/back. *Wings darker than back*,
grey on ♀, *black* on adult ♂ and grey with some blacker coverts and tertials on
1st-winter. (Not illustrated.)

Song Sparrow *Melospiza melodia*   **V***

L 13–15 cm. North American species. A few spring and summer records in
Britain and Norway. – Medium-sized, *long-tailed*. Short, rounded wings. Bill
size variable, often rather narrow and pointed. *Brown lateral crown-stripes* and a
grey, *paler median stripe*, also *grey head-side* with *brown stripes backwards from
eye, along lower edge of ear-coverts and on throat-side*. Distinct dark streaks below,
on upper breast usually coalescing to form *a small dark patch*.

Song Sparrow

White-crowned Sparrow *Zonotrichia leucophrys*   **V***

L 15–16 cm. North American species. Vagrant in Britain, France, Iceland, Ireland and Netherlands. – Close relative of White-throated Sparrow (p. 406), but
differs as follows: lacks clearly defined white throat; has *very broad white median
crown-stripe* with *bold black border*; white supercilium *never yellow in front of eye*;
cheeks and neck-sides plain ash-grey, form *broad pale neck-collar*; upperparts
grey, not brown. 1st-winter has brown and buffish-white head markings.

White-crowned Sparrow

Black-faced Bunting *Emberiza spodocephala*   **V***

L 14–15½ cm. Breeds in Siberia in waterlogged forest or on overgrown bogs.
Vagrant in Britain, Denmark, Finland, Germany and Netherlands. – A streaked
bunting with slightly '*soiled' yellowish-white, grey and brown* plumage. Adult ♂
has *olive-grey head and breast* and swarthy '*face*'. Autumn birds are more nondescript; note *uniform* (or only diffusely spotted) *brownish-grey rump, olive-grey
tone to cheek and neck-side*, dirty and *rather indistinct paler supercilium, yellow
tinge to underparts*, straight culmen and partially pink lower mandible. Call a
sharp 'zit', similar to that of Song Thrush.

Black-faced Bunting   ad. ♂

DZ

Accidentals

The species listed below have been recorded only 1–3 times (in a few cases more) within the region treated in this book (roughly corresponding to the Western Palearctic; see definition on p. 8). Apart from English and scientific names, the origin of each species is given (main or closest breeding range) and where and when they have been recorded. Records which in all likelihood refer only to escapes from captivity, or the authenticity of which may be questionable, have been excluded from the list. Published records have been covered until the end of June 2009.

SPECIES		ORIGIN	RECORDED
Ostrich	*Struthio camelus*	Africa	Egypt (old records in Sinai), Israel (occasional up to 1920s), Jordan (1930s?)
Fulvous Whistling Duck	*Dendrocygna bicolor*	America, Africa, India	Morocco (1977; Sep 1980)
Lesser Whistling Duck	*Dendrocygna javanica*	South Asia	Israel (Nov 1966–Mar 1967)
Spur-winged Goose	*Plectropterus gambensis*	Africa	Morocco (Jun 1984)
Cotton Pygmy Goose	*Nettapus coromandelianus*	South Asia, Australia	Jordan (Apr 1997)
Red-billed Teal	*Anas erythrorhyncha*	Africa	Israel (Jun–Jul 1958)
Southern Pochard	*Netta erythrophthalma*	Africa, South America	Israel (Apr–May 1998)
Yellow-nosed Albatross	*Thalassarche chlororhynchos*	South Atlantic	Britain (Jun 2007), Norway (Apr 1994), Sweden (Jul 2007)
Shy Albatross	*Thalassarche cauta*	South Atlantic, Indian Ocean	Egypt and Israel (Feb–Mar 1981)
Wandering Albatross	*Diomedea exulans*	South Atlantic, Indian Ocean	Italy (Oct 1957)
Tristan Albatross	*Diomedea dabbenena*	South Atlantic	Italy (Oct 1957)
Southern Giant Petrel	*Macronectes giganteus*	Antarctica	Italy (Sep 1991)
Cape Petrel	*Daption capense*	Southern Oceans	Spain (Gibraltar, Jun 1979)
Soft-plumaged Petrel	*Pterodroma mollis*	South Atlantic	Israel and Jordan (Mar 1997)
Atlantic Petrel	*Pterodroma incerta*	South Atlantic	Israel (May 1982; Apr 1989), Jordan (Apr 1989; Mar 1997)
Black-capped Petrel	*Pterodroma hasitata*	West Indies	Britain (spring 1850; Dec 1984), Spain (Apr 2002)
Streaked Shearwater	*Calonectris leucomelas*	Pacific & Indian Oceans	Israel and Jordan (Jun–Sep 1992; May–Jul 1993)
Flesh-footed Shearwater	*Puffinus carneipes*	Indian & Pacific Oceans	Israel and Jordan (Aug 1980)
Wedge-tailed Shearwater	*Puffinus pacificus*	Indian & Pacific Oceans	Egypt (Mar 1988)
Audubon's Shearwater	*Puffinus lherminieri*	Indian & Pacific Oceans, West Indies	Egypt (Sep 1991), Israel (Jan–Feb 1985; Dec 1989)
Double-crested Cormorant	*Phalacrocorax auritus*	North America	Britain (Jan–Apr 1989), France (Oct 2000), Ireland (Nov 1995–Jan 1996)
African Darter	*Anhinga rufa*	Africa	Israel (occasional 1950s), Morocco (Aug 1985), Turkey (bred 1933–*c.* 1950)
Ascension Frigatebird	*Fregata aquila*	South Atlantic	Britain (Jul 1953)
Lesser Frigatebird	*Fregata ariel*	Tropical Oceans	Israel and Jordan (Dec 1997; May 1999)
Least Bittern	*Ixobrychus exilis*	America	Iceland (Sep 1970)
Schrenck's Bittern	*Ixobrychus eurhythmus*	East Asia	Italy (Nov 1912)
Indian Pond Heron	*Ardeola grayii*	Asia	Egypt (Apr 2004)
Little Blue Heron	*Egretta caerulea*	America	Ireland (Sep–Oct 2008)
Tricoloured Heron	*Egretta tricolor*	America	Canary Islands (Nov 2007–Jun 2008)
Black Heron	*Egretta ardesiaca*	Africa	Israel (Oct 1982)

SPECIES		ORIGIN	RECORDED
Black-headed Heron	*Ardea melanocephala*	Africa	France (c. 1845; Nov 1971), Israel and Jordan (Oct–Dec 1987)
Great Blue Heron	*Ardea herodias*	America	Britain (Dec 2007), Canary Islands (Dec 1998–Jan 1999), France (Apr 1996)
Marabou Stork	*Leptoptilos crumeniferus*	Africa	Israel (May 1951; Apr–May 1957)
Sacred Ibis	*Threskiornis aethiopicus*	Africa, Iraq	Azerbaijan (1944), Canary Islands (Mar 1991), Egypt (1889; 1891). – (See also p. 423.)
Bald Eagle	*Haliaeetus leucocephalus*	North America	Britain (Oct 1978), Ireland (Jan 1973; Nov 1987),
Shikra	*Accipiter badius*	Asia, Africa	Azerbaijan (Jun 1933; bred), Israel (Apr 1987)
Striped Crake	*Porzana marginalis*	Africa	Algeria (Jan 1867), Italy (Jan 1997), Libya (Feb 1970), Malta (Mar 1981; Apr 2004)
African Crake	*Crex egregia*	Africa	Canary Islands (Nov 2001; Nov 2006; Jan 2007)
Black Crake	*Amaurornis flavirostra*	Africa	Madeira (Jan 1895)
Hooded Crane	*Grus monacha*	Asia	Russia (date?)
Three-banded Plover	*Charadrius tricollaris*	Africa	Egypt (six records Mar 1993–Nov 2007)
Oriental Plover	*Charadrius veredus*	Asia	Finland (May 2003)
Black-headed Lapwing	*Vanellus tectus*	Africa	Israel (1869; 1995), Jordan (1995)
American Woodcock	*Scolopax minor*	America	France (Oct 2006)
Eskimo Curlew	*Numenius borealis*	North America	Britain (5 autumn records 19th cent.), Ireland (Oct 1870)
Grey-tailed Tattler	*Tringa brevipes*	Siberia	Britain (Oct–Nov 1981; Nov–Dec 1994), Sweden (Jul 2003)
South Polar Skua	*Stercorarius maccormicki*	Antarctica	Faeroes (Sep 1889), Israel (Jun 1983), Israel and Jordan (Jun 1992),
Brown-headed Gull	*Chroicocephalus brunnicephalus*	Central Asia	Israel (May 1985)
Relict Gull	*Larus relictus*	Asia	Russia (May 2000)
Glaucous-winged Gull	*Larus glaucescens*	W North America, N Pacific Ocean	Britain (Dec 2006 and spring 2007; Dec 2008), Morocco (Jan 1996)
Kelp Gull	*Larus dominicanus*	coasts of S Hemisphere	Canary Islands (Apr 2001), Morocco (Aug 2006, Feb 2008)
Slaty-backed Gull	*Larus schistisagus*	Asia	Lithuania (Nov 2008)
Aleutian Tern	*Onychoprion aleutica*	Bering Sea	Britain (May 1979)
Least Tern	*Sternula antillarum*	America	Britain (summers 1983–92; same bird)
Brown Noddy	*Anous stolidus*	West Indies, South Red Sea, Persian Gulf	Germany (Oct 1912)
African Skimmer	*Rynchops flavirostris*	Africa	Israel (c. 1934), Morocco (Jan 1987)
Long-billed Murrelet	*Brachyramphus perdix*	Pacific Ocean	Britain (Nov 2006), Romania (2006), Switzerland (Dec 1997)
Ancient Murrelet	*Synthliboramphus antiquus*	Pacific Ocean	Britain (springs 1990, 1991 and 1992; prob. same bird)
Crested Auklet	*Aethia cristatella*	Bering Sea	Iceland (Aug 1912)
Parakeet Auklet	*Aethia psittacula*	Bering Sea	Sweden (Dec 1860)
Tufted Puffin	*Fratercula cirrhata*	Pacific Ocean	Britain (Sep 2009), Sweden (Jun 1994)
Yellow-eyed Dove	*Columba eversmanni*	Asia	Russia (date?)
Didric Cuckoo	*Chrysococcyx caprius*	Africa	Cyprus (Jun 1982), Israel (Mar 1994)
Indian Roller	*Coracias benghalensis*	Iraq, South Asia	Syria (date?)
Northern Flicker	*Colaptes auratus*	North America	Denmark (May 1972)
Eastern Phoebe	*Sayornis phoebe*	North America	Britain (Apr 1987)

SPECIES		ORIGIN	RECORDED
Acadian Flycatcher	*Empidonax virescens*	North America	Iceland (Nov 1967)
Least Flycatcher	*Empidonax minimus*	North America	Iceland (Oct 2003)
Alder Flycatcher	*Empidonax alnorum*	North America	Iceland (Oct 2003)
Fork-tailed Flycatcher	*Tyrannus savana*	North America	Spain (Oct 2002)
Chestnut-headed Sparrow-lark	*Eremopterix signatus*	East Africa	Israel (May1983)
Hume's Short-toed Lark	*Calandrella acutirostris*	E Central Asia, Himalayas	Israel (Feb 1986)
Tree Swallow	*Tachycineta bicolor*	North America	Britain (Jun 1990; May 2002)
Purple Martin	*Progne subis*	North America	Britain (Sep 2004)
Ethiopian Swallow	*Hirundo aethiopica*	Africa	Israel (Mar 1991)
Northern Mockingbird	*Mimus polyglottos*	North America	Britain (Aug 1882; Feb–Mar 1996), Netherlands (Oct 1988)
Brown Thrasher	*Toxostoma rufum*	North America	Britain (Nov 1966–Feb 1967)
Rufous-tailed Robin	*Luscinia sibilans*	Siberia	Britain (Oct 2004), Poland (Dec 2005)
Eversmann's Redstart	*Phoenicurus erythronotus*	Central Asia, NW Mongolia	Israel (Nov 1988), Russia (Oct 1881; Oct 1888)
Variable Wheatear	*Oenanthe picata*	Central Asia	Israel (Feb 1986)
Varied Thrush	*Ixoreus naevius*	North America	Britain (Nov 1982), Iceland (May 2004)
Wood Thrush	*Hylocichla mustelina*	North America	Britain (Oct 1987), Iceland (Oct 1967)
Tickell's Thrush	*Turdus unicolor*	W Himalayas	Germany (Oct 1932)
Gray's Grasshopper Warbler	*Locustella fasciolata*	E Siberia	Denmark (Sep 1955), France (Sep 1913)
Oriental Reed Warbler	*Acrocephalus orientalis*	E Asia	Israel (Feb–Apr 1988; May 1990)
Plain Leaf Warbler	*Phylloscopus neglectus*	Central Asia	Sweden (Oct 1991)
Ruby-crowned Kinglet	*Regulus calendula*	North America	Iceland (Nov 1987)
Red-breasted Nuthatch	*Sitta canadensis*	North America	Britain (Oct 1989–May 1990), Iceland (Jan 1970)
Northern Shrike	*Lanius borealis*	North America, Siberia	Finland (Mar 2000), Norway (Nov 1881)
Daurian Starling	*Sturnus sturninus*	Asia	Netherlands (Oct 2005), Norway (Sep 1985)
Yellow-throated Vireo	*Vireo flavifrons*	North America	Britain (Sep 1990), Germany (Sep 1998)
Philadelphia Vireo	*Vireo philadelphicus*	North America	Britain (Oct 1987), Ireland (Oct 1985)
Pallas's Rosefinch	*Carpodacus roseus*	Siberia	Denmark (Mar–Apr 1995), Germany (Oct 1987), Hungary, Russia, Ukraine
Long-tailed Rosefinch	*Uragus sibiricus*	Siberia	Russia (records in W Europe are probably all escapes.)
Golden-winged Warbler	*Vermivora chrysoptera*	North America	Britain (Jan–Apr 1989)
Blue-winged Warbler	*Vermivora pinus*	North America	Ireland (Oct 2000)
Chestnut-sided Warbler	*Dendroica pensylvanica*	North America	Britain (Sep 1985; Oct 1995)
Cerulean Warbler	*Dendroica cerulea*	North America	Iceland (Oct 1997)
Black-throated Blue Warbler	*Dendroica caerulescens*	North America	Iceland (Sep 1988)
Black-throated Green Warbler	*Dendroica virens*	North America	Germany (Nov 1858), Iceland (Oct 2003)
Blackburnian Warbler	*Dendroica fusca*	North America	Britain (Oct 1961), Iceland (Nov 1987)
Cape May Warbler	*Dendroica tigrina*	North America	Britain (Jun 1977)
Magnolia Warbler	*Dendroica magnolia*	North America	Britain (Sep 1981), Iceland (Sep–Dec 1995; Oct 1995)
Palm Warbler	*Dendroica palmarum*	North America	Iceland (Oct 1997)
Bay-breasted Warbler	*Dendroica castanea*	North America	Britain (Oct 1995)
Louisiana Waterthrush	*Seiurus motacilla*	North America	Canary Islands (Nov 1991), Morocco (Jan 1999)

SPECIES		ORIGIN	RECORDED
Hooded Warbler	*Wilsonia citrina*	North America	Britain (Sep 1970; Sep 1992)
Wilson's Warbler	*Wilsonia pusilla*	North America	Britain (Oct 1985)
Canada Warbler	*Wilsonia canadensis*	North America	Iceland (Sep 1973)
Summer Tanager	*Piranga rubra*	North America	Britain (Sep 1957)
Eastern Towhee	*Pipilo erythrophthalmus*	North America	Britain (Jun 1966)
Lark Sparrow	*Chondestes grammacus*	North America	Britain (Jun–Jul 1981; May 1991)
Savannah Sparrow	*Passerculus sandwichensis*	North America	Britain (Apr 1982; Sep–Oct 1987)
Fox Sparrow	*Passerella iliaca*	North America	Iceland (Nov 1944), Ireland (Jun 1961)
Chestnut-eared Bunting	*Emberiza fucata*	Asia	Britain (Oct 2004)
Dickcissel	*Spiza americana*	North America	Norway (Jul 1981)
Brown-headed Cowbird	*Molothrus ater*	North America	Britain (Apr 1988), Norway (Jun 1987)
Yellow-headed Blackbird	*Xanthocephalus xanthocephalus*	North America	Iceland (Jul 1983), Netherlands (May–Jun 1982) (records in Britain, France, Norway and Sweden probably not genuine)

Introduced breeding species and species recorded only as escapes

A number of species which are not part of the region's natural fauna but which have been introduced or become established with human assistance and now *breed* locally in the wild *without continued dependence on humans* are described below. The list does not pretend to be complete but aims to include at least the most well-known species.

Also included are some non-breeding species which have been seen in the region, but which in all cases most likely originate from captivity and thus are not genuine wild birds occurring naturally.

Some introduced species are so well known that they have been included in the main section of the book, or have been placed there to facilitate comparison with similar species: Canada Goose (p. 20), Egyptian and Bar-headed Goose (22), Ruddy Duck (44), Pheasant, Golden Pheasant and Lady Amherst's Pheasant (58) and Indigo Bunting (406).

It should be pointed out that a large number of wildfowl from various parts of the world are kept in bird collections. These are sometimes free-flying and can therefore appear at traditional birdwatching sites; not all such species can be listed. There is also a heavy trade in cagebirds, and species from e.g. Asia, Australia and Africa are imported into Europe on a large scale; some of these escape and appear in the wild.

Black Swan *Cygnus atratus* [V***

L 115–140 cm. Australian species. An introduced self-sustaining population exists at least in the Netherlands, and another has been reported from Poland, with odd records in other countries as well. – *When swimming, all black* with *red bill, bill having a white subterminal band across*. Has peculiarly *narrow neck*, short body and oddly shaped *wing-feathers appearing wavy* like roof tiles. In flight shows *white primaries and outer secondaries*. Juveniles are paler, sooty brown-grey. (Not illustrated.)

Ross's Goose *Anser rossii* [V***

L 53–66 cm. Breeds in N Canada. Seen in ones or twos in a number of NW European countries (e.g. Britain, Germany, Netherlands and Sweden), often in the company of Barnacle Geese, but generally regarded as escapes. – Like white morph of Snow Goose, but is *smaller* and has *shorter neck*, *rounder head* and *shorter, 'cuter' bill*. Bill-base greenish-blue. Grey morph exists, but very rare. (Not illustrated.)

Emperor Goose *Anser canagica* [V***

L 66–85 cm. Breeds in Alaska and NE Siberia. Found in a few European countries, with occasional reports of breeding, but overwhelmingly likely that all records refer to escapes. – A handsome, medium large, *compact* and rather *short-necked* goose, *body scaly bluish-grey, head and hindneck white* but *chin and foreneck black*. *Legs orange-yellow*, bill black with large central pinkish-yellow patch. Blue morph of Snow Goose superficially similar at distance but Emperor has shorter neck and scaly plumage. (Not illustrated.)

South African Shelduck *Tadorna cana* —

L 60–68 cm. Breeds in S Africa. Recorded in a few European countries, and escaped or introduced birds have bred occasionally in Sweden. – Quite similar to Ruddy Shelduck, and possible to confuse with that in flight due to very similar wing pattern (black primaries, white 'forearm' and green speculum), but ♂ has *uniform grey head* and ♀ *dark brown head with pure white 'face'*. (Not illustrated.)

Wood Duck *Aix sponsa* [V***

L 43–51 cm. American species, kept in some wildfowl collections in Europe, and escapes occasionally seen, e.g. in Britain and Switzerland. Genuine vagrants reported from Iceland. – Slightly larger than Mandarin Duck. Adult ♂ has *head and upperparts metallic blue and green, head with fine white lines*; chin and upper throat white, sides of neck and breast dark rufous, speckled white. Adult ♀ (and eclipse ♂) like Mandarin, told by having *more white surrounding eye*; *darker head without pale striation on sides of cheeks*; different *outline of base of bill*; dark nail; *finer pale spotting on flanks*.

Wood Duck

♀

♂

KM

Mandarin Duck *Aix galericulata* [r**B**4]

L 41–49 cm, WS 65–75 cm. Breeds in East Asia, feral population established in Britain early 20th century and in Ireland since 1978; also in France and Switzerland. Found in many other European countries. Prefers lakes and rivers with overhanging willows, reedbeds and other sheltering vegetation. – Medium-small and compact, long tail, large head. Sexes markedly different: adult ♂ unmistakable with *red bill, wide whitish band from bill above eye to end of fluffy crest*, orange, 'combed' whiskers and large *orange 'sails' at rear of back*, whereas adult ♀ is dull olive-grey with *white spectacles* and *narrow line back towards nape*; white narrow line inside base of bill, white chin and throat. Flanks boldly spotted pale. Juvenile like ♀ but duller and browner, with more indistinct head markings.

Mandarin Duck

Cinnamon Teal *Anas cyanoptera* [**V*****]

L 38–48 cm. Breeds in western North America and South America. Very rarely seen free-flying in Europe, and then most likely involving escapes only. – Both sexes have *green speculum* and *bright blue upperwing-covert patch*. ♂ is uniformly rufous with dark crown and back. ♀ like Blue-winged Teal (see p. 28), but is more reddish and has *less contrasting head pattern*. (Not illustrated.)

California Quail *Callipepla californica* —

L 24–27 cm. Breeds in North America. Introduced in France, feral stock in Corsica. – *Uniform blue-grey above and on breast*, coarsely spotted white on belly. Both sexes have a *black plume* on forecrown (bigger on ♂). ♂ also has *white-bordered black throat* and *white supercilium*. ♀ has finely spotted head. Display-call a loud, repeated rhythmic 'ka-**kwah**-ko'.

California Quail

Northern Bobwhite *Colinus virginianus* —

L 24–26 cm. Breeds in North America. Introduced in S Europe, breeds in e.g. C France (rare) and N Italy. – Mainly *reddish-brown* with white and black spots and streaks. ♂ has *black head* with *white supercilium and throat*, while ♀ has head patterned dark brown and light rusty-brown. Display-call a clear whistling 'tüh tvei**it**'. (Not illustrated.)

Erckel's Francolin *Francolinus erckelii* —

L 38–43 cm. Breeds in E Africa, in particular in Ethiopia. Local feral population exists in Italy. – Medium-large francolin, brown above and whitish below, whole plumage steaked dark brown. Typically has uniform rufous crown and black forehead and 'mask'. Chin white, unmarked. Yellowish legs very strong, double-spurred. (Not illustrated.)

Reeves's Pheasant *Syrmaticus reevesii* —

L ♂ 140–190 cm, ♀ 60–85 cm. Chinese species introduced in Europe and now breeding in wild in e.g. France and Czech Republic. Forest-dweller. – Large. Mottled brown like Pheasant, but told by ♂'s *extremely long tail* (100–150 cm!), *white head with black band* from bill back to nape, and *white spot beneath eye*. (Not illustrated.)

Sacred Ibis

Sacred Ibis *Threskiornis aethiopicus* —

L 60–85 cm, WS 110–125 cm. Breeds in Africa, building its nest in trees in heronries, just as Glossy Ibis. Introduced in France in 1976, where a free-flying population still breeds. A feral population also in Italy. Occasionally seen elsewhere. – Mainly *white, with black head and neck* (both unfeathered) and *black, strong, downcurved bill*. Tertials blackish, plume-like.

African Spoonbill *Platalea alba* —

L 85–95 cm. Breeds in Africa. Recorded several times in France, a couple of times in Spain and once each in Denmark and Austria. All records probably involve escapes from bird collections. – Same size as Eurasian Spoonbill. *White plumage* and *pinkish-red legs*. Red 'face' and red base and edges to otherwise grey, spoon-shaped bill.

African Spoonbill

KM/DZ

Lesser Flamingo

Chilean Flamingo

ad.

Barbary Dove

♀ ♂

Rose-ringed Parakeet

American Flamingo *Phoenicopterus ruber* —

L 125–145 cm, WS 140–170 cm. Breeds in Central America. Several records in Europe, but these most likely refer to escapes from collections where it is often kept. – *Very large*, as European Flamingo. Plumage differs on having *much stronger and more extensive pink-red colour*. Note that amount of red varies with age and food; some birds are almost as light as European Flamingo, others are *strikingly red on head, neck, breast and tail*. Legs greyish with a little pink on 'knees' (though much less contrast than in Chilean Flamingo), bill like Flamingo but often *deeper pink*. (Not illustrated.)

Lesser Flamingo *Phoenicopterus minor* —

L 80–95 cm, WS 90–110 cm. Breeds in C and E Africa. Several records in France, Morocco and Spain, and some of these may involve wild birds. In 1994 a pair attempted to breed in Camargue, France, and there have been new attempts on later occasions. A number of records also in N Europe, but these most likely refer to escapes from collections. – *Small*. Plumage *pink* with red stripes on back and shoulders. *Legs entirely red, bill deep red with small black tip.*

Chilean Flamingo *Phoenicopterus chilensis* [V***]

L 100–120 cm, WS 120–135 cm. South American species kept in several bird collections in Britain and Europe, and free-flying individuals may be met with. A feral population known from Germany. – Somewhat smaller than Flamingo, has *over half of bill black*, has *buff-grey legs with contrasting red 'knees'* and toes, and has often *pinker-toned plumage*.

Barbary Dove *Streptopelia risoria* (domest.) —

(Alt. names: Ring-necked Dove or Domestic Ringdove.) L 27–29 cm. A long-domesticated form of African Collared Dove (p. 414), sometimes treated as a separate species. A few pairs breed in Tenerife in Canary Islands, originating from escaped cagebirds. – Very like Collared Dove but *smaller*, has slightly *shorter tail*, *paler* buffy grey-white plumage, and is almost white on belly (Collared Dove light grey). Black neck-side marking is often broader, more rounded, not a narrow cross-bar. Is also a touch smaller and paler than the wild ancestral form. Call a pleasing 'ko k'rrooh', repeated a few times.

Rose-ringed Parakeet *Psittacula krameri* [r**B**5]

L 37–43 cm (incl. tail-extension of ad. 18–23 cm). South Asian and sub-Saharan African species, introduced or escaped in treated region, now feral populations in several countries (Belgium, Britain, Egypt, France, Germany, Greece, Israel, Netherlands, Spain, possibly more). – Nearly all *bright green*, with *long, pointed tail* and *long narrow wings* with *darker flight-feathers*. Upper mandible rosehip-red. ♂ has black bib and *narrow black line across side of throat* turning into a rosy-red necklace across neck and nape; ♀ has uniform green head/neck. Pale eye with red orbital ring. Noisy and lively, keeping high up in the canopy of tall trees in larger parks, even in cities.

Alexandrine Parakeet *Psittacula eupatria* [r**B**5]

L 50–62 cm. South Asian species breeding in India and eastward. Small populations in Belgium, Germany, Netherlands and Turkey (Istanbul) emanating from escaped cagebirds. – Like a *large* version of Rose-ringed Parakeet, almost twice as big, long green wings and long pointed tail, told apart from size by proportionately *much heavier red bill* and presence of *rufous 'shoulder patches'* (inner 'forearm' reddish). (Not illustrated.)

Fischer's Lovebird *Agapornis fischeri* —

L 14–16 cm. East African species. Feral population in France, and is spreading. – A quite small, compact parrot with proportionately *large, rounded head* and *short tail*. Easily told by *green body*, *orange head* and *yellow breast*. *Heavy bill is all red*. Dark eye surrounded by *broad white eye-ring*. (Not illustrated.)

Nanday Parakeet *Nandayus nenday* —

, 32–36 cm. South American species. Breeds locally in small colonies in Israel; the birds obviously originate from escaped cagebirds. – A medium-large parrot with *predominantly bright yellow-green* plumage, relieved by *bluish breast-band* and *blackish head*. Amber-coloured eye has white eye-ring. *Long, pointed, green tail*. Legs orangey. (Not illustrated.)

Monk Parakeet

Monk Parakeet *Myiopsitta monachus* —

, 28–31 cm. South American species. Breeds locally in small colonies in Belgium, Canary Islands, Italy, Slovakia and Spain (Barcelona and Baleares); the birds originate from escaped cagebirds. – *Upperparts bright green*, forehead, throat and *breast light grey*, breast finely barred dark. Belly pale green. *Wings tinged bluish*. *Long, pointed, green tail*. Bill orange.

Blue-crowned Parakeet *Aratinga acuticaudata* —

, 33–38 cm. South American species. Reportedly breeds or has bred locally in small colonies in Britain (Kent) and Spain (Barcelona); the birds originate from escaped cagebirds. – Medium-large species with predominantly *green colours*, breast paler green, head darker and tinged blue, especially on crown. *Long and pointed tail mixture of green, golden-brown and reddish*. Rather heavy bill pinkish-brown. Naked orbital skin whitish. (Not illustrated.)

Common Myna

Common Myna *Acridotheres tristis* —

, 22–25 cm. Indian starling which is spreading N and W. Breeds in e.g. Israel, Russia and Tenerife (since 1993), in all cases originating from escaped cagebirds. – About the size of Starling, brown with sooty-black head, *yellow bill* and *bare yellow skin behind eye*, also *yellow legs*. *Large white patches on 'hand' on rounded wings* striking in flight. Fearless, lives near human habitation. Very loud and noisy.

Black-headed Weaver *Ploceus melanocephalus* —

(Alt. name: Yellow-backed Weaver.) L 14–16 cm. African species, breeding in sub-Saharan belt. A free-flying population in Portugal, but undoubtedly referring only to escaped cagebirds. – One in a rather large group of mainly yellowish and *black-headed* weaver species. Told by combination of *lightly streaked grey-brown back*, impression of *yellow neck-collar, yellow breast* lacking orange-red hue, and *darkish eye*. (Not illustrated.)

Streaked Weaver

Streaked Weaver *Ploceus manyar* —

, 12–13 cm. African species. Escaped cagebirds have established colonies in reedbeds in the Nile delta, Egypt. – *Big head* and *sturdy bill*. Tail short and square-cut, legs also rather short. ♂ has *yellow crown, dark head-side* and heavily streaked buffy brownish-white underparts, and during breeding *black bill*. ♀ has *yellow supercilium, yellow neck-side patch* and dark cheeks and crown. Bill ivory-yellow.

♂

Yellow-crowned Weaver *Euplectes afer* —

(Alt. name: Yellow-crowned Bishop.) L 9½–12 cm. Widespread African species. Small free-flying population in Portugal, but undoubtedly referring only to escaped cagebirds. – *A chubby* and *short-tailed*, attractive little sparrow. ♂ is yellow and black, has *yellow crown, back and vent, the remaining parts being black*. ♀ is brown and streaked above, not that unlike a House Sparrow (only considerably smaller). (Not illustrated.)

Red-billed Firefinch

Red-billed Firefinch *Lagonosticta senegala* —

, 9½–11 cm. African species, widespread S of Sahara. Introduced at El Golea, Algeria, but has become rare or extinct lately. Found around human settlements. *Small*. Tiny pale spots on breast-side. ♂ is *red on much of head and breast*, has reddish bill, *plain, brown wings* and black tail. ♀ is grey-brown, with *red lores, bill and rump*. Thin yellow eye-ring around dark eye.

DZ

Common Waxbill

Red Avadavat

♂

Indian Silverbill

DZ

Common Waxbill *Estrilda astrild*

L 11–12 cm. African species, introduced in Iberian peninsula, where it now breeds locally, mainly in Portugal. Sedentary. Breeds in reeds, bulrushes, rushes, spiraea, etc., season protracted (Feb–Nov). – *Small*, grey-brown, finely vermiculated, with *red eye-mask*, a *red patch on belly-centre* and *black* (or dark brown) *vent*. Bill *red* on adult, brownish-black on juvenile. Song a three-syllable, simple verse, two harsh notes followed by a rising, somewhat rolling note, 'chre-chre-srri'.

Red Avadavat *Amandava amandava*

L 9½–10½ cm. Indian species. Introduced in e.g. Spain, Po delta and Nile delta. Breeds in autumn (spring in Nile delta) in lush waterside vegetation, but also in cultivated crops. – *Very small*. In all plumages, grey-brown upperparts with fine *white spots on wing-coverts and tertials*, *red uppertail-coverts* and dark tail. Adult ♂ May–Dec has greater part of *head and underparts saturated red*, *underparts sprinkled with white spots*. ♀ and ♂ winter–spring are off-white below, with buffish-grey on breast and *yellow tone on belly*. Dark lores. ♂ has more red on uppertail-coverts with more distinct white spots, and has *tinge of red on belly*. Flight-call short, muffled 'chick-chick'; when foraging, sharp 'zsi' notes. Song a series of falling, clear, soft notes, often terminating in brief trill.

Indian Silverbill *Lonchura malabarica*

L 11–12 cm. Indian species. Since 1988 has bred in Israel, and from 1990 in Jordan. It also has a free-flying population in Nice, France. Likely origin for all these populations is escaped cagebirds. Social habits, even when breeding. Fearless. – *Small. Stout silvery-grey bill. Uniform light brown upperparts*, whitish underparts with brown-tinged flanks, black primaries and *long, tapering black tail. Uppertail-coverts white*. Sexes alike. Juvenile has shorter, blunter tail, and brownish 'face' and brownish rump.

Black-headed Munia *Lonchura malacca*

L 10½–12 cm. Indian species. Escaped cagebirds have established colonies in Portugal. – Attractive rather small finch-like species with ♂ plumage having *black head and vent* separated by *white across lower breast and onto flanks*, while *upperparts are plain brown*. Heavy conical bill is greyish-white. ♀ is plain greybrown and off-white, lacking features, but usually told by strong, pale grey conical bill. (Not illustrated.)

References

The references listed below have been consulted during the work on this book, many of them repeatedly. Without them our work would have been much more difficult and the end result less good. They can all be recommended. For the

preparation of the distribution maps all known national checklists and atlas surveys referring to the treated region have been used, the most important and up-to-date ones being included in the following list.

GENERAL

Ali, S. & Ripley, S. D. (1987) *Compact Handbook of the Birds of India and Pakistan*. 2nd ed. Oxford Univ. Press, Oxford.

Alström, P., Colston, P. & Lewington, I. (1991) *A Field Guide to the Rare Birds of Britain and Europe*. Domino, Jersey.

Alström, P., Mild, K. & Zetterström, B. (2003) *Pipits & Wagtails of Europe, Asia and North America*. Helm, London.

Alula (1995–2008) Ornithological journal (4 issues/yr). Helsinki.

Baker, K. (1997) *Warblers of Europe, Asia and North Africa*. Helm, London.

Beaman, M. & Madge, S. (1998) *The Handbook of Bird Identification for Europe and the Western Palearctic*. Helm, London.

Birding World (1988–) Ornithological journal (12 issues/yr). Cley-next-the-sea, Holt, England.

Blomdahl, A., Breife, B. & Holmström, N. (2003) *Flight Identification of European Seabirds*. Helm, London.

Brazil, M. (2009) *Birds of East Asia*. Helm, London.

Brewer, D. (2001) *Wrens, Dippers and Thrashers*. Helm, London.

British Birds (1907–) Ornithological journal (12 issues/yr). London.

Brown, L. H., Urban, E. K., Newman, K., Fry, C. H. & Keith, G. S. (ed.) (1982–2004) *The Birds of Africa*. Vol. 1–7. Academic Press/Helm, London.

BWPi (*Birds of the Western Palearctic interactive*). (2004) 1 DVD. Birdguides/Oxford University Press, London/Oxford.

BWP Update. The Journal of the Birds of the Western Palearctic. (1997–2004) Vol. 1–6. Oxford University Press, Oxford.

Campbell, B. & Lack, E. (ed.) (1985) *A Dictionary of Birds*. Poyser, Calton.

Chantler, P. & Driessens, G. (1995) *Swifts*. Pica, Mountfield.

Cleere, N. & Nurney, D. (1998) *Nightjars*. Pica, Mountfield.

Clement, P., Harris, A. & Davis, J. (1993) *Finches and Sparrows*. Helm, London.

Clement, P. & Hathway, R. (2000) *Thrushes*. Helm, London.

Cottridge, D. & Vinicombe, K. (1996) *Rare Birds in Britain & Ireland*. A photographic record. HarperCollins, London.

Cramp, S., Simmons, K. E. L. & Perrins, C. M. (ed.) (1977–94) *The Birds of the Western Palearctic*. Vol. 1–9. Oxford University Press, Oxford.

Curry-Lindahl, K. (ed.) *et al*. (1959–62) *Våra fåglar i Norden*. ('Our Nordic Birds'.) Vol. 1–4. 2nd ed. Natur och Kultur, Stockholm.

Curson, J., Quinn, D. & Beadle, D. (1994) *New World Warblers*. Helm, London.

Delin, H. & Svensson, L. (1988) *Photographic Guide to the Birds of Britain & Europe*. Hamlyn, London.

Delin, H. & Svensson, L. (2007) *Philip's Guide to Birds of Britain & Europe*. Philip's, London.

Doherty, P. (2004) *The Birds of Britain & Europe*. 6 DVD. Bird Images, Sherburn-in-Elmet.

Dunn, J. & Garrett, K. (1997) *A Field Guide to the Warblers of North America*. Houghton Mifflin, Boston.

Dutch Birding (1979–) Ornithological journal (6 issues/yr). Santpoort-Zuid.

Farrand Jr, J. (ed.) *et al*. (1983) *The Audubon Society Master Guide to Birding*. Vol. 1–3. Knopf, New York.

Forsman, D. (1984) *Rovfågelsguiden*. ('The Guide to Raptors'.) Lintutieto, Helsinki.

Forsman, D. (1999) *The Raptors of Europe and The Middle East*. Poyser, London.

Fry, C. H., Fry, K. & Harris, A. (1992) *Kingfishers, Bee-eaters & Rollers*. Helm, London.

Génsbøl, B. (2006) *Rovfåglar i Europa*. ('Raptors in Europe'.) Prisma, Stockholm.

Géroudet, P. (1953–61) *Les Passereaux d'Europe*. I–III. Délachaux et Niestlé, Paris.

Géroudet, P. (1965) *Les Rapaces Diurnes et Nocturnes d'Europe*. Délachaux et Niestlé, Paris.

Géroudet, P. (1965) *Water-birds with webbed feet*. Blandford, London.

Gibbs, D., Barnes, E. & Cox, J. (2001) *Pigeons and Doves*. Pica, Mountfield.

Glutz, U. N. (publ.), Bauer, K., Bezzel, E. *et al*. (1966–98) *Handbuch der Vögel Mitteleuropas*. Vol. 1–14. Aula-Verlag, Wiesbaden.

Grant, P. J. (1986) *Gulls: a guide to identification*. 2nd ed. Poyser, Calton.

Grimmett, R., Inskipp, C. & Inskipp, T. (1998) *Birds of the Indian Subcontinent*. Helm, London.

Hancock, J. & Kushlan, J. (1984) *The Herons Handbook*. Harper & Row, New York.

Harrap, S. & Quinn, D. (1996) *Tits, Nuthatches & Treecreepers*. Helm, London.

Harris, T. & Franklin, K. (2000) *Shrikes & Bush-Shrikes*. Helm, London.

Harrison, P. (1989) *Seabirds of the World*. Helm, London.

Hayman, P., Marchant, J. & Prater, T. (1986) *Shorebirds*. Croom Helm, Beckenham.

Hollom, P. A. D., Porter, R. F., Christensen, S. & Willis, I. (1988) *Birds of the Middle East and North Africa*. Poyser, London.

del Hoyo, J., Elliott, A. & Christie, D. A. (ed.) (1992–2008) *Handbook of the Birds of the World*. Vol. 1–13. Lynx, Barcelona.

Jenni, L. & Winkler, R. (1994) *Moult and Ageing of European Passerines*. Academic Press, London.

Jonsson, L. & Tysse, T. (1992) *Lommar*. ('Loons'.) Suppl. to *Vår Fågelvärld* no. 15. SOF, Stockholm.

Kaufman, K. (1990) *Advanced Birding*. Houghton Mifflin, Boston.

Lefranc, N. & Worfolk, T. (1997) *Shrikes*. Pica, Mountfield.

Limicola (1987–) Ornithological journal (6 issues/yr). Einbeck-Drüber.

Madge, S. & Burn, H. (1988) *Wildfowl*. Helm, London.

Madge, S. & Burn, H. (1991) *Crows and Jays*. Helm, London.

Madge, S. & McGowan, P. (2002) *Pheasants, Partridges & Grouse*. Helm, London.

Mebs, T. & Scherzinger, W. (2000) *Die Eulen Europas*. ('The Owls of Europe'.) Kosmos, Stuttgart.

Mikkola, H. (1983) *Owls of Europe*. Poyser, Calton.

Mitchell, D. & Young, S. (1997) *Rare Birds of Britain and Europe*. New Holland, London.

Naoroji, R. (2006) *Birds of Prey of the Indian Subcontinent*. Helm, London.

O'Brien, M., Crossley, R. & Karlson, K. (2007) *The Shorebird Guide*. Helm, London.

Olsen, K. M. & Larsson, H. (1995) *Terns of Europe and North America*. Helm, London.

Olsen, K. M. & Larsson, H. (1997) *Skuas and Jaegers*. Pica, Mountfield.

Olsen, K. M. & Larsson, H. (2004) *Gulls of Europe, Asia and North America*. 2nd printing. Helm, London.

Olsson, U., Curson, J. & Byers, C. (1995) *Buntings and Sparrows*. Pica, Mountfield.

Palmer, R. S. (ed.) *et al*. (1962–88) *Handbook of North American Birds*. Vol. 1–5. Yale University Press, New Haven.

Porter, R. F., Christensen, S. & Schiermacker-Hansen, P. (1996) *Field Guide to the Birds of the Middle East*. Poyser, London.

Pyle, P *et al*. (1997) *Identification Guide to North American Birds*. Part 1. Slate Creek Press, Bolinas, California.

Rasmussen, P. C. & Anderton, J. (2005) *Birds of South Asia*. The Ripley Guide. Vol. 1–2. Smithonian & Lynx, Barcelona.

Robson, C. (2000) *A Field Guide to the Birds of South-East Asia*. New Holland, London.

Rosair, D. & Cottridge, D. (1995) *Photographic Guide to the Waders of the World*. Hamlyn, London.

Rosenberg, E. (1953) *Fåglar i Sverige*. ('Birds in Sweden'.) Almqvist & Wiksell, Stockholm.

Shirihai, H. (2007) *A Complete Guide to Antarctic Wildlife*. 2nd ed. A. & C. Black, London.

Shirihai, H., Christie, D. & Harris, A. (1996) *Birder's Guide to European and Middle Eastern Birds*. Macmillan, London.

Shirihai, H., Gargallo, G. & Helbig, A. J. (2001) *Sylvia Warblers*. Helm, London.

Sibley, D. A. (2000) *The Sibley Guide to Birds*. Knopf, New York.

Svensson, L. (1992) *Identification Guide to European Passerines*. 4th ed. Stockholm.

Taylor, B. & van Perlo, B. (1998) *Rails*. Pica, Mountfield.
Ticehurst, C. B. (1938) *A Systematic Review of the Genus Phylloscopus*. British Museum, London.
Turner, A. & Rose, C. (1989) *Swallows and Martins of the World*. Helm, London.
Urquhart, E. (2002) *Stonechats*. A Guide to the Genus Saxicola. Ill. by A. Bowley. Helm, London.
Vår Fågelvärld. (1942–) Ornithological journal (8 issues/yr). Stockholm.
Vinicombe, K., Harris, A. & Tucker, L. (1989) *Bird Identification*. Macmillan, London.
Winkler, H., Christie, D. A. & Nurney, D. (1995) *Woodpeckers*. Pica, Mountfield.
Witherby, H. F., Jourdain, F. C. R., Ticehurst, N. F. & Tucker, B. W. (1938–41) *The Handbook of British Birds*. Vol. 1–5. Witherby, London.

ATLASES & CHECKLISTS

Andrews, I. J. (1995) *The Birds of the Hashemite Kingdom of Jordan*. Musselburgh, Scotland.
Bannerman, D. A. & Bannerman, W. M. (1983) *The Birds of the Balearics*. Helm, London.
Baumgart, W. (2003) *Birds of Syria*. 2nd rev. ed. by OSME. (1st ed. in German by Kasparek Verlag, Heidelberg.)
Beaman, M. (1994) *Palearctic Birds*. Stonyhurst.
Clements, J. F. (2000) *Birds of the World*. A Checklist. 5th ed. Pica, Mountfield.
Dickinson, E. C. (ed.) *et al*. (2003) *The Howard & Moore Complete Checklist of the Birds of the World*. 3rd ed. Helm, London.
Dubois, P. J., Le Maréchal, P., Olioso, G., Yésou, P. *et al*. (2000) *Inventaire des Oiseaux de France*. Nathan, Paris.
Estrada, J. *et al*. (ed.) (2004) *Atles dels ocells nidificants de Catalunya 1999–2002*. ('Catalan Breeding Bird Atlas'.) Lynx, Barcelona.
Ferguson-Lees, J, Willis, I. & Sharrock, J. T. R. (1983) The Shell Guide to the Birds of Britain and Ireland. Michael Joseph, London.
Gavrilov, E. & Gavrilov, A. (2005) *The Birds of Kazakhstan*. Tethys, Almaty.
Gjershaug, J. O., Thingstad, P. G., Eldøy, S. & Byrkjeland, S. (ed.) (1994) *Norsk fugleatlas*. NOF, Klæbu.
Goodman, S. M., Meininger, P. L. (ed.) *et al*. (1989) *The Birds of Egypt*. Oxford Univ. Press, Oxford.
Gorman, G. (1996) *The Birds of Hungary*. Helm, London.
Hagemeijer, W. J. M. & Blair, M. J. (ed.) (1997) *The EBCC Atlas of European Breeding Birds*. Poyser, London.
Handrinos, G. & Akriotis, T. (1997) *The Birds of Greece*. Helm, London.
Hartert, E. (1910–22) *Die Vögel der paläarktischen Fauna*. (With Suppl. 1923, 1932–38.) Friedländer, Berlin.
Isenmann, P. & Moali, A. (2000) *Birds of Algeria*. SEOF, Paris.
Isenmann, P. *et al*. (2005) *Birds of Tunisia*. SEOF, Paris.
Kirwan, G. *et al*. (2008) *The Birds of Turkey*. Helm, London.
Kren, J. (2000) *Birds of the Czech Republic*. Helm, London.
Larsson, L., Larsson, E. & Ekström, G. (2002) *Birds of the World*. Interactive electronic checklist, 2 CDs. Väse, Sweden.
Leibak, E., Lilleleht, V. & Veromann, H. (1994) *Birds of Estonia*. Estonian Academy Publ., Tallinn.
Lindell, L. *et al*. (2002) *Sveriges fåglar*. ('The Birds of Sweden'.) 3rd ed. SOF, Stockholm.
Marti, R. & del Moral, J. C. (ed.) (2003) *Atlas de las Aves Reproductoras de España*. ('Atlas of Breeding Birds in Spain'.) Ministerio de Medio Ambiente/ SEO Birdlife, Madrid.
Meschini, E. & Frugis, S. (ed.) (1993) *Atlante degli Uccelli Nidificanti in Italia*. ('Atlas of Breeding Birds in Italy'.) Suppl. Ric. Biol. Selvaggina, Bologna.
Priednieks, J., Strazds, M., Strazds, A. & Petrins, A. (1989) *Latvijas Ligzdojoso Putno Atlants 1980–1984*. Zinatne, Riga. (With English summaries.)
Purroy, F. J. (ed.) *et al*. (1997) *Atlas de las Aves de España*. SEO Birdlife/Lynx, Barcelona.
Rogacheva, H. (1992) *The Birds of Central Siberia*. Husum.
Roselaar, C. S. (1995) *Songbirds of Turkey*. An atlas of biodiversity of Turkish passerine birds. Pica, Mountfield.
Rufino, R. (ed.) *et al*. (1989) *Atlas das Aves que nidificam em Portugal Continental*. Minist. plano e administr. território, Lisbon.
Ryabtsev, V. K. (2001) *Ptitsy Urala*. ('The Birds of Ural'.) Yekaterineburg.
Scott, D. A. *et al*. (ed.) (1983) *The Birds of Iran*. 2nd ed. Dept of Environment, Islamic Republic of Iran, Tehran.
Shirihai, H. *et al*. (1996) *The Birds of Israel*. Academic Press, London.
Stewart, P. F. & Christensen, S. J. (1971) *A Checklist of the Birds of Cyprus*. RAF, Plymouth.
Stresemann, E., Portenko, L. A., Dahte, H., Neufeldt, I. A. *et al*. (1960–88) *Atlas der Verbreitung palaearktischer Vögel*. Vol. 1–15. Akademie-Verlag, Berlin.
Svensson, L., Svensson, M. & Tjernberg, M. (1999) *Svensk fågelatlas*. ('Atlas of Swedish Birds'.) Vår Fågelvärld, Suppl. 31. SOF, Stockholm.

Thévenot, M., Vernon, R. & Bergier, P. (2003) *The Birds of Morocco*. BOU Checklist no. 20. BOU, Tring.
Tomiałojc, L. & Stawarczyk, T. (2003) *Awifauna Polski*. (The Avifana of Poland Distribution, numbers and trends.) 2 vol. PTPP, Wrocław.
Vaurie, C. (1959, 1965) *The Birds of the Palearctic Fauna*. Vol. 1–2. Witherby
Voous, K. H. (1977) *List of Recent Holarctic Bird Species*. BOU, London.
Wassink, A. & Oreel, G. J. (2007) *The Birds of Kazakhstan*. Privately published (Wassink), Texel.
Zink, G. (1973–85) *Der Zug europäischer Singvögel*. Vol. 1–4. Vogelzug-Verlag Möggingen.

SOUND

Andersson, B., Svensson, L. & Zetterström, D. (1990) *Fågelsång i Sverige*. 1 CD and book. Mono Music, Stockholm.
Bergmann, H.-H. & Helb, H.-W. (1982) *Stimmen der Vögel Europas*. BLV, Munich.
Carlsson, S. & Gydemo, P.-E. (1993) *Tättingläten*. ('Passerine calls'.) 1 CD with 99 species.
Chappuis, C. (1987) *Migrateurs et hivernants*. 2 sound cassettes. Grand Couronne, France.
Chappuis, C. (2000) *African Bird Sounds – 1. North-West Africa, Canaria and Cap-Verde Islands*. 4 CD. SEOF/Nat. Sound Arch., London.
Constantine, M. *et al*. (2006) *The Sound Approach to birding*. With 2 CD. Poole.
Cramp, S. (ed.) *et al*. (2006) *BWPi*, version 2.0. (Electronic version of the 9-volume handbook, the Concise edition and the BWP Updates, plus added film footage and sounds.) Birdguides, London.
Gulledge, J. (prod.) *et al*. (1983) *A Field Guide to Bird Songs of Eastern and Central North America*. 2nd ed. 2 sound cassettes. Cornell/Houghton Mifflin, Boston.
Gunn, W. W. H., Kellogg, P. P. (ed.) *et al*. (1975) *A Field Guide to Western Bird Songs*. 3 sound cassettes. Cornell/Houghton Mifflin, Boston.
Jännes, H. (2002) *Calls of Eastern Vagrants*. 1 CD. Earlybird, Helsinki.
Mild, K. (1987) *Soviet Bird Songs*. 2 sound cassettes and booklet. Stockholm.
Mild, K. (1990) *Bird Songs of Israel and the Middle East*. 2 sound cassettes and booklet. Stockholm.
Palmér, S. & Boswall, J. (1981) *A Field Guide to the Bird Songs of Britain and Europe*. 16 sound cassettes. SR Phonogram, Stockholm.
Ranft, R. & Cleere, N. (1998) *A Sound Guide to Nightjars and Related Nightbirds*. 1 CD. Pica/Nat. Sound Arch., Mountfield
Robb, M. S. (2000) *Introduction to vocalizations of crossbills in north-western Europe*. 1 CD and leaflet. Dutch Birding, Amsterdam.
Roché, J. C. *et al*. (1998) *Birds of prey and Owls of Western Europe*. 1 CD. Frémeau, Cervennes.
Roché, J. C. *et al*. (1990) *All the bird songs of Britain and Europe*. 4 CDs. Sittell/ Mens, France.
Roché, J. C., Chevereau, J. (ed.) *et al*. (1998) *A sound guide to the Birds of North West Africa*. 1 CD. Sittelle, Mens, France.
Roché, J. C., Chevereau, J. (ed.) *et al*. (2001) *Guia sonora de las aves de Europa*. 10 CD. Lynx, Barcelona.
Sample, G. (1998) *Bird Call Identification*. Book and 1 CD. HarperCollins, London.
Sample, G. (2003) *Warbler Songs & Calls of Britain and Europe*. Book and 3 CD. HarperCollins, London.
Schubert, M. (1981) *Stimmen der Vögel Zentralasiens*. 2 LP records. Eterna.
Schubert, M. (1984) *Stimmen der Vögel. VII. Vogelstimmen Südosteuropas (2)*. 1 LP record. Eterna.
Schulze, A. (ed.), Roché, J. C., Chappuis, C., Mild, K. *et al*. (2003) *Die Vogelstimmen Europas, Nordafrikas und Vorderasiens*. 17 CDs. Edition Ample, Germering, Germany. [Version with 2 MP3 discs available.]
Strömberg, M. (1994) *Moroccan Bird Songs and Calls*. 1 sound cassett. Sweden.
Svensson, L. (1984) *Soviet Birds*. 1 sound cassette. Stockholm.
Ueda, H. (1998) *283 Wild Bird Songs of Japan*. Yama-kei, Tokyo.
Veprintsev, B. N. *et al*. (1982–86) *'Birds of the Soviet Union: A Sound Guide'*. LP records. Melodia, Moscow.
Veprintsev, B. N., Veprintsev, O. *et al*. (2007) *'Voices of the birds of Russia'*. MP3 disc. Phonotheca, Pushchino, Moscow.
Wahlström, S. (1992) *Från Alfågel till Årtsångare*. ('From Long-tailed Duck to Lesser Whitethroat'.) 2 CDs. Bra Böcker, Viken, Sweden.
Walton, R. K. & Lawson, R. W. (1994) *More Birding by Ear. A Guide to Bird-song Identification*. Eastern/Central North America. 3 CDs and booklet. Houghton Mifflin, Boston.
Wetland Birds – a celebration. (1996) 1 CD. Wildfowl & Wetlands Trust/Bird Watching, England.

Index